AF718031

Root-knot Nematodes

Mixed Sources
Product group from well-managed forests and other controlled sources
www.fsc.org Cert no. SA-COC-1565
© 1996 Forest Stewardship Council
FSC

With thanks to Syngenta for sponsoring the colour plate section in this book.

syngenta

Syngenta is a world-leading agribusiness committed to sustainable agriculture through innovative research and technology. The company is a leader in crop protection as well as in high-value commercial seed markets. Further information is available at www.syngenta.com.

AVICTA®

Seed Treatment **Nematicide**

AVICTA® is the first ever nematicide offered as a seed treatment. AVICTA® is a novel way to protect agricultural crops against damage from most commonly occurring nematodes. Further information is available at www.avicta.com.

Root-knot Nematodes

Edited by

Roland N. Perry

Plant Pathology and Microbiology Department, Rothamsted Research, Harpenden, Hertfordshire, UK and Biology Department, Ghent University, Ghent, Belgium

Maurice Moens

Institute for Agricultural and Fisheries Research, Merelbeke, Belgium and Department of Crop Protection, Ghent University, Ghent, Belgium

and

James L. Starr

Department of Plant Pathology and Microbiology, Texas A&M University, College Station, USA

www.cabi.org

CABI is a trading name of CAB International

CABI Head Office
Nosworthy Way
Wallingford
Oxfordshire OX10 8DE
UK

Tel: +44(0)1491 832111
Fax: +44(0)1491 833508
E-mail: cabi@cabi.org
Website: www.cabi.org

CABI North American Office
875 Massachusetts Avenue
7th Floor
Cambridge, MA 02139
USA

Tel: +1 617 395 4056
Fax: +1 617 354 6875
E-mail: cabi-nao@cabi.org

© CAB International 2009. All rights reserved. No part of this publication may be reproduced in any form or by any means, electronically, mechanically, by photocopying, recording or otherwise, without the prior permission of the copyright owners.

A catalogue record for this book is available from the British Library, London, UK.

Library of Congress Cataloging-in-Publication Data

Root-knot nematodes / edited by Roland N. Perry, Maurice Moens, and James L. Starr.
p. cm.
Includes bibliographical references and index.
ISBN 978-1-84593-492-7 (alk. paper)
1 Root-knot nematodes. 2 Root-knot nematodes--Control. I Perry, R. N. (Roland N.) II Moens, Maurice. III Starr, James L. IV Title.

SB998.M45R655 2010
632'.6257--dc22

2009010917

ISBN-13: 978 1 84593 492 7

Typeset by SPi, Pondicherry, India.
Printed and bound in the UK by the MPG Books Group.

The paper used for the text pages in this book is FSC certified. The FSC (Forest Stewardship Council) is an international network to promote responsible management of the world's forests.

To Clare, Monique and Marylou with thanks for their unfailing tolerance, patience and support during the preparation of this book.

Contents

About the Editors **xix**
Contributors **xxi**
Preface **xxv**

1 *Meloidogyne* species – a Diverse Group of Novel and Important Plant Parasites **1**
Maurice Moens, Roland N. Perry and James L. Starr

1.1 Introduction 1
1.2 Impact 2
1.3 History of the Genus 2
1.4 Current Trends in Species Identification 2
1.5 Life Cycle 3
1.5.1 Incompatible host reactions 6
1.6 Diversity in Biology 6
1.6.1 Concept of host races 7
1.7 Major and Emerging Species 8
1.7.1 *Meloidogyne enterolobii* (= *Meloidogyne mayaguensis*) 9
1.7.2 *Meloidogyne paranaensis* 9
1.7.3 *Meloidogyne fallax* and *Meloidogyne chitwoodi* 10
1.7.4 *Meloidogyne minor* 11
1.8 Interactions with Other Plant Pathogens 11
1.9 Management and Control 11
1.10 Conclusions and Future Directions 12
1.11 References 13

2 General Morphology **18**
Jonathan D. Eisenback and David J. Hunt

2.1 General Morphology 18
2.1.1 Second-stage juvenile 19
2.1.2 Male 21
2.1.3 Female 21
2.1.4 Egg 22

2.2 Body Wall 23
2.2.1 Cuticle 23
2.2.2 Hypodermis 33
2.2.3 Somatic muscles 34
2.3 Nervous System 37
2.3.1 Cephalic sensory structures 39
2.3.2 Caudal sensory structures 40
2.4 Digestive System 40
2.4.1 Stoma and pharynx 40
2.4.2 Intestine 42
2.4.3 Rectum 42
2.5 Secretory–Excretory System 42
2.6 Reproductive System 43
2.6.1 Second-stage juvenile 43
2.6.2 Male 43
2.6.3 Female 44
2.7 Morphological Methods 44
2.8 Minimum Standards for Describing a New Species 44
2.8.1 The text 45
2.8.2 The figures 47
2.9 References 50

3 Taxonomy, Identification and Principal Species **55**
David J. Hunt and Zafar A. Handoo

3.1 Introduction 55
3.1.1 History 55
3.1.2 Major reference sources 60
3.1.3 Rate of species descriptions 60
3.1.4 Recent advances in characterization 61
3.2 Systematic Position 61
3.3 Subfamily and Genus Diagnosis 61
3.4 List of Species and Synonyms 63
3.5 Identification 67
3.5.1 General techniques 67
3.5.2 Perineal pattern 68
3.5.3 Root staining 68
3.5.4 Scanning electron microscopy 68
3.5.5 Diagnostic characters 68
3.5.6 Root-knot or cyst-forming nematode? 70
3.5.7 Differential host test 70
3.5.8 Gall form 70
3.5.9 Isozyme phenotyping 72
3.5.10 Molecular diagnostics 72
3.6 Principal Species 72
3.6.1 *Meloidogyne arenaria* 72
3.6.2 *Meloidogyne hapla* 74
3.6.3 *Meloidogyne incognita* 74
3.6.4 *Meloidogyne javanica* 74
3.6.5 *Meloidogyne acronea* 77
3.6.6 *Meloidogyne chitwoodi* 79
3.6.7 *Meloidogyne enterolobii* 80

3.6.8 *Meloidogyne ethiopica* 80
3.6.9 *Meloidogyne exigua* 80
3.6.10 *Meloidogyne fallax* 84
3.6.11 *Meloidogyne graminicola* 84
3.6.12 *Meloidogyne paranaensis* 84
3.7 Conclusions and Future Directions 88
3.8 Acknowledgements 88
3.9 References 88

4 Biochemical and Molecular Identification **98**
Vivian C. Blok and Thomas O. Powers

4.1 Introduction 98
4.2 Biochemical Methods 100
4.2.1 Isozymes 100
4.2.2 Antibodies 102
4.3 DNA-based Methods 103
4.3.1 DNA extraction 103
4.3.2 Restriction fragment length polymorphisms (RFLPs) 104
4.3.3 Satellite DNA probes and PCR 104
4.3.4 Ribosomal DNA PCR 105
4.3.5 Mitochondrial DNA 106
4.3.6 Sequence characterized amplified regions (SCARs) 108
4.3.7 Random amplified polymorphic DNA (RAPD) 109
4.3.8 Other PCR targets 110
4.3.9 Real-time PCR 110
4.3.10 Microarrays 110
4.4 Conclusions and Future Directions 111
4.5 Acknowledgements 112
4.6 References 112

5 Molecular Taxonomy and Phylogeny **119**
Byron J. Adams, Adler R. Dillman and Camille Finlinson

5.1 Introduction 119
5.2 The History of Reconstructing *Meloidogyne* Phylogenetic History 119
5.3 Molecular Phylogenetics: Genetic Markers and Evolutionary Relationships 120
5.3.1 Nuclear ribosomal DNA sequences 120
5.3.1.1 18S (small ribosomal subunit) 121
5.3.1.2 28S (large ribosomal subunit) 122
5.3.1.3 ITS (internally transcribed spacer region) 123
5.3.2 Orthologous nuclear genes 125
5.3.2.1 Dystrophin 125
5.3.2.2 Major sperm protein (msp) 125
5.3.2.3 Elongation factor 1-alpha (EF1-α) 125
5.3.2.4 RNA polymerase 2 126
5.3.3 Mitochondrial DNA 126
5.3.4 Phylogenomics 127
5.4 A *Meloidogyne* Supertree Analysis 127
5.5 Conclusions and Future Directions 130
5.6 References 135

6 Hatch and Host Location **139**
Rosane H.C. Curtis, A. Forest Robinson and Roland N. Perry

6.1 Introduction 139
6.2 Hatching 139
6.2.1 General hatching response 140
6.2.2 Hatching mechanism 140
6.2.3 Dependence on root exudates 142
6.2.4 Egg numbers and embryogenesis 143
6.3 Movement Through Soil 144
6.3.1 How root-knot juveniles move 144
6.3.2 Factors influencing rate of movement 144
6.3.3 Plant-independent factors influencing the direction of nematode movement 145
6.4 Host Location 147
6.4.1 General considerations 147
6.4.2 Heat 147
6.4.3 Soil gases 147
6.4.4 Uniquely plant-specific compounds 148
6.5 Nematode Changes and Responses at the Root–Soil Interface 149
6.5.1 Chemical communication at the root–soil interface 149
6.5.2 Perturbing chemosensory perception 152
6.5.3 Surface cuticle changes in response to environmental signals 153
6.6 Conclusions and Future Directions 155
6.7 References 155

7 Invasion, Feeding and Development **163**
Pierre Abad, Philippe Castagnone-Sereno, Marie-Noëlle Rosso, Janice de Almeida Engler and Bruno Favery

7.1 Introduction 163
7.2 Root-knot Nematode Life Cycle 164
7.3 Nematode Parasitism 165
7.4 Compatible Interactions with Resistant Plants: the Case of Virulent Root-knot Nematodes 167
7.5 (A)virulence Determinants and Pathogenicity Factors: Root-knot Nematode Effectors with Dual Function? 169
7.6 Tools for Molecular and Functional Analysis of Root-knot Nematode Parasitism 169
7.7 Giant Cell Development 170
7.8 Cytoskeleton Organization and Cell Cycle Progression During Giant Cell Ontogenesis 172
7.9 Extensive Cell Wall Modifications to Build Up Giant Cells 173
7.10 Suppression of Plant Defence Associated with Giant Cell Development 174
7.11 Major Reprogramming of Plant Metabolism and Transport 174
7.12 Comparison between *Meloidogyne* Parasitism and Symbiotic Rhizobia in *Medicago* 175
7.13 Conclusions and Future Directions 176
7.14 Acknowledgements 176
7.15 References 176

8 Reproduction, Physiology and Biochemistry **182**
David J. Chitwood and Roland N. Perry

8.1 Introduction 182
8.2 Reproduction and Moulting 182
8.2.1 Reproduction mechanisms and cytogenetics 182
8.2.1.1 Mode of reproduction 183
8.2.1.2 Sex ratios 183
8.2.1.3 Chromosome complement 183
8.2.1.4 Evolution of *Meloidogyne* species 186
8.2.1.5 Origin and evolution of parthenogenesis 186
8.2.2 Moulting 187
8.3 Physiology 188
8.3.1 Respiration 188
8.3.2 Effects of osmotic and ionic stress 188
8.3.3 Secretory–excretory products 189
8.4 Biochemistry 189
8.4.1 Enzymes 189
8.4.2 Other proteins 190
8.4.3 Amino acids and sugars 190
8.4.4 Neuropeptides 191
8.4.5 Complex carbohydrates and lipids 191
8.4.6 Steroids 191
8.5 Sensory Perception and Neurotransmission 192
8.5.1 Sensory perception 192
8.5.2 Neurotransmission 193
8.6 Conclusions and Future Directions 193
8.7 References 194

9 Survival Mechanisms **201**
Adrian A.F. Evans and Roland N. Perry

9.1 Introduction 201
9.2 Dormancy, Diapause and Quiescence 202
9.3 Embryonation and the Egg Mass Environment 202
9.3.1 The egg mass 202
9.3.2 The effect of soil moisture 203
9.3.3 The effect of soil aeration 204
9.3.4 Other roles for the egg mass 204
9.3.5 The egg mass and dormancy 205
9.4 Temperature Effects on Development of Eggs and Infective Stages 206
9.4.1 Temperature as an isolated factor 207
9.4.2 Low temperature survival 208
9.4.3 The influence of soil type and moisture content on temperature effects 209
9.4.4 A case study investigating factors affecting infectivity of *Meloidogyne javanica* J2 211
9.4.5 Overwintering of adult stages 212
9.4.6 Diapause in *Meloidogyne naasi* 212
9.4.7 A critique of de Guiran's use of 'diapause' as an explanation of late-emerging J2 213

9.5 The Effect of Osmotic Stress on Infective Stages in Soil 213
9.6 Survival Mechanisms Deployed: Life History Strategies in *Meloidogyne* Species 214
9.6.1 *Meloidogyne javanica* 215
9.6.2 *Meloidogyne arenaria* 216
9.6.3 *Meloidogyne incognita* 217
9.6.4 *Meloidogyne hapla* 217
9.7 Conclusions and Future Directions 218
9.8 References 219

10 Interactions with Other Pathogens **223**
Rosa H. Manzanilla-López and James L. Starr

10.1 Introduction 223
10.2 Interactions with Microbial Pathogens 226
10.2.1 Vascular wilt pathogens 226
10.2.2 Root-rot pathogens 228
10.2.3 More recently described disease complexes 230
10.3 Interactions with Other Plant-parasitic Nematodes 230
10.3.1 Interactions and parasitic habits 231
10.3.2 Sequential infections 234
10.3.3 Additive interaction 235
10.3.4 Competition 235
10.3.5 Interactions between *Meloidogyne* species 236
10.3.6 Effect on host 237
10.4 Basis for Interactions 238
10.5 Conclusions and Future Directions 239
10.6 References 240

11 Population Dynamics and Damage Levels **246**
Nicola Greco and Mauro Di Vito

11.1 Introduction 246
11.2 Patterns of Population Dynamics 246
11.3 Factors Affecting Population Dynamics 248
11.3.1 The nematode species 248
11.3.2 Crop and cropping system 249
11.3.3 The season 250
11.3.4 The soil 251
11.4 Modelling Population Dynamics 251
11.5 Damage Levels 253
11.6 Pattern of Nematode Damage to Crop Plants 254
11.7 Factors Affecting Nematode Damage 256
11.7.1 Nematode species and population level 256
11.7.2 Soil and environmental conditions 256
11.7.3 Crop and cropping system 259
11.8 Modelling Damage Levels 259
11.9 Implementing Experiments to Assess Nematode Dynamics and Crop Damage 260
11.9.1 Preparation and type of inoculum 261
11.9.2 Glasshouse experiments 263
11.9.3 Field experiments 263
11.9.4 Microplots 263
11.9.5 Maintenance of experiments 264
11.9.6 Fitting the models to data 266

11.10 Yield Loss Assessment 268
11.11 Importance of Information on Nematode Damage Levels and Dynamics in Management Strategies 268
11.12 Conclusions and Future Directions 269
11.13 Acknowledgements 269
11.14 References 269

12 Sampling Root-knot Nematodes **275**
Larry W. Duncan and Mark S. Phillips

12.1 Introduction 275
12.2 Nematode Spatial Patterns 276
12.3 Characterizing Sample Accuracy and Reliability 278
12.4 Sample Processing 281
12.5 Extracting Nematodes from Soil 281
12.6 Extracting Nematodes from Plant Material 282
12.7 Root Gall Indices 282
12.8 Other Plant Symptoms 283
12.9 Research to Optimize Sampling Programmes for Root-knot Nematodes 283
12.10 Examples of Results from Sampling Programmes 285
12.10.1 Surveys 285
12.10.2 Field experimentation 285
12.11 Conclusions and Future Directions 293
12.12 References 295

13 Mechanisms and Genetics of Resistance **301**
Valerie M. Williamson and Philip A. Roberts

13.1 Introduction 301
13.2 Sources and Inheritance of Root-knot Nematode Resistance 302
13.3 Mechanisms of Resistance to Pathogens in Plants 307
13.4 Structure and Function of the Nematode Resistance Gene *Mi-1* 308
13.5 What is Known About Other Nematode R-Genes 310
13.6 Nematode Virulence and Durability of Resistance 311
13.7 Management of Resistance and Virulence in the Field 315
13.8 Conclusions and Future Directions 317
13.9 References 319

14 Development of Resistant Varieties **326**
James L. Starr and Chris F. Mercer

14.1 Introduction – the Plus Side of Resistance 326
14.2 Introduction – a Look at the Other Side 326
14.3 Successful Use of Resistance – Room for Wider Deployment 327
14.4 Planning a Resistance-breeding Programme 328
14.4.1 Identification of the root-knot nematode species present 329
14.4.2 Establishing pure cultures 329
14.4.3 Nematode variability 329
14.4.4 Screening methods 329
14.4.5 Sources of resistance 330
14.4.6 Mass selection 331
14.4.7 Recurrent selection 331
14.5 Screening Methods, Including Marker-assisted Selection 331
14.6 Quality of Candidate Resistant Material 333

14.7 Engineered Resistance 334
14.8 Conclusions and Future Directions 335
14.9 References 335

15 Plant Biotechnology and Control **338**
Howard J. Atkinson, Peter E. Urwin and Richard S. Hussey

15.1 Introduction 338
15.2 Proteinase Inhibitors 339
15.3 Cry Proteins of *Bacillus thuringiensis* as Biopesticides 340
15.3.1 Cry proteins 340
15.3.2 Activity of Cry proteins against nematodes 340
15.3.3 Activity of Cry6A against *Meloidogyne incognita* 341
15.3.4 Resistance to Cry proteins in nematodes 341
15.4 *In planta* RNAi to Target Plant-parasitic Nematodes 342
15.5 Repellents 345
15.6 The *Mi-1*-mediated Resistance Response 346
15.7 Efficacy and Durability 347
15.7.1 Efficacy 347
15.7.2 Durability 348
15.8 Promoters for Transgenic Control of *Meloidogyne* 349
15.9 Biosafety 349
15.9.1 Food 349
15.9.2 Environment 350
15.10 Developing World Needs 352
15.10.1 The need for biotechnology to control *Meloidogyne* in the developing world 352
15.10.2 Appropriate technology 353
15.11 Conclusions and Future Directions 353
15.11.1 Proteinase inhibitors 353
15.11.2 Cry proteins 354
15.11.3 RNAi 354
15.11.4 Commercial prospects of deployment of transgenic resistance to *Meloidogyne* 354
15.11.5 Prospects of uptake in support of food security 355
15.11.6 Rate of uptake possible 355
15.12 References 356

16 The Complete Sequence of the Genomes of *Meloidogyne incognita* and *Meloidogyne hapla* **363**
Pierre Abad and Charles H. Opperman

16.1 Introduction 363
16.2 *Meloidogyne incognita* Genome 364
16.2.1 A genome constituted by pairs of homologous but divergent segments 365
16.2.2 The gene content of a plant-parasitic nematode 367
16.2.3 Identifying plant parasitism genes 368
16.2.4 A nematode adapted to a privileged plant host environment 370
16.2.5 Does the *Caenorhabditis elegans* genome reflect nematode lifestyle diversity? 370
16.2.6 Exploration of new anti-parasitic drug targets 372

16.3 *Meloidogyne hapla* Genome 372
16.3.1 General characterization of the genome 372
16.3.2 Estimation of gene numbers 373
16.3.3 Gene families 373
16.3.4 Genome organization 373
16.3.5 Pathway conservation with free-living nematodes 374
16.4 Conclusions and Future Directions 375
16.5 Acknowledgements 376
16.6 References 376

17 Biological Control Using Microbial Pathogens, Endophytes and Antagonists **380**
Johannes Hallmann, Keith G. Davies and Richard Sikora

17.1 Introduction 380
17.2 Bacterial Pathogens and Antagonists 381
17.2.1 Endoparasitic bacteria 381
17.2.1.1 *Pasteuria penetrans* 382
17.2.1.2 Mass production, *in vivo* and *in vitro* culturing methods 382
17.2.1.3 Quantification, nematode suppression and the problem of host specificity 383
17.2.1.4 Mechanism of endospore attachment 384
17.2.1.5 Potential for root-knot control 385
17.2.2 Rhizosphere bacteria 385
17.2.3 Endophytic bacteria 387
17.2.4 Other bacteria 387
17.3 Fungal Pathogens and Antagonists 389
17.3.1 Nematophagous fungi 389
17.3.1.1 Predacious fungi 392
17.3.2 Saprophagous fungi 393
17.3.3 Endophytic fungi 394
17.4 Commercialization and Future Directions 395
17.4.1 Commercial products 395
17.4.2 The development of a commercial product 398
17.4.3 Potential markets 398
17.4.4 Enhancement strategies 399
17.4.5 Transgenic approaches 400
17.4.6 Future prospects 400
17.5 References 401

18 Current and Future Management Strategies in Intensive Crop Production Systems **412**
Andrew P. Nyczepir and Stephen H. Thomas

18.1 Introduction 412
18.2 Current Control Practices 413
18.2.1 Chemical control 413
18.2.1.1 Fumigant nematicides 414
18.2.1.2 Non-fumigant nematicides 415
18.2.1.3 Other compounds 416
18.2.2 Cultural control 416
18.2.2.1 Crop and fallow rotation 417

18.2.2.2 Trap crops, cover crops and soil amendments 418
18.2.2.3 Exploitation of phenology 420
18.2.2.4 Sanitation 420
18.2.2.5 Steam heat and solarization 421
18.2.3 Biological control and host plant resistance 422
18.3 Current Management Practices 423
18.3.1 Significance of diagnostic sampling and government regulation 424
18.3.2 Implementation of management strategies 425
18.4 Future Opportunities and Challenges 428
18.4.1 Emerging control options 429
18.4.1.1 Chemical control 429
18.4.1.2 Cultural control 430
18.4.1.3 Technological advances 431
18.4.2 Emerging management options 432
18.4.2.1 Natural resource availability 432
18.4.2.2 Knowledge gaps 432
18.4.2.3 Alternatives to methyl bromide 434
18.5 Conclusions and Future Directions 435
18.6 References 435

19 Current and Future Management Strategies in Resource-poor Farming **444**
Danny L. Coyne, Hendrika H. Fourie and Maurice Moens

19.1 Introduction and Definitions 444
19.2 Options 446
19.3 Correct Diagnosis 446
19.4 Prevention 447
19.4.1 Healthy planting material 447
19.4.2 Seed and seedling supply 448
19.4.3 Heat treatment 448
19.4.4 Tissue culture 450
19.4.5 Quarantine 450
19.5 Cultural Control 451
19.5.1 Removal of infected material 451
19.5.2 Planting date 451
19.5.3 Flooding 451
19.5.4 Mulching and soil amendments 452
19.5.5 Physical methods 453
19.6 Cropping Systems 453
19.6.1 Rotation 453
19.6.2 Fallow 455
19.6.3 Cover crops (improved fallow) 455
19.6.4 Antagonistic or trap crops 456
19.7 Resistance 458
19.8 Biological Control 459
19.9 Chemical Control 459
19.9.1 Past and current nematicide use 460
19.9.2 Bionematicides 461
19.9.2.1 Avermectins 462
19.9.2.2 Neem products 462
19.9.2.3 Glucosinolates in *Brassica* spp. 463

19.9.2.4 Polythienyls in *Tagetes* spp. 463
19.9.2.5 Ricin in *Ricinus communis* 464
19.9.2.6 Velvet bean compounds 464
19.9.2.7 Monocrotaline in *Crotalaria* spp. 464
19.9.2.8 Glucoside in cassava 464
19.9.2.9 Other sources of phytochemicals with nematicidal properties 464
19.10 Conclusions and Future Directions 465
19.11 References 466

Gene Index **477**
Nematode Genus and Species Index **479**
General Index **483**

The colour plate section can be found following p. 262.

The editors, Roland N. Perry (centre), Maurice Moens (left) and James L. Starr (right)

About the Editors

Roland N. Perry Professor Roland Perry is based at Rothamsted Research, UK. He graduated with a BSc (Hons) in Zoology from Newcastle University, UK, where he also obtained a PhD in Zoology on physiological aspects of desiccation survival in *Ditylenchus* spp. After a year's postdoctoral research at Newcastle, he moved to Keele University, UK, for 3 years, where he taught parasitology. He was appointed to Rothamsted Research in 1976. His research interests have centred primarily on plant-parasitic nematodes, especially focusing on nematode hatching, sensory perception, behaviour and survival physiology, and several of his past PhD and postdoctoral students are currently involved in nematology research.

He co-edited *The Physiology and Biochemistry of Free-living and Plant-parasitic Nematodes* (1997) and the textbook *Plant Nematology* (2006). He is author or co-author of over 40 book chapters and refereed reviews, and over 100 refereed research papers. He is co-editor-in-chief of *Nematology* and chief editor of the *Russian Journal of Nematology*. He co-edits the book series *Nematology Monographs and Perspectives*. In 2001, he was elected Fellow of the Society of Nematologists (USA) in recognition of his research achievements, and in 2008 he was elected Fellow of the European Society of Nematologists for outstanding contribution to the science of nematology. He is a visiting professor at Ghent University, Belgium, where he lectures on nematode biology, and also gives regular lectures to the MSc course at Imperial College, London.

Maurice Moens Professor Maurice Moens is Director of Research at the Institute for Agriculture and Fisheries Research (ILVO) at Merelbeke, Belgium and part-time professor at Ghent University, Belgium, where he gives a lecture course on agro-nematology at the Faculty of Bioscience Engineering. He is also director of the Postgraduate International Nematology Course (MSc Nematology), where he gives three lecture courses on plant nematology. The MSc course is organized in the Faculty of Sciences of Ghent University. He graduated as an agricultural engineer from Ghent University and obtained a PhD at the same university on the spread of plant-parasitic nematodes and their management in hydroponic cropping systems. Within the framework of the Belgian Cooperation, he worked from 1972 to 1985 as a researcher in crop protection, including nematology, at two research stations in Tunisia. Upon his return to Belgium, he was appointed as senior nematologist at the Agricultural Research Centre (now ILVO). There, he expanded the research in plant nematology over various areas, covering molecular characterization, biology of host–parasite relationships, biological control, resistance and other forms of non-chemical control. He was appointed head of the Crop Protection Department in 2000 and moved to his present position as Director of Research in 2006.

He supervised 15 PhD students, who are now active in nematology all over the world. He is a partner in research projects being executed in Europe and several developing countries. He is co-author of the text book *Plant Nematology* (2006) and is author or co-author of over 100 refereed research papers. He is a member of the editorial boards of *Nematology*, *Annals of Applied Biology* and the *Russian Journal of Nematology*.

James L. Starr Professor James Starr is a professor in the Department of Plant Pathology and Microbiology at Texas A&M University, USA. He obtained a BS, an MS and a PhD in Plant Pathology from Cornell University. After completing his PhD, he moved to North Carolina State University, where he was successively Research Associate in the Department of Plant Pathology and Chief of the Nematode Advisory Section. He moved to Texas A&M University in 1981, where his main research interests are the identification, characterization and deployment of resistance to nematodes in important crop species.

In 2003 he was elected Fellow of the Society of Nematologists (USA) in recognition of his research achievements, and in 2006 he was elected Fellow of the American Phytopathological Society. He served as editor-in-chief of the *Journal of Nematology* (2003–2006) and was President of the Society of Nematologists (1996–1997). He has published over 80 refereed scientific papers, overseen the release of five germplasm lines and two cultivars resistant to nematodes, and co-edited two books, *Plant Resistance to Parasitic Nematodes* (2002) and *Plant Nematodes of Agricultural Importance* (2007).

Contributors

Pierre Abad INRA, UMR 1301, Interactions Biotiques et Santé Végétale, 400 route des Chappes, F-06903 Sophia Antipolis, France (e-mail: abad@sophia.inra.fr)

Byron J. Adams Microbiology & Molecular Biology Department, and Evolutionary Ecology Laboratory, Brigham Young University, Provo, UT 84602-5253, USA (e-mail: bjadams@byu.edu)

Howard J. Atkinson Centre for Plant Sciences, Faculty of Biological Sciences, University of Leeds, Leeds, West Yorkshire LS2 9JT, UK (e-mail: H.J.Atkinson@leeds.ac.uk)

Vivian C. Blok Plant Pathology Programme, Scottish Crop Research Institute, Invergowrie, Dundee, DD2 5DA, UK (e-mail: Vivian.Blok@scri.ac.uk)

Philippe Castagnone-Sereno INRA, UMR 1301, Interactions Biotiques et Santé Végétale, 400 route des Chappes, F-06903 Sophia Antipolis, France (e-mail: pca@sophia.inra.fr)

David J. Chitwood USDA, ARS, Nematology Laboratory, Bldg 011A, Room 165-B, BARC-West, Beltsville, MD 20705, USA (e-mail: David.Chitwood@ars.usda.gov)

Danny L. Coyne International Institute of Tropical Agriculture (IITA), c/o Lambourn & Co., Carolyn House, 26 Dingwall Road, Croydon CR9 3EE, UK (e-mail: d.coyne@iita.org)

Rosane H.C. Curtis Plant Pathology and Microbiology Department, Rothamsted Research, Harpenden, Hertfordshire AL5 2JQ, UK (e-mail: rosane.curtis@bbsrc.ac.uk)

Keith G. Davies Plant Pathology and Microbiology Department, Rothamsted Research, Harpenden, Hertfordshire AL5 2JQ, UK (e-mail: keith.davies@bbsrc.ac.uk)

Janice de Almeida Engler INRA, UMR 1301, Interactions Biotiques et Santé Végétale, 400 route des Chappes, F-06903 Sophia Antipolis, France (e-mail: Janice.Almeida-Engler@sophia.inra.fr)

Adler R. Dillman Division of Biology, California Institute of Technology, Pasadena, CA 91125, USA (e-mail: adlerd@caltech.edu)

Mauro Di Vito Consiglio Nazionale delle Ricerche, Istituto per la Protezione delle Piante, Via G. Amendola 122-D, 70126 Bari, Italy (e-mail: m.divito@ba.ipp.cnr.it)

Larry W. Duncan University of Florida, Institute of Food and Agricultural Sciences, Citrus Research and Education Center, 700 Experiment Station Road, Lake Alfred, FL 33850, USA (e-mail: lwdn@crec.ifas.ufl.edu)

Jonathan D. Eisenback Department of Plant Pathology, Physiology and Weed Science, 103 Price Hall, Virginia Polytechnic Institute and State University, Blacksburg, VA 24061, USA (e-mail: jon@vt.edu)

Adrian A.F. Evans Silwood Park Campus, Division of Biology, Imperial College London, Sunninghill, Ascot, Berkshire SL5 7PY, UK (e-mail: a.a.f.evans@imperial.ac.uk)

Bruno Favery INRA, UMR 1301, Interactions Biotiques et Santé Végétale, 400 route des Chappes, F-06903 Sophia Antipolis, France (e-mail: favery@sophia.inra.fr)

Camille Finlinson Department of Biology, and Evolutionary Ecology Laboratories, Brigham Young University, Provo, UT 84602-5181, USA (e-mail: finli6@gmail.com)

Hendrika H. Fourie ARC-Grain Crops Institute, Private Bag X1251, Botha Avenue 114, Potchefstroom 2520, South Africa (e-mail: FourieHD@arc.agric.za)

Nicola Greco Consiglio Nazionale delle Ricerche, Istituto per la Protezione delle Piante, Via G. Amendola 122-D, 70126 Bari, Italy (e-mail: n.greco@ba.ipp.cnr.it)

Johannes Hallmann Julius Kühn-Institut, Bundesforschungsinstitut für Kulturpflanzen, Toppheideweg 88, D-48161 Münster, Germany (e-mail: johannes.hallmann@jki.bund.de)

Zafar A. Handoo USDA, ARS, Nematology Laboratory, Bldg 011A, Room 159, BARC-West, Beltsville, MD 20705, USA (e-mail: Zafar.Handoo@ars.usda.gov)

David J. Hunt CABI Europe-UK, Bakeham Lane, Egham, Surrey TW20 9TY, UK (e-mail: d.hunt@cabi.org)

Richard S. Hussey Department of Plant Pathology, University of Georgia, Athens, GA 30602-7274, USA (e-mail: hussey@uga.edu)

Rosa H. Manzanilla-López Plant Pathology and Microbiology Department, Rothamsted Research, Harpenden, Hertfordshire AL5 2JQ, UK (e-mail: rosa.manzanilla-lopez@bbsrc.ac.uk)

Chris F. Mercer AgResearch Grasslands, Private Bag 11-008, Palmerston North, New Zealand (e-mail: chris.mercer@agresearch.co.nz)

Maurice Moens Institute for Agriculture and Fisheries Research, Burg. Van Gansberghelaan 96, 9820 Merelbeke, Belgium (e-mail: maurice.moens@ilvo.vlaanderen.be)

Andrew P. Nyczepir USDA, ARS, Southeastern Fruit and Tree Nut Research Laboratory, Byron, GA 31008, USA (e-mail: Andy.Nyczepir@ars.usda.gov)

Charles H. Opperman Center for the Biology of Nematode Parasitism, Department of Plant Pathology, North Carolina State University, Raleigh, NC 27695-7253, USA (e-mail: warthog@ncsu.edu)

Roland N. Perry Plant Pathology and Microbiology Department, Rothamsted Research, Harpenden, Hertfordshire AL5 2JQ, UK (e-mail: roland.perry@bbsrc.ac.uk)

Mark S. Phillips Plant Pathology Programme, Scottish Crop Research Institute, Invergowrie, Dundee, DD2 5DA, UK (e-mail: Mark.Phillips@scri.ac.uk)

Thomas O. Powers Department of Plant Pathology, University of Nebraska, Lincoln, NE 68583-0722 USA (e-mail: tpowers@unlnotes.unl.edu)

Philip A. Roberts Department of Nematology, University of California, Riverside, CA 92521-0415, USA (e-mail: philip.roberts@ucr.edu)

A. Forest Robinson USDA, ARS, Southern Crops Research Laboratory, 2765 F&B Road, College Station, TX 77845, USA (e-mail: robinson.forest@gmail.com)

Marie-Noëlle Rosso INRA, UMR 1301, Interactions Biotiques et Santé Végétale, 400 route des Chappes, F-06903 Sophia Antipolis, France (e-mail: rosso@sophia.inra.fr)

Richard Sikora INRES-Phytomedicine, Soil-Ecosystem Phytopathology & Nematology, University of Bonn, Nussallee 9, D-53115 Bonn, Germany (e-mail: rsikora@uni-bonn.de)

James L. Starr Department of Plant Pathology and Microbiology, Texas A&M University, College Station, TX 77843-2132, USA (e-mail: j-starr@tamu.edu)

Stephen H. Thomas Department of Entomology, Plant Pathology, and Weed Science, Box 30003, MSC 3BE, New Mexico State University, Las Cruces, NM 88003, USA (e-mail: stthomas@nmsu.edu)

Peter E. Urwin Centre for Plant Sciences, Faculty of Biological Sciences, University of Leeds, Leeds, West Yorkshire LS2 9JT, UK (e-mail: P.E.Urwin@leeds.ac.uk)

Valerie M. Williamson Department of Nematology, University of California, Davis, CA 95616, USA (e-mail: vmwilliamson@ucdavis.edu)

Preface

Among plant-parasitic nematodes, the root-knot nematodes (*Meloidogyne* spp.) are the most economically important group. They have a worldwide distribution and are obligate parasites of the roots, tubers and corms of thousands of plant species, resulting in devastating adverse effects on the quality and yield of crops. They are especially prevalent in the warm temperate and tropical regions where subsistence agricultural systems predominate. Thus, root-knot nematodes have major financial and social impacts and have been the subject of extensive research, including studies on taxonomy, biology, plant–nematode interactions and, especially, control approaches. This research, undertaken over many years, has generated an enormous volume of literature. A summary of the research is needed and, while it is impossible to reference all relevant literature, it is important to condense and focus on the main findings. This is the first aim of the present volume.

The second aim is to reflect the exciting and recent advances in the molecular genetics of root-knot nematodes. With the elucidation in 2008 of the genomic sequences of *M. hapla* and *M. incognita*, it will be possible to perform comparative genomic studies with free-living and animal-parasitic nematode genomes. This will provide not only an insight into aspects of the development and the features of obligate parasitism but also a basis for in-depth analysis of the characteristics implicit in a life strategy that has to cope with the vagaries of the parasitic existence. A comparison of these two Meloidogyne species will also provide information on the differences between an organism reproducing in an asexual manner compared with one reproducing sexually.

A third aim of this book is to highlight the control options and management strategies, especially in the light of an understanding of the biology of the genus as a whole and of the important differences between individual species. The changing control scenarios reflect reduction in the use of chemical control strategies and the concomitant increasing importance of biological management and resistance mechanisms. The increasing relevance of plant biotechnology and other management options will be central to the future control of *Meloidogyne* species. Global warming is likely to result in increased spread of tropical species to regions and crops that hitherto have been unaffected, so targeting of environmentally acceptable control strategies based on a sound knowledge of nematode biology is essential.

The three aims of this book are overall objectives that have to be allied to an understanding of the morphology and identification of the various species of *Meloidogyne*. The last books on this genus were published in 1985 as a two-volume set entitled *An Advanced Treatise on* Meloidogyne. Since then there have been many new species described and major advances in our understanding of the host–parasite interaction. These advances are reflected in the chapters in the present book, while also citing important earlier publications. The second volume of the 1985 publications dealt with methodology. This aspect is not included in the present volume as there have been several books on relevant techniques, and the methodology volume contains protocols that are still in use today.

We are grateful to all the authors of the chapters for their time and effort in compiling their contributions. We are especially indebted to all the scientists who have contributed to the knowledge that this book attempts to summarize. In addition, we wish to thank David Hunt (CABI Bioscience, UK), John Jones (Scottish Crop Research Institute, UK) and Brian Kerry (Rothamsted Research, UK) for their advice and comments on various chapters. The support of Syngenta in funding the colour plates is also gratefully acknowledged.

Editing this book has been an enjoyable challenge, helped greatly by the convivial atmosphere during the undertaking.

Roland N. Perry, Maurice Moens and James L. Starr
January 2009

1 *Meloidogyne* Species – a Diverse Group of Novel and Important Plant Parasites

Maurice Moens[1], Roland N. Perry[2] and James L. Starr[3]
[1]*Institute for Agricultural and Fisheries Research, Merelbeke, Belgium and Ghent University, Ghent, Belgium;* [2]*Rothamsted Research, Harpenden, Hertfordshire, UK and Ghent University, Ghent, Belgium;* [3]*Texas A&M University, College Station, Texas, USA*

1.1	**Introduction**	**1**
1.2	**Impact**	**2**
1.3	**History of the Genus**	**2**
1.4	**Current Trends in Species Identification**	**2**
1.5	**Life Cycle**	**3**
1.6	**Diversity in Biology**	**6**
1.7	**Major and Emerging Species**	**8**
1.8	**Interactions with Other Plant Pathogens**	**11**
1.9	**Management and Control**	**11**
1.10	**Conclusions and Future Directions**	**12**
1.11	**References**	**13**

1.1 Introduction

Root-knot nematodes are members of the genus *Meloidogyne* (Göldi, 1892), *Meloidogyne* is of Greek origin and means 'apple-shaped female'. They are an economically important polyphagous group of highly adapted obligate plant parasites, are distributed worldwide and parasitize nearly every species of higher plant. Typically they reproduce and feed on modified living plant cells within plant roots, where they induce small to large galls or root-knots, hence their vernacular name. The above-ground symptoms are not readily apparent and may be similar to those produced on any plants having a damaged and malfunctioning root system. Hosts may be heavily infected without showing external symptoms on the harvested products, e.g. symptomless potato tubers. The rapid rate of development and reproduction on good hosts results, in the majority of species, in several generations during one cropping season, leading to severe crop damage. Damage may consist of various degrees of stunting, lack of vigour, and wilting under moisture stress. Secondary infection by other pathogens often results in extensive decay of nematode-infected tissues. The common explanation for these above-ground symptoms is that *Meloidogyne* infection affects water and nutrient uptake and upward translocation by the root system. By disrupting the host plant physiology, root-knot nematodes may not only reduce crop yield but also product quality

©CAB International 2009.

(e.g. of potatoes and carrots) and therefore are of great economic and social importance (see Greco and Di Vito, Chapter 11, this volume).

1.2 Impact

Damage and yield losses caused by plant pathogens, including plant-parasitic nematodes, are, on average, greater in tropical than in temperate regions because of greater pathogen diversity, more favourable environmental conditions for pathogen colonization, development, reproduction and dispersal, and lack of human, technical and financial resources to combat infections (De Waele and Elsen, 2007). Severity of damage caused by *Meloidogyne* can be species-specific and also vary by host, crop rotation, season and soil type (Greco *et al.*, 1992; Potter and Olthof, 1993). Similarly, economic thresholds vary, primarily depending on these same factors. Damage thresholds have been established for several crops, where the average is approximately 0.5–2 juveniles/g of soil (or from the lower limits of detection, over 1000 individuals/500 cm^3 of soil; see Greco and Di Vito, Chapter 11, this volume). In addition to direct costs, root-knot nematodes cause indirect costs because of the quarantine status of some species of *Meloidogyne* in several countries or regions. For example, *M. chitwoodi* is increasingly regulated because it is a serious pest of potato and other economically important crops such as carrot, and the known geographical distribution is still relatively limited. It is on the list of prohibited pests of many countries (Canada, the EU, Mexico and other countries in Latin America, and the Far East) (Hockland *et al.*, 2006). *Meloidogyne fallax*, another pest of potato, is recognized as an important pest by fewer countries (e.g. the EU), although it poses a similar economic risk. In the future, additional species of root-knot nematodes that might be added to the list of quarantine species include recently described species such as *M. minor* in Europe and *M. citri* in the USA.

1.3 History of the Genus

A more detailed account of the background literature, progress in descriptions and classification, and authorities for the species of the genus is given in Hunt and Handoo, Chapter 3, this volume; a brief summary is included here as part of the introduction to the genus and its basic biology. The first illustrated report of root-knot disease appeared during the middle of the 19th century when the clergyman Miles Joseph Berkeley (1855) first attributed galls detected on glasshouse cucumber roots to nematodes. The first description of a root-knot nematode was made by Cornu (1879); it was based on nematodes found in root galls that were detected on sainfoin (*Onobrychis sativus* Lam.) in the Loire valley, France. In 1887, Göldi briefly described and illustrated a root-knot nematode from coffee plants in Brazil and named it *M. exigua*. Although the 1887 publication was an advance copy or preprint of the full article subsequently published by Göldi in 1892, the 1887 article meets the requirements to establish the actual publication date for the genus and type species as 1887 (see Hunt and Handoo, Chapter 3, this volume for a full account of this decision).

The name *Heterodera marioni* was widely used for root-knot nematodes until 1949, when Chitwood removed the root-knot nematodes from the genus *Heterodera* because they differed from cyst nematodes. Since the oldest name for the genus was Göldi's *Meloidogyne*, that name had precedence. Chitwood redescribed *M. arenaria*, *M. exigua*, *M. incognita* and *M. javanica*, and described *M. hapla* and a variety of *M. incognita* he termed *M. incognita* var. *acrita*. The species were separated from each other on the basis of perineal pattern morphology, stylet knob shape, and length of stylet and dorsal gland orifice. Since Chitwood's publication, many more *Meloidogyne* species have been described (see Hunt and Handoo, Chapter 3, this volume, for a full list). Species descriptions gradually included increasing numbers of features (observed by light microscopy and/or electron microscopy) of females, males and second-stage juveniles (J2) (see Jepson, 1987, for a review, and Eisenback and Hunt, Chapter 2, this volume).

1.4 Current Trends in Species Identification

Research on cytogenetics, isozymes and the genome of root-knot nematodes, mostly since the

1970s, provided further evidence for the large diversity of species within the genus. Importantly, several new technologies have provided tools to assist in species identification. One of the most important has been the use of isozyme phenotypes, using PAGE of crude protein extracts and histochemical stains for non-specific esterases, superoxide dismutase, malate dehydrogenase, and glutamate oxaloacetate transaminase (Esbenshade and Triantaphyllou, 1985), with the esterases and malate dehydrogenase being the most useful for discriminating the four common *Meloidogyne* species: *M. arenaria*, *M. hapla*, *M. incognita* and *M. javanica*. These enzymes have also been used successfully with less common species such as *M. enterolobii* (= *M. mayaguensis*) (Brito *et al.*, 2004), *M. partityla* (Starr *et al.*, 1996), and *M. trifoliophila* (Mercer *et al.*, 1997). With the appropriate equipment it is possible to make the species identification on individual females. Ibrahim and Perry (1992) showed that PAGE combined with staining for esterases could be used with *Meloidogyne* females in galled roots, thus obviating the need to separate nematodes from the host tissue. It remains to be determined if these procedures will be able to discriminate all of the currently recognized species of this genus.

More recently, much effort has been given to the development of species-specific markers using variation in the genomes of the different species coupled with PCR techniques (see Blok and Powers, Chapter 4, this volume). There are now many reports that describe protocols for distinguishing numerous species using species-specific primers derived from the nuclear genome (Adam *et al.*, 2007) and the mitochondrial genome (Powers and Harris, 1993). A major benefit of these systems is that they can be used on an individual J2, the most common stage of the nematode found in the soil, thus eliminating the need to use mature females from roots of field samples (which is not possible if the host of interest is not actually growing at the time of sample collection) or to establish a culture in the glasshouse and wait for the development of the females. At present, there are no primer pairs available for many species, and multiple assays with different primer pairs are often needed to complete the identification. For some species the PCR amplification product must be digested with a restriction enzyme to complete the analysis (Powers and Harris, 1993). Finally, this approach requires some expensive equipment (the PCR thermocycler) and supplies, and a clean laboratory environment. None the less, these approaches are likely to play an ever-increasing role in species identification (especially by those not well trained or experienced in classical morphometrics) and will be used extensively by regulatory agencies.

1.5 Life Cycle

The root-knot nematode life cycle is summarized in Fig. 1.1 and Plate 36. Females lay eggs into gelatinous masses composed of a glycoprotein matrix, which is produced by rectal glands in the female, keeps the eggs together and protects them against environmental extremes and predation. The egg masses are usually found on the surface of galled roots, although they may also be embedded within the gall tissue. The egg mass is initially soft, sticky and hyaline but becomes firmer and dark brown with age. Surprisingly, there has been only limited analysis of the glycoproteins (Sharon and Spiegel, 1993) or other components of the gelatinous matrix, despite its obvious importance. In addition to providing some protection to the eggs from environmental extremes, it has been demonstrated that the matrix has antimicrobial properties (Orion and Kritzman, 1991).

Within the egg, embryogenesis proceeds to the first-stage juvenile, which moults to the infective J2. Hatch of the J2 is primarily dependent on temperature and sufficient moisture, although other factors, including root diffusate and generation, modify the hatching response (see Curtis *et al.*, Chapter 6, this volume) so that the J2 hatch when conditions are favourable for movement and host location. The ability of *Meloidogyne* to survive is enhanced by several physiological and biochemical adaptations, including delayed embryogenesis, quiescence and diapause, and lipid reserves that prolong viability until the J2 reaches and invades a host; these aspects are discussed in detail by Evans and Perry, Chapter 9, this volume. In the soil, the J2 is vulnerable and needs to locate a host as rapidly as possible. J2 are attracted to roots, and there is evidence that when both resistant and susceptible plant roots are present the susceptible ones are

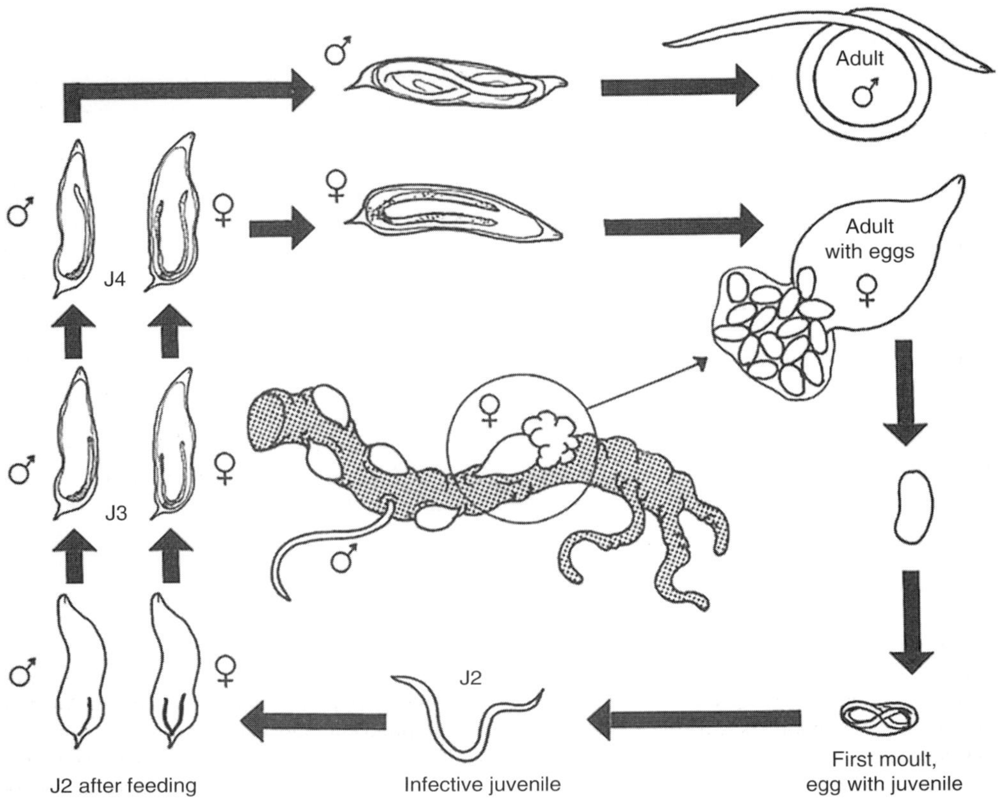

Fig. 1.1. Diagram of the life cycle of the root-knot nematode, *Meloidogyne*. J2: second-stage juvenile; J3: third-stage juvenile; J4: fourth-stage juvenile. (Adapted from Karssen and Moens, 2006.)

more attractive (see Curtis *et al.*, Chapter 6, this volume).

The invasive J2 commences feeding after it has invaded the root, usually behind the root tip, and moved through the root to initiate and develop a permanent feeding site. The feeding of the J2 on protoxylem and protophloem cells induces these cells to differentiate into specialized nurse cells, which are called giant cells. Once a giant cell is initiated, the nematode becomes sedentary and enlarges greatly to assume a 'sausage' shape. Under favourable conditions, the J2 stage moults to the third-stage juvenile (J3) after *c.*14 days, then to the fourth-stage juvenile (J4), and finally to the adult stage. The combined time for the J3 and J4 stages is much shorter than for the J2 or the adult, typically 4–6 days. J3 and J4 lack a functional stylet and do not feed. Males, when present, are vermiform and there is no evidence that they feed. Males may be found in parthenogenetic species when conditions are unfavourable for female development, such as when population densities are very high and presumably there is a limitation of food supply.

The initiation, development and maintenance of the giant cell is the subject of continuing investigation, facilitated by molecular techniques with the impetus of developing novel control strategies based on preventing giant cell formation or, more likely, development. These specialized feeding sites are remarkable for their complexity. They are greatly enlarged from typical phloem and xylem parenchyma, or cortical cells, with final cell volumes nearly 100-fold greater than normal root cells. The giant cells are functionally similar to syncytia induced by other plant-parasitic nematodes that have sedentary adult females, but are distinct in their development. Like syncytia, they are functional transfer cells, based on morphology (Jones and

Northcote, 1972) and because photosynthates pass through the giant cells before being ingested by the nematodes (Bird and Loveys, 1975). Unlike syncytia, each giant cell develops from a single initial cell rather than by coalescence of several adjacent cells. The giant cells are not only multinucleate, containing as many as 80 nuclei each, but individual nuclei within each giant cell are polyploid, some with *c.* eight-fold increase in chromosome number (Huang and Maggenti, 1969; Wiggers *et al.*, 1990). Thus, each giant cell may have up to a 600-fold increase in copy number of each plant gene. Several studies have documented the effects of nematode infection on gene expression, with a variety of genes being upregulated (Gheysen and Jones, 2006; Schaff *et al.*, 2007) and probably a greater number downregulated (Schaff *et al.*, 2007). A few studies on gene expression in the giant cells have reported that the mRNA for some genes can be present in giant cells at levels that are many fold greater than in non-infected root cells (Ramsey *et al.*, 2004; He *et al.*, 2005). These data demonstrate that the giant cells induced by all *Meloidogyne* spp. are unique examples of how parasites can affect normal host development. Most recently (Huang *et al.*, 2006), some progress has been made in the characterization of parasitism genes in the nematode that enable it fundamentally to alter plant growth and development for the benefit of the parasite (see Abad *et al.*, Chapter 7, this volume).

On the bases of cytogenetic studies on about 600 populations (representing 24 species) and in collaboration with the International *Meloidogyne* Project, Triantaphyllou (1985) was able to demonstrate that root-knot nematodes have undergone extensive cytogenetic diversification, probably unparalleled by that of any other animal group. Triantaphyllou concluded that characteristic features are the establishment of meiotic and mitotic parthenogenesis in association with various degrees of polyploidy and aneuploidy. Obligatory cross-fertilization also occurs in some diploid and polyploid forms (e.g. *M. kikuyensis* and *M. megatyla*), whereas facultative meiotic (automixis) (e.g. *M. exigua*, *M. chitwoodi* and *M. graminicola*) and obligatory mitotic parthenogenesis (apomixis) (e.g. *M. incognita*, *M. enterolobii* and *M. oryzae*) prevail in most polyploid and aneuploid forms (see Chitwood and Perry, Chapter 8, this volume).

The trend from amphimictic reproduction to apomixis is generally associated with shorter life cycles, higher reproductive rates and increasing importance as crop pathogens. Only a small number of species reproduce by amphimixis, i.e. with the obligatory fusion of a male and female gamete (e.g. *M. carolinensis*, *M. microtyla*, *M. pini*). These species are considered as minor root-knot nematode species because of their very restricted distribution, host range and economic impact (Jepson, 1987). Automictic root-knot nematode species are facultatively parthenogenetic; apomictic species are obligatory parthenogenetic. The apomictic mode of reproduction is found in the most important species in terms of geographic distribution and agronomic impact. There are two possible explanations for the worldwide distribution of the apomictic root-knot nematodes. Either they are indigenous in much or all of their current range, and therefore are very ancient species, or they are recently evolved and have been widely spread by agriculture (Trudgill and Blok, 2001). The former is widely viewed as unlikely, partly because parthenogenetic species are considered an evolutionary dead end. Most of the amphimictic and automictic species are diploid with a haploid chromosome number of 18. The majority of the apomictic species are polyploid or aneuploid and usually show a wide variation in chromosome number ($2n$ = 30–55 chromosomes) (Karssen and Moens, 2006; see Chitwood and Perry, Chapter 8, this volume). Root-knot nematode species further differ in their male-to-female ratio. Cross-fertilizing species such as *M. carolinensis* and *M. spartinae* usually have a 1:1 ratio. Species that reproduce by facultative or obligatory parthenogenesis such as *M. hapla* and *M. incognita* have variable sex ratios.

The root galling upon which the nematode's common name is based is quite variable among the different species of the genus and plant hosts (Plates 1–12). Some differences in galling among the different species of this genus are well known. *Meloidogyne hapla* is particularly known for the high incidence of adventitious roots that develop from root galls (Sasser, 1954; Plate 4). *Meloidogyne trifoliophila* on clover produces galls that are distinctly elongated, and the egg masses are more typically embedded within the galls than found erupting from the gall surface (Mercer *et al.*, 1997; Plates 6, 7). Other species have a tendency to produce galls at the root terminus. Root galls can be quite small or indistinct on many hosts, which often results in failure to recognize that the plant is being parasitized. Graminaceous hosts rarely form galls. Most plants with fibrous or woody roots will have small or indistinct galls, especially early in a growing season or when nematode population densities are low. Infection sites of *M. partityla* on pecan that contain a single nematode do not form galls; rather, both the mature female and the egg mass are exposed on the root surface (Plate 8). Cotton and groundnut are examples of two highly susceptible crops in which root galls can be difficult to detect early in the growing season but massive galls can be evident at crop maturity. Plants with succulent roots, especially the cucurbits and tomato, develop the readily detectable galls for which the species is named, even with low infection incidence (Plates 2, 3, 9). Under extreme conditions, a plant's root system may be entirely gall tissue with no remaining fibrous roots.

1.5.1 Incompatible host reactions

The induction and maintenance of giant cells by *Meloidogyne* spp. and the associated physiological and molecular changes in a compatible host–parasite interaction are discussed by Bleve-Zacheo and Melillo (1997) and Abad *et al.* (Chapter 7, this volume). Resistance to *Meloidogyne* spp. is a much-researched topic and Veech (1981) provides an excellent summary of the older literature. Williamson and Roberts (Chapter 13, this volume) discuss resistance genes and the genetics of plant resistance to root-knot nematodes. The analysis of histochemical changes and signal transduction pathways using the resistant tomato plant–*Meloidogyne* model system has provided useful information on the changes associated with the incompatible response (Bleve-Zacheo *et al.*, 2007). Many plant defences against pathogens are regulated by signalling pathways in which jasmonic acid, for example, plays a key role. Soriano *et al.* (2004) showed that application of exogenous methyl jasmonate to roots of spinach and oats induced nematode resistance. Jasmonates induce *de novo* ecdysteroid synthesis in roots but Soriano *et al.* (2004) found that the invasion of spinach by *M. javanica* was impaired by the induction of ecdysteroid, indicating movement of ecdysteroid into the rhizosphere and sensitivity of the nematode to its presence; expression of resistance prior to invasion is uncommon. Cooper *et al.* (2005) demonstrated that jasmonic acid induced a systemic defence response that reduced the reproduction of avirulent *Meloidogyne* on susceptible tomato plants. Another resistant response associated with signal transduction during the hypersensitive reaction to pathogen invasion is the stress-induced oxidative burst. This response is complex and involves antioxidants, such as ascorbic acid. Application of ascorbic acid to susceptible tomato plants has been shown to inhibit invasion by *M. incognita*, and resistance has been associated with the ability to synthesize large amounts of ascorbic acid following J2 invasion (Arrigoni *et al.*, 1979). The oxidative burst occurs rapidly after nematode invasion, with a second burst associated with the hypersensitive reaction only detectable in the incompatible tomato–nematode interaction (Melillo *et al.*, 2006).

The responses associated with the resistance genes *Me1* and *Me3* in pepper differed (Bleve-Zacheo *et al.*, 1998). Many fewer J2 of *M. incognita* were able to invade the pepper line HDA149, carrying the *Me3* gene, compared with the line HDA330, carrying the *Me1* gene. The line HDA149 exhibited the typical early hypersensitive response to nematode invasion, while the resistance mechanism in HDA330 involved a delayed plant response after the J2 had set up several imperfect giant cells.

Antioxidant enzymes secreted by root-knot nematodes may be important to overcome the hypersensitive response of resistant roots and the associated generation of reactive oxygen species (Molinari *et al.*, 2008; Molinari, 2009). For example, a selected virulent isolate of *M. incognita* had greater activity of antioxidant enzymes, including catalase, superoxide dismutase and peroxidase, when compared with a near isogenic avirulent isolate and an avirulent field population (Molinari, 2009). However, it is not clear whether enhanced antioxidant activities contribute to the virulent phenotype, or whether they are a side effect.

1.6 Diversity in Biology

Species of root-knot nematodes demonstrate a large diversity in various aspects of their life cycles. With respect to their temperature requirements, root-knot nematodes can be divided into two distinct groups of species, thermophils and cryophils, which can be separated by their ability to survive lipid-phase transitions that occur at 10 °C (Lyons *et al.*, 1975; see Evans and Perry, Chapter 9, this volume). *Meloidogyne chitwoodi*, *M. hapla* and probably *M. naasi* are cryophils and able to survive soil temperatures below 10 °C; *M. arenaria*, *M. javanica* and *M. exigua* are thermophils and do not have extended survival at temperatures below 10 °C. Like survival, hatching is primarily controlled by temperature (see Curtis *et al.*, Chapter 6, this volume). Thermotypes exist within species (e.g. Daulton and Nusbaum, 1961). Root-knot nematode species also differ in the number of generations they can produce per year; this number varies according to species and food availability. Usually there are many generations per year, but in some species (e.g. *M. naasi*) there is only one (Rivoal and Cook, 1993).

Root-knot nematodes demonstrate various degrees of specialization with respect to their host preference. Crops are usually better hosts than weeds (e.g. Mandulu and Trudgill, 1993; Hillocks *et al.*, 1995). Either man, when selecting crop plants, has inadvertently selected for increased susceptibility to root-knot nematode species, or the root-knot nematodes have been selected by repeated exposure to crop plants (Trudgill and Blok, 2001). Most amphimictic species have host ranges confined to a single subclass of plants, on either woody or perennial herbaceous hosts. *Meloidogyne spartinae* seems to be restricted to cordgrass (*Spartina* spp.), both *M. pini* and *M. megatyla* to *Pinus* spp. and *M. subartica* is confined to the Commelinidae (Jepson, 1987). Apart from *M. hapla*, which has a wide host range that does not include graminaceous species, the automictic species tend to have a narrow host range, while the mitotic species have a potential host range containing the majority of the higher plants (Trudgill and Blok, 2001). However, there are exceptions to these generalizations; for example, the apomictic species *M. quericiana* and *M. enterolobii* have restricted host ranges. The majority of apomictic species of root-knot nematodes appear to have a survival strategy based on a wide host range, which enables them to persist whatever the vegetation. They lack specific triggers for hatching, and hatch occurs as soon as the J2 has developed. On good hosts, where generation times are short and fecundity is high, several generations and rapid population increase occur. Consequently, as the growing season progresses, small populations of apomictic root-knot nematodes can become large and very damaging (Trudgill and Blok, 2001).

For those *Meloidogyne* spp. that have host ranges that are large (total number of hosts) and broad (large number of plant families with species susceptible to the nematode), it is somewhat amazing that there are quite distinct differences in their overlapping host ranges. Among the traditional four major species (which were four of the five original species described by Chitwood, 1949) there are some distinct and now classic differences. These four species (*M. arenaria*, *M. hapla*, *M. incognita* and *M. javanica*) have overlapping host ranges as they all infect many common vegetable crops, including tomato, which is often (but inappropriately) considered a universal host for *Meloidogyne* spp. *Meloidogyne hapla* reproduces only poorly or not at all on most grasses and grain crops. Of these four species, only some populations of *M. incognita* parasitize cotton. Groundnut is a good host for *M. hapla* and most *M. arenaria* populations but not for *M. incognita*. Most populations of *M. javanica* in the USA do not reproduce well on groundnut but populations of this species from India and northern Africa generally reproduce well on groundnut. Thus, even though they each have large, broad and overlapping host ranges, there are distinct differences in the host range of each of these species. Perhaps, despite the claims made above, some greater caution is warranted regarding statements of large host ranges. Given the continually increasing number of recognized species, how many of the previous reports on host ranges were based on incorrect species identification and/or having test populations that contained more than one species?

1.6.1 Concept of host races

Sasser (1952) was among the first to report the variation in host range within the original four major species. He later proposed the formal recognition of 'host races' within these species and proposed a standardized set of differential hosts for distinguishing these host races (Hartman and Sasser, 1985). Further, he proposed that the host differential test could be used as an aid to species identification. Here it should be emphasized that Sasser never recommended the use of the host differentials as the sole basis for species identification. Since the 1970s, additional variation among the four major species has been recognized, especially among populations of *M. javanica*, which is now recognized as variable with respect to reproduction on pepper and groundnut, two important differentials for the host race test. This greatly complicates the use of the host race test as an aid to species identification. Perhaps more importantly, there has been a substantial number of new species described since the 1970s. Many of these recently described species have been found in mixed populations with one or more of the traditional major species and attack some of the same important crops. The North Carolina Host Race test does not clearly distinguish most of the newly described species from the original four. Similar variation in host range is reported for other species, with this variation being unrelated to the differential hosts used in the North Carolina Host Differential test.

Although the recognition of variation in host ranges is important, we suggest the formal recognition of the host races (often referred to simply as race) be discontinued. The host race concept has never been universally accepted, in part because it measured only a small portion of the potential variation in parasitic ability. Given the large number of hosts for many species, it is unlikely that the full extent of possible variation will ever be adequately characterized. Assigning each variant population a distinct number is likely to be very unwieldy with time. It had some value in the south-eastern USA because each of the differential hosts was an important and widely grown crop, but this is not always the case. Unfortunately, knowledge of host race rarely allowed one to predict accurately other behavioural characteristics of a population. Race 1 of *M. arenaria* is clearly not as aggressive on soybean as race 2 of that species (Ibrahim and Lewis, 1993) but this correlation of one aspect of behaviour with the host race status is the exception rather than the rule. Perhaps more importantly, the term race, when used in the context of plant disease, generally refers to populations of a pathogen that differ in virulence on host species that carry specific genes for resistance to other populations of that pathogenic species. This terminology is especially prevalent in the USA. The system also mixed variation in host range with variation in virulence. For example, the tobacco cv. NC95 carries specific resistance to *M. incognita*. Thus, variation in reproduction and galling among populations of *M. incognita* on tobacco actually indicates differences in virulence (the ability of some pathogen genotypes to have a compatible interaction with hosts carrying a specific gene for resistance to that pathogen species). Virulence within *M. arenaria*, *M. incognita* and *M. javanica* in relation to the *Mi* resistance gene in tomato is well documented (Semblat *et al.*, 2000). Similarly, variation among *M. incognita* populations with respect to different resistant cotton genotypes has been reported (Zhou *et al.*, 2000). Yet these examples of variation in virulence are unrelated to host race status. Ability to convey clearly information on the behaviour of a population via a proper and correct name is essential. It will be more useful when dealing with a population that has a variant host range to confirm that the population has been adequately identified and then to simply acknowledge the variant host range.

1.7 Major and Emerging Species

Discussion of *Meloidogyne* spp. frequently focuses on the major four species: the three tropical species, *M. incognita*, *M. arenaria* and *M. javanica*, and the temperate species, *M. hapla*. The recognition of these four species as being the 'major species' began with them being four of the five original species (Chitwood, 1949). That each has an extensive host range and that they are each globally distributed further contribute to their recognized importance. Finally, the often-cited publication by Taylor *et al.* (1982), that these four species comprised more than 99% of all species identified from a collection of 662 isolates from 65 countries, further strengthened their status as the major species. However, that survey had a couple of distinct biases that skewed the data. First, most samples were collected from warm temperate or tropical climates, with only 39 of the 662 samples from a climate where the coldest temperature was ≤5 °C. Additionally, although the samples were collected from 121 different crop species, 33% of the samples were from just four crops (tomato, aubergine (= eggplant), tobacco and okra). Probably not surprisingly, tomato alone accounted for nearly 17% of all samples. None the less, the survey greatly strengthened the status of the four major species. Their widespread recognition has probably led to many cases of misidentification, e.g. once *M. hapla* was identified a few times parasitizing a given crop species in a particular region, most further finds of a root-knot nematode attacking that crop would probably be attributed to *M. hapla* without substantial scrutiny. The advent of isozyme phenotyping and species-specific DNA protocols has contributed greatly to our ability to determine rapidly, and with less equivocation than is possible with morphometric analysis, whether a given isolate belongs to one of the well-known species or is possibly a new species.

Notwithstanding the increasing number of new *Meloidogyne* species, the four major ones are undoubtedly of immense economic importance and quite possibly still deserving of their status. They have been the subject of a considerable amount of research, which is reflected in the following chapters. Most knowledge of *Meloidogyne* spp. is based on studies of one or more of these four major species. Further, most of the more fundamental studies on these four species have been based on research with either tomato or tobacco as the host species. In the era of molecular genetics, future studies will depend on model systems for which the genetics of the host are well characterized. *Medicago truncatula* (Dhandaydham *et al.*, 2008) is an example of such a model system and has the advantage of being a legume; many legume crop species are very susceptible to one or more root-knot nematode species. Descriptions and hosts of 12 species, including the four major species, of economic importance in different geographical areas are given by Hunt and Handoo, Chapter 3, this volume, where naming authorities for all species of *Meloidogyne* are provided. Here we describe some species that we consider to be of 'emerging' importance. It is to be expected that changing selection pressures due to evolving cropping and management systems, especially the use of host resistance, will result in a dynamic landscape with regard to important *Meloidogyne* spp.

1.7.1 *Meloidogyne enterolobii* (= *Meloidogyne mayaguensis*)

This nematode is currently considered as one of the most important root-knot nematode species because of its wide geographical distribution, its wide host range, and its ability to overcome the resistance of important crop plants, such as genotypes of tomato, pepper, and some agronomic crops that carry the *Mi-1* gene, which confers resistance to *M. javanica*, *M. arenaria* and *M. incognita* (Fargette, 1987; Fargette *et al.*, 1994). This nematode was originally described from a population collected from aubergine (*Solanum melongena*) in Puerto Rico (Rammah and Hirshmann, 1988). Subsequently, it has been detected in Africa (Fargette *et al.*, 1994; Duponnois *et al.*, 1995; Willers, 1997; Trudgill *et al.*, 2000), the USA (Brito *et al.*, 2004), South and Central America (Decker and Rodriguez-Fuentes, 1989; Carneiro *et al.*, 2000, 2001; Trudgill *et al.*, 2000) and Europe (Blok *et al.*, 2002).

Other recorded hosts include vegetables, and other crops, such as bell pepper (*Capsicum annuum*), soybean (*Glycine max*), sweet potato (*Ipomoea batatas*), tobacco (*Nicotiana tabacum*) and watermelon (*Citrullus lanatus*). A tropical fruit tree, guava (*Psidium guajava*), is also a good host of this nematode (Plate 20). Spanish needle (*Bidens pilosa*), a weed host, has also been identified. In Cuba, reproduction was observed on coffee (*Coffea arabica* cv. Caturra), bean (*Phaseolus vulgaris* cv. Icapijao), beet (*Beta vulgaris*), broccoli (*Brassica oleracea* var. Botrytis), celery (*Apium graveolens* cv. Utah), horsebean (*Cannavalia ensiformis*), parsley (*Petroselynum crispum* cv. Plain), potato (*Solanum tuberosum*) and pumpkin (*Cucurbita* sp.). In Florida, this nematode has been found in roots of angel trumpet (*Brugmansia* cv. Sunray), basil (*Ocimum* sp.), cape honeysuckle (*Tecomaria capensis*), glory bush (*Tibouchina* cv. Compacta and *Tibouchina elegans*), carpet bugleweed (*Ajuga reptans*) and Uganda glorybower (*Clerodendrum ugandense*).

1.7.2 *Meloidogyne paranaensis*

This species was detected on coffee in Paraná state, Brazil, from which it is named (Carneiro *et al.*, 1996). Coffee (*C. arabica*) is the primary host of this species. However, the nematode has also been detected on soybean (Dinnys Roese *et al.*,

2004). In host studies, tobacco, tomato and watermelon were reported as suitable hosts (Carneiro *et al.*, 1996). Decline and dieback of coffee trees, and yield suppression of up to 50%, are associated with nematode infection in Brazil, where the damage occurring on coffee in Paraná state was initially erroneously attributed to *M. incognita* (Carneiro *et al.*, 1996). Currently the nematode has been detected only in the USA (including Hawaii), Central America, the Caribbean, Guatemala and South America.

Specific damage caused by *M. paranaensis* on roots of coffee typically does not involve gall formation. Instead, *M. paranaensis* causes the taproot of coffee to crack and split, as well as damaging other root tissue. Necrosis also occurs where females are embedded and near the giant cells where feeding occurs. Above-ground symptoms generally range from chlorosis and reduced plant growth to death (Carneiro *et al.*, 1996).

Meloidogyne paranaensis may occur by itself or in mixed populations with other *Meloidogyne* spp. (Carneiro *et al.*, 1996). Inserra *et al.* (2003) suggested that this nematode may lower yield potentials by 50%. This estimate is based on information provided by Carneiro *et al.* (1996), but the original authors only suggested that this particular species may '[account] for approximately 52% of all root-knot nematode infestations in Paraná'. Carneiro *et al.* (1996) do not comment on the magnitude of damage when these infestations occur.

1.7.3 *Meloidogyne fallax* and *Meloidogyne chitwoodi*

Meloidogyne chitwoodi and *M. fallax* are closely related species parasitizing monocotyledons and dicotyledons, including several major crop plants such as potatoes, carrots and tomatoes (Santo *et al.*, 1980; O'Bannon *et al.*, 1982; Brinkman *et al.*, 1996; Karssen, 2002 (Plates 10, 11, 22, 25–27)).

Meloidogyne chitwoodi was first described from roots and tubers of potato in a field near Quincy, Washington state, USA. The species has been recorded from Argentina, Belgium, Germany, The Netherlands, Portugal, several states of the USA, Mexico and South Africa (EPPO, 2004). It is not clear whether this is its area of origin. In Europe, the nematode was first detected in the Netherlands, but a review of old illustrations and old specimens of *Meloidogyne* suggests that they may have occurred earlier (EPPO, 1991). The species can begin development when soil temperature rises above 5 °C and requires 600–800 degree-days to complete the first generation; subsequent generations require only 500–600 degree-days (Pinkerton *et al.*, 1991). Intraspecific variation in *M. chitwoodi* is manifest by the occurrence of three biotypes that can be distinguished based on reproduction on lucerne cv. Thor, carrot cv. Red Cored Chantenay, and *Solanum bulbocastanum* Dun. SB22 (Santo and Pinkerton, 1985; Mojtahedi *et al.*, 1988, 1994).

Symptoms caused by *M. chitwoodi* vary according to host, population density of the nematode and environmental conditions. Galls produced on potato tubers are often not easily detected. On carrots, galls appear mainly on the lenticels and reduce the commercial value of the crop (Wesemael and Moens, 2008).

Meloidogyne fallax was detected for the first time in 1992 in a field north of Baexem (The Netherlands) and initially identified as 'a deviating *M. chitwoodi* population' (Karssen, 1994). After this first report, it was recorded on potato at several locations in the southern and south-eastern part of the Netherlands and eventually described as a separate species (Karssen, 1996). Later it was also found in a plastic tunnel house in France (Daher *et al.*, 1996), and in Belgium (Waeyenberge and Moens, 2001) and Germany (Schmitz *et al.*, 1998). The species has been detected outside Europe, in New Zealand (Marshall *et al.*, 2001), Australia (Nobbs *et al.*, 2001) and South Africa (Fourie *et al.*, 2001). Above-ground symptoms of heavily infested plants include stunting and yellowing, while below-ground galling is typical. Variations in host ranges among different populations have not been described for *M. fallax*.

Successful hybridization was not obtained when *M. fallax* and *M. chitwoodi* were crossed; the F1 was viable, but the F2 second-stage juveniles were not viable and showed morphological distortions (van der Beek and Karssen, 1997).

The root galls produced by *M. chitwoodi* and *M. fallax* are comparable to those produced by several other root-knot species, relatively small galls in general, without secondary roots emerging from them (as found in *M. hapla*). On potato tubers, *M. chitwoodi* and *M. fallax* cause numerous small, pimple-like raised areas on the surface (with *M. hapla* these swellings are not evident).

There is a marked contrast in the hatching response of the two species (see Curtis *et al.*, Chapter 6, this volume). Hatching of J2 of *M. chitwoodi* produced on young plants does not require host root diffusate stimulus, whereas at the end of the plant growing season, egg masses contained a percentage of unhatched J2 that require host root diffusate to cause hatch. This form of obligate quiescence at the end of the host growing season was not found in *M. fallax*, which hatched well in water and did not require hatch stimulation from root diffusate, irrespective of the age of the plant on which the egg masses were produced (Wesemael *et al.*, 2006).

1.7.4 *Meloidogyne minor*

Karssen *et al.* (2004) described *M. minor*, which appeared to be the causal agent of yellow patch disease on several golf courses in the British Isles and root-knot symptoms in one potato field in The Netherlands in 2000 (Plates 12–15). A joint pest risk analysis (PRA) by nematologists in The Netherlands and the UK for the EU region established that *M. minor* was present mainly on coastal sand dunes, golf courses and sports grounds in the British Isles, and in The Netherlands *M. minor* was found on several golf courses, sports grounds and pasture fields. *Meloidogyne minor* has been reported on turfgrass in Belgium (Viaene *et al.*, 2007) but it is not known if it is indigenous to Europe and present in other EU countries. Additional surveys are required to determine its distribution and economic importance, but the PRA concluded that, with the current knowledge, *M. minor* was primarily a problem for golf courses, and it is not yet possible to determine whether quarantine measures are appropriate. It is presumed that its spread cannot readily be controlled since it can be carried on footwear and sports equipment. The PRA document will be available on the web sites of the Dutch and British NPPOs (National Plant Protection Organizations; www.minlnv.nl/pd and www.defra.gov.uk).

1.8 Interactions with Other Plant Pathogens

Meloidogyne spp. frequently play a role in disease interactions (Khan, 1993; see Manzanilla-López and Starr, Chapter 10, this volume), especially with other soil-borne pathogens. Plant pathologists often attribute these interactions to wounds made by the nematodes, which ignores the profound effects of parasitism by *Meloidogyne* spp. on plant physiology and gene expression. Regardless of the underlying mechanisms, the numerous interactions with other root and vascular pathogens only serve to exacerbate the ultimate damage to the crop and increase crop losses. Sometimes *Meloidogyne* spp. and ectoparasitic nematodes appear mutually antagonistic. However, interactions between these two groups may be beneficial for one or both of the species (Eisenback, 1993).

1.9 Management and Control

All methods for control of plant pathogens, including parasitic nematodes, can be categorized under one or more principles (Table 1.1). All of the various methods for control of nematodes fit within one of these principles. Management of nematodes (see Nyczepir and Thomas, Chapter 18, and Coyne *et al.*, Chapter 19, this volume) involves the manipulation of nematode densities to non-injurious or sub-economic threshold levels using several measures in relation to the whole production system, whereas control implies the use of a single measure to reduce or eliminate nematode pests, which in most cases is not possible (Thomason and Caswell, 1987). Maintenance of diversity is an objective of management but not of control, and of increasing importance is the additional need to take into consideration the impact of the pest management strategy on biodiversity and the ecological balance in the soil. Biological control (see Hallmann *et al.*, Chapter 17, this volume) is the management of plant diseases and pests with the aid of living organisms, including predators and parasites of organisms that kill or damage their hosts and also microbes that indirectly influence the establishment, function and survival of pathogens and pests. Plant resistance (see Williamson and Roberts, Chapter 13, and Starr and Mercer, Chapter 14, this volume) is biologically based but is considered a distinct approach to control and management.

As with most plant-parasitic nematodes, preventing the introduction and spread of species of

Table 1.1. Principles for control of plant pathogens.

Exclusion	Prohibiting, frequently by governmental regulations, the entry of the pathogen into a region or locale where it does not exist
Eradication	The complete or partial removal of the pathogen from a region, locale, or field (e.g. soil fumigation)
Avoidance	Avoiding environments and conditions that favour pathogen activity (e.g. altering planting date)
Protection	Usually the use of pesticides to inhibit pathogen activity (e.g. use of non-fumigant nematicides)
Resistance	Altering the genetic constitution of the host so it is able to inhibit pathogen activity (for nematodes, this is typically inhibition of reproduction)
Therapy	Action taken after infection has occurred to limit further development of the pathogen (e.g. hot water treatments of infected bulbs or corms)

Meloidogyne is a vital component of management strategies. *Meloidogyne* species may be spread on farm machinery and may be present in planting material, such as corms, bulbs or roots, but are not found in seeds. Cleaning machinery before use is recommended, and planting material can be discarded if infected, or treated with chemicals or hot water to reduce the numbers of *Meloidogyne*. Only seedlings produced in *Meloidogyne*-free seedbeds should be transplanted.

Listing species as quarantine organisms reduces the risks of spread through international trade. In general, root-knot nematodes are not regulated as a group because the major economically important species are already widely distributed (Hockland *et al.*, 2006). However, *M. chitwoodi* is of increasing importance, primarily because it is a serious pest of economically important crops such as potatoes and carrots, and is on the lists of prohibited pests of many countries (Canada, the EU, Mexico, and other countries in Latin America and the Far East). In the future, with the increase in the number of new species being described, initially with limited knowledge of distribution, more species are likely to be of regulatory concern.

Various cultural and physical control methods have been used with varying degrees of success, but often these methods are only of local or regional relevance. For example, soil solarization (Gaur and Perry, 1991) is only of use in regions where sufficient solar energy is available for long periods of time. Similarly, in some climates, ploughing at intervals of 2–4 weeks during the dry season exposes eggs and J2 to desiccation, killing many in the upper layers of soil. Population densities can be reduced by organic amendments, and flooding land to a depth of 10 cm or more for several months is also effective. The adverse effect of *Tagetes* species is highly variable, depending on the combination of *Tagetes* species and cultivar, and the species and population of *Meloidogyne*. It appears that reduction of *M. incognita* by marigold (*Tagetes patula*) is primarily due to an antagonistic or trap crop effect; J2 enter roots but there is neither giant cell formation nor a hypersensitive reaction.

The era of nematicides was the 1950s, 1960s, 1970s and 1980s. Their overall effectiveness is often cited as a reason why other alternative management systems did not receive greater attention for many years. Starting in the late 1970s, the use of some fumigants, initially 1,2-dibromo-3-chloropropane, was greatly restricted or forbidden entirely. That was followed by similar restrictions and eventual suspension (in some cases by corporate rather than governmental decision) of some of the granular nematicides. Currently, the list of available and effective nematicides is very short. Unfortunately, due in part to the small market share of nematicides relative to herbicides or insecticides, there is little prospect for new effective materials in the near future.

1.10 Conclusions and Future Directions

De Waele and Elsen (2007) attribute the general lack of awareness in tropical countries of even the existence of plant-parasitic nematodes to the microscopic nature of nematodes, the lack of characteristic symptoms they cause, and the farmers' limited previous exposure to extension and community information. This is probably universally true. Knowledge of the pest involves identification of the species present and, clearly, the provision of better species descriptions, including all available information on morphology, morphometrics, genetics, phylogenetics, etc., is essential (see Eisenback and Hunt, Chapter 2, this volume). Future research in nematode systematics should comprise well-focused taxonomy based on a combination of classical and molecular methods (Coomans, 2002), and we consider that papers on *Meloidogyne* spp. should include a brief description of the method used to identify the population(s). A relevant example of the need for this was provided when a closer examination of *Meloidogyne* populations associated with coffee in Brazil, combining morphological observations with molecular diagnostics, led to the description of several new *Meloidogyne* species (Plate 17) and the suggestion that *Meloidogyne* spp. populations on coffee from Brazil and other Central and South American countries must frequently have been misidentified (Carneiro *et al.*, 2004). That many of the recently described species have been found in association with more common species suggests that such misidentications may have occurred elsewhere.

There needs to be a uniform method to assess and disseminate information on damage caused by the nematode species. Costs of produce fluctuate greatly and are often not comparable between countries. Thus, economic losses are relevant data only for a particular year. Yield loss data should be in a form that can be easily converted to costs if comparison between years is required. But this presupposes a uniform estimation of damage, which may not be yield loss per se but is likely to include the amount of harvested crop that is unmarketable because of damage by *Meloidogyne* spp.

In this chapter we have recommended abandoning the host race system proposed by Hartman and Sasser (1985) that has been used by many scientists. This is not to dismiss the importance in variation in parasitic abilities but is a recognition that the system is no longer adequate for the greatly expanded genus. The variation in host ranges and parasitic fitness of the more than 90 currently recognized species on individual hosts now appears to be too great to be categorized by a numerical host race designation. With the increasing use of resistant hosts for control of root-knot nematodes, eliminating the use of the host race also avoids potential confusion with races (or pathotypes) that vary in virulence on resistant host genotypes.

This introductory chapter has set the scene for the subsequent chapters, where aspects of nematode biology, host–plant interactions and control will be discussed in depth. Information on the genomes of *M. hapla* and *M. incognita* will enable features of obligate parasitism to be defined, and the genomic information may aid in the identification of novel control targets and the refining of environmentally acceptable management options. Management of *Meloidogyne* in developed and resource-poor regions will be central to the provision of sufficient food for the ever-increasing global population.

1.11 References

Adam, M.A.M., Phillips, M.S. and Blok, V.C. (2007) Molecular diagnostic key for identification of single juveniles of seven common and important species of *Meloidogyne. Plant Pathology* 56, 190–197.

Arrigoni, O., Zacheo, G., Arrigoni-Liso, R., Bleve-Zacheo, T. and Lamberti, F. (1979) Relationship between ascorbic acid and resistance in tomato plants to *Meloidogyne incognita. Phytopathology* 69, 579–581.

Berkeley, M.J. (1855) Vibrio forming cysts on the roots of cucumbers. *Gardeners' Chronicle* 7 April, 220.

Bird, A.F. and Loveys, B.R. (1975) The incorporation of photosynthates by *Meloidogyne javanica. Journal of Nematology* 7, 111–113.

Bleve-Zacheo, T. and Melillo, M.T. (1997) The biology of giant cells. In: Fenoll, C., Grundler, F.M.W. and Ohl, S.A. (eds) *Cellular and Molecular Aspects of Plant–Nematode Interactions.* Kluwer Academic Publishers, Dordrecht, The Netherlands, pp. 65–79.

Bleve-Zacheo, T., Bongiovanni, M., Melillo, M.T. and Castagnone-Sereno, P. (1998) The pepper resistance genes *Me1* and *Me3* induced differential penetration rates and temporal sequences of root cell ultrastructural changes upon nematode infection. *Plant Science* 133, 79–90.

Bleve-Zacheo, T., Melillo, M.T. and Castagnone-Sereno, P. (2007) The contribution of biotechnology to root-knot nematode control in tomato plants. *Plant Technology* 1, 1–16.

Blok, V.C., Wishart, J., Fargette, M., Berthier, K. and Phillips, M.S. (2002) Mitochondrial DNA differences distinguishing *Meloidogyne mayaguensis* from the major species of tropical root-knot nematodes. *Nematology* 4, 773–781.

Brinkman, H., Goossens, J.M. and Van Riel, H.R. (1996) Comparative host suitability of selected crop plants to *Meloidogyne chitwoodi* and *M. fallax*. *Anzeiger für Schädlingskunde Pflanzenschutz Umweltschutz* 96,127–129.

Brito, J., Powers, T.O., Mullin, P.G., Inserra, R.N. and Dickson, D.W. (2004) Morphological and molecular characterization of *Meloidogyne mayaguensis* isolates from Florida. *Journal of Nematology* 36, 232–240.

Carneiro, R.M.D.G., Carneiro, R.G., Abrantes, I.M.O., Santos, M.S.N.A. and Almeida, M.R.A. (1996) *Meloidogyne paranaensis* n. sp. (Nemata: Meloidogynidae), a root-knot nematode parasitizing coffee in Brazil. *Journal of Nematology* 28, 177–189.

Carneiro, R.M.D.G., Almeida, M.R.A. and Quénéhervé, P. (2000) Enzyme phenotypes of *Meloidogyne* spp. isolates. *Nematology* 2, 645–654.

Carneiro, R.M.D.G., Moreira, W.A., Almeida, M.R.A. and Gomes, A.C.M.M. (2001) Primeiro registro de *Meloidogyne mayaguensis* em Goiabeira no Brasil. *Nematologia Brasileira* 25, 223–228.

Carneiro, R.M.D.G., Tigano, M.S., Randig, O., Almeida, M.R.A. and Sarah, J.L. (2004) Identification and genetic diversity of *Meloidogyne* spp. (Tylenchida: Meloidogynidae) on coffee from Brazil, Central America and Hawaii. *Nematology* 6, 287–298.

Chitwood, B.G. (1949) Root-knot nematodes – part I. A revision of the genus *Meloidogyne* Göldi, 1887. *Proceedings of the Helminthological Society of Washington* 16, 90–104.

Coomans (2002) Present status and future of nematode systematics. *Nematology* 4, 573–582.

Cooper, W.R., Jia, L. and Goggin, L. (2005) Effect of jasmonate-induced defenses on root-knot nematode infection of resistant and susceptible tomato cultivars. *Journal of Chemical Ecology* 31, 1953–1967.

Cornu, M. (1879) Etudes sur le *Phylloxera vastatrix*. *Mémoires Divers Savants à l'Académie Royale des Sciences de l'Institut de France* 26, 1–357.

Daher, S., Gillet, S., Mugniéry, D. and Marzin, H. (1996) Discovery in France and characteristics of the Dutch variant of *Meloidogyne chitwoodi*. *Proceedings of the Third International Nematology Congress*, Gosier, Guadeloupe, p. 188.

Daulton, R.A.C. and Nusbaum, C.J. (1961) The effect of soil temperature on the survival of the root-knot nematodes *Meloidogyne javanica* and *M. hapla*. *Nematologica* 6, 280–294.

Decker, H. and Rodriguez-Fuentes, M.E. (1989) The occurrence of root gall nematodes *Meloidogyne mayaguensis* on *Coffea arabica* in Cuba. *Wissenschaftliche Zeitschrift der Wilhelm-Pieck Universität Rostock, Naturwissenschaftliche Reihe* 38, 32–34.

De Waele, D. and Elsen, A. (2007) Challenges in tropical plant nematology. *Annual Review of Phytopathology* 45, 457–485.

Dhandaydham, M., Charles, L., Zhu, H., Starr, J.L., Huguet, T., Cook, D.R., Prosperi, J.-M., and Opperman, C.H. (2008) Characterization of root-knot nematode resistance in *Medicago truncatula*. *Journal of Nematology* 40, 46–54.

Dinnys Roese A., D'Arc De Lima Oliveira R. and De Lanes, F.F. (2004) Reação de cultivares de soja (*Glycine max* L. Merril) a *Meloidogyne paranaensis*. *Nematologia Brasileira* 28, 131–135.

Duponnois, R., Mateille, T. and Gueye, M. (1995) Biological characterization and effects of two strains of *Arthrobotrys ologospora* from Senegal on *Meloidogyne* species parasitizing tomato plants. *Biocontrol Science and Technology* 5, 517–525.

Eisenback, J.D. (1993) Interaction between nematodes in cohabitance. In: Khan, M.W. (ed.) *Nematode Interactions*. Chapman & Hall, London, pp. 143–174.

EPPO (1991) *Meloidogyne chitwoodi* present in Netherlands. *FAO Plant Protection Bulletin* 39, 187.

EPPO (2004) Diagnostic protocols for regulated pests *Meloidogyne chitwoodi* and *Meloidogyne fallax*. *Bulletin OEPP/EPPO Bulletin* 34, 315–320.

Esbenshade, P.R. and Triantaphyllou, A.C. (1985) Identification of major *Meloidogyne* species employing enzyme phenotypes as differentiating characters. In: Sasser, J.N. and Carter, C.C. (eds) *An Advanced Treatise on* Meloidogyne. *Volume I Biology and Control*. North Carolina State University Graphics, Raleigh, North Carolina, pp. 135–140.

Fargette, M. (1987) Use of esterase phenotype in the taxonomy of the genus *Meloidogyne*. 2. Esterase phenotypes observed in West African populations and their characterization. *Revue de Nématologie* 10, 45–56.

Fargette, M., Davies, K.G., Robinson, M.P. and Trudgill, D.L. (1994) Characterization of resistance breaking *Meloidogyne incognita*-like populations using lectins, monoclonal antibodies and spores of *Pasteuria penetrans. Fundamental and Applied Nematology* 17, 537–542.

Fourie, H., Zijlstra, C. and McDonald, A.H. (2001) Identification of root-knot nematode species occurring in South Africa using the SCAR-PCR technique. *Nematology* 3, 675–689.

Gaur, H.S and Perry, R.N. (1991) The use of soil solarization for control of plant parasitic nematodes. *Nematological Abstracts* 60, 153–167.

Gheysen, G. and Jones, J.T. (2006) Molecular aspects of plant–nematode interactions. In: Perry, R.N. and Moens, M. (eds). *Plant Nematology.* CAB International, Wallingford, UK, pp. 234–254.

Göldi, E.A. (1892) Relatoria sôbre a molestia do cafeiro na provincial da Rio de Janeiro. *Archuivos do Museu Nacional do Rio de Janeiro* 8, 1–112.

Greco, N., Vovlas, N., Di Vito, M. and Inserra, R.N. (1992) *Meloidogyne artiellia*: a root-knot nematode parasite of cereals and other field crops. *Nematology Circular* No. 201. Florida Department of Agriculture & Consumer Services, Division of Plant Industry, Gainsville, Florida.

Hartman, K.M. and Sasser, J.N. (1985) Identification of *Meloidogyne* species on the basis of differential host test and perineal pattern morphology. In: Barker, K.R., Carter, C.C. and Sasser, J.N. (eds) *An Advanced Treatise on* Meloidogyne. *Volume II: Methodology.* North Carolina State University Graphics, Raleigh, North Carolina, pp. 69–77.

He, B., Magill, C. and Starr, J.L. (2005) Laser capture microdissection and real-time PCR for measuring mRNA in giant cells induced by *Meloidogyne javanica. Journal of Nematology* 37, 308–312.

Hillocks, R.J., Stokes, S. and Jones, M (1995) Reproduction of *Meloidogyne javanica* on legume crops and some weeds associated with their cultivation in Malawi. *Nematologica* 41, 505–515.

Hockland, S., Inserra, R.N., Millar, L. and Lehman, P.S. (2006) International plant health – putting legislation into practice. In: Perry, R.N. and Moens, M. (eds) *Plant Nematology.* CAB International, Wallingford, UK, pp. 327–345.

Huang, C.S. and Maggenti, A.R. (1969) Mitotic aberration and nuclear changes of developing giant cells in *Vicia faba* caused by *Meloidogyne javanica. Phytopathology* 59, 447–455.

Huang, G., Dong, R., Allen, R., Davis, E.L., Baum, T.J., and Hussey, R.S. (2006) A root-knot nematode secretory peptide functions as a ligand for a plant transcription factor. *Molecular Plant Microbe Interactions* 19, 463–470.

Ibrahim, I.K.A. and Lewis, S.A. (1993) Pathogenicity and reproduction of *Meloidogyne arenaria* races 1 and 2 and *M. incognita* race 3 on soybean. *Journal of Nematology* 23, 159–166.

Ibrahim, S.K. and Perry, R.N. (1992) Use of esterase patterns of females and galled roots for the identification of species of *Meloidogyne. Fundamental and Applied Nematology* 16, 187–190.

Inserra, R.N., Brito, J.A., Dong, K., Handoo, Z.A., Lehman, P.S., Powers, T. and Millar, L. (2003) Exotic nematode plant pests of agricultural and environmental significance to the United States: *Meloidogyne paranaensis. Society of Nematologists.* Available online at http://nematode.unl.edu/wgroup.htm.

Jepson, S.B. (1987) *Identification of Root-knot Nematodes* (Meloidogyne *species*). CAB International, Wallingford, UK.

Jones, M.G.K. and Northcote, D.H. (1972) Multinucleate transfer cells induced in *Coleus* roots by the root-knot nematode *Meloidogyne javanica. Protoplasms* 75, 381–395.

Karssen, G. (1994) The use of isozyme phenotypes for the identification of root-knot nematodes (*Meloidogyne* spp.). *Annual Report 1992 Diagnostic Centre.* Plant Protection Service, Wageningen, The Netherlands, pp. 85–88.

Karssen, G. (1996) Description of *Meloidogyne fallax* n. sp., a root-knot nematode from the Netherlands. *Fundamental and Applied Nematology* 19, 593–599.

Karssen, G. (2002) *The Plant-parasitic Nematode Genus* Meloidogyne *Göldi, 1892 (Tylenchida) in Europe.* Brill, Leiden, The Netherlands.

Karssen, G. and Moens, M. (2006) Root-knot nematodes. In: Perry, R.N. and Moens, M. (eds) *Plant Nematology.* CAB International, Wallingford, UK, pp. 59–90.

Karssen, G., Bolk, R.J., van Aelst, A.C., van den Beld, I., Kox, L.F.F., Korthals, G., Molendijk, L., Zijlstra, C., Van Hoof, R. and Cook, R. (2004) Description of *Meloidogyne minor* n. sp. (Nematoda: Meloidogynidae), a root-knot nematode associated with yellow patch disease in golf courses. *Nematology* 6, 59–72.

Khan, M.W. (1993) Mechanisms of interactions between nematodes and other plant pathogens. In: Khan, M.W. (ed.) *Nematode Interactions.* Chapman & Hall, London, pp. 55–78.

Lyons, J.M., Keith, A.D. and Thomason, I.J. (1975) Temperature-induced phase transitions in nematode lipids and their influence on respiration. *Journal of Nematology* 7, 98–104.

Mandulu, J. and Trudgill, D.L. (1993) Weed hosts of *Meloidogyne javanica* in Tanzania. *Pakistan Journal of Nematology* 11, 61–64.

Marshall, J.W., Zijlstra, C. and Knight, K.W.L. (2001) First record of *Meloidogyne fallax* in New Zealand. *Australasian Plant Pathology* 30, 283–284.

Melillo, M.T., Leonetti, P., Bongiovanni, M., Castagnone-Sereno, P. and Bleve-Zacheo, T. (2006) Modulation of reactive oxygen species activities and H_2O_2 accumulation during compatible and incompatible tomato–root-knot nematode interactions. *New Phytologist* 170, 501–512.

Mercer, C.F., Starr, J.L. and Miller, K.J. (1997) Host-parasite relations of *Meloidogyne trifoliophilia* isolates from New Zealand. *Journal of Nematology* 29, 55–64.

Molinari, S. (2009) Antioxidant enzymes in (a)virulent populations of root-knot nematodes. *Nematology* 11, 703–711.

Molinari, S., Rosso, L. and Ornat Longaron, C. (2008) Manganese superoxide dismutase in *Meloidogyne incognita* isolates selected for virulence in tomato. *Proceedings of the 5th International Congress of Nematology, 13–18 July*. Brisbane, Australia, p. 248 (abstract).

Mojtahedi, H., Santo, G.S. and Wilson, J.H. (1988) Host test to differentiate *Meloidogyne chitwoodi* races 1 and 2 and *M. hapla*. *Journal of Nematology* 20, 468–473.

Mojtahedi, H., Santo, G.S., Brown, C.R., Ferris, H. and Williamson, V. (1994) A new host race of *Meloidogyne chitwoodi* from California. *Plant Disease* 78, 1010.

Nobbs, J.M., Liu, Q., Hartley, D., Handoo, Z., Williamson, V.M., Taylor, S., Walker, G. and Curran, J. (2001) First record of *Meloidogyne fallax*. *Australasian Plant Pathology* 30, 373.

O'Bannon, J.H., Santo, G.S. and Nyczepir, A.P. (1982) Host range of the Columbia root-knot nematode. *Plant Disease* 66, 1045–1048.

Orion, D. and Kritzman, G. (1991) Antimicrobial activity of *Meloidogyne javanica* gelatinous matrix. *Nematologica* 14, 481–483.

Pinkerton, J.N., Santo, G.S. and Mojtahedi, H. (1991) Population dynamics of *Meloidogyne chitwoodi* on Russet Burbank potatoes in relation to degree-day accumulation. *Journal of Nematology* 23, 283–290.

Potter, J.W. and Olthof, T.H.A. (1993) Nematode pests of vegetable crops. In: Evans, K., Trudgill, D.L. and Webster, J.M. (eds) *Plant-parasitic Nematodes in Temperate Agriculture*, CAB International, Wallingford, UK, pp. 171–207.

Powers, T.D.and Harris, T.S. (1993) A polymerase chain reaction method for identification of five major *Meloidogyne* species. *Journal of Nematology* 25, 1–6.

Rammah, A. and Hirshmann, H. (1988) *Meloidogyne mayaguensis* n. sp. (Meloidogynidae), a root-knot nematode from Puerto Rico. *Journal of Nematology* 20, 58–69.

Ramsey, K., Wang, Z. and Jones, M.G.K. (2004) Using laser capture microdissection to study gene expression in early stages of giant cells induced by root-knot nematodes. *Molecular Plant Pathology* 5, 587–592.

Rivoal, R. and Cook, R. (1993) Nematode pests of cereals. In: Evans, K., Trudgill, D.L. and Webster, J.M. (eds) *Plant-parasitic Nematodes in Temperate Agriculture*. CAB International, Wallingford, UK, pp. 259–303.

Santo, G.S. and Pinkerton, J.N. (1985) A second race of *Meloidogyne chitwoodi* discovered in Washington State. *Plant Disease* 69, 361.

Santo, G.S., O'Bannon, J.H., Finley, A.M. and Golden, A.M. (1980) Occurrence and host range of a new root-knot nematode (*Meloidogyne chitwoodi*) in the Pacific Northwest. *Plant Disease* 64, 951–952.

Sasser, J.N. (1952) Identification of root-knot nematodes (*Meloidogyne* spp.) by host reaction. *Plant Disease Reporter* 36, 84–86.

Sasser, J.N. (1954) *Identification and host-parasite relationships of certain root-knot nematodes* (Meloidogyne *spp.*) Maryland Agriculture Experiment Station Bulletin, A-77.

Schaff, J.E., Neilsen, D.M., Smith, C.P., Scholl, E.H., and Bird, D.McK. (2007) Comprehensive transcription profiling in tomato reveals a role for glycosyltransferase in *Mi*-mediated nematode resistance. *Plant Physiology* 144, 1079–1092.

Schmitz, B., Burgermeister, W. and Braasch, H. (1998) Molecular genetic classification of Central European *Meloidogyne chitwoodi* and *M. fallax* populations. *Nachrichtenblatt des Deutschen Pflanzenschutzdienstes* 50, 310–317.

Semblat, J.P., Bongiovanni, M., Wajnberg, E., Dalmasso, A., Abad, P., and Castagnone-Sereno, P. (2000) Virulence and molecular diversity of parthenogenic root-knot nematodes. *Heredity* 1, 81–89.

Sharon, E., and Spiegel, Y. (1993) Glycoprotein characterization of the gelatinous matrix in the root-knot nematode *Meloidogyne javanica*. *Journal of Nematology* 25, 585–589.

Soriano, I.R., Riley, I.T., Potter, M.J. and Bowers, W.S. (2004) Phytoecdysteroids: a novel defence against plant-parasitic nematodes. *Journal of Chemical Ecology* 30, 1885–1899.

Starr, J.L., Tomaszewski, E.K., Mundo-Ocampo, M. and Baldwin, J.G. (1996) *Meloidogyne partityla* on pecan: isozyme phenotypes and other hosts. *Journal of Nematology* 28, 565–568.

Taylor, L.R., Sasser, J.N. and Nelson, L.A. (1982) Relationships of climate and soil characteristics to geographical distribution of *Meloidogyne* species in agricultural soils. Cooperative Publication, Department of Plant Pathology, North Carolina State University and US Agency for International Development, Raleigh, North Carolina.

Thomason, I.J. and Caswell, E.P. (1987) Principles of nematode control. In: Brown, R.H. and Kerry, B.R. (eds) *Principles and Practice of Nematode Control*. Academic Press, London, pp. 87–130.

Triantaphyllou, A.C. (1985) Cytogenetics, cytotaxonomy and phylogeny of root-knot nematodes. In: Sasser, J.N. and Carter, C.C. (eds) *An Advanced Treatise on* Meloidogyne. *Volume I Biology and Control*. North Carolina State University Graphics, Raleigh, North Carolina, pp. 113–126.

Trudgill, D.L. and Blok, V.C. (2001) Apomictic, polyphagous root-knot nematodes: exceptionally successful and damaging biotrophic root pathogens. *Annual Review of Phytopathology* 39, 53–77.

Trudgill, D.L., Bolla, G., Blok, V.C., Daudi, A., Davies, K.G., Gowen, S.R., Fargette, M., Madulu, J.D., Mateille, T., Mwagenui, W., Netscher, C., Phillips, M.S., Sawadogo, A., Trivino, C.G. and Voyoukallou, E. (2000) The importance of tropical root-knot nematodes (*Meloidogyne* spp.) and factors affecting the utility of *Pasteuria penetrans* as a biocontrol agent. *Nematologica* 2, 823–845.

van der Beek, J.G. and Karssen, G. (1997) Interspecific hybridization of meiotic parthenogenetic *Meloidogyne chitwoodi* and *M. fallax*. *Phytopathology* 87, 1061–1066.

Veech, J.A. (1981) Plant resistance to nematodes. In: Zuckerman, B.M. and Rohde, R.A. (eds) *Plant Parasitic Nematodes*, Vol III. Academic Press, New York, pp. 377–403.

Viaene, N., Wiseborn, D.B. and Karssen, G. (2007) First report of the root-knot nematode *Meloidogyne minor* on turfgrass in Belgium. *Plant Disease* 91, 908.

Waeyenberge, L. and Moens, M. (2001) *Meloidogyne chitwoodi* and *M. fallax* in Belgium. *Nematologia Mediterranea* 29, 91–97.

Wesemael, W. and Moens, M. (2008) Quality damage on carrots (*Daucus carota* L.) caused by the root-knot nematode *Meloidogyne chitwoodi*. *Nematology* 10, 261–270.

Wesemael, W., Perry, R. and Moens, M. (2006) The influence of root diffusate and host age on hatching of the root-knot nematodes, *Meloidogyne chitwoodi* and *M. fallax*. *Nematology* 8, 895–902.

Wiggers, R.J., Starr, J.L. and Price, H.J. (1990) DNA content variation and chromosome number in plant cells affected by *Meloidogyne incognita* and *M. arenaria*. *Phytopathology* 80, 1391–1395.

Willers, P. (1997) First record of *Meloidogyne mayaguensis* Rammah & Hirschmann, 1988: Heteroderidae on commercial crops in the Mpumalanga province, South Africa. *Inligtingsbulletin-Instituut vir Tropiese en Subtropiese Gewasse* 294, 19–20.

Zhou, E., Wheeler, T.A. and Starr, J.L. (2000) Root galling and reproduction of *Meloidogyne incognita* populations from Texas on resistant cotton genotypes. *Supplement to Journal of Nematology* 32, 513–518.

2 General Morphology

Jonathan D. Eisenback[1] and David J. Hunt[2]
[1]*Virginia Polytechnic Institute and State University, Blacksburg, Virginia, USA;*
[2]*CABI Europe-UK, Egham, Surrey, UK*

2.1 General Morphology **18**
2.2 Body Wall **23**
2.3 Nervous System **37**
2.4 Digestive System **40**
2.5 Secretory–Excretory System **42**
2.6 Reproductive System **43**
2.7 Morphological Methods **44**
2.8 Minimum Standards for Describing a New Species **44**
2.9 References **50**

2.1 General Morphology

After their first discovery on the roots of cucumber in an English glasshouse (Berkeley, 1855), the root-knot nematodes were soon recognized as important pathogens on numerous host plants all around the world (Greeff, 1864; Licopoli, 1875; Cornu, 1879; Jobert, 1878; Örley, 1880; Bellati and Saccardo, 1881; Müller, 1884; Treub, 1885; Atkinson, 1889; Neal, 1889; Cobb, 1890; Palumbo, 1892; Viala, 1893; van Breda de Haan, 1899; Lavergne, 1901). Near the end of the 19th century, recognition of the economic importance of root-knot nematodes stimulated several detailed morphological studies of these nematodes (Müller, 1884; Göldi, 1887; Atkinson, 1889; Neal, 1889; Cobb, 1890). Although additional contributions on the morphology of all stages in the life cycle of root-knot nematodes were made by Nagakura (1930), the most significant study in this respect was by Chitwood (1949), who revealed that the root-knot nematodes comprised several different species, including *Meloidogyne incognita* (Kofoid & White, 1919) Chitwood, 1949, *M. arenaria* (Neal, 1889) Chitwood, 1949, *M. javanica* (Treub, 1885) Chitwood, 1949 and *M. hapla* Chitwood, 1949.

Histological studies (Elsea, 1951; Maggenti and Allen, 1960) added important details to our understanding of the morphology of the root-knot nematodes, and, shortly thereafter, the use of the transmission electron microscope (TEM) greatly enhanced our knowledge of the fine structure of these nematodes (Bird, 1958, 1959, 1979b; Bird and Rogers, 1965a,b; Bird and Saurer, 1967; Bird and Soeffky, 1972; Baldwin and Hirschmann, 1973, 1975, 1976; Dropkin and Acedo, 1974; Bird and McClure, 1976; Johnson and Graham, 1976; McClure and Bird, 1976; Wergin and Endo, 1976; Baldwin *et al.*, 1977; Endo and Wergin, 1977; Goldstein and Triantaphyllou, 1980; Shepherd and Clark, 1983), as well as the changes in morphology that occur during the onset of parasitism (Bird 1967, 1968a,b, 1969, 1971a).

Additional observations on the external morphology of the various life stages were illuminated by the use of the scanning electron micro-

 ©CAB International 2009. *Root-knot Nematodes* (eds R.N. Perry, M. Moens and J.L. Starr)

scope (SEM). This instrument clarified our understanding of the perineal pattern (Yik and Birchfield, 1978), the morphology of the second-stage juvenile (J2) (Eisenback and Hirschmann, 1979a,b), male (Eisenback and Hirschmann, 1980, 1981) and female heads, and excised stylets (Eisenback *et al.*, 1980). SEM of excised stylets of the J2 revealed minute differences among the four best-known species (Eisenback, 1982).

The morphology of the root-knot nematodes has been reviewed by several authors (Whitehead, 1968; Bird, 1971b, 1979b; Franklin, 1971, 1978; Esser *et al.*, 1976; Taylor and Sasser, 1978; de Guiran and Ritter, 1979; Jepson, 1983a,b,c, 1987; Eisenback, 1985a,b; Hirschmann, 1985a,b; Eisenback and Triantaphyllou, 1991; Kleynhans, 1991; Karssen, 2002; Karssen and Moens, 2006).

The morphological details of root-knot nematodes are important for the identification of species (Chitwood, 1949; Esser *et al.*, 1976; Franklin, 1978; Taylor and Sasser, 1978; Eisenback and Hirschmann, 1981; Eisenback *et al.*, 1981; Jepson, 1983a,b,c, 1987; Hirschmann, 1985b; Kleynhans, 1991; Karssen, 2002; Karssen and Moens, 2006) and for identifying phylogenetic relationships (Franklin, 1971; Wouts, 1979; Hirschmann, 1985a). In addition, these morphological details are often used to determine physiological function (Elsea, 1951; Maggenti and Allen, 1960; Bird and Saurer, 1967; Bird, 1968a,b, 1969, 1979c; Baldwin and Hirschmann, 1973, 1975, 1976; Dropkin and Acedo, 1974; McClure and Bird, 1976; Wergin and Endo, 1976; Dropkin and Bird, 1978; Viglierchio, 1979; Shepherd and Clark, 1983).

Morphological observations are useful in interpreting the interaction of the environment with the nematode (Papadopoulou and Triantaphyllou, 1982), and also provide insight into the intricate host–parasite relationship (Bird, 1979a; de Guiran and Ritter, 1979); this interaction may have an influence on nematode morphology (Bird, 1967, 1968a,b, 1971a).

During the complex life cycle of the root-knot nematodes (Fig. 1.1), the morphology changes from a one-celled zygote to a small, vermiform, first-stage juvenile, which moults once in the egg to become the infective J2, the stage that subsequently hatches from the egg. The infective juveniles may move freely out from the egg mass into the surrounding soil and search for the root tip of a suitable host plant or, in the case of second and further generations of an established infection,

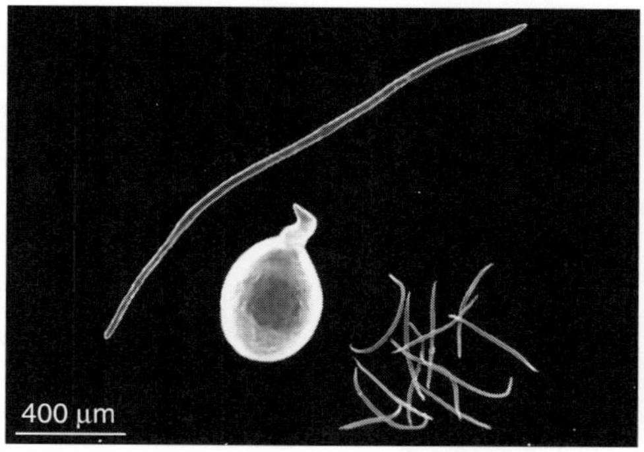

Fig. 2.1. Scanning electron micrograph of root-knot nematode male (left), female (centre) and 13 second-stage juveniles (right), showing general body shape and relative dimensions of the stages. After Eisenback and Triantaphyllou (1991), courtesy of Marcel Dekker, Inc.

may migrate entirely within the root tissue before establishing a new feeding site. The root cortex is penetrated near the zone of differentiation, the J2 migrating predominantly intercellularly toward the root tip and then making a 180° turn back towards the zone of differentiation (Wyss *et al.*, 1992). Once inside the vascular cylinder the J2 initiate a series of three to eight giant cells, where they feed for 3–8 weeks and greatly increase in size by swelling into a sausage-shaped juvenile with a characteristically spicate tail region. The swollen J2 rapidly moult into short-lived third- and fourth-stage juveniles. Fourth-stage juveniles that will develop into males become vermiform after the third moult, but juveniles destined to become females remain swollen. Both types of fourth-stage juveniles moult once more to become either a mature male or a female. The morphology of the J2 and the adult male and female (see Fig. 2.1 for comparison of the general appearance of these stages) will now be described according to body system.

2.1.1 Second-stage juvenile (Figs 2.1–2.9)

Depending upon certain environmental signals, some J2 may enter into a diapause and remain in the egg, where they overwinter (de Guiran and Ritter, 1979; see Evans and Perry, Chapter 9, this volume). Those that hatch from the egg are quite mobile and capable of moving long distances

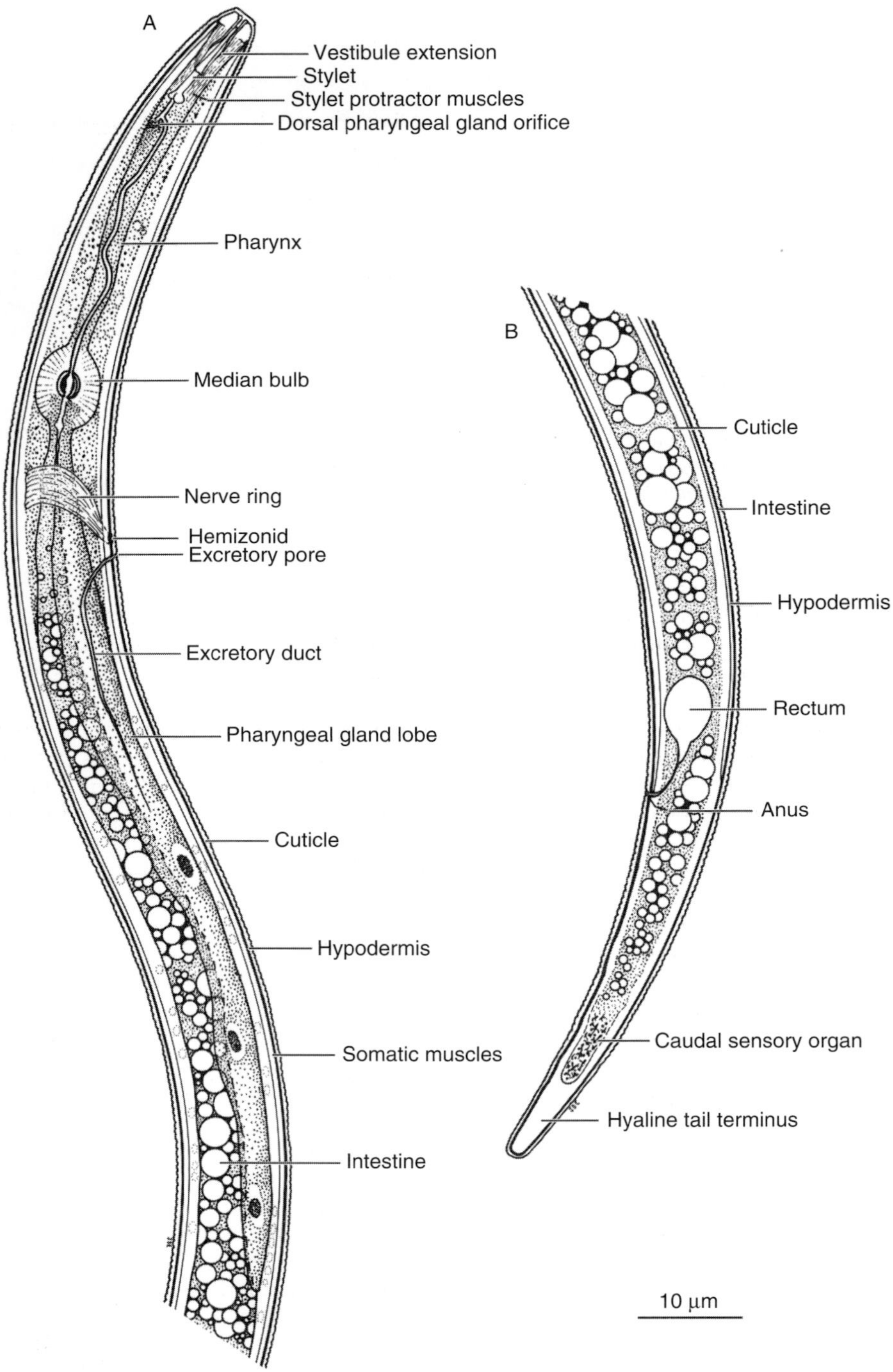

Fig. 2.2. Drawings of second-stage juvenile root-knot nematode. A: anterior region; B: posterior region. After Eisenback (1985a), courtesy of N.C. State University Graphics.

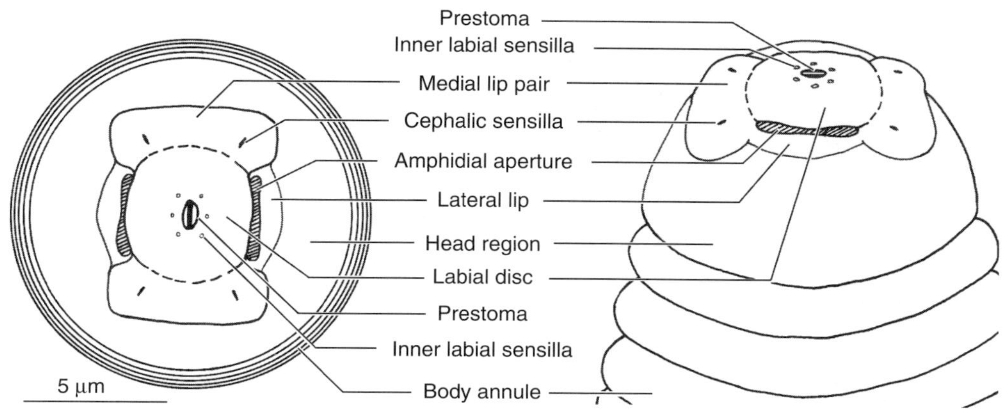

Fig. 2.3. Drawing of anterior end of second-stage juvenile of root-knot nematode, as revealed by scanning electron microscopy, in face and lateral views. After Eisenback and Hirschmann (1979b), courtesy of *Journal of Nematology*.

(40–100 cm) vertically within the soil profile when soil moisture levels are optimum. Mobility allows the J2 to find a suitable host root tip, to penetrate into the cortex and then move to the preferred feeding site. The body wall, together with its protective cuticle and somatic muscles, is controlled by the nervous system, which allows the nematode to respond to environmental cues (see Curtis *et al.*, Chapter 6, this volume) that enable it to move to a suitable site for establishing a host–parasite relationship. The digestive system (Fig. 2.2) initiates the formation of a feeding site, and the nutrients that the nematode absorbs from the plant are stored in the intestine, which becomes enlarged, resulting in a swollen juvenile that is no longer capable of movement. The energy stored in the intestine is eventually transferred to the small genital primordium, which subsequently develops into the adult reproductive systems (Triantaphyllou, 1960, 1962, 1979; Papadopoulou and Triantaphyllou, 1982).

2.1.2 **Male** (Figs 2.1, 2.9–2.16)

Males become vermiform after the third moult and remain vermiform during the fourth moult into the adult stage, the nematode being enclosed within the shed cuticles of the previous stages. They become mobile as adults and eventually emerge from the retained cuticles before leaving the gall and entering the soil phase. There they migrate and attempt to find a female for sexual reproduction. As with the J2, the body wall and nervous system allow the male nematode to respond to environmental cues and to move through the soil in its search for a mate. In species that reproduce by amphimixis males are quiet common, but in parthenogenetic species males may be very rare and are unnecessary for reproduction (Triantaphyllou, 1979). Unlike the J2 and female, the male does not feed, all of the energy required for the development of its reproductive system being obtained while it was a J2. As a consequence, the male pharyngeal glands are degenerate and probably not functional, while the intestine serves as a storage organ for the food reserves obtained as a J2 and supplies energy to the reproductive system for the production of sperm.

2.1.3 **Female** (Figs 2.1, 2.17–2.23)

Females are pear-shaped and sedentary (Figs 2.17, 2.18), although the neck region remains musculated and allows the nematode to change head position so that she can feed on one of several giant cells. The increase in body size and change in shape adds volume to the reproductive system, which is in close contact with the large amorphous intestine. While the digestive system is specialized for the maintenance of the giant cells and the withdrawal of nutrients from the plant, the intestine is less specialized and serves primarily as a storage organ for the nutrients

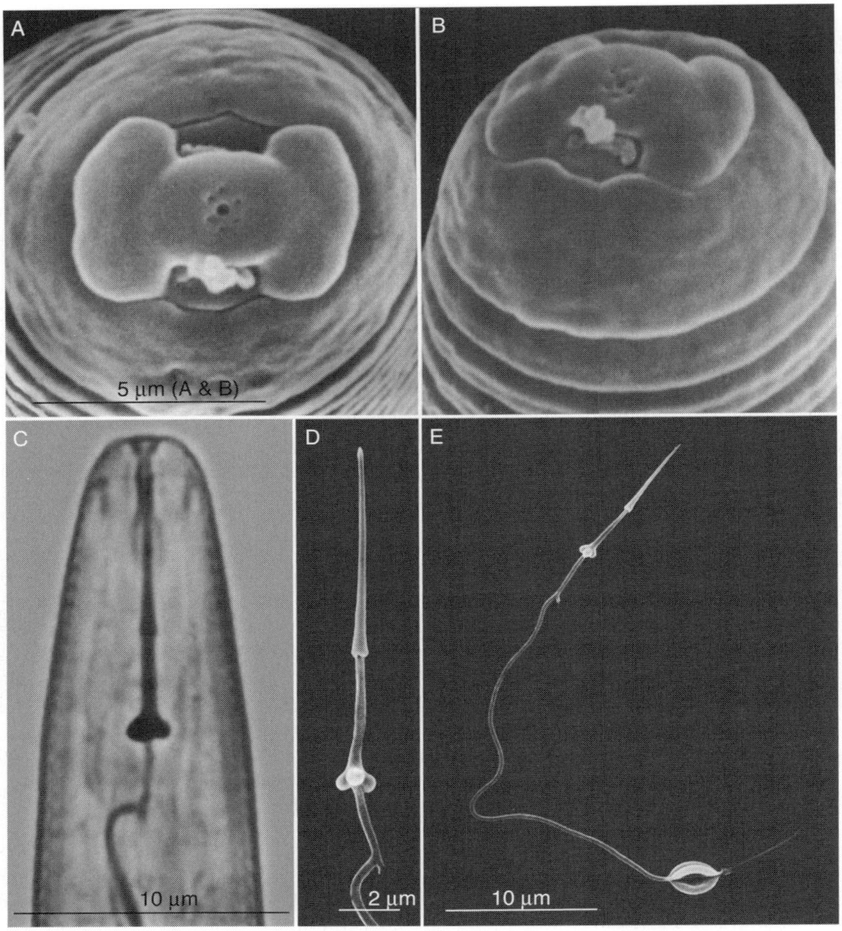

Fig. 2.4. Micrographs of second-stage juvenile root-knot nematode. A: scanning electron micrograph (SEM) of anterior end in face view; B: SEM of anterior end in lateral view; C: light micrograph of anterior, showing shape of head and stylet morphology; D: SEM of excised stylet; E: SEM of excised stylet with attached cuticular lumen lining of entire pharynx. After Eisenback (1982) and Eisenback and Hirschmann (1979a), courtesy of *Journal of Nematology* and *Scanning Electron Microscopy*.

ingested from the plant. Oddly, the intestine is not attached to the rectum and therefore has a blind ending. Instead, six large rectal gland cells empty their contents through the anus to form the gelatinous matrix that surrounds and protects the eggs as they are deposited through the vulva. The two ovaries of the female genital system are in direct contact with the intestine so that nutrients stored there simply diffuse across a few cell membranes to become available for growth of the oogonia and oocytes into mature eggs.

2.1.4 Egg

Eggs of root-knot nematodes vary in size and shape but are typically approximately 95 μm long and 40 μm in diameter (Saigusa, 1957; Bird and McClure, 1976; McClure and Bird, 1976). The eggshell consists of an outer vitelline layer approximately 30 nm thick, a middle chitinous layer about 400 nm thick and an inner glycolipid layer of varying thickness. The glycolipid layer makes the egg very resistant to harsh chemicals

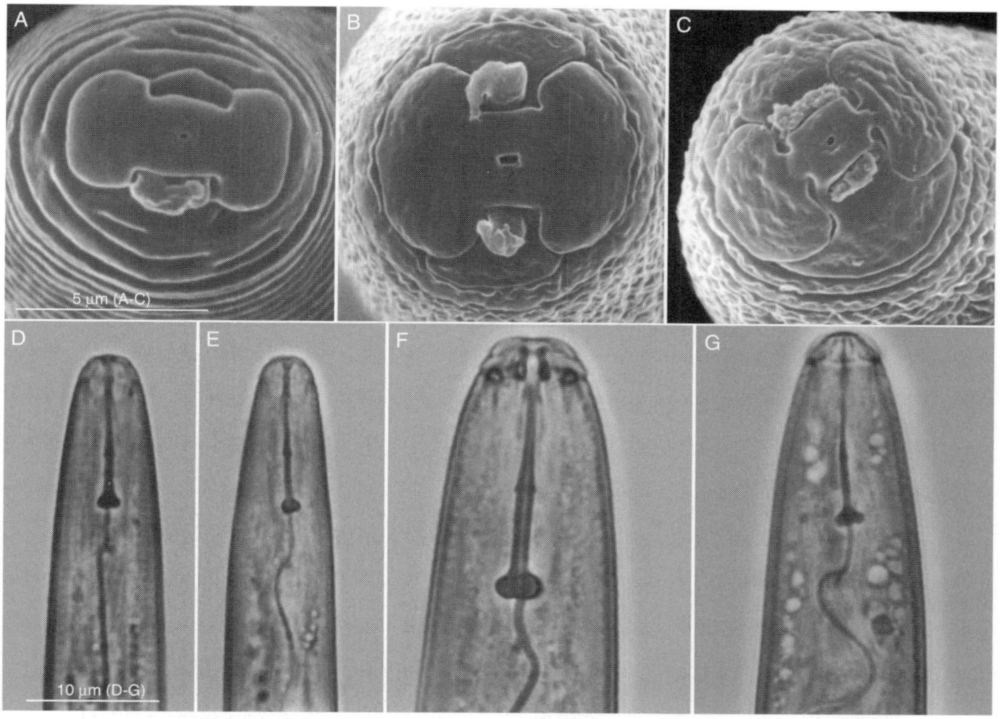

Fig. 2.5. Micrographs of anterior end of second-stage juvenile root-knot nematode. Scanning electron micrographs. A: *Meloidogyne incognita*; B: *M. brevicauda*; and C: *M. nataliae*. Light micrographs. D: *M. arenaria*; E: *M. hapla*; F: *M. brevicauda*; G: *M. nataliae*. After Eisenback (1988), courtesy of Plenum Press.

and, as a consequence, this stage is not sensitive to toxins such as common nematicides.

2.2 Body Wall

In the female, the body wall protects her from the outside environment, but in the J2 and male it also enables them to move through the soil. The body wall has three major layers: the cuticle, hypodermis and somatic muscles (Bird, 1979b). Much of the nervous system is contained within the hypodermis.

2.2.1 Cuticle

The non-cellular, elastic cuticle is secreted by the hypodermis and covers the entire body and all of the openings, including the lining of the pharynx, the amphidial canals, the secretory/excretory duct, the phasmidial ducts, the male cloaca, the J2 rectum and the female vulva (Bird, 1979b). It functions as an interface between the organism and the harsh environment found in the soil, and protects the mobile juvenile and male from numerous biological, chemical and physical hazards. The diffusion of liquids through the body wall is controlled by the structure of the cuticle. Furthermore, the cuticle restricts the body from increasing in diameter and, in concert with the high turgor pressure inside the pseudocoelom, acts as an exoskeleton that enables the nematode to move freely through the soil or plant tissue.

Like most nematodes, the cuticle of the J2 and male has three layers: cortical, medial and basal. The female, however, only has a cortical and basal layer (Bird, 1979b). These layers vary in thickness according to the life stage. In the J2 the cuticle is 0.3–0.4 µm thick, whereas it is 1.5 µm thick in the male and 4–6 µm in the female, where

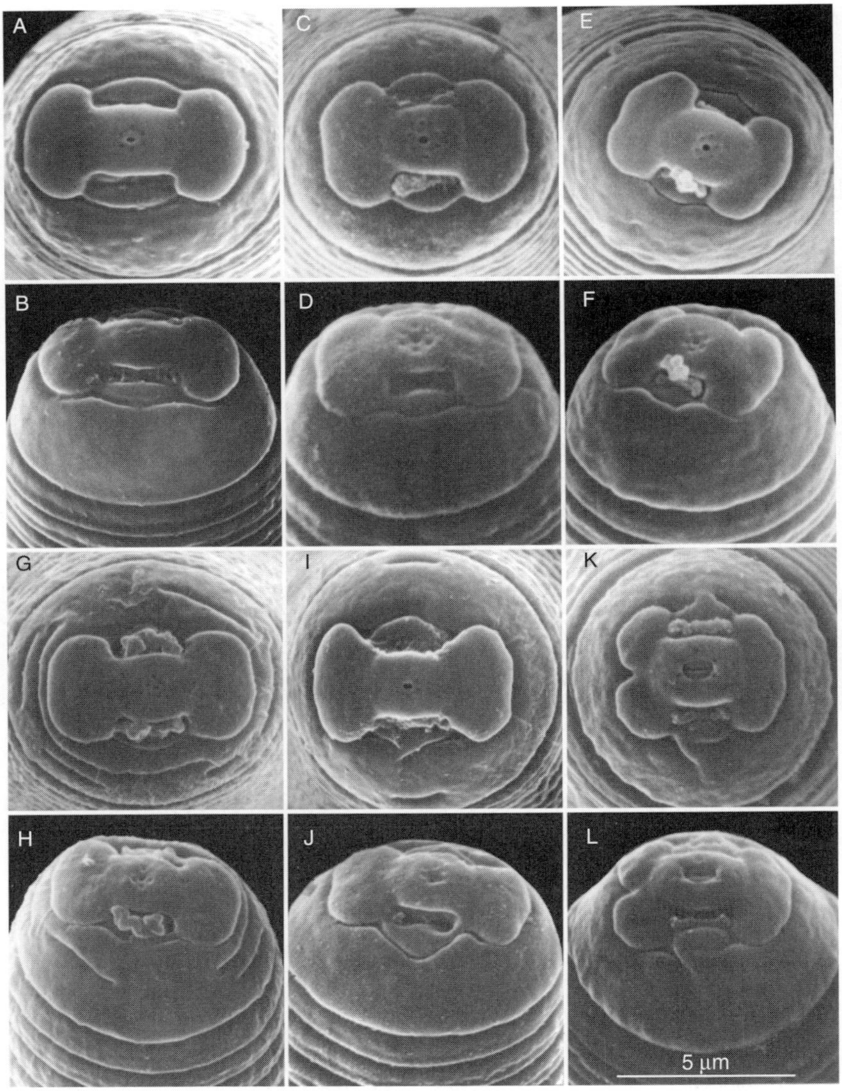

Fig. 2.6. Scanning electron micrographs of face and lateral views of anterior end of second-stage juveniles of several species of root-knot nematodes. A, B: *Meloidogyne arenaria*; C, D: *M. exigua*; E, F: *M. hapla*; G, H: *M. incognita*; I, J: *M. javanica*; K, L: *M. megatyla*. After Eisenback and Hirschmann (1979a), courtesy of *Scanning Electron Microscopy*.

it is thickest in the perineal region and thinnest near the anterior end (Bird, 1979b).

The cortical layer in all three stages is divided into a tri-part external cortical layer composed of two thin, dark-staining regions with a thin, lightly staining inner region, and a thicker internal cortical layer. In the J2, the external cortical layer varies from 35 to 40 nm thick, while in the male it is 100 nm thick and in the females it varies from 100 to 125 nm (Baldwin and Hirschmann, 1975; Johnson and Graham, 1976; Bird, 1979b). The internal cortical layer is much thicker than the external cortical layer, being 50–100 nm thick in the J2, 400–500 nm thick in the male and 700–900 nm thick in the female

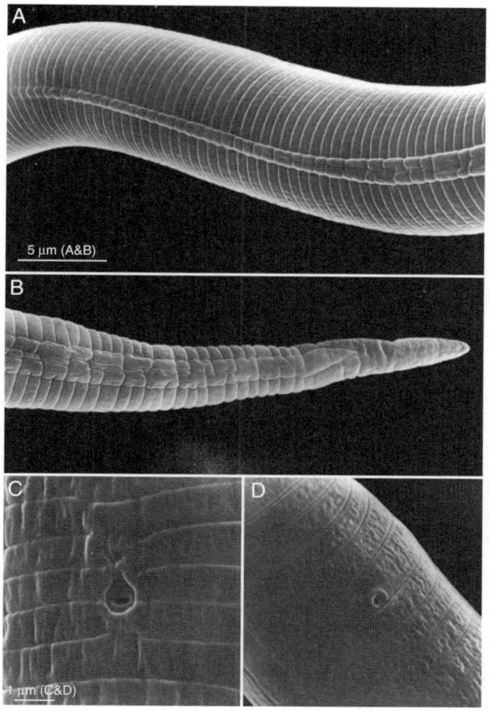

Fig. 2.7. Scanning electron micrographs of second-stage juvenile root-knot nematode. A: anterior portion of body, showing the regular body annulation and beginning of the lateral field; B: posterior end of body, showing end of lateral field and tail terminus; C: secretory/excretory pore; D: anal opening. After Eisenback and Hirschmann (1979b), courtesy of *Journal of Nematology*.

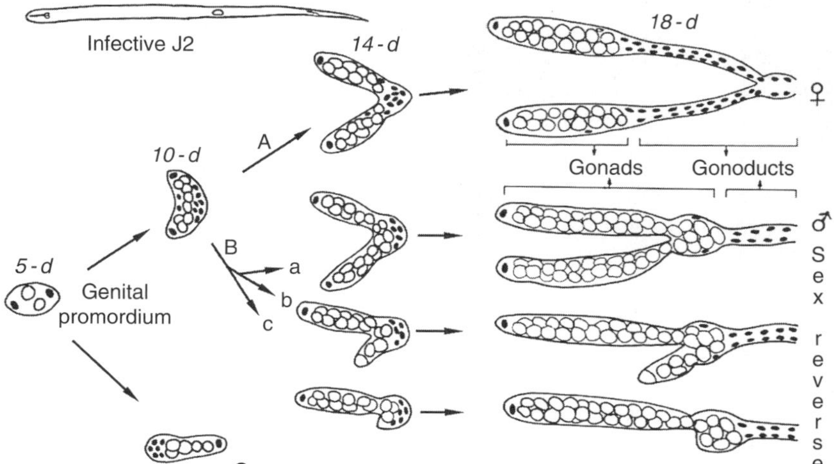

Fig. 2.8. Development of genital primordium into a normal female gonad with two ovaries, sex-reversed gonads of males with two testes and a normal male with only one testis. After Papadopoulou and Triantaphyllou (1982), courtesy of *Journal of Nematology*.

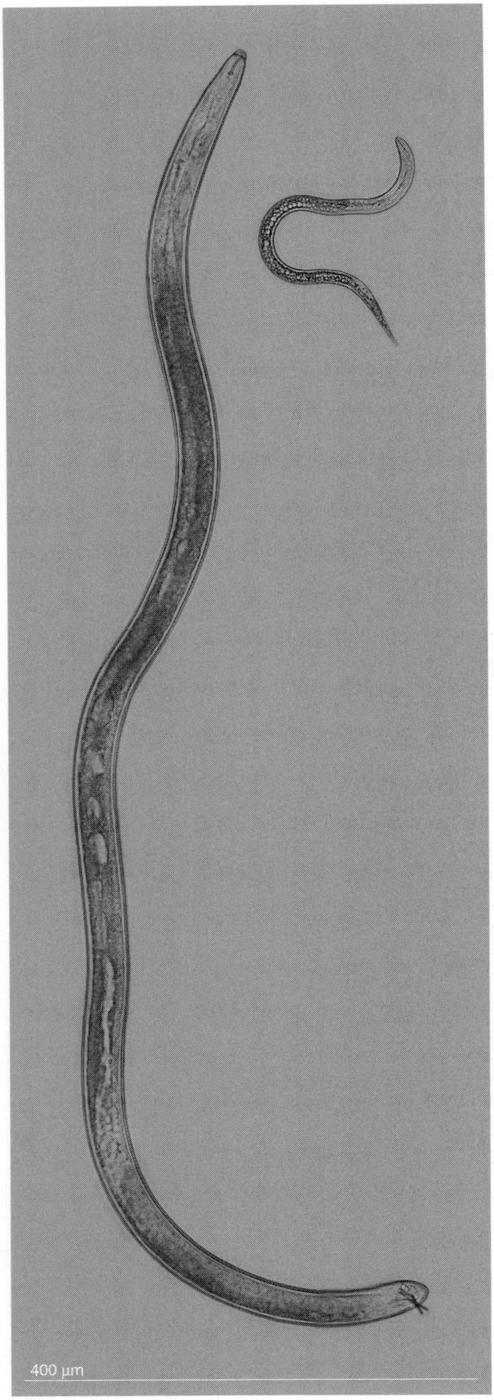

Fig. 2.9. Light micrograph of male and second-stage juvenile root-knot nematode (original).

(Baldwin and Hirschmann, 1975; Johnson and Graham, 1976; Bird, 1979b). Variation in the thickness of this layer is caused by the indentations of the body annulations, the layer being thinner beneath each annulation and thicker under each body annule; this layer is fibrillar. Unlike the other layers of the cuticle, which are broken down and recycled by the hypodermis, the external cortical layer must be cast off during the moulting process.

Although the medial layer is present in the vermiform, mobile J2 and male, it is lacking in the female. This layer may be continuous with the substrate of the internal cortical layer, but it contains electron-dense globules that may be filled with a liquid in the living nematodes. In the J2, the medial layer varies greatly in thickness where it occurs under the body annules (100 nm thick) but disappears completely under the annulations. In the male, this layer is present beneath the annulations and annules but varies in thickness from 300 to 400 nm where the globules are more dense between the annules, and in the lateral fields near the anterior end (Baldwin and Hirschmann, 1975; Johnson and Graham, 1976; Bird, 1979b).

The basal layer is the thickest of the three layers. It is 100–125 nm thick in the J2, moderately thick in the male (400–500 nm) and very thick in the female (3000–4000 nm). In both the male and pre-parasitic juvenile this layer is striated by two sets of laminae orientated at right angles to one another (Popham and Webster, 1978). However, in the female these striae are absent. The orientation of the parallel striae, which occur approximately 22 nm from each other, forms crystalline laminae, which may be either structurally important in protecting the nematode from the harsh external environment or a necessary structural component vital for movement through the soil. A few days after the J2 establishes a host–parasite relationship with the plant, the striae break down and disappear. They are not found in any of the other juvenile stages and reappear only in the cuticle of the mobile male (Bird, 1971b). The laminae appear to be modified beneath the lateral fields where the striae fork, and are replaced by obliquely orientated fibrillar layers at the edges of the lateral field. They are absent from the anterior end of the nematode beyond the normal body annules. Likewise, in the tail tip of the J2 the striae become very disorganized (Bird, 1979c).

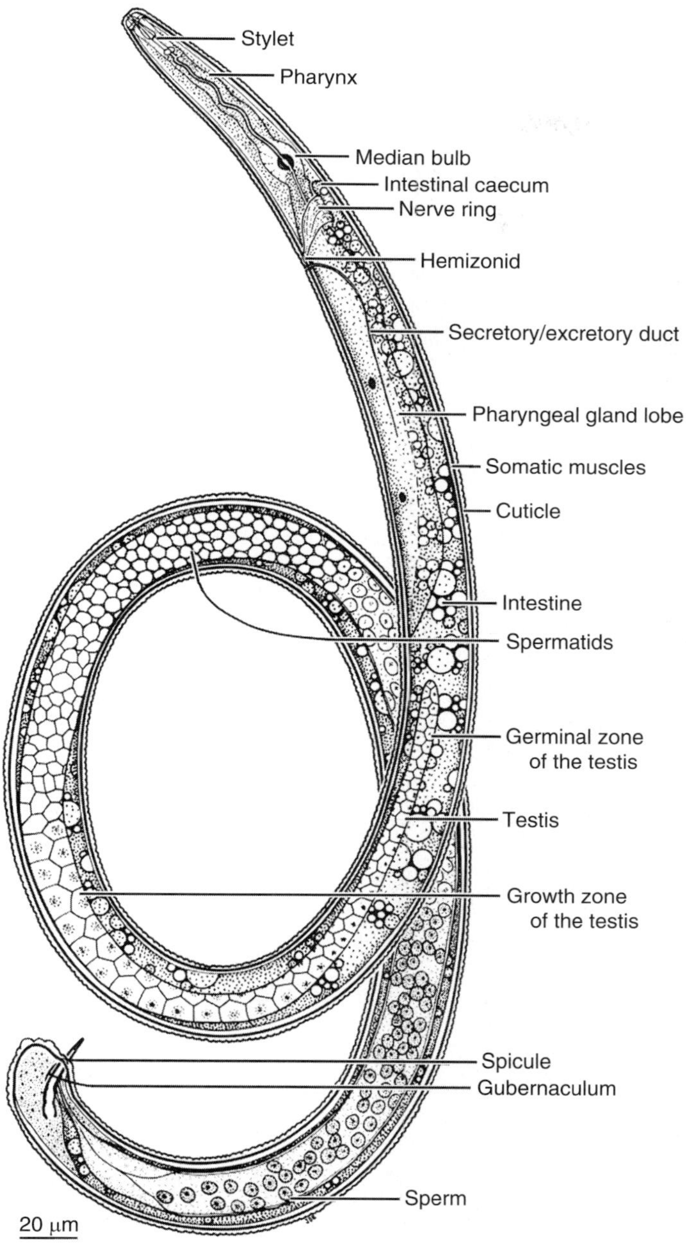

Fig. 2.10. Drawing of entire specimen of male root-knot nematode. After Eisenback (1985a), courtesy of N.C. State University Graphics.

The J2, male and female have similar external cuticular markings, but these differ in their expression (Eisenback and Hirschmann, 1979a,b, 1980, 1981; Eisenback *et al.*, 1980; Jepson, 1987; Kleynhans, 1991; Karssen and Moens, 2006). In all three life stages the slit-like stomatal opening is situated within an ovoid, prestomatal depression. The prestoma is surrounded by six small,

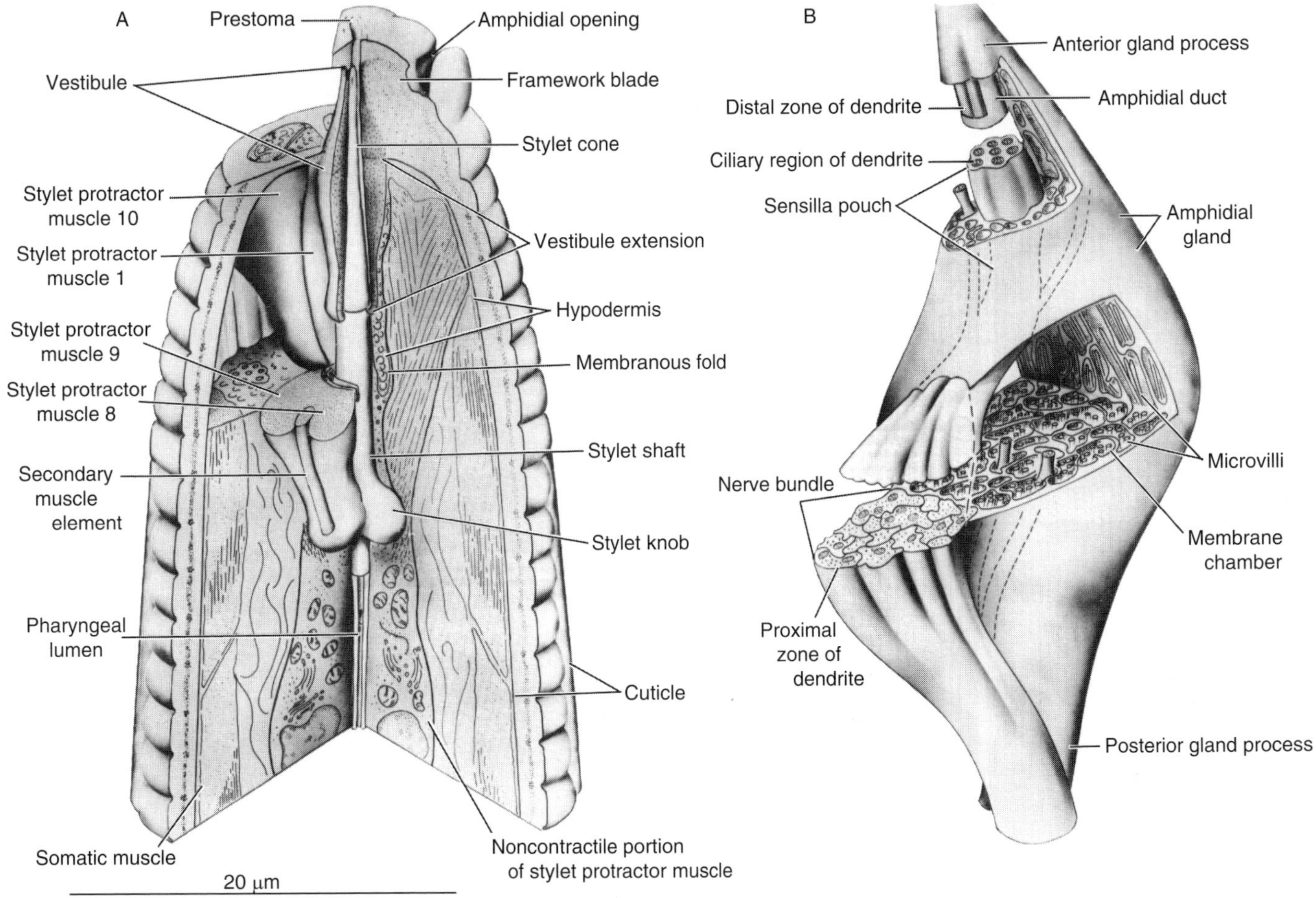

Fig. 2.11. 3-D reconstructions of male root-knot nematode as revealed by transmission electron microscopy. A: anterior end; B: amphid. After Baldwin and Hirschmann (1973, 1976), courtesy of *Journal of Nematology*.

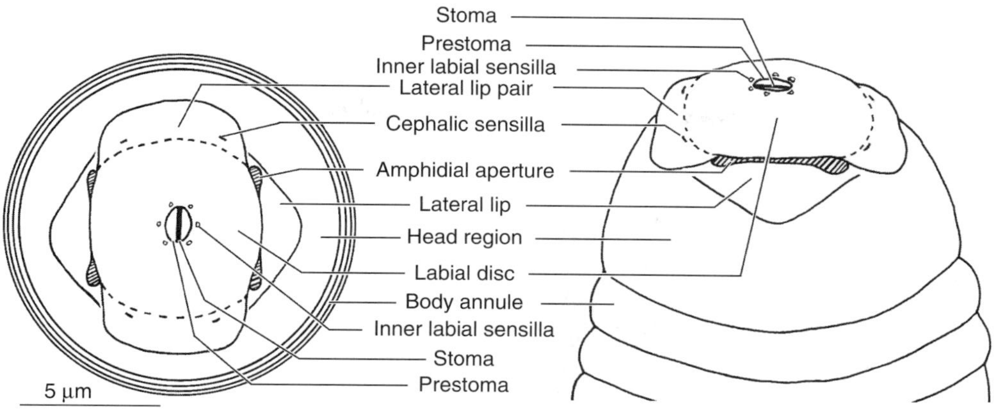

Fig. 2.12. Drawings of anterior end of male root-knot nematode in face and lateral views, as revealed by scanning electron microscope. After Eisenback and Hirschmann (1981), courtesy of *Journal of Nematology*.

pit-like openings of the inner labial sensilla (see Figs 2.6 and 2.13F, for example). In both male and female these sensilla sometimes occur on the edge of the prestoma, causing it to appear hexagonal. The prestoma, stoma and inner labial sensilla are located on the labial disc. In the J2 this disc is moderately sized (Fig. 2.3), but in the male it is relatively large (Fig. 2.12), while in the female it is relatively small (Fig. 2.19). This structure is generally fused with the medial lip pairs and slightly covers the amphidial apertures in face view, to form the head cap. Each medial lip pair contains the nerve endings of two cephalic sensilla, which are located just beneath the cuticle, their positions being sometimes visible as slight depressions. The lateral lips occur posterior to the amphidial apertures and can be either fused with the medial lip pairs or completely or partially fused with the head region. The lips are not a part of the head cap but instead form part of the head region.

Located between the labial disc and lateral lips, the amphidial apertures are elongated ovals in the J2 (Figs 2.3, 2.4A,B, 2.5A–C, 2.6) and female (Figs 2.19, 2.22A,B), but are more slit-like in the male (Figs 2.12, 2.13D–G, 2.14A–C). Depending on the speed of fixation, these openings are often plugged with secretions from the amphidial gland. The secretions are thought to be necessary for cleansing the nerve endings of the amphids.

A large head annule (= the head region) occurs posterior to the head cap and lateral lips (Figs 2.3, 2.12, 2.19). This region may be completely smooth or marked by one to seven complete or incomplete annulations (see Figs 2.13E, 2.14B, for example). Following the head region, regular, evenly spaced body annules continue along the entire length of the J2 and male. They are spaced 0.8–1.0 μm apart in the J2 (Fig. 2.7A) and 2.0–2.5 μm in the male (Fig. 2.15F,G). In the female, the head region is usually less distinct and may sometimes be difficult to distinguish from the regular body annulations (Fig. 2.22A,B). As the body annulations continue along the body of the female, they become less and less set-off, until they become more closely spaced and more distinct where they encircle the perineal area (Figs 2.22E,F, 2.23). The regular body annulations are interrupted near the base of the stylet by a wide, raised lateral field on each side of the body in the J2 and male. Furthermore, the body annulations are interrupted ventrally in all three life stages by the small, ovoid opening of the secretory/excretory pore and by the cloacal opening in the male and anus in the J2 and female (Eisenback and Hirschmann, 1979a,b; Eisenback and Triantaphyllou, 1991).

Commencing as a single ridge about 8–15 body annules from the anterior end (Fig. 2.7A), the lateral field runs the entire length of the nematode. Additional incisures appear along the length of the lateral field, forming three or four incisures in total. In some species the number of lateral lines may increase to more than four, but the number varies along the body and sometimes additional

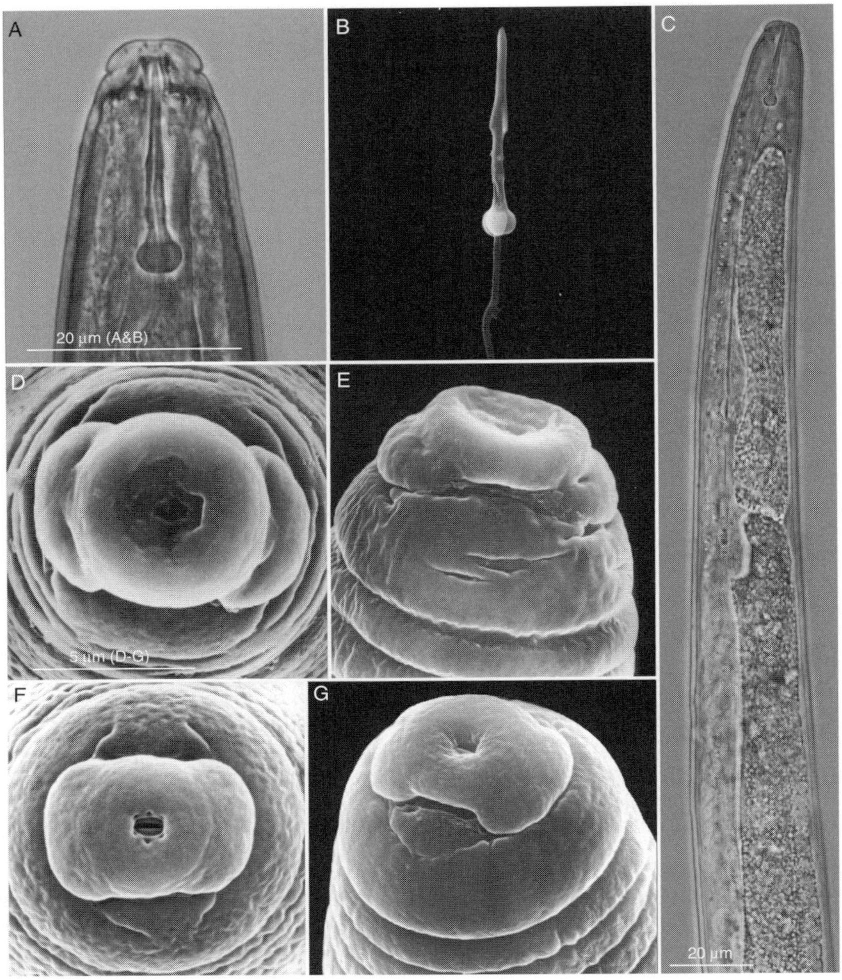

Fig. 2.13. Photomicrographs of male root-knot nematode. A: light micrograph of anterior end, showing head shape and stylet morphology; B: scanning electron micrograph (SEM) of excised male stylet with attached cuticular lumen lining of pharynx; C: light micrograph of anterior end, showing most of pharynx, including stylet, procorpus, metacorpus, basal gland lobe, intestinal caecum and intestine D: SEM of anterior end of *Meloidogyne incognita* in face view; E: SEM of anterior end of *M. incognita* in lateral view; F: SEM of anterior end of *M. hapla* in face view; G: SEM of anterior end of *M. hapla* in lateral view. After Eisenback and Hirschmann (1981), courtesy of *Journal of Nematology*.

folds can form between the lines, thereby making an exact count difficult. In the J2, the lateral fields narrow in the tail region where the two middle incisures merge, resulting in just three incisures. The middle incisure so formed then gradually disappears, the two outer incisures finally merging towards the end of the tail (Fig. 2.7B). In the male, however, the lateral fields appear either to encircle the tail tip or to become very wide and gradually merge with the tail terminus (Fig. 2.15D). The two inner incisures either disappear near the cloacal opening or occur almost all the way to the tail tip. The lateral field is areolated by transverse markings that correspond to the regular body annulation, although they are often incomplete and irregular (Fig. 2.7A,B).

In the female, regular body annules mark the neck region (Fig. 2.22D), where they become

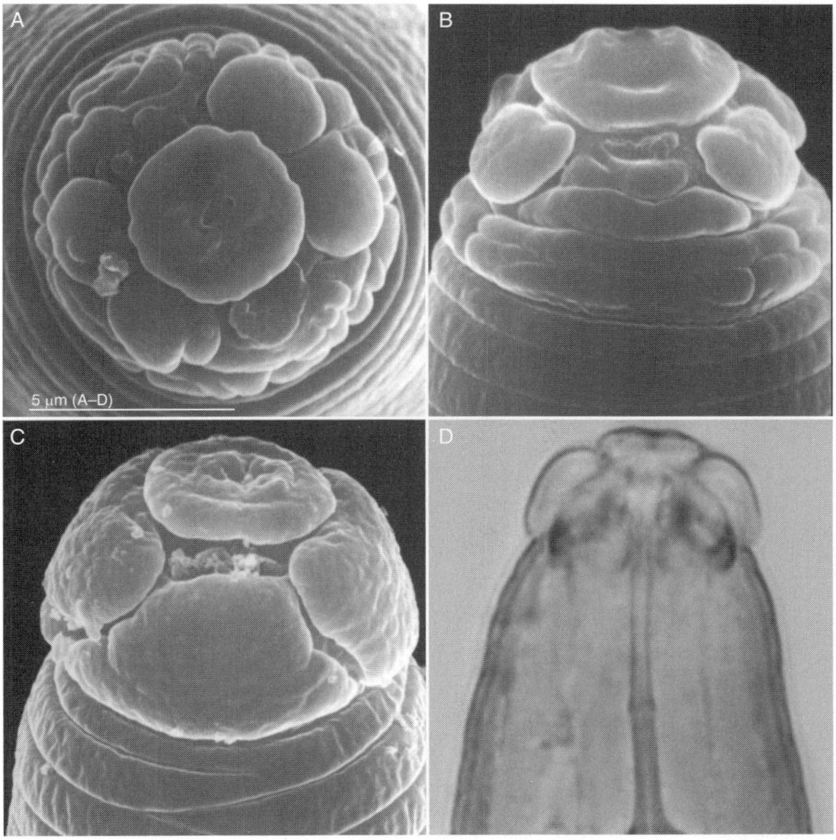

Fig. 2.14. Scanning electron micrographs (SEM) and light micrographs (LM) of male root-knot nematodes. A: SEM of *Meloidogyne nataliae* face view; B: SEM of *M. nataliae* in nearly lateral view; C: SEM of *M. brevicauda* in nearly lateral view. D: LM of *M. brevicauda*. After Eisenback (1988), courtesy of Plenum Press.

more shallow as they proceed down the body, until they nearly disappear altogether in the vicinity of the mid-body region (Eisenback *et al.*, 1980). Near the posterior end, these annulations reappear and become spaced closer together with deeper striae, thereby forming the characteristic, fingerprint-like perineal pattern, which encompasses the vulva, anus and tail tip (Figs 2.18, 2.22E,F, 2.23). The lateral fields are irregular and incomplete in the female, a feature that makes counting the number of incisures uncertain. These fields disappear in the middle of the body but may reappear posteriorly in a few species that have distinct lateral ridges in the perineal pattern (Eisenback *et al.*, 1981).

The phasmidial openings of the J2 and female may be obscured by other features within the lateral field, but in the male they appear as small slit-like openings within the inner incisures at the level of the cloacal opening (Fig. 2.15D) (Eisenback and Hirschmann, 1979a,b; Jepson, 1983c). Likewise, the secretory–excretory pore of the J2 (Fig. 2.7C), male (Fig. 2.15F,G) and female appears as a small, simple, rounded to ovoid ventral opening that may be located in a depression within the cuticle (Fig. 2.15G). Regular body annulations may be interrupted by the occurrence of this pore and these disruptions may be most severe in the female. The anus in the J2 is a small, rounded opening (Fig. 2.7D). In the female it is covered by a flap of cuticle and in the male the cloacal opening is transversely elongate and devoid of regular body annulations, both anteriorly and posteriorly (Fig. 2.15D).

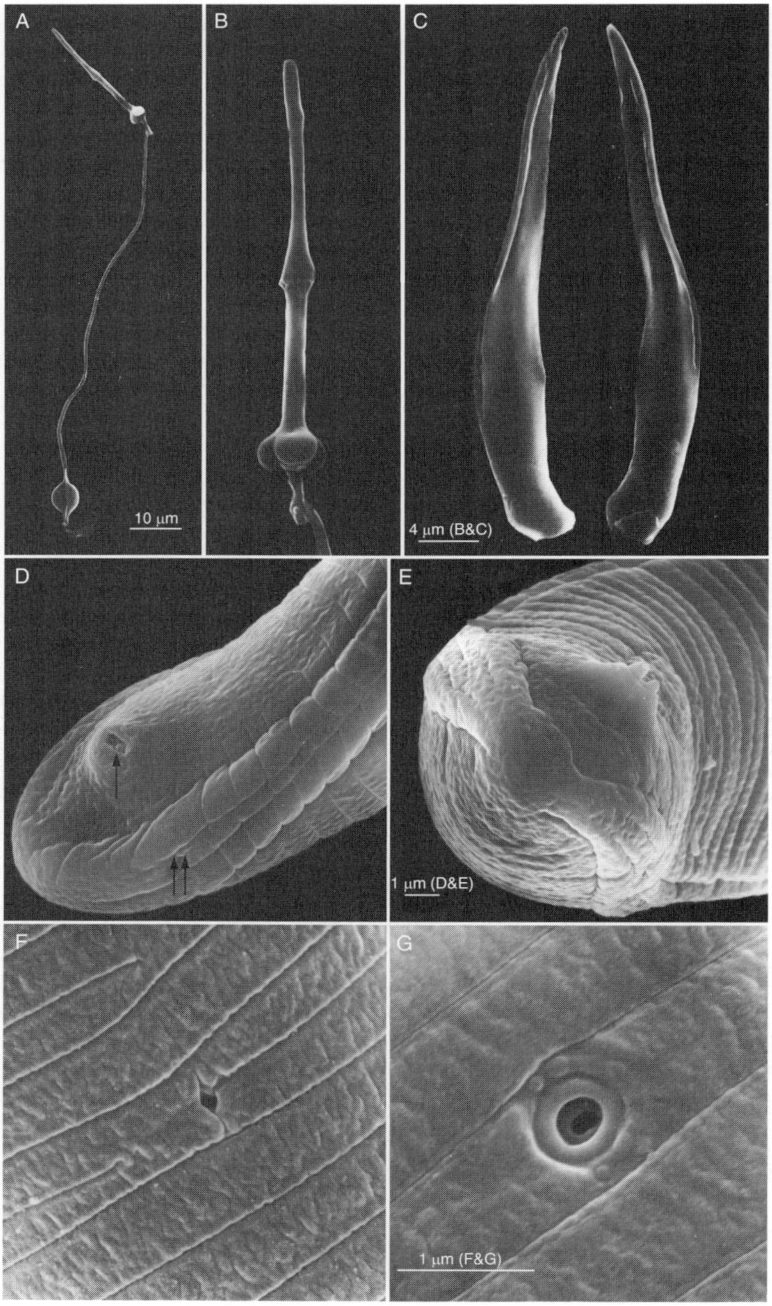

Fig. 2.15. Scanning electron micrographs (SEM) of male root-knot nematodes. A: excised stylet and cuticular lumen lining of pharynx; B: excised stylet; C: pair of spicules; D: posterior end, showing lateral field wrapping around tail, phasmidial opening (double arrows) and cloacal opening (single arrow) with tips of spicules protruding; E: posterior end of tail, showing lateral field and tips of spicules protruding through cloacal opening. F, G; secretory–excretory pore of a typical root-knot nematode and recessed opening of *Meloidogyne javanica*, respectively. After Eisenback (1985a), courtesy of N.C. State University Graphics.

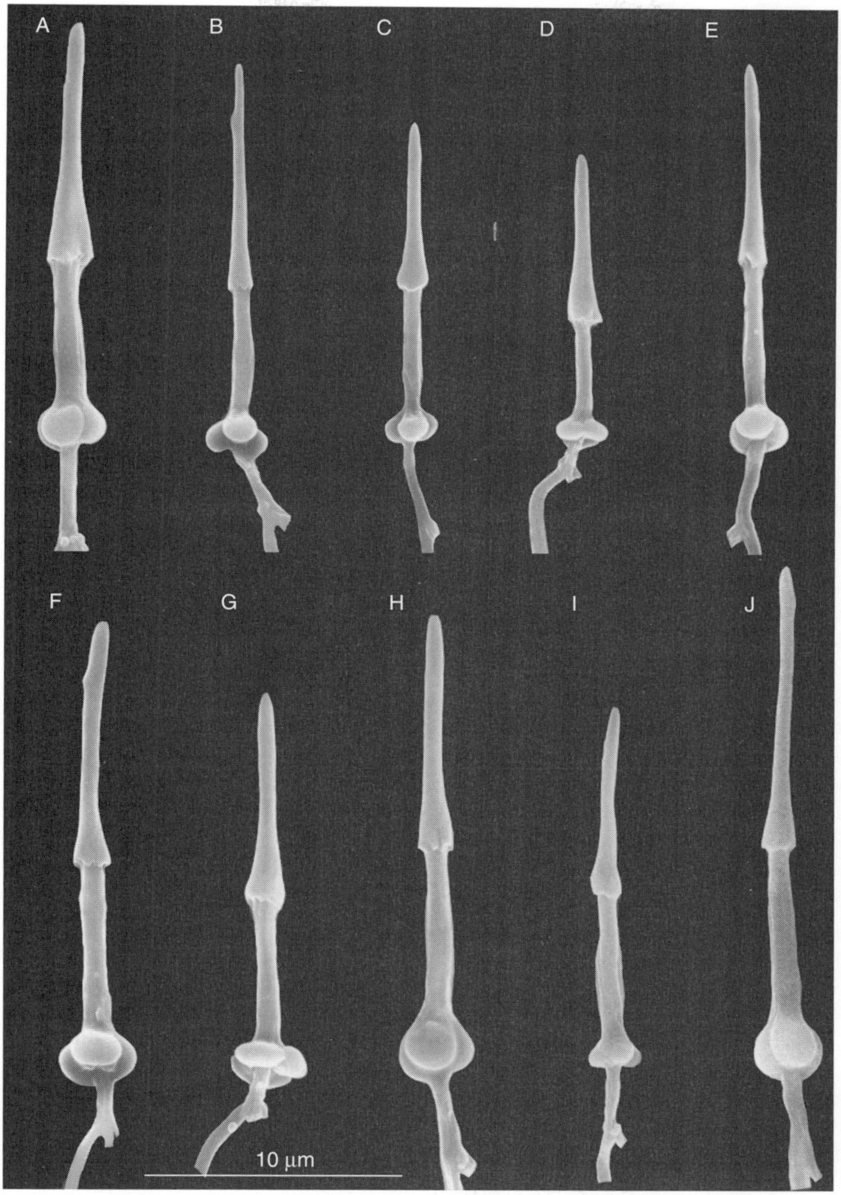

Fig. 2.16. Scanning electron micrographs of excised stylet of male root-knot nematodes. A: *Meloidogyne arenaria*; B: *M. carolinensis*; C: *M. exigua*; D: *M. graminicola*; E: *M. hapla*; F: *M. incognita*; G: *M. javanica*; H: *M. megatyla*; I: *M. naasi*; J: *M. nataliae*. After Eisenback and Hirschmann (1982), courtesy of *Scanning Electron Microscopy*.

2.2.2 Hypodermis

The hypodermis is a thin, living, plasma-membrane-bound syncytium between the cuticle and the somatic muscles. It secretes the cuticle and serves as the interface between the somatic muscles and the cuticle. The hypodermis lines the stomatal cavity, cephalic framework, sensory

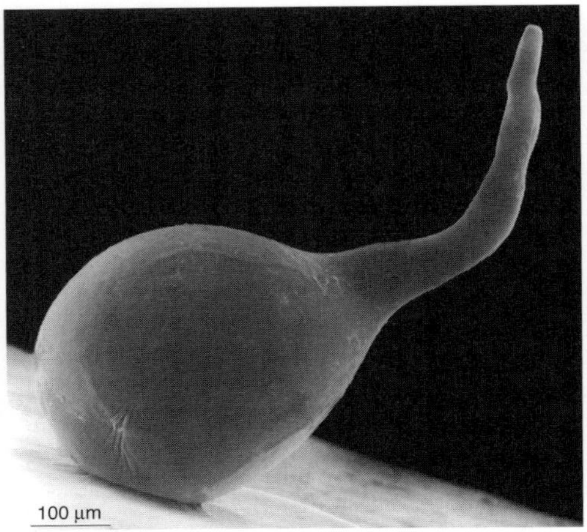

Fig. 2.17. Scanning electron micrograph of entire root-knot nematode female. After Eisenback and Triantaphyllou (1991), courtesy of Marcel Dekker, Inc.

structures and stylet shaft (Fig. 2.11A) (Elsea, 1951; Baldwin and Hirschmann, 1975; Johnson and Graham, 1976; Wergin and Endo, 1976). It extends between the stylet protractor muscle elements and expands between the somatic muscles to form four hypodermal chords: two lateral, one ventral and one dorsal. The lateral chords lie beneath the lateral fields and contain the secretory–excretory duct, longitudinal nerves, transverse nerves and many cell organelles, including mitochondria, ribosomes, Golgi bodies, smooth and rough endoplasmic reticulum, lipid globules and other organelles (Elsea; 1951; Dropkin and Acedo, 1974; Baldwin and Hirschmann, 1975; Johnson and Graham, 1976; Wergin and Endo, 1976). However, the interchordal hypodermis contains none of these organelles, having only nerve processes and numerous hemidesmosomes that serve as attachment points for the somatic muscles to the cuticle (Baldwin and Hirschmann, 1975; Bird, 1979a). The interchordal hypodermis is of varying thickness in the female and approximately 100 nm thick in the J2 and male (Elsea, 1951; Baldwin and Hirschmann, 1975; Bird, 1979a).

2.2.3 Somatic muscles

Somatic muscles are necessary for movement. In the J2, somatic muscles allow the nematode to move inside the egg and thereby facilitate hatch. They also enable it to move through the soil to find a suitable host (Bird, 1967). These muscles help the infective juvenile to penetrate the root and to move to a suitable site for establishment of the host–parasite relationship (Bird, 1967, 1971a). Likewise, somatic muscles in the males allow them to move out of the root tissues where they developed, and to move into the soil in the pursuit of an available mate. In the female, the somatic muscles occur only in the head region, where they allow the nematode to move her head in order to feed on one of several available giant cells (Elsea, 1951; Bird, 1979a,b). The rest of the body lacks these muscles as the female has abandoned her mobile lifestyle in exchange for a protected habitat inside the root system and an enlarged body that significantly increases her reproductive capacity.

Somatic muscle cells are spindle-shaped and grouped into four rows between the hypodermal chords in cross-section. In the J2 and male, longitudinally orientated muscles increase in number as they progress posteriorly down the body wall. There are two interchordal cells in the anterior end, three to four at the base of the stylet, four to five in the pharyngeal region and four to five for the remainder of the body wall (Baldwin and Hirschmann, 1975). The J2 has between 50 and 100 somatic muscle cells, whereas the male has more. In the J2 the somatic muscles end near the level of the anus, whereas they end near the level of the cloacal opening or at the base of the metacorpus in the male and female, respectively.

The contractile portion of a somatic muscle lies adjacent to the hypodermis and cuticle, whereas the non-contractile portion protrudes into the pseudocoelom. In cross-section, the contractile portion is obliquely striated with five to six cycles of I, A, H, A and I bands; however, the tips of the cells contain only fine filaments. Fine filaments only occur in the I bands; A bands contain thick filaments surrounded by a hexagonal pattern of 10–15 fine filaments, while H bands contain only thick filaments. The thin filaments of actin are 6 nm in diameter and the thick filaments of myosin are 22–24 nm in diameter. The muscles contract when the actin filaments slide past the myosin filaments, using the energy that is provided by the high-energy phosphate bonds of myosin.

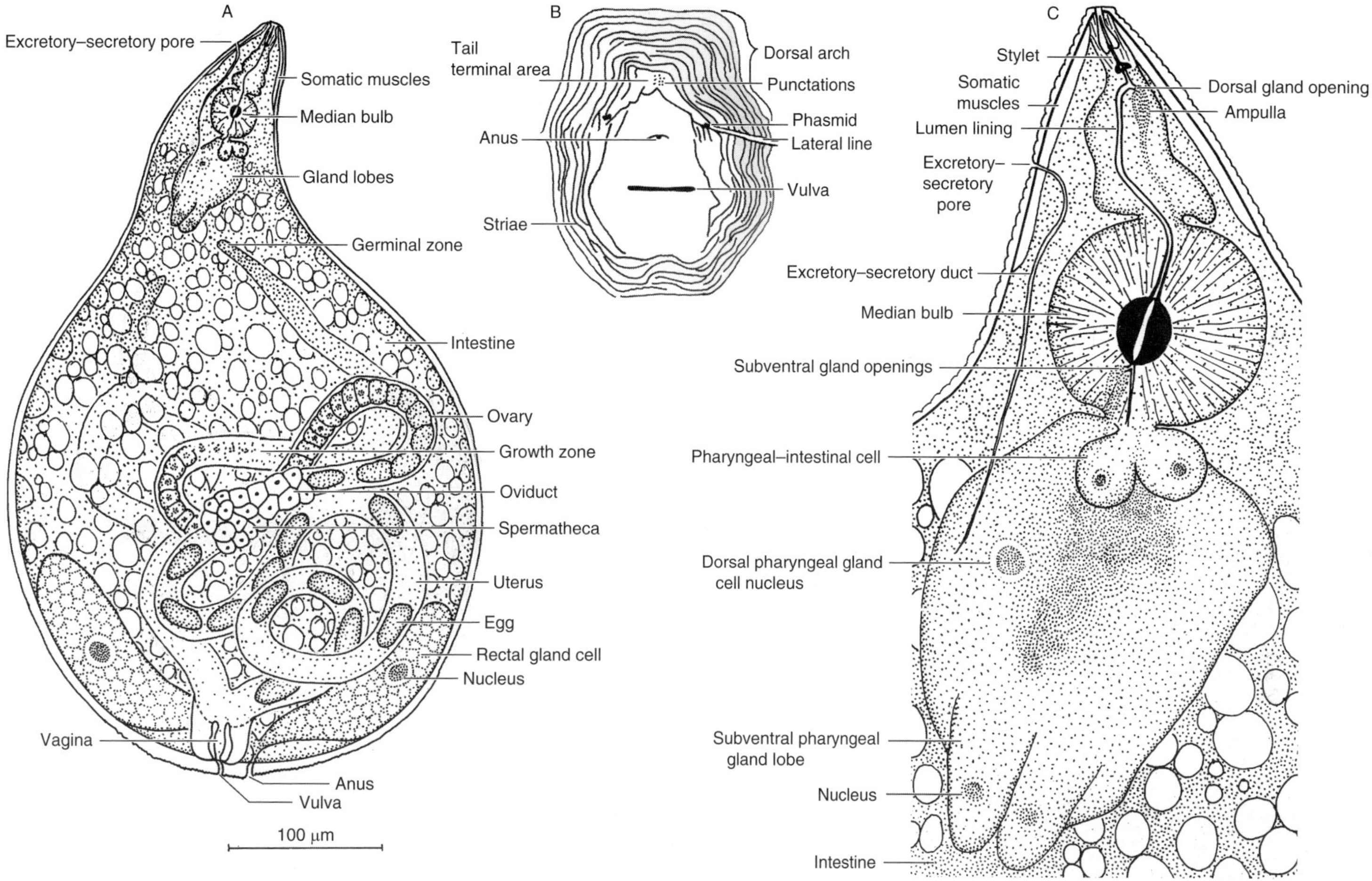

Fig. 2.18. Drawings of female root-knot nematode. A: entire nematode; B: typical perineal pattern; C: detailed view of anterior end. After Eisenback (1985a), courtesy of N.C. State University Graphics.

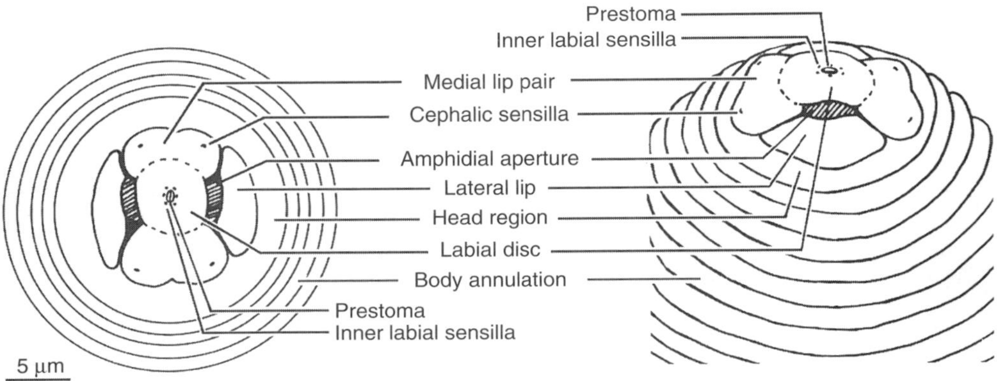

Fig. 2.19. Drawings of anterior end of female root-knot nematode in face and lateral views, as revealed by scanning electron microscope. After Eisenback *et al.* (1980), courtesy of *Journal of Nematology*.

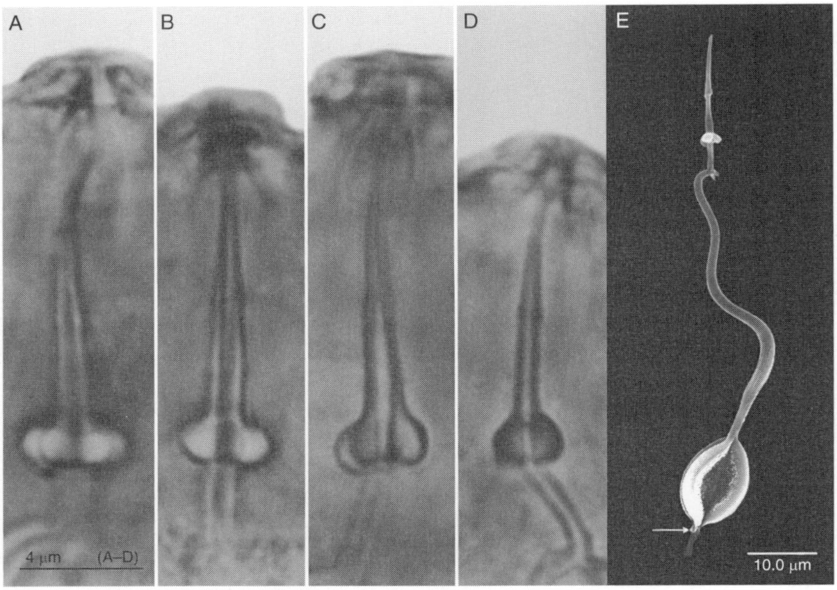

Fig. 2.20. Light micrographs of female stylet. A: *Meloidogyne incognita*; B: *M. javanica*; C: *M. arenaria*; D: *M. hapla*. E: scanning electron micrograph of stylet and cuticular lumen lining of entire pharynx (arrow marks the openings of the subventral pharyngeal glands). After Eisenback *et al.* (1980), courtesy of *Journal of Nematology*.

The non-contractile portion of the muscle cell is innervated by a muscle process that taps into one of the major bundles of longitudinal nerves (Baldwin and Hirschmann, 1973). Muscles in the two subventral quadrants are innervated by the ventral nerve, whereas those in the two subdorsal quadrants are innervated by the dorsal nerve chord. The cell nucleus, numerous mitochondria, lipid globules, glycogen and smooth endoplasmic reticulum occur in the non-contractile portion of the muscle, distinctly separating it from the contractile portion (Bird, 1971b; Baldwin and Hirschmann, 1975).

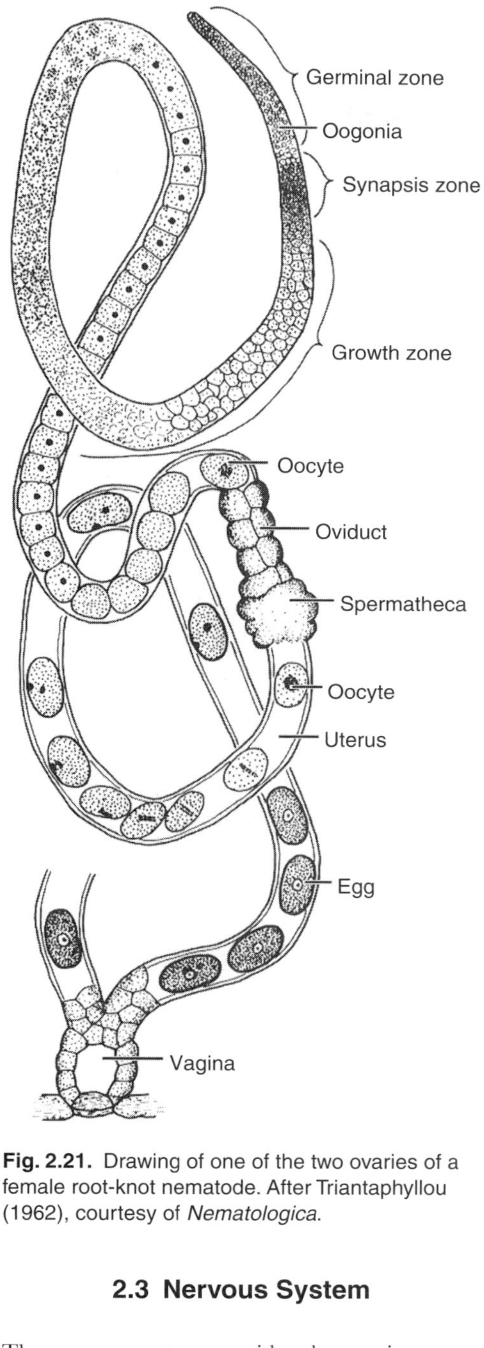

Fig. 2.21. Drawing of one of the two ovaries of a female root-knot nematode. After Triantaphyllou (1962), courtesy of *Nematologica*.

2.3 Nervous System

The nervous system provides the precise motor control that allows the nematode to move around in its environment and the sensory organs that enable the individual to respond to environmental cues. The J2 uses this system to move about inside the egg, to hatch from the egg and move through the soil, to sense the presence of a suitable host root, and to penetrate that host and migrate to the correct position for the development of the host–parasite relationship. In the male, the nervous system is used when migrating out of the root system and moving through the soil to find a suitable mate, while in the female it is utilized to move the head end from one giant cell to another both by coordinating movement of the head and by sensing the correct site for feeding to occur.

Little information is known about the nervous system of root-knot nematodes, the majority of observations relating to the anterior end of the J2 and male. Although the tail of the J2 has been studied, the nervous system in the male tail has not been investigated. Likewise, detailed studies of the female nervous system are not available.

The nerve ring, hemizonid, amphids and phasmids may be visible with the light microscope. The primitive coordinating centre of the nervous system, known as the nerve ring, occurs just posterior to the median bulb. It is formed from numerous nerve processes with their nucleated cell bodies lying anteriorly and posteriorly to the ring itself (Bird, 1971b). The hemizonid is probably a major lateroventral commissure of the nervous system and appears in lateral view as a transparent semicircle on the ventral surface of the body, usually near the secretory–excretory pore in the J2 (Fig. 2.2A) and male (Fig. 2.10).

Six longitudinal nerves extend from the nerve ring anteriorly and innervate the cephalic sensory structures, including the labial and cephalic sensilla, as well as the paired amphids. At least four nerves run posteriorly from the nerve ring, where they turn away from the pharynx and become embedded in the hypodermal chords: two lateral, one dorsal and one ventral. Transverse nerves interconnect all four of the longitudinal nerves. The somatic muscles in the subventral quadrants are innervated by the ventral chord and those in the subdorsal quadrants are innervated by the dorsal chord. The lateral nerves are responsible for innervating the sensory structures in the tail, including the phasmids and spicules in the male, and the phasmids and caudal sensory organ in the J2 (Wergin and Endo, 1976; Endo and Wergin, 1977).

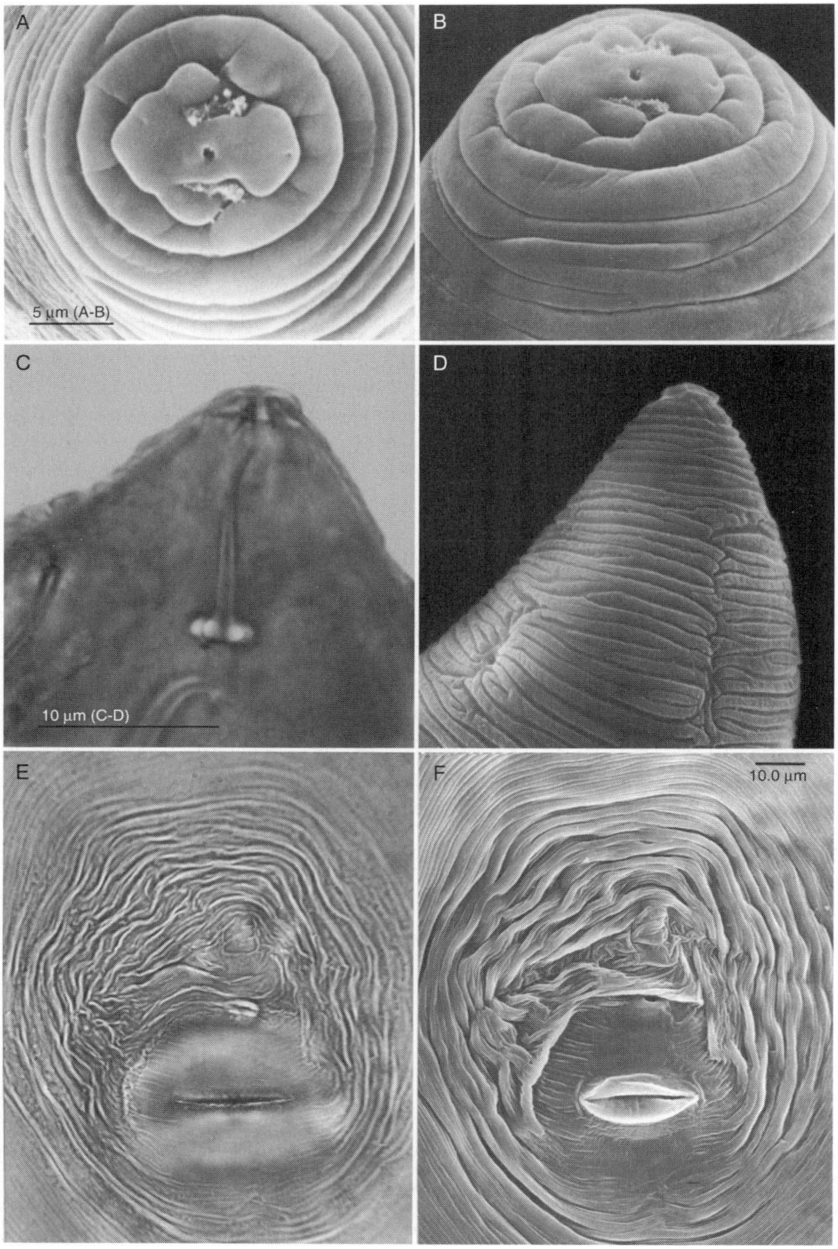

Fig. 2.22. Photomicrographs of a female root-knot nematode. A: scanning electron micrograph (SEM) of anterior end in face view; B: SEM of anterior end in lateral view; C: light micrograph (LM) of anterior end, showing stylet and secretory–excretory duct; D: SEM of anterior end, showing body annulations and secretory/excretory pore; E: LM of perineal pattern; F: SEM of perineal pattern. After Eisenback *et al.* (1980), courtesy of *Journal of Nematology*.

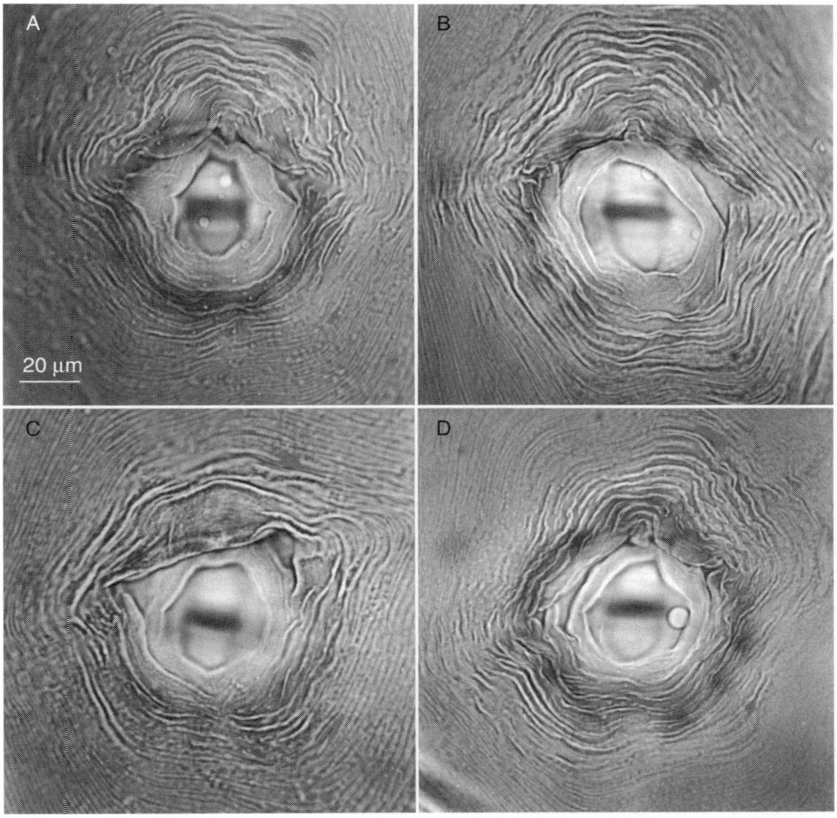

Fig. 2.23. Light micrographs of perineal patterns of female *Meloidogyne pini*. After Eisenback *et al.* (1985), courtesy of *Journal of Nematology*.

2.3.1 Cephalic sensory structures

Sensory structures in the anterior end of the J2, male and female are innervated by nerves extending anteriorly from the nerve ring. These structures include inner labial sensilla, cephalic sensilla, accessory sensory structures and the amphids (Figs 2.3, 2.12, 2.19). In all three life stages, the six inner labial sensilla open around the ovoid prestoma, where they function as chemoreceptors and probably assist during processes such as stylet insertion, feeding and stylet withdrawal. The pore-like openings of the inner labial sensilla are lined with cuticle secreted by the surrounding hypodermal tissue. Each pore contains the endings of two receptor cilia of different lengths, which probably function as chemoreceptors. The four cephalic sensilla are sometimes visible as slight depressions in the cuticle on the medial lip pairs. They are structurally similar to the inner labial sensilla but have just one ciliary receptor. Because they do not open to the external environment, these sensilla probably function as mechanoreceptors.

In the J2, several sensilla are contained in the anterior region of the nematode and may be analogous to the outer labial sensilla (Wergin and Endo, 1976; Endo and Wergin, 1977). These cilia end beneath the cuticle, are not surrounded by cuticle and may function as tactoreceptors. These accessory sensilla occur in the two subdorsal and two subventral sectors of the head region, and are associated with the inner labial sensilla and amphidial canals. Similar sensory structures are present in both male and female, but details of their morphology have not been presented.

Amphidial openings occur on the J2, male and female, and are located between the labial disc and lateral lips. With SEM, these openings are often observed to be plugged by an exudate secreted by the amphidial gland (Figs 2.4A, B, 2.5A–C, 2.6). The amphids are made up of specialized support cells, nerve cells and secretory gland cells (Fig. 2.11B). The amphidial canals are lined with cuticle and surrounded by a support cell (the socket cell), which extends from the amphidial gland cell to the anterior portion of the canal. Seven modified cilia innervate the amphid and lie in the canal, ending near the aperture. The large, irregularly shaped amphidial gland cell (sheath cell) contains numerous mitochondria, microvilli, and intercellular spaces filled with ducts and granules (Wergin and Endo, 1976; Endo and Wergin, 1977). The amphids probably function as chemoreceptors and are essential for the nematode to interface with its environment.

2.3.2 Caudal sensory structures

In the J2, the phasmids are located between the anal opening and the tail tip, whereas in the male they occur near the level of the cloacal opening and in the female are located anterior to the tail terminus and within the perineal pattern. As revealed by TEM studies of the J2, the phasmids are characterized by a cuticle-lined duct, a nerve process and a gland cell (Bird, 1979c). Nerve tissue fills the hypodermis that surrounds the phasmidial gland cell. A nerve process extends from this nerve tissue into the basal layer of the cuticle and a single cilium innervates the phasmidial duct. A gland cell secretes an exudate into the phasmidial canal. Details of the phasmid structure of the male and female stages are probably similar to that of the J2, but they have not been studied in detail.

A caudal sensory organ (Fig. 2.2B) has been described occurring in the J2 tail directly anterior to the hyaline tail terminus (Bird, 1979c). It is 5–10 μm long and is made up of numerous branched nerve processes that originate from the median caudal nerve. This sensory structure is probably a tactoreceptor since it is completely contained within the hypodermal tissues.

In the male, the spicules are innervated by nerves contained within each hollow, cuticularized, organ (Eisenback and Hirschmann, 1980). These nerves open via two small pores at the tip of each spicule. Additional details of these nerves are lacking, but they clearly play a role in finding a mate and/or in copulation. They open to the environment, so these sensory structures probably function as chemoreceptors.

2.4 Digestive System

The digestive system is responsible for obtaining nutrients from a food source in order to support the life of the nematode, including normal metabolic activities, growth and development, movement and fuelling the reproductive system. In the J2 the digestive system is necessary to help the nematode hatch from the egg, penetrate a host root and establish and maintain a host–parasite relationship (Bird, 1967, 1968a,b, 1969; Bird and Saurer, 1967). All of the energy necessary for the three additional moults and development into an adult is obtained by the J2. Additional feeding occurs in the mature female to aid growth of the reproductive system, but the vermiform male does not feed at all (Triantaphyllou and Hirschmann, 1960).

The digestive system is composed of a stoma, pharynx, intestine and rectum (Figs 2.2, 2.10). The stoma is armed with a hollow, protrusible, hypodermic-needle-like stylet that serves as an interface between the nematode and the plant. The pharynx contains three specialized gland cells responsible for several functions in the host–parasite relationship, and a metacorpus that pumps substances from the gland cells into the plant and from the plant into the intestine. Because the food taken in by the nematode (at the expense of the plant) is so highly refined, digestion and absorption are unnecessary. The intestine serves simply as a storage organ and, in the female, is not even connected to the anal opening. Instead, six large rectal gland cells open through this orifice, where they secrete a voluminous gelatinous matrix that serves to protect the eggs as they are deposited to form the egg sac.

2.4.1 Stoma and pharynx

The ovoid prestoma is located on the labial disc where the slit-like stomatal opening guides the stylet as it is protruded into plant tissues. The hypodermic-needle-like stylet lies within a cuticle-lined stomatal cavity surrounded by the hexaradiate cephalic framework (Baldwin and Hirschmann, 1973; Wergin and Endo, 1976; Endo and Wergin, 1977). This framework includes a basal ring, six blades and the vestibule containing the anterior end of the stylet (Fig. 2.11A). The vestibule continues posteriorly as the vestibule extension and connects to the stylet at the junction of the cone and shaft. The external layer of cuticle forming the body wall is continuous with the vestibule and its extension. The blades of the cephalic framework comprise one dorsal, one ventral, two subdorsal and two subventral blades. The lateral sectors are slightly larger than the others. These blades add structural support to the anterior end of the nematode, where they extend from the vestibule to the body wall. The basal ring of the cephalic framework is continuous with the basal layer of cuticle that makes up the body wall (Baldwin and Hirschmann, 1973; Wergin and Endo, 1976; Endo and Wergin, 1977).

Three groups of stylet protractor muscles attach to the body wall and cephalic framework anteriorly and to the stylet knobs posteriorly (Fig. 2.11A). The vestibule extension holds the stylet in place where it attaches at the junction of the shaft and cone. The stylet opens on the ventral side of the cone near the tip in the J2 and female, and approximately 2–3 μm from the tip in the male, at a point that is often marked by a slight protuberance (Fig. 2.16B,F,J) (Eisenback, 1982). The stylet is made up of three distinct parts: cone, shaft and three knobs, one dorsal and two subventral (Figs 2.4C–E, 2.5D–G, 2.13A,B, 2.15B, 2.16, 2.20). The cone is more stable and less soluble than the shaft and knobs, as attested by the fact that these two latter structures often disappear quite rapidly in permanently prepared specimens (Baldwin and Hirschmann, 1973; Wergin and Endo, 1976; Endo and Wergin, 1977).

The stylet protractor muscles force the anterior end of the stylet out through the stoma and into the cell wall of the host plant, where it is used to inject saliva into the cell. This saliva modifies the cell by changing it into a metabolic sink for the nematode, from which the nematode receives nutrients and energy (see Abad *et al.*, Chapter 7, this volume). The vestibule folds backward, and the flexible hypodermal tissues surrounding the stylet allow it to move back and forth. Since stylet retractor muscles are lacking, these same tissues enable the stylet to return to its normal position when the stylet protractor muscles relax, the process being facilitated by the release of tension in the anterior extended portion of the alimentary tract (Baldwin and Hirschmann, 1973; Wergin and Endo, 1976; Endo and Wergin, 1977).

The non-contractile portion of the stylet protractor muscles is contained within the anterior end of the procorpus (Baldwin and Hirschmann, 1973; Wergin and Endo, 1976; Endo and Wergin, 1977). Except in the metacorpus, the lumen lining of the pharynx is circular in cross-section. A short distance from the base of the stylet, at approximately 2–8 μm, the opening of the dorsal pharyngeal gland branches into several channels. The enlarged, muscular metacorpus serves as a pumping mechanism. Along with two strategically placed sphincter muscles, the metacorpus can pump substances either out through the stylet or back into the nematode. The lumen lining of the metacorpus is triradiate in cross-section and heavily sclerotized. The outer edges of this lumen lining are not very flexible but the inner sections are thin and flexible. When the muscles of the metacorpus contract, they pull the thin inner section apart and produce a very strong pumping action. The branched openings of the subventral pharyngeal glands are located in the metacorpus just posterior to the triradiate lumen lining (Baldwin *et al.*, 1977; Eisenback *et al.*, 1980; Eisenback and Hirschmann, 1981; Eisenback, 1982).

In the male, the pharynx is somewhat degenerate and probably not functional. The triradiate lumen lining of the metacorpus is thin and the muscles are poorly developed, sparse in number and disorganized. The lumen ends as a series of folded membranes that are surrounded by two to three pharyngeal–intestinal cells. Three digestive glands overlap the intestine ventrally. The dorsal gland cell is uninucleate and lies anterior to the two subventral gland cells. The dorsal gland cells form an ampulla at the base of the stylet, and the subventral gland cells form an ampulla at the base of the metacorpus (Baldwin *et al.*, 1977).

The three pharyngeal gland cells overlap each other and wrap around the intestine in the J2 (Fig. 2.2A). In the female, these gland cells form one large dorsal lobe and two smaller subventral lobes that are quite variable in morphology (Fig. 2.18B), but in the male they are often reduced and degenerate. Two pharyngo-intestinal cells are present at the junction of the pharynx and intestine (Garcia-Martinez, 1982). They are small, rounded and uninucleate. In the male, however, these two cells are completely contained within the isthmus (Baldwin *et al.*, 1977).

2.4.2 Intestine

Few details of the intestine of root-knot nematodes have been resolved because its morphology is not very distinct. Numerous lipid globules fill the intestine of the J2 (Fig. 2.2), male (Fig. 2.10) and female (Fig. 2.18A). In the J2, a clearly defined lumen with microvilli is absent, but a partially defined lumen may be present in both male and female (Elsea, 1951; Bird, 1979a,b). In the male, an intestinal caecum commonly extends from the base of the isthmus to the base of the metacorpus (Baldwin *et al.*, 1977). In both male and female the intestine is syncytial. In the female, the connection to the rectum has been lost and the intestine functions primarily as a storage organ.

2.4.3 Rectum

The J2 anal opening is a small, round pore, usually contained within one body annule (Fig. 2.7D) (Eisenback and Hirschmann, 1979b). The lumen of the rectum is lined with cuticle that becomes thinner anteriorly until it merges with the plasma membrane. The diameter of the tail is nearly 8 μm, whereas the rectum is 6 μm in diameter (Bird, 1979c). The rectum, in its dilated state, contains a matrix resembling that extruded from the adult female rectal gland cells (Bird, 1979c).

In the J2 and female, the rectal lumen is surrounded by rectal gland cells. The intestine in the female is not connected to the rectum, and the same situation may be true for the J2. As soon as the J2 begins to feed, the rectal gland cells enlarge and become visible as six distinct cells (Fig. 2.18A).

In the male, the intestine and testis share a common duct, the cloaca, which opens about 6–8 μm from the tail tip. Additional details about the morphology of the male tail are lacking.

The anal opening of the female is located between the tail tip and the vulva. As the female matures, the rectal lumen decreases in size. Two subdorsal, two subventral and two lateral rectal glands are connected to the rectum. The uterus separates the two subventral glands, but in general the six glands are equally spaced around the posterior end of the body, where they are closely adpressed to the hypodermis. The rectal gland cells increase in size until they are nearly contiguous (Bird, 1968a,b, 1979b).

The rectal gland cells (Fig. 2.18A) are very large, approximately 100 μm long by 25 μm in diameter, and contain a very large nucleus, approximately 25 μm long by 15 μm in diameter (Maggenti and Allen, 1960; Bird and Rogers, 1965b). These are the largest nuclei in the female. The rectal gland cells contain dense cytoplasm filled with many Golgi bodies, mitochondria, anastomosing endoplasmic reticulum with many ribosomes, and multivesicular lamellae with a repeat periodicity of 75 Å. These cells with giant nuclei and numerous cell organelles are vital for the secretion of the gelatinous matrix. Visible sinus canals are present in each gland cell where they merge to form a single long duct that opens into the anterior end of the rectal lumen. The gelatinous matrix is secreted via the anal opening, which is opened by the anal depressor muscles attached to the rectum and body wall.

When the adult female is mature, copious quantities of gelatinous matrix are secreted, the combined volume of which may be larger than the female herself. The matrix contains proteins, a mucopolysaccharide and various enzymes (see Evans and Perry, Chapter 9, this volume). It forms an irregular, mesh-like structure that physically prevents the J2 from hatching during extreme periods of drought (Maggenti and Allen, 1960; Bird and Rogers, 1965b). It also contains antimicrobial compounds that protect the eggs from attack by soil-borne organisms. The numerous resources that the female uses to produce the gelatinous matrix demonstrate the value that it plays in the long-term survival of this highly evolved sedentary parasite (Bird, 1971a; Bird and Soeffky, 1972; Dropkin and Bird, 1978).

2.5 Secretory–Excretory System

The secretory–excretory system is marked on the body by a small, rounded opening located ventrally in the cuticle where the body annulations are interrupted or slightly deviated. In the J2 (Fig. 2.7C) and male, the pore is located near the metacorpus and 2–8 μm from the hemizonid. However, in the female it is located anterior to the median bulb, often near the stylet base (Fig. 2.18A). The pore is approximately 0.2 μm in diameter in the J2 and 0.6 μm in the female. The duct of the secretory–excretory system extends posteriorly through the hypodermis and becomes a canal that is not lined with cuticle. In the female, the anterior end of this duct is surrounded by numerous vesicles (Bird, 1979b). In the J2, male and female stages, the posterior end of the secretory–excretory canal is closely related to a large sinus gland cell. Although the function of this system is not completely understood, it probably has an excretory function for removing toxic wastes and perhaps a secretory function as well (Bird, 1971b, 1979b).

2.6 Reproductive System

The reproductive system is closely associated with the digestive system, from which it receives all of its nutritional requirements. This system ensures that new individuals will be produced in the struggle for survival of the species. The genital primordium is very small in the pre-parasitic juveniles, but rapidly increases in size as soon as feeding commences (Fig. 2.8) (Papadopoulou and Triantaphyllou, 1982). In the pyriform female, it develops into two very long and convoluted ovaries (Figs 2.18A, 2.21). The shape of the female (Fig. 2.17) allows for this increase in the length of the ovaries, a feature that greatly enhances reproductive capacity. However, such a huge increase in fecundity comes at the expense of mobility. The male, in contrast, remains vermiform and mobile, although the reproductive system is of normal size (Elsea, 1951).

2.6.1 Second-stage juvenile

The genital primordium of the pre-parasitic J2 is formed from four cells, i.e. two small, flattened somatic cells surrounding two large, spherical germinal cells. The primordium lies parallel with the body wall at about 65% of the body length from the anterior end. As soon as feeding begins, it starts to increase in size and develops into either the ovaries or a testis (Fig. 2.8) (Papadopoulou and Triantaphyllou, 1982).

2.6.2 Male

Usually the male has just one testis, but when the environment affects sex expression some individuals may develop two testes. This change in sex, which is caused by various environmental effects, is called sex reversal (Fig. 2.8) (Triantaphyllou, 1979). It is thought to be advantageous to the survival of these obligate parasites because the mechanism reduces the population-induced stresses on the host plant. Males require less energy to produce; they do not feed as adults; they do not add progeny to the burden of the host; and, when present, they may increase genetic variation in the population as a result of sexual reproduction (Triantaphyllou, 1979; Papadopoulou and Triantaphyllou, 1982).

The male gonad is formed from the testis proper (which constitutes approximately one half of the length of the entire tube) and the vas deferens. A cap cell at the distal end of the testis is present in the undifferentiated genital primordium in the J2 (Triantaphyllou, 1979). All of the spermatogonia are derived from this cell in the germinal zone of the testis. The spermatogonia increase in size in the remaining growth zone of the testis proper. Following the growth zone, the glandular vas deferens empties ventrally into the cloaca. The entire gonad is covered by a single layer of epithelial cells (Shepherd and Clark, 1983).

Paired, cuticularized spicules are located within a pouch in the cloaca (Figs 2.10, 2.15C). As these structures are protruded through the cloacal opening they form a tube that facilitates an efficient transfer of sperm into the female vagina. The spicules have a head and shaft that consists of a hollow cytoplasmic core surrounding one or more nerves that open to the exterior via two small pores at the tip of each spicule. A gubernaculum is located dorsally to the spicules and serves to guide them out through the cloacal opening during protrusion and retraction. Spicule protractor and retractor muscles are attached to the spicule head and body wall.

Spermatozoa often pack the vas deferens. They are approximately 6 μm in diameter and 12 μm long. Spermatozoa are divided into two parts: one part contains the nucleus and is surrounded by mitochondria and fibrillar bodies; the other contains most of the cytoplasm, which is used to form several large pseudopodia necessary for crawling through the uterus to the spermatheca. Numerous filopodia are sometimes formed in the region of the spermatozoon that contains the nucleus (Goldstein and Triantaphyllou, 1980; Shepherd and Clark, 1983).

2.6.3 Female

The saccate female body (Fig. 2.17) is filled with two highly convoluted gonads that are in very close proximity to the digestive system, particularly to the nutrient-laden intestine. Approximately 60% of the female gonad is made up of the ovaries proper (Fig. 2.21). The anterior end of each ovary contains a cap cell, which initiates the germinal zone of the organ and produces all of the oogonia. The oogonia are arranged radially around a rachis, to which they are attached by a cytoplasmic bridge. The growth zone of the ovary follows as a region where the oogonia increase in size. The boundaries between the oocytes become more distinct in this zone and the rachis gradually disappears. The oocytes accumulate glycogen, refringent bodies and lipid globules as they pass in single file through the growth zone into the oviduct (Elsea, 1951; Triantaphyllou and Hirschmann, 1960; Triantaphyllou, 1962, 1979; McClure and Bird, 1976).

Each of the two staggered rows of cells in the oviduct (Fig. 2.21) contains four tightly packed cells with large cytoplasmic invaginations and large, irregularly shaped nuclei. The oocytes stretch and flatten these cells as they pass through the narrow valve in the lumen of the oviduct, although they become ovoid afterwards (Elsea, 1951; Triantaphyllou and Hirschmann, 1960; Triantaphyllou, 1962; 1979; McClure and Bird, 1976).

Immediately posterior to the oviduct, the spermatheca (Fig. 2.21) consists of 14–20 rounded, lobe-like cells that have a deeply invaginated margin of plasmalemma. Densely vesiculate bodies located between the lobes of the spermathecal wall have numerous microtubules that project into the lumen and assist in the formation of the protein membrane of the eggshell (McClure and Bird, 1976).

The uterus (Fig. 2.21) occurs posterior to the spermatheca and can be divided into distinct regions: (i) cells at the ovarial end, possessing large intracytoplasmic spaces lined with endoplasmic reticulum and producing the chitin layer; (ii) cells in the middle of the uterus, forming the glycolipid layer; and (iii) cells in the posterior end, consisting of dense cytoplasm with large areas of compact endoplasmic reticulum (Elsea, 1951; Triantaphyllou and Hirschmann, 1960; Triantaphyllou, 1962; 1979; McClure and Bird, 1976).

Near the posterior end of the gonad the two uteri join together to form a common duct (Fig. 2.21). Numerous muscles are attached to the vagina and radiate outward to attach to the body wall. These muscles contract to dilate the vagina during egg laying. The vagina itself is lined with thick cuticle. The vulva is located transversely on the posterior end of the body wall and is surrounded by two, slightly elevated vulval lips (Elsea, 1951).

2.7 Morphological Methods

Preparation of nematodes for SEM and TEM has been reviewed previously by Wergin (1981), Eisenback (1985c, 1991) and Carter (1991). Making perineal patterns for species identification has been outlined by Hartman and Sasser (1985), and additional methods for collecting and preparing nematodes for optical microscopy have been evaluated by Fortuner (1991).

2.8 Minimum Standards for Describing a New Species

With the increasing number of described species in the genus *Meloidogyne*, it is important that species descriptions conform to a general standard in order to facilitate accurate comparisons and differential diagnoses. Because of the relatively conserved morphology in the group, there is an increasing emphasis on techniques such as isozyme phenotyping and molecular sequences. No new species description should be published without at least one of these techniques, a consideration that may require collaboration with workers from other laboratories where specialist methodologies such as molecular characterization are more readily available. New descriptions and other taxonomic work should be submitted to a journal of suitable standing in order to ensure appropriate peer scrutiny – there is little to be gained scientifically by 'hiding' such papers in either obscure journals or those that do not regularly publish manuscripts of a taxonomic nature.

Goodey (1959) provided an excellent 'master-class' on the data to be presented when describing new species. Although not specific to root-knot nematodes, his paper is still recommended reading for anyone embarking on such a task. Authors should also be careful not to cite a new species name in a paper that may be published before the 'official' description – a truly embarrassing and unnecessary calamity, culminating in the creation of a *nomen nudum*.

The description format presented here is a blend of traditional and modern methodologies. The intention is to provide an up-to-date guide to achieve a robust standard of description – and one that should be aimed at by all authors. Journal editors should also find the protocol useful when deciding whether descriptions submitted for publication are adequately supported by appropriate information.

2.8.1 The text

The account below is based, purely as an example, on the style of *Nematology*. It may easily be adapted to that required by alternative journals.

***Meloidogyne ??? *n. sp.**

(Figs ?–?)

2.8.1.1 Measurements

See Table 2.1 for a list of characters that should be measured. At least 20 specimens each of the J2, male and mature female stages should be measured. State whether the material is fresh or has been fixed and processed to glycerine when

Table 2.1. Morphometrics of *Meloidogyne* ??? n. sp. All measurements in μm and in the form: mean ± standard deviation (range).

	Female		Male	J2
Character/stage	Holotype	Paratypes	Paratypes	Paratypes
n	–	20	20	20
L	√	√	√	√
a	√	√	√	√
c	–	–	√	√
T	–	–	√	–
Max. body diameter	√	√	√	√
Neck length	√	√	–	–
Stylet length	√	√	√	√
Stylet knob height	√	√	√	√
Stylet knob width	√	√	√	√
DGO	√	√	√	√
Excretory pore to anterior end	√	√	√	√
Interphasmidial distance	–	√	–	–
Vulva length	–	√	–	–
Vulva–anus distance	–	√	–	–
Tail length	–	–	√	√
Spicule length (median line)	–	–	√	–
Gubernaculum length	–	–	√	–
Testis length	–	–	√	–
Hyaline tail terminus (h)	–	–	–	√
h% (h/tail length × 100)	–	–	–	√

measured. Cite data as: mean ± standard deviation (range). The coefficient of variation may also be given. Measurements are usually expressed in μm, although body length of the male and mature female may be in mm. The majority of measurements are best rounded to the nearest micrometer. Spicules should be measured along the curved median line rather than the chord. When measuring the stylet take particular care that the conus tip is accurately determined (i.e. do not assume that the stylet ends at the anterior extremity of the labial region). Measurements should be made at an appropriate magnification. For smaller structures (e.g. stylet, spicules, J2 hyaline region, perineal pattern) this usually equates to using a ×100 oil immersion objective, although body length of mature females and males may be recorded using a ×10 objective. Ensure that the optical combination used during the measurement process has been properly calibrated. Be aware that adding components such as a drawing tube or NIC (Nomarski Interference Contrast) prism will alter the focal length of the system and hence the magnification factor, thereby necessitating recalibration. Measurements should ideally be taken using either a drawing tube or computer equipped with measurement software. For reasons of convenience and accuracy, the use of a calibrated eyepiece graticule is best restricted to acquiring the length of short, straight structures, such as the stylet.

2.8.1.2 Description

FEMALE. Body shape, presence or absence of terminal cone, form of head region, including annulation, general appearance of stylet and shape of knobs, DGO (dorsal gland orifice) position, position of excretory pore, perineal pattern (type of striae – smooth, wavy, etc., overall shape and proportion of parts dorsal and ventral to vulva, presence or absence of dorsal arch, presence of wings, lateral field development, other pertinent peculiarities, such as punctations, etc.).

MALE. Body form, head shape and form of annules, presence of labial disc, general appearance of stylet and shape of knobs, overlap of pharyngeal glands, spicule shape, lateral field development.

J2. Body form, head shape and form of annules, general appearance of stylet and shape of knobs, excretory pore position, hemizonid position in relation to excretory pore, overlap of pharyngeal glands, form of rectum (inflated or not), tail shape and form of tip, development of hyaline region.

EGG. Length and diameter and the ratio between the two are the main criteria.

2.8.1.3 Type host and locality

The original host of the species should be recorded, together with detailed data concerning the type locality. Altitude and/or GPS (global positioning system) coordinates should be cited if at all possible. If the type material stems from a culture on a host other than the type, then this should also be indicated and appropriate details supplied.

2.8.1.4 Type material

A mature female should be designated as the holotype, the other material examined by the author(s)being paratypes (note that the term ‘allotype’ is unregulated by the International Code of Zoological Nomenclature (ICZN) and has no meaning under the Code; it is best avoided). Type material should be processed to, and mounted in, glycerine as permanent mounts (J2, males, mature females and perineal patterns plus corresponding neck regions). Ensure that the cover slip is supported appropriately (glass rods or beads of slightly larger diameter than specimen, wax ring, etc.) to avoid squashing or pressing the nematodes. Labels must record all pertinent data (stage, paratype/holotype, host, locality, collection number, etc.). Deposition of fixed, but unmounted, material is also desirable, plus material preserved in a manner that enables subsequent molecular studies. Material should be deposited in at least two internationally recognized collections and preferably not restricted to author or laboratory collections. Ideally the number of paratypes, their life cycle stage and collection numbers designated by the depository should be recorded.

2.8.1.5 Diagnosis and relationships

The diagnostic characters of the proposed species should be listed stage by stage. This forms the *diagnosis* of the species and may be repeated in the *Summary* or *Abstract*. The relationships of the new species should then be established by means of detailed comparison with similar species, using comparative data so that the former can be easily distinguished from its congeners. Try not to use expressions that may be ambiguous or subjective, e.g. avoid the use of unsupported phrases such as 'larger than' or 'smaller than' – much better to cite and compare the actual measurements, preferably by providing for each species the mean size and range of the structure in question so that the reader may additionally judge the utility of the comparison.

2.8.1.6 Etymology

The derivation of the specific epithet should be given so that is clear as to how the name was formed and what was the intention of the author (e.g. in case the name is wrongly formed and needs to be emended subsequently). Formation of names and endings must conform to the requirements of the Code. Note: if naming a species in honour of a man, the ending of the patronym must be *-i*; if after a woman, *-ae*.

2.8.1.7 Isozyme phenotype

Isozyme profiles such as esterase and malate dehydrogenase are necessary. A photograph of a good-quality gel with clear bands should be provided. A profile of a known species, such as *M. javanica* or *M. incognita*, should also be run and included next to that of the new species. The number of strong and weak bands should be recorded, together with the migration value (Rm).

2.8.1.8 Molecular characterization

Current molecular data include RAPD (random amplified polymorphic DNA) profiles and sequence data for the ITS (internal transcribed spacer) regions and D2/D3 (dopamine receptor) region. Sequences should be as complete as possible and must be deposited in GenBank or an equivalent open-access database. Accession number(s) must be published with the description. It is recognized that not all laboratories will have access to molecular techniques, but this lack should not prevent the retention of suitably preserved material that can be used for this purpose by other scientists, either as collaborators or subsequent to any published description. If molecular studies cannot be performed at the time of description it is advisable to preserve material to facilitate subsequent DNA analysis. Yoder *et al.* (2006) recommend placing live nematodes directly into DESS solution. Alternative methods include placing live nematodes into saturated NaCl solution, preferably stored at −20 °C, or placing several J2 into a small tube with minimal water and allowing the worms to desiccate before sealing and storing at −20 °C (S. Subbotin, California, 2008, personal communication).

2.8.1.9 Cytogenetics and karyology

Method of reproduction (e.g. meiotic parthenogenesis) and chromosome number should be stated, if known.

2.8.1.10 Bionomics

Host range, differential hosts and gall form are all useful items to record.

2.8.1.11 Remarks

Any other relevant information supporting the description.

2.8.2 The figures

It is imperative that the description of any new species is supported by appropriate line drawings – light micrographs on their own are not sufficient. Line drawings should ideally be supplemented by good-quality light micrographs and scanning electron micrographs.

1. Line drawings: although there is an increasing trend towards molecular-based characterization, it is still vital that good-quality line drawings of the J2, male and mature female are provided. Important features must be drawn at an appropriate magnification to ensure clarity. The primary function of the line drawings is to convey information in as unambiguous a manner as possible. They need not be 'artistic' (some computer-generated toning can be a positive nuisance when reproduced, for example) but should clearly indicate the salient features of the new species and also cover major variation in, for example, perineal pattern and stylet form. A good example of a clearly drawn figure is shown in Fig. 2.24. One could quibble about the fact that only a single perineal pattern is drawn, but in general all the features are deftly presented with

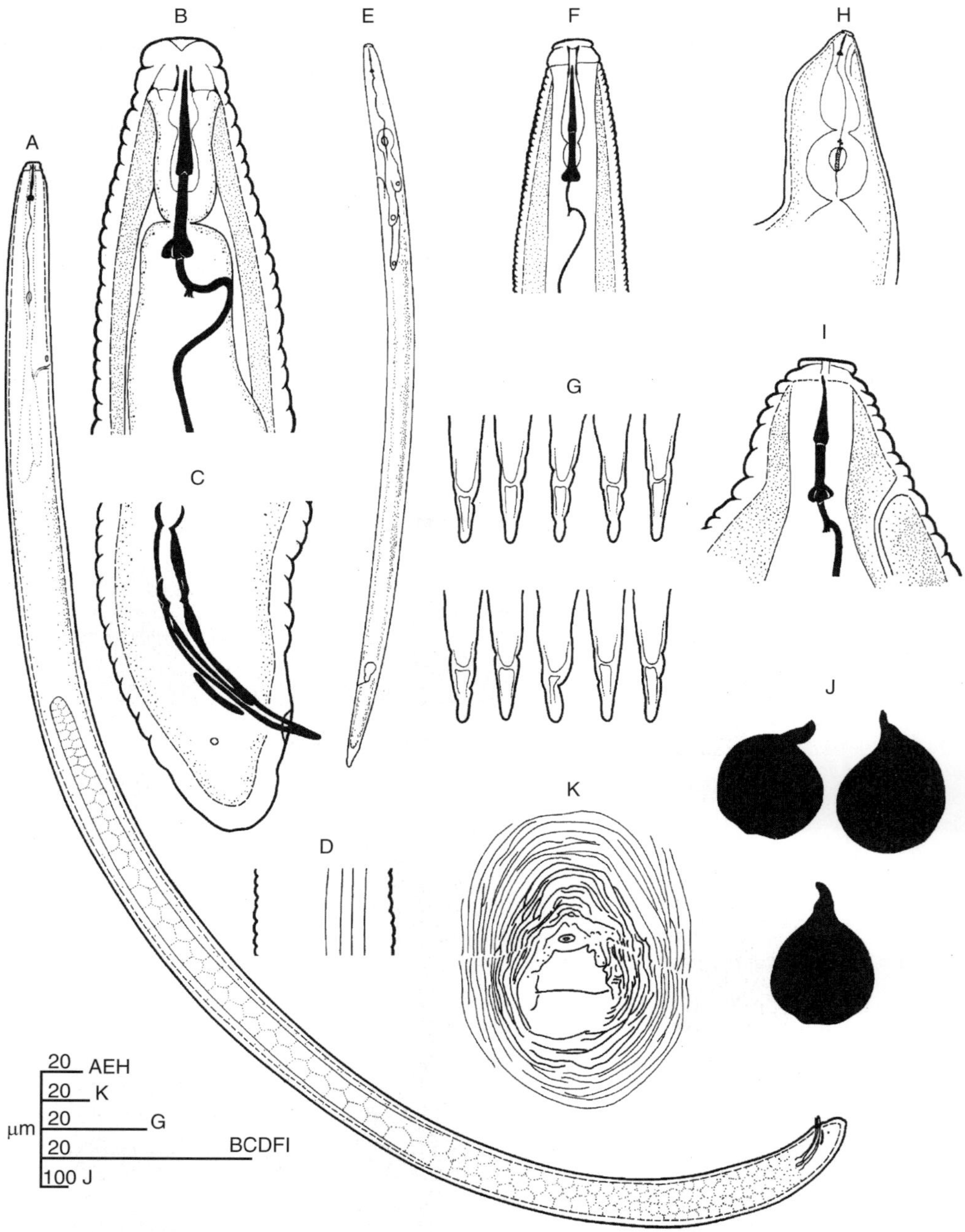

Fig. 2.24. Example of line drawings (of *Meloidogyne chitwoodi*) appropriate to illustrate the description of a new species. A: entire male; B: male anterior region; C: male tail region; D: male lateral field: E: entire J2; F: J2 anterior region; G: J2 tail regions; H, I: female anterior region, lateral view; J: entire females; K: perineal pattern.After Jepson (1985), courtesy of CAB International.

admirable simplicity. Relevant (and accurately calibrated) scales are, of course, an essential addition to any figure.

Suggested features to be reported upon are as follows, although others must be depicted if deemed by the author(s) to be diagnostic:

- J2: anterior region detail, stylet form and basal knob shape, pharyngeal region, including gland overlap, tail region and hyaline area.
- Male: anterior region detail, including that of the annules forming the labial region, stylet form and basal knob shape, pharyngeal region, including gland overlap, tail region, shape of spicules and gubernaculum, lateral field.
- Mature female: variation in general body form (in outline), anterior region detail, stylet form and basal knob shape, pharyngeal region, excretory pore position, perineal pattern form plus major variations encountered.

2. Light micrographs: provide a useful supplement to line drawings, particularly for perineal patterns. Good-quality photographs of the female anterior region, stylet and perineal pattern variation are useful, as are photographs of the male labial region, stylet and knobs and J2 labial and tail region.

3. Scanning electron micrographs: particularly useful for depicting the labial region (*en face* and lateral views needed) of male, female and J2, excised stylets of male and female and perineal pattern of mature female (see Fig. 2.25 for an example of suitable images).

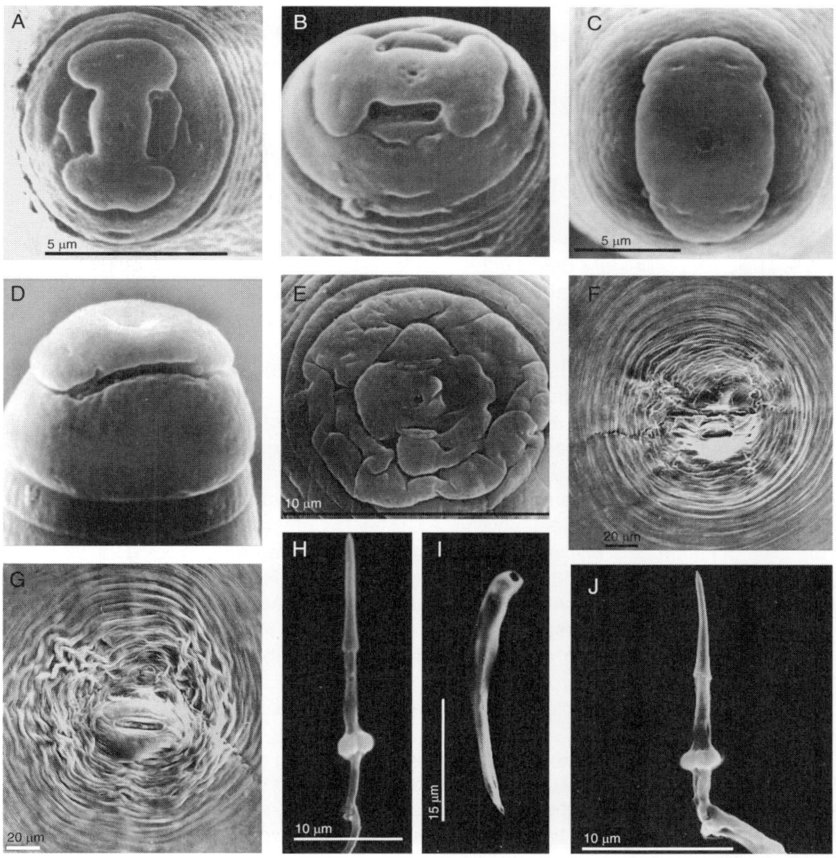

Fig. 2.25. Example of scanning electron micrographs appropriate to support the description of a new species. A, B: cephalic region of second-stage juvenile; C, D: cephalic region of male; E: cephalic region of female; F, G: female perineal pattern; H: male stylet; I: male spicule; J: female stylet. After Charchar *et al.* (2008), courtesy of *Nematology*.

2.8.2.1 Digital files

Very few original figures are now submitted to journals, the illustrations being scanned to digital files by the authors or their associates. This effectively bypasses much of the skill of the publisher's lithographer in delivering a high-quality print. When submitting digital files of figures it is important, therefore, that the correct format and resolution are employed. Black and white line figures are best scanned using the <sharp black and white line drawing> setting (or equivalent) of a professional-quality scanner. This will produce a very-high-quality bitmap file, which at 600 dpi resolution will only occupy about 500 kb when resized to page size and saved as a TIF with the LZW compression algorithm engaged. The 'common or garden' scanners, remarkable beasts that they are, seem to lack this scan option and the best that can be achieved is a greyscale image. Unfortunately, greyscale scans of black and white line drawings tend to generate abundant noise in the form of grey-hued pixels, which are either scattered over the white background of the figure or cluster around the lines, softening and degrading the image as a result. If a greyscale setting has to be employed then an acceptable compromise is to scan at a reasonably high resolution and then simultaneously reduce the file to page size and a resolution of 600 dpi. This file can be saved as a TIF with LZW compression or as a high-quality JPEG. Such images can be processed in a professional image package such as Photoshop® to reduce the effect of noise and thereby arrive at an acceptable result. Greyscale images and SEM micrographs should have a resolution of at least 300 dpi and be saved as high-quality JPEG. Be aware that many 'budget' scanners will automatically scan greyscale images (and also line drawings) as a colour RGB file – be sure to dump this redundant information before submitting the image as otherwise file size will be needlessly bloated by superfluous coding.

2.9 References

Atkinson, G.F. (1889) A preliminary report upon the life history and metamorphoses of a root-gall nematode, *Heterodera radicicola* (Greeff) Müll., and the injuries caused by it upon the roots of various plants. *Scientific Contributions Agricultural Experiment Station, Alabama Polytechnic Institute* 1, 177–226.

Baldwin, J.G. and Hirschmann, H. (1973) Fine structure of the cephalic sense organs in *Meloidogyne incognita* males. *Journal of Nematology* 5, 285–302.

Baldwin, J.G. and Hirschmann, H. (1975) Body wall fine structure of the anterior region of *Meloidogyne incognita* and *Heterodera glycines* males. *Journal of Nematology* 7, 175–193.

Baldwin, J.G. and Hirschmann, H. (1976) Comparative fine structure of the stomatal region of males of *Meloidogyne incognita* and *Heterodera glycines. Journal of Nematology* 8, 1–17.

Baldwin, J.G., Hirschmann, H. and Triantaphyllou, A.C. (1977) Comparative fine structure of the oesophagus of males of *Heterodera glycines* and *Meloidogyne incognita. Nematologica* 23, 239–252.

Bellati, G.B. and Saccardo, P.A. (1881) Sopra rigonfiamneti non filloserici osservati sulle radici di viti europee e cagionati invece dall'*Anguillara radicicola* Greeff in Alano di Piave (Distretto di Feltre). *Atti del Reale Istituto Veneto di Scienze, Lettere ed Arti Ser. V, Vol. VII* 8, 18 pp. + 6 figures.

Berkeley, M.J. (1855) Vibrio forming cysts on the roots of cucumbers. *Gardener's Chronicles* 7 April, p. 220.

Bird, A.F. (1958) The adult female cuticle and egg sac of the genus *Meloidogyne* Goeldi, 1887. *Nematologica* 3, 205–212.

Bird, A.F. (1959) Development of the root-knot nematodes *Meloidogyne javanica* (Treub) and *Meloidogyne hapla* Chitwood in the tomato. *Nematologica* 4, 31–42.

Bird, A.F. (1967) Changes associated with parasitism in nematodes. I. Morphology and physiology of preparasitic larvae of *Meloidogyne javanica. Journal of Parasitology* 53, 768–776.

Bird, A.F. (1968a) Changes associated with parasitism in nematodes. III. Ultrastructure of the egg shell, larval cuticle, and contents of the subventral esophageal glands in *Meloidogyne javanica*, with some observations on hatching. *Journal of Parasitology* 54, 475–489.

Bird, A.F. (1968b) Changes associated with parasitism in nematodes. IV. Cytochemical studies on the ampulla of the dorsal esophageal gland of *Meloidogyne javanica* and on exudations of the buccal stylet. *Journal of Parasitology* 54, 879–890.

Bird, A.F. (1969) Changes associated with parasitism in nematodes. V. Ultrastructure of the stylet exudation and dorsal esophageal gland contents of female *Meloidogyne javanica. Journal of Parasitology* 55, 337–345.

Bird, A.F. (1971a) Specialized adaptations of nematodes to parasitism. In: Zuckerman, B.M., Mai, W.F. and Rohde, R.A. (eds) *Plant Parasitic Nematodes*, Vol. II. Academic Press, New York, pp. 35–49.

Bird, A.F. (1971b) *The Structure of Nematodes*. Academic Press, New York.

Bird, A.F. (1979a) Histopathology and physiology of syncytia. In: Lamberti, F. and Taylor, C.E. (eds) *Rootknot Nematodes (*Meloidogyne *species); Systematics, Biology and Control.* Academic Press, London & New York, pp. 155–171.

Bird, A.F. (1979b) Morphology and ultrastructure. In: Lamberti, F. and Taylor, C.E. (eds) *Rootknot Nematodes (*Meloidogyne *species); Systematics, Biology and Control.* Academic Press, London & New York, pp. 59–84.

Bird, A.F. (1979c) Ultrastructure of the tail region of the second-stage preparasitic larva of the root-knot nematode. *International Journal of Parasitology* 9, 357–370.

Bird, A.F. and McClure, M.A. (1976) The tylenchoid (Nematoda) egg shell: structure, composition and permeability. *Parasitology* 72, 19–28.

Bird, A.F. and Rogers, G.E. (1965a) The ultrastructure of the cuticle and its formation *Meloidogyne javanica. Nematologica* 11, 224–230.

Bird, A.F. and Rogers, G.E. (1965b) Ultrastructural and histochemical studies of the cells producing the gelatinous matrix in *Meloidogyne. Nematologica* 11, 231–238.

Bird, A.F. and Saurer, W. (1967) Changes associated with parasitism in nematodes. II. Histochemical and microspectrophotometric analyses of preparasitic and parasitic larvae of *Meloidogyne javanica. Journal of Parasitology* 53, 1262–1269.

Bird, A.F. and Soeffky, A. (1972) Changes in the ultrastructure of the gelatinous matrix of *Meloidogyne javanica* during dehydration. *Journal of Nematology* 4, 166–169.

Carter, L.K. (1991) Preparation of nematodes for transmission electron microscopy. In: Nickle, W.R. (ed.) *Manual of Agricultural Nematology*. Marcel Dekker, New York, pp. 97–106.

Charchar, J.M., Eisenback, J.D., Charchar, M.J. and Boiteux, M.E.N.F. (2008) *Meloidogyne pisi* n. sp. (Nematoda: Meloidogynidae), a root-knot nematode parasitising pea in Brazil. *Nematology* 10, 479–493.

Chitwood, B.J. (1949) 'Root-knot nematodes' – Part I. A revision of the genus *Meloidogyne* Goeldi, 1887. *Proceedings of the Helminthological Society of Washington* 16, 90–104.

Cobb, N.A. (1890) '*Tylenchus* and root-gall'. *Agricultural Gazette of New South Wales* 1, 155–184.

Cornu, M. (1879) Études sur le *Phylloxera vastatrix. Mémoires Présentés par Divers Savants à l'Académie des Sciences, Institut France* 26, 163–175, 328, 339–341.

de Guiran, G. and Ritter, M. (1979) Life cycle of *Meloidogyne* species and factors influencing their development. In: Lamberti, F. and Taylor, C.E. (eds) *Rootknot Nematodes (*Meloidogyne *species); Systematics, Biology and Control.* Academic Press, London & New York, pp. 172–191.

Dropkin, V.H. and Acedo, J. (1974) An electron microscopic study of the glycogen and lipid in female *Meloidogyne incognita* (root-knot nematode). *Journal of Parasitology* 60, 1013–1021.

Dropkin, V.H. and Bird, A.F. (1978) Physiological and morphological studies on secretion of a protein–carbohydrate complex by a nematode. *International Journal of Parasitology* 8, 225–232.

Eisenback, J.D. (1982) Morphological comparison of head shape and stylet morphology of second-stage juveniles of *Meloidogyne* species. *Journal of Nematology* 14, 339–343.

Eisenback, J.D. (1985a) Detailed morphology and anatomy of second-stage juveniles, males, and females of the genus *Meloidogyne* (root-knot nematodes). In: Sasser, J.N. and Carter, C.C. (eds) *An Advanced Treatise on* Meloidogyne. *Volume I. Biology and Control.* A cooperative publication of the Department of Plant Pathology and the United States Agency for International Development, North Carolina State University Graphics, Raleigh, North Carolina, pp. 47–77.

Eisenback, J.D. (1985b) Diagnostic characters useful in the identification of the four most common species of root-knot nematodes. In: Sasser, J.N. and Carter, C.C. (eds) *An Advanced Treatise on* Meloidogyne. *Volume I. Biology and Control.* A cooperative publication of the Department of Plant Pathology and the United States Agency for International Development, North Carolina State University Graphics, Raleigh, North Carolina, pp. 95–112.

Eisenback, J.D. (1985c) Techniques for preparing nematodes for scanning electron microscopy. In: Barker, K.R., Carter, C.C. and Sasser, J.N. (eds) *An Advanced Treatise on* Meloidogyne. *Volume II. Methodology.* A cooperative publication of the Department of Plant Pathology and the United States Agency for International Development, North Carolina State University Graphics, Raleigh, North Carolina, pp. 79–105.

Eisenback, J.D. (1988) Identification of meloidogynids. In: Fortuner, R. (ed.) *Nematode Identification and Expert System Technology.* Plenum Press, New York, pp. 123–137.

Eisenback, J.D. (1991) Preparation of nematodes for scanning electron microscopy. In: Nickle, W.R. (ed.) *Manual of Agricultural Nematology.* Marcel Dekker, New York, pp. 87–96.

Eisenback, J.D. and Hirschmann, H. (1979a) Morphological comparison of second-stage juveniles of several *Meloidogyne* species (root-knot nematodes) by scanning electron microscopy. *Scanning Electron Microscopy* 3, 223–230.

Eisenback, J.D. and Hirschmann, H. (1979b) Morphological comparison of second-stage juveniles of six populations of *Meloidogyne hapla* by SEM. *Journal of Nematology* 11, 5–16.

Eisenback, J.D. and Hirschmann, H. (1980) Morphological comparison of *Meloidogyne* males by scanning electron microscopy. *Journal of Nematology* 12, 23–32.

Eisenback, J.D. and Hirschmann, H. (1981) Identification of *Meloidogyne* species on the basis of head shape and stylet morphology. *Journal of Nematology* 13, 513–521.

Eisenback, J.D. and Hirschmann, H. (1982) Morphological comparison of stylets of male root-knot nematodes (*Meloidogyne* spp.). *Scanning Electron Microscopy* 2, 837–843.

Eisenback, J.D. and Triantaphyllou, H. (1991) Root-knot nematodes: *Meloidogyne* species and races. In: Nickle, W.R. (ed.) *Manual of Agricultural Nematology.* Marcel Dekker, New York, pp. 191–274.

Eisenback, J.D., Hirschmann, H. and Triantaphyllou, A.C. (1980) Morphological comparison of *Meloidogyne* female head structures, perineal patterns, and stylet. *Journal of Nematology* 12, 300–313.

Eisenback, J.D., Hirschmann, H., Sasser, J.N. and Triantaphyllou, A.C. (1981) *A Guide to the Four Most Common Species of Root-knot Nematodes (*Meloidogyne *species), with a Pictorial Key.* A Cooperative Publication of the Departments of Plant Pathology and Genetics, North Carolina State University and US Agency for International Development, Raleigh, North Carolina.

Eisenback, J.D., Yang, B. and Hartman, K.M. (1985) Description of *Meloidogyne pini* n. sp., a root-knot nematode parasitic on sand pine (*Pinus clausa*), with additional notes on the morphology of *M. megatyla*. *Journal of Nematology* 17, 206–219.

Elsea, J.R. (1951) The histological anatomy of the nematode *Meloidogyne hapla* (Heteroderidae). *Proceedings of the Helminthological Society of Washington* 18, 53–63.

Endo, B.Y. and Wergin, W.P. (1977) Ultrastructure of anterior sensory organs of the root-knot nematode, *Meloidogyne incognita*. *Journal of Ultrastructure Research* 59, 231–249.

Esser, R.P., Perry, V.G. and Taylor, A.L. (1976) A diagnostic compendium of the genus *Meloidogyne* (Nematoda: Heteroderidae). *Proceedings of the Helminthological Society of Washington* 43, 138–150.

Fortuner, R. (1991) Methods for collection and preparation of nematodes. Part 1. Field sampling and preparation of nematodes for optic microscopy. In: Nickle, W.R. (ed.) *Manual of Agricultural Nematology.* Marcel Dekker, New York, pp. 75–87.

Franklin, M.T. (1971) Taxonomy of Heteroderidae. In: Zuckerman, B.M., Mai, W.F. and Rohde, R.A. (eds) *Plant Parasitic Nematodes*, Vol. II. Academic Press, New York, pp. 139–162.

Franklin, M.T. (1978) *Meloidogyne*. In: Southey, J.F. (ed.) *Plant Nematology.* Her Majesty's Stationery Office, London, pp. 98–124.

Garcia-Martinez, R. (1982) Post-infection development and morphology of *Meloidogyne cruciani*. *Journal of Nematology* 14, 332–338.

Göldi, E.A. (1887) Relatório sôbre a moléstia do cafeeiro na provincia do Rio de Janeiro. Extrahido do VIII Vol. dos *Archivos do Museu Nacional*, Rio de Janeiro, Imprensa Nacional, pp. 1–121 + 4 plates and a map.

Goldstein, P. and Triantaphyllou, A.C. (1980) The ultrastructure of sperm development in the plant-parasitic nematode *Meloidogyne hapla*. *Journal of Ultrastructure Research* 71, 143–153.

Goodey, J.B. (1959) Data to be considered, observed and, where possible, reported upon when presenting descriptions of new species. *Nematologica* 4, 211–216.

Greeff, R. (1864) *Verhandlungen des Naturhistorischen Vereins des Preussischen Rheinlande und Westfalens* XXI, pp. 112–113.

Hartman, K.M. and Sasser, J.N. (1985) Identification of *Meloidogyne* species on the basis of differential host test and perineal pattern morphology. In: Barker, K.R., Carter, C.C. and Sasser, J.N. (eds) *An*

Advanced Treatise on Meloidogyne. *Volume II. Methodology.* A cooperative publication of the Department of Plant Pathology and the United States Agency for International Development, North Carolina State University Graphics, Raleigh, North Carolina, pp. 69–77.

Hirschmann, H. (1985a) The classification of the family *Meloidogyne.* In: Sasser, J.N. and Carter, C.C. (eds) *An Advanced Treatise on* Meloidogyne. *Volume I. Biology and Control.* A cooperative publication of the Department of Plant Pathology and the United States Agency for International Development, North Carolina State University Graphics, Raleigh, North Carolina, pp. 35–45.

Hirschmann, H. (1985b) The genus *Meloidogyne* and morphological characters differentiating its species. In: Sasser, J.N. and Carter, C.C. (eds) *An Advanced Treatise on* Meloidogyne. *Volume I. Biology and Control.* A cooperative publication of the Department of Plant Pathology and the United States Agency for International Development, North Carolina State University Graphics, Raleigh, North Carolina, pp. 79–93.

Jepson, S.B. (1983a) The use of second-stage juvenile tails as an aid in the identification of *Meloidogyne* species. *Nematologica* 29, 11–28.

Jepson, S.B. (1983b) Identification of *Meloidogyne* species; a comparison of female stylets. *Nematologica* 29, 132–143.

Jepson, S.B. (1983c) Identification of *Meloidogyne*: a general assessment and a comparison of male morphology using light microscopy, with a key to 24 species. *Revue de Nématologie* 6, 291–309.

Jepson, S.B. (1985) *Meloidogyne chitwoodi.* In: *CIH Descriptions of Plant-parasitic Nematodes*, Set 8, No. 106, Commonwealth Agricultural Bureaux, Farnham Royal, UK.

Jepson, S.B. (1987) *Identification of Root-knot Nematodes (*Meloidogyne *species).* Commonwealth Agricultural Bureaux, Farnham Royal, UK.

Jobert, C. (1878) Sur un maladie du cafeier observée au Bresil. *Compte Rendue Hebdomodaire des Séances del'Académie des Sciences, Paris* 87, 941–943.

Johnson, P.W. and Graham, W.G. (1976) Ultrastructural studies on the cuticle of the second-stage larvae of four root-knot nematodes: *Meloidogyne hapla*, *M. javanica*, *M. incognita*, and *M. arenaria. Canadian Journal of Zoology* 54, 96–100.

Karssen, G. (2002) *The Plant-parasitic Nematode Genus* Meloidogyne *Göldi, 1892 (Tylenchida) in Europe.* Brill Academic Publishers, Leiden, The Netherlands.

Karssen, G. and Moens, M. (2006) Root-knot nematodes. In: Perry, R.N. and Moens, M. (eds) *Plant Nematology.* CAB International, Wallingford, UK, pp. 59–90.

Kleynhans, K.P.N. (1991) *The Root-knot Nematodes of South Africa.* Plant Protection Research Institute, Pretoria. Technical Communication, Department of Agricultural Development, South Africa, No. 231.

Lavergne, G. (1901) [The *Anguillula* in South America.] *Revista Chilena de Historia Natural* 5, 85–91.

Licopoli, G. (1875) Sopra alcuni tubercoli radicellari continente anguillole. *Rendiconti dell'Accademia delle Scienze, Fisiche e Mathematiche, Napoli* 14, 41–42.

Maggenti, A.R. and Allen, M.W. (1960) The origin of the gelatinous matrix in *Meloidogyne. Proceedings of the Helminthological Society of Washington* 27, 4–10.

McClure, M.A. and Bird, A.F. (1976) The tylenchoid (Nematoda) egg shell: formation of the egg shell in *Meloidogyne javanica. Parasitology* 72, 29–39.

Müller, C. (1884) Mittheilungen über die unseren Kulturpflanzen schädlichen, das Gesschlecht *Heterodera* bildenden Würmer. *Landwirthschaftliche Jahrbücher* 13, 1–42.

Nagakura, K. (1930) [Über den Bau and the life cycle of *Heterodera radicicola* (Greeff) Müller.] *Japanese Journal of Zoology* 3, 95–160.

Neal, J.C. (1889) *The Root-knot Disease of the Peach, Orange and Other Plants in Florida, Due to the Work of* Anguillula. Bulletin 20, Division of Entomology, US Department of Agriculture.

Örley, L. (1880) Az Anguillulidák Magánrajza [Monograph of the Anguillulida.] *Természetraji Füzetek, Budapest* 4, 16–150 + 7 plates and 34 figures.

Palumbo, M. (1892) *Anguillula* delle radici della vite. *L'Agricoltura Italiana*, pp. 392–397.

Papadopoulou, J. and Triantaphyllou, A.C. (1982) Sex differentiation in *Meloidogyne incognita* and anatomical evidence of sex reversal. *Journal of Nematology* 14, 549–566.

Popham, J.D. and Webster, J.M. (1978) An alternative interpretation of the fine structure of the basal zone of the cuticle of the dauer-larva of the nematode *Caenorhabditis elegans* (Nematoda). *Canadian Journal of Zoology* 56, 1556–1563.

Saigusa, T. (1957) On the egg development and its morphological observations of the root-knot nematode, *Meloidogyne* spp. *Japanese Journal of Applied Entomology and Zoology* 1, 238–243.

Shepherd, A.M. and Clark, S.A. (1983) Spermatogenesis and sperm structure in some *Meloidogyne* species (Heteroderoidea, Meloidogynidae) and a comparison with those in some cyst nematodes (Heteroderoidea, Heteroderidae). *Revue de Nématologie* 6, 17–32.

Taylor, A.L. and Sasser, J.N. (1978) *Biology, Identification and Control of Root-knot Nematodes (*Meloidogyne *species).* A Cooperative Publication of the Department of Plant Pathology, North Carolina State University and the US Agency for International Development, Raleigh, North Carolina.

Treub, M. (1885) Onderzoekingen over Sereh-Ziek Suikkeriet gedaan in s'Lands Plantentium te Buitenzorg. *Mededeelingen uit's Lands Plantentium, Batavia*, 2, 1–39.

Triantaphyllou, A.C. (1960) Sex determination in *Meloidogyne incognita* Chitwood, 1949 and intersexuality in *M. javanica* (Treub, 1885) Chitwood, 1949. *Annals of the Institute of Phytopathology, Benaki* 3, 12–31.

Triantaphyllou, A.C. (1962) Oogenesis in the root-knot nematode *Meloidogyne javanica. Nematologica* 7, 105–113.

Triantaphyllou, A.C. (1979) Cytogenetics of root-knot nematodes. In: Lamberti, F. and Taylor, C.E. (eds) *Rootknot Nematodes (*Meloidogyne *species); Systematics, Biology and Control.* Academic Press, London & New York, pp. 85–109.

Triantaphyllou, A.C. and Hirschmann, H. (1960) Post infection development of *Meloidogyne incognita* Chitwood, 1949. *Annals of the Institute of Phytopathology, Benaki* 3, 1–11.

van Breda de Haan, J. (1899) Levensgeschiedenis en Bestrijding van het Tabaksaaltje (*Heterodera radicicola*) in Deli. *Mededeelingen uit's Lands Plantentium, Batavia* 35.

Viala, P. (1893) *Les Maladies de la Vigne.* George Masson, Paris, pp. 577–580.

Viglierchio, D.R. (1979) Selected aspects of root-knot nematode physiology. In: Lamberti, F. and Taylor, C.E. (eds) *Rootknot Nematodes (*Meloidogyne *species); Systematics, Biology and Control.* Academic Press, London & New York, pp. 115–154.

Wergin, W.P. (1981) Scanning electron microscopic techniques and applications for use in nematology. In: Zuckerman, B.M. and Rohde, R.A. (eds) *Plant Parasitic Nematodes*, Vol. III. Academic Press, New York, pp. 175–204.

Wergin, W.P. and Endo, B.Y. (1976) Ultrastructure of a neurosensory organ in a root-knot nematode. *Journal of Ultrastructure Research* 56, 258–276.

Whitehead, A.G. (1968) Taxonomy of *Meloidogyne* (Nematoda: Heteroderidae) with descriptions of four new species. *Transactions of the Zoological Society, London* 31, 263–401.

Wouts, W.M. (1979) Characterization of the family Meloidogynidae with a discussion on its relationship to other families of the suborder Tylenchina based on gonad morphology. In: Lamberti, F. and Taylor, C.E. (eds) *Rootknot Nematodes (*Meloidogyne *species); Systematics, Biology and Control.* Academic Press, London & New York, pp. 21–35.

Wyss, U., Grundler, F.M.W. and Munch, A. (1992) The parasitic behaviour of second stage juveniles of *Meloidogyne incognita* in roots of *Arabidopsis thaliana. Nematologica* 38, 98–111.

Yik, C.P. and Birchfield, W. (1978) Scanning electron microscopy of perineal patterns of three species of *Meloidogyne. Journal of Nematology* 10, 118–122.

Yoder, M., Tandingan De Ley, I., King, I.W., Mundo-Ocampo, M., Mann, J., Blaxter, M., Poiras, L. and De Ley, P. (2006) DESS: a versatile solution for preserving morphology and extractable DNA of nematodes. *Nematology* 8, 367–376.

3 Taxonomy, Identification and Principal Species

David J. Hunt[1] and Zafar A. Handoo[2]
[1]*CABI Europe-UK, Egham, Surrey, UK;* [2]*USDA, ARS, BARC-West, Beltsville, Maryland, USA*

3.1 Introduction	**55**
3.2 Systematic Position	**61**
3.3 Subfamily and Genus Diagnosis	**61**
3.4 List of Species and Synonyms	**63**
3.5 Identification	**67**
3.6 Principal Species	**72**
3.7 Conclusions and Future Directions	**88**
3.8 Acknowledgements	**88**
3.9 References	**88**

3.1 Introduction

3.1.1 History

'On closer examination the root was found to be covered with excrescences varying from the size of a small pin's head to that of a little Bean or Nutmeg.' This observation, from one of the first accounts of root-knot nematodes on plants, was made by Berkeley (1855), an eminent Victorian scientist, on publishing his discovery of galls produced by nematodes on the roots of cucumbers growing in a garden frame at Nuneham, England. Miles Joseph Berkeley FRS (1803–1889), an ordained minister, expert draughtsman and pioneering zoologist, plant pathologist and mycologist (he authored over 6000 species), went on to describe the symptoms thus: 'The tubercles were of a dirty cream colour, nearly globose, obscurely furfuraceous, and in almost every case were developed on one side of the root…'. Berkeley (Fig. 3.1) noted the enormous development of the vascular tissues within the galls and recorded the presence of 'Vibrio':

> It appeared that these cysts were regular membranous sacs, exactly resembling the sporangia of Truffles, and filled with a multitude of minute elliptic or slightly cymbiform eggs, averaging not more than 1/250th of an inch in length with a breadth of 1/600th. In many of these the nucleus already showed the form of a Vibrio, folded up once or twice, and several of the animals were free, though still of small size, having escaped from the eggs by a little circular aperture at one extremity.

Berkeley illustrated his account with two rather nice drawings showing the symptoms on the roots and a section through one of the galls (Fig. 3.2).

The occurrence of galls on plant roots was also recorded in some detail by Licopoli (1875), who described tubercles on the roots of *Sempervivum tectorum* L. and other Crassulaceae in Italy. Licopoli dissected the tubercles and recorded the presence of 'miriadi di Anguillole simili a quelle che talvolta rinvengonsi nel frutto del

©CAB International 2009. *Root-knot Nematodes* (eds R.N. Perry, M. Moens and J.L. Starr)

Fig. 3.1. The Rev. Miles Joseph Berkeley FRS (1803–1889). Berkeley, a pioneering mycologist and plant pathologist, was the first to publish a paper on root-knot nematodes.

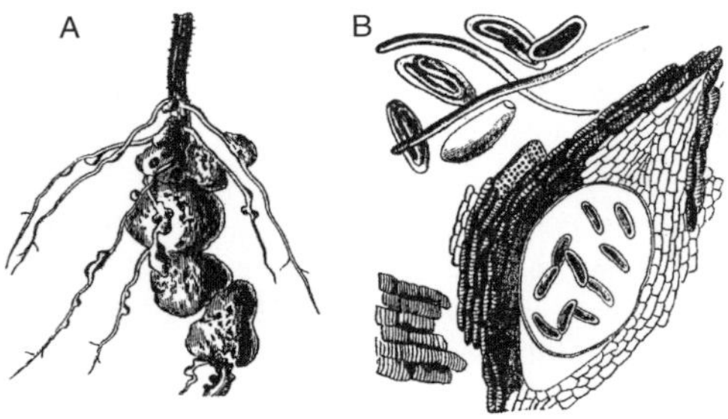

Fig. 3.2. The first illustration of root-knot nematodes on plant roots. A: galled roots; B: section through gall, showing nematodes and eggs. After Berkeley (1855).

formento e del fico non ben maturato' ['myriads of little anguillole, like those sometimes found in the fruit of wheat and in not well-matured fig.']. Similarly, Jobert (1878) described galls on the roots of coffee trees from Rio de Janeiro state, Brazil, and referred to the presence in the galls of eggs containing juveniles.

None of these early workers actually named the nematodes that they found in the galls, this being left to a French botanist, Maxime Cornu, who referred to the nematodes from galls on sainfoin as *Anguillula marioni* Cornu, 1879. Cornu (1843–1901) did a thorough study of the phenomenon, comparing the root-knot galls with the nodules on leguminous plants and those formed by *Phylloxera vastatrix* on vine roots (Cornu, 1879). Carl Müller (1884) was the first to illustrate a perineal pattern while describing root-knot nematodes, which he erroneously referred to as *Heterodera radicicola* (Greeff, 1872) Müller 1884, confusing nematodes previously described as *Anguillula radicicola*[1] by Greeff (1872) with the root-knot nematodes that he, Müller, had found in the same host. This inadvertent error would have unfortunate repercussions as Cobb (1924) subsequently used Greeff's name as the type of his new genus, *Caconema*.

The Dutch botanist Melchior Treub (1851–1910), who worked in the Dutch East Indies from 1880 to 1909, described the next species of root-knot nematode when he proposed *Heterodera javanica* Treub, 1885. This nematode was isolated from the roots of sugarcane from the Buitenzorg Botanical Gardens in Java, Indonesia (Treub, 1885). The description was rather short on detail and lacked figures.

Although two species of root-knot nematode had so far been named, the actual genus name of *Meloidogyne* was not proposed until 1887 when Göldi (Fig. 3.3) described *Meloidogyne exigua* Göldi, 1887 from galls on coffee roots in Rio de Janeiro state, Brazil. Émil August Göldi was born in Switzerland on 28 August 1859 and emigrated to Brazil in 1880, where he worked as a zoologist at the Museu Nacional do Rio de Janeiro, and later at the Museu Paraense, an institution that was renamed in his honour as the Museu Paraense Emílio Goeldi in 1902. In Brazil, he spelled his name as Emílio Augusto Goeldi, hence the two surname variants seen in the literature. Göldi returned to Switzerland in 1905, dying in Zurich on 5 July 1917 at the age of 58. In his proposal of the genus *Meloidogyne*, Göldi (1887) provided a full-page plate of line drawings of the nematodes that he had found, but, although the figures clearly show that it was a root-knot nematode, there are few meaningful data by today's standards to establish its precise identity (Fig. 3.4).

In the literature, the publication date of Göldi's description of *M. exigua* is variously cited as 1887 or 1892. While it is clear that Volume VIII of *Archivos do Museu Nacional*, the journal in which the

[1] Now known as *Subanguina radicicola*.

Fig. 3.3. Emílio Augusto Goeldi (1859–1917), the proposer of the genus *Meloidogyne*. Image from http://pt.wikipedia.org/wiki/Imagem:Goeldi_Emilio_Augusto_1859-1917.jpg. Accessed September 2007.

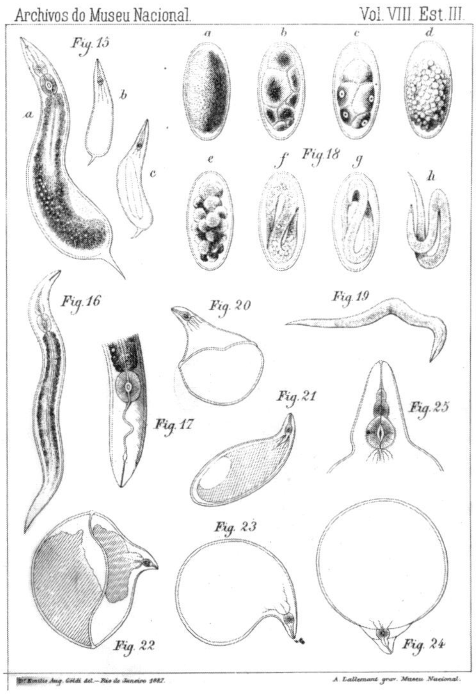

Fig. 3.4. Plate of original line drawings of *Meloidogyne exigua*. Although lacking much of the detail required for species diagnosis, the tail spike characteristic of a developing root-knot nematode can be clearly seen in Fig. 15. After Göldi (1887), courtesy of *Archivos do Museo Nacional, Rio de Janeiro*.

description was printed, was actually published in 1892 (see Lordello, 1951; Wouts and Sher, 1971; Fortuner, 1984; Karssen, 2002, for example), an advance copy or preprint of the article exists, clearly imprinted with 1887 as the date of publication and with different pagination to the 1892 version (Göldi, 1887, 1892). A preprint may be defined as 'a work published, with its own specified date of publication (imprint date), in advance of its later reissue as part of a collective or cumulative work. Preprints may be published works for the purposes of zoological nomenclature' (see Glossary and Article 21.8 in International Commission on Zoological Nomenclature, 1999).

Whitehead (1968) referred to this 'advance copy' in his bibliography, and accordingly used the earlier year of 1887 as the date of publication for the genus and type species. Other authors using the earlier date include Chitwood (1949), who, however, cited the 1892-published journal as the reference source, Sasser (1960), Franklin (1957, 1965a, 1976) and Jepson (1987), although most later authors, including Siddiqi (1986, 2000), Eisenback (1997), Karssen and van Hoenselaar (1998), Karssen (2002) and Karssen and Moens (2006), use the 1892 date.

We have examined an archive copy of the preprint from the United States Department of Agriculture (USDA) National Agricultural Library at Beltsville, Maryland. The title page clearly carries the imprint date of '1887'[2] (compare this with the journal volume title page, which has the imprint 1892) and is paginated from 1 to 121 with an additional index page followed by four plates of figures and a map entitled 'Zona affectada pela Molestia do Cafeeiro, Agosto, 1887'. A typed note (authorship unknown) taped to the inside cover of the preprint declares that 'The material [in the 1892 paper] is the same as in this preprint, 1887.' An additional note indicates that Göldi published essentially the same information, although in a 'somewhat condensed' form, in 1888 (Göldi, 1888). The 1888 paper was published in the March issue of *Revista Agricola* and includes the four plates and map found in the 1887 and 1892 publications. On p. 42 of this article, Göldi discusses the proposal of the new name '*Meloidogyne exigua*' and, in a footnote, cites the new genus name as '*Meloidogyne* nov. gen. Göldi (1887)', and

[2] Thereby distinguishing it from a separate.

provides an etymology and diagnosis of the genus. Göldi also published an article entitled 'Biologische Miscellen aus Brasilien. VII. Der Kaffeenematode Brasiliens (*Meloidogyne exigua* G.)' in 1889 (Göldi, 1889). In this article he discusses '*Meloidogyne exigua* G.', provides some morphological data of the various stages and compares it with *Heterodera*. It is clear, therefore, that the description of the genus and type species was validly published before the 1892 date commonly cited, the earlier preprint of 1887 satisfying the requirements of Article 21.8 of the Code for a nomenclatural act, and thus qualifying as the actual publication date for the genus and binomen. In this chapter, therefore, we accept 1887 as the actual date of publication for the genus and type species.

Shortly after Göldi's proposal of the genus *Meloidogyne*, Neal (1889), clearly unaware of the former publication, proposed the gall-forming nematode, *Anguillula arenaria* Neal, 1889. Neal produced a comprehensive, highly detailed and nicely illustrated paper on the root galls of plants, including radishes, and peach, fig and orange trees, in Florida, USA (Fig. 3.5; Plate 1). The quality of this paper easily surpassed all that preceded it, notwithstanding the fact that Neal was almost certainly dealing with more than one species of root-knot nematode (see Chitwood, 1949). Neal referred to reports of this root-knot disease being recognized as far back as 'the earliest settlement of the South Atlantic and Gulf states by white people' and stated 'In 1869 I found the root-knot prevalent over Florida, and learned from old residents that as far back as 1805 it had been known'.

In the same year that Neal published his work on nematode galls in Florida, Atkinson (1889) found 'giant cells' in cross-sections of root-knot nematode-infected roots, although he interpreted these as dead females rather than nutritional devices. Atkinson gave an account of the life history of these nematodes, which he referred to as '*Heterodera radicicola* (Greeff) Müller', thereby repeating Müller's mistaken identity of some 5 years previously (Müller, 1884).

Other reports of root-knot nematodes include those of Cobb (1890), from New South Wales, Australia, and Lavergne (1901a,b). The latter report is of some interest, as Gaston Lavergne described *Anguillula vialae* Lavergne, 1901 from the roots of vines in Chile; the recent discovery of *Meloidogyne ethiopica* Whitehead, 1968 from vines and kiwi fruit in Chile and other South American countries has led to the suggestion that Lavergne's record may actually refer to *M. ethiopica* (Carneiro *et al.*, 2007), although proof is lacking.

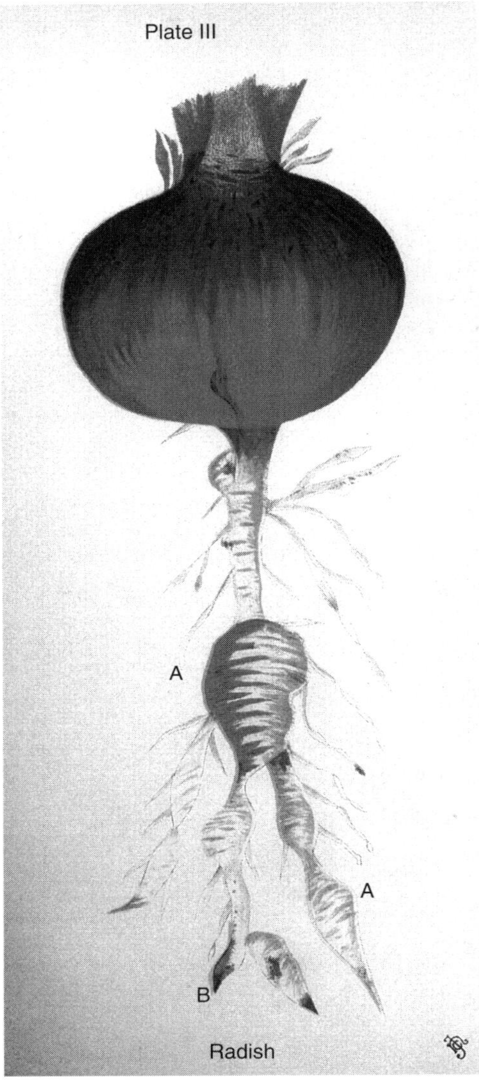

Fig. 3.5. Root galling on radish caused by *Anguillula* [= *Meloidogyne*] '*arenaria*'. After Neal (1889), courtesy of USDA.

In the first decade of the 20th century, Kati Marcinowski (1909) differentiated in detail the differences between cyst and root-knot nematodes. She recognized only one species of root-knot nematode, however, synonymizing all the other nominal species (including *arenaria*, *exigua* and *javanica*) to *H. radicicola*.

One of the more controversial aspects of taxonomy in this genus occurred in 1919 when Kofoid and White, working in the USA, described a new species of oxyurid from 'Man'. They had isolated numerous viable nematode eggs from the faeces of troops stationed in Texas and other military units from Oklahoma, New Mexico and Arizona, yet these eggs did not correspond in dimension to any of the helminths known to be parasitic in humans. They named their species *Oxyuris incognita* Kofoid & White, 1919, the specific epithet reflecting their uncertainty as to the status of this 'new nematode infection of Man'. As is usual for helminth identification, the dimensions of the eggs were recorded, length varying from 68 to 133 μm and diameter from 33 to 43 μm. The large range in length may indicate that more than one nematode species was represented, although the generic identity of these eggs is uncertain.

Sandground (1923), commenting on the occurrence of similar eggs found in the course of hookworm campaigns in the USA and other countries, stated:

> The fact that the eggs were found sporadically in the stools and that their occurrence was especially noticeable in the summer, a period when vegetable salads are a significant article in the diet, made it seem feasible to the writer that they originated in plant parasitic nematodes and were introduced with the food.

His research indicated that when bean roots infected by root-knot nematodes (which he referred to as *H. radicicola*) were ingested by humans the eggs could pass through the body and be recovered from the stools, the eggs being of similar appearance to those previously attributed to *O. incognita*. He therefore concluded that the eggs recorded by Kofoid and White (1919) were the product of a root-knot nematode rather than being laid by an unknown helminth as previously surmized.

Cobb (1924), recognizing that there were differences between cyst-forming and root-knot nematodes, proposed the genus *Caconema* Cobb, 1924 to contain the latter. The type species of this new genus was *H. radicicola*, itself a confused appellation because of the misidentification dating back to Müller (1884). Tom Goodey (1932) did not accept Cobb's proposal of *Caconema* as he regarded root-knot and cyst-forming nematodes as congeneric. Goodey (1932) regarded *A. marioni* Cornu, 1879, the oldest name applied to root-knot nematodes, as belonging to the genus *Heterodera*, a combination that had previously been proposed by Marcinowski (1909). What is now known as *Meloidogyne marioni* is currently regarded as a *species inquirenda* due to a lack of informative morphological detail.

Although Nagakura (1930) published an extensive study on root-knot nematodes, differentiating them in numerous ways from cyst nematodes and making observations on their morphology and life cycle, he still referred to them as *H. radicicola*.

The root-knot nematodes received their first major revision when Chitwood (1949) published a defining overview of the group. Chitwood resurrected the genus *Meloidogyne* as proposed by Göldi (1887) with *M. exigua* as type, even though there was no type material available of this species and the description itself was too poor to enable species identification (yet good enough to establish the genus). Chitwood placed three other species in the genus, making the new combinations *M. incognita*, *M. javanica* and *M. arenaria*. The identity of at least *M. incognita* and *M. javanica* was at best equivocal due to their inadequate original descriptions, *O. incognita* being based solely on a series of rather variable egg dimensions that may well have represented more than one species, while '*Heterodera javanica*' did not even figure in the description by Treub (1885). Although these actions established the fundamental framework for root-knot nematode taxonomy, some authors, notably Gillard (1961) and Whitehead (1968), criticized the resurrection of a genus name perhaps best left forgotten and the recognition of old species names whose true identity was highly dubious – not to mention the fact that Chitwood, perhaps mischievously, cited the type host of *M. incognita* as 'Man', on the basis that the eggs were first isolated from the stools of troops! Chitwood, however, made it very clear that his objective was to establish firmly the older names and thereby stabilize the existing nomenclature. Accordingly he provided redescriptions of the four old names (*arenaria*, *exigua*, *incognita*, *javanica*), based on new material that was often allocated in a somewhat arbitrary way, e.g. his description of *M. exigua*, originally described from Brazil, was based on nematodes from coffee plants in the New York

Botanical Garden, while his description of *M. incognita* used material from carrots in Texas, on the basis that this was the commonest root-knot in the state and therefore the species that the troops, from whom Kofoid and White (1919) had obtained the eggs of *O. incognita*, would probably have ingested. Chitwood also established the use of perineal patterns as a useful diagnostic aid and described *Meloidogyne hapla* Chitwood, 1949. Whatever the veracity of the arguments concerning the validity of the genus and the status of its species, there is no denying that Chitwood produced a seminal paper that clearly demarcated the differences between root-knot and cyst-forming nematodes, and simultaneously laid the foundations for future research.

Sledge and Golden (1964) proposed the genus *Hypsoperine* Sledge & Golden, 1964 for species of *Meloidogyne* where the mature female was characterized by a thicker cuticle and an elevated, cone-like posterior region. The history of this genus has been somewhat chequered, with many authors (e.g. Whitehead, 1968; Hirschmann, 1985; Jepson, 1987) regarding it as a junior synonym of *Meloidogyne*, while Siddiqi (1986) not only recognized it as a valid genus but split it into two subgenera, *Hypsoperine* (*Hypsoperine*) and *Hypsoperine* (*Spartonema*). However, in the second edition of his *magnum opus*, Siddiqi (2000) synonymized *Hypsoperine* (*Hypsoperine*) under *Meloidogyne*, but raised the former subgenus *Spartonema* to genus rank. Plantard *et al.* (2007), on the basis of 18S rDNA sequences, refuted the generic status of *Spartonema*, a decision which is accepted herein.

3.1.2 Major reference sources

The major reference sources for root-knot nematode taxonomy start with the monograph by Chitwood (1949). Although some of his nomenclatural decisions were considered to be at fault by other workers, the paper has rightly become the benchmark for all subsequent work. Intraspecific variability, long the bugbear of root-knot nematode diagnostics, was studied by a number of authors, including Allen (1952) and Dropkin (1953). The importance of the perineal pattern in identification was stressed by Sasser (1954), Taylor *et al.* (1955) and Triantaphyllou and Sasser (1960), among others. Mention must also be made of the monograph by Whitehead (1968), a neat piece of research that drew together much useful information and in which he proposed four new species and recognized 23 as valid. The introduction of that paper is both cogent and comprehensive, and covers the history of the group in far more detail than has been given in the current chapter. Franklin (1971) reviewed the genus and included some 32 species, whilst Esser *et al.* (1976) provided a compendium to facilitate identification of 32 species. In a later work, Franklin (1979) again reviewed the genus, recognizing 36 valid species. Five years later, this total had risen to 54 species and two subspecies (Hirschmann, 1985). Other major reference sources include the monograph by Lamberti and Taylor (1979), the compendium by Hewlett and Tarjan (1983), the two-volume treatise edited by Sasser and Carter (1985) and Barker *et al.* (1985), the insightful, abundantly illustrated monograph by Jepson (1987) and the comprehensive and highly practical root-knot nematode taxonomic database compiled on CD-ROM by Eisenback (1997). The latest monographs are by Karssen and van Hoenselaar (1998) and Karssen (2002), both of which cover the European species of the genus, and Karssen and Moens (2006), where 89 valid species are listed. Karssen (2002) provides an interesting and detailed account of the history of the group.

3.1.3 Rate of species descriptions

Between 1880 and 1960, only eight valid species had been described, many of the intervening decades passing with none or only a single new species being named. No doubt this was in part a reflection of the conserved morphology of this fascinating group of plant parasites, the mature females, for example, showing few useful characters apart from those of the anterior region (labial annulation, stylet form and excretory pore position) and the perineal pattern around the vulva–anus region. Another consideration is the fact that the genus is predominantly tropical or subtropical in distribution, areas where there were few nematologists to take an interest in taxonomy. In the five decades since 1959, the pace of species description has been spectacular, with 18 being described in the 1960s, six in the 1970s, 30 in the 1980s, 22 in the 1990s and 12 since the millennium. By June 2009 there were 97 valid species in the genus.

3.1.4 Recent advances in characterization

The perineal pattern was initially thought to be rather more useful as a diagnostic character than eventually proved to be the case, intraspecific variation and a large increase in the number of nominal species taking their toll on its utility, particularly for the inexperienced eye. Although various characters from the male and second-stage juvenile (J2) were pressed into service, it was not until the advent of isozyme and molecular methodologies that a new window on species concept within the group opened, with the result that species otherwise camouflaged by intraspecific variability of the perineal pattern (the *incognita*-type pattern is particularly common, for example) can now be targeted with greater precision than hitherto.

3.2 Systematic Position

Because of certain similarities in morphology and biology, root-knot nematodes and cyst-forming nematodes have often been thought to be closely related. As a consequence, in many systematic schemes both groups were often placed in a single family or subfamily, the Heteroderidae or Heteroderinae, respectively, closely related to the hoplolaimids. A growing suspicion indicated, however, that the two groups had probably evolved separately and had achieved their similarities via the process of convergent evolution. According to this view, the root-knot nematodes justify their own family or subfamily and are closer to the pratylenchids than to the hoplolaimids.

Root-knot nematodes were first placed in their own subfamily when Skarbilovich (1959) proposed the Meloidogyninae, thereby emphasizing the differences between root-knot and cyst-forming nematodes. The Meloidogyninae was originally regarded as a subfamily under the Heteroderidae, although subsequent workers (e.g. Jepson, 1987; Siddiqi, 1986, 2000; Karssen, 2002) have recognized it at family level, as a sister taxon to either the cyst-forming nematodes or, latterly, the pratylenchids.

The advent of molecular methodologies has facilitated a better understanding of the phylogeny of the Nematoda. As a result, the morphology-based systematics schemes have now been largely replaced by hierarchies based on molecular phylogeny (see De Ley and Blaxter, 2002, 2004). In their attempt to unify the systematics, De Ley and Blaxter introduced the infraorder. The consequence of this action is automatically to cascade a reduction in rank to all subsidiary taxa so that, for example, the former ranks of superfamily, family and subfamily become family, subfamily and tribe, respectively. In this account we follow the scheme outlined in De Ley and Blaxter (2004), with the result that the former family Meloidogynidae is reduced in rank to a subfamily within the Hoplolaimidae. By adopting this scheme, taxa previously regarded as subfamilies should either be reduced in rank to tribe or discarded – we have followed the latter course.

3.3 Subfamily and Genus Diagnosis

The following diagnoses of the family and genus are based on those of Siddiqi (2000) but have been updated to reflect the fact that we reject *Hypsoperine*, including the subgenus *Spartonema*, as valid taxa. *Hypsoperine* and *Spartonema* were also rejected as valid genera by Plantard *et al.* (2007).

Subfamily Meloidogyninae Skarbilovich, 1959
= Meloidogynini Skarbilovich, 1959
= Meloidoderellinae Husain, 1976
= Meloidoderellini Husain, 1976

Diagnosis: Hoplolaimidae. Root-gall inciting, female feeding inciting multinucleate nurse cells. Marked sexual dimorphism. Cuticle striated. Lateral fields bearing four or five incisures. Labial region low, with one to four annules. Under SEM (scanning electron microscopy), female labial disc dorso-ventrally elongate, dumb-bell-shaped with oral opening a small, round pore surrounded by six inner labial pits (sensilla). Framework moderately sclerotized, hexaradiate; lateral sectors equal to, or wider than, submedian sectors. Stylet moderately strong, male stylet longer and more robust than that of female. Orifice of dorsal pharyngeal gland located just posterior to stylet base. Median pharyngeal bulb oval or round, with large refractive thickenings. Pharyngeal glands elongate, extending over intestine mostly ventrally but also laterally; subventrals asymmetrical, extending past dorsal gland, SVN (subventral gland nucleus) always posterior to DN (dorsal gland nucleus). Excretory pore in female opposite or anterior to median bulb, in male usually posterior to median bulb. No preadult vermiform female stage (except in *Meloinema*).

Mature female: Swollen, sedentary, round, oval to pear-shaped with a projecting neck. Cuticle moderately thick, striated, generally forming typical, fingerprint-like perineal pattern terminally. No cyst stage. Vulva subterminal or terminal. Anus located near vulval lip; tail rudimentary or absent. Stylet under 25 μm long in *Meloidogyne* (but 30–35 μm long in *Meloinema*). Median bulb oval or rounded, usually offset, with large refractive thickenings. Didelphic–prodelphic; ovaries coiled. Most eggs not retained in body but laid. Large rectal glands present, gelatinous matrix present.

Male: Vermiform, migratory, generally non-feeding, over 1 mm long, posterior end twisted through 90–180°, developing by metamorphosis within a saccate juvenile. Labial region rather low and continuous; amphidial apertures large transverse slits; labial cap large, prominent; framework moderately sclerotized, lateral sectors wider than submedian sectors. Stylet strong, usually over 20 μm long, basal knobs prominent. Tail short or absent, lacking a bursa (except *Bursadera*). Spicules large (25–64 μm), distally pointed. Gubernaculum linear to trough-shaped, not protrusible. Cloacal lips non-tuboid, generally with hypoptygma.

Juveniles: Second stage migratory and infective. Third- and fourth-stage juveniles swollen, without stylet in type genus. In *Meloinema*, third- and fourth-stage juveniles vermiform. Labial region low, anteriorly flattened or rounded. Lateral sectors wider than submedian sectors, labial disc distinct in type genus. Stylet weak to moderately developed, less than 20 μm long in type genus but strongly developed in *Meloinema*. Tail elongate–conoid, with minutely rounded tip and conspicuous terminal hyaline portion. Phasmids dot-like, located on tail, usually anterior to middle.

Remarks: Meloidogyninae differs from Heteroderinae by its members inciting root-galls and having the lateral sectors of the labial framework wider than the submedian sectors, weaker labial sclerotization and stylet, excretory pore of mature female located opposite or anterior to median bulb, and third- and fourth-stage juveniles lacking a stylet in the type genus.

Type genus:

Meloidogyne Göldi, 1887

Other genera:

Meloinema Choi & Geraert, 1974
= *Nacobbodera* Golden & Jensen, 1974
Bursadera Ivanova & Krall, 1985

Genus *Meloidogyne* Göldi, 1887

= *Hypsoperine* Sledge & Golden, 1964
= *Hypsoperine* (*Hypsoperine*) Sledge & Golden, 1964 (Siddiqi, 1986)
= *Hypsoperine* (*Spartonema*) (Siddiqi, 1986)
= *Spartonema* Siddiqi, 1986

Diagnosis (modified after Siddiqi, 2000)**:** Meloidogyninae. Root-gall inciting.

Mature female: Round to pear-shaped with short projecting neck, white, sedentary. No cyst stage. Vulva and anus located close together, terminal; perineum with a fingerprint-like cuticular pattern, usually flattened, rarely elevated. Phasmids dot-like, slightly anterior to, and on either side of, anus. Cuticle striated. Stylet slender, generally 12–15 μm long, with small basal knobs. Excretory pore anterior to median bulb, often just posterior to base of stylet. Genital tracts paired, prodelphic, convoluted. Six large rectal glands secreting gelatinous material in which eggs are deposited;[3] eggs not retained in body.

Male: Vermiform, up to 2 mm long, tail end twisted, developing by metamorphosis within a swollen juvenile. Cuticle strongly annulated; lateral field with four incisures. Labial region not sharply offset, with distinct labial disc and few (1–3) annules; lateral sectors wider than submedian sectors, appearing as 'cheeks'. Stylet robust (18–25 μm), with large basal knobs. Pharyngeal glands lying mostly ventral to intestine. Spicules slender, generally 25–33 μm long, gubernaculum 7–11 μm long. Testis single, but paired when sex reversal occurs. Tail rounded. Phasmids dot-like, located near cloacal aperture, which is subterminal. Bursa absent.

Juveniles: First stage with a blunt tail tip, moulting within egg; second and third moults occurring within cuticle of second stage. Second stage vermiform, migratory, infective, straight to arcuate habitus upon death. Labial region with coarse annules (1–4), a distinct labial disc, framework lightly sclerotized, lateral sectors wider than submedian sectors, stylet slender, under 20 μm, excretory pore posterior to hemizonid. Median bulb with large oval refractive thickenings. Tail with conspicuous hyaline region, tip narrow, irregular in outline. Third stage sedentary, swollen, sausage-shaped with a short blunt

[3] Rectal glands not developed and no egg mass formed in *M. spartinae* and *M. kikuyensis*.

tail. Stylet absent. Fourth stage sedentary, swollen, with terminal anus. Stylet absent.

Note: The genus *Caconema* Cobb, 1924 has been regarded as a junior synonym of *Meloidogyne*, an action accepted by many taxonomists including Siddiqi (1986). However, after discussion with Andrássy (*in litt.*), Siddiqi (2000) pointed out that *Caconema* was based on *H. radicicola* Greeff as type, Müller's misattribution of his root-knot species to that described by Greeff finally coming home to roost. Greeff's species (*sensu* Greeff, not Müller) is also the type of *Subanguina* Paramonov, 1967 and, therefore, *Caconema* and *Subanguina* are, both being based on the same type species, objective synonyms. *Caconema* is, of course, the senior name, but Siddiqi (2000) argued the case that it should be regarded as an invalid senior synonym and suppressed by the International Code of Zoological Nomenclature (ICZN) on the basis that it was a *nomen oblitum*, whereas *Subanguina* had been widely accepted and used as a valid taxon. Such a ruling has yet to be made by the ICZN.

3.4 List of Species and Synonyms

The following list is based on that of Siddiqi (2000) but includes those species that he regarded as belonging to the genus *Hypsoperine*. A full synonymy is provided together with bibliographic references for all taxonomic authorities. The Principle of Coordination (Article 43.1, International Code of Zoological Nomenclature, 1999) is followed for authorities.

Type species:
Meloidogyne exigua Göldi, 1887
= *Heterodera exigua* (Göldi, 1887) Marcinowski, 1909

Other species:
Meloidogyne acronea Coetzee, 1956
= *Hypsoperine acronea* (Coetzee, 1956) Sledge & Golden, 1964
= *Hypsoperine* (*Hypsoperine*) *acronea* (Coetzee, 1956) Sledge & Golden, 1964 (Siddiqi, 1986)
Meloidogyne actinidiae Li & Yu, 1991
Meloidogyne africana Whitehead, 1960
Meloidogyne aquatilis Ebsary & Eveleigh, 1983
Meloidogyne arabicida López & Salazar, 1989
Meloidogyne ardenensis Santos, 1968
= *Meloidogyne deconincki* Elmiligy, 1968
= *Meloidogyne litoralis* Elmiligy, 1968
Meloidogyne arenaria (Neal, 1889) Chitwood, 1949
= *Anguillula arenaria* Neal, 1889
= *Tylenchus arenarius* (Neal, 1889) Cobb, 1890
= *Heterodera arenaria* (Neal, 1889) Marcinowski, 1909
= *Meloidogyne arenaria arenaria* (Neal, 1889) Chitwood, 1949
= *Meloidogyne arenaria thamesi* Chitwood *in* Chitwood, Specht & Havis, 1952
= *Meloidogyne thamesi* Chitwood *in* Chitwood, Specht & Havis, 1952 (Goodey, 1963)
= *Meloidogyne thamesi gyulai* Amin, 1993
= *Meloidogyne gyulai* Amin, 1993
Meloidogyne artiellia Franklin, 1961
Meloidogyne baetica Castillo, Vovlas, Subbotin & Troccoli, 2003
Meloidogyne brasilensis Charchar & Eisenback, 2002
Meloidogyne brevicauda Loos, 1953
Meloidogyne californiensis Abdel-Rahman & Maggenti, 1987
= *Meloidogyne californiensis* Abdel-Rahman, 1981 (= *nomen nudum*)
Meloidogyne camelliae Golden, 1979
Meloidogyne caraganae Shagalina, Ivanova & Krall, 1985
Meloidogyne carolinensis Eisenback, 1982
= *Meloidogyne carolinensis* Fox, 1967 (=*nomen nudum*)
Meloidogyne chitwoodi Golden, O'Bannon, Santo & Finley, 1980

Meloidogyne chosenia Eroshenko & Lebedeva, 1992
Meloidogyne christiei Golden & Kaplan, 1986
Meloidogyne cirricauda Zhang & Weng, 1991
Meloidogyne citri Zhang, Gao & Weng, 1990
Meloidogyne coffeicola Lordello & Zamith, 1960
= *Meloidodera coffeicola* (Lordello & Zamith, 1960) Kirjanova, 1963
Meloidogyne cruciani García-Martinez, Taylor & Smart, 1982
Meloidogyne cynariensis Pham,[4] 1990
Meloidogyne decalineata Whitehead, 1968
Meloidogyne donghaiensis Zheng, Lin & Zheng, 1990
Meloidogyne dunensis Paolomaes Rius, Vovlas, Troccoli, Liebanas, Landa & Castillo, 2007
Meloidogyne duytsi Karssen, van Aelst & van der Putten, 1998
Meloidogyne enterolobii Yang & Eisenback, 1983
= *Meloidogyne mayaguensis* Rammah & Hirschmann, 1988
Meloidogyne ethiopica Whitehead, 1968
Meloidogyne fallax Karssen, 1996
Meloidogyne fanzhiensis Chen, Peng & Zheng, 1990
Meloidogyne floridensis Handoo, Nyczepir, Esmenjaud, van der Beek, Castagnone-Sereno, Carta, Skantar & Higgins, 2004
Meloidogyne fujianensis Pan, 1985
= *Meloidogyne fujianensis* Pan, Ling & Wang, 1988 (= objective junior homonym)
Meloidogyne graminicola Golden & Birchfield, 1965
Meloidogyne graminis (Sledge & Golden, 1964) Whitehead, 1968
= *Hypsoperine graminis* Sledge & Golden, 1964
= *Hypsoperine (Hypsoperine) graminis* Sledge & Golden, 1964 (Siddiqi, 1986)
Meloidogyne hainanensis Liao & Feng, 1995
Meloidogyne hapla Chitwood, 1949
Meloidogyne haplanaria Eisenback, Bernard, Starr, Lee & Tomaszewski, 2004
Meloidogyne hispanica Hirschmann, 1986
Meloidogyne ichinohei Araki, 1992
Meloidogyne incognita (Kofoid & White, 1919) Chitwood, 1949
= *Oxyuris incognita* Kofoid & White, 1919
= *Heterodera incognita* (Kofoid & White, 1919) Sandground, 1923
= *Meloidogyne incognita incognita* (Kofoid & White, 1919) Chitwood, 1949
= *Meloidogyne acrita* Chitwood, 1949
= *Meloidogyne incognita acrita* Chitwood, 1949
= *Meloidogyne kirjanovae* Terenteva, 1965
= *Meloidogyne elegans* da Ponte, 1977
= *Meloidogyne incognita grahami* Golden & Slana, 1978
= *Meloidogyne grahami* Golden & Slana, 1978
= *Meloidogyne incognita wartellei* Golden & Birchfield, 1978
= *Meloidogyne wartellei* Golden & Birchfield, 1978
Meloidogyne indica Whitehead, 1968
Meloidogyne inornata Lordello, 1956a
= *Meloidogyne incognita inornata* Lordello, 1956a
Meloidogyne izalcoensis Carneiro, Almeida, Gomes & Hernández, 2005
Meloidogyne javanica (Treub, 1885) Chitwood, 1949
= *Heterodera javanica* Treub, 1885
= *Tylenchus (Heterodera) javanicus* (Treub, 1885) Cobb, 1890
= *Anguillula javanica* (Treub, 1885) Lavergne, 1901a

[4] Cited as 'Pham Thanh Binh' after the species name in the original paper.

= *Meloidogyne javanica javanica* (Treub, 1885) Chitwood, 1949
= *Meloidogyne javanica bauruensis* Lordello, 1956b
= *Meloidogyne bauruensis* Lordello, 1956b
= *Meloidogyne lordelloi* da Ponte, 1969
= *Meloidogyne lucknowica* Singh, 1969
Meloidogyne jianyangensis Yang, Hu, Chen & Zhu, 1990
= *Meloidogyne jianyangensis* Zhu, Lan, Hu, Yang & Wang, 1991[5] (= objective junior homonym)
Meloidogyne jinanensis Zhang & Su, 1986
Meloidogyne kikuyensis De Grisse, 1961
= *Spartonema kikuyense* (De Grisse, 1961) Siddiqi, 2000
Meloidogyne konaensis Eisenback, Bernard & Schmitt, 1995
Meloidogyne kongi Yang, Wang & Feng, 1988
Meloidogyne kralli Jepson, 1984
Meloidogyne lini Yang, Hu & Xu, 1988
Meloidogyne lusitanica Abrantes & Santos, 1991
Meloidogyne mali Itoh, Ohshima & Ichinohe, 1969
Meloidogyne maritima Jepson, 1987
Meloidogyne marylandi Jepson & Golden *in* Jepson, 1987
Meloidogyne megadora Whitehead, 1968
Meloidogyne megatyla Baldwin & Sasser, 1979
Meloidogyne mersa Siddiqi & Booth, 1991
= *Meloidogyne* (*Hypsoperine*) *mersa* Siddiqi & Booth, 1991
Meloidogyne microcephalus Cliff & Hirschmann, 1984
Meloidogyne microtyla Mulvey, Townshend & Potter, 1975
Meloidogyne mingnanica Zhang, 1993
Meloidogyne minor Karssen, Bolk, van Aelst, van den Beld, Kox, Korthals, Molendijk, Zijlstra, van Hoof & Cook, 2004
Meloidogyne morocciensis Rammah & Hirschmann, 1990
Meloidogyne naasi Franklin, 1965a
Meloidogyne nataliae Golden, Rose & Bird, 1981
Meloidogyne oryzae Maas, Sanders & Dede, 1978
Meloidogyne oteifai[6] Elmiligy, 1968
Meloidogyne ottersoni (Thorne, 1969) Franklin, 1971
= *Hypsoperine ottersoni* Thorne, 1969
= *Hypsoperine* (*Hypsoperine*) *ottersoni* Thorne, 1969 (Siddiqi, 1986)
Meloidogyne ovalis Riffle, 1963
Meloidogyne panyuensis Liao, Yang, Feng & Karssen, 2005
= *Meloidogyne panyuensis* Liao, 2001 (= *nomen nudum*)
Meloidogyne paranaensis Carneiro, Carneiro, Abrantes, Santos & Almeida, 1996
Meloidogyne partityla Kleynhans, 1986b
Meloidogyne petuniae Charchar, Eisenback & Hirschmann, 1999
Meloidogyne phaseoli Charchar, Eisenback, Charchar & Boiteau, 2008b
Meloidogyne pini Eisenback, Yang & Hartman, 1985
Meloidogyne piperi Sahoo, Ganguly & Eapen, 2000
Meloidogyne pisi Charchar, Eisenback, Charchar & Boiteau, 2008a
Meloidogyne platani Hirschmann, 1982
Meloidogyne propora Spaull, 1977
= *Hypsoperine propora* (Spaull, 1977) Siddiqi, 1986

---

5 The authors cite this name as a new species, thereby creating, albeit unintentionally, a junior objective homonym.

6 Original spelling *oteifae*.

= *Hypsoperine* (*Hypsoperine*) *propora* (Spaull, 1977) Siddiqi, 1986
Meloidogyne querciana Golden, 1979
Meloidogyne salasi López, 1984
Meloidogyne sasseri Handoo, Huettel & Golden, 1994
Meloidogyne sewelli Mulvey & Anderson, 1980
Meloidogyne silvestris Castillo, Vovlas, Troccoli, Liébanas, Palomares Rivs & Landa, 2009
Meloidogyne sinensis Zhang, 1983
Meloidogyne spartinae (Rau & Fassuliotis, 1965) Whitehead, 1968
= *Hypsoperine spartinae* Rau & Fassuliotis, 1965
= *Hypsoperine* (*Spartonema*) *spartinae* Rau & Fassuliotis, 1965 (Siddiqi, 1986)
= *Spartonema spartinae* (Rau & Fassuliotis, 1965) Siddiqi, 1986
Meloidogyne subarctica Bernard, 1981
Meloidogyne suginamiensis Toida & Yaegashi, 1984
Meloidogyne tadshikistanica Kirjanova & Ivanova, 1965
Meloidogyne thailandica Handoo, Skantar, Carta & Erbe, 2005
Meloidogyne trifoliophila Bernard & Eisenback, 1997
Meloidogyne triticoryzae Gaur, Saha & Khan, 1993
Meloidogyne turkestanica Shagalina, Ivanova & Krall, 1985
Meloidogyne ulmi Marinari-Palmisano & Ambrogioni, 2000
Meloidogyne vandervegtei Kleynhans, 1988

***Species inquirendae*:**
Meloidogyne marioni (Cornu, 1879) Chitwood & Oteifa, 1952
= *Anguillula marioni* Cornu, 1879
= *Heterodera marioni* (Cornu, 1879) Marcinowski, 1909
Meloidogyne megriensis (Poghossian, 1971) Esser, Perry & Taylor, 1976
= *Hypsoperine megriensis* Poghossian, 1971
= *Hypsoperine* (*Hypsoperine*) *megriensis* Poghossian, 1971 (Siddiqi, 1986)
Meloidogyne poghossianae Kirjanova, 1963
= *Meloidogyne acronea apud* Poghossian, 1961 *nec M. acronea* Coetzee, 1956
Meloidogyne vialae (Lavergne, 1901b) Chitwood & Oteifa, 1952
= *Anguillula vialae* Lavergne, 1901b
= *Heterodera vialae* (Lavergne, 1901b) Marcinowski, 1909

***Nomina nuda*:**
Meloidogyne californiensis Abdel-Rahman, 1981
Meloidogyne carolinensis Fox, 1967
Meloidogyne goeldii Santos, 1997
Meloidogyne panyuensis Liao, 2001 *nec M. panyuensis* Liao *et al.*, 2005
Meloidogyne zhanjiangensis Liao, 2001

Notes

1. Siddiqi (1986, 2000) cited '*Meloidogyne goeldi* Lordello, 1951' as being proposed as a *nomen novum* for *M. marioni*. This was repeated by Karssen and Moens (2006) in their species list, albeit under the genus *Heterodera*, but is incorrect, apparently stemming from Lordello's citation of the genus and authority being misinterpreted as a binomen (see Lordello, 1951, English Summary, p. 250).
2. In the Chinese literature a species name may be cited as 'n. sp.' subsequent to the original proposal – and with different authors (e.g. *M. fujianensis* and *M. jianyangensis*). Each such subsequent citation appears to be a junior objective homonym and is probably also a *nomen nudum*.
3. The Index of Organism Names (see http://www.organismnames.com/, accessed 3 April 2008) lists the combination '*Meloidogyne zhanjiangensis*'. The publication authority and date for this name are not cited therein, the binomen presumably being culled from other published sources. There is, however, a web reference to a Chinese PhD thesis entitled 'Study on the identification and polymorphism of

rootknot nematodes (*Meloidogyne*)' by J.L. Liao, where the combination '*Meloidogyne zhanjiangensis* n. sp.' occurs; '*Meloidogyne panyuensis* n. sp.' is also mentioned, although this species was subsequently formally described by Liao *et al.* (2005). There is also a web reference to a paper entitled 'Description of *Meloidogyne zhanjiangensis* n. sp. from China' by Liao and Feng. This was supposedly published in the *Russian Journal of Nematology*, volume 11(2) in 2003 (www.nsfc.gov.cn/nsfc/cen/00/kxb/sm/wuchu04.doc, accessed 3 April 2008). However, the only paper by these authors in that issue of the journal is an abstract of a description of a new species of *Meloidogyne* from pea in Guangdong province, China (see Liao and Feng, 2003). No mention is made of a specific name in either the title or text in this abstract (although it seems to be the same article as mentioned above), reference to the specific epithet probably being removed by the editor to avoid creating a *nomen nudum*.

According to Chen Mian-Cai of South China Agricultural University (C. Mian-Cai, China, April 2008, personal communication), the species name was originally proposed in a PhD thesis submitted by J.L. Liao to the university in 2001 (Liao, 2001), although to date a description has not been published in a journal. The binomen was cited by Rui *et al.* (2005), along with a brief set of measurements, pictures of the perineal pattern, J2 and female anterior region, together with Est and Mdh isozyme profiles, but with no formal proposal/description. At the present time, the status of this name is unclear and it is prudent to regard it as a *nomen nudum*.

3.5 Identification

Accurate identification of root-knot nematodes is crucial for effective disease control and depends on rapid and accurate classification of the pathogens involved so that appropriate control measures may be taken. In addition, sound decisions regarding quarantine of imported and exported plant material and commodities also demand timely and accurate diagnostics. None the less, the identification of root-knot nematodes to species level is fraught with difficulty. Conserved morphology, variable morphometrics, host effects, intraspecific variation, the parthenogenetic mode of reproduction, existence of cryptic species or species swarms, and the ever-increasing number of described species – the diagnosis and relationships of many of which vary from less than ideal to dubious – all serve to obfuscate the clear interspecific boundaries that we yearn for. To add to the confusion, there is the not-inconsiderable problem of 'species concept' in organisms that predominantly rely on a parthenogenetic reproductive strategy.

Verification of mixed populations and/or detection of rare species requires identification techniques other than the North Carolina differential host test (see section 3.5.7 below), including morphological (perineal pattern of adult females; male, female and J2 labial region shape, and stylet morphology; length and shape of J2 tail) and, in some cases, biochemical or molecular methodologies. Detailed diagnostic characters differentiating *Meloidogyne* species have been given by authors such as Eisenback *et al.* (1981), Eisenback (1985), Hirschmann (1985), Jepson (1987), Taylor (1987) and Eisenback and Triantaphyllou (1991). They are also covered in Eisenback and Hunt, Chapter 2, this volume.

For many years the form of the perineal pattern of the mature female and various morphometric and morphological features of the J2 were relied upon in species determination. To these were added features of the (often only rarely produced) male, such as the form of the labial region, including the annulation, and form of the stylet and basal knobs. With the increasing number of described species, however, the value of many of these characters, themselves showing often large intraspecific variation, was eroded almost to the point where robust identification tended to involve a fair measure of serendipity.

As an example, what may be termed the *incognita*-type of perineal pattern is now known to occur in a substantial number of species, some of which were commonly misidentified as *M. incognita* (see the work on South and Central American coffee nematodes by Carneiro *et al.*, 2004b, 2005). Isozyme electrophoresis has discriminated a number of these otherwise cryptic species but it is PCR-based molecular methodologies that currently carry the torch and our hopes for the future.

3.5.1 General techniques

For morphological observation, J2 and males can be recovered from fresh infected roots or egg masses incubated in Petri dishes with a small amount of water. They may also be recovered from soil by sieving and Baermann funnel techniques. Females are dissected from infected roots after fixation overnight in 3% formaldehyde. Procedures for measuring and preparing specimens are as given in Golden and Birchfield (1972), except that some females have the anterior and posterior ends cut with a sharp knife, cleaned with a dental root canal file, and mounted permanently on a glass slide in a drop of lactophenol solution. Photomicrographs of perineal patterns, J2s and males can be done with a 35 mm or digital camera attached to a dissecting microscope.

For more details on killing, fixing, processing nematodes to glycerine, preparing temporary and permanent slide mounts, and preserving nematode structures in a life-like manner, the reader is referred to Whitehead (1968), Hooper (1970, 1986, 1990), Golden (1990) and Carta (1991).

3.5.2 Perineal pattern

The character most frequently used for *Meloidogyne* species identification is the morphology of the perineal pattern, which is located in the posterior body region of adult females. This area comprises the vulva–anus area (perineum), tail terminus, phasmids, lateral lines and surrounding cuticular striae. Preparation of perineal patterns (Fig. 3.6) for observation and identification has been covered by Taylor *et al.* (1955), Eisenback (1985), Franklin (1965b), Sasser and Carter (1982), Hartman and Sasser (1985), Hirschmann (1985), Jepson (1987), Riggs (1990) and Charchar and Eisenback (2000). A more detailed account on root-knot perineal pattern development was given by Karssen (2002). Figure 3.7 summarizes the form of perineal pattern in the 12 species of *Meloidogyne* that are considered in greater detail in this chapter.

3.5.3 Root staining

Many methods have been developed for staining and clearing nematode-infected root tissues. Staining with acid fuchsin-lactophenol or lactoglycerol are the most widely used methods. In addition, a method that utilizes chlorine bleach as a prestaining treatment has proven to be very reliable and is relatively simple to use (Byrd *et al.*, 1983). For more detail, see McBeth *et al.* (1941), Hooper (1986, 1990) and Hussey (1990).

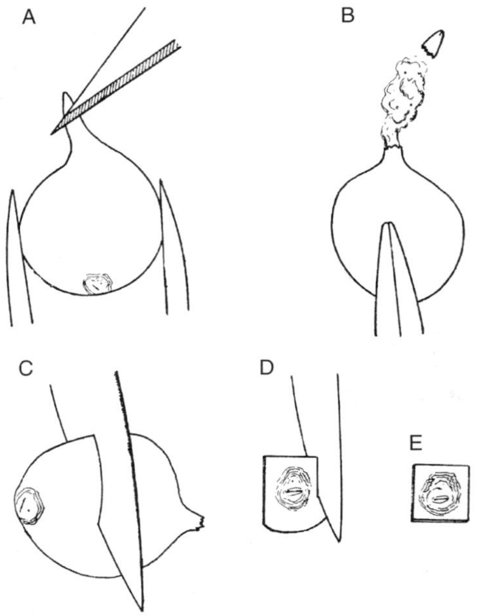

Fig. 3.6. How to cut perineal patterns. A, B: excised female with neck region removed and body contents gently expelled; C: posterior body with perineal pattern removed; D: trimming surplus cuticle around perineal pattern; E: trimmed perineal pattern ready for mounting. After Hartman and Sasser (1985).

3.5.4 Scanning electron microscopy

For scanning electron microscopy, living specimens are fixed in 3% glutaraldehyde solution buffered with 0.05 M phosphate (pH 6.8), dehydrated in a graded series of ethanol, critical-point dried from liquid CO_2 and sputter-coated with a 20–30 nm layer of gold–palladium. For more detail, the papers by Eisenback (1991) and Charchar and Eisenback (2000) are recommended; see also Eisenback and Hunt, Chapter 2, this volume.

3.5.5 Diagnostic characters

The most important diagnostic features used for identification of *Meloidogyne* spp. include: **Female.** Shape of body, labial region, stylet length, shape of stylet cone, shaft and basal knobs, nature of perineal pattern, including form of dorsal arch, lateral field, striae and tail terminus (see Fig. 3.7), and excretory pore/stylet length ratios (EP/ST). **Male.** Size, height and shape of labial cap, the number of annulations, diameter of the labial region as

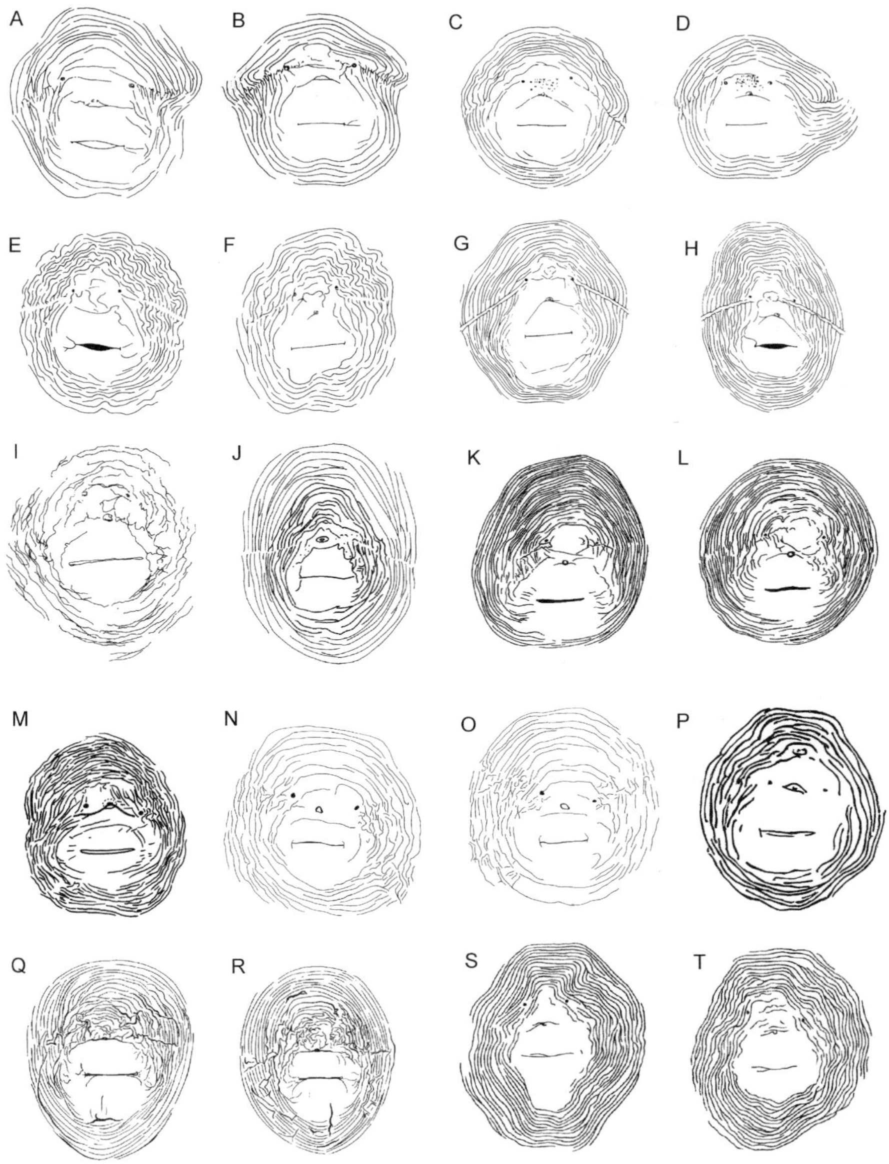

Fig. 3.7. Comparison of perineal patterns for 12 major species of *Meloidogyne*. A, B: *M. arenaria*; C, D: *M. hapla*; E, F: *M. incognita*; G, H: *M. javanica*; I: *M. acronea*; J: *M. chitwoodi*; K, L: *M. enterolobii*; M: *M. ethiopica*; N, O: *M. exigua*; P: *M. fallax*; Q, R: *M. graminicola*; S, T: *M. paranaensis*. Drawings not to scale. A–H, after Orton Williams (1972, 1973, 1974, 1975); I, after Page (1985); J, after Jepson (1985); K, L, after Rammah and Hirschmann (1988); M, after Whitehead (1968); N, O, courtesy of Janet Machon; P, after Karssen (1996); Q, R, after Mulk (1976); S, T, after Carneiro *et al.* (1996).

compared with the first body annule, stylet length, form of stylet cone, shaft and basal knobs, distance of the dorsal gland orifice (DGO) from the stylet base and length and form of spicule. **J2.** Body and stylet length, form of labial region and shape of stylet knobs, location of the hemizonid in relation to the excretory pore, distance of DGO from stylet base, number of lines in the lateral field and shape and length of the tail and hyaline terminus. For more details about these and other differentiating characters, see Whitehead (1968), Esser *et al.* (1976), Eisenback *et al.* (1981), Hirschmann (1985), Kleynhans (1986a), Jepson (1987), Eisenback and Triantaphyllou (1991), Karssen (2002) and Eisenback and Hunt, Chapter 2, this volume.

3.5.6 Root-knot or cyst-forming nematode?

Differentiating root-knot nematodes from cyst-forming nematodes is usually an easy task, regardless of developmental stage. However, in the interest of completeness these differences are summarized and compared in Table 3.1 and Fig. 3.8.

3.5.7 Differential host test

Variations in host range are known to occur in some species of root-knot nematodes, and attempts have been made to characterize these on the basis of differential host range. Sasser (1954) proposed a simple method, based on responses to a series of differential hosts and the amount of galling induced, to identify four of the five species of *Meloidogyne* recognized by Chitwood (1949). Subsequently, this test was often a component in descriptions of new species, purporting to demonstrate a 'unique' host reaction (see also Sasser, 1979). This test became known as the 'North Carolina differential host test' and has been used to detect host races within the 'Chitwood species' (Sasser and Carter, 1982). The 'International *Meloidogyne* project' (IMP) summarized the responses of about 1000 populations of the four most common species of *Meloidogyne* and their races to differential hosts, and more details were provided by Taylor and Sasser (1978), Eisenback *et al.* (1981), Sasser and Carter (1982) and Hartman and Sasser (1985). As discussed by Moens *et al.* (Chapter 1, this volume), the differential host range test is currently out of favour as a diagnostic tool and and has several drawbacks.

3.5.8 Gall form

Species of *Meloidogyne* typically cause galls on plant roots and other below-ground organs. However,

Table 3.1. Comparison between root-knot and cyst-forming nematodes.

Stage	Character	Root-knot	Cyst-forming
Second-stage infective juveniles	Body form	Slender, anterior body tapering	Robust, anterior body more parallel
	Labial region	Weakly cuticularized, not offset	Strongly cuticularized, offset
	Stylet	Slender, < 19 μm long	Robust, 20–30 μm long
	Hyaline region	Short, starting near tail tip	Long, well developed
Sedentary juveniles	Tail spike	Present	Absent
Male	Labial region	Weakly cuticularized, 2 annules	Strongly cuticularized, 4–5 annules
	Stylet	Slender	Robust
Mature female	Excretory pore	Anterior to median bulb valve plates	Posterior to median bulb valve plates
	Cuticle	Thin, white, not tanning to brownish colour on death	Thick, tanning to brownish colour on death
	Eggs	Deposited in external gelatinous egg mass, few retained in body	Mostly retained within body, occasionally a few laid into small egg sac
Biology	Host symptoms	Root galls almost always formed	No root galls formed
	Parasitic habit	Mature female usually endoparasitic	Mature female semi-endoparasitic
	Trophic system	Giant cells	Syncytia

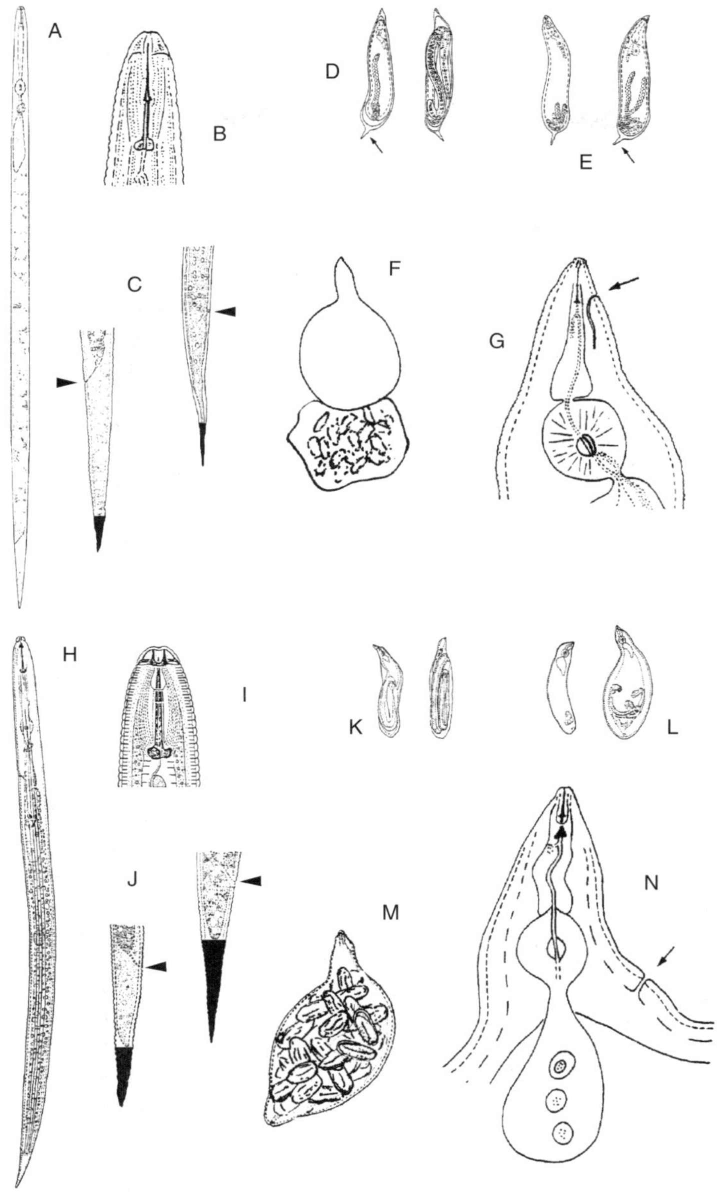

Fig. 3.8. Comparative morphology between root-knot and cyst-forming nematodes. A–G: root-knot nematode. H–N: cyst-forming nematode. A, H: infective J2; Note that the root-knot J2 is more slender. B, I: anterior region of J2, showing weaker labial sclerotization and spear development in B compared with I. C, J: tail region of J2, showing difference in hyaline region (indicated by black infill) as proportion of tail length (arrowhead points at anus in each case). D, K: development of male parasitic stages; Note presence of tail spike in D (arrow). E, L: development of female parasitic stages; Note presence of tail spike in E (arrow). F, M: mature female, showing eggs mostly laid into external gelatinous mass (F) or retained in body (M). G, N: anterior region of female, showing relative position of excretory pore (arrow), which is either anterior (G) or posterior (N) to the median bulb valve plates. Drawings not to scale and adapted from various sources.

they may also parasitize stems, leaves or flowers and incite galls in these tissues in several genera of plants (Lehman, 1985). *Meloidogyne* spp. are the most common and best-known nematodes that cause extensive root galls or 'root-knots', although a few species do not produce galls at all, e.g. *M. sasseri* (Handoo *et al.*, 1994).

The physical appearance and position of galls on roots can be of some assistance in diagnostics. For example, *M. javanica* and *M. incognita* tend to form large and irregular galls some distance from the root tip, whereas *M. exigua* galls on coffee are small, more or less spherical and located at the root tip. Galls of *M. graminicola* on rice are elongate and usually located just behind the root tip, affected roots assuming a characteristic hook-shape, while the relatively small and irregular galls of *M. hapla* often sport several lateral roots.

3.5.9 Isozyme phenotyping

Isozyme electrophoretic profiles, often using esterase and malate dehydrogenase, have been established for a number of species and can provide a useful, routine diagnostic test, particularly for morphologically variable species like *M. incognita* and *M. javanica* (see Esbenshade and Triantaphyllou, 1985a,b, 1987; Karssen *et al.*, 1995; Carneiro and Almeida, 2001; Carneiro *et al.*, 1998, 2000; Hernández *et al.*, 2004; Blok and Powers, Chapter 4, this volume). Some species, such as *M. arenaria*, show several different profiles, although this may be an indication of the existence of cryptic species. Although isozyme electrophoresis is perhaps the current diagnostic method of choice, with profiles accompanying the description of many new species, for example, it seems likely that PCR-based methodologies will soon usurp this method for many applications where finer resolution, particularly of intraspecific variation, is paramount.

3.5.10 Molecular diagnostics

There is no denying that PCR-based methodologies are of ever-increasing importance in species diagnostics and phylogeny within the genus *Meloidogyne*. Techniques include RFLP (restriction fragment length polymorphism) profiles of the ITS (internal transcribed spacer) region of rDNA, RAPD (random amplified polymorphic DNA) fragments, 18S rDNA sequences, satellite DNA probes and species-specific primers. Recent papers include De Ley *et al.* (2002) and Tigano *et al.* (2005), both using sequences of the 18S rDNA to construct phylogenies, while the latter also employed IGS (intergenic spacer) mitochondrial rDNA sequences. Carta *et al.* (2006) recommended molecular protocols for identification of root-knot nematodes on potato. Molecular approaches are covered in greater detail in Blok and Powers, Chapter 4 and Adams *et al.*, Chapter 5, this volume.

3.6 Principal Species

The following 12 species have been selected for further discussion either because they are common, economically important and of worldwide distribution (e.g. *M. arenaria*, *M. hapla*, *M. incognita*, *M. javanica*), or because they represent more regionally constrained species that may be damaging to specific major crops (e.g. *M. chitwoodi*, *M. ethiopica*, *M. fallax*, *M. graminicola*) or have potential as emergent pests. To facilitate comparisons, the four commonest species are presented first, the remainder following in alphabetical order. Each species is illustrated (Figs 3.9–3.20) and perineal patterns are juxtaposed in Fig. 3.7. The data presented here were obtained from various sources, including Jepson (1987), Karssen and Moens (2006), and original descriptions and/or redescriptions.

3.6.1 *Meloidogyne arenaria* (Fig. 3.9)

Morphology: **Female.** Pear-shaped, no posterior terminal protuberance. Stylet 13–17 μm long, cone curved dorsally, gradually tapering to blunt tip anteriorly; shaft broad, cylindrical, gradually widening posteriorly; basal knobs rounded to tear-drop-shaped, offset. Perineal pattern variable, rounded to ovoid with fine to coarse striae. Dorsal arch low, flattened with striae smooth or slightly wavy, continuous or broken, slightly bent towards tail tip at lateral line; generally forming shoulders on lateral portion of arch. Dorsal and ventral striae often meeting at an angle at lateral lines; lateral field distinct, slightly irregular. **Male.** Labial region not offset, smooth, rarely with one or two incomplete annulations, labial disc more or less rounded, slightly raised above level of medial lips, lateral lips usually absent (remnants occasionally present). Stylet 20–28 μm long, basal knobs offset, angular or more amalgamated, DGO = 4–8 μm. **J2.** L (body length) = 392–605 μm, hemizonid one to three annules anterior to excretory pore, tail = 44–69 μm with rounded to pointed tail tip and indistinct 6–13 μm long hyaline region.

Hosts: Extremely polyphagous, attacking both monocotyledons and dicotyledons.

Distribution: Worldwide; found in most of the warmer regions of the world and frequently encountered in glasshouses in cooler climates.

Isozymes: Populations of *M. arenaria* are variable in isozyme phenotype (Esbenshade and Triantaphyllou, 1985a,b). Three phenotypes of esterase activity commonly occur – namely, A1, A2, and A3 – and several other phenotypes occur less commonly. The phenotypes A1 and A2

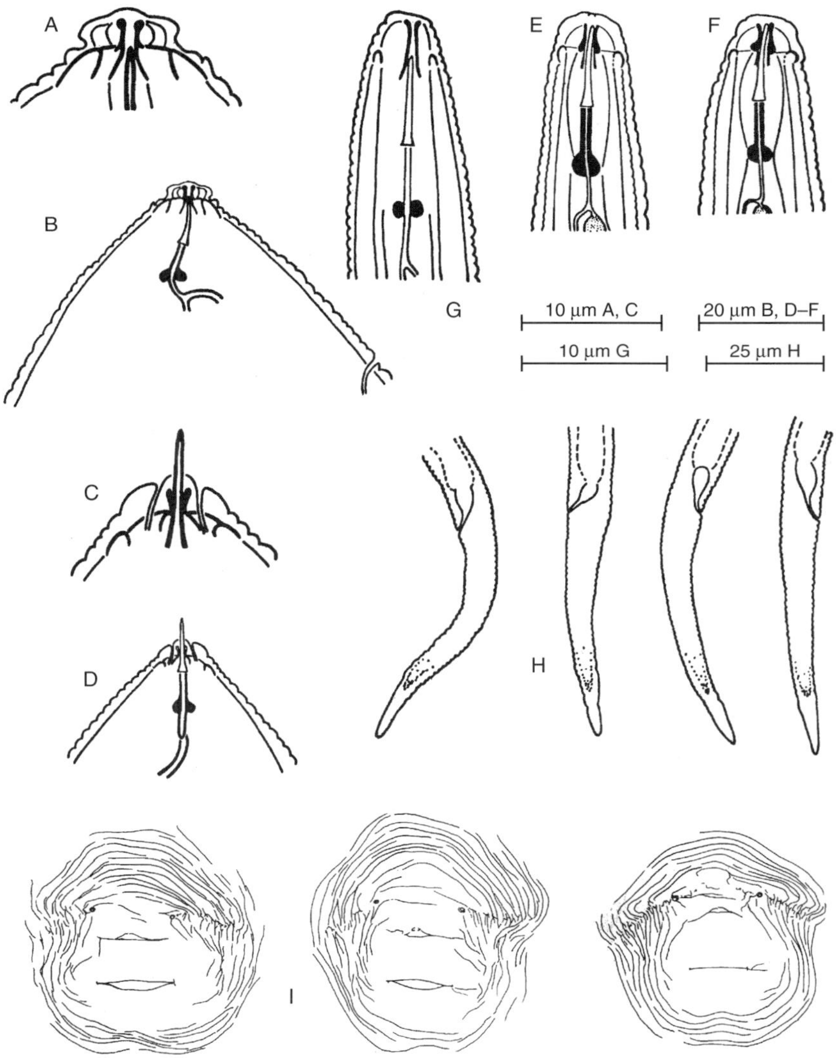

Fig. 3.9. *Meloidogyne arenaria.* A–D: female anterior region; E, F: male anterior region; G: J2 anterior region; H: J2 tail regions; I: perineal patterns. A–H, after Whitehead (1968), courtesy of *Transactions of the Zoological Society of London*; I, after Orton Williams (1975), courtesy of CAB International.

include several cytological forms of *M. arenaria*, whereas phenotype A3 includes only the most typical cytological form with a somatic chromosome number of 51–56. Some populations of *M. arenaria* have atypical esterase phenotypes: namely, S1-M1, S2-M1 and M3-F1. The malate dehydrogenase phenotype is either N1 or N3, according to population.

Remarks: One of the four commonest species worldwide. Typically inciting large, irregular galls. Part of a confusing species complex.

3.6.2 *Meloidogyne hapla* (Fig. 3.10)

Morphology: **Female.** Pear-shaped, terminal protuberance absent. Stylet 13–17 µm long, basal knobs small, rounded, offset. Perineal pattern rounded, low dorsal arch, characteristic punctations usually present near anus, fine striae, lateral field present. **Male.** Labial region offset, labial disc not usually elevated, lateral lips present. Stylet 19–22 µm long, basal knobs small, rounded, offset. DGO = 4–5 µm. **J2.** L = 360–500 µm, hemizonid anterior to excretory pore, tail = 48–70 µm, hyaline region often irregular in form, tail tip finely rounded.

Hosts: Mainly dicotyledonous plants.

Distribution: Common in temperate areas and at higher altitudes in the tropics.

Isozymes: Most populations show the H1 esterase phenotype (Esbenshade and Triantaphyllou, 1985a,b). One population from France does not have any major esterase activity, and another population from Minnesota, USA, has an A1 esterase phenotype identical to *M. arenaria*. The malate dehydrogenase H1 phenotype is unique for this species.

Remarks: One of the four commonest species worldwide. Galls are usually relatively small and may bear subsidiary roots.

3.6.3 *Meloidogyne incognita* (Fig. 3.11)

Morphology: **Female.** Pear-shaped, no posterior terminal protuberance. Stylet 15–16 µm long, basal knobs rounded, offset. Perineal pattern oval to rounded, typically with high, squared, dorsal arch, striae usually wavy, lateral field absent or weakly demarcated by forked striae. **Male.** Labial region not offset, labial disc elevated, lateral lips usually absent. Stylet 23–26 µm long, basal knobs offset, rounded to transversely elongate. DGO = 2–4 µm. **J2.** L = 350–450 µm, hemizonid anterior or adjacent to excretory pore, tail = 43–65 µm with 6–14 µm long hyaline region, rounded tail tip.

Hosts: Extremely polyphagous, attacking both monocotyledons and dicotyledons.

Distribution: Worldwide; restricted to protected cultivation in temperate regions.

Isozymes: The unique esterase I1 type and the malate dehydrogenase N1 type were described by Esbenshade and Triantaphyllou (1985a,b). Variability of esterase activity is small, only one population not having the typical phenotype. The malate dehydrogenase phenotype N1 is similar to that of *M. javanica*, *M. exigua* and some populations of *M. arenaria*.

Remarks: One of the four commonest species worldwide. Typically inciting large, usually irregular, galls. Member of a confusing species complex.

3.6.4 *Meloidogyne javanica* (Fig. 3.12)

Morphology: **Female.** Pear-shaped, no posterior terminal protuberance. Stylet 14–18 µm long, basal knobs ovoid, offset. Perineal pattern rounded, low dorsal arch; striae smooth, tail whorl often distinct, lateral field distinct, clearly demarcated from striae by more or less parallel lines. **Male.** Labial region not offset, labial disc not elevated, lateral lips absent. Stylet 19–24 µm long, basal knobs ovoid, offset. DGO = 3–5.5 µm. **J2.** L = 400–560 µm, hemizonid anterior or adjacent to excretory pore, tail = 47–60 µm with 9–18 µm long hyaline region, finely rounded tail tip.

Hosts: Extremely polyphagous, attacking both monocotyledons and dicotyledons.

Distribution: Worldwide; restricted to protected cultivation in temperate regions.

Isozymes: The esterase phenotype is J3, J2 or J2a (Esbenshade and Triantaphyllou, 1985a,b; Tomaszewski *et al.*, 1994; Castro *et al.*, 2003) and the malate dehydrogenase is of the N1 type (Esbenshade and Triantaphyllou, 1985a,b). Most populations of *M. javanica* have a malate dehydrogenase phenotype N1, similar to that of *M. incognita*, *M. exigua* and some populations of *M. arenaria*. One population of *M. javanica* from Bangladesh and one from Korea have the N3 phenotype, similar to that of some populations of *M. arenaria*.

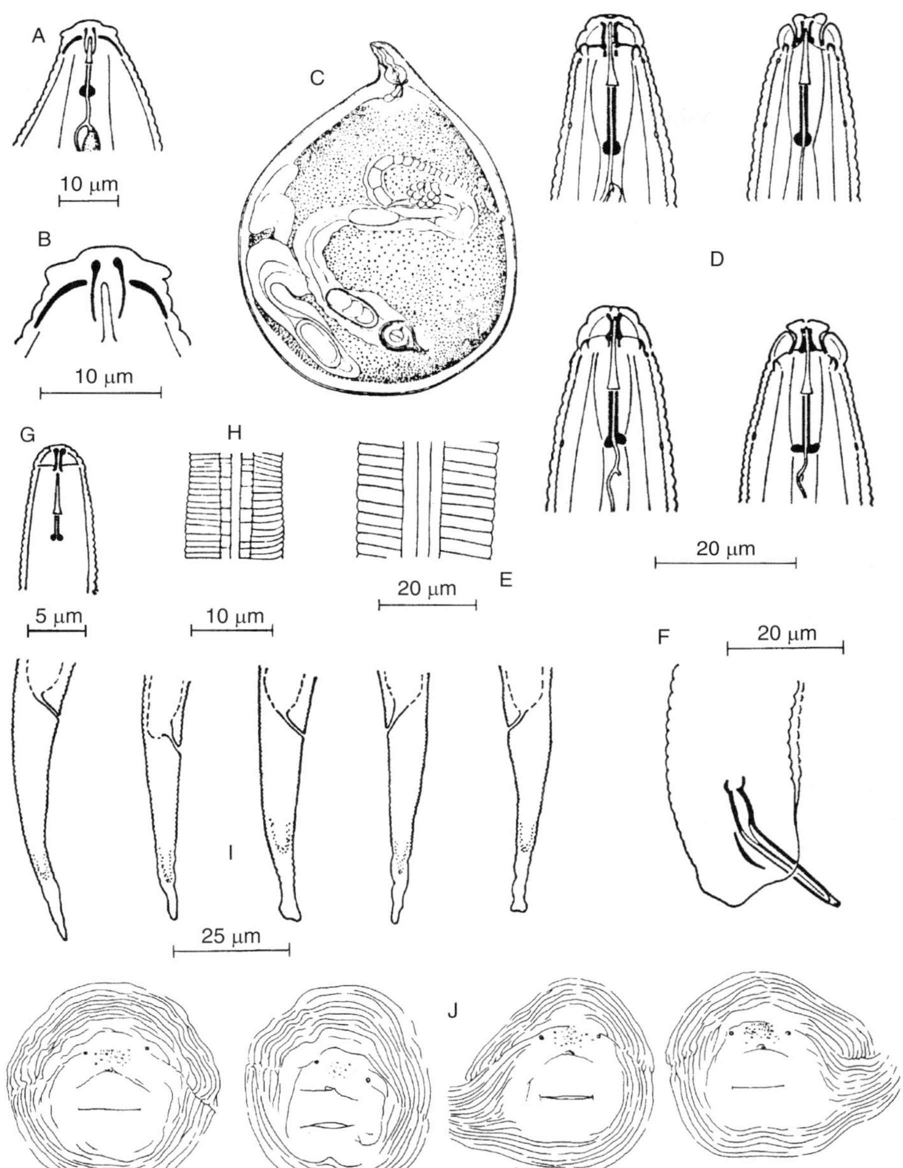

Fig. 3.10. *Meloidogyne hapla.* A, B: female anterior region; C: entire female; D: male anterior region; E: male lateral field; F: male tail; G: J2 anterior region; H: J2 lateral field at mid-body; I: J2 tail regions; J: perineal patterns. A–I, after Whitehead (1968), courtesy of *Transactions of the Zoological Society of London*; J, after Orton Williams (1974), courtesy of CAB International.

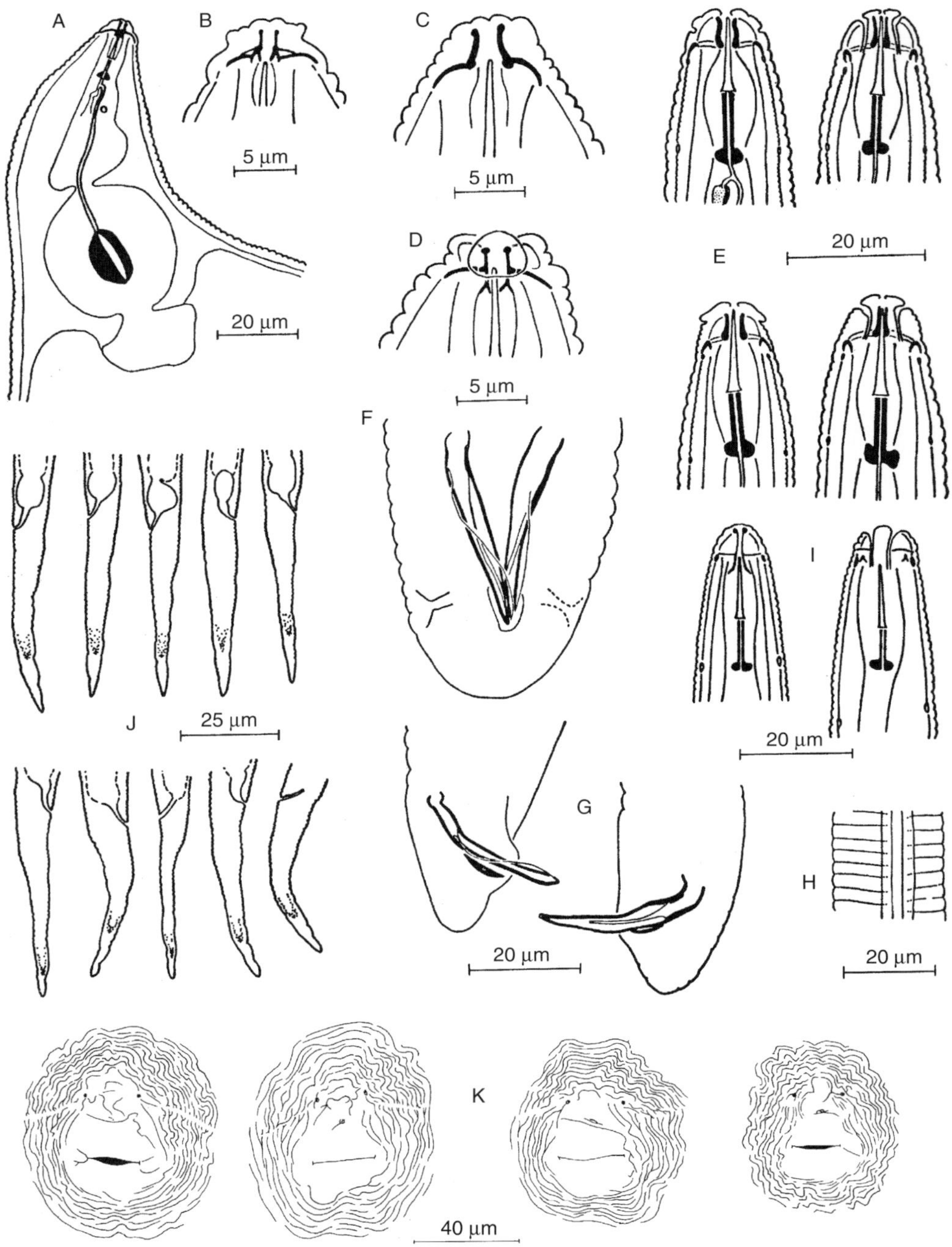

Fig. 3.11. *Meloidogyne incognita*. A: female pharyngeal region; B–D: female anterior region; E: male anterior region; F–G: male tail region; H: male lateral field; I: J2 anterior region; J: J2 tail regions; K: perineal patterns. A–J, after Whitehead (1968), courtesy of *Transactions of the Zoological Society of London*; K, after Orton Williams (1973), courtesy of CAB International.

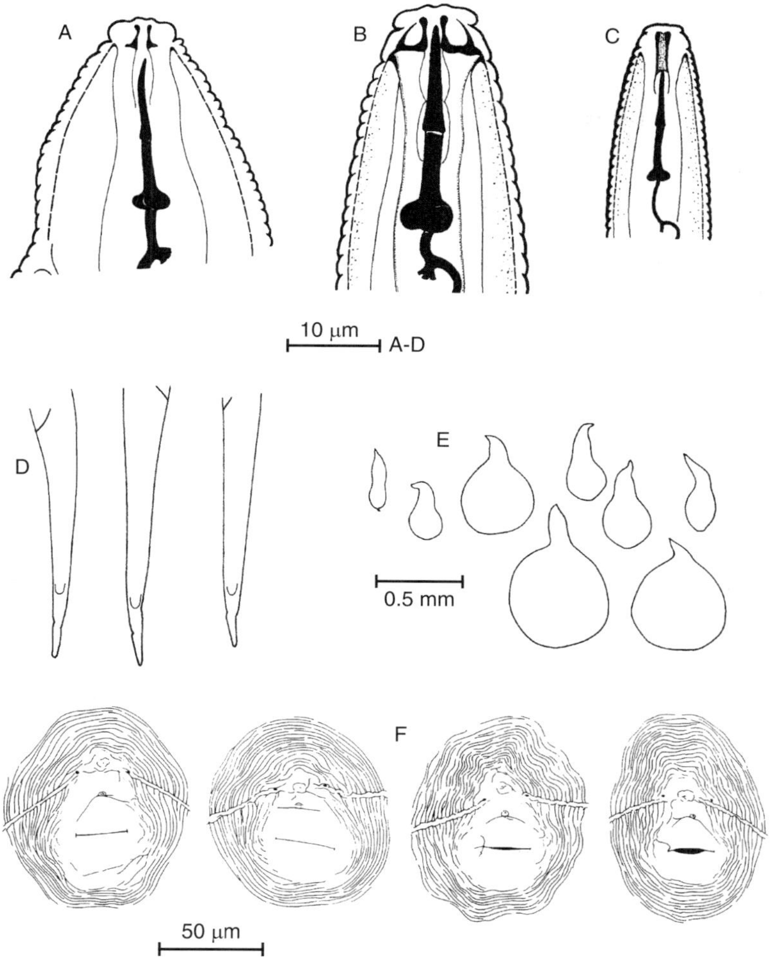

Fig. 3.12. *Meloidogyne javanica*. A: female pharyngeal region, lateral view; B: male anterior region; C: J2 anterior region; D: J2 tail regions; E: entire females; F: perineal pattern variation. A–D after Jepson (1987) and E, F after Orton Williams (1972), courtesy of CAB International.

Remarks: One of the four commonest species worldwide. Typically inciting large, irregular galls.

3.6.5 *Meloidogyne acronea* (Fig. 3.13)

Morphology: **Female.** Oval to spherical, perineal region situated on terminal protuberance. Stylet 10–14 μm long, basal knobs rounded, offset. Perineal pattern rounded, dorsal arch low, striae faint, intermittent, often broken on one side of vulval slit, lateral field absent. **Male.** Labial region not offset, labial disc not elevated, lateral lips usually present. Stylet 16–20 μm long, basal knobs pyriform, offset. DGO = 2–7 μm. **J2.** L = 340–490 μm, hemizonid anterior or adjacent to excretory pore, tail = 33–49 μm, short hyaline region (4–7 μm), tail tip rounded.

Hosts: Cotton, pigeon pea, okra, tomato, sorghum, bulrush millet and grasses.

Distribution: Southern Africa.

Isozymes: Phenotype unknown.

Remarks: Restricted distribution in Africa.

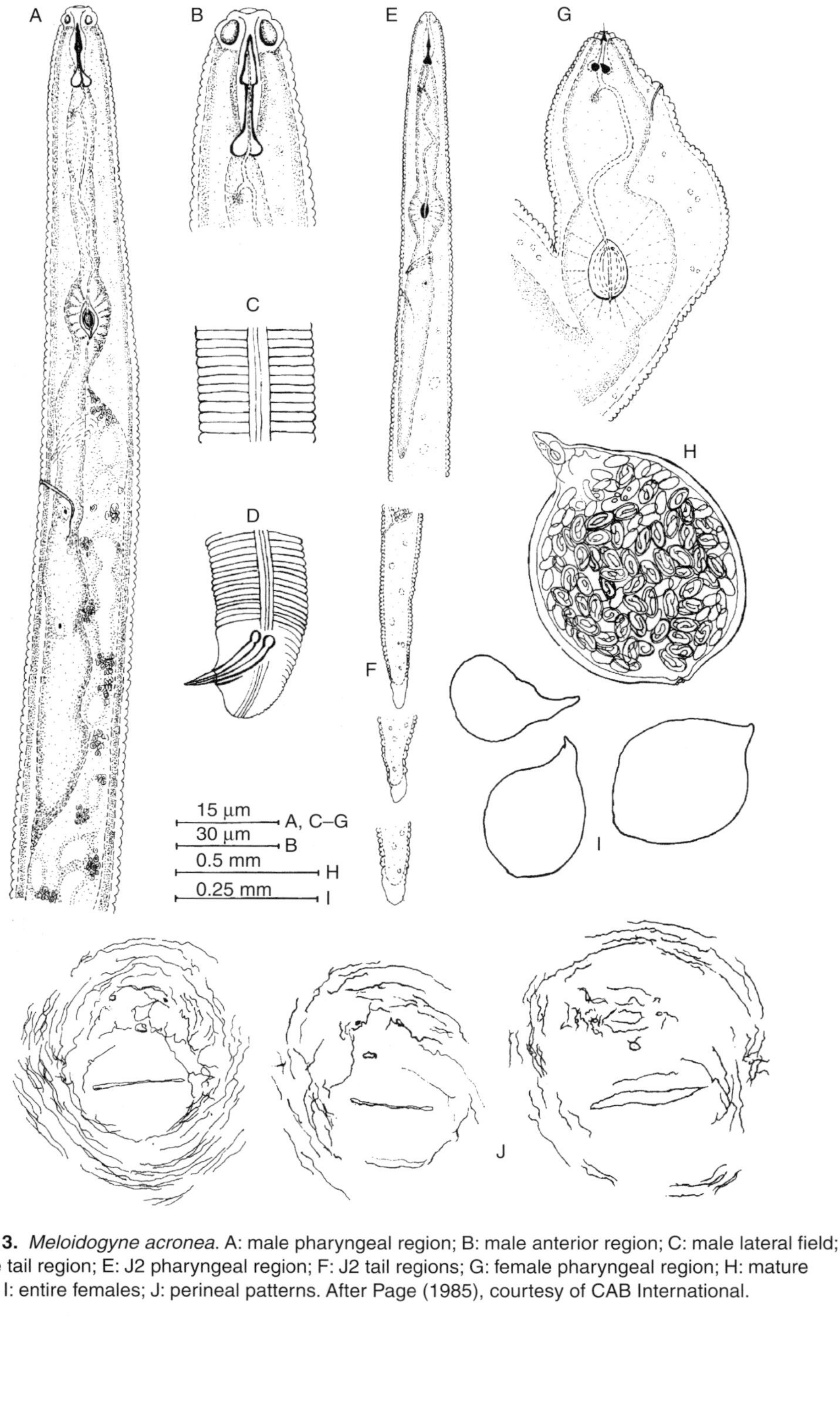

Fig. 3.13. *Meloidogyne acronea*. A: male pharyngeal region; B: male anterior region; C: male lateral field; D: male tail region; E: J2 pharyngeal region; F: J2 tail regions; G: female pharyngeal region; H: mature female; I: entire females; J: perineal patterns. After Page (1985), courtesy of CAB International.

3.6.6 *Meloidogyne chitwoodi* (Fig. 3.14)

Morphology: **Female.** Pear-shaped, with slight posterior protuberance. Stylet relatively small, 11–12.5 µm long, basal knobs small, rounded to irregularly shaped, posteriorly sloping. Several vesicle-like structures usually present within median bulb, clustered around lumen anterior to valve plates of median bulb (Golden *et al.*, 1980). Perineal pattern rounded to oval, dorsal arch low and rounded to high and angular; striae near perineal area broken, curved, twisted; lateral lines weakly visible. **Male.** Labial region not offset, labial disc elevated, lateral lips present. Stylet 18–19 µm long, basal knobs small, irregularly shaped, posteriorly sloping, offset, DGO = 2.2–3.4 µm. **J2.** L = 336–417 µm,

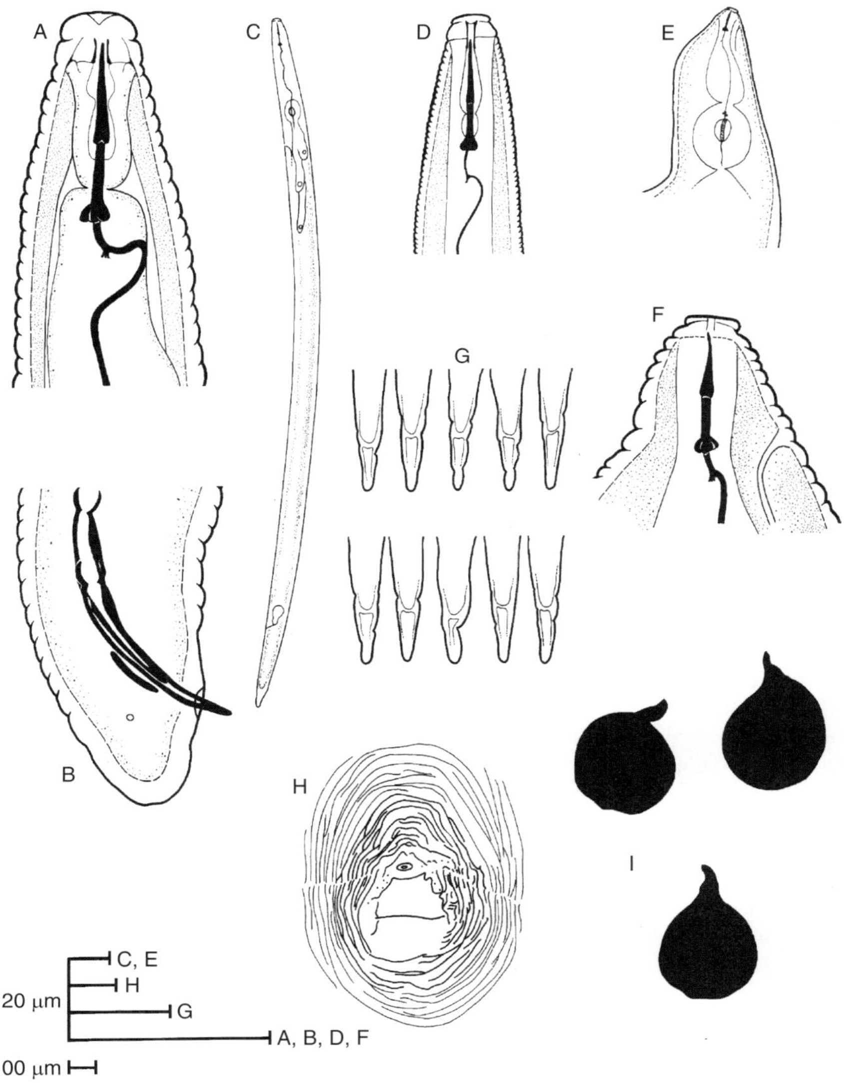

Fig. 3.14. *Meloidogyne chitwoodi.* A: male anterior region; B: male tail region; C: entire J2; D: J2 anterior region; E, F: female anterior region, lateral view; G: J2 tail regions; H: perineal pattern; I: entire females. After Jepson (1985), courtesy of CAB International.

hemizonid anterior or adjacent to excretory pore, tail = 39–47 µm with bluntly rounded tail tip and 9–14 µm long hyaline region.

Hosts: Potato and tomato are good hosts. Wide host range among several plant families, including crop plants and common weed species (barley, carrots, maize, sugarbeet, peas, wheat and various Poaceae). Attacking monocotyledons and dicotyledons.

Distribution: North America: Mexico, USA (California, Colorado, Idaho, Nevada, New Mexico, Oregon, Texas, Utah, Virginia, Washington). South America: Argentina. Europe: Belgium, Germany, The Netherlands, Portugal. Africa: South Africa.

Isozymes: The esterase phenotype is S1 and the malate dehydrogenase phenotype is N1a.

Remarks: Major species of economic importance attacking potato. Typically inciting galls similar to several other root-knot species, i.e. usually relatively small galls. Surface of infected potato tubers with numerous small, pimple-like, raised areas. It was added to the European list of quarantine organisms in 1998 to try to prevent further distribution within Europe.

3.6.7 *Meloidogyne enterolobii* (Fig. 3.15)

Morphology: **Female.** Pear-shaped, no posterior protuberance. Stylet 14–17 µm long, basal knobs reniform, indented, offset. Perineal pattern round to ovoid, dorsal arch rounded; striae fine, widely spaced; lateral field absent or with single line occurring at junction of dorsal and ventral arches. **Male.** Labial region not offset, labial disc not elevated, lateral lips absent. Stylet 18–25 µm long, basal knobs rounded, sloping posteriorly, offset, DGO = 3–5 µm. **J2.** L = 377–528 µm, hemizonid two annules anterior to excretory pore, tail = 43–63 µm with bluntly pointed tail tip and 5–15 µm long hyaline region.

Hosts: Aubergine, basil, bell pepper, coffee, soybean, sweet potato, tobacco, tomato, watermelon, guava, Spanish needle, bean, beet, broccoli, celery, horsebean, parsley, potato, pumpkin, American black nightshade, wild poinsettia, angel trumpet, glory bush, ajuga, glory flower.

Distribution: Brazil, China, Cuba, France, Guadeloupe, Malawi, Martinique, Puerto Rico, Senegal, South Africa, Switzerland, The Netherlands (intercept), Trinidad & Tobago, USA (Florida), Venezuela, West Africa (Ivory Coast and Burkina Faso).

Isozymes: Two major bands (VS1-S1 phenotype) of esterase activity and one strong malate dehydrogenase band (N1a).

Remarks: Widely distributed, emergent pest species with potential to cause great economic damage. Virulent on tomato cultivars with the *Mi*1 gene for resistance to other *Meloidogyne* spp. and on soybean cv. Forrest and sweet potatoes that are resistant to *M. incognita*. Damaging to coffee in Cuba and reproducing on tomato with *Mi* resistance gene. Typically inciting large, irregular galls. The nematode formerly known as *M. mayaguensis* is now regarded as a junior synonym of *M. enterolobii* (see Xu *et al.*, 2004 and EPPO, 2008).

3.6.8 *Meloidogyne ethiopica* (Fig. 3.16)

Morphology: **Female.** Elongate to pyriform, terminal protuberance absent. Stylet 12–15 µm long, basal knobs rounded, tapering gradually into shaft. Perineal pattern oval to squarish; dorsal arch moderately high to high; striae coarse, widely separated, smooth to wavy; lateral field indistinct. **Male.** Labial region not offset, labial disc distinct, lateral lips present. Stylet 23–27 µm long, basal knobs rounded to pear-shaped, offset. DGO = 3–5 µm. **J2.** L = 326–510 µm, hemizonid anterior to excretory pore, tail = 52–72 µm, hyaline region distinct (12–15 µm), tail tip finely rounded to pointed.

Hosts: Grapevine, kiwi, soybean and sugarcane.

Distribution: Mainly in East and Southern Africa and South America (Brazil, Chile) but also known from Slovenia.

Isozymes: The esterase phenotype E3 (Ki3) is species-specific and this is the most useful character for differentiating *M. ethiopica* from other species. Esterase phenotype (E3, Rm = 0.9, 1.05, 1.20); malate dehydrogenase N1 type.

Molecular: RAPD profiles were used by Carneiro *et al.* (2004a) to identify conspecific populations from Brazil, Chile and Kenya.

Remarks: Damaging species on grapevine and kiwi in South America (Brazil, Chile).

3.6.9 *Meloidogyne exigua* (Fig. 3.17)

Morphology: **Female.** Pear-shaped, no posterior terminal protuberance. Stylet 12–14 µm long, basal knobs rounded, offset. Perineal pattern

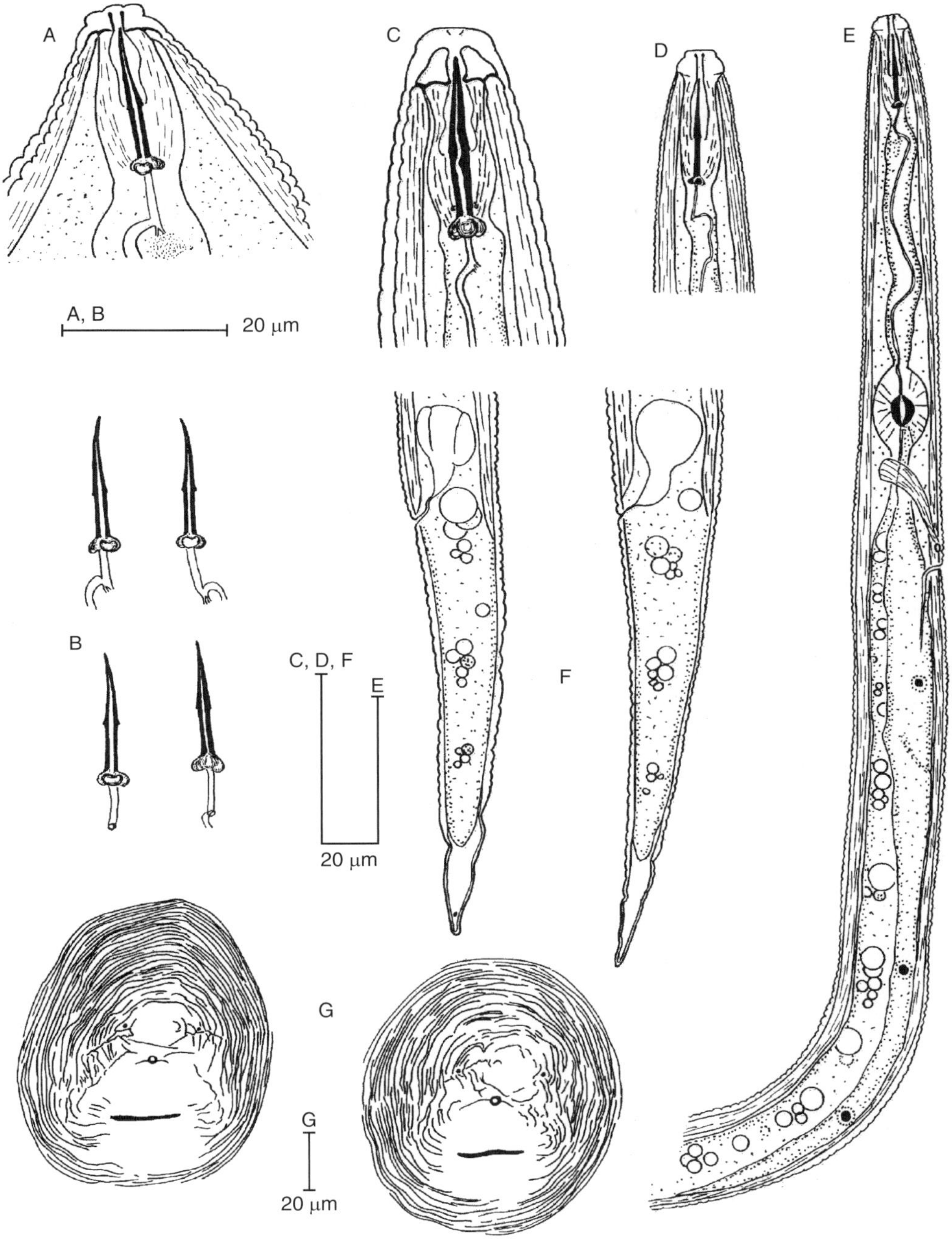

Fig. 3.15. *Meloidogyne enterolobii.* A: female anterior region; B: female stylets; C: male anterior region; D: J2 anterior region: E: J2 pharyngeal region; F: J2 tail regions; G: perineal patterns. After Rammah and Hirschmann (1988), courtesy of *Journal of Nematology.*

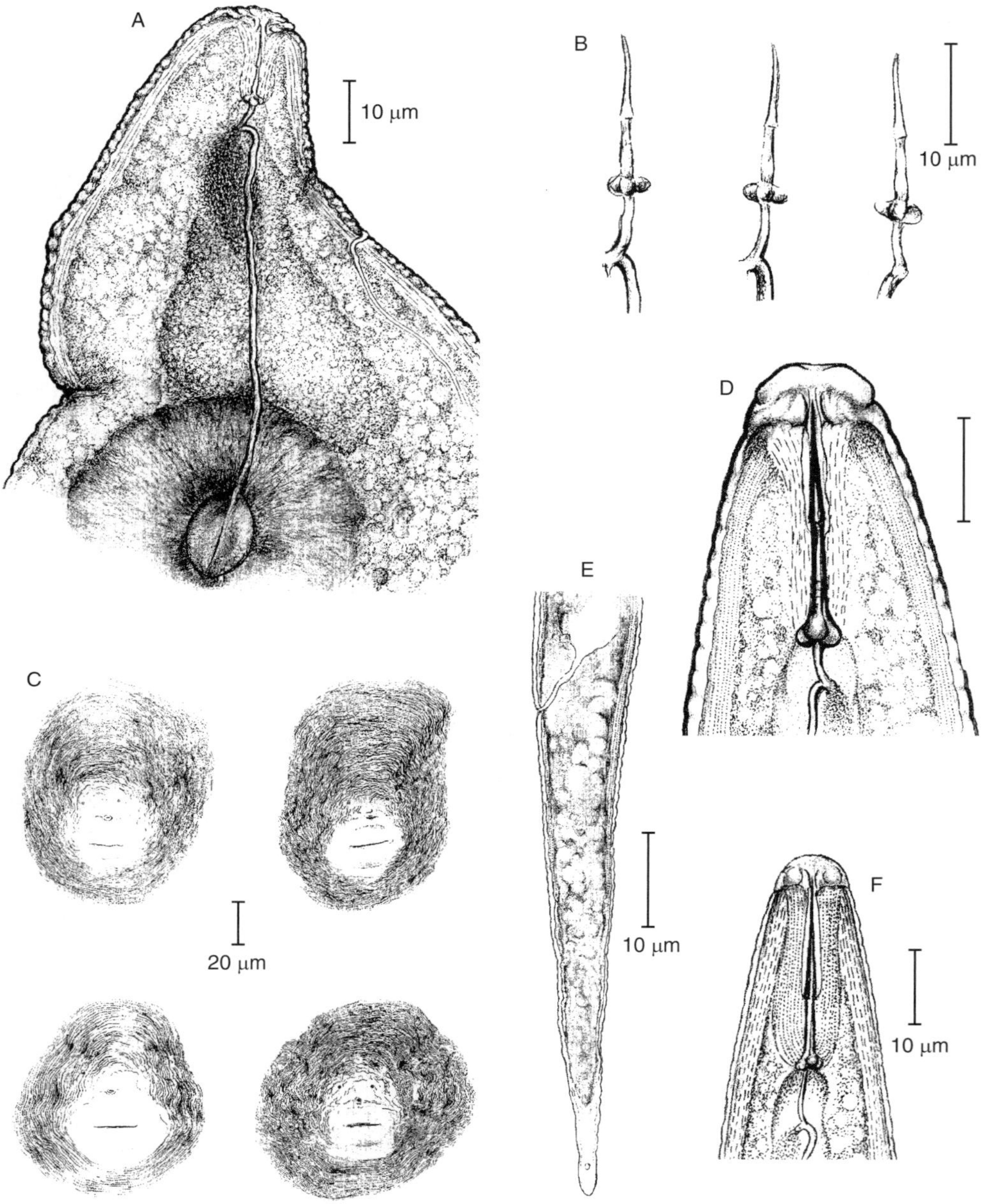

Fig. 3.16. *Meloidogyne ethiopica*. A: female pharyngeal region, lateral view; B: female stylets; C: perineal patterns; D: male labial region, lateral view; E: J2 tail, lateral view, F: J2 anterior region, lateral view. Modified after Carneiro *et al.* (2004a), courtesy of *Nematology*.

rounded/oval, low dorsal arch; striae smooth, widely spaced, coarse, broken and folded in lateral regions; lateral field absent. **Male.** Labial region slightly offset, with distinct elevated labial disc and one annule, lateral lips present, trapezoidal or almost triangular. Stylet 18–20 μm long, basal knobs rounded, posteriorly sloping, offset. DGO = 0.0–3.0 μm. **J2.** L = 290–370 μm, excretory pore opposite to posterior end of isthmus, tail = 39–50 μm with narrowly rounded tip.

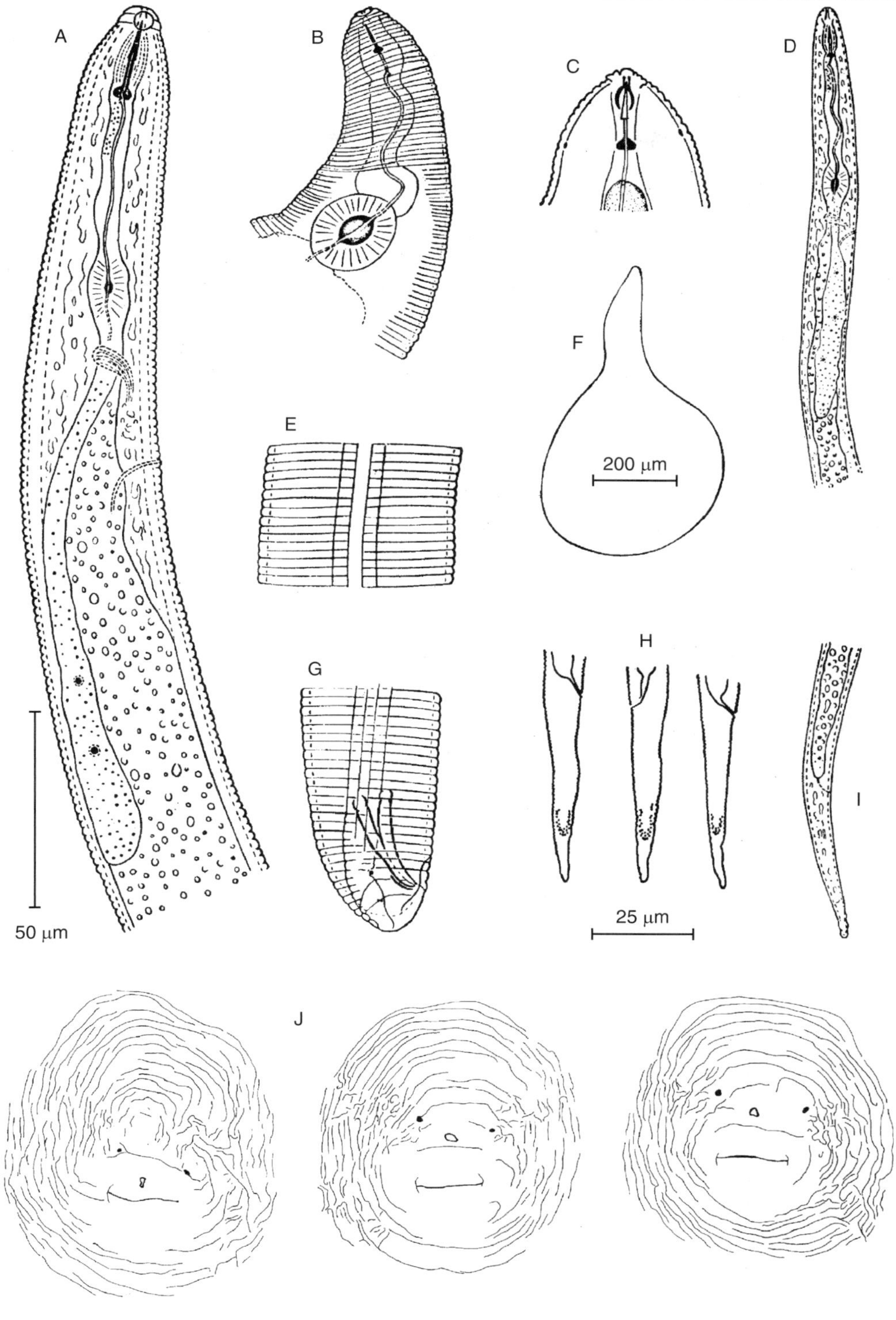

Fig. 3.17. *Meloidogyne exigua*. A: male pharyngeal region; B: female pharyngeal region; C: female anterior region; D: J2 pharyngeal region; E: male lateral field; F: entire female; G: male tail region; H, I: J2 tail regions; J: perineal patterns. A, B, D–G, after Lordello and Zamith (1958), courtesy of *Proceedings of the Helminthological Society of Washington*; C, H, after Whitehead (1968), courtesy of *Transactions of the Zoological Society of London*; J, courtesy of Janet Machon.

Hosts: Mostly attacking coffee, but also common weeds in coffee plantations. Fairly wide host range, including tomato, rice, sugarcane, banana, citrus, onion, etc. Race 3 attacks rubber trees (but not coffee) in Brazil.

Distribution: Widely distributed in humid coffee-producing areas of Central and South America. Occurs on coffee in Brazil, Guatemala, Peru, Suriname, Colombia, Venezuela, Costa Rica, El Salvador, Dominican Republic, Martinique. In Asia: India. Also found on other crops in Trinidad, French West Indies, China. In Europe: Greece, Italy.

Isozyme: The VF1 esterase phenotype of *M. exigua* is shared with *M. naasi*, although it can be differentiated from the latter by its N1 malate dehydrogenase phenotype.

Remarks: Attacks coffee in Central and South America. Outside this region it is a threat to all coffee-growing areas and is targeted in regulatory programmes. Typically inciting severe root galling in heavily attacked coffee and other plants.

3.6.10 *Meloidogyne fallax* (Fig. 3.18)

Morphology: **Female.** Globular to pear-shaped, with slight posterior protuberance. Stylet 14–15 µm long, basal knobs large, rounded to transversally ovoid, offset. One or two large and several smaller vesicle-like structures located along lumen lining. Perineal pattern ovoid to oval-shaped, or rectangular; dorsal arch low to moderately high, striae coarse; tail terminus and lateral field indistinct, resulting in a relatively large area without striae. **Male.** Labial region slightly offset, labial disc rounded, elevated, fused with medial lips, lateral lips present. Stylet 19–21 µm long, basal knobs offset, large rounded, DGO = 3–6 µm. **J2.** L = 381–435 µm, hemizonid at same level as excretory pore, tail = 46–56 µm with broadly rounded tail tip and 12–16 µm long hyaline region.

Hosts: Potato, tomato, oyster plant, carrot. Attacks both monocotyledons and dicotyledons.

Distribution: Australia; Europe (Belgium, France, Germany, The Netherlands); New Zealand.

Isozymes: Characterized by a unique malate dehydrogenase N1b phenotype and the lack of any major esterase band. All populations share this rare malate dehydrogenase phenotype and 'null' esterase phenotype. Prolonged esterase staining (60 min) revealed a very weak, three-banded pattern named F3.

Remarks: Morphologically closely related to *M. chitwoodi* and a pest of potato and cereals in Europe. Typically inciting small galls. Infected potato tubers with blister-like or raised swellings on surface. It was added to the European list of quarantine organisms to try to prevent further distribution within Europe.

3.6.11 *Meloidogyne graminicola* (Fig. 3.19)

Morphology: **Female.** Elongate, slight terminal protuberance present. Stylet 12–15 µm long, basal knobs ovoid, offset. Perineal pattern rounded/oval, striae smooth, lateral field absent. **Male.** Labial region not offset, labial disc not elevated, lateral lips usually present. Stylet 15–20 µm long, basal knobs ovoid, offset. DGO = 3–4 µm. **J2.** L = 410–480 µm, hemizonid anterior or adjacent to excretory pore, tail = 60–80 µm, tail tip finely rounded.

Hosts: Rice and many grasses. Dicotyledonous weeds may also act as good hosts.

Distribution: Common in rice-growing areas.

Isozymes: Esterase VS1 phenotype with one slow band with a large drawn-out area of enzymatic activity (Esbenshade and Triantaphyllou, 1985a,b). The malate dehydrogenase N1a phenotype is similar to that of *M. chitwoodi* and *M. salasi*.

Remarks: Major species attacking rice. Typically incites large galls, often at the root tip, which may become hooked.

3.6.12 *Meloidogyne paranaensis* (Fig. 3.20)

Morphology: **Female.** Ovoid/pear-shaped, no posterior terminal protuberance. Stylet 15–17.5 µm long, basal knobs broad, offset. Perineal pattern rectangular to oval, high dorsal arch; striae fine/coarse, smooth/wavy, lateral field absent. **Male.** Labial region not offset, labial disc elevated, lateral lips absent. Stylet 20–27 µm long, basal knobs round to transversally elongate, offset, DGO = 3.5–5.0 µm. **J2.** L = 389–513 µm,

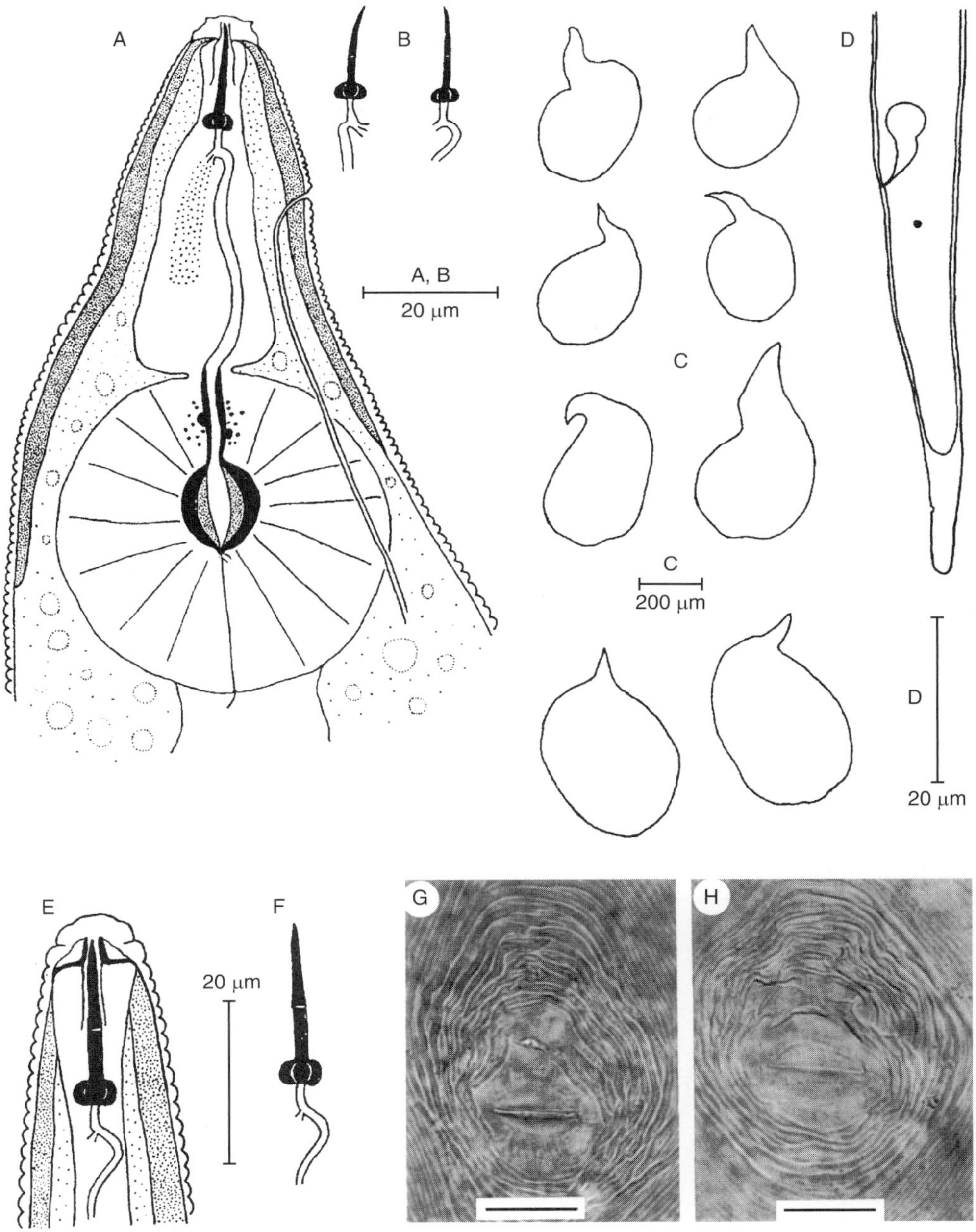

Fig. 3.18. *Meloidogyne fallax*. A: female pharyngeal region, lateral view; B: female stylets; C: entire females; D: J2 tail region, lateral view; E: male anterior region; F: male stylets; G, H: perineal patterns (scale bar = 25 μm). Modified after Karssen (1996), courtesy of *Fundamental and Applied Nematology*.

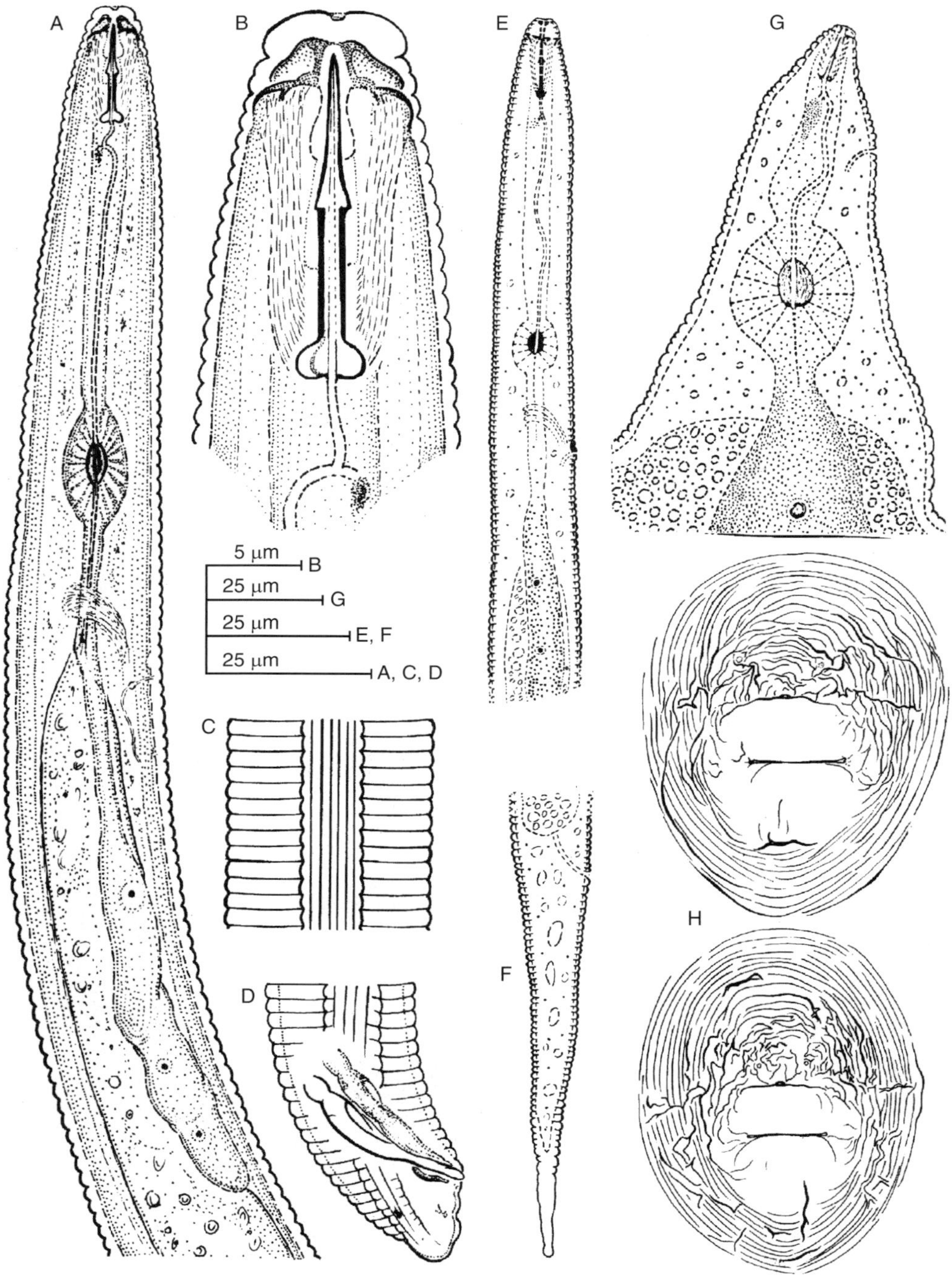

Fig. 3.19. *Meloidogyne graminicola.* A: male pharyngeal region, B: male anterior region; C: male lateral field; D: male tail region; E: J2 pharyngeal region; F: J2 tail region; G: female anterior region; H: perineal patterns. A–D, H, after Mulk (1976), courtesy of CAB International; E–G, after Golden and Birchfield (1965), courtesy of *Proceedings of the Helminthological Society of Washington.*

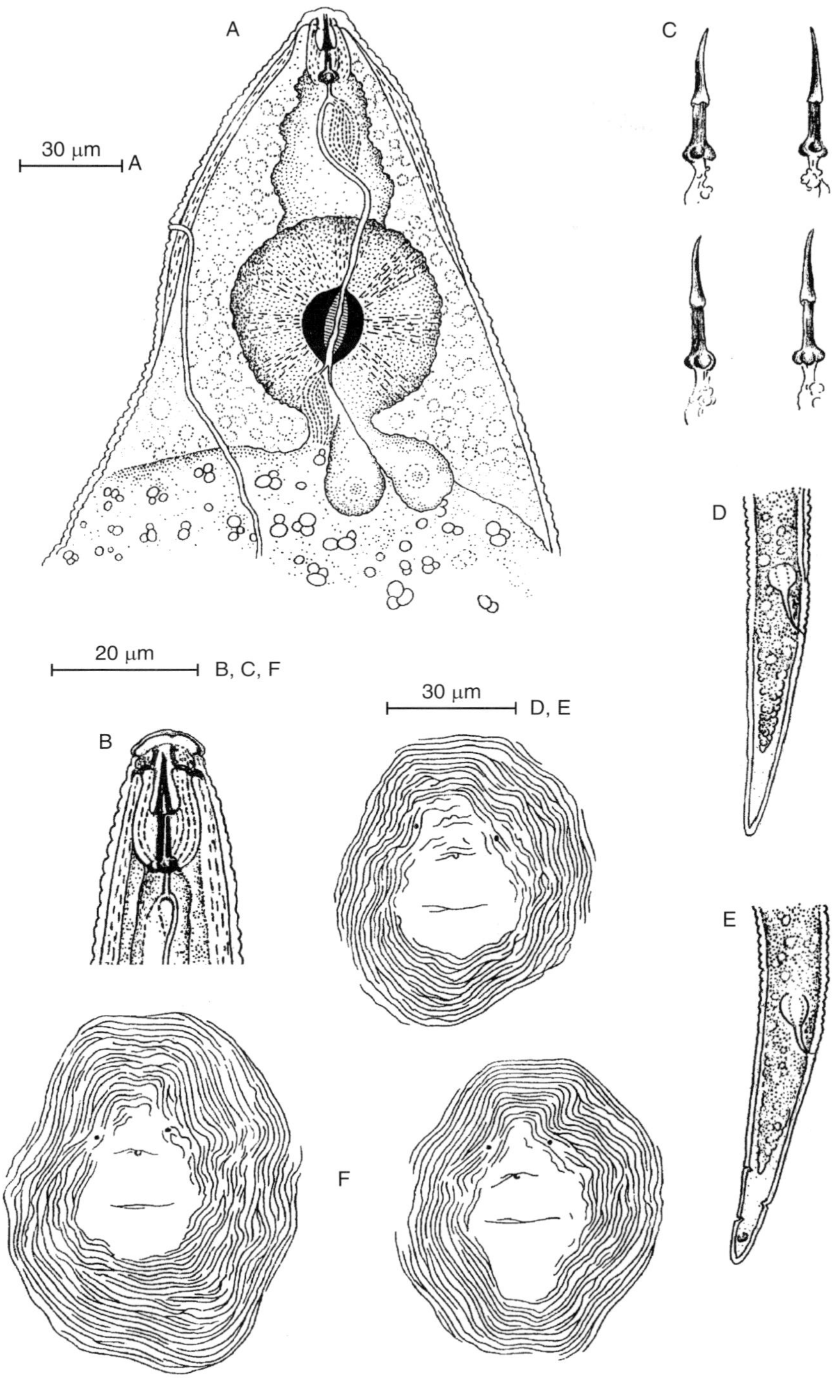

Fig. 3.20. *Meloidogyne paranaensis*. A: female pharyngeal region, lateral view; B: male anterior region, lateral view; C: female stylets; D, E: J2 tail region; F: perineal patterns. Modified after Carneiro *et al.* (1996), courtesy of *Journal of Nematology*.

hemizonid anterior to excretory pore, tail = 48–51 μm with rounded tail tip and 9–10 μm long hyaline region.

Hosts: Coffee, soybean. Also reproduces on tomato, tobacco, watermelon, mate (*Ilex paraguariensis*) and species of Solanaceae, Cucurbitaceae and Aquifoliaceae.

Distribution: Brazil, Guatemala.

Isozymes: The esterase phenotype has one fast-migrating band (F1) and the malate dehydrogenase phenotype is N1.

Remarks: Has potential to adversely affect threatened and endangered plant species. Symptoms include a cracking and splitting of the tap root of coffee but without gall formation.

3.7 Conclusions and Future Directions

There can be no doubt that there is a need for a minimum descriptive standard for any future proposals of new species in this genus. Such a protocol should take cognizance of the need for a blend of morphological (including SEM), morphometric, isozyme and molecular data. Naturally, opinion may vary as to which characters are essential and which are less so, yet still desirable, but with the genus rapidly approaching 100 'valid' species, the implementation of such a protocol cannot be long delayed if we are to avoid utter confusion in one of the most economically important groups of plant-parasitic nematodes. Carta *et al.* (2006) made certain recommendations in this regard, and in Eisenback and Hunt, Chapter 2, this volume, a protocol is put forward which it is hoped will become widely accepted as a minimum standard.

The future prospects in root-knot nematode taxonomy and diagnostics are dependent on molecular-based methodologies that will discriminate not only at the species level but also at the level of host races, thereby opening up opportunities for more focused management strategies. Such techniques offer the possibility of rapid, unequivocal diagnostics and should help resolve the present problems associated with relatively morphologically conserved organisms that reproduce, for the most part, parthenogenetically. Once such techniques are widely employed no doubt a number of the current nominal species will be shown to be junior synonyms, while others, conversely, will be shown to be species complexes, possibly of sibling species. It seems likely that molecular methodologies will replace isozymes as the preferred diagnostic tool because of their inherently higher resolution and the opportunity to develop DNA chips for rapid and reliable field identification. Molecular characterization will also enhance our understanding of the phylogeny of the genus and its relationship with other plant-parasitic nematodes.

3.8 Acknowledgements

DJH is grateful to Ailton Monteiro and Regina Carneiro for assistance in clarifying nomenclatural matters relating to Lordello's 1951 paper and for biographical information concerning Emílio Goeldi.

3.9 References

Abdel-Rahman, F. (1981) Taxonomy, morphology, and biology of two new species of nematodes parasitic on sedge, *Scirpus robustus* Pursh. PhD thesis, University of California, Davis, California.

Abdel-Rahman, F. and Maggenti, A.R. (1987) *Meloidogyne californiensis* n. sp. (Nemata: Meloidogyninae), parasitic on bulrush, *Scirpus robustus* Pursh. *Journal of Nematology* 19, 207–217.

Abrantes, I.M. de O. and Santos, S.N. de A. (1991) *Meloidogyne lusitanica* n. sp. (Nematoda: Meloidogynidae), a root-knot nematode parasitizing olive tree (*Olea europaea* L.). *Journal of Nematology* 23, 210–224.

Allen, M.W. (1952) Observations on the genus *Meloidogyne* Goeldi, 1887. *Proceedings of the Helminthological Society of Washington* 19, 44–51.

Amin, A.W. (1993) A new local race of the root-knot nematode *Meloidogyne thamesi* Chitwood in Chitwood, Specht & Havis, 1952 in Hungary. *Opuscula Zoologica* (*Budapest*) 26, 3–8.

Araki, M. (1992) Description of *Meloidogyne ichinohei* n. sp. (Nematoda: Meloidogynidae) from *Iris laevigata* in Japan. *Japanese Journal of Nematology* 22, 11–20.

Atkinson, G.F. (1889) A preliminary report upon the life history and metamorphoses of a root-gall nematode, *Heterodera marioni* (Greeff) Müll., and the injuries caused by it upon the roots of various plants. *Scientific Contributions of the Agricultural Experiment Station, Alabama Polytechnic Institute* 1, 177–266.

Baldwin, J.G. and Sasser, J.N. (1979) *Meloidogyne megatyla* n. sp., a root-knot nematode from loblolly pine. *Journal of Nematology* 11, 47–56.

Barker, K.R., Carter, C.C. and Sasser, J.N. (eds) (1985) *An Advanced Treatise on* Meloidogyne. *Volume II: Methodology.* A cooperative publication of the Department of Plant Pathology and the United States Agency for International Development, North Carolina State University Graphics, Raleigh, North Carolina.

Berkeley, M.J. (1855) Vibrio forming cysts on the roots of cucumbers. *Gardener's Chronicle and Agricultural Gazette* 14, 220.

Bernard, E.C. (1981) Three new species of Heteroderidae (Nematoda) from the Aleutian Islands. *Journal of Nematology* 13, 499–513.

Bernard, E.C. and Eisenback, J.D. (1997) *Meloidogyne trifoliophila* n. sp. (Nemata: Meloidogynidae), a parasite of clover from Tennessee. *Journal of Nematology* 29, 43–54.

Byrd, D.W. Jr, Kirkpatrick, T. and Barker, K.R. (1983) An improved technique for clearing and staining plant tissues for detection of nematodes. *Journal of Nematology* 15, 142–143.

Carneiro, R.M.D.G. and Almeida, M.R.A. (2001) Técnica de eletroforese usada no estudo de enzimas dos nematóides de galhas para identificação de espécies. *Nematologia Brasileira* 25, 555–560.

Carneiro, R.M.D.G., Carneiro, R.G., Abrantes, I.M.O., Santos, M.S.N.A. and Almeida, M.R.A. (1996) *Meloidogyne paranaensis* n. sp. (Nemata: Meloidogynidae), a root-knot nematode parasitizing coffee in Brazil. *Journal of Nematology* 28, 177–189.

Carneiro, R.M.D.G., Castagnone-Sereno, P. and Dickson, D.W. (1998) Variability among four populations of *Meloidogyne javanica* from Brazil. *Fundamental and Applied Nematology* 21, 319–326.

Carneiro, R.M.D.G., Almeida, M.R.A. and Quénéhervé, P. (2000) Enzyme phenotypes of *Meloidogyne* spp. populations. *Nematology* 2, 645–654.

Carneiro, R.M.D.G., Randig, O., Almeida, M.R.A. and Gomes, A.C.M.M. (2004a) Additional information on *Meloidogyne ethiopica* Whitehead, 1968 (Tylenchida: Meloidogynidae), a root-knot nematode parasitising kiwi fruit and grape-vine from Brazil and Chile. *Nematology* 6, 109–123.

Carneiro, R.M.D.G., Tigano, M.S., Randig, O., Almeida, M.R.A. and Sarah, J.-L. (2004b) Identification and genetic diversity of *Meloidogyne* spp. (Tylenchida: Meloidogynidae) on coffee from Brazil, Central America and Hawaii. *Nematology* 6, 287–298.

Carneiro, R.M.D.G., Almeida, M.R.A., Gomes, A.C.M. and Hernández, A. (2005) *Meloidogyne izalcoensis* n. sp. (Nematoda: Meloidogynidae), a root-knot nematode parasitising coffee in El Salvador. *Nematology* 7, 819–832.

Carneiro, R.M.D.G., Almeida, M.R.A., Cofcewicz, E.T., Magunacelaya, J.C. and Aballay. E. (2007) *Meloidogyne ethiopica*, a major root-knot nematode parasitising *Vitis vinifera* and other crops in Chile. *Nematology* 9, 635–641.

Carta, L.K. (1991) Part 3. Preparation of nematodes for transmission electron microscopy. In: Nickle, W.R. (ed.) *Manual of Agricultural Nematology.* Marcel Dekker, New York, pp. 97–106.

Carta, L.K., Handoo, Z.A., Powers, T.O., Miller, S.A., Pérez-Zubiri, R. and Ramírez-Suárez, A. (2006) Guidelines for isolation and identification of regulated nematodes of potato. *Revista Méxicana de Fitopatología* 23 (2005), 211–222.

Castillo, P., Vovlas, N., Subbotin, S. and Troccoli, A. (2003) A new root-knot nematode, *Meloidogyne baetica* n. sp. (Nematoda: Heteroderidae), parasitizing wild olive in Southern Spain. *Phytopathology* 93, 1093–1102.

Castillo, P., Vovlas, N., Troccoli, A., Liébanas, 6., Palomares Rivs, J.E. and Landa, B.B. (2009) A new root-knot nematode, *Meloidogyne silvestris* n.sp. (Nematoda: Meloidogynidae), parasitizing European holly in northern Spain. *Plant Pathology* 58, 606–619.

Castro, J.M.C., Lima, R.D. and Carneiro, R.M.D.G. (2003) Variabilidade isoenzimática de populações de *Meloidogyne* spp. provenientes de regiões brasileiras produtoras de soja. *Nematologia Brasileira* 27, 1–13.

Charchar, J.M. and Eisenback, J.D. (2000) An improved technique to prepare perineal patterns of root-knot nematodes for SEM. *Nematologia Brasileira* 24, 245–247.

Charchar, J.M. and Eisenback, J.D. (2002) *Meloidogyne brasilensis* n. sp. (Nematoda: Meloidogynidae), a root-knot nematode parasitising tomato cv. Rossol in Brazil. *Nematology* 4, 629–643.

Charchar, J.M., Eisenback, J.D. and Hirschmann, H. (1999) *Meloidogyne petuniae* n. sp. (Nemata: Meloidogynidae), a root-knot nematode parasitic on petunia in Brazil. *Journal of Nematology* 31, 81–91.

Charchar, J.M., Eisenback, J.D., Charchar, M.J. and Boiteux, M.E.N.F. (2008a) *Meloidogyne pisi* n. sp. (Nematoda: Meloidogynidae), a root-knot nematode parasitising pea in Brazil. *Nematology* 10, 479–493.

Charchar, J.M., Eisenback, J.D., Charchar, M.J. and Boiteux, M.E.N.F. (2008b) *Meloidogyne phaseoli* n. sp. (Nematoda: Meloidogynidae), a root-knot nematode parasitising bean in Brazil. *Nematology* 10, 525–538.

Chen, P.S., Peng, D.L. and Zheng, J.W. (1990) [The discovery and description of *Meloidogyne fanzhiensis* n. sp. on potato.] *Journal of Shaxi Agricultural University* 10, 55–60.

Chitwood, B.G. (1949) Root-knot nematodes – Part 1. A revision of the genus *Meloidogyne* Goeldi, 1887. *Proceedings of the Helminthological Society of Washington* 16, 90–104.

Chitwood, B.G. and Oteifa, B.A. (1952) Nematodes parasitic on plants. *Annual Review of Microbiology* 6, 151–184.

Chitwood, B.G., Specht, A.W. and Havis, L. (1952) Root-knot nematodes. III. Effects of *Meloidogyne incognita* and *M. javanica* on some peach rootstocks. *Plant & Soil* 4, 77–95.

Cliff, G.M. and Hirschmann, H. (1984) *Meloidogyne microcephala* n. sp. (Meloidogynidae), a root-knot nematode from Thailand. *Journal of Nematology* 16, 183–193.

Cobb, N.A. (1890) *Tylenchus* and root-gall. *Agricultural Gazette of New South Wales* 1, 155–184.

Cobb, N.A. (1924) The amphids of *Caconema* (*nom. nov.*) and other nemas. *Journal of Parasitology* 11, 118–120.

Coetzee, V. (1956) *Meloidogyne acronea*, a new species of root-knot nematode. *Nature, London* 177, 899–900.

Cornu, M. (1879) Études sur le *Phylloxera vastatrix. Mémoires Présentés par Divers Savants à l'Académie des Sciences, Institut France* 26, 163–175, 328, 339–341.

da Ponte, J.J. (1969) *Meloidogyne lordelloi* n. sp., a nematode parasite of *Cereus macrogonus* Salm. Dick. *Boletin Cearense da Agronomia* 10, 59–63.

da Ponte, J.J. (1977) Nematoides das Galhas: especies ocorrentes no Brasil e seus hospedeiros. *Colecao Mossoroense* 54, 21–25.

De Grisse, A. (1961) *Meloidogyne kikuyensis* n. sp., a parasite of kikuyu grass (*Pennisetum clandestinum*) in Kenya. *Nematologica* 5 (1960), 303–308.

De Ley, P. and Blaxter, M.L. (2002) Systematic position and phylogeny. In: Lee, D.L. (ed.) *The Biology of Nematodes.* Taylor & Francis, London, pp. 1–30.

De Ley, P. and Blaxter, M.L. (2004) A new system for Nematoda: combining morphological characters with molecular trees, and translating clades into ranks and taxa. In: Cook, R. and Hunt, D.J. (eds) *Proceedings of the Fourth International Congress of Nematology, Tenerife, Spain, 8–13 June 2002, Nematology Monographs and Perspectives, Volume 2.* Brill, Leiden, The Netherlands, pp. 633–653.

De Ley, I.T., De Ley, P., Vierstraete, A., Karssen, G., Moens, M. and Vanfleteren, J. (2002) Phylogenetic analyses of *Meloidogyne* small subunit rDNA. *Journal of Nematology* 34, 319–327.

Dropkin, V.H. (1953) Studies on the variability of the anal plate patterns in pure lines of *Meloidogyne* spp., the root-knot nematode. *Proceedings of the Helminthological Society of Washington* 20, 32–39.

Ebsary, B.A. and Eveleigh, E.S. (1983) *Meloidogyne aquatilis* n. sp. (Nematoda: Meloidogynidae) from *Spartina pectinata* with a key to the Canadian species of *Meloidogyne. Journal of Nematology* 15, 349–353.

Eisenback, J.D. (1982) Description of the blueberry root-knot nematode, *Meloidogyne carolinensis* n. sp. *Journal of Nematology* 14, 303–317.

Eisenback, J.D. (1985) Diagnostic characters useful in the identification of the four most common species of root-knot nematodes (*Meloidogyne* spp.). In: Sasser, J.N. and Carter, C.C. (eds) *An Advanced Treatise on* Meloidogyne. *Volume I: Biology and Control.* A cooperative publication of the Department of Plant Pathology and the United States Agency for International Development, North Carolina State University Graphics, Raleigh, North Carolina, pp. 95–112.

Eisenback, J.D. (1991) Part. 2. Preparation of nematodes for scanning electron microscopy. In: Nickle, W.R. (ed.) *Manual of Agricultural Nematology.* Marcel Dekker, New York, pp. 87–96.

Eisenback, J.D. (ed.) (1997) *Root-knot Nematode Taxonomic Database.* CAB International, Wallingford, UK, CD-ROM.

Eisenback, J.D. and Triantaphyllou, H.H. (1991) Root-knot nematode: *Meloidogyne* spp. and races. In: Nickle, W.R. (ed.) *Manual of Agricultural Nematology.* Marcel Dekker, New York, pp. 191–274.

Eisenback, J.D., Hirschmann, H., Sasser, J.N. and Triantaphyllou, A.C. (1981) *A Guide to the Four Most Common Species of Root-knot Nematodes (*Meloidogyne *species) with a Pictorial Key.* A cooperative Publication of the Departments of Plant Pathology and Genetics, North Carolina State University and the United States Agency for International Development.

Eisenback, J.D., Yang, B. and Hartman, K.M. (1985) Description of *Meloidogyne pini* n. sp., a root-knot nematode parasitic on sand pine (*Pinus clausa*), with additional notes on the morphology of *M. megatyla. Journal of Nematology* 17, 206–219.

Eisenback, J.D., Bernard, E.C. and Schmitt, D.P. (1995) Description of the Kona coffee root-knot nematode, *Meloidogyne konaensis* n. sp. *Journal of Nematology* 26, 363–374.

Eisenback, J.D., Bernard, E.C., Starr, J.L., Lee, T.A. Jr and Tomaszewski, E.K. (2004) *Meloidogyne haplanaria* sp. nov. (Nematoda: Meloidogynidae), a root-knot nematode parasitizing peanut in Texas. *Journal of Nematology* 35 (2003), 395–403.

Elmiligy, I.A. (1968) Three new species of the genus *Meloidogyne* Goeldi, 1887 (Nematoda: Heteroderidae). *Nematologica* 14, 577–590.

EPPO (2008) *Meloidogyne enterolobii* http://www.eppo.org/QUARANTINE/Alert_List/nematodes/meloidogyne_enterolobii.htm (accessed June 2008).

Eroshenko, A.S. and Lebedeva, E.V. (1992) [The description of a new species of root-knot eelworms, *Meloidogyne chosenia* sp. n. (Nematoda: Meloidogynidae), a parasite of willow in Kamchatka.] *Parazitologiya (St. Petersburg)* 26, 340–344.

Esbenshade, P.R. and Triantaphyllou, A.C. (1985a) Use of enzyme phenotypes for identification of *Meloidogyne* species (Nematoda: Tylenchida). *Journal of Nematology* 17, 6–20.

Esbenshade, P.R. and Triantaphyllou, A.C. (1985b) Identification of major *Meloidogyne* species employing enzyme phenotypes as differentiating characters. In: Sasser, J.N. and Carter, C.C. (eds) *An Advanced Treatise on* Meloidogyne. *Volume I: Biology and Control.* A cooperative publication of the Department of Plant Pathology and the United States Agency for International Development, North Carolina State University Graphics, Raleigh, North Carolina, pp. 135–140.

Esbenshade, P.R. and Triantaphyllou, A.C. (1987) Enzymatic relationships and evolution in the genus *Meloidogyne* (Nematoda: Tylenchida). *Journal of Nematology* 19, 8–18.

Esser, R.P., Perry, V.G. and Taylor, A.L. (1976) A diagnostic compendium of the genus *Meloidogyne* (Nematoda: Heteroderidae). *Proceedings of the Helminthological Society of Washington* 43, 138–150.

Fortuner, R. (1984) List and status of the genera and families of plant-parasitic nematodes. *Helminthological Abstracts* 53, 87–133.

Fox, J.A. (1967) Biological studies of the blueberry root-knot nematode (*Meloidogyne carolinensis* n. sp.). PhD thesis, North Carolina State University, Raleigh, North Carolina.

Franklin, M.T. (1957) Review of the genus *Meloidogyne. Nematologica* 2 (Suppl.), 387–397.

Franklin, M.T. (1961) A British root-knot nematode, *Meloidogyne artiellia* n. sp. *Journal of Helminthology, R.T. Leiper Supplement* 85–92.

Franklin, M.T. (1965a) *Meloidogyne* – root-knot eelworm. In: Southey, J.F. (ed.) *Plant Nematology.* Technical Bulletin of the Ministry of Agriculture, Fisheries and Food, 2nd edition, No. 7, Her Majesty's Stationery Office, London, pp. 59–88.

Franklin, M.T. (1965b) A root-knot nematode, *Meloidogyne naasi* n. sp., on field crops in England and Wales. *Nematologica* 11, 79–86.

Franklin, M.T. (1971) Taxonomy of Heteroderidae. In: Zuckerman, B.M., Mai, W.F. and Rohde, R.A. (eds) *Plant Parasitic Nematodes. Volume I. Morphology, Anatomy, Taxonomy and Ecology.* Academic Press, New York, pp. 139–162.

Franklin, M.T. (1976) *Meloidogyne.* In: Southey, J.F. (ed.) *Plant Nematology.* Her Majesty's Stationery Office, London, pp. 98–124.

Franklin, M.T. (1979) Taxonomy of the genus *Meloidogyne.* In: Lamberti, F. and Taylor, C.E. (eds) *Rootknot Nematodes (*Meloidogyne *species); Systematics, Biology and Control.* Academic Press, London & New York, pp. 37–54.

García-Martinez, R., Taylor, A.L. and Smart, G.C. Jr (1982) *Meloidogyne cruciani* n. sp., a root-knot nematode from St. Croix (U.S. Virgin Islands) with observations on morphology of this and two other species of the genus. *Journal of Nematology* 14, 292–303.

Gaur, H.S., Saha, M. and Khan, E. (1993) *Meloidogyne triticoryzae* sp. n. (Nematoda: Meloidogynidae) a root-knot nematode damaging wheat and rice in India. *Annals of Plant Protection Sciences* 1, 18–26.

Gillard, A. (1961) Onderzoekingen omtrent de biologie, de verspreiding en de bestrijding van wortelknobbelaaltjes (*Meloidogyne* spp.). *Mededelingen van de Landbouwhogeschool, Gent* 26, 515–646.

Golden, A.M. (1979) Descriptions of *Meloidogyne camelliae* n. sp. and *M. querciana* n. sp. (Nematoda: Meloidogynidae), with SEM and host-range observations. *Journal of Nematology* 11, 175–189.

Golden, A.M. (1990) Preparation and mounting nematodes for microscopic observation. In: Zuckerman, B.M., Mai, W.F. and Krusberg, L.R. (eds) *Plant Nematology Laboratory Manual.* University of Massachusetts Agricultural Experiment Station, Amherst, Massachusetts, pp. 197–205.

Golden, A.M. and Birchfield, W. (1965) *Meloidogyne graminicola* (Heteroderidae) a new species of root-knot nematode from grass. *Proceedings of the Helminthological Society of Washington* 32, 228–231.

Golden, A.M. and Birchfield, W. (1972) *Heterodera graminophila* n. sp. (Nematoda: Heteroderidae) from grass with key to closely related species. *Journal of Nematology* 4, 147–154.

Golden, A.M. and Birchfield, W. (1978) *Meloidogyne incognita wartellei* n. subsp. (Meloidogynidae) a root knot nematode on resistant soybeans in Louisiana. *Journal of Nematology* 10, 269–277.

Golden, A.M. and Kaplan, D.T. (1986) Description of *Meloidogyne christiei* n. sp. (Nematoda: Meloidogynidae) from oak with SEM and host-range observations. *Journal of Nematology* 18, 533–540.

Golden, A.M. and Slana, L.J. (1978) *Meloidogyne grahami* n. sp. (Meloidogynidae), a root-knot nematode on resistant tobacco in South Carolina. *Journal of Nematology* 10, 355–361.

Golden, A.M., O'Bannon, J.H., Santo, G.S. and Finley, A.M. (1980) Description and SEM observations of *Meloidogyne chitwoodi* n. sp. (Meloidogynidae), a root-knot nematode on potato in the Pacific Northwest. *Journal of Nematology* 12, 319–327.

Golden, A.M., Rose, L.M. and Bird, G.W. (1981) Description of *Meloidogyne nataliei* n. sp. (Nematoda: Meloidogynidae) from grape (*Vitis labrusca*) in Michigan, with SEM observations. *Journal of Nematology* 13, 393–400.

Göldi, E.A. (1887) Relatório sôbre a moléstia do cafeeiro na provincia do Rio de Janeiro. Extrahido do VIII Vol. dos. *Archivos do Museu Nacional, Rio de Janeiro, Imprensa Nacional*, pp. 1–121 + 4 plates and a map.

Göldi, E.A. (1888) Resumo do relatório sôbre a moléstia do cafeeiro na provincia do Rio de Janeiro. *Revista Agricola do Imperial Instituto Fluminense de Agricultura* 19, 3–65 + 4 plates and a map.

Göldi, E.A. (1889) Biologische Miscellen aus Brasilien. VII. Der Kaffeenematode Brasiliens (*Meloidogyne exigua* G.). *Zoologische Jahrbücher Jena. Systematik, Ökologie und Geographie der Tiere* 4, 262–267.

Göldi, E.A. (1892) Relatório sôbre a moléstia do cafeeiro na provincia do Rio de Janeiro. *Archivos do Museo Nacional, Rio de Janeiro* 8, 7–123 + 4 plates and a map.

Goodey, J.B. (1963) *Soil and Freshwater Nematodes*, 2nd edn. Methuen & Co., London.

Goodey, T. (1932) On the nomenclature of root-gall nematodes. *Journal of Helminthology* 10, 21–28.

Greeff, R. (1872) Über Nematoden in Wurzelanschwellungen (Gallen) verschiedener Pflanzen. *Sitzungsberichte der Gesellschaft zür Beförderung der Gesamten Naturwissenschaften zu Marburg* 11, 172–174.

Handoo, Z.A., Huettel, R.N. and Golden, A.M. (1994) Description and SEM observations of *Meloidogyne sasseri* n. sp. (Nematoda: Meloidogynidae), parasitizing beachgrasses. *Journal of Nematology* 25 (1993), 628–641.

Handoo, Z.A., Nyczepir, A.P., Esmenjaud, D., van der Beek, J.G., Castagnone-Sereno, P., Carta, L.K., Skantar, A.M. and Higgins, J.A. (2004) Morphological, molecular, and differential-host characterization of *Meloidogyne floridensis* n. sp. (Nematoda: Meloidogynidae), a root-knot nematode parasitizing peach in Florida. *Journal of Nematology* 36, 20–35.

Handoo, Z.A., Skantar, A.M., Carta, L.K. and Erbe, E.F. (2005) Morphological and molecular characterization of a new root-knot nematode, *Meloidogyne thailandica* n. sp. (Nematoda: Meloidogynidae), parasitizing ginger (*Zingiber* sp.). *Journal of Nematology* 37, 343–353.

Hartman, K.M. and Sasser, J.N. (1985) Identification of *Meloidogyne* species on the basis of differential host test and perineal pattern morphology. In: Barker, K.R., Carter, C.C. and Sasser, J.N. (eds) *An Advanced Treatise on* Meloidogyne. *Volume II: Methodology.* A cooperative publication of the Department of Plant Pathology and the United States Agency for International Development, North Carolina State University Graphics, Raleigh, North Carolina, pp. 69–77.

Hernández, A., Fargette, M. and Sarah, J.-L. (2004) Characterization of *Meloidogyne* spp. (Tylenchida: Meloidogynidae) from coffee plantations in Central America and Brazil. *Nematology* 6, 193–204.

Hewlett, T.E. and Tarjan, A.C. (1983) Synopsis of the genus *Meloidogyne* Goeldi, 1987. *Nematropica* 13, 79–102.

Hirschmann, H. (1982) *Meloidogyne platani* n. sp. (Meloidogynidae), a root-knot nematode parasitizing American sycamore. *Journal of Nematology* 14, 84–95.

Hirschmann, H. (1985) The genus *Meloidogyne* and morphological characters differentiating its species. In: Sasser, J.N. and Carter, C.C. (eds) *An Advanced Treatise on* Meloidogyne. *Volume I: Biology and Control.* A cooperative publication of the Department of Plant Pathology and the United States Agency for International Development, North Carolina State University Graphics, Raleigh, North Carolina, pp. 79–93.

Hirschmann, H. (1986) *Meloidogyne hispanica* n. sp. (Nematoda: Meloidogynidae), the 'Seville root-knot nematode'. *Journal of Nematology* 18, 520–532.

Hooper, D.J. (1970) Handling, fixing, staining and mounting nematodes. In: Southey, J.F. (ed.) *Laboratory Methods for Work with Plant and Soil Nematodes*, 5th edn. Her Majesty's Stationery Office, London, pp. 39–54.

Hooper, D.J. (1986) Preserving and staining nematodes in plant tissues. In: Southey, J.F. (ed.) *Laboratory Methods for Work with Plant and Soil Nematodes*, 6th edn. *Reference book 402*. Her Majesty's Stationery Office, London, pp. 81–85.

Hooper, D.J. (1990) Extraction and processing of plant and soil nematodes. In: Luc, M., Sikora, R.A. and Bridge, J. (eds) *Plant Parasitic Nematodes in Subtropical and Tropical Agriculture*. CAB International, Wallingford, UK, pp. 45–68.

Hussey, R.S. (1990) Staining nematodes in plant tissues. In: Zuckerman, B.M., Mai, W.F. and Krusberg, L.R. (eds) *Plant Nematology Laboratory Manual*. University of Massachusetts Agricultural Experiment Station, Amherst, Massachusetts, pp. 190–193.

International Commission on Zoological Nomenclature (1999) *International Code of Zoological Nomenclature*, 4th edn. International Trust for Zoological Nomenclature, London, http://www.iczn.org/iczn/index.jsp (accessed 9 April 2008).

Itoh, Y., Ohshima, Y. and Ichinohe, M. (1969) A root-knot nematode, *Meloidogyne mali* n. sp., on apple-tree from Japan (Tylenchida: Heteroderidae). *Applied Entomology and Zoology, Tokyo* 4, 194–202.

Jepson, S.B. (1984) *Meloidogyne kralli* n. sp. (Nematoda: Meloidogynidae) a root-knot nematode parasitising sedge (*Carex acuta* L.). *Revue de Nématologie* 6(1983), 239–245.

Jepson, S.B. (1985) *Meloidogyne chitwoodi*. In: *CIH Descriptions of Plant-parasitic Nematodes*, Set 8, No. 106, Commonwealth Agricultural Bureaux, Farnham Royal, UK.

Jepson, S.B. (1987) *Identification of Root-knot Nematodes (*Meloidogyne *species)*. Commonwealth Agricultural Bureaux, Farnham Royal, UK.

Jobert, C. (1878) Sur un maladie du cafeier observée au Bresil. *Compte Rendue Hebdomodaire des Séances del'Académie des Sciences, Paris* 87, 941–943.

Karssen, G. (1996) Description of *Meloidogyne fallax* n. sp. (Nematoda: Heteroderidae), a root-knot nematode from The Netherlands. *Fundamental and Applied Nematology* 19, 593–599.

Karssen, G. (2002) *The Plant-parasitic Nematode Genus* Meloidogyne *Göldi, 1892 (Tylenchida) in Europe*. Brill Academic Publishers, Leiden, The Netherlands.

Karssen, G. and Moens, M. (2006) Root-knot nematodes. In: Perry, R.N. and Moens, M. (eds) *Plant Nematology*. CAB International, Wallingford, UK, pp. 59–90.

Karssen, G. and van Hoenselaar, T. (1998) Revision of the genus *Meloidogyne* Göldi, 1892 (Nematoda: Heteroderidae) in Europe. *Nematologica* 44, 713–788.

Karssen, G., van Hoenselaar, T., Verkerk-Bakker, B. and Janssen, R. (1995) Species identification of cyst and root-knot nematodes from potato by electrophoresis of individual females. *Electrophoresis* 16, 105–109.

Karssen, G., van Aelst, A. and van der Putten, W.H. (1998) *Meloidogyne duytsi* n. sp. (Nematoda: Heteroderidae), a root-knot nematode from Dutch coastal foredunes. *Fundamental and Applied Nematology* 21, 299–306.

Karssen, G., Bolk, R.J., van Aelst, A.C., van den Beld, I., Kox, L.F.F., Korthals, G., Molendijk, L., Zijlstra, C., van Hoof, R. and Cook, R. (2004) Description of *Meloidogyne minor* n. sp. (Nematoda: Meloidogynidae), a root-knot nematode associated with yellow patch disease in golf courses. *Nematology* 6, 59–72.

Kirjanova, E.S. (1963) [Collection and taxonomy of root nematodes of the family Heteroderidae (Skarbilovich, 1947) Thorne, 1949.] In: [*Methods of Investigating Nematodes of Plants, Soil and Insects*.] USSR Academy of Sciences, Zoological Institute, Leningrad, pp. 6–32.

Kirjanova, E.S. and Ivanova, T.S. (1965) [Eelworm fauna of *Pelargonium roseum* L. in the Tadzhik SSR.] *Izvestiya Akademii Nauk Tadzhikskoĭ SSR* 1(18), 24–31.

Kleynhans, K.P.N. (1986a) Useful new characters for the identification of four *Meloidogyne* species. *Phytophylactica* 18, 93–94.

Kleynhans, K.P.N. (1986b) *Meloidogyne partityla* sp. nov. from pecan nut [*Carya illinoensis* (Wangenh.) C. Koch] in the Transvaal lowveld (Nematoda: Meloidogynidae). *Phytophylactica* 18, 103–106.

Kleynhans, K.P.N. (1988) *Meloidogyne vandervegtei* sp. nov. from a subtropical coastal forest in Natal (Nemata: Heteroderidae). *Phytophylactica* 20, 263–267.

Kofoid, C.A. and White, W.A. (1919) A new nematode infection of man. *Journal of the American Medical Association* 72, 567–569.

Lamberti, F. and Taylor, C.E. (eds) (1979) *Rootknot Nematodes (*Meloidogyne *species); Systematics, Biology and Control*. Academic Press, London & New York.

Lavergne, G. (1901a) La Anguillula en Sud-América. *Revista Chilena de Historia Natural* 5, 85–91.

Lavergne, G. (1901b) L'anguillule du Chili (*Anguillula vialae*). *Revue de Viticulture* 16, 445–452.

Lehman, P.S. (1985) Galls on aboveground plants caused by root-knot nematodes. Florida Department of Agriculture and Consumer Services, Division of Plant Industry. *Nematology Circular No. 125.*

Li, S.J. and Yu, Z. (1991) [A new species of root-knot nematode (*Meloidogyne actinidiae*) on *Actinidia chinensis* in Henan Province.] *Acta Agriculturae Universitatis Henanensis* 25, 251–253.

Liao, J.L. (2001) [Study on the identification and polymorphism of rootknot nematodes (*Meloidogyne*).] PhD thesis, South China Agricultural University, Guangzhou, P.R. China.

Liao, J.L. and Feng, Z.X. (1995) [A root-knot nematode *Meloidogyne hainanensis* sp. nov. (Nematoda: Meloidogynidae) parasitizing rice in Hainan, China.] *Journal of South China Agricultural University* 16, 34–39.

Liao, J. and Feng, Z. (2003) A new species of *Meloidogyne* from China. *Russian Journal of Nematology* 11, 142 (abstr.).

Liao, J.L., Yang, W.C., Feng, Z.X. and Karssen, G. (2005) Description of *Meloidogyne panyuensis* sp. n. (Nematoda: Meloidogynidae), parasitic on peanut (*Arachis hypogaea* L.) in China. *Russian Journal of Nematology* 13, 107–114.

Licopoli, G. (1875) Sopra alcuni tubercoli radicellari continente anguillole. *Rendiconti dell'Accademia delle Scienze, Fisiche e Mathematiche, Napoli* 14, 41–42.

Loos, C.A. (1953) *Meloidogyne brevicauda* n. sp., a cause of root-knot of mature tea in Ceylon. *Proceedings of the Helminthological Society of Washington* 20, 83–91.

López, R. (1984) *Meloidogyne salasi* sp. n. (Nematoda: Meloidogynidae), a new parasite of rice (*Oryza sativa* L.) from Costa Rica and Panama. *Turrialba* 34, 275–286.

López, R. and Salazar, L. (1989) *Meloidogyne arabicida* sp. n. (Nemata: Heteroderidae) nativo de Costa Rica: un nuevo y severo patogeno del cafeto. *Turrialba* 39, 313–323.

Lordello, L.G.E. (1951) Sôbre o estado atual da taxonomia dos Nematoídes causadores das galhas nas raíces. *Revista de Agricultura, Piracicaba* 26, 247–250.

Lordello, L.G.E. (1956a) *Meloidogyne inornata* sp. n. a serious pest of soybean in the state of São Paulo, Brazil (Nematoda, Heteroderidae). *Revista Brasil Biologica* 16, 65–70.

Lordello, L.G.E. (1956b) Nematóides que parasitanica soja na região de Bauru. *Bragantia* 15, 55–64.

Lordello, L.G.E. and Zamith, A.P.L. (1958) On the morphology of the coffee root-knot nematode, *Meloidogyne exigua* Goeldi, 1887. *Proceedings of the Helminthological Society of Washington* 24, 133–137.

Lordello, L.G.E. and Zamith, A.P.L. (1960) *Meloidogyne coffeicola* sp. n., a pest of coffee trees in the State of Paraná, Brazil (Nematoda: Heteroderidae). *Revista Brasil Biologica* 20, 375–379.

Maas, P.W.T., Sanders, H. and Dede, J. (1978) *Meloidogyne oryzae* n. sp. (Nematoda, Meloidogynidae) infesting irrigated rice in Surinam (South America). *Nematologica* 24, 305–311.

Marcinowski, K. (1909) Parasitisch und semiparasitisch an Pflanzen lebende Nematoden. *Arbeiten aus der Kaiserlichen Biologischen Bundesanstalt für Land- und Forstwirtschaft, Berlin* 7, 1–192.

Marinari-Palmisano, A. and Ambrogioni, L. (2000) *Meloidogyne ulmi* sp. n., a root-knot nematode from elm. *Nematologia Mediterranea* 28, 279–293.

McBeth, C.W., Taylor, A.L. and Smith, A.L. (1941) Notes on staining nematodes in root tissue. *Proceedings of the Helminthological Society of Washington* 8, 26.

Mulk, M.M. (1976) *Meloidogyne graminicola.* In: *CIH Descriptions of Plant-parasitic Nematodes*, Set 6, No. 87. Commonwealth Agricultural Bureaux, Farnham Royal, UK.

Müller, C. (1884) Mittheilungen über die unseren Kulturpflanzen schädlichen, das Geschlecht *Heterodera* bildenden Wurmer. *Landwirtschaftliche Jarhbücher, Berlin* 13, 1–42.

Mulvey, R.H. and Anderson, R.V. (1980) Description and relationships of a new root-knot nematode, *Meloidogyne sewelli* n. sp. (Nematoda: Meloidogynidae) from Canada and a new host record for the genus. *Canadian Journal of Zoology* 58, 1551–1556.

Mulvey, R.H., Townshend, J.L. and Potter, J.W. (1975) *Meloidogyne microtyla* sp. nov. from southwestern Ontario, Canada. *Canadian Journal of Zoology* 53, 1528–1536.

Nagakura, K. (1930) Über den Bau und die lebensgeschichte der *Heterodera radicicola* (Greeff) Müller. *Japanese Journal of Zoology* 3, 95–160.

Neal, J.C. (1889) *The Root-knot Disease of the Peach, Orange and Other Plants in Florida, Due to the Work of* Anguillula. Bulletin 20, Division of Entomology, US Department of Agriculture.

Orton Williams, K.J. (1972) *Meloidogyne javanica.* In: *CIH Descriptions of Plant-parasitic Nematodes*, Set 1, No. 3. Commonwealth Agricultural Bureaux, Farnham Royal, UK.

Orton Williams, K.J. (1973) *Meloidogyne incognita.* In: *CIH Descriptions of Plant-parasitic Nematodes*, Set 2, No. 18. Commonwealth Agricultural Bureaux, Farnham Royal, UK.

Orton Williams, K.J. (1974) *Meloidogyne hapla*. In: *CIH Descriptions of Plant-parasitic Nematodes*, Set 3, No. 31. Commonwealth Agricultural Bureaux, Farnham Royal, UK.
Orton Williams, K.J. (1975) *Meloidogyne arenaria*. In: *CIH Descriptions of Plant-parasitic Nematodes*, Set 5, No. 62. Commonwealth Agricultural Bureaux, Farnham Royal, UK.
Page, S.L.J. (1985) *Meloidogyne acronea*. In: *CIH Descriptions of Plant-parasitic Nematodes*, Set 8, No. 114. Commonwealth Agricultural Bureaux, Farnham Royal, UK.
Pan, C.S. (1985) [Studies on plant-parasitic nematodes on economically important crops in Fujian. III. Description of *Meloidogyne fujianensis* n. sp. (Nematoda: Meloidogynidae) infesting *Citrus* in Nanjing County.] *Acta Zoologica Sinica* 31, 263–268.
Pan, C.S., Ling, J. and Wang, S.Y. (1988) [Studies on parasitic nematodes in crops of economical importance in Fujian. IV. *Meloidogyne fujianensis*.] *Acta Zoologica Sinica* 34, 305–309.
Paolomaes Rius, J.E., Vovlas, N., Troccoli, A., Liebanas, G., Landa, B.B. and Castillo, P. (2007) A new root-knot nematode parasitizing sea rocket from Spanish Mediterranean coastal dunes: *Meloidogyne dunensis* n. sp. (Nematoda: Meloidogynidae). *Journal of Nematology* 39, 190–202.
Pham, T.B.[7] (1990) [Gall nematodes of vegetables and potatoes in Da Lat (Tei Nguen Plateau, Vietnam) and description of *Meloidogyne cynariensis*, a parasite of artichokes.] *Zoologichesky Zhurnal* 69 (4), 128–131.
Plantard, O., Valette, S. and Gross, M.F. (2007) The root-knot nematode producing galls on *Spartina alterniflora* belongs to the genus *Meloidogyne*: rejection of *Hypsoperine* and *Spartonema* spp. *Journal of Nematology* 39, 127–132.
Poghossian, E.E. (1961) [A root-knot nematode new for the USSR from Armenia.] *Izvestiya Akademii Nauk Armyanskoĭ SSR, Biologicheskie i Sel'skokhozyaĭstvennye Nauki* 14, 95–97.
Poghossian, E.E. (1971) [*Hypsoperine megriensis* n. sp. (Nematoda: Heteroderidae) in the Armenian SSR.] *Doklady Akademii Nauk Armyanskoĭ SSR* 53, 306–312.
Rammah, A. and Hirschmann, H. (1988) *Meloidogyne mayaguensis* n. sp. (Meloidogynidae), a root-knot nematode from Puerto Rico. *Journal of Nematology* 20, 58–69.
Rammah, A. and Hirschmann, H. (1990) *Meloidogyne morocciensis* n. sp. (Meloidogyninae), a root-knot nematode from Morocco. *Journal of Nematology* 22, 279–291.
Rau, G.J. and Fassuliotis, G. (1965) *Hypsoperine spartinae* sp. n., a gall forming nematode on the roots of smooth cordgrass. *Proceedings of the Helminthological Society of Washington* 32, 159–162.
Riffle, J.W. (1963) *Meloidogyne ovalis* (Nematoda: Heteroderidae), a new species of root-knot nematode. *Proceedings of the Helminthological Society of Washington* 30, 287–292.
Riggs, R.D. (1990) Making perineal patterns of root-knot nematodes and vulval cones of cyst nematodes. In: Zuckerman, B.M., Mai, W.F. and Krusberg, L.R. (eds) *Plant Nematology Laboratory Manual*. University of Massachusetts Agricultural Experiment Station, Amherst, Massachusetts, pp. 103–106.
Rui, K., Chen, M.C., Guo, A.P., Xiao, T.B., Xie, S.H., Wu, F.Z. and Liao, J.L. (2005) The identification of guava root-knot nematodes in Hainan Province. *Journal of South China Agricultural University* 26, 53–58.
Sahoo, N.K., Ganguly, S. and Eapen, S.J. (2000) Description of *Meloidogyne piperi* sp. n. (Nematoda: Meloidogynidae) isolated from the roots of *Piper nigrum* in India. *Indian Journal of Nematology* 30, 203–209.
Sandground, J.H. (1923) '*Oxyuris incognita*' or *Heterodera radicicola*? *Journal of Parasitology* 10, 92–94.
Santos, J.M. (1997) Estudo das principais espécies de *Meloidogyne* Goeldi que infectam o cafeeiro no Brasil com descrição de *Meloidogyne goeldii* sp. n. PhD thesis, Faculdade de Ciências Agronômicas/UNESP, Botucatu, Brazil.
Santos, M.S.N. de A. (1968) *Meloidogyne ardenensis* n. sp. (Nematoda: Heteroderidae), a new British species of root-knot nematode. *Nematologica* 13 (1967), 593–598.
Sasser, J.N. (1954) Identification and host parasite relationships of certain root-knot nematodes (*Meloidogyne* spp.). *Maryland Agricultural Experiment Station Bulletin* A-77.
Sasser, J.N. (1960) The genera *Meloidogyne* and *Meloidodera*. In: Sasser, J.N. and Jenkins, W.R. (eds) *Nematology*. University of North Carolina Press, Chapel Hill, North Carolina, pp. 216–219.
Sasser, J.N. (1979) Pathogenicity, host ranges and variability in *Meloidogyne* species. In: Lamberti, F. and Taylor, C.E. (eds) *Root-knot Nematodes (*Meloidogyne *species)*. Academic Press, New York, pp. 257–268.

[7] Cited as 'Pham Thanh Binh' after the species name in the original paper.

Sasser, J.N. and Carter, C.C. (1982) Root-knot nematodes (*Meloidogyne* spp.): identification, morphological and physiological variation, host-range, ecology, and control. In: Riggs, R.D. (ed.) *Nematology in the Southern Region of the United States.* Southern Cooperative Series Bulletin 276, Arkansas Agricultural Experiment Station, Fayetteville, Arkansas, pp. 21–32.

Sasser, J.N. and Carter, C.C. (eds) (1985) *An Advanced Treatise on* Meloidogyne. *Volume I: Biology and Control.* A cooperative publication of the Department of Plant Pathology and the United States Agency for International Development, North Carolina State University Graphics, Raleigh, North Carolina.

Shagalina, L., Ivanova, T. and Krall, E. (1985) [Two new gall nematodes from the genus *Meloidogyne* (Nematoda: Meloidogynidae), parasites of trees and shrubs.] *Eesti NSV Teaduste Akadeemia Toimetised, Bioloogialne* 34, 279–287.

Siddiqi, M.R. (1986) *Tylenchida: Parasites of Plants and Insects.* Commonwealth Agricultural Bureaux, Farnham Royal, UK.

Siddiqi, M.R. (2000) *Tylenchida: Parasites of Plants and Insects*, 2nd edn. CAB International, Wallingford, UK.

Siddiqi, M.R. and Booth, W. (1991) *Meloidogyne* (*Hypsoperine*) *mersa* sp. n. (Nematoda: Tylenchina) attacking *Sonneratia alba* trees in mangrove forest in Brunei Darussalam. *Afro-Asian Journal of Nematology* 1, 212–220.

Singh, S.P. (1969) A new plant parasitic nematode *Meloidogyne lucknowica* n. sp. from the root galls of *Luffa cylindrica* (sponge gourd) in India. *Zoologischer Anzeiger* 182, 259–270.

Skarbilovich, T.S. (1959) On the structure of systematics of nematodes order Tylenchida Thorne, 1949. *Acta Parasitologica* 7, 117–132.

Sledge, E.B. and Golden, A.M. (1964) *Hypsoperine graminis* (Nematoda: Heteroderidae), a new genus and species of plant-parasitic nematode. *Proceedings of the Helminthological Society of Washington* 31, 83–88.

Spaull, V.W. (1977) *Meloidogyne propora* n. sp. (Nematoda: Meloidogynidae) from Aldabra atoll, western Indian ocean, with a note on *M. javanica* (Treub). *Nematologica* 23, 177–186.

Taylor, A.L. (1987) *Identification and Estimation of Root-knot Nnematode Species in Mixed Populations.* Bulletin 12. Florida Department of Agriculture and Consumer Services, Division of Plant Industry, Contribution Number 335, pp. 1–73.

Taylor, A.L. and Sasser, J.N. (1978) *Biology, Identification and Control of Root-knot Nematodes (*Meloidogyne *species).* A cooperative publication of the Department of Plant Pathology, North Carolina State University and the United States Agency for International Development. North Carolina State Graphics, Raleigh, North Carolina.

Taylor, A.L., Dropkin, V.H. and Martin, G.C. (1955) Perineal patterns of root-knot nematodes. *Phytopathology* 45, 26–34.

Terenteva, T.G. (1965) [*Meloidogyne kirjanovae* n. sp. (Nematoda: Heteroderidae).] *Materialy Nauchnoi Konferentsii Vesouyuznoi Obscheĭ Gel'mint* 4, 277–281.

Thorne, G. (1969) *Hypsoperine ottersoni* sp. n. (Nematoda: Heteroderidae) infesting Canary Grass, *Phalaris arundinaceae* (L.) Reed in Wisconsin. *Proceedings of the Helminthological Society of Washington* 36, 98–102.

Tigano, M.S.T., Carneiro, R.M.D.G., Jeyaprakash, A., Dickson, D.W. and Adams, B.J. (2005) Phylogeny of *Meloidogyne* spp. based on 185 rNA and the intergenic region of mitochondrial DNA sequences. *Nematology* 7, 851–862.

Toida, Y. and Yaegashi, T. (1984) Description of *Meloidogyne suginamiensis* n. sp. (Nematoda: Meloidogynidae) from mulberry in Japan. *Japanese Journal of Nematology* 14, 49–57.

Tomaszewski, E.K., Khalil, A.M.M., El-Deeb, A.A., Powers, T.O. and Starr, J.L. (1994) *Meloidogyne javanica* parasitic on peanut. *Journal of Nematology* 26, 436–441.

Treub, M. (1885) Onderzoekingen over Sereh-Ziek Suikkeriet gedaan in s'Lands Plantentium te Buitenzorg. *Mededeelingen uit's Lands Plantentium, Batavia*, 2, 1–39.

Triantaphyllou, A.C. and Sasser, J.N. (1960) Variation in perineal patterns and host specificity of *Meloidogyne incognita. Phytopathology* 50, 724–735.

Whitehead, A.G. (1960) The root-knot nematodes of East Africa. I. *Meloidogyne africana* sp. n., a parasite of arabica coffee (*Coffea arabica* L.). *Nematologica* 4, 272–278.

Whitehead, A.G. (1968) Taxonomy of *Meloidogyne* (Nematoda: Heteroderidae) with descriptions of four new species. *Transactions of the Zoological Society of London* 31, 263–401.

Wouts, W.M. and Sher, S.A. (1971) The genera of the subfamily Heteroderinae (Nematoda: Tylenchoidea) with a description of two new genera. *Journal of Nematology* 3, 129–144.

Xu, J., Liu, P., Meng, Q. and Long, H. (2004) Characterisation of *Meloidogyne* species from China using isozyme phenotypes and amplified mitochondrial DNA restriction fragment length polymorphism. *European Journal of Plant Pathology* 110, 309–315.

Yang, B.J. and Eisenback, J.D. (1983) *Meloidogyne enterolobii* n. sp. (Meloidogynidae), a root-knot nematode parasitizing pacara earpod tree in China. *Journal of Nematology* 15, 381–391.

Yang, B.J., Wang, Q.L. and Feng, R.Z. (1988) [*Meloidogyne kongi* n. sp. (Nematoda: Meloidogynidae) a root-knot nematode parasitizing *Citrus* sp. in Guangxi, China.] *Journal of Jiangsu Agricultural College* 7, 1–9.

Yang, B.J., Hu, K.J. and Xu, B.W. (1988) [A new species of root-knot nematode *Meloidogyne lini* n. sp. parasitizing rice.] *Journal of Yunnan Agricultural University* 3, 11–17.

Yang, B.J., Hu, K.J., Chen, H. and Zhu, W.S. (1990) [A new species of root-knot nematode *Meloidogyne jianyangensis* n. sp. parasitizing mandarin orange.] *Acta Phytopathologica Sinica* 20, 259–264.

Zhang, S.S. (1993) [*Meloidogyne mingnanica* n. sp. (Meloidogynidae) parasitizing citrus in China.] *Journal of Fujian Agricultural University* 22 (Suppl.), 69–76.

Zhang, S.S. and Weng, Z.M. (1991) [Identification of root-knot nematode species in Fujian.] *Journal of Fujian Agricultural College* 20, 158–160.

Zhang, S.S., Gao, R.X. and Weng, Z.M. (1990) [*Meloidogyne citri* n. sp. (Meloidogynidae), a new root-knot nematode parasitizing citrus in China.] *Journal of Fujian Agricultural College* 19, 305–311.

Zhang, Y.M. (1983) [A root-knot nematode, *Meloidogyne sinensis* n. sp., from potatoes in China.] *Acta Scientiarum Naturalium Universitatis Shandong* 2, 88–95.

Zhang, Y.M. and Su, C.D. (1986) [A new species of the genus *Meloidogyne* from Shandong Province, China (Tylenchida: Meloidogynidae).] *Journal of Shandong University* 21, 95–103.

Zheng, L., Lin, M. and Zheng, M. (1990) [Occurrence and identification of a new disease of the citrus, Donghai root-knot nematode – *Meloidogyne donghaiensis* sp. nov. in coast soil of Fujian, China.] *Journal of Fujian Academy of Agricultural Sciences* 5, 56–63.

Zhu, W.S., Lan, X.Y., Hu, K.J., Yang, B.J. and Wang, Q.L. (1991) [Preliminary report of the survey on the occurrence of nematodes in citrus orchards in Sichuan Province.] *China Citrus* 20, 12–13.

4 Biochemical and Molecular Identification

Vivian C. Blok[1] and Thomas O. Powers[2]
[1]*Scottish Crop Research Institute, Invergowrie, Dundee, UK;* [2]*University of Nebraska, Lincoln, Nebraska, USA*

4.1 Introduction **98**
4.2 Biochemical Methods **100**
4.3 DNA-based Methods **103**
4.4 Conclusions and Future Directions **111**
4.5 Acknowledgements **112**
4.6 References **112**

4.1 Introduction

Meloidogyne identification has always presented challenges to the diagnostician. Conservative morphology, life stages in different habitats, wide host ranges, indistinct species boundaries or species complexes, sexual dimorphism, species with a potential hybrid origin, polyploidy, and over a century of human-aided dispersal are just some of the complicating features in the identification of *Meloidogyne* spp. Consider the infective stage: if *Meloidogyne* is present in a field, a soil nematode extraction typically recovers the small (< 0.6 mm) infective juvenile stage. Trained nematologists using a dissecting microscope can readily recognize members of the genus based on the fine stylet, the characteristically tapering tail, body movement or body shape if the juvenile is not moving. Yet even the most seasoned diagnostician would hesitate to assign an individual juvenile to a species. Morphometrics of juveniles can provide a relatively reliable assessment for species assignation (Hirshmann, 1985; Jepson, 1987; Karssen, 2002), but species-level identification, in practice, is complicated by genetic, climatic and anthropogenic factors associated with the dynamic nature and global scope of present-day agricultural production. In other words, there is no guarantee that an agricultural field contains only a single species of *Meloidogyne* or that the diagnostic descriptions currently available cover all of the diversity in the genus and will permit reliable identification. Are seed potatoes, for example, which are routinely shipped across international borders, responsible for the widespread distribution of *M. chitwoodi*? The planting of infected seed potatoes may occur in a field already infested with another *Meloidogyne* species, as it is now recognized that soils containing multiple *Meloidogyne* species are fairly common. Furthermore, *M. enterolobii* (= *M. mayaguensis*) was recognized as a problem in Florida when galls appeared on root-knot-resistant tomatoes, grown in response to persistent populations of *M. incognita* (Brito *et al.*, 2004), and *M. floridensis* was recognized as a distinct species following its discovery on root-knot-resistant peach rootstock (Carneiro *et al.*, 2000; Handoo *et al.*, 2004). *Meloidogyne parenaensis*, *M. izalcoensis* and *M. mayaguensis* (now

 ©CAB International 2009. *Root-knot Nematodes* (eds R.N. Perry, M. Moens and J.L. Starr)

M. enterolobii; see Hunt and Handoo, Chapter 3, this volume) were probably misdiagnosed on coffee for over 20 years due to over-reliance on perineal patterns and differential host tests (Carneiro *et al.*, 1996a,b, 2004a, 2005a).

Agricultural fields with multiple *Meloidogyne* species are not the only diagnostic challenge. A more basic concern is the actual genetic nature of our diagnostic target. As techniques have increased our ability to resolve more finely *Meloidogyne* genetics, it has become clear that many of our 'species' are collections of lineages that may or may not share a recent common ancestry. Common ancestry and descent provide the framework for the species concepts and the recognition of species boundaries. For *Meloidogyne*, this framework is still in the early stages of development (see Adams *et al.*, Chapter 5, this volume). In this chapter we briefly review the historical development of biochemical and molecular-based identification methods for root-knot nematodes. The application of these methods needs to be considered in terms of the cost and the accuracy that they provide, and will vary depending on the application, such as for routine quarantine or ecological studies, or for functional and evolutionary studies. Our lack of knowledge of the biogeography and evolutionary history of these organisms, and their genetics, is overlaid by complications that have been introduced through dispersal with agriculture, and this must also be remembered when using identification methods.

De Waele and Elsen (2007) noted that by 2006 about half (47) of the 92 nominal species (now 97, see Hunt and Handoo, Chapter 3, this volume) of *Meloidogyne* that they listed were described in the last 20 years; 29 of these were from Central and South America, Africa or Asia, with 14 of the new species from China. Thus, the possibility is high of encountering a new *Meloidogyne* species, particularly in tropical regions, where species diversity is rich. Where species identification is critical for intercepting and deploying appropriate quarantine steps, for the detection of emerging nematode threats and for appropriate nematode management, accurate identification is fundamental. While new species-continue to be described and identification methods improve, it is also important to recognize that often the expertise and facilities are lacking in parts of the world where *Meloidogyne* spp. are most prevalent and problematic. Basic education, training and appropriate infrastructure and funding are required for *Meloidogyne* spp. diagnostics to be utilized where they are most needed and for the benefit of the international community.

'Emerging threats' have been highlighted through extensive surveys and the use of a range of diagnostic tools to aid species identification. Distributions and host ranges, and morphological and molecular descriptions, in addition to revealing these new threats, also aid in defining the most stable diagnostic features and those that can be most practically utilized. For example, *M. mayaguensis*, first described in 1988 by Rammah and Hirschmann, is now recorded from West Africa (Senegal, Ivory Coast, Burkina Faso), South Africa, Malawi, the Caribbean (Puerto Rico, Cuba, Dominican Republic, Guadeloupe, Martinique, Trinidad), Brazil, the USA (Florida), and a glasshouse in France (De Waele and Elsen, 2007). It is now considered to be conspecific with *M. enterolobii*, described by Yang and Eisenback (1983) in China (Xu *et al.*, 2004), based on the identical sequence of a mitochondrial DNA region obtained for both species (see Eisenback and Hunt, Chapter 2, this volume). The elevation in status of *M. enterolobii* as one of the most economically important root-knot nematode (RKN) species has arisen through surveys that have established its wide geographic distribution and which have utilized biochemical and molecular diagnostics for identification (Fargette and Braaksma, 1990; Fargette *et al.*, 1996; Blok *et al.*, 2002). The recognition of its wide host range, combined with its virulence characteristics, makes it a major threat. The overlap of its morphometric characters with those of the most common tropical species (Brito *et al.*, 2004) has probably led to this species being misidentified in the past. Another species, *M. paranaensis*, which parasitizes coffee, has recently been shown to be widely distributed in coffee-growing regions in Central and South America and was probably confused with *M. incognita* until isozyme esterase phenotyping and RAPDs (random amplified polymorphic DNAs) were used in its identification Carneiro *et al.*, 1996a,b, 2004b; Hervé *et al.*, 2005). More species are likely to emerge as threats as further surveys are conducted and combined with more reliable diagnostic methods.

Before introducing biochemical and molecular diagnostics, it is worth considering the sample types that may be used; the various life stages (egg, juvenile, female, male), root tissue or soil may preclude or limit the suitability of a particular diagnostic method and influence the level of specificity and sensitivity that can be achieved. Bioassays involving nematicide testing or germplasm screening in which defined inocula are used will have different requirements from field surveys and diversity studies. Assays that distinguish between biotypes require another level of discrimination. Traditional methods that require laborious extraction techniques and microscope observation combined with manual enumeration are still used frequently and may be the most efficient and cost-effective method available for some applications. Biochemical or molecular methods that combine identification with quantification with the potential for automation are still under development and beyond the resources of many organizations. However, examples are provided that illustrate both the practical benefits that biochemical and molecular diagnostics are bringing and how they are improving our understanding of *Meloidogyne* spp.

4.2 Biochemical Methods

4.2.1 Isozymes

One of the earliest examples of the use of isozyme phenotypes to distinguish *Meloidogyne* spp. was published by Esbenshade and Triantaphyllou (1985), who reported esterase patterns from 16 *Meloidogyne* species, with the most common phenotypes being A2 and A3 (*M. arenaria*), H1 (*M. hapla*), I1 (*M. incognita*) and J3 (*M. javanica*). In 1990, Esbenshade and Triantaphyllou used isozymes in their landmark survey involving approximately 300 populations originating from 65 countries and various continents. In later surveys, Carneiro *et al.* (2000) found 18 esterase phenotypes among 111 populations of *Meloidogyne* spp. in Brazil and other South American countries, and Zu *et al.* (2004) examined 46 populations from 14 provinces in China and found five esterase phenotypes. Isozymes continue to be widely used for studies of *Meloidogyne* despite some of their limitations, and isozyme phenotypes for a large number of species have been published (Table 4.1). Schematic diagrams of isozyme patterns based on surveys, including those conducted in the International *Meloidogyne* project, have been published (Bergé and Dalmasso, 1975; Dalmasso and Bergé, 1978; Fargette, 1978; Janati *et al.*, 1982; Esbenshade and Triantaphyllou, 1985, 1990; Carneiro *et al.*, 2000; Hernandez *et al.*, 2004) and provide important references.

Several isozyme systems have been used, with carboxylesterase/esterase EST (EC 3.1.1.1) proving to be most useful for discriminating *Meloidogyne* species, with others such as malate dehydrogenase MDH (1.1.1.37), superoxide dismutase SOD (1.15.1.1) and glutamate-oxaloacetate transaminase GOT (EC 2.6.1.1) also often included to confirm species identifications (Esbenshade and Triantaphyllou, 1985). Enzyme phenotypes are designated, indicating the *Meloidogyne* species that each specifies and the number of bands detected. Phenotypes with the same number of bands are differentiated by small letters (Esbenshade and Triantaphyllou, 1985, 1990). Enzyme patterns are usually compared with a known standard, frequently from *M. javanica*, which is included in the electrophoresis to determine migration distances. Isozymes are used primarily with the female egg-laying stage, using single individuals (Dalmasso and Bergé, 1978), although the use of galled root tissue has also been reported (Ibrahim and Perry, 1992). Miniaturization and automation of the electrophoresis systems and the use of precast polyacrylamide gels (i.e. PhastSytem, Pharmacia Ltd, Uppsala, Sweden) has made isozyme phenotyping a widely used technique (Esbenshade and Triantaphyllou, 1985; Karssen *et al.*, 1995; Chen *et al.*, 1998; Molinari, 2001). These systems are not technically sophisticated and more than one enzyme system can be stained on the same gel. Aside from the initial expense of equipment, the consumables required are relatively inexpensive and so isozymes are often used for field surveys and have been used for routine screening of glasshouse cultures to assure their species stability.

The relative stability of the isozyme phenotypes within *Meloidogyne* species (De Waele and Elsen, 2007) makes them an attractive system, although there are some complications. The occurrence of intraspecific variants and the difficulty in resolving size variants between species (e.g. the esterases of *M. incognita* and *M. hapla*) has necessitated the use of more than one enzyme system to confirm the identity of some isolates. Malate dehydrogenase separates *M. hapla* from

Table 4.1. Esterase and malate dehydrogenase (Mdh) isozyme phenotypes of *Meloidogyne* spp., including atypical esterase patterns.

Species	Esterase phenotype	Atypical esterase patterns	Mdh
M. arabicida	AR2[7], M1F1b[22]		N1[22]
M. ardenensis			N1a[23]
M. arenaria	A1[18], A2[18], A3[18]	S1-M1[18], S2-M1[18], M3-F1[18]	N1[18], N3[18]
M. artiella	M2-VF1[23]		N1b[23]
M. baetica	Rm 0.31[12]		
M. carolinensis	VS1-S1a[18]		H1[18]
M. chitwoodi	S1[18]		N1a[18]
M. coffeicola	C2[5]		C1[5]
M. cruciani	M3a[18]		N1[18]
M. duytsi	VS1[23]		N2[23]
M. dunensis	VS1[28]		N1c[28]
M. enterolobii	VS1-S1[18]		N1a[18]
M. ethiopica	E3[9]		N1[9]
M. exigua	VS1[20], E1 (VF1)[5], E1b (VF1)[5], E2[11], E2a[11], E3[11]		N1[20]
M. fallax	F3[33]		N1b[23]
M. floridensis	P3[5, 21]		N1[5]
M. graminicola	VS1[18]		N1a[18]
M. graminis	VS1[19] G1[3]		N4[19], N1a[3]
M. hapla	H1[18]	A1[18]	H1[18]
M. haplanaria	Rm 0.61[17]		Rm 0.44[17]
M. hispanica	S2-M1[18], Hi3[6]		N1[18]
M. incognita	I1[18], I2[5], M1a[22]	S1[18]	N1[18]
M. inornata	I3[10]		N1[5]
M. izalcoensis	I4=S4[8, 22]		N1[22]
M. javanica	J3[18], J2a[13], J2[32]		N1[18, 15]
M. jianyangensis	Rm 0.41, 0.45, 0.48[1]		
M. konaensis	F1[16], K3[5], I1[31,] F1-I1[31]		N1[5]
M. kralli			N1c[23]
M. lusitanica	P1[27], A1[23]		P3[27], N1c[23]
M maritima	VS1-S1[23]		N1c[23]
M. marylandi	VS1[26]		N1c[26]
M. enterolobii	VS1-S1[29,] M2[5]		N1a[5], N3c[29]
M. microcephala	A1[18]		N1[18]
M. microtyla	M1[18]		H1[18]
M. minor	VS1[24]		N1a[24]
M. morocciensis	A3[30]		N1[30]
M. naasi	VF1[18]		N1a[18]
M. oryzae	VS1[18]		N1a[18]
M. panyuensis	S1-F1[25]		N1b[25]
M. paranaensis	P1[4, 5, 7], F1[5], P2[7]		N1[5]
M. partityla	Mp3[2]		N1a[2]
M. petuniae	VS1-S1[14]		N1[14]
M. plantani	S1[18]		N1a[18]
M. querciana	F1[18]		N3a[18]

[1]Baojun *et al.* (1990); [2]Brito *et al.* (2008); [3]Brito, pers. comm.; [4]Carneiro *et al.* (1996b); [5]Carneiro *et al.* (2000); [6]Carneiro *et al.* (2004a); [7]Carneiro *et al.* (2004b); [8]Carneiro *et al.* (2005a); [9]Carneiro *et al.* (2007); [10]Carneiro *et al.* (2008); [11]Carneiro, pers. comm.; [12]Castillo *et al.* (2003); [13]Castro *et al.* (2003); [14]Charchar *et al.* (1999); [15]Cofcewick *et al.* (2005); [16]Eisenback *et al.* (1994); [17]Eisenback *et al.* (2003); [18]Esbenshade and Triantaphyllou (1985); [19]Esbenshade and Triantaphyllou (1987); [20]Esbenshade and Triantaphyllou (1990); [21]Handoo *et al.* (2004); [22]Hernandez *et al.* (2004); [23]Karssen and van Hoenselaar (1998); [24]Karssen *et al.* (2004); [25]Liao *et al.* (2005), [26]Oka *et al.* (2003); [27]Pais and Abrantes (1989); [28]Palomares Rius *et al.* (2007); [29]Rammah and Hirschmann (1988); [30]Rammah and Hirschmann (1990); [31]Sipes *et al.* (2005); [32]Tomaszewski *et al.* (1994); [33]van der Beek and Karssen (1997).

M. incognita, *M. arenaria* and *M. javanica*, whereas glutamate dehydrogenase separates *M. incognita* from *M. javanica*, *M. arenaria* and *M. hapla* (Esbenshade and Triantaphyllou, 1985). Poor signal intensity can also necessitate the use of several females (e.g. with *M. exigua* (Carneiro *et al.*, 2000)).

In surveys concerning *Meloidogyne* biodiversity and nature conservancy, isozymes are a convenient first stage in species identification and have enabled species diversity and the frequency of particular species and their abundance to be determined. Females recovered after allowing multiplication of field samples on a generally susceptible host such as *Solanum lycopersicum* can be tested for their isozyme phenotype and the associated egg mass reserved for further characterization if necessary. Lima *et al.* (2005) used this approach in their study of the nematofauna of the Atlantic forest in Brazil, and Hernandez *et al.* (2004) in their survey of coffee-growing areas in Central America. Novel isozyme phenotypes have been frequently encountered in these surveys of biodiverse regions, adding to the understanding of the species ecology and biogeography of *Meloidogyne* spp.

The literature gives many examples of atypical isozyme phenotypes or those from undescribed species, some of which are resolved as new species in due course. Some examples of these are given here, but it remains one of the challenges of using isozymes where novel phenotypes are obtained, to relate these to previous examples in the literature. Esbenshade and Triantaphylllou (1985) listed F1, VS1, VS1-S1, VS1-M2, S1-M1, M3, A2 as undescribed phenotypes; Cenis *et al.* (1992) reported an atypical esterase pattern for *M. incognita* from Spain; Hernandez *et al.* (2004) found M1F1a (Rm 73.5, 78.0), M1F1b (Rm 73.5, 82.0) and Sa4 (Rm 73.5, 78.0, 53.0, 59.0) esterase phenotypes with isolates from coffee in Central America; and Adam *et al.* (2005) described an S2 phenotype for an *M. incognita* isolate from Libya. Molinari *et al.* (2005) reported atypical EST patterns in their survey of populations from India, Venezuela, Cuba and Egypt. Lima *et al.* (2005) found an MC4 phenotype in their survey of montane forest in Brazil; Carneiro *et al.* (2005b) lists unknown populations, including esterase phenotype Br2 (Rm 0.92, 1.02), in a survey of coffee in Brazil; Medina *et al.* (2007) found Est S1, Est F2b and Est F2a in 20% of the samples from fig trees in Brazil; and Carneiro *et al.* (2007) found atypical esterase patterns L3 (Rm 1.0, 1.1, 1.3) and V3/V4 with a minor band (Rm 1.3) and three major bands (Rm 0.9, 1.2, 1.3) in their survey of vineyards in Chile. Clearly many novel esterase patterns are still being discovered, and to determine whether these represent novel or aberrant patterns additional information from host range, geographic distributions and other biochemical, molecular or morphological features are needed. Intraspecific diversity or differences in the patterns obtained from different laboratories may also contribute to slight variations in phenotypes, as highlighted by Hernandez *et al.* (2004).

4.2.2 Antibodies

Polyclonal and monoclonal antibodies have been produced for root-knot nematode identification purposes, as well as for investigations of the nematode surface and secretions, interactions with the host and other parasites, for localization studies, development of plantibodies and behavioural studies (Tastet *et al.*, 2001). Qualitative and quantitative features of immunoassays using poly- or monoclonal antibodies have determined their utility for diagnostic purposes. The sample type from which the antigen will be extracted and whether cross-reaction may occur, the life-stage and the antibody sensitivity and specificity all contribute to whether an immunoassay is appropriate for the particular application. In addition, the process of developing an antiserum requires a considerable investment and hence applications must justify these costs. Polyclonal antibodies tend to be highly sensitive; however, they may also be cross-reactive and lack the specificity required, and different batches may vary in their binding characteristics. Production of a polyclonal antibody to a diagnostic protein can overcome some of the problems with cross-reactivity but producing sufficient pure antigen can be challenging. Monoclonal antibodies (Mabs) produced from cell lines can give high specificity and better reproducibility between batches but their production is expensive and cell lines can be unstable. Screening existing libraries for an antibody that has the required specificity is another alterative, but requires technical expertise. For routine testing, such as for nematicide or germplasm screening where a defined nematode species is involved, an ELISA may be the most appropriate assay. However, when dealing with unknowns, such as in surveys or in quarantine situations, immunoassays are usually not the most appropriate technique to use.

The use of antibodies as diagnostic tools for *Meloidogyne* spp. is limited to a few examples, having mainly been superseded by DNA-based diagnostics, which generally have greater sensitivity and specificity. Davies *et al.* (1996) selected three Mabs that could distinguish females of *M. incognita*, *M. javanica* and *M. arenaria* by ELISA and dot blots; however, cross-reactivity was found when used in Western blots. Antisera raised to purified species-specific esterase bands did permit differentiation of *M. incognita* from *M. javanica* but the Mabs cross-reacted with other species of root-knot nematodes (Davies *et al.* 1996; Ibrahim *et al.* 1996). Ibrahim *et al.* (1996) raised a Mab to purified esterase from *M. incognita* and were able to distinguish *M. incognita* from *M. javanica* in crude extracts of non-denatured protein. Tastet *et al.* (2001) used two-dimensional electrophoresis to identify a major protein of *M. chitwoodi* and *M. fallax* that was not found in several other *Meloidogyne* species, and following internal amino-acid sequencing, a peptide was synthesized and used to raise antisera in rabbits. They were able to distinguish *M. chitwoodi* and *M. fallax* from eight other *Meloidogyne* species in a dot-blot hybridization with soluble proteins extracted from a single female.

Quantification of root-knot nematodes directly in soil using antibodies has not proved successful, and some level of nematode extraction has been required (Davies *et al.*, 1996). However, immunocapture to recover particular nematodes from mixtures has been achieved. Antiserum-coated magnetized beads (Dynabeads) were used to recover *M. arenaria* from mixtures with other species of nematodes (Chen *et al.*, 2001). Combining an enrichment approach with highly specific antibodies may provide a fruitful avenue for the future. Targets that are unique to particular species may be identified from the considerable sequence information that is being generated, and synthetic peptides that are based on unique sequence regions could be used to raise antibodies and generate a new source of diagnostic antibodies.

4.3 DNA-based Methods

4.3.1 DNA extraction

Many methods have been reported for the extraction of DNA from bulk samples of second-stage juveniles (J2) as well as from single J2, females and males. Methods for the extraction of DNA from plant roots and galls infected with *Meloidogyne* and from soil samples are also available.

For a single J2, DNA extraction methods include crushing the nematode on a glass slide with a pipette tip (http://nematode.unl.edu/nemaid.pdf), treatment of intact nematodes with NaOH (Stanton *et al.*, 1998), and proteinase K treatment following cutting in worm lysis buffer (Castagnone-Sereno *et al.*, 1995). A systematic diagnostic key for the identification of seven of the common and economically important *Meloidogyne* spp. by Adam *et al.* (2007) provides a logical process for molecular identification of individual nematodes in, at most, three steps. The extraction method used yields sufficient DNA for 15 PCR reactions and the key can be readily expanded to include more species. Multiple displacement amplification (MDA) of total genomic DNA from *Meloidogyne* spp. is also possible to increase the amount of template for molecular analyses from small samples (Skantar and Carta, 2005). For larger samples of juvenile nematodes or egg masses, extraction of DNA using phenol:chloroform (Blok *et al.*, 1997a) or DNA extraction kits such as those of Qiagen are suitable.

Nematodes extracted from soil using a Baermann funnel can be individually isolated for diagnostic analyses. Examples of the application of molecular diagnostics to DNA extracted from the total soil nematode communities are limited. Methods for the extraction of DNA directly from soil, including proteinase K digestion followed by phenol:chloroform extraction, NaOH extraction, and bead beating combined with a commercial kit for DNA recovery, have been compared by Donn *et al.* (2008), but use of these methods for detection of *Meloidogyne* spp. was not reported.

4.3.2 Restriction fragment length polymorphisms (RFLPs)

Initially the use of restriction fragment length polymorphisms (RFLPs) to distinguish species and isolates of root-knot nematodes involved the extraction and purification of genomic DNA, restriction digestion and visualization of banding patterns following gel electrophoresis. An early example of the application of RFLPs to *Meloidogyne* spp. was reported by Curran *et al.* (1985, 1986). The DNA isolated from large numbers of eggs was digested and then subjected to electrophoresis in an agarose gel, followed by visualization of the DNA banding patterns with ethidium bromide. The patterns representing highly repeated regions of DNA allowed samples to be distinguished, but required large amounts of DNA and, hence, prior culturing of the isolates. The patterns were often not clearly seen against the background smear of DNA. However, the advantage of the removal of dependency on a particular stage in the life cycle and the inclusion of the whole genome was apparent with this approach. Later, RFLPs were combined with DNA hybridization and the use of either probes labelled radioactively, or a non-radioactive detection system using randomly selected clones from genomic DNA, mitochondrial DNA or satellite DNA sequences as probes (Curran and Webster, 1987; Castagnone-Sereno *et al.*, 1991; Gárate *et al.*, 1991; Cenis *et al.*, 1992; Piotte *et al.*, 1992, 1995; Xue *et al.*, 1992; Baum *et al.*, 1994; Hiatt *et al.*, 1995). Although interspecific discrimination was demonstrated in these experiments, the lack of sensitivity, i.e. the requirement for DNA from multiple individuals, the use of radioactivity and the relative complexity of the technique, limited its application. The development of PCR has largely supplanted hybridization-based approaches to RFLP analysis for nematode species identification.

4.3.3 Satellite DNA probes and PCR

Satellite DNAs (satDNAs) are highly repeated tandem arrays of short sequences (~70–2000 bp in length) that are associated with heterochromatin, centromeric and telomeric regions of chromosomes. The detection of satDNAs in nematode tissue squashed on to a membrane and then hybridized with a satellite probe is an attractive diagnostic approach as it requires limited molecular equipment or expertise and can be used efficiently where there are large numbers of samples to screen, such as from field surveys. This method usually does not require DNA extraction or PCR amplification of the nematode DNA and, when used with the non-radioactive detection system DIG (which uses digoxigenin-labelled DNA), is safe, stable and reusable (Castagnone-Sereno *et al.*, 1999). SatDNAs have different signature sequences and can differ in their copy number, length and polymorphic regions in *Meloidogyne* spp. (Meštrović *et al.*, 2006), and satDNAs assays have been described for several species of *Meloidogyne*. The highly repetitive nature of satDNAs aids in their ease of detection (satDNA comprises 2.5% of the genome of *M. incognita* (Piotte *et al.*, 1994) and 20% of *M. fallax* (Castagnone-Sereno *et al.*, 1998)), and the discovery that some satDNAs are divergent between different species has been exploited to develop various diagnostic probes for RFLPs, dot blots and for designing PCR assays. The distribution of these sequences in the genome and the mechanisms involved in their evolution are not well understood; however, with the determination of the genomic sequences of *M. incognita* and *M. hapla* (see Abad and Oppermann, Chapter 16, this volume), the number of different types and their location in the genome is being revealed and may help us to understand how satDNA might be further exploited in the future for diagnostic purposes.

Examples of the use of satDNA as a diagnostic probe include repeat sequences from *M. hapla* that were radioactively labelled and had sufficient sensitivity to detect DNA from individual females of *M. hapla*, including those in root tissue (Piotte *et al.*, 1995; Dong *et al.*, 2001a); this probe detected *M. hapla* but not *M. chitwoodi* or *M. incognita*. Castagnone-Sereno *et al.* (2000) isolated a conserved Sau3A satDNA from *M. arenaria*, which was subsequently also described in *M. javanica* (Meštrović *et al.*, 2005). Randig *et al.* (2002a) cloned a BglII satellite from *M. exigua* and used it as a radioactively labelled probe to detect single individuals (J2, females, egg masses and galls squashed on to nylon membrane) and showed it to be specific to *M. exigua* when tested with eight other *Meloidogyne* spp. Similarly, single J2 of *M. chitwoodi* or *M. fallax* could be distinguished from one of *M. hapla* in a simple squash blot using DIG-labelled probes from AluI satDNA pMcCo and pMfFd, and, conversely, *M. hapla* was detected and distinguished from *M. chitwoodi* or *M. fallax* with the pMhM satDNA probe isolated from *M. hapla* (Castagnone-Sereno *et al.*, 1998). Conversion of satellite DNA probes into a PCR-based detection system has provided an alternative approach for sensitive detection of *Meloidogyne* spp. This was demonstrated for *M. hapla* by Castagnone-Sereno *et al.* (1995).

4.3.4 Ribosomal DNA PCR

The ribosomal DNA (rDNA) repeating unit, including 18S, 28S, and 5.8S coding genes and the internal transcribed spacer (ITS), external transcribed spacer (ETS) and intergenic spacer (IGS) regions, has been used extensively for both phylogenetic studies and diagnostic purposes. The ITS regions are possibly the most widely used genetic markers among living organisms and the most common species-level marker used for plants, protists and fungi (Hajibabaei *et al.*, 2007). The multi-copy basis of rDNA provides ample target for PCR amplification, and sufficient variation and stability occurs within it for reliable discrimination of most species, although intraspecific variation has been found (Zijlstra *et al.*, 1995; Hugall *et al.*, 1999; Adam *et al.*, 2007) and there is evidence for intra-individual variation (Blok *et al.*, 1997b; Powers *et al.*, 1997; Zijlstra *et al.*, 1997; Hugall *et al.*, 1999). Differences in sequence variation occur between the regions of the rDNA cistron, with regions coding for structural RNAs (18S, 28S, 5.8S) showing greater conservation than the transcribed and non-transcribed intergenic regions (ITS, ETS, IGS). For diagnostic purposes, rDNA PCR amplification products that are polymorphic in size, with or without subsequent restriction enzyme digestion, have been used to identify many *Meloidogyne* spp. For example, PCR–RFLP of the ITS regions has been used to identify *M. arenaria*, *M. camelliae*, *M. mali*, *M. marylandi*, *M. suginamiensis* (Orui, 1999), *M. incognita*, *M. javanica*, *M. hapla*, *M. chitwoodi*, *M. fallax* (Zijlstra *et al.*, 1995) and *M. naasi* (Schmitz *et al.*, 1998). Size polymorphisms of rDNA amplification products, where products are amplified from more than one species but the size is characteristic of a particular species, are used in the scheme of Adam *et al.* (2005). Sequence analysis of rDNA is, however, increasingly being used for identification of *Meloidogyne* spp. (Powers, 2004), and this approach is useful when the resources are available and when supported with a sound phylogenetic basis for distinguishing species, which is validated with many isolates (see Adams *et al.*, Chapter 5, this volume). These analyses have also led to a published patent which describes primers based on sequence polymorphisms in rDNA for distinguishing *M. incognita*, *M. javanica*, *M. arenaria*, *M. hapla*, *M. microtyla*, *M. ardenensis*, *M. maritima*, *M. duytsi*, *M. chitwoodi*, *M. fallax*, *M. minor*, *M. naasi*, *M. oryzae* and *M. graminicola* (Helder *et al.*, 2008).

Various sets of primers are reported in the literature for amplifying different rDNA regions. The primers designed by Vrain *et al.* (1992) have been widely used to amplify the ITS region for *Meloidogyne* spp. and produce a product of ~800 bp, which can then be sequenced to produce species-specific primers or restriction enzyme digestions. For example, this approach was used by Zijlstra (1997) and Zijlstra *et al.* (2004), who sequenced rDNA ITS of *M. naasi*, *M. chitwoodi*, *M. fallax*, *M. hapla*, *M. minor* and *M. incognita*, and then designed specific primers for each species to produce products unique to each species. The ITS–RFLP approach, as well as producing characteristic digestion patterns, has been used to determine the composition of species in mixtures by comparing the intensity of bands produced for each species. This was demonstrated for mixtures of *M. hapla*, *M. chitwoodi*, *M. incognita* and *M. fallax* by Zijlstra *et al.* (1997).

Most reports have concluded that there is limited sequence polymorphism in the ITS sequences of the most common species – *M. incognita*, *M. javanica* and *M. arenaria* – to distinguish them, although Hugall *et al.* (1999), in their detailed sequence analyses, did reveal polymorphisms in the ITS region, which they suggested are indicative of genetic lineages shared in these species and illustrate the potential for misidentification if ITS sequence is used exclusively for identification of these species. Because of the limited sequence polymorphism in ITS rDNA to distinguish *M. incognita*, *M. javanica* and *M. arenaria* reliably, specific sequenced characterized amplified region (SCAR) primers have been developed for these species. In Table 4.2, examples of species-specific primers are given, some of which are based on rDNA sequences and others developed from RAPDs. Although these primers are described as 'species specific', they must be considered in relation to the species and isolates that have been used for comparison.

Other species combinations that have proved difficult to distinguish using morphological and biological features, such as *M. hispanica* from *M. incognita* and *M. arenaria*, have been differentiated by comparing their sequences from ITS, 18S and D2/D3, a variable region within the 28S gene. However, *M. hispanica* has also been reported to have an identical ITS sequence to *Meloidogyne ethiopica*, suggesting caution is needed with this approach too, although these species can be differentiated by their D2/D3 sequences (Landa *et al.*, 2008). The IGS region has been found to contain repeated sequences and sequence polymorphisms that have been exploited to distinguish *M. chitwoodi* and *M. fallax* (Blok *et al.*, 1997a, 2002; Petersen *et al.*, 1997; Wishart *et al.*, 2002) from other species, including *M. enterolobii*, *M. hapla* and *M. incognita*/*M. javanica*/*M. arenaria*. Distinguishing species based on size polymorphisms of the amplification products has the additional advantage that the products act as positive controls, in contrast to species-specific primer sets, where a product is only obtained from the species that the primers are specific for and the negative results cannot be distinguished from failed reactions.

4.3.5 Mitochondrial DNA

From the perspective of identification, the mitochondrial genome (mitochondrial DNA, mtDNA) provides a rich source of genetic markers for identification (Rubinoff and Holland, 2005; Hu and Glasser, 2006). Multiple copies of the circular mitochondrial genome are contained within each cell, providing ample template for PCR assays. Uniparental inheritance and a low level of recombination facilitate the construction of phylogenies that can be used to address questions of species boundaries and variation among populations. Rates of evolution of the mitochondrial genome are generally higher than rates for corresponding nuclear genes, creating sufficient nucleotide variation for species-level analyses (Brown *et al.*, 1979). The conserved gene content among mitochondrial genomes of animals allows investigators to compare similar experimental approaches across widely divergent phyla. For example, postglacial recolonization patterns in Europe have been inferred by an examination of the mitochondrial cytochrome b gene of wood rats and their nematode parasites (Nieberding *et al.*, 2005). The Consortium for the Barcode of Life (http://www.barcoding.si.edu/) has exploited these mitochondrial features and proposed a worldwide initiative in which all known species are 'bar coded' by DNA sequence from the cytochrome oxidase subunit I (COI) gene. One objective of this initiative is to develop a rapid method to identify all known animal species. The initiative has generated considerable debate, particularly on theoretical and philosophical issues, with critics claiming that proponents of the approach oversell the advantages and insights to be gained from a 'one-gene-fits-all' approach (Moritz and Cicero, 2004). Advocates point to a growing literature of empirical taxonomic studies that employ the COI bar code (Vogler and Monaghan, 2007). Several nematode taxa have been bar coded by COI, such as *Bursaphelenchus* (Ye *et al.*, 2007); however, studies on *Meloidogyne* are yet to be published.

A structural map of the mitochondrial genome of *Meloidogyne* was published by Okimoto *et al.* (1991), although not the full sequence. The map showed the location of 12 protein-coding genes, the large and small rRNA genes, and tRNA (transfer RNA) genes (Fig. 4.1). In gene content and overall structure, the *Meloidogyne* mitochondrial genome resembled other animal mtDNAs. It is a circular molecule with genes colinearly arranged without intervening non-coding DNA sequences. However, several unique characteristics of the genome highlighted features that subsequently were incorporated into diagnostic assays. Gene order in *Meloidogyne* mitochondria differed from that of two other nematodes, *Ascaris suum* and *Caenorhabditis elegans* (Okimoto *et al.*, 1991). The differences in gene order allowed for the development of PCR-based diagnostic assays with reduced probability of false-positive amplifications. This could be accomplished by placing primer pairs in two genes that were not adjacent to each other in non-target mitochondrial genomes. A second feature, relatively rare in animal mitochondrial genomes, was the presence of non-coding, repeated sequences. Three sets of different-sized

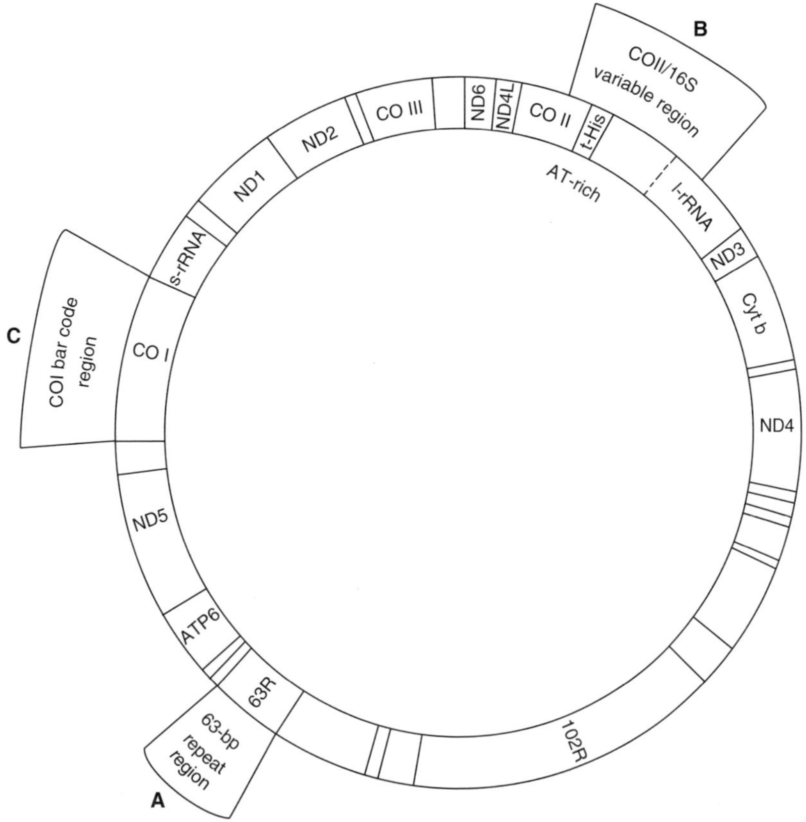

Approximate amplification product sizes

A 63-bp repeat region:
322 bp – *Meloidogyne enterolobii* (*M. mayaguensis*)

B COII/16S variable region:
Using primer set C_2F_3/1108
(C_2F_3 5'-GGTCAATGTTCAGAAATTTGTGG-3')
(1108 5'-TACCTTTGACCAATCACGCT-3')
1.5–1.6 kb – *M. arabicida*, *M. arenaria**, *M. ethiopica*, *M. incognita*, *M. javanica*
1.2 kb – *M. paranaensis*
1.1 kb – *M. arenaria**, *M. floridensis*, *M. morocciensis*, *M. thailandica*
750 bp – *M. enterolobii* (*M. mayaguensis*)
520–540 bp – *M. chitwoodi*, *M. fallax*, *M. graminicola*, *M. graminis*, *M. hapla*, *M. haplanaria*, *M. mali*, *M. marylandi*, *M. microtyla*, *M. naasi*, *M. oryzae*, *M. partityla*, *M. suginamiensis*, *M. trifoliophila*

C COI bar code region

*Hugall *et al.*, 1994

Fig. 4.1. *Meloidogyne* mitochondrial genome structure, showing regions used for diagnostics. (After Okimoto *et al.*, 1991 and sequences from NCBI (US National Center for Biotechnology Information).)

repeats, 102, 63 and 8 nucleotides in length, are clustered apart from the protein-coding genes. Blok *et al.* (2002) discovered that amplification of the 63 bp repeating region by flanking primers produced a discrete 320 bp product with *M. enterolobii*, whereas other *Meloidogyne* species produced either a multi-banded pattern or no amplification product. Hyman and Whipple (1996) and Lunt *et al.* (2002) have explored the possibility of using the repeated region as a marker to examine population dynamics. The extreme variability of this region within and among offspring of the same parent makes this an intriguing target for genealogical studies, but amplification properties make it procedurally difficult to analyse (Lunt *et al.*, 2002). A second region of the *Meloidogyne* genome amenable to diagnostic development is the portion of the genome flanked by the COII gene and the large (16S) ribosomal gene. Between these two genes is the tRNA-His gene (53 bp) and, in the mitotically parthenogenetic species, non-coding sequences that include a stem and loop structure characteristic of the AT-rich region or control region of the mitochondrial molecule (Hugall *et al.*, 1994, 1997; Jeyaprakash *et al.*, 2006). This region was originally targeted as a potential means for differentiating the five common *Meloidogyne* species of different-sized amplified products generated by primers positioned in the 3′ portion of COII and the 5′ portion of 16S rRNA (Powers and Harris, 1993). Three size classes were recognized: (i) an approximately 530 bp amplification product was observed in *M. hapla*, which included the flanking portions of COII and 16S rDNA and the complete tRNA-His, but no AT-rich region; (ii) a 1.1 kb amplification product was found in *M. arenaria*, which included an approximately 570 bp AT-rich region; and (iii) *M. incognita* and *M. javanica* had the largest amplification products (~ 1.6 kb) due to an AT-rich region of approximately 1.0 kb. Today, more than 15 years later, many additional *Meloidogyne* species have been examined, resulting in numerous size classes (Fig. 4.1). These size classes result primarily from insertions and deletions in the AT-rich region. A large group of species fall into the smallest size class, those lacking an AT-rich region in the amplified product. Together with *M. hapla*, these include *M. chitwoodi*, *M. fallax*, *M. graminicola*, *Meloidogyne graminis*, *M. mali*, *M. marylandi*, *M. microtyla*, *M. naasi*, *M. oryzae*, *M. suginamiensis* and *M. trifolliophila*. Presumably this is the ancestral state for *Meloidogyne* since non-*Meloidogyne* spp., such as *Nacobbus aberrans*, share this trait (Powers, unpublished observation). *M. mayaguensis* and *M. enterolobii* share a 167 bp AT-rich region, identical in size and sequence, a key feature which led to their synomyzation (Blok *et al.*, 2002; Xu *et al.*, 2004). *M. arenaria* and *M. floridensis* share an intermediate-sized AT-rich region of 573 and 603 bp respectively, and *M incognita*, *M. javanica* and other mitotically parthenogenetic species possess AT-rich regions that range from 963 to 1100 bp in size (Jeyaprakash *et al.*, 2006). Size classes of amplification will probably diminish in diagnostic value as more species are examined and distinctions among groups based on size alone are blurred. However, sequence polymorphism among species remains sufficient to construct diagnostic assays, keeping in mind that all diagnostic assays must be grounded in an understanding of species boundaries. The disparities among phylogenetic trees generated from 18S ribosomal DNA and mitochondrial DNA suggest a full understanding of *Meloidogyne* species boundaries is yet to be obtained (Tigano *et al.*, 2005).

4.3.6 Sequence characterized amplified regions (SCARs)

Specific primers have been developed to PCR-amplify diagnostic repetitive regions of sequence: sequence characterized amplified regions (SCARs). Typically, characteristic repetitive sequences have been identified following an analysis of a panel of isolates from several *Meloidogyne* species with short RAPD primers of eight to ten nucleotides; the differential bands are isolated, sequenced and long specific primers designed. Examples of 'species-specific' primer sets based on RAPD product and rDNA sequences are shown in Table 4.2 for ten species. For several species there are choices of primer sets. The sensitivity and the specificity of these primer sets will vary and depend on the number of species and isolates that they have been tested with. There are also examples where several sets of SCAR primers have been used together in multiplex reactions, which allows several species

Table 4.2. 'Species-specific' primers for *Meloidogyne* identification. See references for species and isolates used for validation of these primers.

Species	Primer set (5′–3′)	Amplicon length	Reference
M. arenaria	TCGGCGATAGAGGTAAATGAC TCGGCGATAGACACTACAACT	420 bp	Zijlstra *et al.*, 2000
	TCGAGGGCATCTAATAAAGG GGGCTGAATAATCAAAGGAA	950 bp	Dong *et al.*, 2001b
M. chitwoodi	CCAATGATAGAGATAGGAAC CTGGCTTCCTCTTGTCCAAA	400 bp	Williamson *et al.*, 1997
	GATCTATGGCAGATGGTATGGA AGCCAAAACAGCGACCGTCTAC	900 bp	Petersen *et al.*, 1997
	TGGAGAGCAGCAGGAGAAAGA GGTCTGAGTGAGGACAAGAGTA	800 bp	Zijlstra, 2000
M. exigua	CATCCGTGCTGTAGCTGCGAG CTCCGTGGGAAGAAAGACTG	562 bp	Randig *et al.*, 2002a
M. fallax	TGGGTAGTGGTCCCACTCTG AGCCAAAACAGCGACCGTCTAC	1100 bp	Petersen *et al.*, 1997
	CCAAACTATCGTAATGCATTATT GGACACAGTAATTCATGAGCTAG	515 bp	Zijlstra, 2000
M. hapla	CAGGCCCTTCCAGCTAAAGA CTTCGTTGGGGAACTGAAGA	960 bp	Williamson *et al.*, 1997
	TGACGGCGGTGAGTGCGA TGACGGCGGTACCTCATAG	610 bp	Zijlstra, 2000
	GGCTGAGCATAGTAGATGATGTT ACCCATTAAAGAGGAGTTTTGC	1500 bp	Dong *et al.*, 2001b
	GGATGGCGTGCTTTCAAC AAAAATCCCCTCGAAAAATCCACC	440 bp	Wishart *et al.*, 2002
M. incognita	CTCTGCCCAATGAGCTGTCC CTCTGCCCTCACATTAGG	1200 bp	Zijlstra *et al.*, 2000
	TAGGCAGTAGGTTGTCGGG CAGATATCTCTGCATTGGTGC	1350 bp	Dong *et al.*, 2001b
	GGGATGTGTAAATGCTCCTG CCCGCTACACCCTCAACTTC	399 bp	Randig *et al.*, 2002a
	GTGAGGATTCAGCTCCCCAG ACGAGGAACATACTTCTCCGTCC	955 bp	Meng *et al.*, 2004
M. javanica	CCTTAATGTCAACACTAGAGCC GGCCTTAACCGACAATTAGA	1650 bp	Dong *et al.*, 2001b
	GGTGCGCGATTGAACTGAGC CAGGCCCTTCAGTGGAACTATAC	670 bp	Zijlstra *et al.*, 2000
	ACGCTAGAATTCGACCCTGG GGTACCAGAAGCAGCCATGC	517 bp	Meng *et al.*, 2004
M. enterolobii	GAAATTGCTTTATTGTTACTAAG TAGCCACAGCAAAATAGTTTTC	322 bp	Blok *et al.*, 2002
M. naasi	CTCTTTATGGAGAATAATCGT CCTCCGCTTACTGATATG	433 bp	Zijlstra *et al.*, 2004
M. paranaensis	GCCCGACTCCATTTGACGGA CCGTCCAGATCCATCGAAGTC	208 bp	Randig *et al.*, 2002b

to be identified in a single reaction (Zijlstra, 2000; Randig *et al.*, 2002b). Interference between primers can be a problem in multiplexing so that specificity is compromised, and usually multiplexing only works with a limited number of primers.

4.3.7 Random amplified polymorphic DNA (RAPD)

Random amplified polymorphic DNAs (RAPDs) have been developed to examine intra- and interspecific relationships of *Meloidogyne* spp. (Blok *et al.*, 1997b), from which SCAR primers for species identification have been developed (see section 4.3.6), and they have been used directly to assist with species identification. Characteristic amplification patterns that are obtained with certain RAPD primers are used to distinguish individuals. Species-specific diagnostic primers are preferred for identification purposes as the relatively high annealing temperatures that are used with species-specific primers enhance their specificity. However, occasionally ambiguous results are obtained with specific primers, possibly due to a polymorphism within the binding site of the primers or a deletion within the amplification region, which leads to an atypical size of amplification product. In these instances RAPDs have been used, even with individual nematodes, to assist with identifications (Adam *et al.*, 2007). Orui (1999) used RAPD amplification with DNA extracted from single J2 or males to distinguish ten *Meloidogyne* spp., and Randig *et al.* (2001) observed stable RAPD profiles from single females and showed that they remained stable for three subsequent generations using DNA equivalent to a quarter of a female nematode in each reaction. Adam *et al.* (2007) also found consistent amplification patterns from individual J2, females and males of *M. javanica* using RAPDs. Obtaining reproducible amplification patterns with RAPDs requires rigorous application of procedures; however, they are useful in certain circumstances.

4.3.8 Other PCR targets

The potential for using RKN pathogenicity and avirulence factors for diagnostic purposes remains largely unexplored and may provide rational bases for deployment of resistance and cropping regimes in the future. For example, the pharyngeal gland protein SEC 1 sequence was used by Tesařová *et al.* (2003) to distinguish *M. incognita* from *M. javanica*, *M. arenaria*, *M. hapla*, *M. chitwoodi* and *M. fallax*, although the molecular bases for the differentiation was not explained.

4.3.9 Real-time PCR

Few examples have been published using real-time PCR for identification and quantification of root-knot nematodes. Increased sensitivity compared with conventional PCR, simultaneous detection of more than one species and the absence of post-PCR processing steps are advantages; however, real-time PCR does require specialized equipment and reagents. The use of probes can increase the specificity of real-time PCR assays, and minor sequence polymorphisms can be exploited with novel chemistries in the probes that maximize sequence discrimination, particularly when size differences in the products cannot be distinguished reliably or when heteroduplex formation may confound interpretations. Applications include ecological studies involving species mixtures or the examination of quarantine samples where closely related species may be present. Zijlstra and van Hoof (2006) reported a real-time multiplex test for *M. chitwoodi* and *M. fallax*, two species that are sympatric and of economic and quarantine importance in a number of countries. Ciancio *et al.* (2005) and Toyota *et al.* (2008) have reported real-time PCR primers for *M. incognita*, and Berry *et al.* (2008) have reported real-time PCR primers for *M. javanica*. Stirling *et al.* (2004) describe the use of real-time PCR to evaluate a risk assessment of *Meloidogyne* spp. damage to tomato using 400 g soil samples; however, primer sequences are not provided.

4.3.10 Microarrays

The potential of microarray technology for diagnosis of plant-parasitic nematodes in complex samples is a new approach being developed. The principle has already been demonstrated for the detection of human and plant pathogens with oligonucleotide spotted arrays. Microarrays can circumvent some of the limitations of multiplex PCR where several optimized primer sets are used in a reaction and interference/competition/loss in specificity in the amplification reactions, as well as problems in discriminating the products, can be problematic. An attraction of microarrays is the potential to monitor a large number of possible targets simultaneously, a feature that is important for plant protection organizations with responsibility for many different organisms, as well as for those conducting ecological studies involving complex communities. The specificity of microarrays is dependent on unique signature sequences being available for each species.

However, the large number of sequences (probes for specific targets) that can be screened simultaneously allows for more than one capture probe to be used for each species, thus increasing the confidence in the results. Improvements have been made in the sensitivity and specificity of arrays with shorter and more sequence-specific oligonucleotides, as well as in the chemistries of the probes. Major issues that remain are the amplification of unknowns from complex samples and the non-expected behaviour of some probes, in which sequences that have mismatches with the target hybridize better than the perfectly matched target (Frederique Pasquer, personal communication), as well as cost. Examples of published microarray results that include *Meloidogyne* spp. are still limited (Szemes *et al.*, 2005; François *et al.*, 2006; van Doorn *et al.*, 2007), but they illustrate the potential of the technology. To obtain the sensitivity that is required for the detection of nematodes, amplification of the target DNA of the nematode is necessary. Multiplex amplification strategies involve either amplification with generic primers or multiple primer sets that target a genomic region containing species-specific information (to be recognized on the microarray); both approaches face serious limitations. Targeting a conserved genome region limits the analysis to a taxonomically defined group, while combining several primer sets may present a significant technical challenge. François *et al.* (2006) generated PCR products from *M. chitwoodi* using specific primers labelled with Cyanine 3 or Cyanine 5 fluorescent dyes, and hybridized them overnight to the microarray. They were able to detect *M. chitwoodi* in pure and mixed samples (i.e. when *M. chitwoodi* DNA was mixed with DNA from a congeneric nematode species), and they found that simultaneous hybridization of the microarray with two amplified targets labelled with different dyes gave no significant competition between the targets. Padlock probes offer a means of combining pathogen-specific molecular recognition and universal amplification. In combination with a microarray, padlock probe technology has been shown to enable the sensitive simultaneous detection of ten different plant pathogens, among them *M. hapla* (Szemes *et al.*, 2005). More recently, a similar approach using OpenArrays enabled the quantitative multiplex detection of 13 plant pathogens, among them *M. hapla* (van Doorn *et al.*, 2007). Microarrays do offer the possibility of a uniform and standardized detection system for a wide range of pathogens, and further developments are expected in the future.

4.4 Conclusions and Future Directions

Biochemical and molecular methods for identification of *Meloidogyne* spp. are now widely used and, in some cases, essential for species diagnosis. They cannot, however, be used with confidence to identify all *Meloidogyne* spp. A clear understanding of species boundaries and adequate sampling of known species across their geographic range are lacking (see Adams *et al.*, Chapter 5, this volume). Particularly noteworthy are the recent conclusions of Lunt (2008), which strongly suggest that the tropical apomictic *Meloidogyne* species result from interspecific hybridizations; this is also indicated in the genome sequence of *M. incognita* (Abad *et al.*, 2008). Depending on the nature of the interspecific hybidization and the parental species involved, these hybrids pose special difficulties for diagnostics based on single genetic loci. Several species, such as *M. chitwoodi* and *M. enterolobii*, are well characterized by multiple genetic markers and have been sampled across much of their known range. Other species, such as *M. floridensis* and *M. fallax*, have been characterized molecularly but are currently known from a relatively limited geographic region. The so-called major species – *M. arenaria*, *M. hapla*, *M. incognita* and *M. javanica* – have been extensively studied biochemically and molecularly, resulting in an increasingly large set of 'atypical' diagnostic characters (see above). The unexpectedly high levels of intraspecific variation within the clonal, mitotically parthenogenetic species mirror their observed cytological and physiological variation (Castagnone-Sereno, 2006). *Meloidogyne hapla* is readily differentiated morphologically and molecularly from the other three species; none the less, documentation of nuclear and mitochondrial variation in *M. hapla* has steadily accumulated (Peloquin *et al.*, 1993; Piotte *et al.*, 1995; Hugall *et al.*, 1997; Handoo *et al.*, 2005; Powers *et al.*, 2005). In addition to the continued evaluation of intraspecfic variation, there is a pressing need to incorporate newly discovered tropical and Asian species of *Meloidogyne* into current identification protocols (De Waele and Elsen, 2007). Validation and adaptation of these methods for different geographic regions and different working conditions is a challenging goal. To date, most studies employing molecular diagnostic methods have been conducted at academic or national institutions, and few large-scale surveys, such as that conducted by Powers *et al.* (2005), have employed molecular diagnostics and taken the theory into practice. If routine use of molecular identification to meet regulatory demands or to enhance management decisions is a goal of diagnostics, then it will be necessary to emphasize methods that are robust, reliable and inexpensive. Given the current concerns in relation to climate change, food security and the global transport of agricultural commodities, the use of diagnostics for *Meloidogyne* spp. is highly relevant.

4.5 Acknowledgements

The authors acknowledge the assistance of Drs Janet Brito and Regina Carneiro in the collation of the isozyme data, Rebecca S. Higgins for generating the mtDNA figure, and Keith Davies, Frédérique Pasquer and Carolien Zijlstra for assistance in the preparation of this chapter.

4.6 References

Abad, P., Gouzy, J., Aury, J.-M. *et al.* (2008) Genome sequence of the metazoan plant-parasitic nematode *Meloidogyne incognita. Nature Biotechnology* 26, 909–915.

Adam, M.A.M., Phillips, M.S. and Blok, V.C. (2005) Identification of *Meloidogyne* spp. from north east Libya and comparison of their inter- and intro-specific genetic variation using RAPDs. *Nematology* 7, 599–609.

Adam, M.A.M., Phillips, M.S. and Blok, V.C. (2007) Molecular diagnostic key for identification for single juveniles of seven common and economically important species of root-knot nematode (*Meloidogyne* spp.). *Plant Pathology* 56, 190–197.

Baojun, Y., Kaiji, H., Hui, C. and Weisheng, Z. (1990). A new species of root-knot nematode *Meloidogyne jianyangensis* n. sp. parasitizing mandarin orange. *Acta Phytopathologica Sinica* 20, 259–264.

Baum, T.J., Lewis, S.A. and Dean, R.A. (1994) Isolation, characterization and application of DNA probes specific to *Meloidogyne arenaria. Phytopathology* 84, 489–494.

Bergé, J.-B. and Dalmasso, A. (1975) Biochemical characteristics of several populations *of Meloidogyne hapla* and *Meloidogyne* spp. *Cahiers O.R.S.T.O.M. Serie Biologie, Nematologie* 10, 263–271.

Berry, S.D., Fargette, M., Spaull, V.W., Morand, S. and Cadet, P. (2008) Detection and qunatification of root-knot nematode (*Meloidogyne javanica*), lesion nematode (*Pratylenchus zeae*) and adagger nematode (*Xiphinema elongatum*) parasites of sugarcane using real-time PCR. *Molecular and Cellular Probes* 22, 168–176.

Blok, V.C., Phillips, M.S. and Fargette, M. (1997a) Comparison of sequences from the ribosomal DNA intergenic region of *Meloidogyne mayaguensis* and other major tropical root knot nematodes. *Journal of Nematology* 29, 16–22.

Blok, V.C., Phillips, M.S., McNicol, J.W. and Fargette, M. (1997b) Genetic variation in tropical *Meloidogyne* spp. as shown by RAPDs. *Fundamental and Applied Nematology* 20, 127–133.

Blok, V.C., Wishart, J., Fargette, M., Berthier, K. and Phillips, M.S. (2002) Mitochondrial differences distinguishing *Meloidogyne mayaguensis* from the major species of tropical root-knot nematodes. *Nematology* 4, 773–781.

Brito, J., Powers, T.O., Mullin, P.G., Inserra, R.N. and Dickson, D.W. (2004) Morphological and molecular characterisation of *Meloidogyne mayaguensis* isolates from Florida. *Journal of Nematology* 36, 232–240.

Brito, J.A., Kaur, R., Cetintas, R., Stanley, J.D., Mendes, M.L., McAvoy, E.J., Powers, T.O. and Dickson, D.W. (2008) Identification and isozyme characterization of *Meloidogyne* spp. infecting horticultural and agronomic crops, and weed plants in Florida. *Nematology* 10, 757–766.

Brown, W.M., George, M. Jr and Wison, A.C. (1979) Rapid evolution of animal mitochondrial DNA. *Proceedings of the National Academy of Sciences of the USA* 76, 1967–1971.

Carneiro, R.M.D.G., Almeida, M.R.A. and Carneiro, R.G. (1996a) Enzyme phenotypes of Brazilian populations of *Meloidogyne* spp. *Fundamental and Applied Nematology* 19, 555–560.

Carneiro, R.M.D.G., Carneiro, R.G., Abrantes, I.M.O., Santos, M.S.N.A. and Almeida, M.R.A. (1996b) *Meloidogyne paranaensis* n. sp. (Nemata: Meloidogynidae), a root-knot nematode parasitizing coffee in Brazil. *Journal of Nematology* 28, 177–189.

Carneiro, R.M.D.G., Almeida, M.R.A. and Quénéhervé, P. (2000) Enzyme phenotypes of *Meloidogyne* spp. populations. *Nematology* 2, 645–654.

Carneiro, R.M.D.G., Tigano, M.S., Randig, O., Almeida, M.R.A. and Sarah, J.L. (2004a) Identification and genetic diversity of *Meloidogyne* spp. (Tylenchida: Meloidogynidae) on coffee from Brazil, Central America and Hawaii. *Nematology* 6, 287–298.

Carneiro, R.M.D.G., Ritta, M., Almeida, A. and Gomes, A.C.M.M. (2004b) First record of *Meloidogyne hispanica* Hirschmann, 1986 on squash in state of Bahia, Brazil. *Nematologia Brasileira* 28, 215–218.

Carneiro, R.M.D.G., Almeido, M.R.A., Gomes, A.C.M.M. and Hernandez, A. (2005a) *Meloidogyne izalcoensis* n. sp. (Nematoda: Meloidogynidae), a root-knot nematode parasitizing coffee in El Salvador. *Nematology* 7, 819–832.

Carneiro, R.M.D.G., Randig, O., Almeida, M.R.A. and Gonçalves, W. (2005b) Identification and characterization of *Meloidogyne* species on coffee from São Paulo and Minas Gerais states of Brazil using esterase phenotypes and SCAR-PCR multiplex. *Nematologia Brasileira* 29, 233–241.

Carneiro, R.M.D.G., Almeida, M.R.A., Cofcewicz, E.T., Magunacelaya, J.C. and Aballay, E. (2007) *Meloidogyne ethiopica*, a major root-knot nematode parasitising *Vitis vinifera* and other crops in Chile. *Nematology* 9, 635–641.

Carneiro, R.M.D.G., Mendes, M. De L., Almeida, M.R.A., dos Santos, M.F.A., Gomes, A.C.M.M. and Karssen, G. (2008) Additional information on *Meloidogyne inornata* Lordello, 1056 (Tylenchida: Meloidogynidae) and its characterisation as a valid species. *Nematology* 10, 123–136.

Castagnone-Sereno, P. (2006) Genetic variability and adaptive evolution in parthenogenetic root-knot nematodes. *Heredity* 96, 282–289.

Castagnone-Sereno, P., Piotte, C., Abad, P., Bongiovanni, M. and Dalmasso, A. (1991) Isolation of a repeated DNA probe showing polymorphism among *Meloidogyne incognita* populations. *Journal of Nematology* 23, 316–320.

Castagnone-Sereno, P., Esparrago, G., Abad, P., Leroy, F. and Bongiovanni, M. (1995) Satellite DNA as a target for PCR-specific detection of the plant-parasitic nematode *Meloidogyne hapla*. *Current Genetics* 28, 566–570.

Castagnone-Sereno, P., Semblat, J.P., Leroy, F. and Abad, P. (1998) A new *Alu*I satellite DNA family in the root-knot nematode *Meloidogyne fallax*: relationships with satellites from the sympatric species *M. hapla* and *M. chitwoodi*. *Molecular Biology and Evolution* 15, 1115–1122.

Castagnone-Sereno, P., Leroy, F., Bongiovanni, M., Zijlstra, C. and Abad, P. (1999) Specific diagnosis of two root-knot nematodes, *Meloidogyne chitwoodi* and *M. fallax*, with satellite DNA probes. *Phytopathology* 89, 380–384.

Castagnone-Sereno, P., Leroy, F. and Abad, P. (2000) Cloning and characterization of an extremely conserved satellite DNA family from the root-knot nematode *Meloidogyne arenaria*. *Genome* 43, 346–353.

Castillo, P., Vovlas, N., Subbotin, S. and Troccoli, A. (2003) A new root-knot nematode, *Meloidogyne baetica* n. sp. (Nematoda: Heteroderidae), parasitizing wild olive in southern Spain. *Phytopathology* 93, 1093–1102.

Castro, J.M D.C.E., De Lima, R.D. and Carneiro, R.M.D.G. (2003) Isoenzymatic variability in Brazilian populations of *Meloidogyne* spp. from soybean. *Nematologia Brasileira* 27, 1–12.

Cenis, J.L., Opperman, C.H. and Triantaphyllou, A.C. (1992) Cytogenetic, enzymatic and restriction fragment length polymorphism variation of *Meloidogyne* spp. from Spain. *Phytopathology* 82, 527–531.

Charchar, J.M., Eisenback, J.D. and Hirschmann, H. (1999) *Meloidogyne petuniae* n. sp. (Nemata: Meloidogynidae), a root-knot nematode parasitic on petunia in Brazil. *Journal of Nematology* 31, 81–91.

Chen, Q., Robertson, L., Jones, J.T., Blok, V.C., Phillips, M.S. and Brown, D.J.F. (2001) Capture of nematodes using antiserum and lectin-coated magnetised beads. *Nematology* 3, 593–601.

Chen, Y.F., Wu, J.Y., Hu, X.O. and Yu, S.F. (1998) Using PhastSystem for rapid identification of root-knot nematodes. *Acta Phytopathology Sinica* 28, 73–77.

Ciancio, A., Loffredo, A., Paradies, F., Turturo, C. and Sialer, M.F. (2005) Detection of *Meloidogyne incognita* and *Pochonia chlamydosporia* by fluorogenic molecular probes. *EPPO Bulletin* 35, 157–164.

Cofcewick, E.T., Carneiro, R.M.D., Randing, O., Chabrier, C. and Quénéhervé, P. (2005) Diversity of *Meloidogyne* spp. on *Musa* in Martinique, Guadeloup, and French Guiana. *Journal of Nematology* 37, 313–322.

Curran, J. and Webster, J.M. (1987) Identification of nematodes using restriction fragment length differences and species-specific DNA probes. *Canadian Journal of Plant Pathology* 9, 162–166.

Curran, J., Baillie, D.L. and Webster, J.M. (1985) Use of genomic DNA restriction fragment length differences to identify nematode species. *Parasitology* 90, 137–144.

Curran, J., McClure, M.A. and Webster, J.M. (1986) Genotypic differentiation of *Meloidogyne* populations by detection of restriction fragment length difference in total DNA. *Journal of Nematology* 18, 83–86.

Dalmasso, A. and Bergé, J.-B. (1978) Molecular polymorphism and phylogenetic relationship in some *Meloidogyne* spp.: application to the taxonomy of *Meloidogyne. Journal of Nematology* 10, 323–332.

Davies, K.G., Curtis, R.H. and Evans, K. (1996) Serologically based diagnostic and quantification tests for nematodes. *Pesticide Science* 47, 81–87.

De Waele, D. and Elsen, A. (2007) Challenges in tropical plant nematology. *Annual Review of Phytopathology* 45, 457–485.

Dong, K., Dean, R.A., Fortnum, B.A. and Lewis, S.A. (2001a) A species-specific DNA probe for the identification of *Meloidogyne hapla. Nematropica* 31, 17–23.

Dong, K., Dean, R.A., Fortnum, B.A. and Lewis, S.A. (2001b) Development of PCR primers to identify species of root-knot nematodes: *Meloidogyne arenaria, M. hapla, M. incognita* and *M. javanica. Nematropica* 31, 271–280.

Donn, S., Griffiths, B.S., Neilson, R. and Daniell, T.J. (2008) DNA extraction from soil nematodes for multi-sample community studies. *Applied Soil Ecology* 38, 20–26.

Eisenbeck, J.D., Bernard, E.C. and Schmitt, D.P. (1994) Description of the Kona coffee root-knot nematode, *Meloidogyne konaensis* n. sp. *Journal of Nematology* 26, 363–374.

Eisenback, J.D., Bernard, E.C., Starr, J.L., Lee, T.A. Jr and Tomaszewski, E.K. (2003) *Meloidogyne haplanaria* n. sp. (Nematoda: Meloidogynidae), a root-knot nematode parasitizing peanut in Texas. *Journal of Nematology* 35, 395–403.

Esbenshade, P.R. and Triantaphyllou, A.C. (1985) Use of enzyme phenotypes for identification of *Meloidogyne* species. *Journal of Nematology* 17, 6–20.

Esbenshade, P.R. and Triantaphyllou, A.C. (1987) Enzymatic relationships and evolution in the genus *Meloidogyne* (Nematoda: Tylenchida). *Journal of Nematology* 19, 8–18.

Esbenshade, P.R. and Triantaphyllou, A.C. (1990) Isozyme phenotypes for the identification of *Meloidogyne* species. *Journal of Nematology* 22, 10–15.

Fargette, M. (1978) Use of the esterase phenotype in the taxonomy of the genus *Meloidogyne*. 2. Esterase phenotypes observed in West African populations and their characterisation. *Revue de Nématologie* 10, 45–56.

Fargette, M. and Braaksma, R. (1990) Use of the esterase phenotype in taxonomy of the genus *Meloidogyne*. 3. A study of some 'B' race lines and their taxonomic position. *Revue de Nématologie* 13, 375–386.

Fargette, M., Phillips, M.S., Blok, V.C., Waugh, R. and Trudgill, D.L. (1996) An RFLP study of relationships between species, populations and resistance-breaking lines of tropical species of *Meloidogyne. Fundamental and Applied Nematology* 19,193–200.

François, C., Kebdani, N., Barker, I., Tomlinson, J., Boonham, N. and Castagnone-Sereno, P. (2006) Towards specific diagnosis of plant-parasitic nematodes using DNA oligonucleotide microarray technology: a case study with the quarantine species *Meloidogyne chitwoodi. Molecular and Cellular Probes* 20, 64–69.

Gárate, T., Robinson, M.P., Chacón, M.R. and Parkhouse, R.M.E. (1991) Characterization of species and races of the genus *Meloidogyne* by DNA restriction enzyme analysis. *Journal of Nematology* 23, 414–420.

Hajibabaei, M., Singer, G.A.C., Hebert, P.D.N. and Hickey, D.A. (2007) DNA barcoding: how it complements taxonomy, molecular phylogenetics and population genetics. *Trends in Genetics* 23, 167–172.

Handoo, Z.A., Nyczepir, A.P., Esmenjaud, D., van der Beek, J.G., Castagnone-Sereno, P., Carta, L.K., Skantar, A.M. and Higgins, J.A. (2004) Morphological, molecular, and differential-host characterization of *Meloidogyne floridensis* n. sp. (Nematoda: Meloidogynidae), a root-knot nematode parasitizing peach in Florida. *Journal of Nematology* 36, 20–35

Handoo, Z.A., Skantar, A.M., Carta, L.K. and Schmitt, D.P. (2005) Morphological and molecular evaluation of a *Meloidogyne hapla* population damaging coffee (*Coffea arabica* L.) in Maui, Hawaii. *Journal of Nematology* 37, 136–145.

Helder, J., Karssen, G., Van Den, E.S.J.J., Holterman, M.H.M., Veenhuizen, P.T.M., Landeweert, R., Hekman, H. and Bakker, J. (2008) Methods of detecting root knot nematodes. www.freepatentsonline.com/EP1878803.html (1/16/08).

Hernandez, A., Fargette, M. and Sarah, J.L. (2004) Characterisation of *Meloidogyne* spp. (Tylenchida: Meloidogynidae) from coffee plantations in Central America and Brazil. *Nematology* 6, 193–204.

Hervé, G., Bertand, B., Villain, L., Licardie, D. and Cilas, C. (2005) Distribution analyses of *Meloidogyne* spp. and *Pratylenchus coffeae sensu lato* in coffee plots in Costa Rica and Guatemala. *Plant Pathology* 54, 471–475.

Hiatt, E.E., Georgi, L., Huston, S., Harshman, D.C., Lewis, S.A. and Abbott, A.G. (1995) Intra- and inter-population genome variation in *Meloidogyne arenaria*. *Journal of Nematology* 27, 143–152.

Hirschmann, H. (1985) The genus *Meloidogyne* and morphological characters differentiating its species. In: Sasser, J.N. and Carter, C.C. (eds) *An Advanced Treatise on* Meloidogyne. *Volume 1, Biology and Control*. North Carolina State University Graphics, Raleigh, North Carolina, pp. 79–93.

Hu, M. and Gasser, R.B. (2006) Mitochondrial genomes of parasitic nematodes – progress and perspectives. *Trends in Parasitology* 22, 78–84.

Hugall, A., Moritz, C., Stanton, J. and Wolstenholme, D.R. (1994) Low, but strongly structured mitochondrial DNA diversity in root-knot nematodes (*Meloidogyne*). *Genetics* 136, 903–912.

Hugall, A., Stanton, J. and Moritz, C. (1997) Evolution of the AT-rich mitochondrial DNA of the root knot nematode, *Meloidogyne hapla*. *Molecular and Biological Evolution* 14, 40–48.

Hugall, A., Stanton, J. and Moritz, C. (1999) Reticulate evolution and the origins of ribosomal internal transcribed spacer diversity in apomictic *Meloidogyne*. *Molecular and Biological Evolution* 16, 157–164.

Hyman, B.C. and Whipple, L.E. (1996) Application of mitochondrial DNA polymorphism to *Meloidogyne* molecular population biology. *Journal of Nematology* 20, 268–276.

Ibrahim, S.K. and Perry, R.N. (1992) Use of esterase patterns of females and galled roots for the identification of species of Meloidogyne. *Fundamental and Applied Nematology* 16, 187–190.

Ibrahim, S.K., Davies, K.G. and Perry, R.N. (1996) Identification of the root-knot nematode, *Meloidogyne incognita*, using monoclonal antibodies raised to non-specific esterases, *Physiological and Molecular Plant Pathology* 49, 79–88.

Janati A., Bergé, J.-B., Triantaphyllou, A.C. and Dalmasso, A. (1982) Nouvelles données sur l'utilisation des isoestérases pour l'identification des *Meloidogyne*. *Revue de Nématologie* 5, 147–154.

Jepson, S.B. (1987) *Identification of Root-knot Nematodes* (Meloidogyne *species*). CAB International Publishing, Wallingford, UK.

Jeyaprakash, A., Tigano, M.S., Brito, J., Carneiro, R.M.D.G. and Dickson, D.W. (2006) Differentiation of *Meloidogyne floridensis* from *M. arenaria* using high-fidelity PCR amplified mitochondrial AT-rich sequences. *Nematropica* 36, 1–12.

Karssen, G. (2002) *The Plant-parasitic Nematode Genus* Meloidogyne *Göldi, 1892 (Tylenchida) in Europe*. Brill Academic Publishers, Leiden, The Netherlands.

Karssen, G. and van Hoenselaar, T. (1998) Revision of the genus *Meloidogyne* Göldi, 1892 (Nematoda: Heteroderidae) in Europe. *Nematologica* 44, 713–788.

Karssen, G., van Hoenselaar, T., Verkerk-Bakker, B. and Janssen, R. (1995) Species identification of cyst and root-knot nematodes from potato by electrophoresis of individual females. *Electrophoresis* 16,105–109.

Karssen, G., Bolk, R.J., van Aelst, A.C., van den Beld, I., Kox, L.F.F., Korthals, G., Molendijk, L., Zijlstra, C., van Hoof, R. and Cook, R. (2004) Description of *Meloidogyne minor* n. sp. (Nematoda: Meloidogynidae), a root-knot nematode associated with yellow patch disease in golf courses. *Nematology* 6, 59–72.

Landa, B.B., Palomares Rius, J.E., Vovlas, N., Carneiro, R.M.D.G., Maleita, C.M.N. and Abrantes, I.M. de O. (2008) Molecular characterization of *Meloidogyne hispanica* (Nematoda, Meloidogynidae) by phylogenetic analysis of genes within the rDNA in *Meloidogyne* spp. *Plant Disease* 92, 1104–1110.

Liao, J., Yang, W., Feng, Z. and Karssen, G. (2005). Description of *Meloidogyne panyuensis* sp. n. (Nematoda: Meloidogynidae), parasitic on peanut (*Arachis hypogea* L.) in China. *Russian Journal of Nematology* 13, 107–114.

Lima, I.M., Souza, R.M., Silva, C.P. and Carneiro, R.M.D.G. (2005) *Meloidogyne* spp. from preserved areas of Atlantic forest in the state of Rio de Janeiro, Brazil. *Nematologia Brasileira* 29, 31–38.

Lunt, D.H. (2008) Genetic tests of ancient asexuality in root knot nematodes reveal recent hybrid origins. *BMC Evolutionary Biology* 8, 194.

Lunt, D.H., Whipple, L.E. and Hyman, B.C. (2002) Mitochondrial DNA variable number tandem repeats (VNTRs): utility and problems in molecular ecology. *Molecular Ecology* 7, 1441–1455.

Medina, I.L., Gomes, C.B., Rossi, C. and Carneiro, R.M.D.G. (2007) Characterization and identification of root-knot nematodes (*Meloidogyne* spp.) from fig (*Ficus carica* L.) trees in Rio Grande do Sul and São Paulo states of Brazil. *Nematologia Brasileira* 30, 179–187.

Meng, Q.-P., Long, H. and Xu, J.H. (2004) PCR assays for rapid and sensitive identification of three major root-knot nematodes, *Meloidogyne incognita*, *M. javanica* and *M. arenaria*. *Acta Phytopathologcia Sinica* 34, 204–210.

Meštrović, N., Randig, O., Abad, P., Plohl, M. and Castagnone-Sereno, P. (2005) Conserved and variable domains in satellite DNAs of mitotic parthenogenetic root-knot nematode species. *Gene* 362, 44–50.
Meštrović, N., Catagnone-Sereno, P. and Plohl, M (2006) High conservation of the differentially amplified MPA2 satellie DNA family in parthenogenetic root-knot nematodes. *Gene* 376, 260–267.
Molinari, S. (2001) Polymorphism of esterase isozyme zymograms of *Meloidogyne* populations detected by PhastSystem. *Nematologica Mediterranea* 29, 63–66.
Molinari, S., Lamberti, F., Crozzoli, R., Sharma, S.B. and Sánchez Portales, L. (2005) Isozyme patterns of exotic *Meloidogyne* spp. populations. *Nematologia Mediterranea* 33, 61–65.
Moritz, C. and Cicero, C. (2004) DNA barcoding: promise and pitfalls. *PLoS Biology* 10, 1529–1531.
Nieberding, C., Libois, R., Douady, C.J., Morand, S. and Michaux, J.R. (2005) Phylogeography of a nematode (*Heligmosomoides polygyrus*) in the Western Palearctic region: persistence of northern cryptic populations during ice ages? *Molecular Ecology* 14, 765–779.
Oka, Y., Karssen, G. and Mor, M. (2003) Identification, host range and infection process of *Meloidogyne marylandi* from turf grass in Israel. *Nematology* 5, 727–734.
Okimoto, R., Chamberlin, H.M., Macfarlane, J.L. and Wolstenholme, D.R. (1991) Repeated sequence sets in mitochondrial DNA molecules of root knot nematodes (*Meloidogyne*): nucleotide sequences genome location and potential for host-race identification. *Nucleic Acids Research* 19, 1619–1626.
Orui, Y. (1999) Species identification of *Meloidogyne* spp. (Nematoda: Meloidogynidae) in Japan by random amplified polymorphic DNA (RAPD-PCR). *Japanese Journal of Nematology* 29, 7–15.
Pais, C.S. and Abrantes, I.M. De O. (1989) Esterase and malate dehydrogenase phenotypes for identification of *Meloidogyne* species. *Journal of Nematology* 21, 342–346.
Palomares Rius, J.E., Volvas, N., Troccoli, A., Liébanas, G., Land, B.B. and Castillo, P. (2007) A new root-knot nematode parasitizing sea rocket from Spanish Mediterranean coastal dunes: *Meloidogyne dunensis* n. sp. (Nematoda: Meloidogynidae). *Journal of Nematology* 39, 190–202.
Peloquin, J.J., Bird, M.D., Kaloshian, I. and Matthews, C. (1993) Isolates of *Meloidogyne hapla* with distinct mitochondrial genomes. *Journal of Nematology* 25, 239–243.
Petersen, D.J., Zijlstra, C., Wishart, J., Blok, V. and Vrain, T.C. (1997) Specific probes efficiently distinguish root-knot nematode species using signature sequences in the ribosomal intergenic spacer. *Fundamental and Applied Nematology* 20, 619–626.
Piotte, C., Castagnone-Sereno, P., Uijthof, J., Abad, P., Bongiovanni, M. and Dalmasso, A. (1992) Molecular characterization of species and populations of *Meloidogyne* from various geographic origins with repeated-DNA homologous probes. *Fundamental and Applied Nematology* 15, 271–276.
Piotte, C., Castagnone-Sereno, P., Bongiovanni, M., Dalmasso, A. and Abad, P. (1994) Cloning and characterization of two satellite DNAs in the low-C-value genome of *Meloidogyne* spp. *Gene* 138,175–180.
Piotte, C., Castagnone-Sereno, P., Bongiovanni, M., Dalmasso, A. and Abad, P. (1995) Analysis of a satellite DNA from *Meloidogyne hapla* and its use as a diagnostic probe. *Phytopathology* 85, 458–462.
Powers, T. (2004) Nematode molecular diagnostics: from bands to barcodes. *Annual Review of Phytopathology* 42, 367–383.
Powers, T.O. and Harris, T.S. (1993) A polymerase chain reaction for the identification of five major *Meloidogyne* species. *Journal of Nematology* 25, 1–6.
Powers, T.O., Todd, T.C., Burnell, A.M., Murray, P.C.B., Fleming, C.C., Szalanski, A.L., Adams, B.A. and Harris, T.S. (1997) The rDNA internal transcribed spacer region as a taxonomic marker for nematodes. *Journal of Nematology* 29, 441–450.
Powers, T.O., Mullin, P.G., Harris, T.S., Sutton, L.A. and Higgins, R.S. (2005) Incorporating molecular identification of *Meloidogyne* spp. into large-scale regional nematode survey. *Journal of Nematology* 37, 226–235.
Rammah, A. and Hirschmann, H. (1988) *Meloidogyne mayaguensis* n. sp. (Meloidogynidae), a root-knot nematode from Puerto-Rico. *Journal of Nematology* 20, 65–68.
Rammah, A. and Hirschmann, H. (1990) *Meloidogyne morocciensis* n.sp. (Meloidogyninae), a root-knot nematode from Morocco. *Jouranl of Nematology* 22, 279–292.
Randig, O., Leroy, F., Bongiovanni, M. and Castagnone-Sereno, P. (2001) RAPD characterisation for single females of the root-knot nematodes, *Meloidogyne* spp. *European Journal of Plant Pathology* 107, 639–643.
Randig, O., Bongiovanni, M., Carneiro, R.M.D.G., Sarah, J.-L. and Castagnone-Sereno, P. (2002a) A species-specific satellite DNA family in the genome of the coffee root-knot nematode *Meloidogyne exigua*: application to molecular diagnostics of the parasite. *Molecular Plant Pathology* 3, 431–437.

Randig, O., Bongiovanni, M., Carneiro, R.M.D.G. and Castagnone-Sereno, P. (2002b) Genetic diversity of root-knot nematodes from Brazil and development of SCAR markers specific for the coffee-damaging species. *Genome* 45, 862–870.

Rubinoff, D. and Holland, B.S. (2005) Between two extremes: mitochondrial DNA is neither the panacea nor the nemesis of phylogenetic and taxonomic inference. *Systematic Biology* 54, 952–961.

Schmitz, B., Burgermeister, W. and Braasch, H. (1998) Molecular genetic classification of Central European *Meloidogyne chitwoodi* and *M. fallax*. *Nachrichtenblatt des Deutschen Pflanzenschutzdienstes* 50, 310–317.

Sipes, B.S., Schmitt, D.P., Xu, K. and Serracin, M. (2005) Esterase polymorphism in *Meloidogyne konaensis*. *Journal of Nematology* 37, 438–443.

Skantar, A.M. and Carta, L.K. (2005) Multiple displacement amplification (MDA) of total genomic DNA from *Meloidogyne* spp. and comparison to crude DNA extracts in PCR of ITS1, 28S D2-D3 rDNA and Hsp90. *Nematology* 7, 285–293.

Stanton, J.M., McNicol, C.D. and Steele, V. (1998) Non-manual lysis of second-stage *Meloidogyne* juveniles for identification of pure and mixed samples based on the polymerase chain reaction. *Australian Plant Pathology* 27, 112–115.

Stirling, G.R., Griffin, D., Ophel-Keller, K., McJay, A., Hartley, D., Curran, J., Stirling, A.M., Monsour, C., Winch, J. and Hardie, B. (2004) Combining an intial risk assessment process with DNA assays to improve prediction of soilborne diseases caused by root-knot nematode (*Meloidogyne* spp.) and *Fusarium oxysporum* f. sp. *lycopersici* in the Queensland tomato industry. *Australasian Plant Pathology* 33, 285–293.

Szemes, M., Bonants, P., de Weerdt, M., Baner, J., Landegren, U. and Schoen, C.D. (2005) Diagnostic application of padlock probes – multiplex detection of plant pathogens using universal microarrays. *Nucleic Acids Research* 33, e70.

Tastet, C., Val, F., Lesage, M., Renault, L., Marché, L., Bossis, M. and Mugniéry, D. (2001) Application of a putative fatty-acid binding protein to discriminate serologically the two European quarantine root-knot nematodes, *Meloidogyne chitwoodi* and *M. fallax*, from other *Meloidogyne* species. *European Journal of Plant Pathology* 107, 821–832.

Tesařová, B., Zouhar, M. and Ryšánek, P. (2003) Development of PCR for specific determination of root-knot nematode *Meloidogyne incognita*. *Plant Protection Sciences* 39, 23–28.

Tigano, M.S., Carneiro, R.M.D.G., Jeyaprakash, A., Dickson, D.W. and Adams, B.J. (2005) Phylogeny of *Meloidogyne* spp. based on 18S rDNA and the intergenic region of mitochondrial DNA sequences. *Nematology* 7, 851–862.

Tomaszewski, E.K., Khalil, M.A.M., El-Deeb, A.A., Powers, T.O. and Starr, J.L. (1994) *Meloidogyne javanica* parasitic on peanut. *Journal of Nematology* 26, 436–441.

Toyota, K., Shirakashi, T., Sato, E. and Wada, S. (2008) Development of a real-time PCR method for the potato-cyst nematode *Globodera rostochiensis* and the root-knot nematode *Meloidogyne incognita*. *Soil Science and Plant Nutrition* 54, 72–76.

van der Beek, J.G. and Karssen, G. (1997) Interspecific hybridisation of meiotic parthenogenetic *Meloidogyne chitwoodi* and *M. fallax*. *Phytopathology* 87, 1061–1066.

van Doorn, R., Szemes, M., Bonants, P., Kowalchuk, G.A., Salles, J.F., Ortenberg, E. and Schoen, C.D. (2007) Quantitative multiplex detection of plant pathogens using a novel ligation probe-base system coupled with universal, high-throughput real-time PCR and OpenArrays. *BMC Genomics* 8, 276.

Vogler, A.P. and Monaghan, M.T. (2007) Recent advances in DNA taxonomy. *Journal Zoological Systematics and Evolutionary Research* 45, 1–10.

Vrain, T.C., Wakarchuk, D.A., Levesque, A.C. and Hamilton, R.L. (1992) Intraspecific rDNA restriction fragment length polymorphism in the *Xiphinema americanum* group. *Fundamental and Applied Nematology* 15, 563–573.

Williamson, V.M., Caswell Chen, E.P., Westerdahl, B.B., Wu, F.F. and Caryl, G. (1997) A PCR assay to identify and distinguish single juveniles of *Meloidogyne hapla* and *M. chitwoodi*. *Journal of Nematology* 29, 9–15.

Wishart, J., Phillips, M.S. and Blok, V.C. (2002) Ribosomal intergenic spacer: a PCR diagnostic for *Meloidogyne chitwoodi*, *M. fallax* and *M. hapla*. *Phytopathology* 92, 884–892.

Xu, J., Liu, P., Meng, Q. and Long, H. (2004) Characterisation of *Meloidogyne* species from China using isozyme phenotypes and amplified mitochondrial DNA restriction fragment length polymorphism. *European Journal of Plant Protection* 110, 309–315.

Xue B., Baillie D.L., Beckenbach, K. and Webster, J.M. (1992) DNA hybridisation probes for studying the affinities of three *Meloidogyne* populations. *Fundamental and Applied Nematology* 15, 35–41.

Yang, B. and Eisenback, J.D. (1983) *Meloidogyne enterolobii* n. sp. (Meloidogynidae), a root-knot nematode parasitizing pacara earpod tree in China. *Journal of Nematology* 15, 381–391.

Ye, W., Giblin-Davis, R.M., Braasch, H., Morris, K. and Thomas, W.K. (2007) Phylogenetic relationships among *Bursaphelenchus* species (Nematoda: Parasitaphelenchidae) inferred from nuclear ribosomal and mitochondrial DNA sequence data. *Molecular and Phylogenetic Evolution* 43, 1185–1197.

Zijlstra, C. (1997) A fast PCR assay to identify *Meloidogyne hapla, M. chitwoodi* and *M. fallax* and to sensitively differentiate them from each other and from *M. incognita* in mixtures. *Fundamental and Applied Nematology* 20, 505–511.

Zijlstra, C. (2000) Identification of *Meloidogyne chitwoodi, M. fallax* and *M. hapla* based on SCAR-PCR: a powerful way of enabling reliable identification of populations or individuals that share common traits. *European Journal of Plant Pathology* 106, 283–290.

Zijlstra, C. and van Hoof, R.A. (2006) A multiplex real-time polymerase chain reaction (TaqMan) assay for the simultaneous detection of *Meloidogyne chitwoodi* and *M. fallax. Phytopathology* 96, 1255–1262.

Zijlstra, C., Lever, A.E.M., Uenk, B.J. and van Sifhout, C.H. (1995) Differences between ITS regions of isolates of root-knot nematodes *Meloidogyne hapla* and *M. chitwoodi. Phytopathology* 85, 1231–1237.

Zijlstra, C., Uenk, B.J. and van Silfhout, C.H. (1997) A reliable, precise method to differentiate species of root-knot nematodes in mixtures on the basis of ITS-RFLPs. *Fundamental and Applied Nematology* 20, 59–63.

Zijlstra, C., Donkers-Venne, D.T.H.M. and Fargette, M. (2000) Identification of *Meloidogyne incognita, M. javanica* and *M. arenaria* using sequence characterised amplified region (SCAR) based PCR assays. *Nematology* 2, 847–853.

Zijlstra, C., van Hoof, R. and Donkers-Venne, D. (2004) A PCR test to detect the cereal root-knot nematode *Meloidogyne naasi. European Journal of Plant Pathology* 110, 855–860.

Zu, J., Liu, P., Meng, Q. and Long, H. (2004) Characterisation of *Meloidogyne* species from China using isozyme phenotypes and amplified mitochondrial DNA restriction fragment length polymorphism. *European Journal of Plant Pathology* 110, 309–315.

5 Molecular Taxonomy and Phylogeny

Byron J. Adams[1], Adler R. Dillman[2] and Camille Finlinson[1]
[1]Brigham Young University, Provo, Utah, USA; [2]California Institute of Technology, Pasadena, California, USA

5.1 Introduction 119
5.2 The History of Reconstructing *Meloidogyne* Phylogenetic History 119
5.3 Molecular Phylogenetics: Genetic Markers and Evolutionary Relationships 120
5.4 A *Meloidogyne* Supertree Analysis 127
5.5 Conclusions and Future Directions 130
5.6 References 135

5.1 Introduction

The genus *Meloidogyne* contains over 90 described species and each of these species typically has an extremely broad host range (as many as 3000 plant species; Trudgill and Blok, 2001). In addition to their host-range diversity, they also exhibit tremendous cytogenetic variation (aneuploidy and polyploidy) and mode of reproduction (from obligatory amphimixis to meiotic and mitotic parthenogenesis) (Triantaphyllou, 1985; see Chitwood and Perry, Chapter 8, this volume). In current practice, identification of species is based primarily on the morphological features of females, males and second-stage juveniles (Eisenback and Triantaphyllou, 1991; see Hunt and Handoo, Chapter 3, this volume), as well as esterase and malate dehydrogenase isozyme profiles derived from single females by polyacrylamide gel electrophoresis (Esbenshade and Triantaphyllou, 1985, 1990; Carneiro *et al.*, 2000) and DNA-based bar codes (Powers and Harris, 1993; Powers, 2004; Powers *et al.*, 2005; see Blok and Powers, Chapter 4, this volume). Historically, the diagnostic features deemed most valuable for identification commonly preceded their use as important characters for taxonomic statements (such as new species descriptions) and, subsequently, phylogenetic analyses. As molecular markers increasingly demonstrated improved resolving power, they became more commonplace as diagnostic tools, eventually becoming more prominent as parts of formal taxonomic statements (including descriptions of new species) and phylogenetic analyses (e.g. Castillo *et al.*, 2003; Landa *et al.*, 2008). With the incorporation of molecular sources of characters and refinements to phylogenetic theory, the fields of taxonomy and evolutionary biology have now become more completely integrated as a discipline, such that the terms molecular taxonomy and phylogenetics are (or should be) subsumed as a single research programme (systematics). In this chapter we present a summary of early and contemporary research on the molecular systematics of *Meloidogyne*.

5.2 The History of Reconstructing *Meloidogyne* Phylogenetic History

Some of the earliest work on evolutionary relationships among species of *Meloidogyne* was based on morphological characteristics and relied heavily on many of the characters used for identification

(Eisenback and Triantaphyllou, 1991). Subsequent efforts involved cytogenetics (Triantaphyllou, 1966, 1985), producing evidence that supported hypotheses consistent with the idea that mitotic parthenogens evolved from meiotic parthenogenetic ancestors, following suppression of meiotic processes and the establishment of various ploidy levels. Triantaphyllou also hypothesized that the amphimictic species in the genus, such as *M. exigua*, are highly specialized parasites and should not be considered as ancestral forms (Triantaphyllou, 1985). Cytogenetic studies, followed by protein and DNA analyses, implied a unique origin of, and monophyly among, the ameiotic species (Dickson *et al.*, 1971; Dalmasso and Bergé, 1978; Esbenshade and Triantaphyllou, 1987; Castagnone-Sereno *et al.*, 1993; Baum *et al.*, 1994; van der Beek *et al.*, 1998). Later, studies based on mitochondrial genes soon revealed that these genes can be hypervariable, both in patterns of sequence substitution and in gene content and arrangement (Powers and Sandall, 1988; Powers *et al.*, 1993). These properties are desirable for diagnostic or population genetic markers, or for resolving phylogenetic relationships among closely related species (see Blok and Powers, Chapter 4, this volume). More recent phylogenetic analyses have utilized small ribosomal subunit (18S) rDNA (ribosomal DNA) sequences (De Ley *et al.*, 2002), large subunit (28S) rDNA (Castillo *et al.*, 2003), and mitochondrial DNA (mtDNA) sequences (Tigano *et al.*, 2005). Lunt (2008) performed separate analyses of four genes – those for dystrophin, elongation factor 1-alpha, major sperm protein, and RNA polymerase 2 – as part of a clever study to elucidate whether the origin of the asexual *Meloidogyne* lineages was ancient or recent (Adams and Powers, 1996; Hugall *et al.*, 1999). A phenomenon that emerges from each of these studies is the close relationships among the three major mitotic parthenogenetic species: *M. arenaria*, *M. javanica* and *M. incognita*. Regardless of the type of phylogenetic analysis performed, or the genetic locus examined, the preponderance of evidence from single gene analyses suggests that the mitotic parthenogens are clearly evolutionarily distinct from either the meiotic or obligatory amphimictic species. However, gene trees are not always concordant with the evolutionary history of independently evolving species, and the discordance between the two different histories can confound phylogenetic inference. Discordance between gene trees and species trees is most commonly explained as lineage sorting among mitochondrial haplotypes, but can also arise through paralogous genes, as would be expected for the rDNA tandem array if intraspecific concerted evolution were non-uniform, or non-orthologous genes go undetected (Maddison, 1997; Maddison and Knowles, 2006). Phylogenomic analyses hold the promise of resolving problematic phylogenies by swamping the data sets with signal, despite high noise, by including character information from numerous loci (Eisen, 1998; Eisen and Fraser, 2003), but see Longhorn *et al.* (2007). Although phylogenomic analyses that could exhaustively sample all *Meloidogyne* species are premature (if not unnecessary), in a preliminary effort of this kind Scholl and Bird (2005) identified numerous putative homologues and used them to generate a phylogeny for a subset of *Meloidogyne* species. Although this effort was based on a small sample of taxa, the major contribution was the elucidation of relationships among three mitotic parthenogens (*M. hapla*, *M. incognita*, *M. javanica*) that had been poorly resolved in previous phylogenetic analyses. Subsequent refinements to *Meloidogyne* phylogeny have consisted primarily of analyses that have added new or previously unsampled taxa to existing databases (Castillo *et al.*, 2003; Landa *et al.*, 2008).

5.3 Molecular Phylogenetics: Genetic Markers and Evolutionary Relationships

5.3.1 Nuclear ribosomal DNA sequences

Nuclear ribosomal DNA is currently the most extensively employed molecular marker for *Meloidogyne* molecular systematics. Variation in mutation rates observed among different genes and spacers within an rDNA transcription unit results in regions of adjacent DNA segments in the cistron that are useful across a wide range of taxonomic hierarchical levels (Hillis and Dixon, 1991). This includes conserved and variable regions of the 18S and 28S subunits, and the more highly variable ITS (internal transcribed spacer) region. These three rRNA gene regions are the most commonly used genetic markers for nematode molecular systematics, and each of these regions has been employed for *Meloidogyne* phylogenetics (Landa *et al.*, 2008).

While rRNA genes may provide optimal levels of variation for investigating *Meloidogyne* phylogeny, they are not without significant theoretical and analytical drawbacks. The single biggest obstacle in using rRNA genes is that, unlike protein-coding genes, they are not constrained to maintain codon fidelity or even an open reading frame. Whereas the length and composition of protein-coding genes are generally subject to selection by codon usage, rRNA genes are not. Thus, for some rDNA regions, insertion and deletion events (indels) can be as frequent as transitions and transversions, often involving blocks of multiples of nucleotides (Powers *et al.*, 1997; Powers, 2004). Indel events can result in substantial rDNA size differences between sequences (taxa), which complicates the process of generating multiple sequence alignments and reduces confidence in the homology statements for each nucleotide in the multiple sequence alignment. In our experience, and as shown by others, there is usually more variation in tree topology due to differences in the multiple sequence alignment than there is among the different methods used to generate the trees (i.e. parsimony, maximum likelihood, Bayesian and distance methods) (Morrison and Ellis, 1997). Approaches to addressing this problem require thoughtful consideration of the mechanics of how multiple sequence alignments (homology statements) are constructed. These involve the nuts and bolts of how computer algorithms generate multiple sequence alignments, removing the alignment-ambiguous regions based on an *a priori* metric (i.e. remove ambiguous indels that lie between a predetermined number of invariant nucleotides (Nguyen *et al.*, 2001)), direct optimization (Terry and Whiting, 2005), comparison of secondary structure based on minimum energy models (Subbotin *et al.*, 2006), and minimum posterior probabilities among alternative placements of nucleotides (characters) in the alignment (Loytynoja and Milinkovitch, 2003). These problems are not unique to rDNA, as alignment ambiguity can also arise where protein-coding genes have undergone tremendous divergence, or for other non-coding sequences (such as non-coding regions of mitochondrial DNA; see section 5.3.3).

5.3.1.1 18S (small ribosomal subunit)

Although 18 S sequences for several *Meloidogyne* species had long been available in public databases, De Ley *et al.* (2002) were the first to use this locus as part of a rigorous reconstruction of *Meloidogyne* phylogeny. Their analysis included 12 species of *Meloidogyne* and four outgroup taxa subject to phylogenetic analyses generated from three different multiple sequence alignment methodologies and three different tree-building optimality criteria (distance, parsimony and maximum likelihood). Calculations of phylogenetic signal (skewness of tree length distribution) were high and intraspecific sequence polymorphism low, suggesting that the locus was appropriately robust for resolving relationships among the sampled species but with nodal support strongest at the deeper nodes. The De Ley *et al.* (2002) analysis showed strong support for three clades, which they designated: clade I, conscribing the mitotic parthenogens (*M. incognita*, *M. arenaria* and *M. javanica*); clade II, including the obligatory amphimictic, meiotic and mitotic parthenogens (*M. hapla* races A and B, respectively) as well as *M. duytsi* and *M. maritima*; and clade III, containing three meiotic parthenogens (*M. exigua*, *M. graminicola* and *M. chitwoodi*). As with earlier phylogenetic efforts (Triantaphyllou, 1985; Castagnone-Sereno *et al.*, 1993), their analysis supported the location of the ameiotic species as distantly related to either the obligate amphimictic or meiotic species (Fig. 5.1).

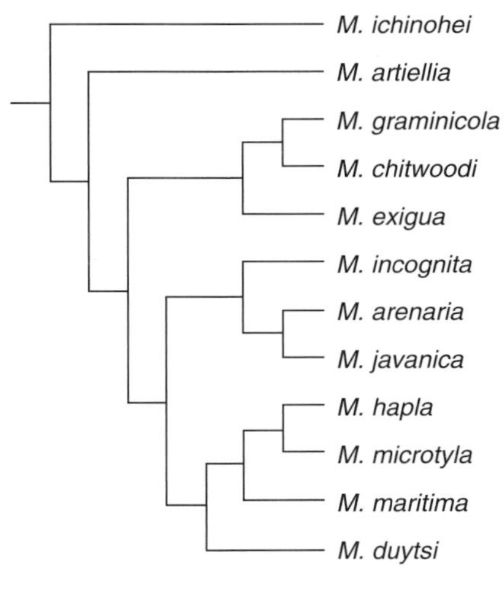

Fig. 5.1. 18 S rDNA *Meloidogyne* phylogeny of De Ley *et al.* (2002). The tree was generated from a secondary-structure-based multiple sequence alignment and resolved via maximum parsimony. (Adapted from De Ley *et al.*, 2002.)

The subsequent 18S analyses of Tigano *et al.* (2005) included 19 additional sequences, representing 12 nominal species, as well as several unknown isolates from disparate geographic locations, in an effort that revealed interesting comparisons between the 18S rDNA sequences and other tools for diagnosing species, including morphological and isozyme phenotypes. The Tigano *et al.* (2005) 18S rDNA analysis proceeded from sequences profile-aligned to the optimal sequence alignment of De Ley *et al.* (2002) by distance, parsimony and maximum likelihood tree-building algorithms, all of which produced congruent topologies (Fig. 5.2). Their results revealed polymorphisms between isolates of the same species (as per morphological and isozyme diagnoses), including *M. arenaria*, *M. incognita*, *M. javanica*, *M. exigua* and *M. hapla*. Some of the differences were slight, but several were substantial: *M. arenaria* of De Ley *et al.* (2002) differed from the Tigano *et al.* (2005) sequences by two substitutions and four insertions; two *M. incognita* isolates differed by one insertion and ten substitution events; two *M. javanica* sequences differed by 24 substitutions, 18 ambiguities and four insertions. However, some 18S rDNA sequences were highly conserved between species. For example, *M. hispanica* and *M. ethiopica* yield identical 18S sequences, so only *M. hispanica* is listed in the analysis of Landa *et al.* (2008) (Fig. 5.3).

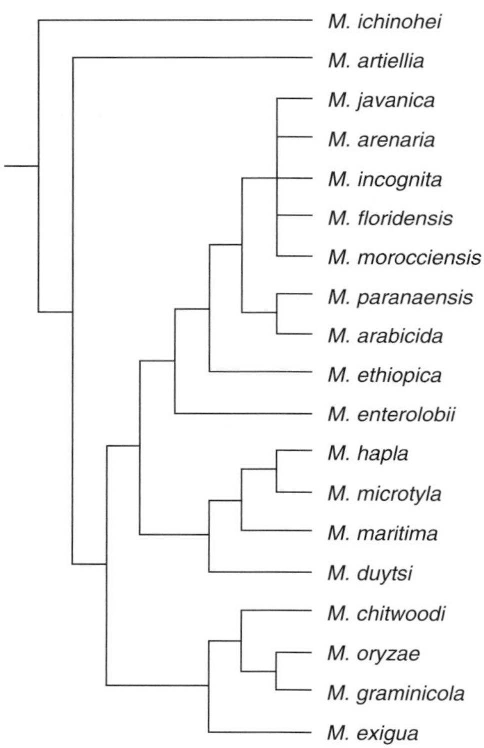

Fig. 5.2. 18S rDNA *Meloidogyne* phylogeny of Tigano *et al.* (2005). The tree was generated from sequences profile-aligned to the secondary-structure-based alignment of De Ley *et al.* (2002) and resolved via maximum likelihood. (Adapted from Tigano *et al.*, 2005.) Only nominal taxa were retained. Taxa with multiple representative sequences were represented by a single semaphoront and relationships collapsed to their most inclusive clade.

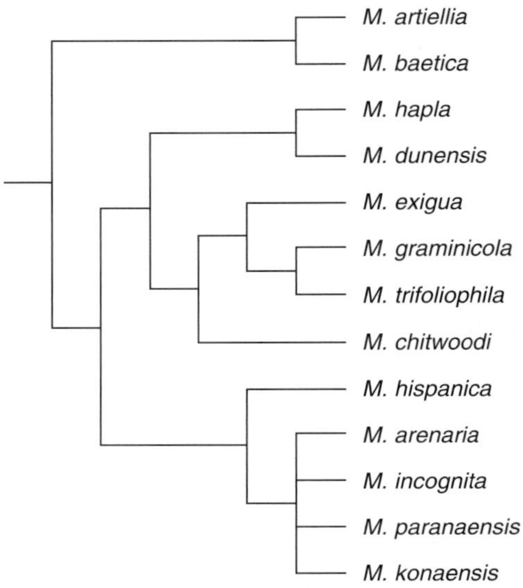

Fig. 5.3. 18S rDNA *Meloidogyne* phylogeny of Landa *et al.* (2008). The tree was generated from sequences aligned using Bionumerics software ver. 4.5 (Applied Maths, Kortrijk, Belgium) and resolved via maximum parsimony (also by Bionumerics). (Adapted from Landa *et al.*, 2008.)

5.3.1.2 28S (large ribosomal subunit)

The first effort to use the 28S rDNA region to resolve phylogenetic relationships among *Meloidogyne* spp. was that of Tenente *et al.* (2004). Their analysis included nine species represented by 12 sequences, and their most robust phylogenetic analyses included exhaustive, unrooted maximum parsimony searches and maximum likelihood searches based on various ClustalX alignments. The data for each of the generated alignments revealed strong phylogenetic signal as based on g1 statistics, and most nodes, particularly for the deeper clades, were strongly supported by bootstrap resampling estimates. The Tenente *et al.* (2004) analysis found fairly strong support for two monophyletic clades that are compatible with clades I and III of the 18 S analysis of De Ley *et al.* (2002) (but did not include any of the clade II taxa) (Fig. 5.4). The weakest supported nodes, and the ones that differed the most by alignment and tree search strategy, involved the relationships among the mitotic parthenogenetic species *M. arenaria*, *M. incognita*, *M. konaensis*, and *M. paranaensis*. They conclude that the D2/D3 (dopamine receptor) region of this marker, which showed high degrees of variation between two species of *Acrobleoides* that are morphologically virtually indistinguishable (De Ley *et al.*, 1999), were 'simply too conserved for the phylogenetic analysis of mitotic parthenogenetic *Meloidogyne* species' (Tenente *et al.*, 2004). A contemporary analysis by Castillo *et al.* (2003) also reflected these sentiments; their analysis, which included the Tenente *et al.* (2004) sequences, also included several other unpublished *Meloidogyne* and outgroup (*Pratylenchus*) sequences, and was based on simple ClustalX default alignment parameters under the maximum parsimony optimality criterion. The Castillo *et al.* (2003) results are consistent with those of Tenente *et al.* (2004), whereby monophyletic groups I and III form a clade with the more ancestral lineage comprising *M. artiellia* and *M. baetica* (representatives of clade II were not included in the analysis). Like Tenente *et al.* (2004), Castillo *et al.* (2003) and Landa *et al.* (2008) were unable to resolve relationships among species of clade I but resolved deeper nodes with much greater support (Figs 5.5 and 5.6).

5.3.1.3 ITS (internal transcribed spacer region)

Early work by Hugall *et al.* (1999) on *Meloidogyne* ITS rDNA sequences revealed a somewhat surprising phenomenon: although amphimictic species exhibited only a single ITS lineage, the ameiotic species *M. hapla*, *M. arenaria* and *M. incognita* exhibited numerous lineages, even within individuals. In fact, Hugall *et al.* (1999) showed that up to 90% of the total ITS diversity could be found within an individual nematode, which contained as many as 9–13 different sequence variants. Such variation, and the way it is partitioned across the genus, poses challenges to using it to infer phylogenetic relationships, but does provide strong evidence for the hybrid origins of *M. hapla*, *M. arenaria* and *M. incognita* (Hugall *et al.*, 1999).

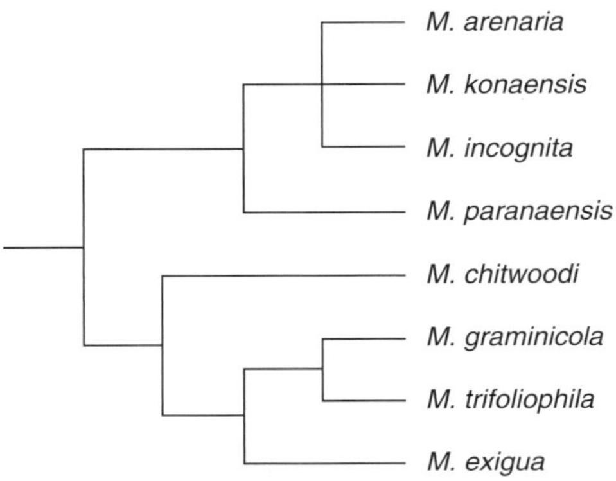

Fig. 5.4. 28 S rDNA *Meloidogyne* phylogeny of Tenente *et al.* (2004). The tree was generated from sequences aligned using ClustalX (Thompson *et al.*, 1997) default parameters and resolved using maximum parsimony. (Adapted from Tenente *et al.*, 2004.)

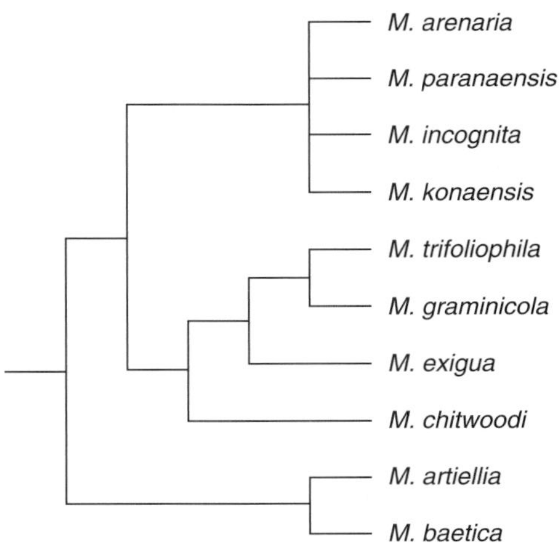

Fig. 5.5. 28 S rDNA *Meloidogyne* phylogeny of Castillo *et al.* (2003). The tree is a strict consensus of five equally parsimonious trees generated from sequences aligned using ClustalX default parameters (Thompson *et al.*, 1997) and resolved via maximum parsimony (Swofford, 2002). (Adapted from Castillo *et al.*, 2003.)

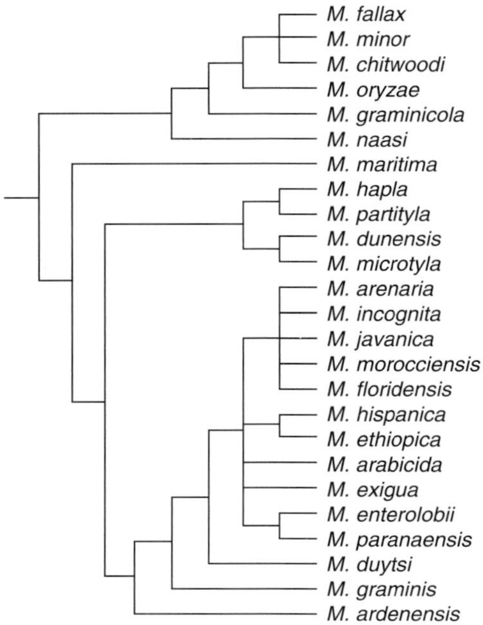

Fig. 5.6. 28S rDNA Meloidogyne phylogeny of Landa *et al.* (2008). The tree was generated from sequences aligned using Bionumerics software ver. 4.5 (Applied Maths, Kortrijk, Belgium) and resolved via maximum parsimony (also by Bionumerics). (Adapted from Landa *et al.*, 2008.)

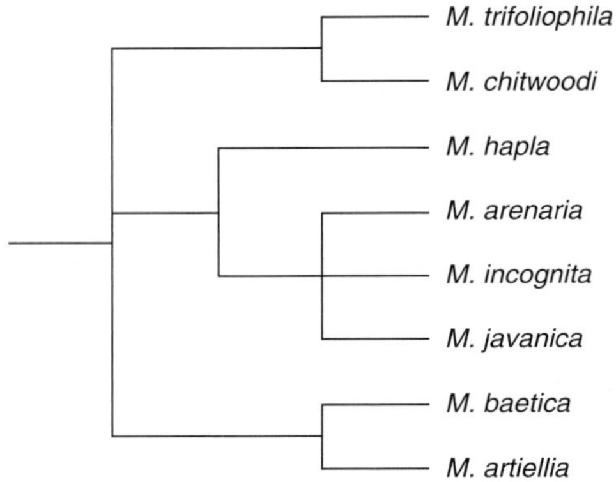

Fig. 5.7. ITS rDNA *Meloidogyne* phylogeny of Castillo *et al.* (2003). The tree was generated from sequences aligned using ClustalX default parameters (Thompson *et al.*, 1997) and resolved via maximum parsimony (Swofford, 2002). The original solution was presented as unrooted; here we root it with *M. artiellia* and *M. baetica*. (Adapted from Castillo *et al.*, 2003.)

One of the by-products of exploring ITS sequence variation in search of diagnostic markers is a large, publicly available database of sequences. In an early effort, Castillo *et al.* (2003) generated an unrooted maximum parsimony tree from ITS sequences for nine species. Subsequently, Landa *et al.* (2008) used these and additional sequences to perform a phylogenetic analysis of 16 species from 29 different isolates. Although details of their sequence alignment process are not explicit, the maximum parsimony tree they generated depicts a completely resolved tree, but with varying levels of support, particularly through the intermediate nodes. Clades III and I are monophyletic. Clade II is depicted as paraphyletic, but the discordant nodes are also those that are most weakly supported. Also evident in the resulting tree is the apparent paraphyletic nature of the ITS lineages within and among the ameiotic species (Figs 5.7 and 5.8).

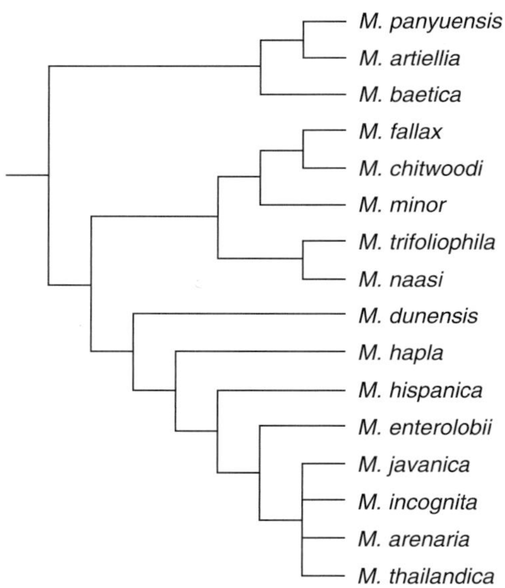

Fig. 5.8. ITS rDNA *Meloidogyne* phylogeny of Landa *et al.* (2008). The tree was generated from sequences aligned using Bionumerics software ver. 4.5 (Applied Maths, Kortrijk, Belgium) and resolved via maximum parsimony (also by Bionumerics). (Adapted from Landa *et al.*, 2008.)

5.3.2 Orthologous nuclear genes

As with gene families, concerted evolution does not always work fast enough to homogenize all copy variants in the rDNA cistron. The result is that paralogous gene genealogies can contradict actual phylogenetic relationships of species. This is evidenced by lineage sorting events for mtDNA (discussed below) and also by species that arose via hybridization events, such as the mitotic parthenogenetic species of *Meloidogyne* (Hugall *et al.*, 1999). Thus, an optimal phylogenetic analysis will include as many nuclear, single-copy genes in the transformation series as possible. In an effort to distinguish the relative timing of the hybridization events involved in the origin of the ameiotic lineages, Lunt (2008) explored the evolution of four different single-copy nuclear genes. Prior to this effort, Scholl and Bird (2005) undertook a phylogenomic approach in order to tease out the relationships between the mitotic parthenogenetic *Meloidogyne* species. Each of these efforts is discussed below.

5.3.2.1 Dystrophin

Dystrophin is a muscle protein that connects the cytoskeleton of a muscle fibre to the surrounding extracellular matrix. It is coded by the longest gene in the human genome but exists in single-copy form in invertebrates (Roberts *et al.*, 1995; Roberts and Bobrow, 1998). Lunt (2008) designed primers that amplified a 670–770bp product of three exons and two introns from multiple populations of five *Meloidogyne* spp. (*M. incognita*, *M. javanica*, *M. arenaria*, *M. enterolobii* (= *M. mayaguensis*) and *M. hapla*) and an outgroup taxon (*Globodera pallida*) for phylogenetic analyses using maximum likelihood. The results for this gene are consistent with the clade designations of De Ley *et al.* (2002) but fail to resolve unambiguously relationships among the members of clade I (*M. enterolobii*, *M. arenaria*, *M. incognita*, *M. javanica*). However, none of the paraphyletic nodes is well supported by approximate likelihood ratio tests (Fig. 5.9).

5.3.2.2 Major sperm protein (msp)

Major sperm protein is the most abundant protein in nematode sperm and is responsible for

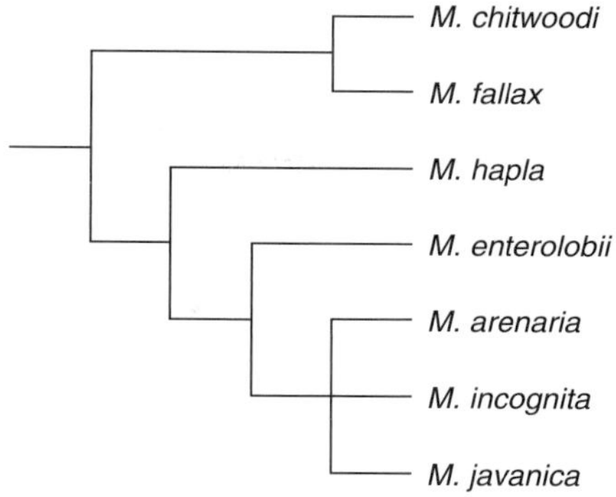

Fig. 5.9. Dystrophin exon gene genealogy of *Meloidogyne* (Lunt, 2008). The tree was generated from exon sequences aligned with Clustal X (Thompson *et al.*, 1997) (unambiguous alignment due to codon structure) and resolved using maximum likelihood as implemented in PhyML (Guindon and Gascuel, 2003).

the cell's motility (Roberts, 2005). Lunt (2008) studied *Meloidogyne* msp gene genealogies in order to see if the gene underwent an accelerated rate of mutation after the evolution of mitotic parthenogenesis and was, therefore, putatively no longer under selection pressure to maintain its function; it did not. Maximum parsimony analysis from multiple populations of five *Meloidogyne* spp. (*M. incognita*, *M. javanica*, *M. arenaria*, *M. enterolobii*, *M. hapla*) and an outgroup taxon (*G. pallida*) yielded a phylogenetic tree that is consistent with the three clade designation of De Ley *et al.* (2002), except that the positions of clade II and III are reversed relative to clade I, but the node involved in this reversal is not well supported (Fig. 5.10).

5.3.2.3 Elongation factor 1-alpha (EF1-α)

Elongation factor 1-alpha (EF-1α), the GTP binding protein involved in catalysing the binding of the aminoacyl-transfer RNAs to the ribosome, is an essential component of eukaryotic translation (Watson, 2008). Lunt (2008) pursued this gene as an independent estimate of phylogenetic relationships but, upon inspection of the resulting phylogenetic tree, suspected that instead of it being single copy, a gene duplication event might have occurred, resulting in paralogous loci. In support of this, Lunt identified two copies of

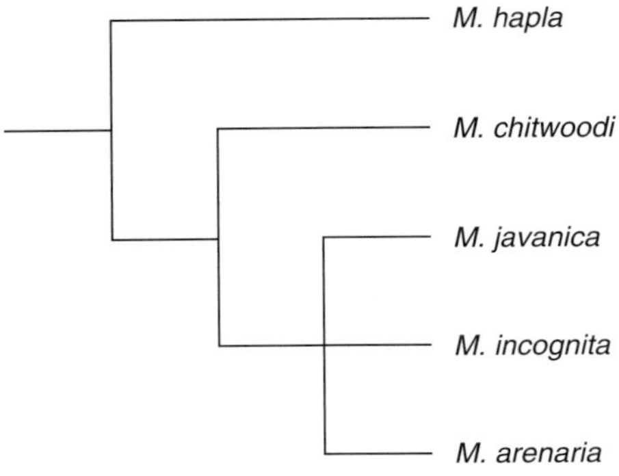

Fig. 5.10. Major sperm protein exon gene genealogy of *Meloidogyne* (Lunt, 2008). The tree was generated from exon sequences aligned with Clustal X (Thompson *et al.*, 1997) (unambiguous alignment due to codon structure) and resolved using maximum likelihood as implemented in PhyML (Guindon and Gascuel, 2003).

the gene in the *Caenorhabditis elegans* genome, yet the maximum likelihood solution among his sampled species was still congruent with the clade designations of De Ley *et al.* (2002). However, the EF-1α gene genealogy among the ameiotic species is paraphyletic, poorly supported and too conserved to distinguish *M. fallax* from *M. chitwoodi* (Fig. 5.11).

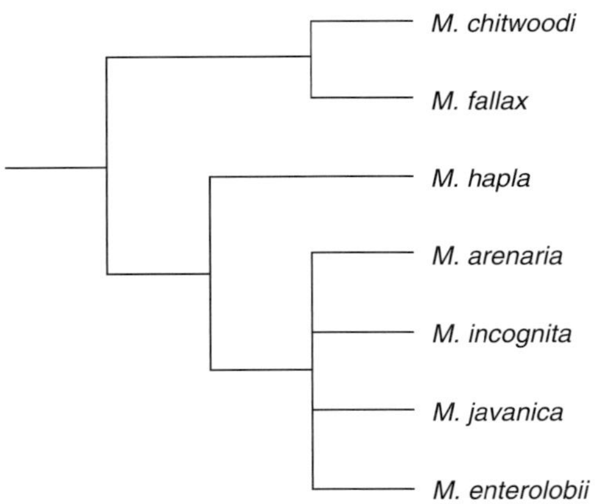

Fig. 5.11. Elongation factor 1-alpha exon gene genealogy of *Meloidogyne* (Lunt, 2008). The tree was generated from exon sequences aligned with Clustal X (Thompson *et al.*, 1997) (unambiguous alignment due to codon structure) and resolved using maximum likelihood as implemented in PhyML (Guindon and Gascuel, 2003).

5.3.2.4 RNA polymerase 2

RNA polymerase 2 is the eukaryotic enzyme responsible for synthesis of mRNA during transcription (Kornberg, 2007). Lunt (2008) designed primers that amplified an approximately 710 bp fragment that included coding sequence from two exons and one intron. Using the exon sequences only resulted in the maximum likelihood solution in Fig. 5.12. The tree is consistent with the clade designations of De Ley *et al.* (2002) but fails to resolve *M. chitwoodi* and *M. fallax* lineages within clade III, and where they are resolved, the *M. javanica*, *M. incognita* or *M. arenaria* relationships within clade I are paraphyletic.

5.3.3 Mitochondrial DNA

Despite apparent low genetic diversity (or because of it; see Blok and Powers, Chapter 4, this volume) among populations of *M. arenaria*, *M. javanica* and *M. incognita* (Hugall *et al.*, 1994, 1997; Stanton *et al.*, 1997), mtDNA sequences spanning the COII (cytochrome oxidase subunit II) through lRNA (16S rRNA) genes have been intensively studied for *Meloidogyne* molecular diagnostics (Blok *et al.*, 2002; Powers, 2004; Brito *et al.*, 2004; Powers *et al.*, 2005). The diagnostic utility of the marker persuaded Tigano *et al.* (2005) to explore

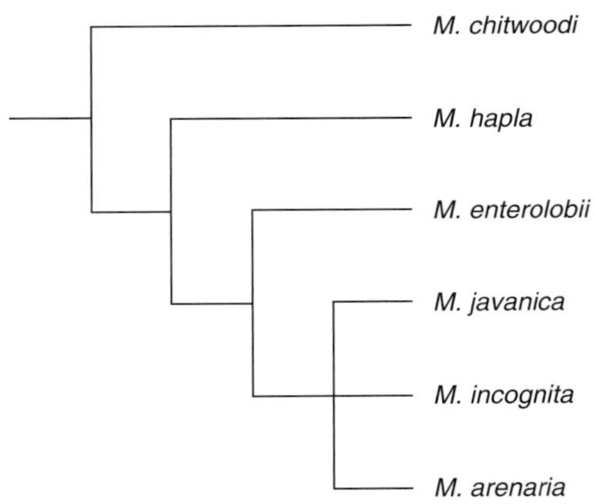

Fig. 5.12. RNA polymerase 2 exon gene genealogy of *Meloidogyne* (Lunt, 2008). The tree was generated from exon sequences aligned with Clustal X (Thompson *et al.*, 1997) (unambiguous alignment due to codon structure) and resolved using maximum likelihood as implemented in PhyML (Guindon and Gascuel, 2003).

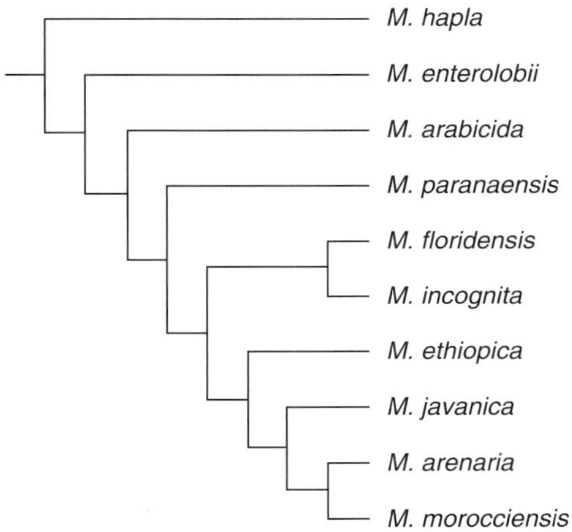

Fig. 5.13. Mitochondrial DNA *Meloidogyne* phylogeny of Tigano *et al.* (2005). The tree was generated from DNA sequences that span the COII through lRNA region, including the complete sequence for tRNA-His, and the AT-rich region. Sequences were aligned using Clustal X (Thompson *et al.*, 1997), adjusted by eye using MacClade (Maddison and Maddison, 2002), and resolved by maximum parsimony as implemented in PAUP* (Swofford, 2002).

its ability to resolve phylogenetic relationships among the very closely related clade I taxa. By parsimony, distance and maximum likelihood approaches they analysed the complete region, including partial COII and partial lRNA sequence, the complete sequence for tRNA-His and the AT-rich region, and obtained a single optimal solution. The concatenation of the different gene regions was justified by failure to reject shared evolutionary histories by way of an incongruence length difference test. Although there is some discordance between this tree and the Tigano *et al.* (2005) rDNA solution, the relationship among the ameiotic species is congruent, if poorly supported, by both the 18S rDNA analyses of Tigano *et al.* (2005) and De Ley *et al.* (2002) (Fig. 5.13).

5.3.4 Phylogenomics

Optimally, phylogenetic construction proceeds from consideration of as many independently evolving, heritable characters as possible (Farris, 1983; Kluge, 1997). In order to address the possible problems associated with gene tree/species tree discordance, a reasonable expectation might be that if one simply looks at enough characters, or in this case enough DNA sequences from enough genes, eventually the phylogenetic signal will swamp noise. Such is the idea of using genomic data to resolve phylogenetic relationships, where it has been shown that using small numbers of genes can produce support for incorrect phylogenies, but that support and resolution become optimized at about 20 genes. Taking this into consideration, Scholl and Bird (2005) sampled 47 orthologous genes from several tylenchid nematodes, including *M. chitwoodi*, *M. hapla*, *M. arenaria*, *M. javanica* and *M. incognita*, with the goal of resolving the relationships between the apomicts (*M. arenaria*, *M. javanica* and *M. incognita*) where previous analyses appeared to show conflicting relationships: based on mtDNA, Powers and Sandall (1988) suggested the relationship to be (*M. arenaria* (*M. javanica* + *M. incognita*), whereas the 18S best estimate of De Ley *et al.* (2002) was (*M. incognita* (*M. javanica* + *M. arenaria*)). To construct their phylogenetic trees, Scholl and Bird (2005) first performed rigorous screens to identify orthologous genes from EST (expressed sequence tag) sequence databases that had a homologue in *C. elegans* (and were thus conserved across a large phylogenetic distance, reflecting evolutionary constraint). The orthologues were aligned based on their inferred amino acid sequence and then back-translated to their DNA sequence to maintain open reading frame fidelity. Phylogenies were constructed using multiple alignments of the individual genes and concatenated full-length data sets containing all the genes, or different subsets of genes, via Bayesian inference, maximum likelihood and minimum evolution (via neighbour joining). The optimal solution based on all of the analyses favoured the hypothesis of Powers and Sandall (1988) (Fig. 5.14).

5.4 A *Meloidogyne* Supertree Analysis

It would be excellent if we could just take all the DNA sequences for all of the *Meloidogyne* taxa ever generated, compile them into a single giant multiple sequence alignment, crunch it through some tree-building algorithms and confidently

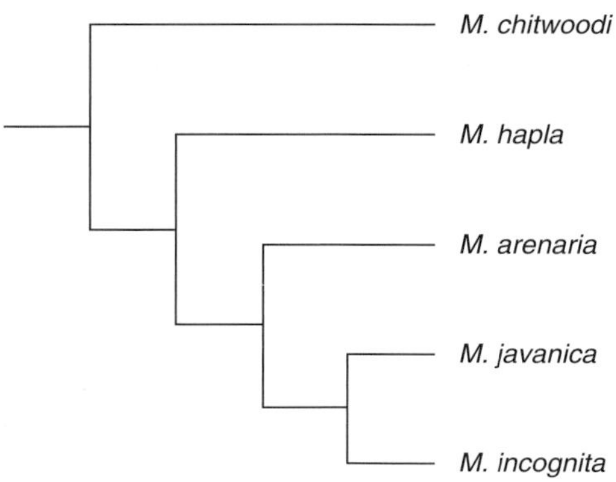

Fig. 5.14. *Meloidogyne* phylogeny of Scholl and Bird (2005). The tree was generated from a concatenated matrix of 47 putative orthologous genes aligned using Clustal W (Thompson *et al.*, 1994) and resolved using Bayesian analysis as implemented in MRBAYES (Huelsenbeck and Ronquist, 2001).

report the one true tree of *Meloidogyne* spp. evolutionary history. While perhaps optimal, such an exercise is obviated by the range of sequence variation that spans the gamut of evolutionary rates of change on the loci under study. For example, for many of the mtDNA and ITS sequences, divergence and length differences prohibit making unambiguous multiple sequence alignments among all taxa in the genus. Similarly, not all taxa are present in all of the data sets, and the inclusion of large amounts of missing characters can result in spurious phylogenetic inference (Maddison, 1993; Wiens, 1998). One solution to this problem is to generate a supertree – an evolutionary tree that is assembled from a bunch of smaller trees that share some, but not necessarily all, common taxa (Bininda-Emonds, 2004).

Although it may sound straightforward to take several phylogenetic trees and spin them into a single tree that is the sum of all the parts, it is not. What should not be surprising is that the quality of the obtained supertree is a function of the quality of the phylogenies used to build it. Even if the phylogenies used to build the supertree are robust, there are still several important aspects of building a supertree to consider; we will only touch on a few here that are relevant to *Meloidogyne* (but see Bininda-Emonds *et al.*, 2004). In the case for building a *Meloidogyne* supertree one must consider not only the quality of the different phylogenies used to build the tree but also whether the phylogenies are independent. As opposed to using morphological or molecular characters to build phylogenies, the raw data for generating supertrees are lifted from the topological arrangements of two or more partially overlapping phylogenetic trees. So just as using the character 'male stylet length' five times in a character matrix would be redundant, using three different 18S trees to build a supertree that was to include seven other genetic loci could artificially bias the overall supertree topology in favour of the 18S topology. In other words, the three 18S trees are not independent estimates of phylogenetic relationships and must be dealt with somehow so as not to bias the analysis unfairly. Similarly, what about different trees that are generated by the same author? If Lunt (2008) used the same general methodology, from alignment to tree-building strategy, to construct all of his phylogenetic trees, is there an element of non-independence among his different gene trees? In fact, the majority of *Meloidogyne* phylogenetic analyses reported to date employed several different approaches in the same publication, from alignment strategy to tree-building algorithms and optimality criteria. More often than not, these different approaches generated several different hypotheses of relationships for each data set. Of these, which do we choose to use as source trees to generate a supertree?

With the above caveats in mind, we generated a supertree based on the best estimates of phylogenetic relationships from the most robust and comprehensive molecular phylogenetic analyses we could find in the literature. In order to account for duplication of gene trees (three 18S, three 28S, and two ITS gene trees), we generated 'mini-supertrees' from each set of redundant gene trees and then used the 'mini-supertree' as the source tree for the main analysis (Bininda-Emonds *et al.*, 1999). In choosing among the alternative phylogenetic arrangements published in most papers (i.e. the maximum parsimony tree versus the maximum likelihood tree), we did what any good taxonomist would do – appeal to authority. We simply chose the hypothesis that was most favoured by the authors.

To construct the supertree we mined the optimal topologies from the following publications: De Ley *et al.* (2002), Castillo *et al.* (2003), Tenente *et al.* (2004), Scholl and Bird (2005), Tigano *et al.* (2005), Landa *et al.* (2008) and Lunt (2008). To account for non-independent estimates we generated mini-supertrees from the data sets that were represented more than once (18S, 28S and ITS). The mtDNA data set was treated as a single estimate, even though it comprised a concatenation of several genes. While potentially misleading, we justify this because Tigano *et al.* (2005) tested for shared evolutionary history and common inheritance of each of the individual genes. The phylogenomic analysis, of Scholl and Bird (2005) presents an interesting case because their analysis consisted of 47 concatenated genes. In theory, for the present analysis, we should be analysing each of the 47 gene tree topologies independently and then using each one of those topologies as an independent estimate of relationships (source tree) to construct the supertree. When Scholl and Bird (2005) performed their analyses on the concatenated data set of 47 genes, the data matrix was treated as if it were a single gene, assuming a single model of evolution (general time-reversible with four categories of gamma-distributed rate heterogeneity). In reality, since it is unlikely that all 47 genes evolved under the same model of sequence evolution, it would have been more appropriate to partition the concatenated data set by gene, each partition with its own most appropriate model of evolution, but that approach was still under development at the time of the Scholl and Bird (2005) analyses (Huelsenbeck *et al.*, 2008); however, Scholl and Bird did do independent analyses of subsets of single genes and reported that they were congruent with the overall concatenated gene phylogeny. Thus, as with the mtDNA topology, for our supertree analysis we used the Scholl and Bird (2005) topology as a single, independent estimate of relationships, acknowledging that were we to use all 47 gene topologies (which were identical) that there would be overwhelming support for a (*M. arenaria* (*M. javanica* + *M. incognita*)) clade in the present analysis. Similarly, by using the mtDNA data as a single source tree, we are probably underestimating the overall support for its topology in the supertree.

We used two supertree construction methods: matrix representation of parsimony (MRP; Purvis, 1995) with the matrix generated in RadCon 1.1.6 (Thorley and Page, 2000) and implemented in PAUP* (Swofford, 2002), and the most similar supertree method (MSSA, or dfit – distance fit) as implemented in CLANN ver. 3.0.0 (Creevey and McInerney, 2005) (Fig. 5.15). Even though these reconstructions involved the analysis of only nine different topologies, their sum of possible unrooted solutions is 2.92156×10^{40}, presenting a fairly computationally intensive effort. Each of the two supertree analyses yielded multiple equally parsimonious (MRP) or costly (dfit) trees (1736 and 7, respectively), which we represent here using two different consensus approaches (combinable components and majority rule) (Fig. 5.15). For both types of supertree construction methods, the consensus trees differed only in terms of resolution (combinable components being more conservative and less resolved). Overall, the obtained topologies among the different supertree methods are quite similar. Both approaches identified *M. ichinohei* as the lineage that shares a most recent common ancestry with the remaining members of the genus, with the next lineage to branch being the monophyletic clade of (*M. baetica* (*M. panyuensis* + *M. artiellia*)). Membership in clade III is congruent between the two analyses, with the exception that the dfit solution suggests sister relationships for (*M. oryzae* + *M. graminicola*) and (*M. trifoliophila* + *M. naasi*). Clade II membership is monophyletic in the dfit topology, differing from the MRP solution, which suggests that *M. dunensis* and *M. microtyla* are sister taxa. Also, the MRP topology differs radically from the dfit solution, as in the latter, *M. graminis* and *M. ardenensis* nest within clade II, but in the MRP solution, they are ancestral to clades I, II and III. clade I membership is congruent between both analyses, and lack of resolution is completely understandable given that these relationships were poorly resolved in most of the source trees. Both approaches favour *M. floridensis* and *M. incognita* as sister taxa (Fig. 5.15B,C), but beyond this relationship there is only discord. The most fundamental difference is that of the relationship between *M. arenaria*, *M. incognita* and *M. javanica*. The MRP solution favours (*M. incognita* (*M. arenaria* + *M. javanica*)), whereas the dfit solution is (*M. arenaria* (*M. javanica* + *M. incognita*)). Because the actual number of 'characters' (tree topologies) is so small ($n = 9$), bootstrap support for any of the relationships is virtually non-existent (data not shown). However, the phylogenetic signal was significantly better than random (permutation tail probability test, $P < 0.01$).

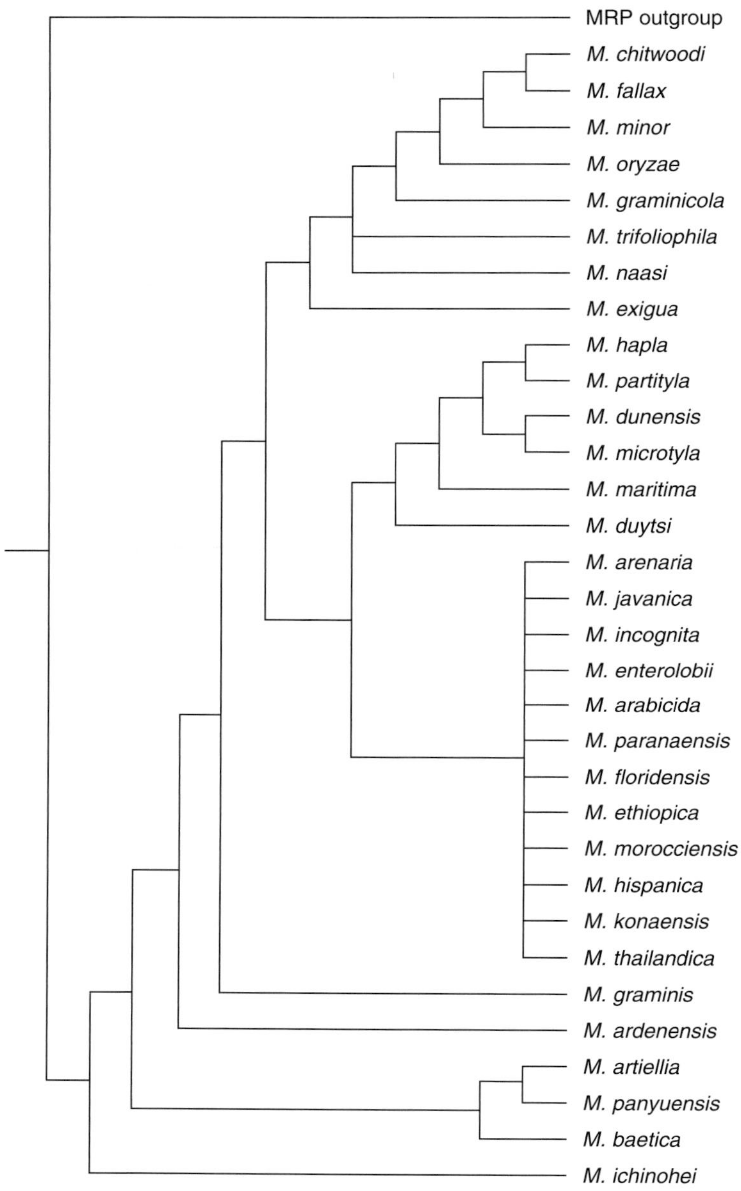

Fig. 5.15A. Supertree solutions for seven independent *Meloidogyne* source trees. MRP (matrix representation of parsimony) combinable components consensus.

5.5 Conclusions and Future Directions

Overall, the topologies of the phylogenies we used as source trees are remarkably similar. Following the clade designations of De Ley *et al.* (2002), clade I includes the mitotic parthenogens, *M. arenaria*, *M. incognita* and *M. javanica*, as well as *M. enterolobii*, *M. morociensis*, *M. ethiopica*, *M. hispanica*, *M. konaensis*, *M. paranaensis*,

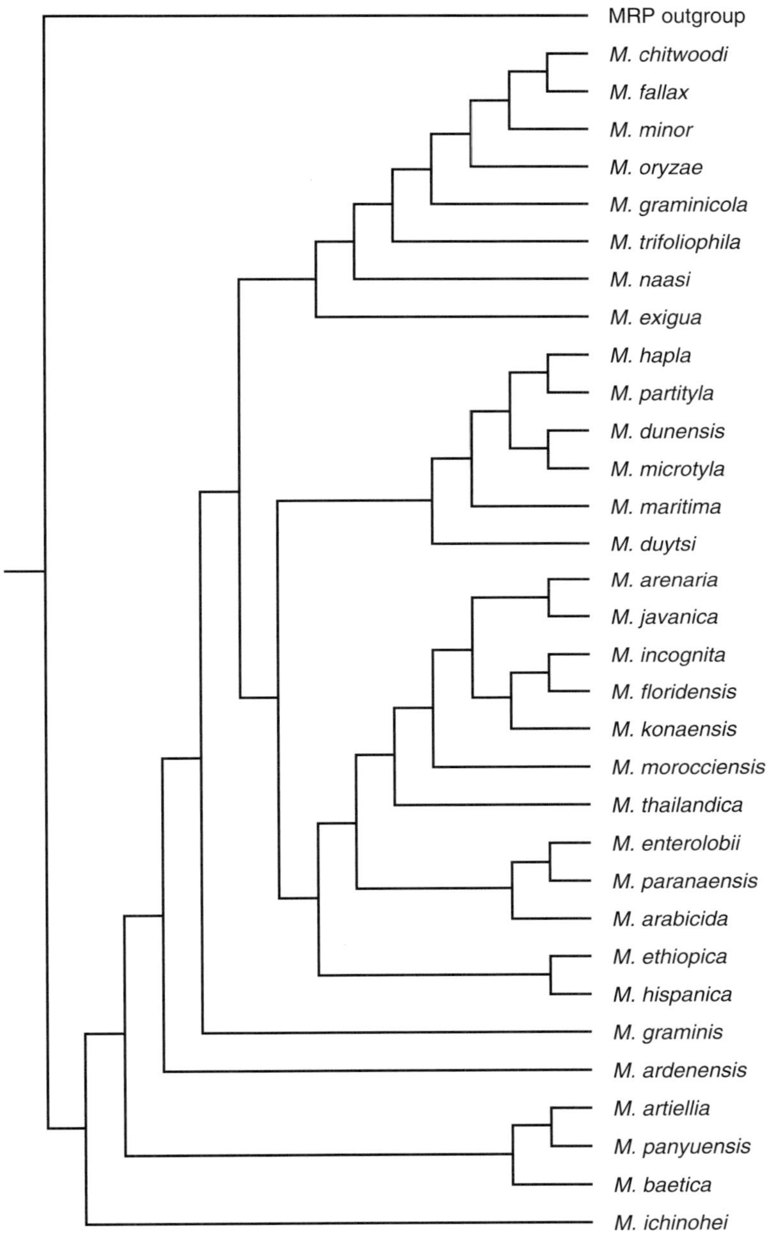

Fig. 5.15B. MRP (matrix representation of parsimony) majority rule consensus tree.

M. thailandica and *M. arabicida*. The only known exceptional member of the clade is *M. floridensis*, which is described as a meiotic parthenogen (Handoo *et al.*, 2004) but counter-intuitively nests as sister taxon to *M. incognita*. Clade I is sister to clade II, which contains *M. hapla*, *M. partityla*, *M. dunensis*, *M. microtyla*, *M. maritima* and *M. duytsi*. The dfit analysis also includes in this group *M. graminis* and *M. ardenensis*. Clades I and II form a clade with respect to clade

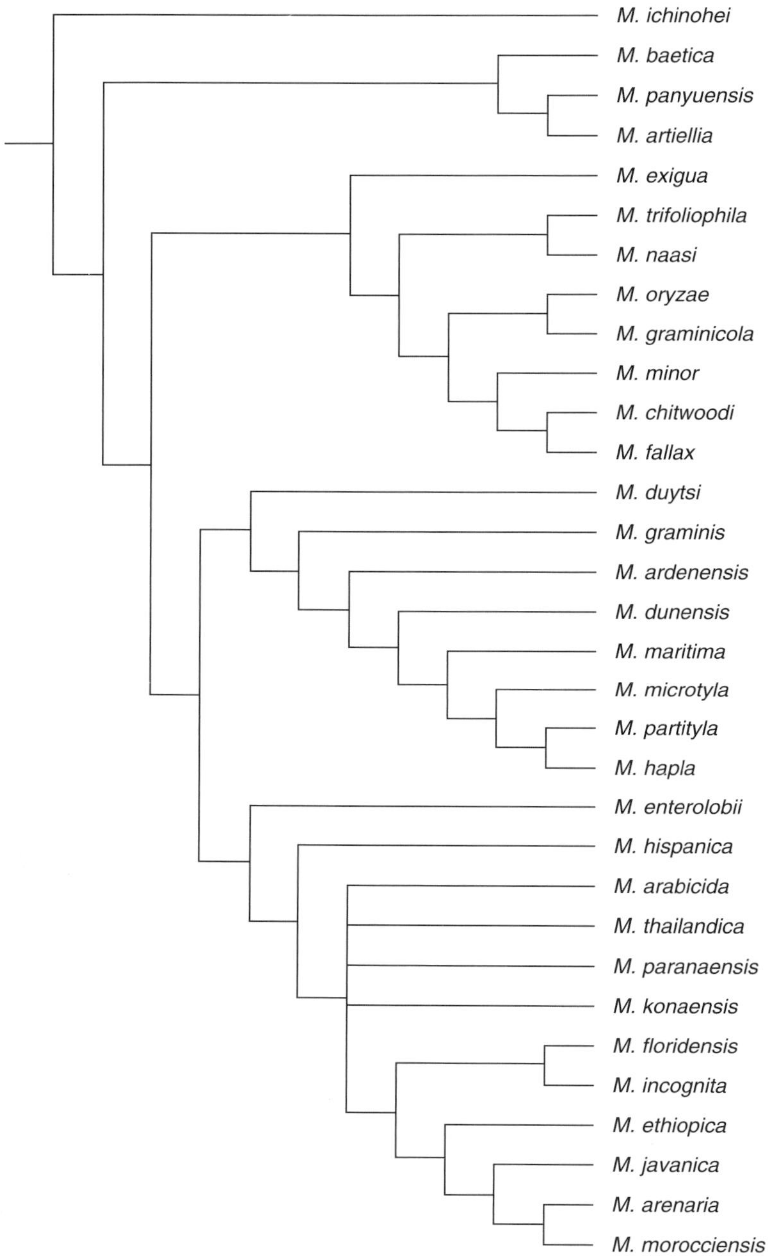

Fig. 5.15C. Dfit (most similar supertree, or distance fit) combinable components consensus.

III, which contains *M. graminicola*, *M. chitwoodi*, *M. exigua*, *M. trifoliophila*, *M. naasi*, *M. minor*, *M. fallax* and *M. oryzae*.

Tenente *et al.* (2004) showed that there was considerable variation in the 28S region among their sampled taxa (52 differences between *M. chitwoodi* and its nearest taxon). The analyses of Tigano *et al.* (2005) and De Ley *et al.* (2002) reveal even more amazing variation at the 18S locus. At the conservative end of this spectrum, *M. hispanica*

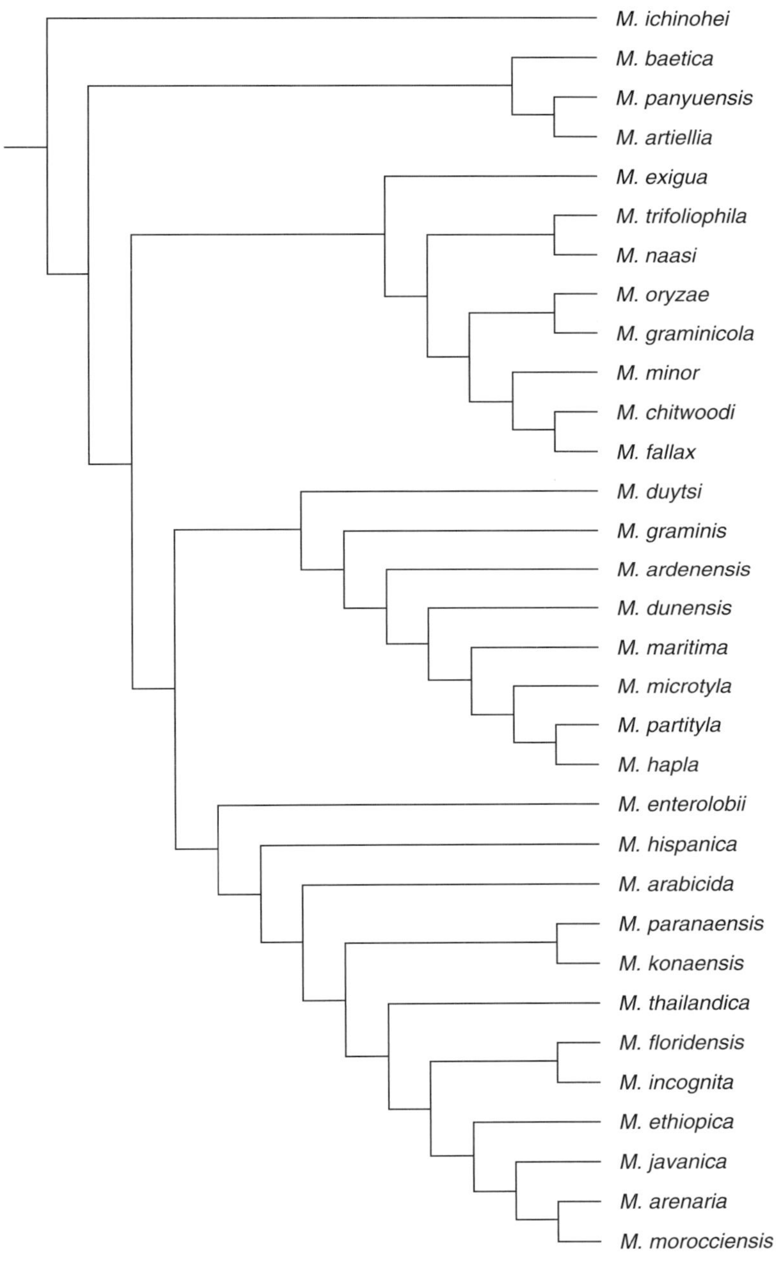

Fig. 5.15D. Dfit (most similar supertree, or distance fit) majority rule consensus tree.

and *M. ethiopica* sequences are identical (Landa *et al.*, 2008). At the other end is the fact that there can be more variation between two species of *Meloidogyne* (say, *M. maritima* and *M. artiellia*) than between a human and a platypus (63 differences). Many of the differences are attributable to base-call ambiguity that could be resolved with more persistent sampling and sequencing effort (probably attributable to intraspecific and intra-individual variation). Still, such discrepancies are astonishing when one considers that two organisms identified by nematode taxonomists as the same species (i.e. two isolates of *M. javanica*) have far greater 18S rDNA sequence divergence (24 substitutions, four indels) than an Australian wombat and a North American possum (three substitutions, two indels). Certainly, nematodes are reported to have high rates of sequence divergence relative to morphological evolution (Stein *et al.*, 2003; Sudhaus and Kiontke, 2007), and *Meloidogyne* spp. are part of a rapidly evolving clade (Holterman *et al.*, 2006), but such high divergence at the 18S locus among closely related species, and even among individuals within a species, is sufficient cause for further investigation into the origin(s) of *Meloidogyne* species and the morphological and molecular characters considered informative for diagnosis and species delimitation.

There are several reasons why the published 18S and mtDNA analyses might have returned different topologies for the clade I mitotic parthenogens. The first, and most obvious, is that they represent discordant evolutionary histories. This notion was first empirically tested by Tigano *et al.* (2005), who performed an incongruence length difference test (Farris *et al.*, 1994, 1995) (but see Hipp *et al.*, 2004; Barker and Lutzoni, 2002; Darlu and Lecointre, 2002) on their mtDNA and 18S rDNA sequences and found significant differences, sufficient to conclude that the two data sets did not share a common evolutionary history. This observation could simply be due to the fundamental nature of the two markers: 18S sequences reside within a tandemly repeated cistron that is subject to mutation and gene conversion through concerted evolution, which could result in paralogous, and not orthologous, gene sequences (Slowinski and Page, 1999). Alternatively, it is possible that lineage sorting of mitochondrial haplotypes, mutation rate heterogeneity or sampling error from the small number of phylogenetically informative nucleotide bases examined could result in discordance between gene and species trees (Maddison, 1997; Funk and Omland, 2003; Avise, 2007).The problem of resolving phylogenetic relationships among the mitotic parthenogenetic species goes far beyond differential lineage sorting and gene conversion, and is most certainly compounded by their probable hybrid origins. It has long been suspected that *M. incognita*, *M. arenaria* and *M. javanica* arose through hybridization events between sexual or meiotic parthenogenic taxa (Triantaphyllou, 1985; Castagnone-Sereno *et al.*, 1993).

What will it take to achieve a fully resolved, robust *Meloidogyne* phylogeny with near complete representation of all its species? First, and perhaps most importantly, it will take a rigorous sampling effort of both genes and taxa. This Herculean step requires thorough field sampling of genetic variation across the globe, collaborative research involving experts in both morphological and molecular identification, and the resources and will to generate enormous amounts of DNA sequence data for each species. Second, it must be recognized that because the relative amount of DNA sequence divergence is so varied between taxa (some extremely high, some extremely low), phylogenetic analysis of the whole group will require use of suites of genes that evolve very slowly for deep nodes and very rapidly for shallow nodes of the tree. It is likely that genes appropriate for resolving relationships among closely related species will be inappropriate, if not completely alignment ambiguous, among distantly related species. Thus, phylogenomic and total evidence analyses using concatenated data sets will probably be highly informative for resolving relationships among deeper nodes, but unless they can also sample variation within and between populations, meta-analyses and supertree construction are likely to be required to assemble a tree with the greatest explanatory power.

The observed inability to obtain monophyletic relationships among ITS rDNA sequences (Hugall *et al.*, 1999) and/or 'alleles' of putative single-copy nuclear loci (Lunt, 2008) from hybridogenic lineages is completely consistent with the expected fate of such genes upon phylogenetic analysis. In fact, it is highly likely that even the resolution of those genealogies exhibited by the phylogenomic analysis of Scholl and Bird (2005) would dissolve upon further sampling of 'allelic' variation among additional individuals of these species from disparate populations. Simply put, the evolutionary lineages that comprise these lineages may not have unique evolutionary origins or fates and thus are not only intractable phylogenetically but also ontologically (Ghiselin, 1997; Adams, 2001). Regardless, the resolution of historical relationships among the genes that comprise the 'species' remains the single most powerful tool in the arsenal of comparative methods for understanding the origin and evolution of what are arguably the most perplexing, and vexing, nematodes on earth.

5.6 References

Adams, B.J. (2001) The species delimitation uncertainty principle. *Journal of Nematology* 33, 153–160.

Adams, B.J. and Powers, T.O. (1996) An rDNA phylogenetic analysis of the genus *Meloidogyne*: a recent origin of mitotic parthenogenetic species. *Nematropica* 26, 193–194 (abstract).

Avise, J.C. (2007) *On Evolution*. The Johns Hopkins University Press, Baltimore, Maryland.

Barker, F.K. and Lutzoni, F.M. (2002) The utility of the incongruence length difference test. *Systematic Biology* 51, 625–637.

Baum, T.J., Gresshoff, P.M., Lewis, S.A. and Dean, R.A. (1994) Characterization and phylogenetic analysis of 4 root-knot nematode species using DNA amplification fingerprinting and automated polyacrylamide gel electrophoresis. *Molecular Plant-Microbe Interactions* 7, 39–47.

Bininda-Emonds, O.R.P. (2004) *Phylogenetic Supertrees: Combining Information to Reveal the Tree of Life*. Kluwer Academic Publishers, Dordrecht, The Netherlands.

Bininda-Emonds, O.R.P., Gittleman, J.L. and Purvis, A. (1999) Building large trees by combining phylogenetic information: a complete phylogeny of the extant Carnivora (Mammalia). *Biological Reviews* 74, 143–175.

Bininda-Emonds, O.R.P., Jones, K.E., Price, S.A., Cardillo, M.G.R. and Purvis, A. (2004) Garbage in, garbage out: data issues in supertree construction. In: Bininda-Emonds, O.R.P. (ed.) *Phylogenetic Supertrees: Combining Information to Reveal the Tree of Life*. Kluwer Academic Publishers, Dordrecht, The Netherlands, pp. 267–280.

Blok, V.C., Wishart, J., Fargette, M., Berthier, K. and Phillips, M.S. (2002) Mitochondrial DNA differences distinguishing *Meloidogyne mayaguensis* from the major species of tropical root-knot nematodes. *Nematology* 4, 773–781.

Brito, J., Powers, T.O., Mullin, P.G., Inserra, R.N. and Dickson, D.W. (2004) Morphological and molecular characterization of *Meloidogyne mayaguensis* isolates from Florida. *Journal of Nematology* 36, 232–240.

Carneiro, R.M.D.G., Almeida, M.R.A. and Quénéhervé, P. (2000) Enzyme phenotypes of *Meloidogyne* spp. populations. *Nematology* 2, 645–654.

Castagnone-Sereno, P., Piotte, C., Uijthof, J., Abad, P., Wajnberg, E., Vanlerberghe Masutti, F., Bongiovanni, M. and Dalmasso, A. (1993) Phylogenetic relationships between amphimictic and parthenogenetic nematodes of the genus *Meloidogyne* as inferred from repetitive DNA analysis. *Heredity* 70, 195–204.

Castillo, P., Vovlas, N., Subbotin, S. and Troccoli, A. (2003) A new root-knot nematode, *Meloidogyne baetica* n. sp (Nematoda : Heteroderidae), parasitizing wild olive in Southern Spain. *Phytopathology* 93, 1093–1102.

Creevey, C.J. and McInerney, J.O. (2005) Clann: investigating phylogenetic information through supertree analyses. *Bioinformatics* 21, 390–392.

Dalmasso, A. and Bergé, J.B. (1978) Molecular polymorphism and phylogenetic relationship in some *Meloidogyne* spp.: application to taxonomy of *Meloidogyne*. *Journal of Nematology* 10, 323–332.

Darlu, P. and Lecointre, G. (2002) When does the incongruence length difference test fail? *Molecular Biology and Evolution* 19, 432–437.

De Ley, I.T., De Ley, P., Vierstraete, A., Karssen, G., Moens, M. and Vanfleteren, J. (2002) Phylogenetic analyses of *Meloidogyne* small subunit rDNA. *Journal of Nematology* 34, 319–327.

De Ley, P., Felix, M.A., Frisse, L.M., Nadler, S.A., Sternberg, P.W. and Thomas, W.K. (1999) Molecular and morphological characterisation of two reproductively isolated species with mirror-image anatomy (Nematoda : Cephalobidae). *Nematology* 1, 591–612.

Dickson, D.W., Huisingh, D. and Sasser, J.N. (1971) Dehydrogenases, acid and alkaline phosphatases, and esterases for chemotaxonomy of selected *Meloidogyne, Ditylenchus, Heterodera,* and *Aphelenchus* spp. *Journal of Nematology* 3, 1–16.

Eisen, J.A. (1998) Phylogenomics: improving functional predictions for uncharacterized genes by evolutionary analysis. *Genome Research* 8, 163–167.

Eisen, J.A. and Fraser, C.M. (2003) Phylogenomics: intersection of evolution and genomics. *Science* 300, 1706–1707.

Eisenback, J.D. and Triantaphyllou, H.H. (1991) Root-knot nematodes: *Meloidogyne* species and races. In: Nickle, W.R. (ed.) *Manual of Agricultural Nematology*. Marcel Dekker, New York, pp. 191–274.

Esbenshade, P.R. and Triantaphyllou, A.C. (1985) Use of enzyme phenotypes for identification of *Meloidogyne* species. *Journal of Nematology* 17, 6–20.

Esbenshade, P.R. and Triantaphyllou, A.C. (1987) Enzymatic relationships and evolution in the genus *Meloidogyne* (Nematoda, Tylenchida). *Journal of Nematology* 19, 8–18.

Esbenshade, P.R. and Triantaphyllou, A.C. (1990) Isozyme phenotypes for the identification of *Meloidogyne* species. *Journal of Nematology* 22, 10–15.

Farris, J.S. (1983) The logical basis of phylogenetic analysis. In: Platnick, N.I. and Funk, V.A. (eds) *Advances in Cladistics*. Columbia University Press, New York, pp. 570–572.

Farris, J.S., Kallersjo, M., Kluge, A.G. and Bult, C. (1994) Testing significance of incongruence. *Cladistics* 10, 315–319.

Farris, J.S., Källersjö, M., Kluge, A.G. and Bult, C. (1995) Constructing a significance test for incongruence. *Systematic Biology* 44, 570–572.

Funk, D.J. and Omland, K.E. (2003) Species-level paraphyly and polyphyly: frequency, causes, and consequences, with insights from animal mitochondrial DNA. *Annual Review of Ecology Evolution and Systematics* 34, 397–423.

Ghiselin, M.T. (1997) *Metaphysics and the Origin of Species*. State University of New York Press, Albany, New York.

Guindon, S. and Gascuel, O. (2003) A simple, fast, and accurate algorithm to estimate large phylogenies by maximum likelihood. *Systematic Biology* 52, 696–704.

Handoo, Z.A., Nyczepir, A.P., Esmenjaud, D., van der Beek, J.G., Castagnone-Sereno, P., Carta, L.K., Skantar, A.M. and Higgins, J.A. (2004) Morphological, molecular, and differential-host characterization of *Meloidogyne floridensis* n. sp (Nematoda : Meloidogynidae), a root-knot nematode parasitizing peach in Florida. *Journal of Nematology* 36, 20–35.

Hillis, D.M. and Dixon, M.T. (1991) Ribosomal DNA: molecular evolution and phylogenetic inference. *The Quarterly Review of Biology* 66, 411–453.

Hipp, A.L., Hall, J.C. and Sytsma, K.J. (2004) Congruence versus phylogenetic accuracy: revisiting the incongruence length difference test. *Systematic Biology* 53, 81–89.

Holterman, M., van der Wurff, A., van den Elsen, S., van Megen, H., Bongers, T., Holovachov, O., Bakker, J. and Helder, J. (2006) Phylum-wide analysis of SSU rDNA reveals deep phylogenetic relationships among nematodes and accelerated evolution toward crown clades. *Molecular Biology and Evolution* 23, 1792–1800.

Huelsenbeck, J.P. and Ronquist, F. (2001) MRBAYES: Bayesian inference of phylogenetic trees. *Bioinformatics* 17, 754–755.

Huelsenbeck, J.P., Joyce, P., Lakner, C. and Ronquist, F. (2008) Bayesian analysis of amino acid substitution models. *Philosophical Transactions of the Royal Society B – Biological Sciences* 363, 3941–3953.

Hugall, A., Moritz, C., Stanton, J. and Wolstenholme, D.R. (1994) Low, but strongly structured mitochondrial-DNA diversity in root-knot nematodes (*Meloidogyne*). *Genetics* 136, 903–912.

Hugall, A., Stanton, J. and Moritz, C. (1997) Evolution of the AT-rich mitochondrial DNA of the root knot nematode, *Meloidogyne hapla*. *Molecular Biology and Evolution* 14, 40–48.

Hugall, A., Stanton, J. and Moritz, C. (1999) Reticulate evolution and the origins of ribosomal internal transcribed spacer diversity in apomictic *Meloidogyne*. *Molecular Biology and Evolution* 16, 157–164.

Kluge, A.G. (1997) Testability and the refutation and corroboration of cladistic hypotheses. *Cladistics – the International Journal of the Willi Hennig Society* 13, 81–96.

Kornberg, R.D. (2007) The molecular basis of eukaryotic transcription. *Proceedings of the National Academy of Sciences of the USA* 104, 12955–12961.

Landa, B.B., Rius, J.E.P., Vovlas, N., Carneiro, R., Maleita, C.M.N., Abrantes, I.M.D. and Castillo, P. (2008) Molecular characterization of *Meloidogyne hispanica* (Nematoda, Meloidogynidae) by phylogenetic analysis of genes within the rDNA in *Meloidogyne* spp. *Plant Disease* 92, 1104–1110.

Longhorn, S.J., Foster, P.G. and Vogler, A.P. (2007) The nematode-arthropod clade revisited: phylogenomic analyses from ribosomal protein genes misled by shared evolutionary biases. *Cladistics* 23, 130–144.

Loytynoja, A. and Milinkovitch, M.C. (2003) A hidden Markov model for progressive multiple alignment. *Bioinformatics* 19, 1505–1513.

Lunt, D.H. (2008) Genetic tests of ancient asexuality in root knot nematodes reveal recent hybrid origins. *Bmc Evolutionary Biology* 8, 194.

Maddison, W.P. (1993) Missing data versus missing characters in phylogenetic analysis. *Systematic Biology* 42, 576–581.

Maddison, W.P. (1997) Gene trees in species trees. *Systematic Biology* 46, 523–536.

Maddison, W.P. and Knowles, L.L. (2006) Inferring phylogeny despite incomplete lineage sorting. *Systematic Biology* 55, 21–30.

Maddison, W.P. and Maddison, D.R. (2002) *MacClade version 4.0.5*. Sinauer, Sunderland, Massachusetts.

Morrison, D.A. and Ellis, J.T. (1997) Effects of nucleotide sequence alignment on phylogeny estimation: a case study of 18S rDNAs of Apicomplexa. *Molecular Biology and Evolution* 14, 428–441.

Nguyen, K.B., Maruniak, J. and Adams, B.J. (2001) Diagnostic and phylogenetic utility of the rDNA internal transcribed spacer sequences of *Steinernema. Journal of Nematology* 33, 73–82.

Powers, T.O. (2004) Nematode molecular diagnostics: from bands to barcodes. *Annual Review of Phytopathology* 42, 367–383.

Powers, T.O. and Harris, T.S. (1993) A polymerase chain reaction method for identification of five major *Meloidogyne* species. *Journal of Nematology* 25, 1–6.

Powers, T.O. and Sandall, L.J. (1988) Estimation of genetic divergence in *Meloidogyne* mitochondrial DNA. *Journal of Nematology* 20, 505–511.

Powers, T.O., Harris, T.S. and Hyman, B.C. (1993) Mitochondrial DNA sequence divergence among *Meloidogyne incognita*, *Romanomermis culicivorax*, *Ascaris suum*, and *Caenorhabditis elegans. Journal of Nematology* 25, 564–572.

Powers, T.O., Todd, T.C., Burnell, A.M., Murray, P.C.B., Fleming, C.C., Szalanski, A.L., Adams, B.A. and Harris, T.S. (1997) The rDNA internal transcribed spacer region as a taxonomic marker for nematodes. *Journal of Nematology* 29, 441–450.

Powers, T.O., Mullin, P.G., Harris, T.S., Sutton, L.A. and Higgins, R.S. (2005) Incorporating molecular identification of *Meloidogyne* spp. into a large-scale regional nematode survey. *Journal of Nematology* 37, 226–235.

Purvis, A. (1995) A modification to Baum and Ragan's method for combining phylogenetic trees. *Systematic Biology* 44, 251–255.

Roberts, R.G. and Bobrow, M. (1998) Dystrophins in vertebrates and invertebrates. *Human Molecular Genetics* 7, 589–595.

Roberts, R.G., Nicholson, L., Bone, Q. and Bobrow, M. (1995) Phylogenetic analysis of vertebrate and invertebrate dystrophins and utrophins. *Journal of Medical Genetics* 32, 142–142.

Roberts, T.M. (2005) Major sperm protein. *Current Biology* 15, R153.

Scholl, E.H. and Bird, D.M. (2005) Resolving tylenchid evolutionary relationships through multiple gene analysis derived from EST data. *Molecular Phylogenetics and Evolution* 36, 536–545.

Slowinski, J.B. and Page, R.D.M. (1999) How should species phylogenies be inferred from sequence data? *Systematic Biology* 48, 814–825.

Stanton, J., Hugall, A. and Moritz, C. (1997) Nucleotide polymorphisms and an improved PCR-based mtDNA diagnostic for parthenogenetic root-knot nematodes (*Meloidogyne* spp.). *Fundamental and Applied Nematology* 20, 261–268.

Stein, L.D., Bao, Z., Blasiar, D. *et al.* (2003) The genome sequence of *Caenorhabditis briggsae*: a platform for comparative genomics. *Public Library of Science – Biology* 1, 166–192.

Subbotin, S.A., Sturhan, D., Vovlas, N., Castillo, P., Tanyi Tambe, L., Moens, M. and Baldwin, J.G. (2006) Application of secondary structure model of rRNA for phylogeny: D2-D3 expansion segments of the LSU gene of plant-parasitic nematodes from the family Hoplolaimidae Filipjev, 1934. *Molecular Phylogenetics and Evolution* 43, 881–890.

Sudhaus, W. and Kiontke, K. (2007) Comparison of the cryptic nematode species *Caenorhabditis brenneri* sp n. and *C. remanei* (Nematoda : Rhabditidae) with the stem species pattern of the *Caenorhabditis elegans* group. *Zootaxa*, 45–62.

Swofford, D.L. (2002) *PAUP*: Phylogenetic Analysis Using Parsimony (*and Other Methods).* Sinauer Associates, Sunderland, Massachusetts.

Tenente, G.C.M.V., De Ley, P., De Ley, I.T., Karssen, G. and Vanfleteren, J.R. (2004) Sequence analysis of the D2/D3 region of the large subunit RDNA from different *Meloidogyne* isolates. *Nematropica* 34, 1–12.

Terry, M.D. and Whiting, M.F. (2005) Behavior of POY versus CLUSTALL across a broad parameter landscape. *Cladistics* 21, 272–281.

Thompson, J.D., Higgins, D.G. and Gibson, T.J. (1994) Clustal-W – improving the sensitivity of progressive multiple sequence alignment through sequence weighting, position-specific gap penalties and weight matrix choice. *Nucleic Acids Research* 22, 4673–4680.

Thompson, J.D., Gibson, T.J., Plewniak, F., Jeanmougin, F. and Higgins, D.G. (1997) The CLUSTAL X windows interface: flexible strategies for multiple sequence alignment aided by quality analysis tools. *Nucleic Acids Research* 25, 4876–4882.

Thorley, J.L. and Page, R.D M. (2000) RadCon: phylogenetic tree comparison and consensus. *Bioinformatics* 16, 486–487.

Tigano, M.S., Carneiro, R., Jeyaprakash, A., Dickson, D.W. and Adams, B.J. (2005) Phylogeny of *Meloidogyne* spp. based on 18S rDNA and the intergenic region of mitochondrial DNA sequences. *Nematology* 7, 851–862.

Triantaphyllou, A.C. (1966) Polyploidy and reproductive patterns in the root-knot nematode *Meloidogyne hapla*. *Journal of Morphology* 118, 403–414.

Triantaphyllou, A. (1985) Cytogenetics, cytotaxonomy and phylogeny of root-knot nematodes. In: Sasser, J.N. and Carter, C.C. (eds) *An Advanced Treatise on* Meloidogyne, Volume 1. North Carolina State University Graphics, North Carolina, Raleigh, North Carolina, pp. 113–126.

Trudgill, D.L. and Blok, V.C. (2001) Apomictic, polyphagous root-knot nematodes: exceptionally successful and damaging biotrophic root pathogens. *Annual Review of Phytopathology,* 39, 53–77.

van der Beek, J.G., Folkertsma, R., Zijlstra, C., Van Koert, P.H.G., Poleij, L.M. and Bakker, J. (1998) Genetic variation among parthenogenetic *Meloidogyne* species revealed by AFLPs and 2D-protein electrophoresis contrasted to morphology. *Fundamental and Applied Nematology* 21, 401–411.

Watson, J.D., Baker, T.A., Bell, S.P., Gann, A., Levine, M. and Losick, R. (2008) *Molecular Biology of the Gene.* Benjamin-Cummings, Upper Saddle River, New Jersey.

Wiens, J.J. (1998) Does adding characters with missing data increase or decrease phylogenetic accuracy? *Systematic Biology* 47, 625–640.

6 Hatch and Host Location

Rosane H.C. Curtis[1], A. Forest Robinson[2] and Roland N. Perry[1,3]
[1]*Rothamsted Research, Harpenden, Hertfordshire, UK;* [2]*USDA, ARS, Southern Crops Research Laboratory, College Station, Texas, USA;* [3]*Ghent University, Ghent, Belgium*

6.1 Introduction **139**
6.2 Hatching **139**
6.3 Movement Through Soil **144**
6.4 Host Location **147**
6.5 Nematode Changes and Responses at the Root–Soil Interface **149**
6.6 Conclusions and Future Directions **155**
6.7 References **155**

6.1 Introduction

As an obligate plant endoparasite, the infective second-stage juvenile (J2) of *Meloidogyne* has to locate a host as soon as possible after hatching in order to feed and develop. Optimum conditions for hatching are also ideal for movement through the soil. Thus, hatching and movement occur under favourable conditions of temperature, moisture and aeration to maximize the probability of host location. However, the dynamics of the soil ecosystem make analysis of the importance of individual factors and their interactions difficult, and most research has been based on *in vitro* analysis of individual features. Several factors influence the movement and attraction of J2 towards roots and, once at the root surface, changes to the J2 occur as a preparation for invasion. These aspects of the life cycle are discussed in this chapter.

The terms 'root diffusate', 'root leachate' and 'root exudate' have been used interchangeably by plant nematologists to describe the various compounds that emanate from plant roots that may affect nematode behaviour. Leachate refers more to the *in vitro* method of obtaining the solution rather than the solution itself. Diffusate conveys the idea of non-volatile and, especially, volatile substances diffusing through the soil for some distance, perhaps many centimetres, and establishing a concentration gradient; it is an especially apposite term in relation to hatching and attraction of nematodes. The word exudate is more properly restricted to liquids that gradually ooze out of a source. It is more relevant to localized changes at the root surface. However, to avoid confusion, in this chapter the term 'root exudate' will be used throughout, with the understanding that it relates to volatile and non-volatile components emanating from the plant root.

6.2 Hatching

The hatching behaviour of many species of parasitic nematodes is an essential component of the life cycle for optimizing the probabilities of successful infection by synchronization with host availability (Perry, 2002). The host influence on hatching, manifest primarily by the hatch-inducing effects of host root exudate, is most evident in species of cyst nematodes, such as *Globodera rostochiensis* and *G. pallida*, with narrow host ranges (Perry and Clarke, 1982). For species of *Meloidogyne* with relatively wide host ranges, the influence of exudate is less apparent, with most J2 hatching in water without the need for exudate stimulation. However, as will be discussed below, root exudates influence hatch of some species of *Meloidogyne*.

6.2.1 General hatching response

A simplistic general description of the hatching of *Meloidogyne* is that J2 hatch when environmental conditions, such as appropriate temperature, oxygen availability and soil moisture levels, are suitable, and there is an absence of physiological barriers, such as diapause. Diapause is a state of arrested development whereby development does not occur until specific requirements have been satisfied, even if favourable conditions return. Diapause is a well-documented phenomenon in insects and was described in detail for plant-parasitic nematodes by de Guiran (1979), working with *M. incognita*. Subsequent work has demonstrated that certain species of *Meloidogyne* exhibit diapause, although the percentage of unhatched J2 in diapause varies considerably from 10% in the predominantly tropical *M. arenaria* to 94% in *M. naasi*, which has a more temperate distribution (de Guiran and Villemin, 1980). Diapause as a physiological adaptation is an important component of the life cycle in relation to hatching and survival; diapause in nematodes has been reviewed by Jones *et al.* (1998) and Wright and Perry (2006) and is assessed in detail by Evans and Perry, Chapter 9, this volume.

6.2.2 Hatching mechanism

Females of *Meloidogyne* do not retain eggs internally or form a cyst, but lay eggs into a gelatinous matrix secreted through the anus by six large rectal glands. The gelatinous matrix consists of an irregular meshwork of glycoprotein material (Sharon and Spiegel, 1993). Each female may lay 30–40 eggs per day, and in a favourable host several hundred eggs are produced by each female (Karssen and Moens, 2006). Starr (1993) found a mean of 770 ± 190 eggs per egg mass of *M. incognita* on cotton. If the female is exposed on the root surface, the outer layers of the gelatinous matrix may become dry, which results in shrinkage and hardening of the matrix, thus exerting mechanical pressure on the eggs to inhibit hatching of J2 during desiccating conditions (Wallace, 1968b; Bird and Soeffky, 1972). Wallace (1968b) also considered that the swelling and shrinking properties of the gelatinous matrix regulate hatch so that under low moisture conditions it is reduced, thus ensuring that hatch occurs mainly when conditions are favourable for subsequent J2 movement through the soil. Conditions of low aeration of soils may reduce hatch of J2 by inducing quiescence (Wallace, 1968a), a dormant state induced by unfavourable conditions that is readily reversible when favourable conditions return. There is a decrease in percentage hatch with increase in the number of eggs in egg masses of *M. incognita* (Ishibashi *et al.*, 1960), indicating the accumulation of an inhibitor as hatching occurs.

Hatched J2 are vulnerable to environmental stresses and they are viable in the soil for periods much shorter than if they had remained unhatched. In addition to the gelatinous matrix, the eggshell affords protection to the enclosed J2. The egg of *Meloidogyne* spp. is cylindrical with three layers: an outer vitelline, a middle chitinous and an inner lipid. Chitin has been found only in the eggshells of nematodes, and this has been demonstrated experimentally in *M. javanica* using histological tests and examination of electron diffraction patterns (Bird and Self, 1995). Bird and McClure (1976) estimated that the eggshell of *Meloidogyne* is 30% chitin. Veronico *et al.* (2001) and Fanelli *et al.* (2004) found that a single chitin synthase gene is responsible for chitin production in eggs of *M. artiellia*. Hatching of juveniles of *Meloidogyne* is primarily temperature driven. Species of *Meloidogyne* have been separated into two groups, thermophils and cryophils, based on their ability to survive lipid-phase transitions that occur at 10°C (Van Gundy, 1985). This grouping also relates to hatching; for example, *M. chitwoodi*, *M. hapla* and *M. naasi* are cryophils and can hatch at temperatures below 10°C, whereas *M. javanica* and *M. exigua* are thermophils and do not hatch at temperatures below 15°C.

Although there is considerable information about the cascade of events after hatch stimulation of cyst nematodes, especially *Globodera* spp., there is less about the hatching biology of species of *Meloidogyne* and information is fragmentary and derives from work on several species. The hatching process of cyst nematodes has been divided into three phases: changes in the eggshell, activation of the J2, and eclosion (Perry, 2002). In *Meloidogyne*, activation of the J2 precedes, and probably causes, changes in eggshell structure, in contrast to cyst nematodes, such as *G. rostochiensis*, where changes in the eggshell initiate the hatching sequence (Perry, 2002; Table 6.1).

The changes in the eggshell structure of *Meloidogyne* immediately before hatching are manifested by a marked change in permeability. Eggs of *M. javanica* are initially impermeable to osmium tetroxide (Bird and Bird, 1991), but immediately prior to hatch the eggshell becomes permeable. When eggshells are permeable, the unhatched J2 are susceptible to toxic compounds, including plant extracts that may have potential as control agents. For example, Meyer *et al.* (2006) found that the extracts of *Plantago lanceolata* and *P. rugelii* were toxic to unhatched J2 of *M. incognita* eggs. Other compounds, such as the biogenic amines serototin and octopamine, reduced hatch of *M. incognita* (Masler, 2008).

Bird (1968) suggested that, in *M. javanica*, enzymes from the pharyngeal glands of the J2 caused hydrolysis and flexibility of the eggshell. Using Coomassie Brilliant Blue G-250 dye, Premachandran *et al.* (1988) showed that unhatched J2 of *M. incognita* produced protein secretions from the amphidial and secretory–excretory systems. Subsequently, Perry *et al.* (1992) demonstrated that the hatch of *M. incognita* was positively correlated with lipase activity in the hatching fluid (Fig. 6.1), and proteinase, including collagenase, and chitinase activity also were detected; these enzymes are likely to erode layers of the eggshell, resulting in increased flexibility and softening of the eggshell prior to eclosion. Wallace (1968b) noted increased permeability to water of *M. javanica* eggs before hatching occurs. The flexible egg of *M. incognita* has about 30% internal free space, which allows the J2 to be fully hydrated and active by the time of eclosion (Ellenby, 1974). In *M. javanica*, the anterior end of the J2 projects into the flexible eggshell, and stylet thrusts cause a tear through which the J2 escapes. *M. arenaria* made 70–90 thrusts a minute and repeatedly returned to the same place on the eggshell (Dropkin *et al.*, 1958). In this overall scenario, hatching of *Meloidogyne* occurs once the J2 has fully developed. In contrast to many cyst nematodes, where root diffusates are required to change the eggshell permeability characteristics and initiate hatching, it is the *Meloidogyne* J2 itself that changes the eggshell structure prior to hatching.

Thus, in general, hatch of *Meloidogyne* occurs without requiring specific cues from host roots (Perry, 1987). However, there are exceptions, and responses to root exudates may be more important than previously realized (Perry and Wesemael, 2008). Root exudates can affect the rate of hatch (i.e. the number of J2 that hatch per unit time), and some J2 within a population depend on root exudate to initiate the hatching process.

Table 6.1. Comparison between the hatching process of *Meloidogyne* spp. and *Globodera* spp. Information from various sources summarized by Perry (2002).

Meloidogyne spp.	*Globodera* spp.
Majority of J2 hatch in water without stimulation from host root exudates	Majority of J2 need stimulation by host root exudates
Permeability change of the eggshell is caused by enzyme secretion and, possibly, J2 activity	Eggshell permeability change caused by Ca^{2+}-induced alteration; there is no evidence of enzyme secretion
J2 activity occurs before eggshell permeability change	J2 activity occurs after eggshell permeability change
Egg becomes flexible before J2 hatches	Egg remains rigid during the hatching process
Stylet used to puncture eggshell; J2 extends tear as it hatches	Coordinated stylet thrusts cut subpolar slit in eggshell
J2 fully hydrated before hatching	J2 takes up water after hatching

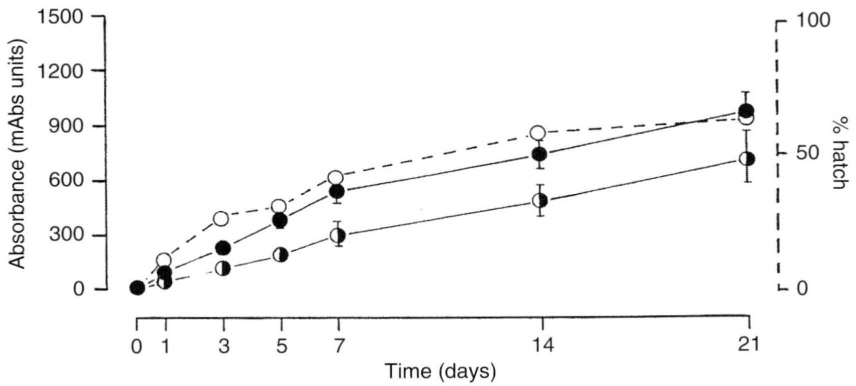

Fig. 6.1. Cumulative lipase activity, assayed at pH 5 (●) and 8 (◑), and percentage hatch (○) of *Meloidogyne incognita* in glass distilled water over a 3-week period. Enzyme activity is expressed in absorbance units as mean ± standard error of three separate analyses. (From Perry *et al.*, 1992.)

6.2.3 Dependence on root exudates

Early work indicated that root exudates can enhance the rate of hatch and the total percentage hatch of some species of *Meloidogyne*. For example, hatch of *M. hapla*, *M. incognita* and *M. javanica* was enhanced when stimulated by host root exudates (Viglierchio and Lownsbery, 1960; Brzeski and Hendricks, 1971). However, Shepherd and Clarke (1971), reviewing the early work on observations on increased hatch in *in vitro* experiments, concluded that the increase was small and barely significant. Much of the early work on enhancement of the rate of hatch is limited and results are often conflicting, perhaps because the age (generation) of egg masses and host from which they were obtained were not defined. More recent information illustrates the importance of the changes in hatch in relation to host age (Perry and Wesemael, 2008).

It is clear that, for some species of *Meloidogyne*, root exudates enhance hatch. Hydroponic culture media in which tomato, okra, cucumber and bean had been grown were tested for their *in vitro* hatching effect on *M. incognita*; tomato and okra culture media increased the numbers of J2 that hatched as compared with those in water or fresh culture medium (Oka and Mizukubo, 2009). The sophisticated host–parasite interaction of *Meloidogyne* extends to modifications in hatching response as a reflection of changes in cues from host root exudates. The number of generations per year varies according to species and food availability. Most species have multiple generations during a host growing season but some, such as *M. naasi*, have only one generation. For some species of *Meloidogyne* with multiple generations per year, there is evidence that a large proportion of J2 from later generations do not hatch but remain protected by the egg and gelatinous matrix during the period between crops. This variation in hatching between generations was investigated by Ishibashi (1969), who found that old or poorly nourished females of *M. incognita* produce brown egg masses containing eggs with dormant J2, which are resistant to environmental stresses and nematicides, and which depend on hatch stimulation by host root exudates. By contrast, young and well-nourished females produce white egg masses susceptible to environmental stresses but from which J2 hatch readily in water without needing exudate stimulation. Immediate hatching of J2 from white egg masses ensures more than one generation per host growing season, while the dormant J2 in brown egg masses ensure survival and carry-over of infective J2 from one season to the next. Gaur *et al.* (2000) showed that females of *M. triticoryzae* from India produced three types of eggs: those that hatch in water, those that hatch in host root exudates and those that do not hatch even in the presence of exudate. The proportion of these three types varies with generation, with the final generation produced on senescing plants having a large proportion of unhatched juveniles of the third type, which is likely to equate with diapause. The difference between generations in hatching of J2 from egg masses produced at different stages of plant growth illustrates the influence of the plant on development and subsequent hatching. This reflects a change of priority for the species during the host-plant growing season, from rapid reinfection of young plants to survival after host senescence.

The hatching response of species that coexist may also differ. *Meloidogyne chitwoodi* and *M. fallax* cause severe damage to economically important crops and, in 1998, both species were listed as quarantine pests in Europe (see Moens *et al.*, Chapter 1, this volume). Comparative studies were made on the effects of root exudates and host age on the *in vitro* hatching of *M. chitwoodi* and *M. fallax* (Wesemael *et al.*, 2006). There is a marked contrast in the hatching response of the two species. Hatching of J2 of *M. chitwoodi* produced on young plants did not require host root exudate stimulus, whereas at the end of the plant growing season, egg masses contained a percentage of unhatched J2 that required host root exudate to cause hatch (Fig. 6.2). This form of obligate dormancy at the end of the host growing season was not found in *M. fallax* (Fig. 6.2). This species hatched well in water and did not require hatch stimulation from root exudate, irrespective of the age of the plant on which the egg masses were produced. The factors causing the change in hatching response of species such as *M. triticoryzae* and *M. chitwoodi* are unknown.

In the *in vitro* experiments on *M. chitwoodi* and *M. fallax*, egg masses were exposed to optimum conditions for hatch, including exudates from 6-week-old plants that had maximum hatching activity (Wesemael *et al.*, 2006). Work on cyst nematode hatching has demonstrated that the production of active exudates declines with the onset of plant senescence (Perry, 1997) and, in several plant species, it is confined to a short period of plant growth. This is an additional factor ensuring that J2 do not hatch at a time when their food source is senescing but carry over to the next crop growing season; research on this aspect is needed for *Meloidogyne* species.

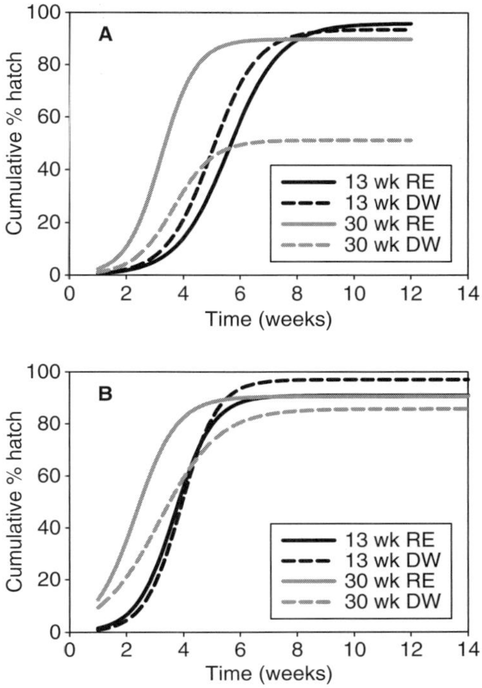

Fig. 6.2. Fitted curves showing cumulative percentage hatch in distilled water (DW), and tomato root exudate (RE) from egg masses of *Meloidogyne chitwoodi* and *M. fallax* collected from 13-week- and 30-week-old tomato plants. (Adapted from Wesemael *et al.*, 2006.)

6.2.4 Egg numbers and embryogenesis

Other differences between *M. chitwoodi* and *M. fallax* were noted by Wesemael *et al.* (2006). The number of eggs per egg mass for *M. fallax* collected on senescing plants was significantly greater than that for *M. chitwoodi*, and the number of eggs per egg mass of *M. chitwoodi* decreased with plant age. Together with the fact that 90% of J2 of *M. fallax* from egg masses on senescing plants hatched without root exudate stimulation, this might indicate that a greater number of J2 of *M. fallax* remains in the soil after crop harvest compared with *M. chitwoodi*.

Although no quiescent unhatched J2 were found in egg masses of *M. fallax* obtained from senescing plants, there was a small proportion (4–14%) of unembryonated eggs that did not contain developed J2. Unembryonated eggs were also found in *M. chitwoodi*. This type of arrested development during embryogenesis of species of *Meloidogyne* was reviewed by Evans (1987), who concluded that it could not be considered as a diapause as there was no element of periodicity linked to season.

The differences between the hatching responses of species of *Meloidogyne* species such as *M. chitwoodi* and *M. fallax* may be linked to different survival strategies. *M. chitwoodi* seems to have a strategy for survival in the intercrop period that is centred on quiescent J2, which hatch only in the presence of root exudates, and on delayed development of unembryonated eggs. By contrast, survival of *M. fallax* seems to be based on delayed development of embryonating eggs and, presumably, the ability of hatched J2 to survive in the soil. The greater number of eggs per egg mass of *M. fallax* collected on senescing plants indicates that a large number of J2 will be in the soil.

6.3 Movement Through Soil

6.3.1 How root-knot juveniles move

Movement of J2 of root-knot nematodes is similar to that of most nematodes. They move by dorsoventral undulations of the body that are propagated backward from the anterior end, resulting in sinusoidal paths on agar (Croll, 1970; Croll and Sukhdeo, 1981; Alexander, 2002; Burr and Robinson, 2004). Body waves are propagated endogenously and rhythmically but more slowly (one wave per 2–5 seconds) than in most nematodes, with continuous wave propagation persisting until an obstacle is encountered. The response then is to stop, reverse one or two wavelengths, probe with the anterior end, and then resume forward movement in a new direction. On agar, the amplitude to wavelength ratio varies directly with external resistance (Robinson and Perry, 2006), increasing purchase against the substrate, and this response is likely to be differentially modulated along the body in response to surface irregularities, as has been shown in other nematode groups (Gans and Burr, 1994). Nematodes have the sensory capacity to detect a wide range of stimuli in their environment, and readers are referred to Perry and Maule (2004), Riga (2004) and reviews of nematode sensory structures and responses (Jones, 2002; Baldwin and Perry, 2004).

6.3.2 Factors influencing rate of movement

J2 of *Meloidogyne* do not exhibit developmentally induced changes in behaviour (see section 6.5) until encountering or penetrating a root, and so most of the factors affecting the rate of movement through soil are extrinsic. It is widely appreciated among nematologists working on *Meloidogyne* spp. that if stored lipids are exhausted sufficiently for J2 to lose optical refractivity, they often move more slowly. Also, the relationship between temperature and developmental rates differs significantly among species of root-knot nematodes, with *M. chitwoodi*, for example, exhibiting a base temperature of 4 °C, compared with *c.* 8, 10 and 14 °C for *M. hapla*, *M. incognita* and *M. javanica*, respectively (Inserra *et al.*, 1983; Lahtinen *et al.*, 1988; Pinkerton *et al.*, 1991; Madulu and Trudgill, 1994; Ploeg and Maris, 1999). It is likely that differences in thermal optima for motility rank similarly to those for nematode development. For example, vertical movement in 55-cm-long soil columns at 12 °C was considerably greater by *M. chitwoodi* than by *M. hapla*, consistent with the better adaptation of *M. chitwoodi* to low temperatures (Pinkerton *et al.*, 1987).

Moisture, porosity, oxygen availability, toxins and temperature can limit or stop movement by root-knot nematodes through soil. All nematodes the size of the root-knot nematode J2 require a film of moisture of a certain thickness for movement. The ideal moisture and porosity shown in the careful studies of Wallace (1958, 1959, 1960, 1968c) can be defined in terms of the effective pore size and the Gibbs free energy (or water potential) of the water. Porosity varies partly with the absolute size of soil particles, partly with the particle aggregate size, and partly with the degree of compaction, which often is expressed as bulk density. Root-knot nematodes are well known for their tendency to be a problem in sandy soils, and their movement also appears to be favoured by low soil compaction (Eo *et al.*, 2007).

The Gibbs free energy of water in soil is a function of the affinity of soil particle surfaces for water (matric potential) and the salt concentration (osmotic potential). Blake (1961) showed that, over ranges that occur in most natural soils, only the matric potential is an important factor. Osmotic potential is not important in most soils because as soil dries, matric effects prevent nematode movement well before water removal elevates salt concentrations to physiologically significant levels. Wallace's most important finding was probably that optimal soil moisture for movement by nematodes occurs at 'field capacity', or the point when saturated soil has drained into equilibrium by gravity, and no further extraction of water by plants has yet occurred (Wallace, 1958). Field capacity occurs at much higher total moisture contents in clay soils than in sandy soils. At field capacity, conditions are optimal for nematode hatching and root growth: soil is moist, friable and permeated by continuous channels of air, through which respiratory gases readily diffuse over considerable distances. Thus, conditions are optimally suitable for the establishment of long-distance gradients of volatiles within the gas phase of soil, and relatively short-distance gradients of solutes. Either might be utilized by nematodes to locate roots.

The data available indicate that, although root-knot nematode J2 lose motility and eventually are killed under strongly reducing, anaerobic conditions, motility seems little or not affected over the concentration ranges of O_2 and CO_2 typically found in soil. Oxygen in soil occurs at *c.* 20% but may drop to nearly zero with a high biological oxygen demand. CO_2 concentrations in soil are about 0.25% in well-aerated soil, but may be as high as 1.0 or 2.0% in poorly aerated soil. Generally, as soil O_2 drops from a maximum concentration in the gas phase of 20% to 0%, CO_2 concentration increases from around 0.25% to 1.0 or 2.0% (Campbell, 1985; Bajracharya *et al.*, 2000). Oxygen appears to limit activity of *Meloidogyne* J2 only at very low concentrations. CO_2 intoxication in *Rotylenchulus reniformis* (Robinson and Heald, 1991) and in *M. incognita* (Robinson, unpublished data), for example, began at about 5%, which is typically encountered only in waterlogged soil. Ammonium also has been shown to be toxic to root-knot J2 and, as will be discussed, may play a role in modulating orientation to CO_2.

6.3.3 Plant-independent factors influencing the direction of nematode movement

Soil is an immense mass of porous material containing water, air and many living organisms and, ignoring deep percolation in some strata, has basically only one (upper) surface across which heat, water and respiratory gases move. Moreover, soil is created by the deposition of horizontal layers over long periods of time, producing textural strata. Daily, the soil surface is markedly heated and then cooled, with a resulting heat wave propagated down through the soil every 24 h. At less frequent intervals, wetting fronts move variable distances downward through soil following rain and irrigation. As soil contains many living organisms, significant respiratory gas exchange occurs at the surface. Thus, with depth, as total resistance to gaseous diffusion increases, O_2 decreases and CO_2 increases in concentration. In sum, just about any measurable substance or quality within soil might be expected to occur as a vertical gradient, and many of those gradients fluctuate over time. The challenge for plant-parasitic nematodes seeking roots is to distinguish these ever-present vertical gradients from the radial gradients around roots. Using micro-moulded substrates for *in vitro* testing of nematode responses to gravity, Eo *et al.* (2008) found that several nematode species, including *M. incognita*, were not geotaxic. Thus, vertical movement in soil is unlikely to be influenced by gravity and must be in response to other factors.

Temperature clearly has a potent influence on the vertical movement of root-knot nematode juveniles. Predicting which way they will go, however, is not simple. When placed on a static temperature gradient, *M. incognita* exhibits a preferred temperature, or intermediate temperature, towards which juveniles move (Wallace, 1961; Diez and Dusenbery, 1989a). They also exhibit adaptation to that temperature (Diez and Dusenbery, 1989a). Since vertical gradients in soil invert twice during the 24-h day, the direction of nematode movement depends on the preferred temperature, the rate of adaptation, the rate the heat wave is propagated down through the soil, and the rate of nematode movement, which was computer-modelled for soil organisms in general by Dusenbery (1988a) to predict various possible outcomes. In the case of *M. incognita*, laboratory experiments imposing natural patterns of temperature gradient fluctuations in sand were found to induce upward movement when J2 were placed at a depth of 10 cm (Robinson, 1994).

Vertical gradients of porosity and moisture result from many events, including soil deposition and evolution, disturbance and compaction by farm implements, earthworm activity, decreases in organic matter content with depth, precipitation and evapotranspiration. From the nematode's perspective, some soil layers at any given time will be more suited for movement than others. Based solely on kinetic effects, nematodes would be expected to accumulate in regions where conditions become unsuitable for movement, like automobiles at a traffic light. However, several experiments have revealed significant net movement opposite to the direction predicted by kinetics, suggesting orientation to gradients.

Effects of gradients of gases and salts can be difficult to distinguish from those of moisture and porosity, as all four factors are interrelated in soil. Gas and salt gradients are also affected by plants, since plant roots remove water, differentially take up salts, take up oxygen, release CO_2, and grow primarily in soil zones where relatively high levels of organic matter have accumulated in past years. However, vertical gradients of salts and respiratory gases occur independently of those extending radially from roots due to evaporation and exchange of respiratory gases at the surface. Those vertical gradients are likely to influence vertical nematode movement. As will be discussed, several simple salts, CO_2, and ammonia can profoundly influence the net direction of movement of nematodes in soil.

Movement by *M. javanica in vitro* towards moisture in tubes of sand was reversed when the concentration of Hoagland's plant nutrient solution and other salts was highest at the wet end (Prot, 1978b, 1979b). However, salts affect different species somewhat differently, and while *M. javanica* and *M. incognita* were repelled by a wide range of salts, *Heterodera oryzae* was repelled only by sodium, and *Scutellonema cavenessi* was unaffected (Prot, 1978a,c, 1979a). Vertical movement away from salts would usually lead nematodes to deeper regions of higher water content. The role of salts as root-finding cues, however, was not clear in these studies because sodium, which occurs in highest concentrations near roots, was repellent. Subsequent experiments by Castro *et al.* (1990, 1991) showed K^+, NH_4^+, Cs^+, NO_3^- and Cl^- to be strongly repellent to *M. incognita in vitro*, and glasshouse experiments indicated that ions could be used to protect plant roots from nematode invasion. In a different study, *M. javanica* also was usually, but not always, repelled by sodium salts (Abou-Setta and Duncan, 1998). Saux and Quénéhervé (2002) noted that the response of *M. incognita* is governed more by the constitutive cation than by the constitutive anion; they showed that calcium salts had no effect on the orientation of J2 of *M. incognita*, while ammonium salts and ammonium nitrate were strongly repellent. Unfortunately, many behavioural experiments testing responses of root-knot nematodes to salts have tested unbalanced salts in an otherwise salt-free water fraction, without confirmation of results in solutions containing the more balanced mixtures of salts present in naturally occurring soils.

Comparisons between responses to CO_2 *in vitro* and gradients measured in an apple orchard predicted that CO_2 gradients may draw root-knot nematodes downward as far as 1 m (Pline and Dusenbery, 1987). In a different study, *M. incognita* accumulated at the high end of linear gradients of CO_2 within moist sand packed inside a plastic tube, with the gradient generated by leaving each end of the tube open and exposing it to a continually purged atmosphere with a controlled CO_2 concentration and high relative humidity to prevent evaporative chilling (Robinson, 1995). Nematodes were deposited on the sand surface on one end, or injected into the centre of the sand mass, after sufficient time had passed to establish a concentration gradient. These experiments confirmed earlier *in vitro* experiments showing that root-knot nematodes were attracted to CO_2 at remarkably low concentrations and low concentration gradients, as long as exposure to toxic concentrations was avoided. Very importantly, sensitivity increased as ambient CO_2 concentration decreased (Pline and Dusenbery, 1987), allowing nematodes to be responsive at very low concentrations a long distance from the source. The maximal distance over which CO_2 can attract root-knot nematodes in soil has not been tested. However, experiments in soil with an insect, the corn root worm (*Diabrotica virgirifera*), which also is attracted to CO_2, have shown not only attraction many centimetres from the source but also disruption of root location in glasshouse and field settings by strategic placement of CO_2-generating granules (Bernklau and Bjostad, 1998; Bernklau *et al.*, 2004).

One other possible role of CO_2 should be noted. Pline and Dusenbery (1987) proposed that CO_2, like temperature, might also serve as a collimating stimulus, providing a vertical directional reference for nematodes, as well as for other soil organisms.

6.4 Host Location

6.4.1 General considerations

As mentioned above, in order for nematodes to find roots in the stimulus-laden soil environment, it would seem they need to respond to stimulus gradients around roots that are unique in character, concentration or combination. Participating stimuli may or may not also occur as vertical gradients or as gradients generated by decaying roots, tubers, fruits or insects within soil. Even though J2 of *Meloidogyne* are notoriously sluggish, controlled studies on long-distance movement frequently have shown them to move vertically more than 15 cm and sometimes 1 m in less than 1 month (Prot, 1980; Dickson and Hewlett, 1986; Mojtahedi *et al.*, 1991; De Rooij-Van der Goes *et al.*, 1998). Under field crops, the relative distribution of *M. chitwoodi*, assessed by monthly sampling over a 2-year period, was consistent in the fields tested and was not influenced significantly by different successions with good, moderate and poor hosts (Wesemael and Moens, 2008). These authors found that the highest final population densities of *M. chitwoodi* were greatest in the soil layers corresponding to the highest root densities, but this did not change the relative vertical distribution.

In order to locate host roots, nematodes respond to gradients of attractants via input from the sensilla (see section 6.5.2 and Eisenback and Hunt, Chapter 2, this volume). Perry (2005) classified attractants as long-distance, short-distance and local attractants. Long-distance attractants are those that attract nematodes to the general root area; short-distance attractants are those that attract nematodes to roots themselves; and local attractants are those that enable endoparasitic nematodes, such as *Meloidogyne*, to orientate to the preferred invasion site. This is a useful generalized framework to envisage attractants, but it must be realized that the classes are not mutually exclusive. Heat, soil gases and plant-specific compounds are among others that have been suggested as long-distance attractants

6.4.2 Heat

When El-Sherif and Mai (1969) originally observed that minute sources of heat attracted *Ditylenchus dipsaci*, *Pratylenchus penetrans* and *Tylenchorhynchus claytoni in vitro*, they speculated that metabolic heat from roots might attract nematodes. The requisite sensitivity is supported by theory (Dusenbery, 1988b) and by experiments (Dusenbery, 1988c; Pline *et al.*, 1988) demonstrating responses by root-knot nematodes to temperature changes of 0.001°C. Laboratory simulations of the heat waves that move through natural soil every day as a result of surface heating and cooling, however, confirmed Dusenbery's predictions about movement in response to fluctuating gradients, and showed such a profound influence on the vertical movement of *M. incognita* as to question how the relatively miniscule gradient around roots could possibly be detected by nematodes except in the deepest soil, probably more that 1 m deep (Robinson, 1994). More research is needed to define the role of metabolic heat in root attraction; Perry (2005) suggested that such heat may only be of importance in orientating the J2 to the preferred site of invasion at the metabolically highly active area just behind the root tip (see section 6.5.1).

6.4.3 Soil gases

CO_2 has been shown to attract *M. hapla*, *M. incognita* and *M. javanica* as well as seven other plant-parasitic and seven non-plant-parasitic nematodes (Robinson, 2004; Robinson and Perry, 2006). Based on behavioural thresholds measured for root-knot nematodes *in vitro*, the predicted theoretical distance from which they might be attracted to CO_2 produced by roots in soil is on the order of 1 m (Dusenbery, 1987). *Meloidogyne incognita* was attracted over distances of several centimetres to minute sources of CO_2 in sand in less than 24 h, and behaviourally relevant release rates of the gas from a point source were shown to be achievable by roots and other biological sources (Robinson, 1995). Indeed, CO_2 could be the most common and potent nematode attractant in nature, since it has been shown to attract a wide range of nematodes (Robinson, 2004).

CO_2 is released abundantly by living and decaying plant and animal tissues, providing an obvious cue to the possible presence of food. However, plant-parasitic nematodes are confronted with the problem of distinguishing living roots from decaying material from which they would be unable to obtain food. It is possible that other substances modulate nematode attraction to CO_2. For example, *M. incognita* is repelled by ammonia and several nitrogenous salts released by decaying material (Castro *et al.*, 1991; C.E. Castro, personal communication).

In a search for both volatile and non-volatile attractants of *M. incognita* to host roots, the only attractive chemical found was CO_2, although the presence of complex unidentified repellent chemicals was demonstrated (Diez and Dusenbery, 1989b; McCallum and Dusenbery, 1992). Robinson and Perry (2006) pointed out that evaluating the significance of CO_2 from roots as an attractant is compromised by the effects of CO_2 on gradients of redox potential, pH, carbonic acid, bicarbonate and carbonate in the soil. The consensus is that CO_2 attracts nematodes, including J2 of *Meloidogyne*, to roots and that either dissolved CO_2 or carbonic acid is the attractive species.

6.4.4 Uniquely plant-specific compounds

Chemicals that cause interaction between organisms are called semiochemicals, which include allelochemicals that mediate interspecific responses. Steiner (1925) and Prot (1980) both concluded that nematodes must be attracted to soluble substances released by plants. Yet, there is still very little direct information to support the idea that *Meloidogyne*, or other root-feeding nematodes, are attracted to allelochemicals that are of uniquely vascular plant origin. Many early experiments *in vitro* demonstrated that nematodes are attracted to roots and even within sand to zones where roots had been growing (Prot, 1980). Moreover, experiments with sand supplemented or not with low concentrations of clay revealed that adding a small amount of clay greatly enhanced root finding by *Meloidogyne* spp. (Prot and Van Gundy, 1980), suggesting that the charged surfaces of clay micelles facilitated establishment of essential gradients of organic compounds released by roots.

The attraction of J2 of *Meloidogyne* spp. to roots was studied by Wang *et al.* (2009) using *in vitro* assays with pluronic gel, a transparent medium that allows three-dimensional movement. J2 moved freely through the gel and were attracted to roots of tomato, *Medicago truncatula*, common bean, and *Arabidopsis*. *Meloidogyne javanica* and *M. incognita* moved to roots much more rapidly than *M. hapla*. Nematodes formed aggregates when in contact with root tips, indicating that a signal from the root is involved in the attraction. It is also possible that the nematodes at the root emit an aggregation pheromone to attract other J2 to the invasion site. There is evidence that this type of pheromone exists in *Caenorhabditis elegans* and some animal-parasitic nematodes (reviewed by Huettel, 1986; Thompson and Geary, 2002), but studies on this aspect are needed for plant-parasitic nematodes. Nematodes aggregated when suspended in pluronic gel without roots, and a coverslip placed on the gel accelerated and served as a focus for the aggregation; Wang *et al.* (2009) suggested that lower oxygen or a volatile attractant is involved in this aggregation behaviour.

Overall, contradictory results have been obtained regarding nematode responses under controlled conditions to most substances known to occur as gradients around roots (Klingler, 1965; Prot, 1980). The only consistent directions of movement seem to be toward CO_2 and to the wet end of a soil moisture gradient (Wallace, 1960). Movement toward moisture vertically would help nematodes avoid desiccation, but horizontally it would take them away from roots. This reinforces the evidence that CO_2 is the prime candidate as a root signal.

As well as limited information about plant-derived attractants, there is also a paucity of research about the extent to which plants may have evolved repellents or feeding deterrents against nematodes. Some plants do appear to repel or fail to attract nematodes, but there is conflicting evidence about whether this is associated with resistance. For example, roots of cucumber plants carrying the *Bi* (bitter) locus for triterpenoid cucurbitacins were less attractive to root-knot nematodes than roots of other cucumber plants (Kaplan and Keen, 1980), but plants with the *Bi* gene, none the less, can become infected and heavily galled in response to nematode feeding. The resistant grass *Aegilops variabilis* was less attractive to J2 of *M. naasi* than susceptible barley (*Hordeum vulgare*) and wheat (*Triticum aestivum*) (Balhadere and Evans, 1994), but susceptible and resistant barley cultivars were similarly attractive. *Meloidogyne javanica*, which is controlled by the resistance gene *Mi-1* in tomato, was equally attracted to tomato cultivars irrespective of whether or not they carried the resistance gene (Wang *et al.*, 2009). By contrast, Griffin (1969) observed that more J2 of *M. hapla* were attracted to susceptible than to resistant seedlings of lucerne. Differential responses may be associated with local exudates. For example, Zhao *et al.* (2000) showed that root border cell exudates affected the behaviour of J2 of *M. incognita* differently: while resistant lucerne cv. Moapa 69 repelled these nematodes, susceptible hosts, such as pea and lucerne cv. Thor, were chemotactically attractive. It is interesting to note that while four species – maize, pea, lucerne and snap bean – are hosts to *Meloidogyne*, the root caps of these plants nevertheless are effectively 'resistant' to nematode attack (Zhao *et al.*, 2000; Hubbard *et al.*, 2005).

6.5 Nematode Changes and Responses at the Root–Soil Interface

Once the nematode is in the root area, short-distance attractants, which may also include CO_2 as well as other components of root exudates, including non-volatile compounds, are thought to enable the nematodes to locate a root. The main emphasis is on host recognition, with soluble and high molecular weight compounds acting as short and local orientation signals, which cause the nematode to move to individual host roots and to the preferred site of invasion in the root tip, respectively. When a root is encountered, its surface is explored for a suitable penetration site, and at this stage plant chemicals in the rhizosphere can influence nematode behaviour.

The term 'rhizosphere', from the Greek words rhizo (root) and sphere (zone of influence), encompasses the soil within millimetres of a plant root where complex biological and ecological processes occur. Compounds produced by microorganisms, which are more abundant around roots than elsewhere in the soil (Russell *et al.*, 1985), might also contribute to bringing nematodes close to roots. The various parameters of the plant rhizosphere that determine the structure of the microbial community in the vicinity of the plant root include plant species and soil type (Garbeva *et al.*, 2008). It is possible that the individuality of the rhizosphere microbial community in relation to the plant species may directly affect the compounds to which different species of nematodes respond. Clearly, more than one compound present in the rhizosphere, from plants and/or microorganisms, might be involved in the below-ground chemical communication between the nematode and its respective host plant.

The rhizosphere can be divided into the endorhizosphere – within the root (Lynch and Whipps, 1990), the rhizoplane – root surface (Clark, 1949), and the ectorhizosphere – outside the root (Lynch and Whipps, 1990). The ectorhizosphere can range from several millimetres in the case of soluble nutrients and volatiles to <1 mm for sparingly soluble nutrients (Curl and Truelove, 1986; Neumann *et al.*, 2006). This section of the chapter will deal with the nature of the rhizosphere as a habitat for nematodes and refers to the root–soil interface, which encompasses the rhizoplane, the actual surface of plant roots and the ectorhizosphere.

6.5.1 Chemical communication at the root–soil interface

In root exudates, plants provide the raw material that feeds the microbial community of the rhizosphere, and chemical components of root exudates released into the rhizosphere are involved in plant–nematode interactions by acting as hatching stimulants, repellents, attractants or inhibitors; some can be toxic to nematodes (Zhao *et al.*, 2000; Robinson, 2004; Wuyts *et al.*, 2006a).

Root exudates include low molecular weight compounds, such as amino compounds, organic acids, carbohydrates, phenolics, flavonoids, enzymes, nucleotides, chalcones, fatty acids, sterols and other miscellaneous compounds, and high molecular weight compounds, such as mucilage (polysaccharides) and proteins. Such compounds may influence nematode behaviour at the rhizoplane.

Most root exudations from young and healthy plants come from the root cap region (here defined as the terminal 1–2 mm at the root apex) and consist of root border cells (RBC), which slough off from the cap as metabolically active RBC and their associated materials (Sievers and Braun, 1996; Hawes *et al.*, 1998).

RBC and root cap exudates from peas increased motility and attracted J2 of *M. incognita*. However, when in contact with root border cells and mucilage, J2 of *M. incognita* entered a reversible state of immobility (Fig. 6.3; Zhao *et al.*, 2000; Rodger *et al.*, 2003; Hubbard *et al.*, 2005). When border cells and associated root cap exudates were washed from pea roots prior to challenge with nematodes, no such accumulation and immobility was observed (Zhao *et al.*, 2000). The same phenomenon has also been noted in *Radopholus similis* and *C. elegans* (Hubbard *et al.*, 2005; Wuyts *et al.*, 2006b). Root cap exudates of legumes, but not of solanaceous plants, induce a reversible immobility in root-knot and free-living nematodes. Nevertheless, plant-parasitic nematodes recover fully, with a dramatic increase in the capacity to find and penetrate root tips within 24 h after recovery from immobility (Hubbard *et al.*, 2005). This ability of the root cap to deliver products that temporarily immobilize nematodes could conceivably play a role in the ability of the root cap to escape invasion, as they can grow away from the threat (the root tip of pea plants moves at a rate of 1 mm/h), avoid invasion and preserve the apical meristem from damage (Zhao *et al.*, 2000; Wuyts *et al.*, 2006b).

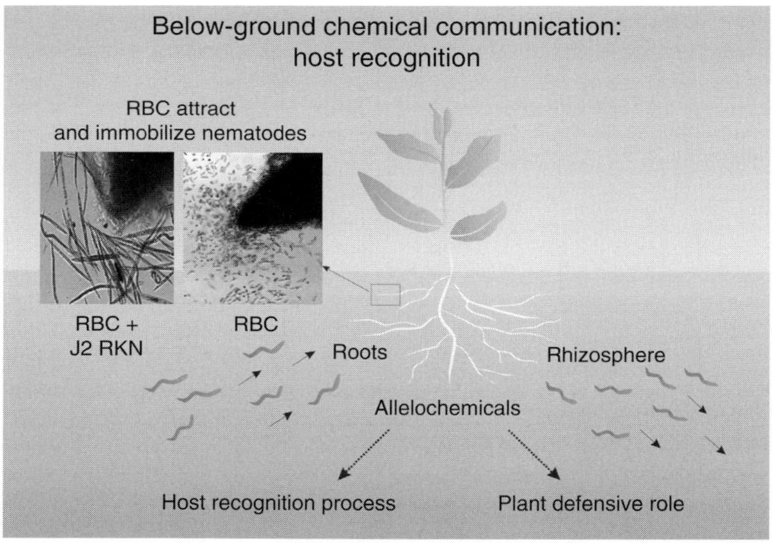

Fig. 6.3. Allelochemicals can be found in parts of the plant and can also be secreted as root exudates or released as volatile compounds. They mediate chemical interactions between two plants as well as plant–pathogen chemical communication. Attraction, penetration and feeding behaviour of plant-parasitic nematodes (in this case, second-stage juvenile, J2, root-knot nematodes, RKN) involve molecular communication between the nematode and respective plants. Root exudates can induce attraction, repellence, inhibition and hatching stimulation. Root border cells (RBC) play an important role in protecting the root tip from infection by acting as a natural trap for pathogenic organisms and can attract and produce a reversible state of immobility in free-living and plant-parasitic nematodes.

The major site for release of root exudation is the zone of elongation (Rovira and Ridge, 1973), and J2 of *Meloidogyne* spp. invade this region behind the root tip, just past the region where border cells are released (von Mende, 1997; Zhao *et al.*, 2000). However, root hairs and lateral roots also produce root exudations (Curl and Truelove, 1986), and other sites of root exudation can occur along main roots during the formation of lateral roots, when tissue damage causes leakage of the contents of damaged cells to the ectorhizosphere at the point of root emergence. Indeed, invasion of nematodes has been observed at young lateral apices and injured tissues (Egeraat, 1975; Prot, 1980; Wyss, 1981) and this might indicate a correlation between the concentration of root exudations and the preferred site of nematode invasion of roots.

Although the functions of most root exudates have not been determined, several compounds play important roles in biological processes (Bais *et al.*, 2006). The allelopathic potential of compounds present in root exudates, such as flavonoids and phenolics, has been analysed by examining the effects of such compounds on the behaviour of J2 of *M. incognita*. These compounds play important roles in plant defence and resistance against pests and diseases (Nicholson and Hammerschmidt, 1992; Plowright *et al.*, 1996; Baldridge *et al.*, 1998; Pegard *et al.*, 2005) and have also been shown to have a direct effect on nematode behaviour *in vitro* and to act as a repellent and as an inhibitor of motility of the J2 of *M. incognita* (Mahajan *et al.*, 1992; Wuyts *et al.*, 2006b). For example, in *in vitro* behavioural assays, flavonols were found to be repellent compounds for J2 of *M. incognita*, while in their degraded form they were motility inhibitors; by contrast, salicylic acid was a strong attractant but was also nematicidal and an irreversible inhibitor of hatch (Wuyts *et al.*, 2006b). Drench application of salicylic acid controlled *M. incognita* (Maheshwari and Anwar, 1990), and foliar application of salicylic acid induces suppression of *M. incognita* infection in plants (Nandi *et al.*, 2003), although in the latter work the effect was likely to be due to the signalling role of salicyclic acid in inducing pathogen resistance. Salicylic acid is an important component of the *Mi-1*-mediated defence response to root-knot nematode in tomato (Branch *et al.*, 2004). J2 of *M. javanica* and *M. hapla* have also been shown to be positively stimulated by ascorbic acid, gibberellin or glutamic acid (Bird, 1959, 1962). Chitwood (2002) has reviewed in detail the potential of phytochemical-based strategies for nematode control. However, it is important to stress that *in vitro* assessment of nematode responses to individual components may be unrealistic in the context of the soil environment; many compounds identified as attractants from *in vitro* results may not have the temporal or spatial attributes required to set up a gradient in the soil (Perry, 2005) and, as pointed out by Wuyts *et al.* (2006b), there is too little information about compounds, including phenylpropanoids, in roots and the rhizosphere of nematode hosts to link *in vitro* studies to actual behaviour in the soil.

Acquisition of nutrients and water by roots creates gradients in the rhizosphere that can also be involved in the root–nematode interaction (Prot, 1980; Claassen and Steingrobe, 1999; Jungk, 2002). Roots rely on chemical strategies for nutrient mobilization by inducing modification of pH and redox potential; these gradients formed along the root axis and in the rhizosphere could influence nematodes at the root–soil interface (Neumann *et al.*, 2006). The driving force for nutrient uptake by roots is H^+ extrusion, mediated by the activity of a plasma-membrane-bound H^+ pumping ATPase, which creates an outward positive gradient in electropotential and pH between the cytosol (pH 7–7.5) and the apoplast (pH 5–6) (Gerendas and Ratcliffe, 2000; Neumann *et al.*, 2006). It has been suggested that *Meloidogyne* spp. are attracted to roots and that contact is maintained by lower redox potentials and also by the acidic nature of the root surface (Bird, 1959; Prot, 1980).

J2 of *Meloidogyne* spp. are attracted to growing root tips and display characteristic nematode exploratory behaviour at the root surface, including stylet thrusting, release of secretions in preparation for root penetration, aggregation and increase in nematode mobility (von Mende, 1997). This exploratory behaviour can also be induced *in vitro* by compounds present in root exudations, and a number of chemicals have been found to induce nematode stylet thrusting and production of secretions, among which are some plant compounds such as cathecol and caffeic acid (McClure and von Mende, 1987; Grundler *et al.*, 1991; Robinson, 2004; Curtis, 2007).

6.5.2 Perturbing chemosensory perception

Host-finding behaviour involves recognition of host signals by the nematode's sensilla (sense organs), which function as the conduit between stimulus, reception and behavioural output. The main nematode chemosensory organs involved in host-recognition processes are two bilaterally symmetric amphids in the nematode head and two paired pore-like phasmids located in the lateral field of the nematode tail (see Eisenback and Hunt, Chapter 2, this volume). Hilliard *et al.* (2002) suggested that nematodes have the ability to chemo-orientate using a combination of head-to-tail chemosensory sensors to compare simultaneously the intensities of the stimulus across their body length. This idea had previously been rejected by Ward (1973) because mutants of *C. elegans* with blisters over the phasmids were still able to orient. It is probable that *Meloidogyne*, in common with other nematodes, compares concentrations successively in time by side-to-side displacement of the anterior end (klinotaxis).

Each amphid consists of three basic cell types: a glandular sheath cell, a supporting socket cell and a number of dendritic processes that are surrounded by secretions. There is a continuity of secretions within the socket cell and the material in the amphidial duct, and Baldwin and Perry (2004) considered that the amphid secretions emanate from the socket cell. Typically, the amphid sheath cell is deeply folded, resulting in a large surface area. In *Meloidogyne* males and, to a lesser extent, in J2 the sheath cell has many extracellular fluid-filled caverns continuous with a larger pouch surrounding the receptors (Baldwin and Hirschmann, 1973; Wergin and Endo, 1976). The compounds present in the amphidial secretions are undoubtedly important in chemoreception, and early work (reviewed by Perry and Aumann, 1998) used lectins to demonstrate the presence of carbohydrate residues in secretions, and some of these were constituents of glycoproteins. Components in amphidial secretions of J2 of *M. incognita* include *N*-acetylgalactosamine and fucose. There is evidence that the composition of amphidial secretions varies between species. Stewart *et al.* (1993a) found a 32 kDa glycoprotein associated with amphidial secretions of J2 of six species of *Meloidogyne*, but it was not present in representatives from eight other genera, including *Globodera* and *Heterodera*. The glycoprotein was found in all active life-cycle stages of *M. javanica*, but was not found in the sedentary adult female, where the amphids may be non-functional.

Nematode signalling components present in the chemosensory organs are in contact with the external environment and therefore can be targeted with antagonistic compounds that could block their interaction with host signals, leading to disruption of host-finding mechanisms. Nematodes would become disorientated in the soil and unable to sense and respond to host signals present in the rhizosphere, and infection of host plants would be adversely affected (Zuckerman, 1983; Fioretti *et al.*, 2002; Perry, 2005). Working with *P. penetrans*, Trett and Perry (1985) showed that the neuronal uptake of nematicides such as aldicarb disrupts chemosensory cells. Low doses of aldicarb (1 μM) disrupted chemoreception in plant-parasitic nematodes by paralysing the nematodes (Liu *et al.*, 2005) and caused a considerable reduction of penetration of host roots by J2 of *M. javanica* (Hough and Thomason, 1975). Winter *et al.* (2002) proposed an uptake pathway for aldicarb by retrograde transport along chemosensory dendrites to their site of action at cell bodies and synapses, and suggested that this may be a general mechanism for the low-dose effects of some nematicides. Chemoreception interference, by blocking specific nematode amphidial secretions using monoclonal and polyclonal antibodies, retarded movement of J2 of *M. javanica* (Stewart *et al.*, 1993b) and *G. pallida* (Fioretti *et al.*, 2002) and significantly reduced infection of plants. However, responses were not permanently blocked as, after a period, turnover of sensilla secretions apparently 'unblocked' the amphids. Lectins have been shown to bind to the surface cuticle and amphids of *Meloidogyne* spp. (Davis *et al.*, 1988), and treatment of infected soil with lectins significantly reduced root galling of *M. incognita* in tomato, probably by blocking the nematode chemoreceptors and interfering with host finding (Marbán-Mendoza *et al.*, 1987). Treatment of J2 of *Heterodera glycines in vitro* with root exudates from transgenic potato plants expressing a chemoreception-disruptive peptide that inhibits acetylcholinesterase reduced root invasion. The establishment of *G. pallida* in these transgenic plants was also suppressed in glasshouse experiments. This work shows that disruption of chemoreception occurs after initial root invasion and suggests that transgenic plants expressing chemoreception-disruptive peptides might suppress parasitism of cyst nematodes (Liu *et al.*, 2005).

6.5.3 Surface cuticle changes in response to environmental signals

The external cuticular layer of nematodes is the epicuticle, covered in many species by a coating material termed 'surface coat' (SC). The SC is composed mainly of proteins, carbohydrates and lipids (Spiegel and McClure, 1995). Nematodes can rapidly change their surface composition in response to environmental signals, which may enable animal-parasitic nematodes to escape host immune responses and free-living nematodes to escape pathogenic infections (Proudfoot *et al.*, 1993; Olsen *et al.*, 2007). Surface composition of an individual stage of the life cycle can also change during entry of parasitic nematodes into a new host or host tissue (Proudfoot *et al.*, 1993; Modha *et al.*, 1995). One of the most interesting features of the nematode SC is its dynamic nature; there is a continuous turnover of the surface-associated antigens, which involves shedding and replacing of the antigens (Blaxter and Robertson, 1998).

The dynamic nature of the SC of plant-parasitic nematodes was demonstrated for the pre-parasitic juveniles of *Meloidogyne* spp., as SC proteins of *M. incognita* were released from the nematode's surface when J2 were incubated in water, indicating that SC proteins may be transitory (Lin and McClure, 1996). Also, when J2 of *M. incognita* were treated with detergents there was a reduction in the binding of red blood cells to the nematode surface, but the binding properties were completely renewed after 24 h, indicating that the sloughing-off process of the nematode's SC is an active metabolic event (Spiegel *et al.*, 1997). The ability of these nematodes to shed and renew the SC continuously may help the nematode avoid recognition, and the induction of resistance in the host plant. Therefore, the SC may help to protect the J2 during its movement in the soil environment and in the host plant. The origin of surface-associated antigens on nematodes may differ for various antigens and is still not clear in most cases (Blaxter and Robertson, 1998). These non-structural proteins can originate from gland cells such as excretory cells, pharyngeal glands, amphids and phasmids, as well as from the hypodermis (Blaxter and Robertson, 1998). The antibodies Misec 3F.4 and Mj PC E2 reactive with the SC of *M. incognita* and *M. javanica* also showed reactivity with the hypodermis (Fig. 6.4), perhaps indicating the origin of these surface antigens (Sharon *et al.*, 2002). Nevertheless, the rectal glands were also suggested as the origin of a surface antigen of *M. incognita* race 3 (Hu *et al.*, 2000).

In vitro, plant signals present in root exudates trigger a rapid alteration of the surface cuticle of *M. incognita* and *G. rostochiensis* (Lopéz de Mendoza *et al.*, 2000; Akhkha *et al.*, 2002), and the same increase of the surface lipophilicity was also induced by phytohormones, in particular indoleacetic acid (IAA) and kinetin (Fig. 6.5), in *M. incognita*, but not in *G. rostochiensis* (Akhkha *et al.*, 2002, 2004; Curtis, 2007). As *M. incognita* can infect a large range of host plants, it was not surprising that it was responding to such a general plant compound. This increase in the lipophilicty of the SC of *M. incognita* J2, induced by plant signals might allow this nematode to adapt to and survive plant defence processes. However, more specific host cues from root exudates of solanaceous plants were responsible for increasing the lipophilicity of the surface cuticle of infective J2 of *Globodera* species (Akhkha *et al.*, 2002). *In vitro*, IAA has also been shown to induce the production of nematode secretions (Duncan *et al.*, 1995), and an increase in nematode motility (Curtis, 2007). *Globodera pallida* (Duncan *et al.*, 1995), but not *Meloidogyne* spp. (R. Curtis, unpublished data), was shown to have an immunological cross-reactivity to the maize auxin-binding protein, indicating that these sedentary plant-parasitic nematodes may respond differently to IAA.

Exogenous application of IAA shifted the response of resistant tomato plants towards susceptibility, indicating that higher levels of auxin in plants favour infection by *Meloidogyne* spp. (Dropkin *et al.*, 1969; Sawhney and Webster, 1975). Nematodes can encounter an IAA gradient inside roots, and free IAA has been detected

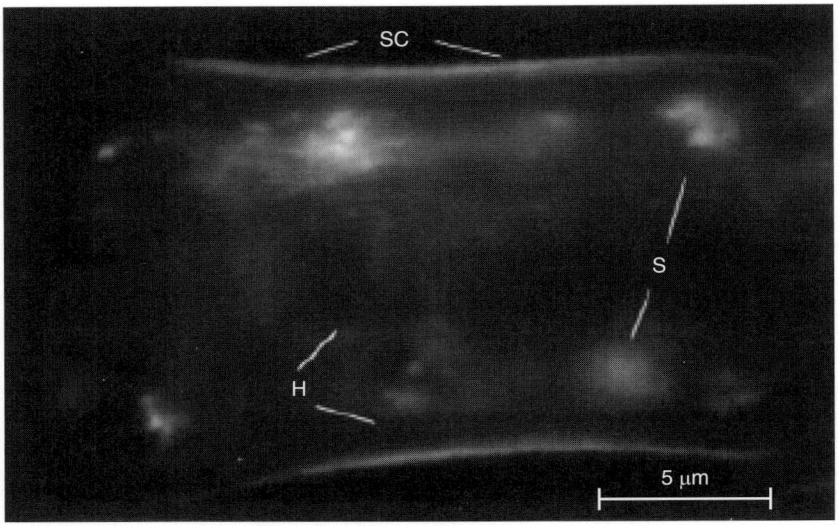

Fig. 6.4. Immunofluorescent labelling of a longitudinal cryosection of *Meloidogyne javanica* infective second-stage juvenile, showing reactivity of the MAb Misec 3F.4 with the surface cuticle (SC) and with 'spots' (S) along the hypodermis (H). (From Sharon *et al.*, 2002.)

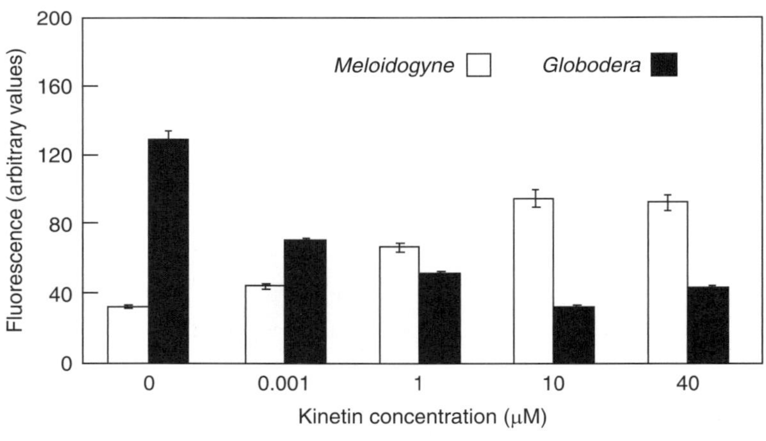

Fig. 6.5. Effect of kinetin on the uptake of AF18 (5-*N*-(octodecanoyl)-aminofluorescein) by J2 of *Meloidogyne incognita* and *Globodera rostochiensis*. J2 were incubated in different kinetin concentrations: 0 μM, 0.001 μM, 1 μM, 10 μM and 40 μM. Relative fluorescence was expressed in arbitrary values; error bars are standard errors ($n = 30$). (From Akhkha *et al.*, 2002.)

at concentrations of up to 1 μM in phloem exudates (Friml, 2003). Auxin conjugates such as IAA methyl-glutamate have been shown to be present in root exudates of *Arabidopsis thaliana* (Narasimhan *et al.*, 2003), and therefore nematodes may also be affected by IAA from plant origin in the rhizosphere.

Caenorhabditis elegans also responds to environmental conditions by modifying its surface, and these environmental signals are detected by the nematodes's chemosensory organs (Grenache *et al.*, 1996; Olsen *et al.*, 2007). These studies suggest that surface switching might also rely on chemosensation, and it can be speculated that free-living and parasitic nematodes use their sensilla to detect environmental signals that lead to changes in the surface composition. This behavioural adaptation may protect the nematodes from biological attack by helping the nematode to evade host responses. Artificially manipulating the signals controlling surface switching could prevent a parasite from evading host responses.

6.6 Conclusions and Future Directions

During co-evolution with the host plant, parasitic nematodes have developed the capacity to recognize and respond to chemical signals of host origin. Signals from roots present in the rhizosphere and bulk soil can specifically influence nematode behaviour, inducing hatching, attraction, surface cuticle changes, root exploratory behaviour and penetration of plant roots, and involve molecular communication between the nematode and respective host plant.

Understanding the complexity of the molecular signal exchange and response during the early stages of the plant–nematode interactions is important to identify vulnerable points in the parasite life cycle that can be targeted to disrupt nematode host recognition. A better understanding of the nature of exudates affecting nematode behaviour will reveal targets for chemical or genetic intervention to control root-knot nematodes.

6.7 References

Abou-Setta, M.M. and Duncan, L.W. (1998) Attraction of *Tylenchulus semipenetrans* and *Meloidogyne javanica* to salts in vitro. *Nematropica* 28, 49–59.

Akhkha, A., Kusel, J., Kennedy, M. and Curtis, R.H.C. (2002) Effects of phytohormones on the surfaces of plant-parasitic nematodes. *Parasitology* 125, 165–175.

Akhkha, A., Curtis, R., Kennedy, M. and Kusel, J. (2004) The potential signalling pathways which regulate surface changes induced by phytohormones in the potato cyst nematode (*Globodera rostochiensis*). *Parasitology* 128, 533–539.

Alexander, R.M. (2002) Locomotion. In: Lee, D.L. (ed.) *The Biology of Nematodes*. Taylor and Francis, London, pp. 345–352.

Bais, H.P., Weir, T.L., Perry, L.G., Gilroy, S. and Vivanco, J.M. (2006) The role of root exudates in rhizosphere interactions with plants and other organisms. *Annual Review of Plant Biology* 57, 233–266.

Bajracharya, R.M., Lal, R. and Kimble, J.M. (2000) Erosion effects on carbon dioxide concentration and carbon flux from an Ohio alfisol. *Soil Science Society of America Journal* 64, 694–700.

Baldridge, G.D., O'Neill, N.R. and Samac, D.A. (1998) Alfalfa (*Medicago sativa* L.) resistance to the root-lesion nematode, *Pratylenchus penetrans*: defense-response gene mRNA and isoflavonoid phytoalexin levels in roots. *Plant Molecular Biology* 38, 999–1010.

Baldwin, J.G. and Hirschmann, H. (1973) Fine structure of cephalic sense organs in *Meloidogyne incognita* males. *Journal of Nematology* 5, 285–302.

Baldwin, J.G. and Perry, R.N. (2004) Nematode morphology, sensory structure and function. In: Cheng, Z., Cheng, S. and Dickson, D.W. (eds) *Nematology – Advances and Perspectives*, Vol. 1. CAB International, Wallingford, UK, pp. 175–256.

Balhadere, P. and Evans, A.A.F. (1994) Characterization of attractiveness of excised root tips of resistant and susceptible plants for *Meloidogyne naasi*. *Fundamental and Applied Nematology* 17, 527–536.

Bernklau, E.J. and Bjostad, L.B. (1998) Reinvestigation of host location by western corn rootworm larvae (Coleoptera: Chrysomelidae): CO_2 is the only volatile attractant. *Ecology and Behavior* 91, 1331–1340.

Bernklau, E.J., Fromm, E.A. and Bjostad, L.B. (2004) Disruption of host location of western corn rootworm larvae (Coleoptera: Chrysomeolidae) with carbon dioxide. *Journal of Economic Entomology* 97, 330–339.

Bird, A.F. (1959) The attractiveness of roots to the plant parasitic nematodes *Meloidogyne javanica* and *M. hapla*. *Nematologica* 4, 322–335.

Bird, A.F. (1962) Orientation of the larvae of *Meloidogyne javanica* relative to roots. *Nematologica* 8, 275–287.

Bird, A.F. (1968) Changes associated with parasitism in nematodes. III. Ultrastructure of the egg shell, larval cuticle, and contents of the subventral oesophageal glands in *Meloidogyne javanica* with some observations on hatching. *Journal of Parasitology* 54, 475–489.

Bird, A.F. and Bird, J. (1991) *The Structure of Nematodes*, 2nd edn. Academic Press, London and New York.

Bird, A.F. and McClure, M.A. (1976) The tylenchid (Nematoda) eggshell: structure, composition and permeability. *Parasitology* 72, 19–28.

Bird, A.F. and Self, P.G. (1995) Chitin in *Meloidogyne javanica. Fundamental and Applied Nematology* 18, 235–239.

Bird, A.F. and Soeffky (1972) Changes in the ultrastructure of the gelatinous matrix of *Meloidogyne javanica* during dehydration. *Journal of Nematology* 14, 166–169.

Blake, C.D. (1961) Importance of osmotic potential as a component of the soil water on the movement of nematodes. *Nature* 191, 144–145.

Blaxter, M.L. and Robertson, W.M. (1998) The cuticle. In: Perry, R.N. and Wright, D.J. (eds) *Free-living and Plant-parasitic Nematodes.* CAB International, Wallingford, UK, pp. 25–48.

Branch, C., Hwang, C.F., Navarre, D.A. and Williamson, V.M. (2004) Salicylic acid is part of the Mi-1-mediated defence response to root-knot nematode in tomato. *Molecular Plant–Microbe Interactions* 17, 351–356.

Brzeski, M.W. and Hendricks, E.K. (1971) The overwintering and hatching of *Meloidogyne hapla* Chitw. *Zeszyty Problemowe Postepow Nauk Rolniczych* 121, 235–244.

Burr, A.H.J. and Robinson, A.F. (2004) Locomotion behaviour. In: Gaugler, R. and Bilgrami, A.L. (eds) *Nematode Behaviour.* CAB International, Wallingford, UK, pp. 25–62.

Campbell, G.S. (1985) *Soil Physics with Basic – Transport Models for Soil–Plant Systems.* Elsevier, Amsterdam.

Castro, C.E., Belser, N.O., McKinney, H.E. and Thomason, I.J. (1990) Strong repellency of the root knot nematode, *Meloidogyne incognita* by specific inorganic ions. *Journal of Chemical Ecology* 16, 1199–1205.

Castro, C.E., McKinney, H.E. and Lux, S. (1991) Plant protection with inorganic ions. *Journal of Nematology* 23, 409–413.

Chitwood, D.J. (2002) Phytochemical based strategies for nematode control. *Annual Review of Phytopathology* 40, 221–249.

Claassen, N. and Steingrobe, B. (1999) Mechanistic simulation models for a better understanding of nutrient uptake from soil. In: Rengel, Z. (ed.) *Mineral Nutrition of Crops: Fundamental Mechanisms and Implications.* Food Product Press, Binghamton, New York, pp. 327–367.

Clark, F.E. (1949) Soil microorganisms and plant roots. *Advances in Agronomy* 1, 241–288.

Croll, N.A. (1970) *The Behaviour of Nematodes: their Activity, Senses and Responses.* St. Martin's Press, New York, pp. 117–132.

Croll, N.A., and Sukhdeo, M.V.K. (1981) Hierarchies in nematode behavior. In: Zuckerman, B.M. and Rohde, R.A. (eds) *Plant Parasitic Nematodes*, Vol. III. Academic Press, New York, pp. 227–251.

Curl, E.A. and Truelove, B. (1986) Root exudates. In: *The Rhizosphere.* Vol. 15 in the Advanced Series in Agricultural Sciences. Springer-Verlag, Berlin, pp. 55–92.

Curtis, R.H.C. (2007) Do phytohormones influence nematode invasion and feeding site establishment? *Nematology* 9, 155–160.

Davis, E.L., Kaplan, D.T., Dickson, D.W. and Mitchell, D.J. (1988) Root-tissue response of 2 near-isogenic soybean cultivars to infection by lectin-treated *Meloidogyne* spp. *Journal of Nematology* 20, 632–633.

de Guiran, D. (1979) A necessary diapause in root-knot nematodes. Observations on its distribution and inheritance in *Meloidogyne incognita. Revue de Nématologie* 2, 223–231.

de Guiran, D. and Villemin, M.A. (1980) Spécificité de la diapause embryonnaire des oeufs de *Meloidogyne incognita* (Nematoda). *Revue de Nématologie* 3, 115–121.

De Rooij-Van der Goes, P.C.E.M., Peter, B.A.M. and van der Putten, W.H. (1998) Vertical migration of nematodes and soil-borne fungi to developing roots of *Ammophila arenaria* (L.) Link after sand accretion. *Applied Soil Ecology* 10, 1–10.

Dickson, D.W., and Hewlett, T.E. (1986) Vertical migration of *Meloidogyne arenaria* in soil columns in the field. *Revue de Nématologie* 9, 295.

Diez, J.A. and Dusenbery, D.B. (1989a) Preferred temperature of *Meloidogyne incognita. Journal of Nematology* 21, 99–104.

Diez, J.A. and Dusenbery, D.B. (1989b) Repellent of root-knot nematodes from exudate of host roots. *Journal of Chemical Ecology* 15, 2445–2455.

Dropkin, V.H., Martin, G.C. and Johnson, R.W. (1958) Effects of osmotic concentrations on the hatching of some plant parasitic nematodes. *Nematologica* 3, 115–126.

Dropkin, V.H., Helgeson, J.P. and Upper, C.D. (1969) Hypersensitivity reaction of tomatoes resistant to *Meloidogyne incognita* – reversal by cytokinins. *Journal of Nematology* 1, 55–61.

Duncan, L.H., Robertson, W.M. and. Kusel, J.R. (1995) Induction of secretions in *Globodera pallida* by phytohormones. *Journal of Nematology* 27, 498.

Dusenbery, D.B. (1987) Theoretical range over which bacteria and nematodes locate plant roots using carbon dioxide. *Journal of Chemical Ecology* 13, 1617–1625.

Dusenbery, D.B. (1988a) Avoided temperature leads to the surface: computer modeling of slime mold and nematode thermotaxis. *Behavioral Ecology and Sociobiology* 22, 219–223.

Dusenbery, D.B. (1988b) Limits of thermal sensation. *Journal of Theoretical Biology* 131, 263–271.

Dusenbery, D.B. (1988c) Behavioral responses of *Meloidogyne incognita* to small temperature changes. *Journal of Nematology* 20, 351–355.

Egeraat, A.W.S.M.v. (1975) Exudation of ninhydrin-positive compounds by pea-seedling roots: a study of the sites of exudation and of the composition of the exudate. *Plant and Soil* 42, 37–47.

Ellenby, C. (1974) Water uptake and hatching in the potato cyst nematode, *Heterodera rostochiensis*, and the root-knot nematode, *Meloidogyne incognita. Journal of Experimental Biology* 61, 773–779.

El-Sherif, M. and Mai, W.F. (1969) Thermotactic response of some plant parasitic nematodes. *Journal of Nematology* 1, 43–48.

Eo, J., Nakamoto, T., Otobe, K. and Mizukubo, T. (2007) The role of pore size on the migration of *Meloidogyne incognita* juveniles under different tillage systems. *Nematology* 9, 751–758.

Eo, J., Otobe, K. and Mizukubo, T. (2008) Absence of geotaxis in soil-dwelling nematodes. *Nematology* 10, 147–149.

Evans, A.A.F. (1987) Diapause in nematodes as a survival strategy. In: Veech, J.A. and Dickson, D.W. (eds) *Vistas on Nematology*. Society of Nematologists, Hyattsville, Maryland, pp.180–187.

Fanelli E., Di Vito, M., Jones, J.T. and De Giorgi, C. (2004) Analysis of chitin synthase function in a plant-parasitic nematode, *Meloidogyne artiellia*, using RNAi. *Gene* 349, 87–95.

Fioretti, L., Porter, A., Haydock, P.J. and Curtis, R. (2002) Monoclonal antibodies reactive with secreted-excreted products from the amphids and the cuticle surface of *Globodera pallida* affect nematode movement and delay invasion of potato roots. *International Journal for Parasitology* 32, 1709–1718.

Friml, J. (2003) Auxin transport – shaping the plant. *Current Opinion in Plant Biology* 6, 7–12.

Gans, C. and Burr, A.H.J. (1994) The unique locomotory mechanism of *Mermis nigrescens*, a large nematode which crawls over soil and climbs through vegetation. *Journal of Morphology* 222, 133–148.

Garbeva, P., van Elsen, J.D. and van Veen, J.A. (2008) Rhizosphere microbial community and its response to plant species and soil history. *Plant Soil* 302, 19–32.

Gaur, H.S., Beane, J. and Perry, R.N. (2000) The influence of root diffusate, host age and water regimes on hatching of the root-knot nematode, *Meloidogyne triticoryzae. Nematology* 2, 191–199.

Gerendas, J. and Ratcliffe, R.G. (2000) Intracellular pH regulation in maize root tips exposed to ammonium at high external pH. *Journal of Experimental Botany* 51, 207–219.

Grenache, D.G., Caldicott, I., Albert, P.S., Riddle, D.L and. Politz, S.M. (1996) Environmental induction and genetic control of surface antigen switching in the nematode *Caenorhabditis elegans. Proceedings of the National Academy of Sciences of the USA* 93, 12388–12393.

Griffin, G.D. (1969) Attractiveness of resistant and susceptible alfalfa to stem and root-knot nematodes. *Journal of Nematology* 1, 9.

Grundler, F., Schnibbe, L. and Wyss, U. (1991) *In vitro* studies on the behaviour of 2nd stage juveniles of *Heterodera schachtii* (Nematoda, Heteroderidae) in response to host plant-root exudates. *Parasitology* 103, 149–155.

Hawes, M.C., Brigham, L.A., Wen, F., Woo, H.H. and Zhu, Z. (1998) Function of root border cells in plant health: pioneers in the rhizosphere. *Annual Review of Phytopathology* 36, 311–327.

Hilliard, M.A., Bargmann, C.I. and Bazzicalupo, P. (2002) *C. elegans* responds to chemical repellents by integrating sensory inputs from the head and the tail. *Current Biology* 12, 730–734.

Hough, A. and. Thomason, I.J. (1975) Effects of aldicarb on behavior of *Heterodera schachtii* and *Meloidogyne javanica. Journal of Nematology* 7, 221–229.

Hu, G.G., McClure, M.A. and Schmitt, M.E. (2000) Origin of a *Meloidogyne incognita* surface coat antigen. *Journal of Nematolology* 32, 174–182.

Hubbard, J.E., Flores-Lara, Y., Schmitt, M., McClure, M.A., Stock, S.P. and Hawes, M.C. (2005) Increased penetration of host roots by nematodes after recovery from quiescence induced by root cap exudate. *Nematology* 7, 321–331.

Huettel, R.N. (1986) Chemical communication in nematodes. *Journal of Nematology* 18, 3–8.

Inserra, R.N., Griffin, G.D. and Sisson, D.V. (1983) Effects of temperature and root leachates on embryonogenic development and hatching of *Meloidogyne chitwoodi* and *M. hapla*. *Journal of Nematology* 15, 123–127.

Ishibashi, N. (1969) Studies on the propagation of the root-knot nematode *Meloidogyne incognita* (Kofoid & White), Chitwood, 1949. *Review of Plant Protection Research* 2, 125–128.

Ishibashi, N., Kegasawa, K. and Kunii, Y. (1960) Studies on the hatching of the root-knot nematode *Meloidogyne incognita* var. *acrita* Chitwood. I. The relation between hatching and the original egg content of egg mass. *Japanese Journal of Applied Entomology* 4, 249–255.

Jones, J.T. (2002) Nematode sense organs In: Lee, D.L. (ed.) *The Biology of Nematodes*. Taylor and Francis, London, pp. 353–368.

Jones, P.W., Tylka, G.L. and Perry, R.N. (1998) Hatching. In: Perry, R.N. and Wright, D.J. (eds) *The Physiology and Biochemistry of Free-living and Plant-parasitic Nematodes*. CAB International, Wallingford, UK, pp. 181–212.

Jungk, A.O. (2002) Dynamics of nutrient movement at the soil–root interface. In: Waisel, Y., Eshel, A. and Kafhafi, V. (eds) *Plant Roots. The Hidden Half.* Marcel Dekker, New York, pp. 587–616.

Kaplan, D.T. and Keen, N.T. (1980) Mechanisms conferring plant incompatibility to nematodes. *Revue de Nématologie* 3, 123–134.

Karssen, G. and Moens, M. (2006) Root-knot nematodes. In: Perry, R.N. and Moens, M. (eds) *Plant Nematology*. CAB International, Wallingford, UK, pp. 59–90.

Klinger, J. (1965) On the orientation of plant nematodes and of some other soil animals. *Nematologica* 11, 4–18.

Lahtinen, A.E., Trudgill, D.L. and Tiilikkala, K. (1988) Threshold temperature and minimum time requirements for the complete life cycle of *Meloidogyne hapla* from northern Europe. *Nematologica* 34, 443–451.

Lin, H.J. and McClure, M.A. (1996) Surface coat of *Meloidogyne incognita*. *Journal of Nematology* 28, 216–224.

Liu, B., Hibbard, J.K., Urwin, P.E. and Atkinson, H.J. (2005) The production of synthetic chemodisruptive peptides in planta disrupts the establishment of cyst nematodes. *Plant Biotechnology Journal* 3, 487–496.

Lopéz de Mendoza, M.E., Modha, J., Roberts, M.C., Curtis, R. and Kusel, J.R. (2000) Changes in the lipophilicity of the surfaces of *Meloidogyne incognita* and *Haemonchus contortus* during exposure to host signals. *Parasitology* 120, 203–209.

Lynch, J.M. and Whipps, J.M. (1990) Substrate flow in the rhizosphere. *Plant and Soil* 129, 1–10.

Madulu, J. and Trudgill, D.L. (1994) Influence of temperature on *Meloidogyne javanica*. *Nematologica* 40, 230–243.

Mahajan, R., Kaur, D.J. and Bajaj, K.L. (1992) Nematicidal activity of phenolic compounds against *Meloidogyne incognita*. *Nematologia Mediterranea* 20, 217–219.

Maheshwari, D.K. and Anwar, M. (1990) Nematocidal activity of some phenolics on root-knot, growth and yield of *Capsicum frutescens* cv California Wonder. *Journal of Phytopathology–Phytopathologische Zeitschrift* 129, 159–164.

Marbán-Mendoza, N., Jeyaprakash, A., Jansson, H.-B., Damon, H.A. and Zuckerman, B.M. (1987) Control of root-knot nematode on tomato by lectins. *Journal of Nematology* 19, 331–335.

Masler, E.P. (2008) Response of *Heterodera glycines* and *Meloidogyne incognita* to exogenously applied biogenic amines. *Nematology* 10, 911–917.

McCallum, M.E. and Dusenbery, D.B. (1992) Computer tracking as a behavioral GC detector: nematode responses to vapor of host roots. *Journal of Chemical Ecology* 18, 585–592.

McClure M.A. and von Mende, N. (1987) Induced salivation in plant-parasitic nematodes. *Phytopathology* 77, 1463–1469.

Meyer, S.L.F., Zasada, I.A., Roberts, D.P., Viviard, B.T., Lakshman, D.K., Lee, J.K., Chitwood, D.J. and Carta, L. (2006) *Plantago lanceolata* and *Plantago rugelii* extracts are toxic to *Meloidogyne incognita* but not to certain microbes. *Journal of Nematology* 38, 333–338.

Modha, J., Kusel, J.R. and Kennedy, M.W. (1995) A role for 2nd-messengers in the control of activation-associated modification of the surface of *Trichinella spiralis* infective larvae. *Molecular and Biochemical Parasitology* 72, 141–148.

Mojtahedi, H., Ingham, R.E., Santo, G.S., Pinkerton, J.N., Reed, G.L. and Wilson, J.H. (1991) Seasonal migration of *Meloidogyne chitwoodi* and its role in potato production. *Journal of Nematology* 23, 162–169.

Nandi, B., Kundu, K., Banerjee, N. and Babu, S.P.S. (2003) Salicylic acid-induced suppression of *Meloidogyne incognita* infestation of okra and cowpea. *Nematology* 5, 747–752.

Narasimhan, K., Basheer, C., Bajic, V.B. and Swarup, S. (2003) Enhancement of plant-microbe interactions using a rhizosphere metabolomics-driven approach and its application in the removal of polychlorinated biphenyls. *Plant Physiology* 132, 146–153.

Neumann, G., Kohls, S., Landsberg, E, Souza, K.S.O., Yamada, T. and Romheld, V. (2006) Relevance of glyphosate transfer to non-target plants via the rhizosphere. *Journal of Plant Diseases and Protection* 20, 963–969.

Nicholson, R.L. and Hammerschmidt, R. (1992) Phenolic-compounds and their role in disease resistance. *Annual Review of Phytopathology* 30, 369–389.

Oka, Y. and Mizukubo, T. (2009) Tomato culture filtrate stimulates hatching and activity of *Meloidogyne incognita* juveniles. *Nematology* 11, 51–61.

Olsen, D.P., Phu, D., Libby, L.J.M., Cormier, J.A., Montez, K.M., Ryder, E.F. and Politz, S.M. (2007) Chemosensory control of surface antigen switching in the nematode *Caenorhabditis elegans. Genes Brain and Behavior* 6, 240–252.

Pegard, A., Brizzard, G., Fazari, A., Soucaze, O., Abad, P. and Djian-Caporalino, C. (2005) Histological characterisation of resistance to different root-knot nematode species related to phenolics accumulation in *Capsicum annuum. Phytopathology* 95, 158–165.

Perry, R.N. (1987) Host induced hatching of phytoparasitic nematode eggs. In: Veech, J.A. and Dickson, D.W. (eds), *Vistas on Nematology.* Society of Nematologists, Hyattsville, Maryland, pp. 159–164.

Perry, R.N. (1997) Plant signals in nematode hatching and attraction. In: Fenoll C., Grundler, F.M.W. and Ohl, S.A. (eds) *Cellular and Molecular Aspects of Plant–Nematode Interactions.* Kluwer Academic Press, Dordrecht, The Netherlands, pp. 38–50.

Perry, R.N. (2002) Hatching. In: Lee, D.L. (ed.) *The Biology of Nematodes.* Taylor and Francis, London, pp.147–169.

Perry, R.N. (2005) An evaluation of types of attractants enabling plant-parasitic nematodes to locate plant roots. *Russian Journal of Nematology* 13, 83–88.

Perry, R.N. and Aumann, J. (1998) Behaviour and sensory responses. In: Perry, R.N. and Wright, D.J. (eds) *Free-living and Plant-parasitic Nematodes.* CAB International, Wallingford, UK, pp. 75–102.

Perry, R.N. and Clarke, A.J. (1982) Hatching mechanisms of nematodes. In: Crompton, D.W.T. and Newton, B.A. (eds) *Trends and Perspectives in Parasitology*, Vol. 2. Cambridge University Press, Cambridge, UK, pp. 63–77.

Perry, R.N. and Maule, A.G. (2004) Physiological and biochemical basis of behaviour. In: Gaugler, R. and Bilgrami, A.L. (eds) *Nematode Behaviour.* CAB International, Wallingford UK, pp. 197–238.

Perry, R.N. and Wesemael, W.M.L. (2008) Host plant effects on hatching of root-knot nematodes. *Russian Journal of Nematology* 16, 1–5.

Perry, R.N., Knox, D.P. and Beane, J. (1992) Enzymes released during hatching of *Globodera rostochiensis* and *Meloidogyne incognita. Fundamental and Applied Nematology* 15, 283–288.

Pinkerton, J.N., Mojtahedi, H., Santo, G.S. and O'Bannon, J.H. (1987) Vertical migration of *Meloidogyne chitwoodi* and *Meloidogyne hapla* under controlled temperature. *Journal of Nematology* 19, 152–157.

Pinkerton, J.N., Santo, G.S. and Mojtahedi, H. (1991) Population dynamics of *Meloidogyne chitwoodi* on Russet Burbank potatoes in relation to degree-day accumulation. *Journal of Nematology* 23, 282–290.

Pline, M. and Dusenbery, D.B. (1987) Responses of plant-parasitic nematode *Meloidogyne incognita* to carbon dioxide determined by video camera-computer tracking. *Journal of Chemical Ecology* 13, 873–888.

Pline, M., Diez, J.A. and Dusenbery, D.B. (1988) Extremely sensitive thermotaxis of the nematode *Meloidogyne incognita. Journal of Nematology* 20, 605–608.

Ploeg, A.T. and Maris, P.C. (1999) Effects of temperature on the duration of the life cycle of a *Meloidogyne incognita* population. *Nematology* 1, 389–393.

Plowright, R.A., Grayer, R.J., Gill, J.R., Rahman, M.L. and Harborne, J.B. (1996) The induction of phenolic compounds in rice after infection by the stem nematode *Ditylenchus angustus. Nematologica* 42, 564–578.

Premachandran, D., von Mende, N., Hussey, R.S. and McClure, M.A. (1988) A method for staining nematode secretions and structures. *Journal of Nematology* 20, 70–78.

Prot, J.-C. (1978a) Influence of concentration gradients of salts on the movement of second stage juveniles of *Meloidogyne javanica. Revue de Nématologie* 1, 21–26.

Prot, J.-C. (1978b) Vertical migration of four natural populations of *Meloidogyne. Revue de Nématologie* 1, 109–112.

Prot, J.-C. (1978c) Behavior of juveniles of *Meloidogyne javanica* in salt gradients. *Revue de Nématologie* 2, 135–142.

Prot, J.-C. (1979a) Influence of concentration gradients of salts on the behaviour of four plant parasitic nematodes. *Revue de Nématologie* 2, 11–16.

Prot, J.-C. (1979b) Horizontal migrations of second-stage juveniles of *Meloidogyne javanica* in sand in concentration gradients of salts and in a moisture gradient. *Revue de Nématologie* 2, 17–21.

Prot, J.-C. (1980) Migration of plant-parasitic nematodes towards plant roots. *Review de Nematologie* 3, 305–318.

Prot, J.-C. and Van Gundy, S.D. (1980) Effect of soil texture and the clay component on migration of *Meloidogyne incognita* second-stage juveniles. *Journal of Nematology* 13, 213–217.

Proudfoot, L., Kusel, J.R., Smith, H.V., Harnett, W., Worms, M.J. and. Kennedy, M.W. (1993) Rapid changes in the surface of parasitic nematodes during transition from pre-parasitic to post-parasitic forms. *Parasitology* 107, 107–117.

Riga, E. (2004) Orientation behaviour. In: Gaugler, R. and Bilgrami, A.L. (eds) *Nematode Behaviour.* CAB International, Wallingford, UK, pp. 63–90.

Robinson, A. (1994) Movement of five nematode species through sand subjected to natural temperature gradient fluctuations. *Journal of Nematology* 26, 46–58.

Robinson, A.F. (1995) Optimal release rates for attracting *Meloidogyne incognita, Rotylenchulus reniformis,* and other nematodes to carbon dioxide in sand. *Journal of Nematology* 27, 42–50.

Robinson, A.F. (2004) Nematode behavior and migrations through soil and host tissue. In: Chen, Z., Chen, S. and Dickson D.W. (eds) *Nematology – Advances and Perspectives, Volume I: Nematode Morphology, Physiology, and Ecology.* Tsinghua University Press, Beijing and CAB International, Wallingford, UK, pp. 330–405.

Robinson, A.F. and Heald, C.M. (1991) Carbon dioxide and temperature gradients in Baermann funnel extraction of *Rotylenchulus reniformis. Journal of Nematology* 23, 28–38.

Robinson, A.F. and Perry, R.N. (2006) Behaviour and sensory perception. In: Perry, R.N. and Moens, M. (eds) *Plant Nematology.* CAB International, Wallingford, UK, pp. 210–233.

Rodger, S., Bengough, A.G., Griffiths, B.S., Stubbs, V. and Young, I.M. (2003) Does the presence of detached root border cells of *Zea mays* alter the activity of the pathogenic nematode *Meloidogyne incognita*? *Phytopathology* 93, 1111–1114.

Rovira, A.D. and Ridge, E.H. (1973) Exudation of ^{14}C-labelled compounds from wheat roots: influence of nutrients, micro-organisms and added organic compounds. *New Phytologist* 72, 1081–1087.

Russell, E., Ingham, J.A., Trofymow, E., Ingham, R. and Coleman, D.C. (1985) Interactions of bacteria, fungi, and their nematode grazers: effects on nutrient cycling and plant growth. *Ecological Monographs* 55, 119–140.

Saux, R.L. and Quénéhervé, P. (2002) Differential chemotactic responses of two plant-parasitic nematodes, *Meloidogyne incognita* and *Rotylenchulus reniformis,* to some inorganic ions. *Nematology* 4, 99–105.

Sawhney, R. and Webster, J.M. (1975) Role of plant-growth hormones in determining resistance of tomato plants to root-knot nematode, *Meloidogyne incognita. Nematologica* 21, 95–103.

Sharon, E. and Spiegel, Y. (1993) Glycoprotein characterization of the gelatinous matrix in the root-knot nematode *Meloidogyne javanica. Journal of Nematology* 25,585–589.

Sharon, E., Spiegel, Y., Solomon, R. and Curtis, R.H.C. (2002) Characterization of *Meloidogyne javanica* surface coat using antibodies and their effect on nematode behaviour. *Parasitology* 125, 177–185.

Shepherd, A.M. and Clarke, A.J. (1971) Molting and hatching stimuli. In: Zuckerman, B.M., Mai, W.F. and Rohde, R.A. (eds) *Plant Parasitic Nematodes. Vol. II. Cytogenetics, Host–Parasite Interactions, and Physiology.* Academic Press, New York and London, pp. 267–287.

Sievers, A. and Braun, M. (1996) The root cap: structure and function. In: Waisel, Y., Eshel, A. and Kafhafi, V. (eds) *Plant Roots. The Hidden Half.* Marcel Dekker, New York, pp. 31–49.

Spiegel, Y. and McClure, M.A (1995) The surface-coat of plant-parasitic nematodes – chemical composition, origin, and biological role – a review. *Journal of Nematology* 27, 127–134.

Spiegel, Y., Kahane, I., Cohen, L. and Sharon, E. (1997) *Meloidogyne javanica* surface proteins: characterisation and lability. *Parasitology* 115, 513–519.
Starr, J.L. (1993) Recovery and longevity of egg masses of *Meloidogyne incognita* during simulated winter survival. *Journal of Nematology* 25, 244–248.
Steiner, G. (1925) The problem of host selection and host specialization of certain plant-infesting nemas and its application in the study of nemic pests. *Phytopathology* 15, 499–534.
Stewart, G.R., Perry, R.N., Alexander, J. and Wright, D.J. (1993a) A glycoprotein specific to the amphids of *Meloidogyne* species. *Parasitology* 106, 405–412.
Stewart, G.R., Perry, R.N. and Wright, D.J. (1993b) Studies on the amphid specific glycoprotein gp32 in different life-cycle stages of *Meloidgyne* species. *Parasitology* 107, 573–578.
Thompson, D.P. and Geary, T.G. (2002) Excretion/secretion, ionic and osmotic regulation. In: Lee, D.L. (ed.) *The Biology of Nematodes*. Taylor and Francis, London, pp. 291–320.
Trett, M.W. and Perry, R.N. (1985) Effects of the carbamoyloxime, aldicarb, on the ultrastructure of the root-lesion nematode, *Pratylenchus penetrans* (Nematoda, Pratylenchidae). *Nematologica* 31, 321–334.
Van Gundy, S.D. (1985) Ecology of *Meloidogyne* spp. – emphasis on environmental factors affecting survival and pathogenicity. In: Sasser, J.N. and Carter, C.C. (eds) *An Advanced Treatise on* Meloidogyne. *Volume 1, Biology and Control*. North Carolina State University Graphics, Raleigh, North Carolina, pp. 177–182.
Veronico, P, Gray, L.J., Jones, J.T., Bazzicalupo, P, Arbucci, S., Cortese, M.R., Di Vito, M. and De Giorgi, C. (2001) Nematode chitin synthases: gene structure, expression and function in *Caenorhabditis elegans* and the plant parasitic nematode *Meloidogyne artiellia*. *Molecular Genetics and Genomics* 266, 28–34.
Viglierchio, D.R. and Lownsbery, B.F. (1960) The hatching response of *Meloidogyne* species to the emanations from the roots of germinating tomatoes. *Nematologica* 5, 153–157.
von Mende, N. (1997) Invasion and migration behaviour of sedentary nematodes. In: Fenol, C., Grundler, F.M.W. and Ohl, S.A. (eds) *Cellular and Molecular Aspects of Plant–Nematode Interactions*. Kluwer Academic Publishers, Dordrecht, The Netherlands, pp. 51–64.
Wallace, H.R. (1958) Movement of eelworms I. The influence of pore size and moisture content of the soil on the migration of larvae of the beet eelworm, *Heterodera schachtii* Schmidt. *Annals of Applied Biology* 46, 74–85.
Wallace, H.R. (1959) Movement of eelworms. IV. The influence of water percolation. *Annals of Applied Biology* 47, 131–139.
Wallace, H.R. (1960) Movement of eelworms. VI. The influences of soil type, moisture gradients and host plant rots on the migration of the potato-root eelworm *Heterodera rostochiensis* Wollenweber. *Annals of Applied Biology* 48, 107–120.
Wallace, H.R. (1961) The orientaton of *Ditylenchus dipsaci* to physical stimuli. *Nematologica* 6, 222–236.
Wallace, H.R. (1968a) The influence of aeration on the survival and hatch of *Meloidogyne javanica*. *Nematologica* 14, 223–230.
Wallace, H.R. (1968b) The influence of soil moisture on the survival and hatch of *Meloidogyne javanica*. *Nematologica* 14, 231–242.
Wallace, H.R. (1968c) The dynamics of nematode movement. *Annual Review of Phytopathology* 6, 91–114.
Wang, C., Lower, S. and Williamson, V.M. (2009) Application of pluronic gel to the study of root-knot nematode behaviour. *Nematology* 11, 453–464.
Ward, S. (1973) Chemotaxis by the nematode *Caenorhabditis elegans*: identification and analysis of response by the use of mutants. *Proceedings of the National Academy of Sciences of the USA* 70, 817–821.
Wergin, W.P. and Endo, B.Y. (1976) Ultrastructure of a neurosensory organ in a root-knot nematode. *Journal of Ultrastructural Research* 56, 258–276.
Wesemael, W.M.L. and Moens, M. (2008) Vertical distribution of the plant-parasitic nematode, *Meloidogyne chitwoodi*, under field crops. *European Journal of Plant Pathology* 120, 249–257.
Wesemael, W.M.L., Perry, R.N. and Moens, M. (2006) The influence of root diffusate and host age on hatching of the root-knot nematodes, *Meloidogyne chitwoodi* and *M. fallax*. *Nematology* 8, 895–902.
Winter, M.D., McPherson, M.J. and Atkinson, H.J. (2002) Neuronal uptake of pesticides disrupts chemosensory cells of nematodes. *Parasitology* 125, 561–565.
Wright, D.J. and Perry, R.N. (2006) Reproduction, physiology and biochemistry. In: Perry, R.N. and Moens, M. (eds) *Plant Nematology*. CAB International, Wallingford, UK, pp. 187–209.

Wuyts, N., Swennen, R. and De Waele, D. (2006a) Effects of plant phenylpropanoid pathway products and selected terpenoids and alkaloids on the behaviour of the plant-parasitic nematodes *Radopholus similis*, *Pratylenchus penetrans* and *Meloidogyne incognita. Nematology* 8, 89–101.

Wuyts, N., Maung, Z.T.Z., Swennen, R. and De Waele, D. (2006b). Banana rhizodeposition: characterization of root border cell production and effects on chemotaxis and motility of the parasitic nematode *Radopholus similis. Plant and Soil* 283, 217–228.

Wyss, U. (1981) Ectoparasitic root nematodes: feeding behaviour and plant cell responses. In: Zuckerman, B.M. and Rhode, R.A (eds) *Plant Parasitic Nematodes*, Vol. III. Academic Press, New York, pp. 325–351.

Zhao, X.W., Schmitt, M. and Hawes, M.C. (2000) Species-dependent effects of border cell and root tip exudates on nematode behaviour. *Phytopathology* 90, 1239–1245.

Zuckerman, B.M. (1983) Hypotheses and possibilities of intervention in nematode chemoresponses. *Journal of Nematology* 15, 173–182.

7 Invasion, Feeding and Development

Pierre Abad[1], Philippe Castagnone-Sereno[1], Marie-Noëlle Rosso[1], Janice de Almeida Engler[1] and Bruno Favery[1]
[1]*INRA/CNRS UMA/Université de Nice Sophia-Antipolis, Sophia Antipolis, France*

7.1	**Introduction**	**163**
7.2	**Root-knot Nematode Life Cycle**	**164**
7.3	**Nematode Parasitism**	**165**
7.4	**Compatible Interactions with Resistant Plants: the Case of Virulent Root-knot Nematodes**	**167**
7.5	**(A)virulence Determinants and Pathogenicity Factors: Root-knot Nematode Effectors with Dual Function?**	**169**
7.6	**Tools for Molecular and Functional Analysis of Root-knot Nematode Parasitism**	**169**
7.7	**Giant Cell Development**	**170**
7.8	**Cytoskeleton Organization and Cell Cycle Progression During Giant Cell Ontogenesis**	**172**
7.9	**Extensive Cell Wall Modifications to Build Up Giant Cells**	**173**
7.10	**Suppression of Plant Defence Associated with Giant Cell Development**	**174**
7.11	**Major Reprogramming of Plant Metabolism and Transport**	**174**
7.12	**Comparison between *Meloidogyne* Parasitism and Symbiotic Rhizobia in *Medicago***	**175**
7.13	**Conclusions and Future Directions**	**176**
7.14	**Acknowledgements**	**176**
7.15	**References**	**176**

7.1 Introduction

Root-knot nematodes, *Meloidogyne* spp., are sedentary, endoparasitic nematodes that interact with their hosts in a fascinating way. These obligate parasites have evolved the ability to manipulate host functions to their own benefit. Root-knot nematodes induce the redifferentiation of parenchyma root cells into multinucleate and hypertrophied feeding cells, named giant cells. These giant cells constitute the exclusive source of nutrients for the developing nematode. Hyperplasia of the surrounding cells leads to the formation of the typical root gall, the primary visible symptom of infection. Like other plant-parasitic nematodes, root-knot nematodes have a stylet, a hollow retractable needle connected to the pharynx, and three unicellular pharyngeal glands. This stylet is used to pierce plant cell walls, to release pharyngeal gland secretions into the host tissue and to take up nutrients

©CAB International 2009.

from the giant cells. Plant nutrient and water uptake are substantially reduced by the resulting damage to the root system, and infested plants are therefore weak and give low yields. Unlike plant-parasitic nematodes that kill the cells from which they feed, it is essential for root-knot nematodes that the feeding cells remain healthy and metabolically active throughout the nematode's life cycle. This chapter reviews progress being made towards understanding this compatible interaction. Recent investigations of nematode signals that trigger plant molecular and developmental changes associated with parasitism are also discussed.

7.2 Root-knot Nematode Life Cycle

Root-knot nematodes undergo a first moult inside the eggs to develop from first-stage juveniles (J1) to second-stage juveniles (J2) before hatching (Fig. 7.1A; Plate 36). Hatched, pre-parasitic J2 then penetrate host plant roots, usually close to the tip, by using their protractible stylets and releasing secretions containing cell-wall-degrading enzymes (Abad *et al.*, 2003) (Fig. 7.1B). They migrate intercellularly between the cortical cells down towards the root tip, lining up parallel to the long axis of the root. The casparian strips may form an insuperable barrier, forcing the nematode to make a U-turn to enter into the vascular cylinder. Only after this migratory phase, which leads the now-parasitic J2 to the vicinity of the vascular tissue, do the nematodes become sedentary and begin feeding. In order to sustain their subsequent sedentary parasitic stages, each J2 then induces the redifferentiation of five to seven parenchymatic root cells into a multinucleate and hypertrophied feeding cell (Fig. 7.2; Plate 37).

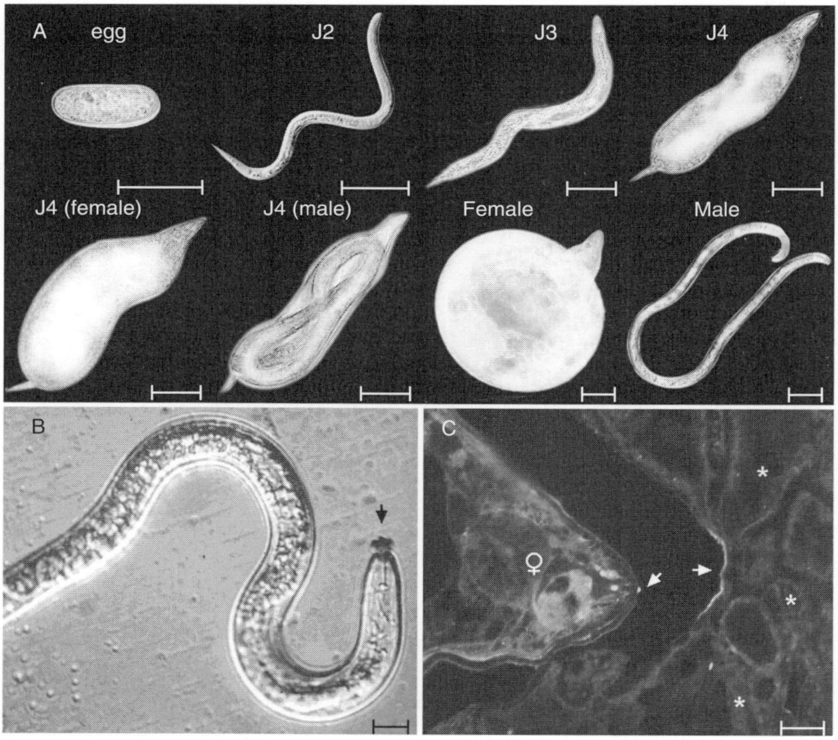

Fig. 7.1. Developmental stages and stylet secretions of *Meloidogyne incognita*. A: developmental stages, from eggs to adult nematodes; B: proteins secreted via the stylet by second-stage juveniles are visualized with Coomassie staining (arrow); C: fluorescence immunolabelling of a calreticulin (arrows) secreted during parasitism. The calreticulin accumulates at the stylet tip of the female and along the cell wall of adjacent giant cells. Asterisks, giant cell; ♀, female. Scale bars = 40 μm (A), 10 μm (B and C). (B and C from Caillaud *et al.*, 2008a.)

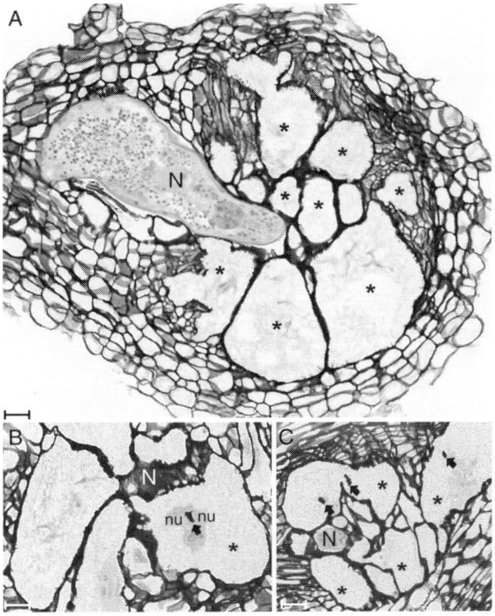

Fig. 7.2. Giant cells induced by *Meloidogyne incognita* in *Arabidopsis thaliana.* A: multinucleate giant cells; B–C: giant cells with mini cell plates (arrows) separating daughter nuclei. Sections through a gall at 10 days post infection stained with toluidine blue. Asterisks, giant cell; N, nematode; nu, nuclei. Scale bars = 20 µm.

These 'giant cells' function as specialized sinks, supplying nutrients to the nematode until reproduction (reviewed by Caillaud *et al.*, 2008a). While the two subventral pharyngeal gland cells are most active during the early stages of parasitism, i.e. root penetration, migration and the early events of giant cell formation, the single dorsal gland cell becomes most active in the later stages of giant cell formation and maintenance, as well as during feeding. Root-knot nematodes exclusively feed from giant cells and undergo three additional moults to reach the reproductive mature adult stage. Most root-knot nematodes reproduce by parthenogenesis (reviewed by Castagnone-Sereno, 2006; see Chitwood and Perry, Chapter 8, this volume). Males migrate out of the plant (Fig. 7.1A); only a few species, e.g. *M. carolinensis*, *M. microtyla* and *M. pini*, reproduce by amphimixis, with the obligatory fusion of a male and female gamete. After the development of the pear-shaped female, eggs are released on the root surface in a protective gelatinous matrix. Embryogenesis within the egg is followed by the first moult, leading to the J2.

7.3 Nematode Parasitism

Secretions from root-knot nematodes contain major effectors of parasitism, and these are the primary signalling molecules at the plant–nematode interface (reviewed by Jasmer *et al.*, 2003; Neveu *et al.*, 2003a; Davis *et al.*, 2004). The molecular dialogue between root-knot nematodes and the host starts at a distance, with modifications of the surface of infective J2 in response to root diffusates (López de Mendoza *et al.*, 2000; Akhkha *et al.*, 2002; see Curtis *et al.*, Chapter 6, this volume). A Nod-like effector secreted by *M. hapla* is suspected to be responsible for the induction at distance of root hair deformation, similar to the early legume response to *Rhizobium* (Weerasinghe *et al.*, 2005). Once the nematode has reached the root, different secretory organs participate in the molecular interaction with the host. As a general feature of parasitic nematodes, secretions from the cuticle build up a surface coat that is likely to hide the nematode from host perception throughout the interaction (reviewed by Curtis, 2007). In addition, and similar to animal-parasitic and cyst nematodes, it is thought that root-knot nematodes secrete, through the cuticle, antioxidant enzymes that are produced in the hypodermis and may protect the nematode from the oxidative response of the host to nematode infection (Robertson *et al.*, 2000; Prior *et al.*, 2001; Jones *et al.*, 2004). The amphids are two chemosensory organs located on the head of the nematode and are involved in the perception of the environment (see Curtis *et al.*, Chapter 6, this volume). Amphidial secretions may act as signalling molecules during the interaction and may elicit defence responses from the plant (Semblat *et al.*, 2001). The role of secretions produced by other secretory organs, such as the phasmids, is so far unknown. By contrast, extensive efforts have been focused on the nematode secretions injected through the stylet into the plant tissue. The nematode proteins produced in, and secreted from, the pharyngeal gland cells into the plant host via the stylet are signal molecules that trigger the manipulation of signalling pathways, leading to host defence suppression and giant cell induction.

A complex panel of proteins secreted through the stylet is probably necessary for the successful establishment of the nematode (Jaubert *et al.*, 2002a; Huang *et al.*, 2003). To date, at least 60 different proteins with various predicted functions have been identified, indicating that several cellular processes are targeted by the nematode for successful manipulation of the host response. Among the identified functions for these secretions is plant cell wall degradation. Some effectors have also been proposed to be involved in the modulation of plant defences and induction of giant cells.

Infective juveniles secrete a battery of cell wall degrading and modifying enzymes that participate in cell wall softening during the penetration and migration in the root tissues. All major constituents of the cell wall are targeted by enzymes secreted by root-knot nematodes: beta-1,4-endoglucanases are active on cellulose (Rosso *et al.*, 1999; Ledger *et al.*, 2006); pectate lyases and polygalacturonases are active on pectin (Jaubert *et al.*, 2002b; Huang *et al.*, 2005); and xylanases are active on hemicellulose (Mitreva-Dautova *et al.*, 2006). In addition, proteins homologous to expansins have been identified, which could weaken the intermolecular bonds between cell wall polysaccharides (Roze *et al.*, 2008). Recently, a survey of the genome of *M. incognita* revealed that all these genes are members of multigene families and that some families, such as cellulases and pectate lyases, underwent considerable expansion (Abad *et al.*, unpublished data). The number and diversity of these genes is unprecedented in any other animal, and the similarity between root-knot nematode and bacterial genes suggests that some of these genes may have been acquired by horizontal gene transfer from bacteria (Scholl and Bird, 2005; Ledger *et al.*, 2006).

Although J2 wound the plant at the penetration point of the root, they subsequently migrate between the cells, causing little damage. However, the host plant responds to the pathogen attack by a rapid and transient production of reactive oxygen species (ROS; Melillo *et al.*, 2006; Gao *et al.*, 2008). The endophyte stages of root-knot nematodes secrete proteins with scavenger activities, which protect the parasite from the damaging effects of ROS (Molinari and Miacola, 1997). In cyst nematodes and animal-parasitic nematodes, secreted antioxidant enzymes are produced in the hypodermis and secreted through the cuticle (Henkle-Dührsen and Kampkötter, 2001; Jones *et al.*, 2004). Surprisingly, a glutathione S-transferase (GST) from class sigma is produced in the pharyngeal glands and secreted by *M. incognita* juveniles. This GST could participate in the detoxification of cytotoxic compounds generated by the oxidative response of the plant (Dubreuil *et al.*, 2007).

Later on during parasitism, the observed downregulation of defence genes from host plants suggests an active modulation of the plant response by root-knot nematodes (Jammes *et al.*, 2005). An active role in the suppression of plant defences has been proposed for a chorismate mutase expressed in the pharyngeal glands of the nematode (Doyle and Lambert, 2003; Long *et al.*, 2006). In bacteria, chorismate mutases are key enzymes of the shikimic acid pathway, which directs the synthesis of aromatic amino acids. In plants, chorismate mutases are also required for the synthesis of several secondary metabolites, including phytohormones such as auxin, and plant defence compounds such as flavonoids, salicylic acid and phytoalexins. The root-knot nematode chorismate mutase may impair the development of the root tissue or suppress host plant defence by affecting the synthesis of chorismate-derived compounds (Doyle and Lambert, 2003). More secreted proteins could interfere with plant signalling pathways during the interaction. A 14-3-3 protein, isoform zeta, expressed in the pharyngeal glands, was isolated from purified stylet secretions (Jaubert *et al.*, 2004). Interestingly, 14-3-3 proteins have been shown to be associated with leucine-rich repeat receptor-like kinases (LRR-RLK), multiprotein complexes in *Arabidopsis*, as well as in mammals (Rienties *et al.*, 2005; Karlova *et al.*, 2006). Membrane-located LRR-RLKs play important roles in plant signalling pathways during plant development and pathogen interactions. For example, LRR-RLK proteins are involved in hormone perception, bacterial PAMP (pathogen-associated molecular pattern) signalling and disease resistance in *Arabidopsis* (reviewed by Torii, 2004). Most likely, these multiprotein complexes require 14-3-3 proteins as adaptor proteins to guide protein–protein interactions. Plant 14-3-3 proteins have also been shown to regulate the activity of transcription factors by sequestering them in cytoplasmic or nuclear compartments (reviewed by Muslin and Xing, 2000). In addition,

14-3-3 proteins are secreted by animal parasites, although their role in parasitism is unclear (Siles-Lucas *et al.*, 2008). A calreticulin has been shown to be secreted into the feeding site by the sedentary stages of the nematode (Fig. 7.1C). Accumulation of the protein was observed at the cell wall of the giant cells (Jaubert *et al.*, 2005). Although a function for the nematode calreticulin in the modulation of the plant responses has still to be determined, calreticulins secreted by animal parasites, including nematodes, platyhelminths and arthropods, have been identified as key modulators of the host immune defences. Interestingly, a calreticulin secreted by the animal-parasitic nematode *Necator americanus* interacts with the surface receptor C1q of the complement system of vertebrates (Kasper *et al.*, 2001; Suchitra and Joshi, 2005). Other calreticulin binding partners have been identified at the surface of animal cells that regulate calcium signalling and the cell cycle (Borisjuk *et al.*, 1998; Ghiran *et al.*, 2003).

Besides the suppression of the plant defences, the successful establishment of the nematode in the plant depends on the differentiation of root parenchyma cells in specialized giant cells, which provide the nematode with the nutrients required for its development during the 3–8 weeks necessary before the production of progeny (depending on the plants and temperature conditions). The induction of giant cell formation is certainly triggered by nematode secretions. An increasing number of genes potentially involved in giant cell induction have been identified (reviewed by Davis *et al.*, 2004; Vanholme *et al.*, 2004; Dubreuil *et al.*, 2007). The characterization of a secreted peptide gave new insights into the way nematodes can manipulate host functions to their own benefit (Huang *et al.*, 2006a). The 13-amino-acid peptide 16D10, expressed in the pharyngeal glands of parasitic J2, stimulates root growth and the generation of extensive lateral roots when overexpressed in tobacco hairy roots. In addition, 16D10 interacts *in planta* with two putative plant SCARECROW-like transcription factors (Huang *et al.*, 2006a). The role of the identified transcription factors during plant development is still to be determined; this work presented the first characterization of plant targets for a nematode-secreted peptide and suggested that nematodes actively intervene in the regulation of plant genes during parasitism. Interestingly, the peptide 16D10 has similarity with plant CLE domains. In plants, CLE peptides control the differentiation of stem cells from the vascular bundle and from the root and shoot apical meristems. The demonstration of the importance of 16D10 for successful development of the nematode, together with the identification of more CLE homologues functionally similar to plant CLE peptides in the closely related cyst nematodes, indicate that root-knot and cyst nematodes have evolved plant mimicry strategies to interfere with the signalling pathways of plant cell development (Huang *et al.*, 2006b; Mitchum *et al.*, 2008).

It is suspected that more plant signalling or regulatory pathways are targeted by nematode secretions, as the secreted proteins identified so far from root-knot nematodes are predicted to act in various cellular compartments of plant cells. The low amount of proteins secreted by the nematode limits their precise localization *in vivo*. *In planta* localization of the secreted calreticulin showed the accumulation of the protein along the cell wall of the giant cells (Jaubert *et al.*, 2005), notwithstanding a possible injection inside the plant cell. Injection by the nematode of the secreted peptide 16D10 into the cytoplasm of plant cells is suggested by its ability to bind cytoplasmic transcription factors (Huang *et al.*, 2006a). Finally, few putative parasitism genes encode proteins with predicted nuclear localization signals (Huang *et al.*, 2003), suggesting that some nematode effectors may be targeted to the nucleus of the host cell.

7.4 Compatible Interactions with Resistant Plants: the Case of Virulent Root-knot Nematodes

Plant resistance is currently an effective and environmentally safe method to control root-knot nematodes. Resistance genes that act by inducing a typical hypersensitive reaction (HR) that prevents the parasite establishment and/or reproduction have been identified in a number of host plants. A detailed review of the mechanisms and genetics of resistance to root-knot nematodes is provided by Williamson and Roberts, Chapter 13, this volume. However, as reported for other plant pathogens, the occurrence of virulent biotypes able to overcome resistance genes has been documented in *Meloidogyne* (Castagnone-Sereno,

2002), and the phenotypic expression resulting from the infection of resistant plants by virulent nematodes is very similar to the compatible interaction described above. Indeed, reproduction of root-knot nematodes has been reported on resistant host species as diverse as tomato, cowpea or grape (Huang *et al.*, 2004; Petrillo *et al.*, 2006; McKenry and Anwar, 2007), which indicates that virulent nematodes are able to complete their life cycle, i.e. to induce a functional feeding site in the roots of plants bearing a resistance gene without triggering HR. Recently, a detailed histological and biochemical analysis of tomato–root-knot nematode interactions involving both avirulent/virulent *M. incognita* lines and susceptible/resistant plants was performed (Melillo *et al.*, 2006), which illustrates this point.

The concept of host resistance genes and nematode avirulence (Avr) genes, and their interaction to specify gene-for-gene relationships will be presented elsewhere (see Williamson and Roberts, Chapter 13, this volume). Simply explained, disease resistance is observed when any particular resistance gene specifically interacts (directly or indirectly) with a particular pathogen *Avr* gene. In contrast to other plant pathogens, knowledge about genes involved in nematode (a)virulence remains scarce and fragmentary. The first candidate gene coding for a nematode Avr protein was isolated in *M. incognita*, by a comparative AFLP (amplified fragment length polymorphism) fingerprinting analysis of virulent and avirulent near-isogenic lines selected on resistant and susceptible tomatoes, respectively (Semblat *et al.*, 2001). This gene, named *map-1*, encoded a putative secreted protein (MAP-1) containing a predicted N-terminal signal peptide. Interestingly, antibodies raised against MAP-1 specifically labelled amphidial secretions from infective J2 (Semblat *et al.*, 2001). Amphids are the primary chemosensory organs of nematodes, and blocking of their secreted products is known to disrupt the plant–nematode interaction (Perry, 1996; Fioretti *et al.*, 2002; see Curtis *et al.*, Chapter 6, this volume). In that respect, MAP-1 may be involved in the early steps of recognition between resistant plants and avirulent nematodes, and polymorphism in MAP-1 sequence or expression may contribute to the development of a compatible interaction between virulent nematodes and resistant plants. A BLASTX search showed some similarity of the C terminus of MAP-1 with the EXPB2 expansin secreted by the potato cyst nematode *Globodera rostochiensis* (Qin *et al.*, 2004), which suggests that MAP-1 may be distantly related to expansins. Further functional analyses will no doubt help to reveal the function of this protein in the plant–nematode interaction.

Using an established model system of avirulent and virulent *M. incognita* near-isogenic lines, a cDNA (complementary DNA)–AFLP-based transcriptomic approach has been developed to monitor differences in gene expression, which resulted in the identification of 22 transcript-derived fragments (from more than 24,000 generated) present in avirulent lines and absent in virulent lines (Neveu *et al.*, 2003a). Analysis of the full-length cDNAs revealed a signal peptide for some of these candidates, and further *in situ* hybridization experiments showed their specific expression in the intestinal or pharyngeal gland cells of infective J2 (Neveu *et al.*, 2003a). Among them, a cysteine protease gene, *Mi-cpl-1*, was shown to be expressed only in the developmental stages that are in close interaction with the root tissues (i.e. juveniles and females), which may indicate that the cysteine protease in *M. incognita* is related to the parasitic aspects of the plant–nematode relationship, e.g. pathogenicity and/or evasion of primary host plant defence systems (Neveu *et al.*, 2003b). Indeed, the effects of knocking-out *Mi-cpl-1* gene function were consistent with a reduction in nematode feeding efficiency, and showed a correlation between transcript abundance, proteinase activity and parasitic success of *M. incognita* (Shingles *et al.*, 2007; Rosso *et al.*, unpublished). More recently, one of the candidate genes was further characterized as an aspartic protease and shown to be specifically expressed in the J2 subventral pharyngeal glands (Neveu *et al.*, unpublished). Evidence is accumulating that a growing list of plant pathogen Avr genes function as proteases that are secreted into plant cells to modify host proteins (e.g. *XopD* and *AvrXv4* in the bacterium *Xanthomonas campestris*, the NIa protease from the potato virus Y, and *AvrPi-ta* in the rice blast fungus; Rathjen and Moffett, 2003; Hotson and Mudgett, 2004; Xia, 2004). These findings reveal that post-translational modification of plant proteins through proteolytic processing is a widely used mechanism in regulating the plant defence response and/or the plant innate immunity. Remarkably, a 'no *a priori*' cloning strategy allowed the identification of two proteases as root-knot nematode candidate Avr genes, and a more profound knowledge of their function will help to reveal the role of these proteins in the molecular events associated with nematode (a)virulence and the activation of plant responses.

7.5 (A)virulence Determinants and Pathogenicity Factors: Root-knot Nematode Effectors with Dual Function?

Nematode genes encoding secretory products are considered members of the 'parasitome', i.e. a set of genes that promote parasitism of plant hosts (see above and Atkinson *et al.*, Chapter 15, this volume). The few putative Avr genes of root-knot nematodes characterized so far have been shown to encode secretory/excretory proteins, which suggests that they do indeed belong to the so-called parasitome. This observation could suggest a possible dual role for such secretions, in line with current accumulating evidence that the avirulence and virulence activities of pathogen effector molecules involved in plant-pathogen recognition events are often linked. Avr genes, including some proteases, may contribute to pathogenicity in the absence of specific R (resistance) components in the host (Rathjen and Moffett, 2003; Alfano and Collmer, 2004; Skamnioti and Ridout, 2005). Thus, the current view that the primary role of many Avr proteins is to target unknown determinants of susceptibility in the host could be applied to root-knot nematodes as well. Together with the pathogenicity components presented earlier in this chapter, the putative Avr proteins of root-knot nematodes belong to the battery of secreted effectors that nematodes deliver into cells of their hosts, and probably also provide non-specific pathogenicity functions in the absence of plant resistance, as demonstrated for other pathogens (Bent and Mackey, 2007).

7.6 Tools for Molecular and Functional Analysis of Root-knot Nematode Parasitism

The difficulty in assigning a role for nematode-secreted proteins has long been due to the lack of a functional analysis tool for root-knot nematode genes. The need for such a tool is particularly striking for the nematode-secreted proteins for which no function could be predicted on the basis of sequence similarity (Huang *et al.*, 2003; Dubreuil *et al.*, 2007). In this respect, the development of RNA interference (RNAi) has been an enormous breakthrough as it allows the knock-down of *Meloidogyne* genes (Rosso *et al.*, 2005). RNAi is the suppression of gene expression induced by double-stranded RNA (dsRNA) molecules homologous to the targeted transcript. The dsRNA molecules synthesized *in vitro* can be delivered to the nematodes by soaking or by artificial stimulation of uptake through the stylet. In addition, nematodes can ingest, by feeding, dsRNA molecules produced by transgenic plants (Huang *et al.*, 2006b). RNAi has allowed the functional analysis of genes involved in various functions, such as development, cuticle biosynthesis, digestion, gene transcription and parasitism (reviewed by Fairbairn *et al.*, 2007; Lilley *et al.*, 2007).

An exhaustive analysis of the compounds secreted by *Meloidogyne* spp. during the parasitic phase of their life cycle is limited by their obligate parasitism. Therefore, technological limitations have led to a focus on protein secretions. However, secreted non-protein compounds probably also play a role in the establishment of the nematode in the plant tissue, as demonstrated by the presence of different cytokinins in exudates from infective juveniles (De Meutter *et al.*, 2003).

The success of the establishment of root-knot and other plant-parasitic nematodes in the host plant involves various functions that highlight similarities with phytoparasitic microbes for root invasion and plant defence modulation. Striking similarities between proteins from root-knot nematodes and bacteria have led to the speculation that some nematode parasitism genes may have been acquired by horizontal gene transfer from bacteria (Veronico *et al.*, 2001; Scholl and Bird, 2005; Ledger *et al.*, 2006). In addition, plant-parasitic nematodes and root-knot nematodes share protection strategies with animal-parasitic nematodes in order to cope with host defences, suggesting convergent evolution for these functions. An additional interesting feature of nematode adaptation to plant parasitism is the ability to mimic plant regulators of cell fate or development. The genome sequences for two species, *M. incognita* and *M. hapla* (see Abad and

Oppermann, Chapter 16, this volume), are undoubtedly unprecedented sources providing a comprehensive view of the strategies *Meloidogyne* has developed for successful manipulation of plant functions.

7.7 Giant Cell Development

Root-knot nematodes are obligate biotrophic pathogens that can only feed on living cells. They establish and maintain an intimate relationship with their host plants. Within the root vascular cylinder, J2s induce the redifferentiation of root cells into giant cells, which represent specialized feeding cells (Fig. 7.2; Plate 37). Fully differentiated giant cells contain more than 100 polyploid nuclei, which have also possibly undergone extensive endoreduplication (Wiggers *et al.*, 1990). Giant cells may reach a final size about 400 times that of individual root vascular cells. In addition, giant cells show an increase in cytoplasmic density and a loss of normal vacuolization. The dense cytoplasm contains a well-developed Golgi apparatus and smooth endoplasmic reticulum, generally organized in swirls, and numerous mitochondria, plastids and ribosomes (Jones and Payne, 1978). Another characteristic feature of these giant cells is the development of cell wall ingrowths, typical of transfer cells (Jones, 1981). These cell wall ingrowths in contact with the xylem elements increase the surface area of the associated membrane and probably enhance solute uptake from the vascular system. Giant cells serve as the sole food source for the subsequent sedentary parasitic stages. The establishment and maintenance of fully differentiated giant cells are essential to fulfil the nematode nutritional demands for growth and reproduction.

One of the first signs of giant cell induction is the formation of vascular binucleate cells (Fig. 7.2A,B). The initial selected cells enlarge considerably and become multinucleate through synchronous repeated nuclear divisions without cell division. Despite the fact that karyokinesis occurs without complete cell division in giant cells, cytokinesis is initiated at the end of the mitosis. Jones and Payne (1978) first described a normal alignment of cell plate vesicles between two daughter nuclei in giant cells, followed by the dispersal of these vesicles and arrest of cytokinesis. Recently, *in vivo* confocal microscopy of gall sections revealed the presence in mitotic giant cells of early synchronous phragmoplast arrays, which do not develop further (Caillaud *et al.*, 2008b). The phragmoplast serves as a scaffold for the growing cell plate assembly, guiding vesicles with cell wall material, and subsequent formation of a new cell wall. In developing giant cells, a restricted out-growth of the phragmoplast leads to the formation of a novel cell plate structure – the giant cell mini cell plate – which does not extend across adjacent faces of the cell. Optical and electron microscopy have confirmed that giant cell mini cell plates were frequently observed between two nuclei in giant cells. Interestingly, this initiation of cytokinesis has been demonstrated to be essential for giant cell ontogenesis (Caillaud *et al.*, 2008b).

It is not yet understood how feeding cells are induced, but it is believed that pathogenicity factors secreted by the nematode might have direct effects on recipient host cells (Davis *et al.*, 2004, Vanholme *et al.*, 2004). Because *Meloidogyne* spp. can induce giant cells in thousands of plant species in a similar manner, they probably interact with and manipulate fundamental host functions to their own benefit. The temporal requirement for an inductive signal is unknown and a transient induction should be sufficient (Bird and Kaloshian, 2003). However, some ongoing interaction between nematode and giant cells is required, as removal of the nematode leads to giant cell destruction (Bird, 1962). Whether this constitutive stimulus is caused by the metabolic sink of feeding or more specifically by a nematode secretion remains unknown. The identification of plant genes required for giant cell ontogenesis remains a major challenge and should greatly improve our understanding of the way nematodes dramatically alter root development to produce and maintain giant cells.

The transformation of root cells into hypertrophied feeding cells with unique morphology and functions requires extensive changes to gene expression in infected root cells (reviewed by Gheysen and Fenoll, 2002; de Almeida Engler *et al.*, 2005; Caillaud *et al.*, 2008a). Molecular and genetic approaches have been developed based on: differential gene expression between healthy and infected root regions, such as cDNA subtraction or differential display; expression analyses of candidate genes by promoter-GUS (β-glucuronidase) reporter fusions; *in situ* hybridization or RT-PCR (reverse transcription-PCR); and promoter trap strategies in which a promoterless GUS construct was introduced randomly into the plant genome. As an example, a screening of 20,000 T-DNA (bacterial plasmid transferred DNA)-tagged *Arabidopsis* lines for GUS expression after *M. incognita* infection enabled the identification of about 200 lines showing GUS induction in galls (Abad *et al.*, 2003; Favery *et al.*, 2004); these lines also presented GUS expression in healthy plants within different cell types and at different developmental times. These results support the hypothesis that plant functions have been recruited to allow pathogen growth, and confirm the complex morphological and physiological changes in cells during their modification into nematode feeding cells. The existence of these strategies has resulted in the characterization of tens of plant genes, mostly upregulated in response to root-knot nematode infection, and these nematode-responsive plant genes have highlighted changes in some key plant development processes, such as cell cycle and cytoskeleton organization regulation (de Almeida Engler *et al.*, 1999, 2004; Favery *et al.*, 2004; Caillaud *et al.*, 2008b). In addition, cell wall modification (Goellner *et al.*, 2001), hormone and defence responses (Lohar *et al.*, 2004; Jammes *et al.*, 2005) and genes involved in the general metabolism have been identified as differentially expressed during giant cell formation. The recent development of plant microarrays has made it possible to generate large-scale patterns of plant gene expression during giant cell formation. Genome-wide expression profiling, using gene-specific CATMA or Affymetrix genechips, has been used to study the response of *Arabidopsis thaliana* to *M. incognita* infection (Hammes *et al.*, 2005; Jammes *et al.*, 2005). In addition, a tomato microarray containing 12,500 cDNAs has been used to profile the response of tomato to *M. javanica* infection (Bar-Or *et al.*, 2005). These studies identified a large number of new genes regulated in response to *Meloidogyne* infection. The proportion (5–15%) of genes displaying differential expression reflects the complexity of nematode feeding site ontogenesis. In total, 3373 *A. thaliana* genes were found to be differentially expressed between uninfected and giant-cell-enriched root tissues, with similar proportions of genes up- and downregulated (Jammes *et al.*, 2005). Thus, microarray experiments have shown that gene downregulation may also be essential for correct gall formation. As might be expected, not all genes from a given pathway are similarly induced or repressed. Moreover, most plant genes in a given family display different patterns of regulation in compatible interaction, possibly accounting for the conflicting results obtained in previous studies. Plant aquaporin genes, for example, may be upregulated, downregulated or unaffected by nematode infection. As 65% of all *A. thaliana* genes belong to gene families, this highlights the importance of the specificity of the probes used.

Characterization of the genes specifically regulated during giant cell development is a first step towards understanding compatible plant–root-knot nematode interactions. Extensive analysis must be coupled with a detailed cellular expression pattern analysis, the characterization of knockout mutants and biochemical investigations, to dissect more accurately gene function during giant cell development. Large genetic screens and knockouts of genes activated in giant cells have led to the characterization of few mutants in which nematode infection was reduced. Five EMS (ethyl methanesulfonate)-induced mutants with altered responses to *M. incognita* (AMi mutants) were isolated from a screen of 5000 M2 seeds (Niebel *et al.*, 1994). Only two genes have been yet identified in gene knockout studies as essential during the early steps of giant cell formation (Favery *et al.*, 1998; Caillaud *et al.*, 2008b). The gene for ribulose-5-phosphate 3-epimerase (*RPE*) is the first plant gene to be described as being required for nematode susceptibility. This gene has been identified by a promoter trap screen for genes upregulated in giant cells. It encodes a key enzyme in the pentose phosphate pathway necessary for metabolic cell reprogramming when turning into giant cells. This pathway plays a crucial role in cells by producing the NADPH required in numerous biosynthetic reactions, and by generating carbohydrate intermediates for the synthesis of nucleotides and cell wall polymers. Recently, Caillaud *et al.* (2008b) described a unique defect in nematode feeding cell formation. In the absence of a microtubule-associated protein, MAP65-3, giant cells started to develop but did not complete their differentiation process and were aborted.

7.8 Cytoskeleton Organization and Cell Cycle Progression During Giant Cell Ontogenesis

The cytoskeleton plays a central role in the cell cycle, differentiation and morphogenesis. The distribution of microtubules (MTs) and microfilaments in giant cells has recently attracted the attention of researchers. Transcriptional activity of actins and tubulins and organization of the actin filaments and MTs have been analysed *in situ* by de Almeida Engler *et al.* (2004). Promoter–GUS fusions of two actin genes, *ACT2* and *ACT7*, expressed in roots and other vegetative tissues, and mRNA *in situ* hybridizations of the three plant tubulins showed high promoter activity of both actin genes and high mRNA levels of the α-, β- and γ-tubulin genes in giant cells and neighbouring cells throughout gall development. Immunolocalization using actin and tubulin antibodies revealed that major rearrangements of the cytoskeleton occur during the formation of nematode feeding cells. *In vivo* observations and immunofluorescence microscopy of galls revealed thick actin cables going through the giant cell cortex. In the cytoplasm, actin bundle segments were shorter, thinner and distributed at random within a slight amorphous actin staining. Chemical blocking of the actin or microtubule cytoskeleton dynamics using cytochalasin D, colchicin or taxol, respectively, resulted in the arrest of proper giant cell development (Wiggers *et al.*, 2002; de Almeida-Engler *et al.*, 2004). These results point to the relevance of the cytoskeleton rearrangements for giant cell initiation and maintenance; therefore, it is of interest to identify interacting/binding proteins. A gene encoding an *A. thaliana* formin homology protein, AtFH6, identified by a promoter trap strategy was the first plant candidate gene found to be implicated in giant cell actin cytoskeleton reorganization (Favery *et al.*, 2004). *AtFH6* is induced at the early stages of giant cell formation and expression persists until their final differentiated state. Formins have been shown to be actin-nucleating proteins that stimulate the *de novo* polymerization of actin filaments (Sagot *et al.*, 2002; Staiger and Blanchoin, 2006). In contrast to AtFH proteins of animals and fungi, AtFH6 has an N-terminal domain with a putative signal peptide or membrane anchor and a transmembrane domain. Subcellular localization analyses showed that AtFH6 is anchored and uniformly distributed throughout the giant cell plasma membrane. Expression analysis of the additional 20 *AtFH* genes showed that two additional *AtFH* genes, *AtFH1* and *AtFH10*, were upregulated in galls 7 and 14 days post-inoculation (dpi). These three formin genes may regulate giant cell isotropic growth by controlling the assembly of actin cables. During giant cell formation, these actin cables might guide the vesicle trafficking needed for extensive plasma membrane and cell wall biogenesis. The identification of forming-interacting proteins in giant cells would allow the study of regulatory mechanisms and signalling molecules responsible for actin cytoskeleton reorganization (Favery *et al.*, 2004).

MT immunolocalization showed giant cell cytoplasm filled with a dense and diffuse fluorescence (de Almeida Engler *et al.*, 2004). A detailed analysis of dynamic changes in the organization of the MT cytoskeleton of giant cells has been performed using the MAP4 microtubule-binding domain (MBD)–GFP reporter protein. *In vivo* confocal microscopy in giant cells revealed MBD–GFP decorated MT cortical arrays with bundling MTs (Caillaud *et al.*, 2008b) (Plate 37). During cell cycle progression, microtubules reorganize into an anastral bipolar spindle, ensuring the accurate segregation of chromosomes during anaphase (Wasteneys, 2002). Multiple spindles were observed in giant cells. The presence of early synchronous phragmoplast was confirmed in mitotic giant cells, whereas late phragmoplast has never been observed (Caillaud *et al.*, 2008b). Detailed functional analyses of the *Arabidopsis* microtubule-associated protein MAP65-3 showed that this protein was associated with the mini cell plates formed between daughter nuclei during cytokinesis initiation in developing giant cells (Fig. 7.2B,C). In the absence of functional MAP65-3, giant cells started to develop but did not complete their differentiation and were eventually destroyed. These giant cell defects impaired the maturation of the infecting nematodes, which are dependent on the nutrients supplied by the giant cells. No fourth-stage juvenile was ever observed in the T-DNA *map65-3* mutants. Thus, MAP65-3 is essential for giant cell ontogenesis and is required for successful pathogen growth and development. Giant cell mini cell plates were never observed in the absence of MAP65-3. Instead, aberrant cell wall stubs were observed in the first steps of giant cell formation in T-DNA *map65-3* mutants. Thus, the giant cell mini cell plate most probably forms a physical barrier separating the two daughter nuclei, required for the multiple rounds of mitosis that occur in developing giant cells, resulting in the formation of a functional feeding site. Defects in mini cell plate formation in the absence of MAP65-3 would lead to the accumulation of mitosis defects (cell wall stubs and connected nuclei) during repeated mitoses. These defects may prevent the development of functional feeding cells, resulting in the death of the nematode. Detailed functional analysis during plant development highlighted the role of MAP65-3 in plant cell division. MAP65-3 plays a key role in the organization of microtubule arrays during both mitosis (spindle morphogenesis) and cytokinesis (phragmoplast expansion) in dividing plant cells. In the early stages of giant cell formation, the activation of MAP65-3 transcription reflects rapid cell cycle reactivation. The expression of this gene then rapidly declines before the development of fully mature giant cells (Caillaud *et al.*, 2008b).

Early transcriptional activation of the genes for cell cycle markers such as cyclin-dependent kinases (*CDC2a* and *CDC2b*) and the mitotic cyclins (*CYC1At* and *CYCA2;1*) has been reported previously (Niebel *et al.*, 1996; de Almeida-Engler *et al.*, 1999). Recently, a detailed expression analysis of 61 core cell cycle genes has been performed, and most genes are shown to be expressed in giant cells (de Almeida Engler, unpublished data). It seems plausible that there should be a large number of other cell cycle regulators that would be found to be upregulated when the cell cycle is restarted during the infection process. In addition, genes involved in endoreduplication, such as *CCS52*, have been shown to be induced in giant cells (Favery *et al.*, 2002). Endoreduplication is the DNA duplication of the genome in the absence of mitosis. This common process in eukaryotes results in an increase in the nuclear DNA content, permitting amplification of the genome of specialized cells. Moreover, the increase in the ploidy level correlates with an increase in nuclear volume and cell size, suggesting that elevated nuclear DNA content is required to maintain larger cells. Indeed, preventing DNA synthesis or blocking the cell cycle at later phases significantly inhibited feeding cell progression (de Almeida-Engler *et al.*, 1999). Further studies to validate the potential relevance of endoreduplication for feeding site formation and maintenance will illustrate the relevance of this process during the interaction between plants and parasitic nematodes.

7.9 Extensive Cell Wall Modifications to Build Up Giant Cells

A characteristic feature of giant cells is their outstanding isotropic growth. Such cell expansion requires extensive and coordinated cell wall remodelling. Plant hydrolases and expansins might play a fundamental role in loosening the cellulose/cross-linking glycan network. Goellner *et al.* (2001) have validated the idea that cell-wall-modifying enzymes of plant origin, such as endo-β-D-glucanases, are implicated in feeding cell formation. The *Arabidopsis* endo-1,4-β-glucanase gene *CEL1* and a gene encoding a pectin acetylesterase have been shown to be upregulated in developing giant cells (Vercauteren *et al.*, 2002; Mitchum *et al.*, 2004). Microarray analysis showed that all genes encoding class A (α) and B (β) expansins regulated upon *Meloidogyne* infestation were activated in *A. thaliana* (Jammes *et al.*, 2005). Expansin proteins rapidly induce extension of plant cell walls by weakening the non-covalent interactions that help to maintain their integrity. In tomato expansin, the gene *LeEXPA5* has been shown to be expressed in gall cells adjacent to the giant cells. A decrease in *LeEXPA5* expression affected the ability of the nematode to complete its life cycle (Gal *et al.*, 2006). In addition, all regulated *A. thaliana* pectate lyases, and most of the polygalacturonases and pectinesterases, are also activated in response to *M. incognita* infestation (Jammes *et al.*, 2005). The concomitant deposition of newly synthesized wall material is associated with this loosening process. Throughout feeding site development, further modifications to the cell wall result in wall thickening and development of wall ingrowths. Several additional genes encoding cell wall proteins (e.g. hydrolases and structural proteins) have been identified as potentially induced or repressed upon infestation (Jammes *et al.*, 2005).

7.10 Suppression of Plant Defence Associated with Giant Cell Development

J2 of *Meloidogyne* cause little damage to the roots during the invasion process as they migrate between the cells. However, wound or defence responses are detected from initial J2 penetration onward. In general, the initial reactions and associated patterns of gene regulation are similar in susceptible and resistant plants. In compatible tomato–root-knot nematode interaction, the generation of reactive oxygen species has been observed at the time of nematode invasion (12 h post inoculation) but was not detectable cytologically 2 dpi, at the time of giant cell induction (Mellilo *et al.*, 2006).

The importance of plant defence suppression during plant–pathogen interactions has been highlighted in recent studies. Indeed, *in planta* development of obligate biotrophic and hemibiotrophic fungi is associated with a phase of active suppression of plant defence (Waspi *et al.*, 2001; Bouarab *et al.*, 2002). Recent data reveal that phytopathogenic bacteria use type III secreted effector proteins, toxins and other factors to interfere with host defence (Abramovitch and Martin, 2004). Thus, successful pathogens seem to have evolved specialized strategies to suppress plant defence response and generate susceptibility in host plants. Defence suppression also appears to play an important role in symbiotic plant–microbe interactions. The NopL effector of *Rhizobium* sp. NGR234 suppresses PR (pathogenesis-related) gene expression when expressed in tobacco or *Lotus japonicus* (Bartsev *et al.*, 2004).

Global analysis showed that the successful establishment of *Meloidogyne* is associated with the suppression of plant defence responses (Jammes *et al.*, 2005). In *A. thaliana*, 70% of the nematode-regulated genes involved in defence were repressed, particularly 14 and 21 dpi. These downregulated defence-related genes included genes that were previously shown to be induced during other plant–pathogen interactions. The suppression of plant defence included resistance genes and resistance-associated genes (*PAD4*, *NHL3*), genes associated with the jasmonic acid/ethylene-dependent pathways (*EIN3*, *ERF1*, *PR4*) and potential antimicrobial genes. Interestingly, no local change was observed in the expression of genes known to be involved in the salicylic acid pathway (*ICS1*, *SID1*, *NPR1*, *WHY1*, *PR5*). In galls, 17 of the 21 *WRKY* genes identified are downregulated, whereas the accumulation of *WRKY* transcripts appears to be a general characteristic of plant defence in response to pathogens. Studies of plant defence in the context of giant cell development would certainly provide useful information on the interplay between defence responses and plant development, about which little is currently known (Whalen, 2005).

7.11 Major Reprogramming of Plant Metabolism and Transport

Giant cells are metabolically hyperactive and form a nutrient sink for the nematode. A major reprogramming of plant metabolism also occurs throughout giant cell formation. Differential display analysis of gene expression in giant cells induced in tomato roots identified transcripts with significant homology to S-adenosylmethionine decarboxylases, cysteine synthases and ribosomal proteins (Wang *et al.*, 2003). These results confirmed the high metabolic turnover in mature giant cells. The specific upregulation in 7 dpi galls of 71 *Arabidopsis* genes encoding 40S and 60S ribosomal proteins clearly suggests increased levels of protein synthesis during giant cell initiation (Jammes *et al.*, 2005). Genome-wide analysis showed that whereas many genes related to metabolism and energy are upregulated in galls, others are downregulated. Similar numbers of genes involved in metabolism were shown to be induced and repressed. The functionality of many overproduced proteins surely benefits from chaperones such as those encoded by *HasHSP17.7* (Escobar *et al.*, 2003). These authors showed that a short fragment of the promoter of the *Hahsp17.7*G4 gene that encodes a small heat-shock protein involved in embryogenesis and stress response is specifically expressed in tobacco galls. Analysis of regulatory sequences and interacting transcription factors should add information on signal transduction pathways essential for feeding cell development.

Large amounts of water and solutes are transported from the xylem through the cell wall ingrowths of the giant cells, probably via water channels that facilitate the passage across biological membranes (Gheysen and Fenoll, 2002). The fine transcriptional regulation of aquaporin genes in giant cells may account for the several functions proposed for these proteins in growth control, water transport and cell osmoregulation (Maurel and Chrispeels, 2001). A previous study reported downregulation of the *Arabidopsis* *γTIP1;1* gene, encoding a tonoplast aquaporin, in galls (Goddijn *et al.*, 1993), whereas later studies reported the upregulation of the tobacco *TobRB7* gene (Opperman *et al.*, 1994) and the *M. truncatula* NIP *NOD26* gene (Favery *et al.*, 2002) in giant cells and galls, respectively. The use of microarrays containing gene-specific sequences showed that genes from the same family may be differently regulated, possibly explaining conflicting results. Among 25 *Arabidopsis* aquaporin genes analysed, three were shown to be upregulated (one *NOD26*-like and two plasma membrane *PIPs*). Seven aquaporin genes, three *PIPs* and four *TIPs*, were shown to be downregulated, including *AtTIP1.1* and *AtPIP1.5* genes. Hammes *et al.* (2005) confirmed by quantitative RT-PCR and promoter GUS fusion that *AtPIP2.5* was specifically upregulated in galls. In a transcriptome analysis of 635 transporter genes, the authors identified 50 genes up- or downregulated in roots infected by *Meloidogyne*. Peptide transport seems to display overall downregulation, whereas an upregulation of amino acid transporters is observed after *M. incognita* inoculation. Two genes encoding amino acid transporters, *AtAAP6* and *AtCAT6*, have been shown to be expressed at higher levels in the galls that contain the giant cells (Hammes *et al.*, 2005, 2006).

The induction, in two independent microarray studies, of *AUX1* and *AtAUX4/LAX3* encoding putative auxin transporters is consistent with a role for plant hormones in the successful establishment of root-knot nematodes. The early, localized and transient activation of the synthetic auxin-responsive promoter element DR5, derived from the soybean promoter GH3, points to a local increase of auxin in feeding sites (Karczmarek *et al.*, 2004). In addition, induction in *Medicago truncatula* giant cells of the *PHAN* and *KNOX* genes, required for normal meristem function, and their involvement in changes in phytohormone levels suggests the implication of these genes in the regulation of auxin distribution during feeding cell development (Koltai *et al.*, 2001). Auxin is probably not the only hormone that plays a role in feeding cell induction. The observed upregulation of a cytokinin-responsive *ARR5* (*Arabidopsis* response regulator) promoter during the early stage of plant–root-knot nematode interaction suggests that a spike of cytokinin is also required during giant cell initiation (Bird, 2004; Lohar *et al.*, 2004). However, the regulation of only a few components of the auxin, abscisic acid, gibberellin and cytokinin pathways has been demonstrated in giant cells. Only one hormone mutant, that of the tomato *diageotropica* gene, has been reported to alter *Meloidogyne* parasitism (Richardson and Price, 1984).

7.12 Comparison between *Meloidogyne* Parasitism and Symbiotic Rhizobia in *Medicago*

Plants have engaged in associations with a wide range of mutualistic and parasitic biotrophic organisms, so it is quite conceivable that these biotrophic interactions might have evolved certain common core components affecting cellular processes (Parniske, 2000). In spite of the fact that the development and physiology of specialized plant cells during biotrophic interactions are specific and differ significantly depending on the type of microorganism, the interactions occurring between plants and endoparasitic nematodes, and in the legume–*Rhizobium* symbioses, are among the most elaborate. For example, as for giant cell formation, the differentiation of the nodule primordium starts by division arrest, and leads to enlarged plant cells with increased DNA. Nod factors secreted by rhizobia play a central role as external mitogenic signals that induce cell division in the root cortex (Foucher and Kondorosi, 2000). Several studies have stressed the existence of common features in the developmental programmes of galls and nitrogen-fixing nodules (Koltai *et al.*, 2001; Favery *et al.*, 2002; Bird, 2004).

Analysis of the expression pattern of ~200 nodule-expressed genes revealed that those such as the early nodulin genes *ENOD40, CCS52a, NOD26* and *CYCD3* (Favery *et al.*, 2002), and those for the regulatory proteins required for the establishment of meristems (Koltai *et al.*, 2001), are significantly upregulated in plant interactions with both rhizobia and parasitic nematodes. These results suggest that cellular processes dealing with endoreduplication, cell cycle regulation, cell to cell communication and water transport might be shared by the complex developmental processes of nodule organogenesis and gall formation. The marker genes induced early in response to *Sinorhizobium meliloti* Nod factors *ENOD11* and *ENOD40* have been shown to be expressed in tissues surrounding the developing giant cells during infections by root-knot nematodes (Favery *et al.*, 2002; Boisson-Dernier *et al.*, 2005). Weerasinghe *et al.* (2005) showed that root-knot nematodes trigger a cytoskeletal response identical to that induced by Nod factors, leading to root-hair waviness and branching in legumes, mediated by a nematode signal acting at distance. In addition, studies in *L. japonicus* mutants have shown that SYMRK, NFR1 and NFR5 are probably involved in the perception of these nematode-derived signals (Weerasinghe *et al.*, 2005).

7.13 Conclusions and Future Directions

The routes by which nematodes manipulate their plant hosts are still not well understood. In the past few years, substantial progress has been made in characterizing the host targets of bacterial, viral and filamentous pathogen virulence factors, providing unique insights into basic plant cellular processes such as gene silencing, vesicle trafficking, hormone signalling and innate immunity (Bray-Speth *et al.*, 2007). The genome sequences of *M. incognita* (P. Abad, http://meloidogyne.toulouse.inra.fr/) and *M. hapla* (C. Opperman, D. Bird and V. Williamson, http://www.hapla.com) will provide a new panorama for studying plant–nematode interactions (see Abad and Opperman, Chapter 16, this volume). The identification of secreted nematode effectors that alter or manipulate plant cell division will enhance our understanding of fundamental cellular mechanisms in plants. Furthermore, determining how a nematode selects particular root cells and modifies them to serve as a feeding cell will enhance our understanding of plant cell development. Thus, the nematode infection process provides a wonderful avenue to explore cell biological events.

7.14 Acknowledgments

Research programmes in our laboratory, initially financed by the European Community (programme BIOTECH NONEMA 'Making plants resistant to plant parasitic nematodes: No access – No feeding'), were supported by INRA and GENOPLANTE contracts AF2001032 ('Identification of plant genes from the feeding sites induced by root-knot nematodes') and ANR05GPLA020 'AFINDIS' (*Arabidopsis* Functions INvolved in DIsease Susceptibility). We thank Marie-Cécile Caillaud for confocal microscopy images and critical reading of the manuscript. We apologize to all colleagues and researchers whose work has not been cited directly.

7.15 References

Abad, P., Favery, B., Rosso, M.N. and Castagnone-Sereno, P. (2003) Root-knot nematode parasitism and host response: molecular basis of a sophisticated interaction. *Molecular Plant Pathology* 4, 217–224.

Abramovitch, R.B. and Martin, G.B. (2004) Strategies used by bacterial pathogens to suppress plant defenses. *Current Opinion in Plant Biology* 7, 356–364.

Akhkha, A., Kusel, J., Kennedy, M. and Curtis, R.H.C. (2002) Effects of phytohormones on the surfaces of plant parasitic nematodes. *Parasitology* 125, 165–175.

Alfano, J.R. and Collmer, A. (2004) Type III secretion system effector proteins: double agents in bacterial disease and plant defense. *Annual Review of Phytopathology* 42, 385–414.

Bar-Or, C., Kapulnik, Y. and Koltai, H. (2005) A broad characterization of the transcriptional profile of the compatible tomato response to the plant parasitic root knot nematode *Meloidogyne javanica. European Journal of Plant Pathology* 111, 181–192.

Bartsev, A.V., Deakin, W.J., Boukli, N.M., McAlvin, C.B., Stacey, G., Malnoe, P., Broughton, W.J. and Staehelin, C. (2004) NopL, an effector protein of *Rhizobium* sp. NGR234, thwarts activation of plant defense reactions. *Plant Physiology* 134, 871–879.

Bent, A.F. and Mackey, D. (2007) Elicitors, effectors and R genes: the new paradigm and a lifetime supply of questions. *Annual Review of Phytopathology* 45, 399–436.

Bird, A.F. (1962) The inducement of giant cells by *Meloidogyne javanica. Nematologica* 8, 1–10.

Bird, D.M. (2004) Signaling between nematodes and plants. *Current Opinion in Plant Biology* 7, 372–376.

Bird, D.M. and Kaloshian, I. (2003) Are roots special? Nematodes have their say. *Physiological and Molecular Plant Pathology* 62, 115–123.

Boisson-Dernier, A., Andriankaja, A., Chabaud, M., Niebel, A., Journet, E.P., Barker, D.G. and de Carvalho-Niebel, F. (2005) *MtENOD11* gene activation during rhizobial infection and mycorrhizal arbuscule development requires a common AT-rich-containing regulatory sequence. *Molecular Plant–Microbe Interactions* 18, 1269–1276.

Borisjuk, N., Sitailo, L., Adler, K., Malysheva, L., Tewes, A., Borisjuk, L. and Manteuffel, R. (1998) Calreticulin expression in plant cells: developmental regulation, tissue specificity and intracellular distribution. *Planta* 206, 504–514.

Bouarab, K., Melton, R., Peart, J., Baulcombe, D. and Osbourn, A. (2002) A saponin-detoxifying enzyme mediates suppression of plant defences. *Nature* 418, 889–892.

Bray-Speth, E., Lee, Y.N. and He, S.Y. (2007) Pathogen virulence factors as potential molecular probes of basic plant cellular functions. *Current Opinion in Plant Biology* 10, 580–586.

Caillaud, M.C., Dubreuil, G., Quentin, M., Perfus-Barbeoch, L., Lecomte, P., de Almeida Engler, J., Abad, P., Rosso, M.-N. and Favery, B. (2008a) Root-knot nematodes manipulate plant cell functions during a compatible interaction. *Journal of Plant Physiology* 165, 104–113.

Caillaud, M.C., Lecomte, P., Jammes, F., Quentin, M., Pagnotta, S., Andrio, E., de Almeida-Engler, J., Marfaing, N., Gounon, P., Abad, P. and Favery, B. (2008b) MAP65-3 microtubule-associated protein is essential for nematode-induced giant cell ontogenesis in *Arabidopsis. Plant Cell* 20, 423–437.

Castagnone-Sereno, P. (2002) Genetic variability of nematodes: a threat to the durability of plant resistance genes? *Euphytica* 124, 193–199.

Castagnone-Sereno, P. (2006) Genetic variability and adaptive evolution in parthenogenetic root-knot nematodes. *Heredity* 96, 282–289.

Curtis, R.H. (2007) Plant parasitic nematode proteins and the host parasite interaction. *Briefings in Functional Genomics and Proteomics* 6, 50–58.

Davis, E.L., Hussey, R.S. and Baum, T.J. (2004) Getting to the roots of parasitism by nematodes. *Trends in Parasitology* 20, 134–141.

de Almeida Engler, J., De Vleesschauwer, V., Burssens, S., Celenza, J.L., Inzé, D., Van Montagu, M., Engler, G. and Gheysen, G. (1999) Molecular markers and cell cycle inhibitors show the importance of the cell cycle progression in nematode-induced galls and syncytia. *Plant Cell* 11, 793–807.

de Almeida Engler, J., Van Poucke, K., Karimi, M., De Groodt, R., Gheysen, G., Engler, G. and Gheysen, G. (2004) Dynamic cytoskeleton rearrangements in giant cells and syncytia of nematode-infected roots. *Plant Journal* 38, 12–26.

de Almeida Engler, J., Favery, B., Engler, G. and Abad, P. (2005) Loss of susceptibility as an alternative for nematode resistance. *Current Opinion in Biotechnology* 16, 112–117.

De Meutter, J., Tytgat, T., Witters, E., Gheysen, G., Van Onckelen, H. and Gheysen, G. (2003) Identification of cytokinins produced by the plant parasitic nematodes *Heterodera schachtii* and *Meloidogyne incognita. Molecular Plant Pathology* 4, 271–277.

Doyle, E.A. and Lambert, K.N. (2003) *Meloidogyne javanica* chorismate mutase 1 alters plant cell development. *Molecular Plant–Microbe Interactions* 16, 123–131.

Dubreuil, G., Magliano, M., Deleury, E., Abad, P. and Rosso, M.N. (2007) Transcriptome analysis of root-knot nematode functions induced in the early stages of parasitism. *New Phytologist* 176, 426–436.

Escobar, C., Barcala, M., Portillo, M., Almoguera, C., Jordano, J. and Fenoll, C. (2003) Induction of the Hahsp17.7G4 promoter by root-knot nematodes: involvement of heat-shock elements in promoter activity in giant cells. *Molecular Plant–Microbe Interactions* 16, 1062–1068.

Fairbairn, D.J., Cavallaro, A.S., Bernard, M., Mahalinga-Iyer, J., Graham, M.W. and Botella, J.R. (2007) Host-delivered RNAi: an effective strategy to silence genes in plant parasitic nematodes. *Planta* 226, 1525–1533.

Favery, B., Lecomte, P., Gil, N., Bechtold, N., Bouchez, D., Dalmasso, D. and Abad, P. (1998) RPE, a plant gene involved in early developmental steps of nematode feeding cells. *EMBO Journal* 17, 6799–6811.

Favery, B., Complainville, A., Vinardell, J.M., Lecomte, P., Vaubert, D., Mergaert, P., Kondorosi, A., Kondorosi, E., Crespi, M. and Abad, P. (2002) The endosymbiosis-induced genes ENOD40 and CCS52a are involved in endoparasitic-nematode interaction in *Medicago truncatula. Molecular Plant–Microbe Interactions* 15, 1008–1013.

Favery, B., Chelysheva, L.A., Lebris, M., Jammes, F., Marmagne, A., de Almeida Engler, J., Lecomte, P., Vaury, C., Arkowitz, R.A. and Abad, P. (2004) *Arabidopsis* formin AtFH6 is a plasma membrane-associated protein upregulated in giant cells induced by parasitic nematodes. *Plant Cell* 16, 2529–2540.

Fioretti, L., Porter, A., Haydock, P.J. and Curtis, R.H. (2002) Monoclonal antibodies reactive with secreted–excreted products from the amphids and the cuticle surface of *Globodera pallida* affect nematode movement and delay invasion of potato roots. *International Journal of Parasitology* 32, 1709–1718.

Foucher, F. and Kondorosi, E. (2000) Cell cycle regulation in course of nodule organogenesis in *Medicago. Plant Molecular Biology* 43, 773–786.

Gal, T.Z., Aussenberg, E.R., Burdman, S., Kapulnik, Y. and Koltai, H. (2006) Expression of a plant expansin is involved in the establishment of root knot nematode parasitism in tomato. *Planta* 224, 155–162.

Gao, X., Starr, J., Göbel, C., Engelberth, J., Feussner, I., Tumlinson, J. and Kolomiets, M. (2008) Maize 9-lipoxygenase ZmLOX3 controls development, root-specific expression of defense genes, and resistance to root-knot nematodes. *Molecular Plant–Microbe Interactions* 21, 98–109.

Gheysen, G. and Fenoll, C. (2002) Gene expression in nematode feeding sites. *Annual Review of Phytopathology* 40, 191–219.

Ghiran I., Klickstein, L.B. and Nicholson-Weller, A. (2003) Calreticulin is at the surface of circulating neutrophils and uses CD59 as an adaptor molecule. *Journal of Biological Chemistry* 278, 21024–21031.

Goddijn, O.J.M., Lindsey, K., Van der Lee, F.M., Klap, J.C. and Sijmons, P.C. (1993) Differential gene expression in nematode-induced feeding structures of transgenic plants harbouring promoter-*gus*A fusion constructs. *Plant Journal* 4, 863–873.

Goellner, M., Wang, X. and Davis, E.L. (2001) Endo-beta-1,4-glucanase expression in compatible plant–nematode interactions. *Plant Cell* 13, 2241–2255.

Hammes, U.Z., Schachtman, D.P., Berg, R.H., Nielsen, E., Koch, W., McIntyre, L.M. and Taylor, C.G. (2005) Nematode-induced changes of transporter gene expression in *Arabidopsis* roots. *Molecular Plant–Microbe Interactions* 18, 1247–1257.

Hammes, U.Z., Nielsen, E., Honaas, L.A., Taylor, C.G. and Schachtman, D.P. (2006) AtCAT6, a sink-tissue-localized transporter for essential amino acids in *Arabidopsis. Plant Journal* 48, 414–426.

Henkle-Dührsen, K. and Kampkötter, A. (2001) Antioxidant enzyme families in parasitic nematodes. *Molecular and Biochemical Parasitology* 114, 129–142.

Hotson, A. and Mudgett, M.B. (2004) Cysteine proteases in phytopathogenic bacteria: identification of plant targets and activation of innate immunity. *Current Opinion in Plant Biology* 7, 384–390.

Huang, G., Gao, B., Maier, T., Allen, R., Davis, E.L., Baum, T.J. and Hussey, R.S. (2003) A profile of putative parasitism genes expressed in the esophageal gland cells of the root-knot nematode *Meloidogyne incognita. Molecular Plant–Microbe Interactions* 16, 376–381.

Huang, G., Dong, R., Allen, R., Davis, E.L., Baum, T.J. and Hussey, R.S. (2005) Developmental expression and molecular analysis of two *Meloidogyne incognita* pectate lyase genes. *International Journal of Parasitology* 35, 685–692.

Huang, G., Dong, R., Allen, R., Davis, E.L., Baum, T.J. and Hussey, R.S. (2006a) A root-knot nematode secretory peptide functions as a ligand for a plant transcription factor. *Molecular Plant–Microbe Interactions* 19, 463–470.

Huang, G., Allen, R., Davis, E.L., Baum, T.J. and Hussey, R.S. (2006b) Engineering broad root-knot resistance in transgenic plants by RNAi silencing of a conserved and essential root-knot nematode parasitism gene. *Proceedings of the National Academy of Sciences of the USA* 103, 14302–14306.

Huang, X., McGiffen, M. and Kaloshian, I. (2004) Reproduction of *Mi*-virulent *Meloidogyne incognita* isolates on *Lycopersicon* spp. *Journal of Nematology* 36, 69–75.

Jammes, F., Lecomte, P., de Almeida Engler, J., Bitton, F., Martin-Magniette, M.L., Renou, J.P., Abad, P. and Favery, B. (2005) Genome-wide expression profiling of the host response to root-knot nematode infection in *Arabidopsis. Plant Journal* 44, 447–458.

Jasmer, D.P., Goverse, A. and Smant, G. (2003) Parasitic nematode interactions with mammals and plants. *Annual Review of Phytopathology* 41, 245–270.

Jaubert, S., Ledger, T.N., Laffaire, J.B., Piotte, C., Abad, P. and Rosso, M.N. (2002a) Direct identification of stylet secreted proteins from root-knot nematodes by a proteomic approach. *Molecular and Biochemical Parasitology* 121, 205–211.

Jaubert, S., Laffaire, J.B., Abad, P. and Rosso, M.N. (2002b) A polygalacturonase of animal origin isolated from the root-knot nematode *Meloidogyne incognita*. *FEBS Letter* 522, 109–112.

Jaubert, S., Laffaire, J.B., Ledger, T.N., Escoubas, P., Amri, E.Z., Abad, P. and Rosso, M.N. (2004) Comparative analysis of two 14-3-3 homologues and their expression pattern in the root-knot nematode *Meloidogyne incognita*. *International Journal of Parasitology* 34, 873–880.

Jaubert, S., Milac, A.L., Petrescu, A.J., de Almeida-Engler, J., Abad, P. and Rosso, M.N. (2005) *In planta* secretion of a calreticulin by migratory and sedentary stages of root-knot nematode. *Molecular Plant–Microbe Interactions* 18, 1277–1284.

Jones, M.G.K. (1981) The development and function of plant cells modified by endoparasitic nematodes. In: Zuckerman, B.M. and Rhode, R.A. (eds) *Plant Parasitic Nematodes*. Academic Press, New York, pp. 225–279.

Jones, M.G.K. and Payne, H.L. (1978) Early stages of nematode-induced giant cell formation in roots of *Impatiens balsamina*. *Journal of Nematology* 10, 70–84.

Jones, J.T., Reavy, B., Smant, G. and Prior, A.E. (2004) Glutathione peroxidases of the potato cyst nematode *Globodera rostochiensis*. *Gene* 324, 47–54.

Karczmarek, A., Overmars, H., Helder, J. and Goverse, A. (2004) Feeding cell development by cyst and root-knot nematodes involves a similar early, local and transient activation of a specific auxin-inducible promoter element. *Molecular Plant Pathology* 5, 343–346.

Karlova, R., Boeren, S., Russinova, E., Aker, J., Vervoort, J. and de Vries, S. (2006) The *Arabidopsis* somatic embryogenesis receptor-like kinase1 protein complex includes brassinosteroid-insensitive1. *Plant Cell* 18, 626–638.

Kasper, G., Brown, A., Eberl, M., Vallar, L., Kieffer, N., Berry, C., Girdwood, K., Eggleton, P., Quinnell, R. and Pritchard, D.I. (2001) A calreticulin-like molecule from the human hookworm *Necator americanus* interacts with C1q and the cytoplasmic signalling domains of some integrins. *Parasite Immunology* 23, 141–152.

Koltai, H., Dhandaydham, M., Opperman, C., Thomas, J. and Bird, D. (2001) Overlapping plant signal transduction pathways induced by a parasitic nematode and a rhizobial endosymbiont. *Molecular Plant–Microbe Interactions* 14, 1168–1177.

Ledger, T.N., Jaubert, S., Bosselut, N., Abad, P. and Rosso, M.N. (2006) Characterization of a new beta-1, 4-endoglucanase gene from the root-knot nematode *Meloidogyne incognita* and evolutionary scheme for phytonematode family 5 glycosyl hydrolases. *Gene* 38, 121–128.

Lilley, C.J., Bakhetia, M., Charlton, W.L. and Urwin, P.E. (2007) Recent progress in the development of RNA interference for plant parasitic nematodes. *Molecular Plant Pathology* 8, 701–711.

Lohar, D.P., Schaff, J.E., Laskey, J.G., Kieber, J.J., Bilyeu, K.D. and Bird, D.M. (2004) Cytokinins play opposite roles in lateral root formation, and nematode and rhizobial symbioses. *Plant Journal* 38, 203–214.

Long, H., Wang, X., Xu, J.H. and Hu, Y.J. (2006) Isolation and characterization of another cDNA encoding a chorismate mutase from the phytoparasitic nematode *Meloidogyne arenaria*. *Experimental Parasitology* 113, 106–111.

López de Mendoza, M., Modha, J., Roberts, C., Curtis, R.H.C. and Kusel, J. (2000) Observations of the changes of the surface cuticle of parasitic nematode using fluorescent probes. *Parasitology* 120, 203–209.

Maurel, C. and Chrispeels, M.J. (2001) Aquaporins. A molecular entry into plant water relations. *Plant Physiology* 125, 135–138.

McKenry, M.V. and Anwar, S.A. (2007) Virulence of *Meloidogyne* spp. and induced resistance in grape rootstocks. *Journal of Nematology* 39, 50–54.

Melillo, M.T., Leonetti, P., Bongiovanni, M., Castagnone-Sereno, P. and Bleve-Zacheo, T. (2006) Modulation of reactive oxygen species activities and H_2O_2 accumulation during compatible and incompatible tomato–root-knot nematode interactions. *New Phytologist* 170, 501–512.

Mitchum, M.G., Sukno, S., Wang, X., Shani, Z., Tsabary, G., Shoseyov, O. and Davis E.L. (2004) The promoter of *Arabidopsis thaliana* cell endo-1,4-β-glucanase gene is differentially expressed in plant feeding cells induced by root-knot and cyst nematodes. *Molecular Plant Pathology* 5, 175–181.

Mitchum M.G., Wang, X. and Davis, E.L. (2008) Diverse and conserved roles of CLE peptides. *Current Opinion in Plant Biology* 11, 75–81.

Mitreva-Dautova, M., Roze, E., Overmars, H., de Graaff, L., Schots, A., Helder, J., Goverse, A., Bakker, J. and Smant, G. (2006) A symbiont-independent endo-1,4-beta-xylanase from the plant-parasitic nematode *Meloidogyne incognita*. *Molecular Plant–Microbe Interactions* 19, 521–529.

Molinari, S. and Miacola, C. (1997) Antioxidant enzymes in phytoparasitic nematodes. *Journal of Nematology* 29, 153–159.

Muslin, A.J. and Xing, H. (2000) 14-3-3 proteins: regulation of subcellular localization by molecular interference. *Cellular Signalling* 12, 703–709.

Neveu, C., Jaubert, S., Abad, P. and Castagnone-Sereno, P. (2003a) A set of genes differentially expressed between avirulent and virulent *Meloidogyne incognita* near-isogenic lines encode secreted proteins. *Molecular Plant–Microbe Interactions* 16, 1077–1084.

Neveu, C., Abad, P. and Castagnone-Sereno, P. (2003b) Molecular cloning and characterization of an intestinal cathepsin L protease from the root-knot nematode *Meloidogyne incognita. Physiological Molecular Plant Pathology* 63, 159–165.

Niebel, A., Barthels, N., de Almeida-Engler, J., Tire, C., Karimi, M., Vercauteren, I., Van Montagu, M. and Gheysen, G. (1994) *Arabidopsis thaliana* as a model host plant to study molecular interactions with root-knot and cyst nematodes. In: Lamberti, F. *et al.* (eds) *Advances in Molecular Plant Nematology.* Plenum Press, New York, pp. 161–170.

Niebel, A., de Almeida-Engler, J., Hemerly, A., Ferreira, P., Inze, D., Van Montagu, M. and Gheysen, G. (1996) Induction of *cdc2a* and *cycAt* expression in *Arabidopsis thaliana* during early phases of nematode-induced feeding cell formation. *Plant Journal* 10, 1037–1043.

Opperman, C.H., Taylor, C.G. and Conkling, M.A. (1994) Root-knot nematode-directed expression of a plant root-specific gene. *Science* 263, 221–223.

Parniske, M. (2000) Intracellular accommodation of microbes by plants: a common developmental program for symbiosis and disease? *Current Opinion Plant Biology* 3, 320–328.

Perry, R.N. (1996) Chemoreception in plant-parasitic nematodes. *Annual Review of Phytopathology* 34, 181–199.

Petrillo, M.D., Matthews, W.C. and Roberts, P.A. (2006) Dynamics of *Meloidogyne incognita* virulence to resistance genes *Rk* and *Rk(2)* in cowpea. *Journal of Nematology* 38, 90–96.

Prior, A., Jones, J.T., Blok, V.C., Beauchamp, J., McDermott, L., Cooper, A. and Kennedy, M.W. (2001) A surface-associated retinol- and fatty acid-binding protein (Gp-FAR-1) from the potato cyst nematode *Globodera pallida*: lipid binding activities, structural analysis and expression pattern. *Biochemical Journal* 356, 387–394.

Qin, L., Kudlav, U., Roze, E. *et al.* (2004) Plant degradation: a nematode expansin acting on plants. *Nature* 427, 30.

Rathjen, J.P. and Moffett, P. (2003) Early signal transduction events in specific plant disease resistance. *Current Opinion in Plant Biology* 6, 300–306.

Richardson, L. and Price, N.S. (1984) Observations on the biology of *Meloidogyne incognita* and the diageotropica tomato mutant. *Revue de Nématologie* 7, 97–99.

Rienties, I.M., Vink, J., Borst, J.W., Russinova, E. and de Vries, S.C. (2005) The *Arabidopsis* SERK1 protein interacts with the AAA-ATPase AtCDC48, the 14-3-3 protein GF14lambda and the PP2C phosphatase KAPP. *Planta* 221, 394–405.

Robertson L., Robertson, W.M., Sobczak, M., Helder, J., Tetaud, E., Ariyanayagam, M.R., Ferguson, M.A., Fairlamb, A. and Jones, J.T. (2000) Cloning, expression and functional characterisation of a peroxiredoxin from the potato cyst nematode *Globodera rostochiensis. Molecular and Biochemical Parasitology* 111, 41–49.

Rosso, M.N., Favery, B., Piotte, C., Arthaud, L., De Boer, J.M., Hussey, R.S., Bakker, J., Baum, T.J. and Abad, P. (1999) Isolation of a cDNA encoding a beta-1,4-endoglucanase in the root-knot nematode *Meloidogyne incognita* and expression analysis during plant parasitism. *Molecular Plant–Microbe Interactions* 12, 585–591.

Rosso, M.N., Dubrana, M.P., Cimbolini, N., Jaubert, S. and Abad, P. (2005) Application of RNA interference to root-knot nematode genes encoding esophageal gland proteins. *Molecular Plant–Microbe Interactions* 18, 615–620.

Roze, E., Hanse, B., Mitreva, M., Vanholme, B., Bakker, J. and Smant, G. (2008) Mining the secretome of the root-knot nematode *Meloidogyne chitwoodi* for candidate parasitism genes. *Molecular Plant Pathology* 9, 1–10.

Sagot, I., Klee, S.K. and Pellman, D. (2002) Yeast formins regulate cell polarity by controlling the assembly of actin cables. *Nature Cell Biology* 4, 42–50.

Scholl, E.H. and Bird, D.M. (2005) Resolving tylenchid evolutionary relationships through multiple gene analysis derived from EST data. *Molecular Phylogenetis and Evolution* 36, 536–545.

Semblat, J.P., Rosso, M.N., Hussey, R.S., Abad, P. and Castagnone-Sereno, P. (2001) Molecular cloning of a cDNA encoding an amphid-secreted putative avirulence protein from the root-knot nematode *Meloidogyne incognita. Molecular Plant–Microbe Interactions* 14, 72–79.

Shingles, J., Lilley, C.J., Atkinson, H.J. and Urwin, P.E. (2007) *Meloidogyne incognita*: molecular and biochemical characterisation of a cathepsin L cysteine proteinase and the effect on parasitism following RNAi. *Experimental Parasitology* 115, 114–120.

Siles-Lucas, M., Merli, M. and Gottstein, B. (2008) 14-3-3 proteins in *Echinococcus*: their role and potential as protective antigens. *Experimental Parasitology* 119, 516–523.

Skamnioti, P. and Ridout, C.J. (2005) Microbial avirulence determinants: guided missiles or antigenic flak? *Molecular Plant Pathology* 6, 551–559.

Staiger, C.J. and Blanchoin, L. (2006) Actin dynamics: old friends with new stories. *Current Opinion in Plant Biology* 9, 554–562.

Suchitra, S. and Joshi, P. (2005) Characterization of *Haemonchus contortus* calreticulin suggests its role in feeding and immune evasion by the parasite. *Biochimical Biophysical Acta* 1722, 293–303.

Torii, K.U. (2004) Leucine-rich repeat receptor kinases in plants: structure, function, and signal transduction pathways. *International Review of Cytolology* 234, 1–46.

Vanholme, B., De Meutter, J., Tytgat, T., Van Montagu, M., Coomans, A. and Gheysen, G. (2004) Secretions of plant-parasitic nematodes: a molecular update. *Gene* 332, 13–27.

Vercauteren, I., de Almeida Engler, J., De Groodt, R. and Gheysen, G. (2002) An *Arabidopsis thaliana* pectin acetylesterase gene is upregulated in nematode feeding sites induced by root-knot and cyst nematodes. *Molecular Plant–Microbe Interactions* 15, 404–407.

Veronico, P., Jones, J., Di Vito, M. and De Giorgi, C. (2001) Horizontal transfer of a bacterial gene involved in polyglutamate biosynthesis to the plant-parasitic nematode *Meloidogyne artiellia. FEBS Letters* 508, 470–474.

Wang, Z., Potter, R.H. and Jones, M.G.K. (2003) Differential display analysis of gene expression in the cytoplasm of giant cells induced in tomato roots by *Meloidogyne javanica. Molecular Plant Pathology* 4, 361–371.

Waspi, U., Schweizer, P. and Dudler, R. (2001) Syringolin reprograms wheat to undergo hypersensitive cell death in a compatible interaction with powdery mildew. *Plant Cell* 13, 153–161.

Wasteneys, G.O. (2002) MT organization in the green kingdom: chaos or self-order? *Journal of Cell Science* 115, 1345–1354.

Weerasinghe, R.R., Bird, D.M. and Allen, N.S. (2005) Root-knot nematodes and bacterial Nod factors elicit common signal transduction events in *Lotus japonicus. Proceedings of the National Academy of Sciences of the USA* 102, 3147–3152.

Whalen, MC. (2005) Host defence in a developmental context. *Molecular Plant Pathology* 6, 347–360.

Wiggers, R.J., Starr, J.L. and Price, H.J. (1990) DNA content and variation in chromosome number in plant cells affected by *Meloidogyne incognita* and *M. arenaria. Phytopathology* 80, 1391–1395.

Wiggers, R.J., Thornton, N.T. and Starr, J.L. (2002) The effects of colchicine on number of giant cell nuclei and nematode development in *Pisum sativum* infected by *Meloidogyne incognita. Nematology* 4, 107–109.

Xia, Y. (2004) Proteases in pathogenesis and plant defence. *Cellular Microbiology* 6, 905–913.

8 Reproduction, Physiology and Biochemistry

David J. Chitwood[1] and Roland N. Perry[2]
[1]*USDA, ARS, BARC-West, Beltsville, Maryland, USA;* [2]*Rothamsted Research, Harpenden, Hertfordshire, UK and Ghent University, Ghent, Belgium*

8.1 Introduction 182
8.2 Reproduction and Moulting 182
8.3 Physiology 188
8.4 Biochemistry 189
8.5 Sensory Perception and Neurotransmission 192
8.6 Conclusions and Future Directions 193
8.7 References 194

8.1 Introduction

The extensive amount of information on the reproduction and cytogenetics of species of *Meloidogyne* contrasts with the limited information on physiology, biochemistry and biochemical pathways. In common with other species of plant-parasitic nematodes, the obligate parasitism and small size of *Meloidogyne* make research on physiology and traditional biochemistry challenging. This chapter aims to discuss the reproductive strategies of species of *Meloidogyne* and to discuss the data available on aspects of their physiology, biochemistry and sensory biology.

8.2 Reproduction and Moulting

The structure of the reproductive system of *Meloidogyne* and the basic features of the development of the gonads are given in Eisenback and Hunt, Chapter 2, this volume. The various reproductive mechanisms of free-living and plant-parasitic nematodes have been reviewed in detail by Evans (1998). Although the plethora of information about the development of *Caenorhabditis elegans* is not matched by data on other species of nematodes, the information available on the cytogenetics of *Meloidogyne* is more extensive than that on any other genus of plant-parasitic nematodes. In the following sections, it is possible only to give a summary of available information.

8.2.1 Reproduction mechanisms and cytogenetics

Among all areas of root-knot nematode biology, none induces admiration from many researchers more than the complex cytogenetics of *Meloidogyne*. A genus with less complicated reproductive cytogenetics would not have attracted the historical or contemporary interest devoted to the root-knot nematodes.

 ©CAB International 2009. *Root-knot Nematodes* (eds R.N. Perry, M. Moens and J.L. Starr)

8.2.1.1 Mode of reproduction

Three types of reproduction exist within the genus: (i) amphimixis, in which sperm from males fertilize oocytes in females and meiosis subsequently occurs; (ii) facultative meiotic parthenogenesis, in which amphimixis occurs in the presence of males but, in their absence, meiosis occurs within oocytes, but two of the nuclei with reduced chromosomal complement (the egg pronucleus and the second polar body) subsequently fuse (automixis); and (iii) obligate mitotic parthenogenesis, where males are not involved (apomixis or amixis), in which one of the two nuclei produced during an initial mitotic division within the oocyte deteriorates and the other becomes the predecessor of the subsequent embryo. Only seven of 37 species of *Meloidogyne* studied to date are amphimictic. Like many soil nematodes, most *Meloidogyne* spp. are parthenogenetic. Some are facultative meiotic parthenogens, and several of the most widespread and economically important species are obligate mitotic parthenogens. Populations of the same *Meloidogyne* species may differ in mode of reproduction; for example, 29 of 32 populations of *M. hapla* reproduced by facultative meiotic parthenogenesis, the others were mitotic parthenogens (Triantaphyllou, 1966). A slightly different form of meiotic parthenogenesis in *M. floridensis* was reported by Handoo *et al.* (2004), where there was a suppression of the second maturation division, indicating that this species has a type of parthenogenesis intermediate between the meiotic form with two maturation divisions and mitotic parthenogenesis.

8.2.1.2 Sex ratios

In *Meloidogyne*, as with *Globodera* and *Heterodera*, sex chromosomes are absent and the sex ratio may be influenced by environmental factors. In the species reproducing by meiotic parthenogenesis, overcrowding, food shortage, temperature extremes or other adverse environmental stresses may lead to the formation of males. These males rarely inseminate females; even when they do, a mitotic division in the oocyte initiates embryogenesis without any fusion with the spermatozoon nucleus. The opportunity for engineering gender reversal in some species is an attractive, although unrealized, crop protection target. In the mitotically parthenogenetic species, the production of males is also induced by unfavourable environmental conditions. As these males were genetically destined to become females, they are produced by a process of sex reversal. Indeed, the timing of the reversal influences the morphology of the males produced: an early reversal results in males with only one testis very similar to normal males; a later reversal induces the production of males with two testes of unequal size; a still later reversal yields males with two testes akin to the two ovaries of females (Papadopoulou and Triantaphyllou, 1982).

8.2.1.3 Chromosome complement

The fascinating cytogenetics of *M. arenaria*, *M. hapla*, *M. incognita* and *Meloidogyne javanica* were elucidated by Triantaphyllou (1962, 1963, 1966, 1981, 1985), whose monumental studies involved the painstaking preparation and microscopic examination of tens of thousands of stained gonads. More recently, the area has been reviewed in detail by Castagnone-Sereno (2006).

The chromosomal complement of *Meloidogyne* spp. reflects the complexity of their reproduction. Perhaps as would be expected in a largely parthenogenic group of often polyploid species, chromosome numbers are not exact multiples of a simpler haploid chromosomal complement. The generally accepted haploid number is $n = 18$ (Triantaphyllou, 1985; Castagnone-Sereno, 2006), although that number might have originated from a chromosomal complement doubling (see next section for discussion). Fertile tetraploid populations of *M. hapla* and *M. microcephalus* have been discovered (Triantaphyllou, 1984; Triantaphyllou and Hirschmann, 1997). The chromosome number is quite variable (Table 8.1) and populations of the same parthenogenetic species may differ greatly in number. For example, various isolates of *M. arenaria* contained 36, 51, 53 or 54 chromosomes (Triantaphyllou, 1966); 220 populations of *M. incognita* had primarily 40–46, although some isolates had only 32–36 chromosomes (Triantaphyllou, 1981), and one female had 88, the sole obviously polyploid specimen in this mammoth study involving the visual counting of approximately 100,000 tiny chromosomes. In 29 facultatively meiotic parthenogenic

Table 8.1. Chromosome number and mode of reproduction of some species of *Meloidogyne*.

Species	Mode of reproduction	Chromosome number	Reference
M. arenaria	obligatory mitotic parthenogenesis	$2n = 30–38, 40–48, 51–56$	18, 22
M. carolinensis	amphimixis	$n = 18$	8
M. chitwoodi	facultative meiotic parthenogenesis	$n = 14–18$	22, 25
M. cruciani	obligatory mitotic parthenogenesis	$2n = 42–44$	22
M. enterolobii	obligatory mitotic parthenogenesis	$2n = 42–44$	16, 22
M. ethiopica	obligatory mitotic parthenogenesis	$2n = 36–44$	2
M. exigua	facultative meiotic parthenogenesis	$n = 18$	15, 22
M. fallax	facultative meiotic parthenogenesis	$n = 18$	25
M. floridensis	facultative meiotic parthenogenesis	$n = 18$	9
M. graminicola	facultative meiotic parthenogenesis	$n = 18$	22
M. graminis	facultative meiotic parthenogenesis	$n = 18$	22
M. hapla cytological race A	facultative meiotic parthenogenesis	$n = 13–17$ (polyploids $n = 28$ & 34)	13, 19, 22
M. hapla cytological race B	obligatory mitotic parthenogenesis	$2n = 30–32, 43–48$	19, 22
M. hispanica	obligatory mitotic parthenogenesis	$2n = 33–36$	19, 22
M. incognita	obligatory mitotic parthenogenesis	$2n = 32–38, 41–46$	21, 22
M. inornata	obligatory mitotic parthenogenesis	$2n = 54–58$	4
M. izalcoensis	obligatory mitotic parthenogenesis	$2n = 44–48$	3
M. javanica	obligatory mitotic parthenogenesis	$2n = 42–48$	22
M. kikuyensis	amphimixis	$n = 7$	23
M. konaensis	obligatory mitotic parthenogenesis	$2n = 44$	Triantaphyllou, pers. comm. in 7
M. megatyla	amphimixis	$n = 18$	8
M. microcephalus	obligatory mitotic parthenogenesis	$2n = 36–38$	22, 24
M. microtyla	amphimixis	$n = 18$	8, 22
M. minor	facultative meiotic parthenogenesis	$n = 17$	van der Beek, pers. comm. in 12
M. morocciensis	obligatory mitotic parthenogenesis	$2n = 47–49$	17
M. naasi	facultative meiotic parthenogenesis	$n = 18$	22

M. oryzae	obligatory mitotic parthenogenesis	$2n = 51–55$	22
M. ottersoni	facultative meiotic parthenogenesis	$n = 18$	22
M. paranaensis	obligatory mitotic parthenogenesis	$3n = 50–52$	1
M. partityla	obligatory mitotic parthenogenesis	$2n = 40–42$	14
M. petuniae	obligatory mitotic parthenogenesis	$2n = 47$	Triantaphyllou, pers. comm. in 5
M. pini	amphimixis	$n = 18$	11, Triantaphyllou, pers. comm. in 6
M. platani	obligatory mitotic parthenogenesis	$2n = 42–44$	22
M. querciana	obligatory mitotic parthenogenesis	$2n = 30–32$	22
M. salasi	obligatory mitotic parthenogenesis	$2n = 36$	22 (as '*Meloidogyne* sp. from rice')
M. spartinae	amphimixis	$n = 7$	20
M. subarctica	amphimixis	$n = 18$	20
M. trifoliophila	facultative meiotic parthenogenesis		10, based on similarity to *M. graminicola*

References: [1]Carneiro *et al.* (1996); [2]Carneiro *et al.* (2004); [3]Carneiro *et al.* (2005); [4]Carneiro *et al.* (2008); [5]Charchar *et al.* (1999); [6]Eisenback *et al.* (1985); [7]Eisenback *et al.* (1994); [8]Goldstein and Triantaphyllou (1982); [9]Handoo *et al.* (2004); [10]Hugall *et al.* (1999); [11]Karssen and Moens (2006); [12]Karssen *et al.* (2004); [13]Liu and Williamson (2006); [14]Marais and Kruger (1991); [15]Muniz *et al.* (2009); [16]Rammah and Hirschmann (1988); [17]Rammah and Hirschmann (1990); [18]Triantaphyllou (1963); [19]Triantaphyllou (1966); [20]Triantaphyllou (1971); [21]Triantaphyllou (1981); [22]Triantaphyllou (1985); [23]Triantaphyllou (1990); [24]Triantaphyllou and Hirschmann (1997); [25]van der Beek and Karssen (1997).

populations of *M. hapla*, the haploid chromosome number was 15–17 (with three presumably polyploid mitotic pathenogens possessing 45 chromosomes) (Triantaphyllou, 1966). Four facultative meotic parthenogenic strains of *M. hapla* had n = 16 (Liu and Williamson, 2006).

8.2.1.4 Evolution of Meloidogyne species

The relationships among species of *Meloidogyne* are examined by Adams *et al.* (Chapter 5, this volume), but the evolution of the genus, especially in relation to the mode of reproduction, has also attracted interest. Triantaphyllou (1985) summarized the speculations about the evolution of *Meloidogyne* spp. based upon cytogenetics. He regarded the obligate amphimictic species (e.g. *M. megatyla*, *M. microtyla* and *M. carolinensis*) with n = 18 or 19 as the current species most closely related to the ancestral predecessors of *Meloidogyne* spp. He also speculated that the low chromosomal numbers in most other nematodes (generally n = 4–12; see Coghlan, 2005) offered support for a polyploid origin of nearly all of the species of *Meloidogyne*. At that time *M. spartinae* was regarded as being in a now-defunct closely related genus, *Hypsoperine*, but its low chromosomal complement (n = 7) was regarded as additional evidence for a condition of tetraploidy in the many species of *Meloidogyne* with n = 14–18. Plantard *et al.* (2007) consider that the n = 7 chromosome number found in only a few species of *Meloidogyne* is a derived character from species with n = 13–19. Triantaphyllou (1985) regarded parthenogenetic species with 30–38 chromosomes as diploids, having arisen from diploid amphmictic species with n ~18, and species with *c.* 54 chromosomes as being triploids produced by the fusion of the chromosomal complements of diploid and haploid forms. The previously discussed existence of naturally occurring polyploid individuals in diploid populations provides additional support for polyploidy as a force in evolution, with aneuploidy or chromosomal fragmentation further modifying the chromosomal complement.

Triantaphyllou (1985) pointed out that as most species of *Meloidogyne* reproduce by mitotic parthenogenesis and have variable chromosome numbers, their status as distinct species may be unclear. The exceptions are *M. javanica* and *M. incognita*, where their distinct biological features represent defined species. Even with the obligate amphimictic species, reproductive isolation tests are extremely difficult because of host specialization of these species.

8.2.1.5 Origin and evolution of parthenogenesis

Evans (1998) pointed out that, although conventional understanding indicated that amphimictic reproduction, with full genetic reassortment, would be the only long-term method enabling species to adapt to environmental change, the most sophisticated and successful genus of plant-parasitic nematodes, *Meloidogyne*, has flourished using, primarily, mitotic parthenogenesis. Such success seems counter to neo-Darwinian wisdom! However, there are advantages to parthenogenesis and, as detailed below, genetic mixing can exist with this mode of reproduction. Parthenogenesis is a reproductive strategy speculated to have advantages in colonizing new ecological niches or in improving the odds of reproductive success in an environment in which the few potential mates may have difficulty in finding females (Ritz and Trudgill, 1999).

The origin of parthenogenesis in *Meloidogyne* spp. and its evolution, have been the subjects of frequent comment, a situation undoubtedly enlivened because of the polyploidy and aneuploidy in the mitotic parthenogenetic species. Parthenogenesis in *Meloidogyne* has been speculated to evolve from two, not necessarily mutually exclusive, pathways: hybridization and mutation (Triantaphyllou, 1985; Trudgill and Blok, 2001; Castagnone-Sereno, 2006; Lunt, 2008), probably with a reticulate evolutionary pattern (Hugall *et al.*, 1999; Trudgill and Blok, 2001). Although van der Beek and Karssen (1997) produced hybrid females from crosses between *M. fallax* and *M. chitwoodi*, these females failed to produce viable second-stage juveniles (J2s). None the less, the exchange of genetic information, as evidenced by hybrid enzyme patterns in the females, would be an important component of speciation.

In a study of AFLP (amplified fragment length polymorphism)-quantified variation in *M. incognita* and facultatively meiotic *M. hapla*, van der Beek and Pijnacker (2008) expectedly discovered almost no variation among *M. incognita* females derived from single juvenile inoculations of tomato roots. By contrast, the variation between sixth-generation descendents of a single egg mass of *M. hapla* was quite large and exceeded the variation between representatives of two different egg masses from the parental population. This variation was regarded as indicative of inverted meiosis, i.e. a process in which chromosomal reduction occurs during the second division instead of the first. The results apparently conflicted with those of Liu *et al.* (2007), who reported a strong tendency to homozygosity in meiotically parthenogenic *M. hapla*.

Although parthenogenetic reproduction would tend to minimize the genetic variation of offspring descending from a single individual, the lack of genetic recombination among different individuals of a parthenogenetic species can result in enhanced accumulation of mutation-initiated divergence of allelic sequences within that species, in comparison with sexually reproducing species (Lunt, 2008). Particularly if only one or a very few parental females were the source of an apomictic species, and if this parental material had originated as a hybrid between two divergent species, the variation between two alleles in the species could be striking. Such individual females might have been the progenitors of parthenogenetic *Meloidogyne* spp. (Hugall *et al.*, 1999; Castagnone-Sereno, 2006; Lunt, 2008). Additional sources of genetic variation in parthenogenetic *Meloidogyne* spp. could result from the abundant transposable elements occurring in these species (Castagnone-Sereno, 2006; Abad *et al.*, 2008; Opperman *et al.*, 2008).

In a nuclear gene sequence study which supported the hybridization ontogeny, Lunt (2008) discovered enhanced sequence divergence within two of three studied genes from mitotic parthenogenetic species (*M. arenaria*, *M. incognita*, *M. javanica* and *M. enterolobii* (=*M. mayaguensis*)), compared with species in which sexual reproduction can occur (*M. chitwoodi*, *M. hapla* and *M. fallax*). Interestingly, Davies *et al.* (2008) reported that cuticular variation in a line derived from a single J2 of mitotically parthenogenetic *M. incognita* was surprisingly high, as reflected by the ability of *Pasteuria penetrans* endospores to attach, and was equal to that of a similar line derived from a single J2 of the facultatively meiotic parthenogen, *M. hapla*. The explanation for the high degree of variation could be that the homologous chromosomes within the polyploid *M. incognita* possess substantial heterozygosity, or that unknown epigenetic mechanisms are responsible for the variation in *M. incognita*.

8.2.2 Moulting

Like most nematodes, *Meloidogyne* moults four times during development to adult. The first moult occurs in the egg, when the small, vermiform, first-stage juvenile (J1) moults to become the infective J2, which subsequently hatches. Details of hatching are given by Curtis *et al.* (Chapter 6, this volume). The infective J2 invade a suitable host plant, initiate a permanent feeding site and feed and grow. The resulting swollen J2 moult into third- and fourth-stage juveniles, which do not feed. Fourth-stage juveniles destined to become males revert to a vermiform shape after the third moult, whereas juveniles destined to become females remain swollen (Fig. 1.1). Both types of fourth-stage juveniles moult once more to become either a mature male or a female.

Moulting of the J1 depends on food reserves stored in the egg. All of the energy required for the three additional moults is contained within the J2. The cuticle of *Meloidogyne* comprises three layers: cortical, medial and basal. The cortical layer is also divided into three layers. At the start of moulting in *M. javanica* the hypodermis becomes thickened and filled with ribosome-like granules and the old cuticle separates from the hypodermis (Bird and Rogers, 1965). The hypodermis first starts to secrete the external cortical layer and then the rest of the new cuticle. The space between the cuticles becomes filled with particles that may be associated with the enzymatic breakdown and reabsorption of the innermost layers of the old cuticle, so that finally only the external cortical layer of the old cuticle is left. After moulting, the new cuticle retains its close cytoplasmic relationship with the hypodermis and increases in thickness. Resorption of the cuticle and recycling its proteins may be an adaptation to endoparasitism, because a sedentary nematode, such as *Meloidogyne*, may have difficulty escaping from a cuticle if it was not absorbed (Lee and Atkinson, 1976).

Having complete sequences of the genomes of *M. hapla* and *M. incognita* (see Abad and Opperman, Chapter 16, this volume) will provide information about the genes involved in moulting in *Meloidogyne*, and this information may aid in the identification of novel control targets. The process of moulting has already attracted attention as a putative target. Soriano *et al.* (2004) examined the effects of the ecdysteroid 20-hydroxyecdysone (20E), a major moulting hormone of insects, on *M. javanica*. Exogenous application of 20E resulted in immobility and death of J2. Furthermore, invasion was partially inhibited and development was halted in spinach with induced high levels of endogenous 20E; however, in the few J2 that invaded, no abnormal moulting was observed. The biosynthesis of ecdysteroids by any nematode has yet to be demonstrated, and specific efforts to detect 20E and its precursor, ecdysone, in *M. arenaria* and *M. incognita* were unsuccessful (Chitwood *et al.*, 1987).

8.3 Physiology

The small size of *Meloidogyne* has limited experimentation on aspects of its physiology. Thus, information on respiration, metabolism and excretion, for example, is limited, and data on the associated biochemical pathways are at best fragmentary and at worst completely lacking.

8.3.1 Respiration

In common with other plant-parasitic nematodes, *Meloidogyne* is sufficiently small for diffusion across the cuticle to provide enough oxygen for aerobic respiration. The limit is likely to be a partial pressure of oxygen of 15 mmHg; below this, nematode activity is adversely affected (Wright and Perry, 2006). Reduced oxygen availability in soils retarded development of *M. javanica* (Van Gundy and Stolzy, 1961), and hatch of J2 from single eggs and from egg masses was reduced at low oxygen concentrations (Baxter and Blake, 1969). Nematodes are more likely to be exposed to low oxygen conditions in soils, especially in waterlogged soil, than in plant tissue, although the root tissues of mangroves and paddy rice may have low oxygen tensions. Robinson and Carter (1986) demonstrated that respiration in J2 of *M. incognita* was essential to survive changes in water potential; when aerobic respiration was prevented, J2 were unable to regulate their volume.

Fumigant nematicides, such as 1,3-D and methyl bromide, are likely to affect biochemical pathways of respiration; those that release methyl isothiocyanate also act on respiration, because once inside the nematode cyanide prevents the utilization of oxygen. The respiration of J2 of *Meloidogyne* spp. treated with 0.5 g methylene bisthiocyanate/ml for 5 min increased significantly but declined when treatment times longer than 5 min were used (Qi *et al.*, 2008). Nordmeyer and Dickson (1989) found that J2 of *M. arenaria* consumed more oxygen than J2 of *M. incognita*, which consumed more than J2 of *M. javanica*. The sensitivity of these three species to nematicides varied in the *in vitro* tests and may relate to differential sensitivity in the field.

8.3.2 Effects of osmotic and ionic stress

Several studies have examined the behaviour of J2 in response to treatment with various solutions (see Curtis *et al.*, Chapter 6, this volume, and Evans and Perry, Chapter 9, this volume), but it is unclear how differences in responses relate to the behaviour of nematodes in soils of varying ionic content. There have been no direct studies on osmotic and ionic regulation by species of *Meloidogyne*. It is probable that they are able to tolerate marked fluctuations in water potentials within plants, particularly at times of drought or nematode-induced stress.

Two complementary osmoregulatory mechanisms are found in animals: isosmotic intracellular regulation, where the osmolarity is adjusted to conform with the extracellular osmotic pressure, and anisosmotic extracellular regulation, where the extracellular fluid is maintained hypo- or hyperosmotic to the external environment (Wright and Perry, 2006). The pseudocoelom is the principal extracellular fluid compartment in nematodes and may act as a primitive circulatory system. In actively moving nematodes, sinusoidal waves of contraction and accompanying internal pressure changes will result in some mixing of the pseudocoelomic fluid. The body wall, the intestine and the secretory–excretory system have been suggested as sites of urine production in nematodes capable of volume regulation in hypo-osmotic environments, but there is no direct evidence for *Meloidogyne*. Regular removal of material by defaecation in actively feeding nematodes suggests that the intestine has an important role in fluid excretion.

The ionic composition of the pseudocoelomic fluid in several animal-parasitic species suggests that ionic regulation must occur. However, there is only limited, largely indirect, physiological and biochemical evidence for the ion channels and pumps that would be required to maintain electrochemical gradients across nematode epidermal and intestinal cells (Thompson and Geary, 2002). There is molecular evidence for K–Cl cotransporter protein in *M. incognita*, which could be involved in ionic and osmotic regulation, and the gene was expressed in both mobile and sedentary stages (Neveu *et al.*, 2002). These authors hypothesized that the gene is involved in the regulation of osmotic pressure of cells in order to maintain nematode body fluids hyperosmotic to the environment.

8.3.3 Secretory–excretory products

In general, nematodes are ammonotelic, with the majority of nitrogenous waste product being ammonia (Wright, 1998). Ammonia is easily soluble in water and poses no problems for nematodes in an aqueous environment. However, ammonia is also toxic, and where water for dilution and diffusion is limited, ammonia is converted to a less toxic end product, such as urea. Information about the excretory products from *Meloidogyne* is lacking, and it is unknown what the excretory end product is once a feeding site has been established. There is no evidence for an excretory function by the secretory–excretory system of *Meloidogyne*, and the role of the system and detailed information of the molecules it secretes remain to be ascertained.

The surface of *Meloidogyne* appears to be covered by glycoprotein and protein, at least some of which appear to be secreted via the nematode secretory–excretory system (Bird *et al.*, 1988) and amphids (Davis *et al.*, 1988; McClure and Stynes, 1988; Davis and Kaplan, 1992). Glycoprotein is also a major component of the gelatinous matrix in the egg mass (Sharon and Spiegel, 1993), and the importance of the gelatinous matrix in the survival of unhatched J2 of *Meloidogyne* is discussed by Evans and Perry, Chapter 9, this volume. Other sources for secretions are the pharyngeal gland cells, and through the cuticle itself. Much of the early work on nematode secretions and their origins has been reviewed by Jones and Robertson (1997). Blaxter and Robertson (1998) reviewed information on the nematode cuticle and pointed out that it plays an important role in nematode physiology, including protection from the environment and excretion, and there may be proteins with potential roles in host recognition.

The secretion of proteins during migration through the host tissue is an essential component of the host–parasite interaction. The source of these proteins and the genes encoding them are being defined, and the functions of secreted proteins coded by the *Meloidogyne* parasitism genes are being elucidated; some are discussed by Atkinson *et al.*, Chapter 15, this volume. The spectrum of proteins associated with species of *Meloidogyne* is discussed in the next section.

8.4 Biochemistry

8.4.1 Enzymes

Investigation of the biochemistry of *Meloidogyne* spp., as in similar studies with other plant-parasitic nematodes, has been hindered by the inability to culture the nematodes independently from their host plants. Much of the early literature on the biochemistry of root-knot nematodes focused on biochemical or histochemical assays for characterizing nematode enzymes or other proteins, or direct analysis of nematode homogenates, extracts or secretions for specific components. Employing electrophoretic or cytochemical techniques, numerous investigators reported discoveries of numerous enzymes in root-knot nematodes, such as acid and alkaline phosphatases, ATPase, catalase, cytochrome oxidase, diaphorase, esterase, β-galactosidase, glucose-6-phosphate dehydrogenase, glucose phosphate isomerase, β-glucosidase, glutamate oxaloacetate transaminase, α-glycerophosphate dehydrogenase, lactate dehydrogenase, lipase, malate dehydrogenase, peroxidase, 6-phosphogluconate dehydrogenase, succinic dehydrogenase, and superoxide dismutase (Ishibashi, 1970; Dickson *et al.*, 1971; Hussey *et al.*, 1972; Dalmasso and Berge, 1978; Starr, 1981; Marwah and Khera, 1988; Esbenshade and Triantaphyllou, 1990; Navas *et al.*, 2001; Molinari *et al.*, 2005). For nearly two decades, a few of these, especially esterase (EST) and malate dehydrogenase, have proved to be useful in the initial molecular identification of species of *Meloidogyne* (see Blok and Powers, Chapter 4, this volume).

In recent years, much of the research on nematode biochemistry has been driven by molecular genetics approaches to elucidate enzymes necessary for nematode-specific functions, or secreted enzymes involved in parasitism or other aspects of the nematode–plant relationship. These molecules provide targets that are hopefully exploitable by potential control strategies (see Atkinson *et al.*, Chapter 15, this volume) and, in *Meloidogyne*, include the following: cathepsin L protease, chorismate mutase, dual oxidase (NADPH oxidase and peroxidase), β-1,4-endoglucanase, pectate lyase, polygalacturonase, serine protease and endo-1,4-β-xylanase (Lambert *et al.*, 1999: Rosso *et al.*, 1999; Doyle and Lambert, 2002, 2003; Jaubert *et al.*, 2002; Neveu *et al.*, 2003; Huang *et al.*, 2004; Bakhetia *et al.*, 2005; Fragoso *et al.*, 2005; Ledger *et al.*, 2006; Long *et al.*, 2006a,b; Mitreva-Dautova *et al.*, 2006; Shingles *et al.*, 2007). Even before the success of the *Meloidogyne* genome projects, early large-scale EST analyses revealed numerous enzymes involved in major biochemical pathways (e.g. McCarter *et al.*, 2003). Indeed, if molecular genetics is regarded as a subset of biochemistry, the entire opus of biochemical literature on *Meloidogyne* is sufficiently voluminous as to be nearly unreviewable. By contrast, the quantity of literature focusing on the analysis of specific compounds in root-knot nematodes and the elucidation of biochemical pathways involving their biosynthesis or metabolism is frighteningly scarce. Although *Meloidogyne* would be expected to share large facets of biochemical machinery with other nematode genera, extrapolation could be dangerous.

8.4.2 Other proteins

Several structural proteins have been detected in *Meloidogyne*. The collagenous nature of the nematode cuticle has been known for decades. Collagen has been localized immunologically in the cuticle of the major species of *Meloidogyne*, and several collagen genes, expression of which is often developmentally correlated, have been discovered (Van der Eycken *et al.*, 1994; Ray *et al.*, 1996a,b; Koltai *et al.*, 1997; Wang *et al.*, 1998; Abrantes and Curtis, 2002). Collagenous proteins isolated from the adult and J2 of *M. javanica* differ in size and amino acid composition: a 76-kDa protein comprises nearly half of the collagen of adult *M. incognita* yet is absent from the J2 (Reddigari *et al.*, 1986). Cuticulins, non-collagenous cuticular proteins that are not readily solubilized, are encoded by at least two distinct genes in *M. artiellia*, and the expression of at least one is highly developmentally regulated (De Georgi *et al.*, 1997).

The eggshells of root-knot nematodes are composed of three major layers: an outer vitelline membrane, a chitin layer also containing protein, and an inner lipid layer, again containing protein. Chemical analysis of eggshells revealed that protein was the most abundant component, and that proline comprised as much as 40% of the total amino acid composition of the eggshell protein of *M. incognita* (Bird and McClure, 1976). Autoradiographic studies with radiolabelled proline indicated that the proline-containing protein was incorporated into both the chitin and lipid layer of *M. javanica* (McClure and Bird, 1976).

8.4.3 Amino acids and sugars

Not unsurprisingly, numerous studies have involved the amino acid composition of root-knot nematodes, but the sole investigation to use a radiolabelled precursor was by Myers and Krusberg (1965), who reported that *M. incognita* was capable of biosynthesizing glutamic acid, glutamine, alanine, asparagine, aspartic acid, glycine, serine and tryptophan, the latter being an essential amino acid in mammals. Wang and Bergeson (1978) demonstrated the presence of 15 different amino acids and six sugars in secretions of J2 of *M. incognita*. Sadly, limited information is available on the biosynthesis of specific sugars by

Meloidogyne; most of the specific information has resulted from the several excellent lectin-based analyses of sugar residues on the *Meloidogyne* surface (e.g. Davis *et al.*, 1988; McClure and Stynes, 1988; Ibrahim, 1991; Davis and Kaplan, 1992; Spiegel *et al.*, 1995; Lin and McClure, 1996).

8.4.4 Neuropeptides

FMRFamide-like peptides (FLPs) are polypeptides containing only a few amino acids and are known neuromodulators of muscular activity in other species of nematodes (Brownlee *et al.*, 2000; Perry and Maule, 2004). Although their presence has not been demonstrated in *Meloidogyne*, numerous genes encoding at least 15 distinct FLPs have been elucidated (Fleming *et al.*, 2007; McVeigh *et al.*, 2008). Nematode FLPs are very attractive targets for development of control techniques because of their potential susceptibility to RNAi (RNA interference)-based disruption.

8.4.5 Complex carbohydrates and lipids

On a dry weight basis, J2 of *M. javanica* contained approximately 7% carbohydrate. The sugar polymer glycogen would be expected to be a major food reserve in root-knot nematodes, and electron microscopy has indicated that the major food reserves of J2 of *M. incognita* are intestinal lipids, with smaller reserves comprising hypodermal lipid and glycogen (Dropkin and Acedo, 1974). As feeding commences and the three moults to parasitic developmental stages ensue, glycogen appears to be the predominant food reserve, although adult females again contain massive quantities of lipids.

Chitin, a polymer of the amino sugar *N*-acetylglucosamine, comprises as much as 30% of the dry weight of the *M. javanica* eggshell (Bird and McClure, 1976). This is apparently the only nematode life stage where chitin exists, as a well-designed electron microscopy study clearly demonstrated that the chitin thought possibly to exist in the gelatinous matrix, if present, was likely to be a product of fungal contamination (Bird and Self, 1995).

Perhaps because of lipid abundance in nematodes (with adult females, J2 and eggs consisting, respectively, of nearly 50%, 40% or 66% lipid on a dry weight basis; Krusberg, 1967; Reversat, 1976), the lipids of *Meloidogyne* spp. have received substantial attention. In addition to obvious roles as food reserves and structural components of membranes, the nematode surface also contains some lipids, and the lipophilicity increases in response to host root exudate (López de Mendoza *et al.*, 2000). About one-sixth of the lipid of females of *M. javanica* is phospholipid; only 2.5% is glycolipid (Chitwood and Krusberg, 1981). The major fatty acid in root-knot nematodes is vaccenic acid, an oleic acid isomer that is common in bacteria; it amounted to over 25% of the dry weight of females and 40% in eggs of *M. incognita* and *M. arenaria* (Krusberg *et al.*, 1973). At least 30 other fatty acids also exist in *Meloidogyne*, including several iso-branched acids. Although the roles of the latter are speculative, one possibility would be that they act as metabolic precursors to an as yet unidentified *Meloidogyne* analogue of the *C. elegans* dauer pheromone, which consists of a short branched-chain fatty acid attached to the sugar ascarylose (Jeong *et al.*, 2005). As much as a third of the individual phospholipids of *M. javanica* contain ether-linked alkyl moieties attached to the C-2 position of the glycerol backbone (Chitwood and Krusberg, 1981). The phospholipid alkyl groups are remarkably undiverse in structure, primarily consisting of saturated 18-carbon moieties.

8.4.6 Steroids

The major roles of sterols in organisms are to modulate membrane fluidity as components of cellular membranes, provide biochemical precursors for steroid hormones, and interact with specific proteins in the regulation of organismal development. Perhaps because *Meloidogyne* membrane phospholipids contain so much polyunsaturated fatty acid that any role of sterol to modulate fluidity would be superfluous, sterols comprise a remarkably small percentage of the nematode, only 0.02% of the eggs of *M. incognita* and *M. arenaria* (Chitwood *et al.*, 1987). In a comparison of the sterols of *M. incognita* and *M. arenaria* with those of their host plant, *Solanum melongena*, the major sterols of nematode eggs were 24-ethylcholesterol, 24-ethylcholestanol,

24-methylcholestanol, 24-ethylcholest-22-enol, cholesterol and cholestanol. The results indicated that saturation of the sterol nucleus was the major metabolic transformation of host sterols in *Meloidogyne*; additionally, the nematodes appeared to remove the C-24 methyl or ethyl substituent in the side chain of typical plant sterols. However, Hedin *et al.* (1995) did not detect the products of sterol nuclear saturation in eggs of *M. incognita* propagated on *Gossypium hirsutum*. Explanations for the analytical variation could include methodological differences, a host-mediated effect on parasite biochemistry, or a true biochemical difference between the two populations of *M. incognita* employed.

Although there have been several reports of toxicity of specific steroids to *Meloidogyne* spp., attempts to demonstrate the biosynthesis of ecdysteroids or other steroids in *Meloidogyne* have been unsuccessful (Chitwood *et al.*, 1987). Indeed, in only one case has the biosynthesis of any steroid with hormonal function in any nematode been conclusively demonstrated, in the case of the 3-ketocholest-4-en-26-oic acid and 3-ketocholest-7-en-26-oic acid, two steroid acids which are involved in the regulation of dauer larva formation in *C. elegans* (Motola *et al.*, 2006).

8.5 Sensory Perception and Neurotransmission

Of the various types of sensory perception, most research on *Meloidogyne* has focused on chemoreception, especially in relation to the amphids. Chemoreception plays an important role in the oriented movement (taxis) of nematodes. Orientation by a nematode can be achieved with one sensor, by sequentially sensing and comparing the stimulus on either side of the path of movement (klinotaxis), or with two or more sensors on different parts of the body, by simultaneously sensing different points within the stimulus field (tropotaxis). Bargmann and Horvitz (1991) demonstrated that killing neurons in either the left or the right chemosensory sensilla of *C. elegans* did not prevent chemotaxis, indicating that the nematode does not perform tropotaxis. Extrapolating this information to *Meloidogyne*, it is likely that repeated side-to-side movements of the head during sinusoidal body movement is responsible for alternating sampling of the stimulus field from either side of the head, and that klinotaxis is involved in orientation. The orientation and attraction of *Meloidogyne* to host roots and their movement in response to edaphic factors are discussed in detail by Curtis *et al.*, Chapter 6, this volume. In the following two sections, we examine the information on amphid functioning and neurotransmission in *Meloidogyne*.

8.5.1 Sensory perception

As in other nematodes, amphids are considered to be the primary chemosensilla of *Meloidogyne*. They are present as paired organs, positioned laterally and with external openings (see Eisenback and Hunt, Chapter 2, this volume). The amphidial cavity contains secretions, apparently produced by the sheath cell, that appear to have multiple roles. Trett and Perry (1985) suggested that the secretions may serve to maintain electrical continuity between the bases and tips of the dendritic processes, and they may also protect the dendritic endings of sensory nerve cells against desiccation and microbial attack (Aumann, 1993). Several investigations have centred on analysing the components of the secretions.

Bird (1966) detected esterases, enzymes that rapidly hydrolyse esters of short-chain fatty acids, in the amphidial ducts of juveniles and adults of *M. javanica* and *M. hapla*. Premachandran *et al.* (1988) found that the protein-specific dye, Coomassie Brilliant Blue R-250, bound to the amphidial secretions of a number of nematodes, including J2 of *M. incognita*, indicating that the secretions contain protein. Lectins have been used to demonstrate the presence of carbohydrate residues in amphidial secretions of species of *Meloidogyne* (McClure and Stynes, 1988). Components of amphidial secretions in J2 of *M. incognita* are thought to include *N*-acetylgalactosamine and fucose (McClure and Stynes, 1988; Spiegel and McClure, 1991).

Davis *et al.* (1992) found differences in the composition of amphidial secretions of *M. incognita* using a monoclonal antibody that reacted with the amphids of adult females but not with the amphids of J2. Stewart *et al.* (1993a) demonstrated the presence of a 32 kDa glycoprotein in the amphidial duct secretions and the sheath cell of J2 of six species of *Meloidogyne*, but it was not found in representatives from eight other genera, including *Globodera* and *Heterodera*, indicating a specialized function for this protein in *Meloidogyne*. The protein was found in all stages of the *Meloidogyne* life cycle, including males of *M. javanica*, but not in the sedentary adult female, where the amphids appear to be non-functional. Electron microscopy indicated a difference in the morphology of amphidial secretions in the J2 and the adult female (Stewart *et al.*, 1993b). In agar plate behavioural assays, prior incubation of *M. javanica* J2 in the antiserum against the protein significantly retarded the ability of the J2 to orientate to host roots (Stewart *et al.*, 1993b). Thus, there are indications that at separate stages of the life cycle the amphids may have a different function or a different combination of functions. Lima *et al.* (2005) used antibodies to immunolocalize secreted–excreted products of species of *Meloidogyne*. The antibodies reacted with antigens present in the amphids of *M. incognita* and *M. arenaria*, and one antibody recognized secretions in the amphidial and phasmidial glands of *M. arenaria*. The functions of the proteins from the amphids identified using antibodies are still unknown. Two of the polyclonal antibodies used by Lima *et al.* (2005) bound to the surface coat of the cuticle as well as to the amphids of *M. incognita* and *M. arenaria*; surface coat components are reviewed by Curtis *et al.*, Chapter 6, this volume.

Semblat *et al.* (2001) cloned a cDNA (complementary DNA) encoding a secretory protein from the amphids of *M. incognita*. The protein (MAP-1) expressed by the *map-1* gene was restricted to the three species of *Meloidogyne*, *M. arenaria*, *M. incognita* and *M. javanica*, controlled by the *Mi* resistance gene, suggesting a specialized function for this protein. The authors speculate that the MAP-1 protein might be involved in the early steps of recognition between (resistant) plants and (avirulent) nematodes.

8.5.2 Neurotransmission

Neurotransmission in nematodes has been summarized by Wright and Perry (1998) and Perry and Maule (2004), and reviewed in detail for *C. elegans* by Rand and Nonet (1997), but there is only fragmentary information on *Meloidogyne*. Typical synaptic transmission involves the arrival of an action potential at a presynaptic nerve ending, causing the opening of voltage-gated Ca^{2+} ion channels and the influx of Ca^{2+} ions into the nerve cell. This results in the secretion of neurotransmitter molecules, which diffuse across the synaptic cleft and bind reversibly to specific receptor proteins on the post-synaptic membrane of a nerve or muscle cell. This causes a conformational change in the receptor proteins that are linked to ion channels. Whether the response is excitatory or inhibitory depends on the type of receptor and, thus, which ion channel is activated. Hence, the same neurotransmitter can be excitatory and inhibitory.

Classical transmitters include acetylcholine, probably the primary excitatory transmitter, several amino acids and various biogenic amines. The amino acid transmitter, γ-amino butyric acid (GABA), and the biogenic amine, dopamine, have been reported in J2 of *M. incognita* (Stewart *et al.*, 1994, 2001).

An essential feature of all neurotransmitter systems is a mechanism for the rapid removal of neurotransmitter from the synaptic cleft. Removal of acetylcholine is enzymatic, and inhibition of acetylcholinesterase is the target for the control of plant-parasitic nematodes, including *Meloidogyne*, by organophosphate and carbamate nematicides. In *C. elegans*, four acetylcholinesterase genes have been isolated, *ace-1* to *ace-4*, coding for biochemical classes of acetylcholinesterase. *Meloidogyne arenaria* and *M. incognita* J2 contain several molecular forms of acetylcholinesterase, which vary in sedimentation coefficient, substrate affinity, thermal inactivation profiles, and/or inhibitor and detergent sensitivity (Nordmeyer and Dickson, 1990; Chang and Opperman, 1991). At least two acetylcholinesterase genes have been identified in *Meloidogyne*. A single gene homologous to the *ace-1* gene of *C. elegans* has been isolated from *M. incognita* and *M. javanica* (Piotte *et al.*, 1999), and Laffaire *et al.* (2003) isolated a new acetylcholinesterase-encoding gene, named *Mi-ace-2*, from *M. incognita*, which is transcribed in J2 before and after hatching, and in females and males.

8.6 Conclusions and Future Directions

There have been rapid advances in experimental techniques, especially in genome sequencing, proteomics and metabolomics, which have generated large amounts of data relating to helminth physiology and biochemistry (Barrett, 2009). Proteomics is the large-scale analysis of proteins in a single cell or tissue, whereas metabolomics is the analysis of the low molecular weight metabolites in the tissue or cell (Barrett, 2009). Instead of focusing on a single gene or protein, proteomics can reveal the relative amounts of protein, the degree of protein modification and turnover, and the interactions between proteins (Barrett *et al.*, 2005; Barrett, 2009). The information that will be generated by the various nematode genome projects, including those of *M. incognita* and *M. hapla* (Abad *et al.*, 2008, Opperman *et al.*, 2008), presages an exciting future for elucidating nematode biological systems. Annotation of the sequences will provide a vast amount of data relating to the physiology and biochemistry of *Meloidogyne*. Although the ability to identify candidate parasitism genes is one obvious outcome of this research, rapid insight should accrue with respect to the basic biochemical mechanisms that, for example, permit reproductive success through varied reproductive strategies mediated by differing cytogenetics.

Thus, genome annotation, proteomics and metabolomics herald an era of enormous opportunities for research workers. However, it will be necessary to interpret and understand the biological relevance of these data and to clarify details of the biological functioning of *Meloidogyne* and other nematodes. The current lack of information on important aspects of physiology and biochemistry reflects, in part, the worrying paucity of researchers trained to use physiological, biochemical and behavioural techniques to investigate gene functioning. The need to interpret components of the host–parasite system is predicated on the existence of these research skills. This is a vital component of future progress, especially as the information gained could lead to much-needed environmentally benign, novel control approaches.

8.7 References

Abad, P., Gouzy, J., Aury, J.-M. *et al.* (2008) Genome sequence of the metazoan plant-parasitic nematode *Meloidogyne incognita. Nature Biotechnology* 8, 909–915.

Abrantes, I.M. de O. and Curtis, R.H.C. (2002) Immunolocalization of a putative cuticular collagen protein in several developmental stages of *Meloidogyne arenaria*, *Globodera pallida* and *G. rostochiensis. Journal of Helminthology* 76, 1–6.

Aumann, J. (1993) Chemosensory physiology of nematodes. *Fundamental and Applied Nematology* 16, 193–198.

Bakhetia, M., Charlton, W., Atkinson, H. J. and McPherson, M. J. (2005) RNA interference of dual oxidase in the plant nematode *Meloidogyne incognita. Molecular Plant–Microbe Interactions* 18, 1099–1106.

Bargmann, C.I. and Horvitz, H.R. (1991) Chemosensory neurons with overlapping functions direct chemotaxis to multiple chemicals in *C. elegans. Neuron* 7, 729–742.

Barrett, J. (2009) Forty years of helminth biochemistry. *Parasitology* 136, 1–10.

Barrett, J., Brophy, P.M. and Hamilton, J.V. (2005) Analysing proteomic data. *International Journal for Parasitology* 35, 543–553.

Baxter, R.I. and Blake, C.D. (1969) Oxygen and the hatch of eggs and migration of larvae of *Meloidogyne javanica. Annals of Applied Biology* 63, 191–203.

Bird, A.F. (1966) Esterases in the genus *Meloidogyne. Nematologica* 12, 359–361.

Bird, A.F. and McClure, M.A. (1976) The tylenchid (Nematoda) egg shell: structure, composition and permeability. *Parasitology* 72, 19–28.

Bird, A.F. and Rogers, G.E. (1965) Ultrastructure of the cuticle and its formation in *Meloidogyne javanica. Nematologica* 11, 224–230.

Bird, A.F. and Self, P.G. (1995) Chitin in *Meloidogyne javanica. Fundamental and Applied Nematology* 18, 235–239.

Bird, A.F., Bonig, I. and Bacic, A. (1988) A role for the ‘excretory’ system in secernentean nematodes. *Journal of Nematology* 20, 493–496.

Blaxter, M.L. and Robertson, W.M. (1998) The cuticle. In: Perry, R.N. and Wright, D.J. (eds) *The Physiology and Biochemistry of Free-living and Plant-parasitic Nematodes.* CAB International, Wallingford, UK, pp. 25–48.

Brownlee, D., Holden-Dye, L. and Walker, R. (2000) The range and biological activity of FMRFamide-related peptides and classical neurotransmitters in nematodes. *Advances in Parasitology* 45, 109–180.

Carneiro, R.M.D.G., Carneiro, R.G., Abrantes, I.M.O., Santos, M.S.N.A. and Almeida, M.R.A. (1996) *Meloidogyne paranaensis* n. sp. (Nemata: Meloidogynidae), a root-knot nematode parasitizing coffee in Brazil. *Journal of Nematology* 28, 177–189.

Carneiro, R.M.D.G., Randig, O., Almeida, M.R.A. and Gomes, A.C.M.M. (2004) Additional information on *Meloidogyne ethiopica* Whitehead, 1968 (Tylenchida: Meloidogynidae), a root-knot nematode parasitising kiwi fruit and grape-vine from Brazil and Chile. *Nematology* 6, 109–123.

Carneiro, R.M.D.G., Almeida, M.R.A., Gomes, A.C.M.M. and Hernandez, A. (2005) *Meloidogyne izalcoensis* n. sp. (Nematoda: Meloidogynidae), a root-knot nematode parasitising coffee in El Salvador. *Nematology* 7, 819–832.

Carneiro, R.M.D.G., Mendes, M.L., Almeida, M.R.A., dos Santos, M.F.A., Gomes, A.C.M.M. and Karssen, G. (2008) Additional information on *Meloidogyne inornata* Lordello, 1956 (Tylenchida: Meloidogynidae) and its characterisation as a valid species. *Nematology* 10, 123–136.

Castagnone-Sereno, P. (2006) Genetic variability and adaptive evolution in parthenogenetic root-knot nematodes. *Heredity* 96, 282–289.

Chang, S. and Opperman, C.H. (1991) Characterization of acetylcholinesterase molecular forms of the root-knot nematode, *Meloidogyne. Molecular and Biochemical Parasitology* 29, 205–214.

Charchar, J.M., Eisenback, J.D. and Hirschmann, H. (1999) *Meloidogyne petuniae* n. sp. (Nemata: Meloidogynidae), a root-knot nematode parasitic on petunia in Brazil. *Journal of Nematology* 31, 81–91.

Chitwood, D.J. and Krusberg, L.R. (1981) Diacyl, alkylacyl, and alkenylacyl phospholipids of *Meloidogyne javanica* females. *Journal of Nematology* 13, 105–111.

Chitwood, D.J., McClure, M.A., Feldlaufer, M.F., Lusby, W.R. and Oliver, J.E. (1987) Sterol composition and ecdysteroid content of eggs of the root-knot nematodes, *Meloidogyne incognita* and *M. arenaria. Journal of Nematology* 19, 352–360.

Coghlan, A. (2005) Nematode genome evolution. In: The *C. elegans* Research Community (eds) *WormBook*, http://www.wormbook.org/chapters/www_genomevol/genomevol.html.

Dalmasso, A. and Berge, J.B. (1978) Molecular polymorphism and phylogenetic relationship in some *Meloidogyne* spp.: application to the taxonomy of *Meloidogyne. Journal of Nematology* 10, 323–332.

Davies, K.G., Rowe, J. and Williamson, V.M. (2008) Inter- and intra-specific cuticle variation between amphimictic and parthenogenetic species of root-knot nematode (*Meloidogyne* spp.) as revealed by a bacterial parasite (*Pasteuria penetrans*). *International Journal for Parasitology* 38, 851–859.

Davis, E.L. and Kaplan, D.T. (1992) Lectin binding to aqueous-soluble and body wall proteins from infective juveniles of *Meloidogyne* species. *Fundamental and Applied Nematology* 15, 243–250.

Davis, E.L., Kaplan, D.T., Permar, T.A., Dickson, D.W. and Mitchell, D.J. (1988) Characterization of carbohydrates on the surface of second-stage juveniles of *Meloidogyne* spp. *Journal of Nematology* 20, 609–619.

Davis, E.L., Aaron, L.M., Pratt, L.H. and Hussey, R.S. (1992) Novel immunization procedures used to develop antibodies that bind to specific structures in *Meloidogyne* spp. *Phytopathology* 82, 1244–1250.

De Giorgi, C., De Luca, F., Di Vito, M. and Lamberti, F. (1997) Modulation of expression at the level of splicing of *cut-1* RNA in the infective second-stage juvenile of the plant parasitic nematode *Meloidogyne artiellia. Molecular and General Genetics* 253, 589–598.

Dickson, D., Huisingh, D. and Sasser, J.N. (1971). Dehydrogenases, acid and alkaline phosphatases, and esterases for chemotaxonomy of selected *Meloidogyne*, *Ditylenchus*, *Heterodera* and *Aphelenchus* spp. *Journal of Nematology* 3, 1–16.

Doyle, E.A. and Lambert, K.N. (2002) Cloning and characterization of an esophageal-gland-specific pectate lyase from the root-knot nematode *Meloidogyne javanica. Molecular Plant–Microbe Interactions* 15, 549–556.

Doyle, E.A. and Lambert, K.N. (2003) *Meloidogyne javanica* chorismate mutase 1 alters plant cell development. *Molecular Plant–Microbe Interactions* 16, 123–131.

Dropkin, V.H. and Acedo, J. (1974) An electron microscopic study of glycogen and lipid in female *Meloidogyne incognita* (root-knot nematode). *Journal of Parasitology* 60, 1013–1021.

Eisenback, J.D., Yang, B. and Hartman, K.M. (1985) Description of *Meloidogyne pini* n. sp., a root-knot nematode parasitic on sand pine (*Pinus clausa*), with additional notes on the morphology of *M. megatyla. Journal of Nematology* 17, 206–219.

Eisenback, J.D., Bernard, E.C. and Schmitt, D.P. (1994) Description of the Kona coffee root-knot nematode, *Meloidogyne konaensis* n. sp. *Journal of Nematology* 26, 363–374.

Esbenshade, P.R. and Triantaphyllou, A.C. (1990) Isozyme phenotypes for the identification of *Meloidogyne* species. *Journal of Nematology* 22, 10–15.

Evans, A.A.F. (1998) Reproductive mechanisms. In: Perry, R.N. and Wright, D.J. (eds) *The Physiology and Biochemistry of Free-living and Plant-parasitic Nematodes*. CAB International, Wallingford, UK, pp. 133–154.

Fleming, C.C., McKinney, S., McMaster, S., Johnston, M.J.G., Donnelly, P., Kimber, M.J. and Maule, A.G. (2007) Getting to the root of neuronal signalling in plant-parasitic nematodes using RNA interference. *Nematology* 9, 301–315.

Fragoso, R. da R., Batista, J.A.N., Neto, O.B.O. and Grossi de Sa, M.F. (2005) Isolation and characterization of a cDNA encoding a serine proteinase from the root-knot nematode *Meloidogyne incognita*. *Experimental Parasitology* 110, 123–133.

Goldstein, P. and Triantaphyllou, A.C. (1982) The synaptonemal complexes of *Meloidogyne*: relationship of structure and evolution of parthenogensis. *Chromosoma* 87, 117–124.

Handoo, Z.A., Nyczepir, A.P., Esmenjaud, D., van der Beek, J.G., Castagnone-Sereno, P., Carta, L.K., Skantar, A.M. and Higgins, J.A. (2004) Morphological, molecular, and differential-host characterization of *Meloidogyne floridensis* n. sp. (Nematoda: Meloidogynidae), a root-knot nematode parasitizing peach in Florida. *Journal of Nematology* 36, 20–35.

Hedin, P.A., Callahan, F.E., Dollar, D.A. and Creech, R.G. (1995) Total sterols in root-knot nematode *Meloidogyne incognita* infected cotton *Gossypium hirsutum* (L.) plant roots. *Comparative Biochemistry and Physiology* 111B, 447–452.

Huang, G., Dong, R., Maier, T., Allen, R., Davis, E.L., Baum, T.J. and Hussey, R.S. (2004) Use of solid-phase subtractive hybridization for the identification of parasitism gene candidates from the root-knot nematode *Meloidogyne incognita*. *Molecular Plant Pathology* 5, 217–222.

Hugall, A., Stanton, J. and Moritz, C. (1999) Reticulate evolution and the origins of ribosomal internal transcribed spacer diversity in apomictic *Meloidogyne*. *Molecular Biology and Evolution* 16, 157–164.

Hussey, R.S., Sasser, J.N. and Huisingh, D. (1972) Disc-electrophoretic studies of soluble proteins and enzymes of *Meloidogyne incognita* and *M. arenaria*. *Journal of Nematology* 4, 183–189.

Ibrahim, S.K. (1991) Distribution of carbohydrates on the cuticle of several developmental stages of *Meloidogyne javanica*. *Nematologica* 37, 275–284.

Ishibashi, N. (1970) Variations of the electrophoretic protein patterns of Heteroderidae (Nematodea: Tylenchida) depending on the developmental stages of the nematode and on the growing conditions of the host plants. *Applied Entomology and Zoology* 5, 23–32.

Jaubert, S., Laffaire, J.-B., Abad, P. and Rosso, M.-N. (2002) A polygalacturonase of animal origin isolated from the root-knot nematode *Meloidogyne incognita*. *FEBS Letters* 522, 109–112.

Jeong, P.-Y., Jung, M., Yim, Y.-H., Kim, H., Park, M., Hong, E., Lee, W., Kim, Y.H., Kim, K. and Paik, Y.-K. (2005) Chemical structure and biological activity of the *Caenorhabditis elegans* dauer-inducing pheromone. *Nature* 433, 541–545.

Jones, J.T. and Robertson, W.M. (1997) Nematode secretions. In: Fenoll, C., Grundler, F.M.W. and Ohl, S.A. (eds) *Cellular and Molecular Aspects of Plant–Nematode Interactions*. Kluwer Academic Publishers, Dordrecht, The Netherlands, pp. 98–106.

Karssen, G. and Moens, M. (2006) Root-knot nematodes. In: Perry, R.N. and Moens, M. (eds) *Plant Nematology*. CAB International, Wallingford, UK, pp. 59–90.

Karssen, G., Bolk, R.J., van Aelst, A.C., van den Beld, I., Kox, L.F.F., Korthals, G., Molendijk, L., Zijlstra, C., van Hoof, R. and Cook, R. (2004) Description of *Meloidogyne minor* n. sp. (Nematoda: Meloidogynidae), a root-knot nematode associated with yellow patch disease in golf courses. *Nematology* 6, 59–72.

Koltai, H., Chejanovsky, N., Raccah, B. and Spiegel, Y. (1997) The first isolated collagen gene of the root-knot nematode *Meloidogyne javanica* is developmentally regulated. *Gene* 196, 191–199.

Krusberg, L.R. (1967) Analysis of total lipids and fatty acids of plant-parasitic nematodes and host tissues. *Comparative Biochemistry and Physiology* 21, 83–90.

Krusberg, L.R., Hussey, R.S. and Fletcher, C.L. (1973) Lipid and fatty acid composition of females and eggs of *Meloidogyne incognita* and *M. arenaria*. *Comparative Biochemistry and Physiology* 45B, 335–341.

Laffaire, J.-B., Jaubert, S., Abad, P. and Rosso, M.-N. (2003) Molecular cloning and life stage expression pattern of a new acetylcholinesterase gene from the plant-parasitic nematode *Meloidogyne incognita*. *Nematology* 5, 213–217.

Lambert, K.N., Allen, K.D. and Sussex, I.M. (1999) Cloning and characterization of an esophageal-gland-specific chorismate mutase from the phytoparasitic nematode *Meloidogyne javanica. Molecular Plant–Microbe Interactions* 12, 328–336.

Ledger, T.N., Jaubert, S., Bosselut, N., Abad, P. and Rosso, M.-N. (2006) Characterization of a new [beta]-1,4-endoglucanase gene from the root-knot nematode *Meloidogyne incognita* and evolutionary scheme for phytonematode family 5 glycosyl hydrolases. *Gene* 382, 121–128.

Lee, D.L. and Atkinson, H.J. (1976) *The Physiology of Nematodes,* 2nd edn. MacMillan Press, London.

Lima, L.M., Grossi-de-Sa, M.F., Pereira, R.A. and Curtis, R.H.C. (2005) Immunolocalisation of secreted–excreted products of *Meloidogyne* spp. using polyclonal and monoclonal antibodies. *Fitopatologia Brasileira* 30, 629–633.

Lin, H.-J. and McClure, M.A. (1996) Surface coat of *Meloidogyne incognita. Journal of Nematology* 28, 216–224.

Liu, Q.L. and Williamson, V.M. (2006) Host-specific pathogenicity and genome differences between inbred strains of *Meloidogyne hapla. Journal of Nematology* 38, 158–164.

Liu, Q.L., Thomas, V.P. and Williamson, V.M. (2007) Meiotic parthenogenesis in a root-knot nematode results in rapid genomic homozygosity. *Genetics* 176, 1483–1490.

Long, H., Wang, X. and Xu, J. (2006a) Molecular cloning and life-stage expression pattern of a new chorismate mutase gene from the root-knot nematode *Meloidogyne arenaria. Plant Pathology* 55, 559–563.

Long, H., Wang, X., Xu, J.H. and Hu, Y.J. (2006b) Isolation and characterization of another cDNA encoding a chorismate mutase from the phytoparasitic nematode *Meloidogyne arenaria. Experimental Parasitology* 113, 106–111.

López de Mendoza, M.E., Modha, J., Roberts, M.C., Curtis, R. and Kusel, J. R. (2000). Changes in the lipophilicity of the surfaces of *Meloidogyne incognita* and *Haemonchus contortus* during exposure to host signals. *Parasitology* 120, 203–209.

Lunt, D.H. (2008) Genetic tests of ancient asexuality in root knot nematodes reveal recent hybrid origins. *BMC Evolutionary Biology* 8, 194.

Marais, M. and Kruger, J.-C. de W. (1991) The cytogenetics of some South African root-knot nematodes (Heteroderidae: Nematoda). *Phytophylactica* 23, 265–272.

Marwah, R. and Khera, S. (1988) Histochemical localization of nucleic acids and some enzymes in females of *Meloidogyne incognita. Helminthologia* 25, 49–60.

McCarter, J.P., Mitreva, M.D., Martin, J. *et al.* (2003) Analysis and functional classification of transcripts from the nematode *Meloidogyne incognita. Genome Biology* 4, R3.

McClure, M.A. and Bird, A.F. (1976) The tylenchid (Nematoda) egg shell: formation of the egg shell in *Meloidogyne javanica. Parasitology* 72, 29–39.

McClure, M.A. and Stynes, B.A. (1988) Lectin binding sites on the amphidial exudates of *Meloidogyne. Journal of Nematology* 20, 321–326.

McVeigh, P., Alexander-Bowman, S., Veal, E., Mousley, A., Marks, N. and Maule, A.G. (2008) Neuropeptide-like protein diversity in phylum Nematoda. *International Journal for Parasitology* 38, 1493–1503.

Mitreva-Dautova, M., Roze, E., Overmars, H., de Graaff, L., Schots, A., Helder, J., Goverse, A., Bakker, J. and Smant, G. (2006) A symbiont-independent endo-1,4-β-xylanase from the plant-parasitic nematode *Meloidogyne incognita. Molecular Plant–Microbe Interactions* 19, 521–529.

Molinari, S., Lamberti, F., Crozzoli, R., Sharma, S.B. and Sanchez Portales, L. (2005) Isozyme patterns of exotic *Meloidogyne* spp. populations. *Nematologia Mediterranea* 33, 61–65.

Motola, D.L., Cummins, C.L., Rottiers, V. *et al.* (2006) Identification of ligands for DAF-12 that govern dauer formation and reproduction in *C. elegans. Cell* 124, 1209–1223.

Muniz, M. de F., Campos, V.P., Almeida, M.R.A., Gomes, A.C.M.M., dos Santos, M.F. and Carneiro, R.M.D.G. (2009) Additional information on an atypical population of *Meloidogyne exigua* Göldi 1887 (Tylenchida: Meloidogynidae) parasitising rubber trees in Brazil. *Nematology* 11, 95–106.

Myers, R.F. and Krusberg, L.R. (1965) Organic substances discharged by plant-parasitic nematodes. *Phytopathology* 55, 429–437.

Navas, A., Castagnone-Sereno, P., Blazquez, J. and Esparrago, G. (2001) Genetic structure and diversity within local populations of *Meloidogyne* (Nematoda: Meloidogynidae). *Nematology* 3, 243–253.

Neveu, C., Semblat, J-P., Abad, P. and Castagnone-Sereno, P. (2002) Characterization of a cDNA related to K–Cl cotransporters in the root-knot nematode *Meloidogyne incognita. DNA Sequence – the Journal of Sequencing and Mapping* 13, 117–121.

Neveu, C., Abad, P. and Castagnone-Sereno, P. (2003) Molecular cloning and characterization of an intestinal cathepsin L protease from the plant-parasitic nematode *Meloidogyne incognita. Physiological and Molecular Plant Pathology* 63, 159–165.

Nordmeyer, D. and Dickson, D.W. (1989) Effect of carbamate, organophosphate and avermectin nematicides on oxygen consumption by three *Meloidogyne* spp. *Journal of Nematology* 21, 472–476.

Nordmeyer, D. and Dickson, D.W. (1990) Molecular forms of cholinesterase in the plant-parasitic nematodes *Meloidogyne incognita* and *Radopholus similis. Revue de Nématologie* 13, 311–316.

Opperman, C.H., Bird, D.McK., Williamson, V.M. *et al.* (2008) Sequence and genetic map of *Meloidogyne hapla*: a compact nematode genome for plant parasitism. *Proceedings of the National Academy of Sciences of the USA* 105, 14802–14807.

Papadopoulou, J. and Triantaphyllou, A.C. (1982) Sex differentiation in *Meloidogyne incognita* and anatomical evidence of sex reversal. *Journal of Nematology* 14, 549–566.

Perry, R.N. and Maule, A.G. (2004) Physiological and biochemical basis of behaviour. In: Gaugler, R. and Bilgrami, A.L. (eds) *Nematode Behaviour.* CAB International, Wallingford, UK, pp. 197–238.

Piotte, C., Arthaud, L., Abad, P. and Rosso, M.-N. (1999) Molecular cloning of an acetylcholinesterase gene from the plant parasitic nematodes, *Meloidogyne incognita* and *Meloidogyne javanica. Molecular and Biochemical Parasitology* 99, 247–256.

Plantard, O., Valette, S. and Gross, M.F. (2007) The root-knot nematode producing galls on *Spartina alterniflora* belongs to the genus *Meloidogyne*: rejection of *Hypsoperine* and *Spartonema* spp. *Journal of Nematology* 39, 127–132.

Premachandran, D., von Mende, N., Hussey, R.S. and McClure, M.A. (1988) A method for staining nematode secretions and structures. *Journal of Nematology* 20, 70–78.

Qi, Z.Q., Chen, C.J., Wang, J.X. and Zhou, M.G. (2008) Mode of action of methylene bisthiocyanate and its decomposition products to *Meloidogyne* spp. juveniles. *Acta Phytopathologica Sinica* 38, 420–424.

Rammah, A. and Hirschmann, H. (1988) *Meloidogyne mayaguensis* n. sp. (Meloidogynidae), a root-knot nematode from Puerto Rico. *Journal of Nematology* 20, 58–69.

Rammah, A. and Hirschmann, H. (1990) *Meloidogyne morocciensis* n. sp. (Meloidogynidae), a root-knot nematode from Morocco. *Journal of Nematology* 22, 279–291.

Rand, J.B. and Nonet, M.L. (1997) Synaptic transmission. In: Riddle, D.L., Blumenthal, T., Meyer, B.J. and Priess, J.R. (eds) *C. elegans II.* Cold Spring Harbor Laboratory Press, Cold Spring Harbor, New York, pp. 611–643.

Ray, C., Wang, T.Y. and Hussey, R.S. (1996a) Identification and characterization of the *Meloidogyne incognita col*1 cuticle collagen gene. *Molecular and Biochemical Parasitology* 83, 121–124.

Ray, C., Reddigari, S.R., Jansma, P.L., Allen, R. and Hussey, R.S. (1996b) Immunocytochemical analysis of the stage-specific distribution of collagen in the cuticle of *Meloidogyne incognita. Fundamental and Applied Nematology* 19, 71–78.

Reddigari, S.R., Jansma, P.L., Premachandran, D. and Hussey, R.S. (1986) Cuticular collagenous proteins of second-stage juveniles and adult females of *Meloidogyne incognita. Journal of Nematology* 18, 294–302.

Reversat, G. (1976) Etude de la composition biochimique globale des juvéniles des nématodes *Meloidogyne javanica* et *Heterodera oryzae. Cahiers ORSTOM, Série Biologie* 11, 225–234.

Ritz, K. and Trudgill, D.L. (1999) Utility of nematode community analysis as an integrated measure of the functional state of soils: perspectives and challenges. *Plant and Soil* 212, 1–11.

Robinson, A.F. and Carter, W.W. (1986) Effects of cyanide ion and hypoxia on the volumes of second-stage juveniles of *Meloidogyne incognita* in polyethylene glycol solutions. *Journal of Nematology* 18, 563–570.

Rosso, M.-N., Favery, B., Piotte, C., Arthaud, L., De Boer, J.M., Hussey, R.S., Bakker, J., Baum, T.J., and Abad, P. (1999) Isolation of a cDNA encoding a β-1,4-endoglucanase in the root-knot nematode *Meloidogyne incognita* and expression analysis during plant parasitism. *Molecular Plant–Microbe Interactions* 12, 585–591.

Semblat, J.-P., Rosso, M.-N., Hussey, R.S., Abad, P. and Castagnone-Sereno, P. (2001) Molecular cloning of a cDNA encoding an amphid-secreted putative avirulence protein from the root-knot nematode *Meloidogyne incognita. Molecular Plant–Microbe Interactions* 14, 72–79.

Sharon, E. and Spiegel, Y. (1993) Glycoprotein characterization of the gelatinous matrix in the root-knot nematode *Meloidogyne javanica. Journal of Nematology* 25, 585–589.

Shingles, J., Lilley, C.J., Atkinson, H.J. and Urwin, P.E. (2007) *Meloidogyne incognita*: molecular and biochemical characterisation of a cathepsin L cysteine proteinase and the effect on parasitism following RNAi. *Experimental Parasitology* 115, 114–120.

Soriano, I.R., Riley, I.T., Potter, M.J. and Bowers, W.S. (2004) Phytoecdysteroids: a novel defense against plant-parasitic nematodes. *Journal of Chemical Ecology* 30, 1885–1899.

Spiegel, Y. and McClure, M. (1991) Stage-specific differences in lectin binding to the surface of *Anguina tritici* and *Meloidogyne incognita*. *Journal of Nematology* 23, 259–263.

Spiegel, Y., Inbar, J., Kahane, I. and Sharon, E. (1995) Carbohydrate-recognition domains on the surface of phytophagous nematodes. *Experimental Parasitology* 80, 220–227.

Starr, J.L. (1981) Beta-glycosidases from *Meloidogyne incognita* and *M. javanica*. *Journal of Nematology* 13, 413–414.

Stewart, G.R., Perry, R.N., Alexander, J. and Wright, D.J. (1993a) A glycoprotein specific to the amphids of *Meloidogyne* species. *Parasitology* 106, 405–412.

Stewart, G.R., Perry, R.N. and Wright, D.J. (1993b) Studies on the amphid specific glycoprotein gp32 in different life cycle stages of *Meloidogyne* species. *Parasitology* 107, 573–578.

Stewart, G.R., Perry, R.N. and Wright, D.J. (1994) Immunocytochemical studies on the occurrence of gamma-aminobutyric acid in the nervous system of the nematodes *Panagrellus redivivus*, *Meloidogyne incognita* and *Globodera rostochiensis*. *Fundamental and Applied Nematology* 17, 433–439.

Stewart, G.R., Perry, R.N. and Wright, D.J. (2001) Occurrence of dopamine in *Panagrellus redivivus* and *Meloidogyne incognita*. *Nematology* 3, 843–848.

Thompson, D.P. and Geary, T.G. (2002) Excretion/secretion, ionic and osmotic regulation. In: Lee, D.L. (ed.) *The Biology of Nematodes*. Taylor and Francis, London, pp. 291–320.

Trett, M.W. and Perry, R.N. (1985) Functional and evolutional implications of the anterior sensory anatomy of species of root-lesion nematodes (genus *Pratylenchus*). *Revue de Nématologie* 8, 106–110.

Triantaphyllou, A.C. (1962) Oogenesis in the root-knot nematode *Meloidogyne javanica*. *Nematologica* 7, 105–113.

Triantaphyllou, A.C. (1963) Polyploidy and reproductive patterns in the root-knot nematode *Meloidogyne arenaria*. *Journal of Morphology* 113, 489–499.

Triantaphyllou, A.C. (1966) Polyploidy and reproductive patterns in the root-knot nematode *Meloidogyne hapla*. *Journal of Morphology* 118, 403–413.

Triantaphyllou, A.C. (1971). Genetics and cytology. In: Zuckerman, B.M., Mai W.F. and Rohde, R.A. (eds) *Plant Parasitic Nematodes*, Vol. II. Academic Press, New York, pp. 1–34.

Triantaphyllou, A.C. (1981) Oogenesis and the chromosomes of the parthenogenetic root-knot nematode *Meloidogyne incognita*. *Journal of Nematology* 13, 95–104.

Triantaphyllou, A.C. (1984) Polyploidy in meiotic parthenogenetic populations of *Meloidogyne hapla* and a mechanism of conversion to diploidy. *Revue de Nématologie* 7, 65–72.

Triantaphyllou, A.C. (1985) Cytogenetics, cytotaxonomy and phylogeny of root-knot nematodes. In: Sasser, J.N. and Carter, C.C. (eds) *An Advanced Treatise on* Meloidogyne. *Vol. I, Biology and Control*. North Carolina State University Graphics, Raleigh, North Carolina, pp. 113–126.

Triantaphyllou, A.C. (1990) Cytogenetic status of *Meloidogyne kikuyensis* in relation to other root-knot nematodes. *Revue de Nématologie* 13, 175–180.

Triantaphyllou, A.C. and Hirschmann, H. (1997) Evidence of direct polyploidization in the mitotic parthenogenetic *Meloidogyne microcephala*, through doubling of its somatic chromosome number. *Fundamental and Applied Nematology* 20, 385–391.

Trudgill, D.L and Blok, V.C (2001) Apomictic, polyphagous root-knot nematodes: exceptionally successful and damaging biotrophic root pathogens. *Annual Review of Phytopathology* 39, 53–77.

van der Beek, J.G. and Karssen, G. (1997) Interspecific hybridization of meiotic parthenogenetic *Meloidogyne chitwoodi* and *M. fallax*. *Phytopathology* 87, 1061–1066.

van der Beek, J.G. and Pijnacker, L.P. (2008) Maintenance of genetic variation in automictic root-knot nematodes. *Nematology* 10, 313–321.

Van der Eycken, W., de Almeida Engler, J., Van Montagu, M. and Gheysen, G. (1994) Identification and analysis of a cuticular collagen-encoding gene from the plant-parasitic nematode *Meloidogyne incognita*. *Gene* 151, 237–242.

Van Gundy, S.D. and Stolzy, L.H. (1961) Influence of soil oxygen concentrations on the development of *Meloidogyne javanica*. *Science* 134, 665–666.

Wang, E.L.H. and Bergeson, G.B. (1978) Amino acids and carbohydrates secreted by *Meloidogyne incognita*. *Journal of Nematology* 10, 367–368.

Wang, T., Deom, C.M. and Hussey, R.S. (1998) Identification of a *Meloidogyne incognita* cuticle collagen gene and characterization of the developmental expression of three collagen genes in parasitic stages. *Molecular and Biochemical Parasitology* 93, 131–134.

Wright, D.J. (1998) Respiratory physiology, nitrogen excretion and osmotic and ionic regulation. In: Perry, R.N. and Wright, D.J. (eds) *The Physiology and Biochemistry of Free-living and Plant-parasitic Nematodes.* CAB International, Wallingford, UK, pp. 103–131.

Wright, D.J. and Perry, R.N. (1998) Musculature and neurobiology. In: Perry, R.N. and Wright, D.J. (eds) *Free-living and Plant-parasitic Nematodes.* CAB International, Wallingford, UK, pp. 49–73.

Wright, D.J. and Perry, R.N. (2006) Reproduction, physiology and biochemistry. In: Perry, R.N. and Moens, M. (eds) *Plant Nematology.* CAB International, Wallingford, UK, pp. 187–209.

9 Survival Mechanisms

Adrian A.F. Evans[1] and Roland N. Perry[2]
[1]*Imperial College London, Sunninghill, Ascot, Berkshire, UK;* [2]*Rothamsted Research, Harpenden, Hertfordshire, UK and Ghent University, Ghent, Belgium*

9.1 Introduction	**201**
9.2 Dormancy, Diapause and Quiescence	**202**
9.3 Embryonation and the Egg Mass Environment	**202**
9.4 Temperature Effects on Development of Eggs and Infective Stages	**206**
9.5 The Effect of Osmotic Stress on Infective Stages in Soil	**213**
9.6 Survival Mechanisms Deployed: Life History Strategies in *Meloidogyne* Species	**214**
9.7 Conclusions and Future Directions	**218**
9.8 References	**219**

9.1 Introduction

The majority of species of *Meloidogyne* have wide host ranges and are able to persist in fields between the main host crops. Persistence of *Meloidogyne*, and even increase in populations, during rotations with resistant crops or during fallow periods can occur because weeds act as hosts. For example, Kutywayo and Been (2006) found that three common weed species were good hosts for *M. chitwoodi*, a quarantine species in Europe. Cover crops are used to suppress weeds and increase soil organic matter, but they may also act as hosts for *Meloidogyne* spp.; for example, substantial reproduction of *M. incognita* occurred on winter cover crops in the USA and could lead to damaging populations in a subsequent cotton crop (Timper *et al.*, 2006). The large reproductive rates of species of *Meloidogyne* still enable populations to persist despite high mortality rates for eggs and second-stage juveniles (J2). However, there are situations where the persistence of *Meloidogyne* populations involves survival in the absence of food, or the ability to withstand adverse environmental conditions, including temperature extremes and desiccation. The processes of, first, surviving in soil long enough to find and infect a host and, second, ensuring the best chances for survival of progeny when the host is no longer supportive are the essential non-parasitic tasks of the life cycle, which must be integrated into a successful life history strategy appropriate to the environmental selection pressures.

This account follows the reproductive propagule – the egg – of the *Meloidogyne* female, deposited initially with siblings in a maternally provided gelatinous matrix of the egg mass, and the subsequent progression of the J2 to parasitizing a host where, if successful, it may recapitulate the reproductive process.

©CAB International 2009. *Root-knot Nematodes* (eds R.N. Perry, M. Moens and J.L. Starr)

9.2 Dormancy, Diapause and Quiescence

In this chapter, the terms dormancy, diapause and quiescence are used. They describe arrested development, either at the embryonic stage or at the juvenile stages, which enables many organisms, including *Meloidogyne* spp., to survive adverse conditions. The research leading up to the acceptance of these terms in plant nematology will be discussed, and the following definitions will enable previous work and the early use of these terms to be put in context. Dormancy involves a reduction in metabolism to conserve energy and is usually divided into quiescence and diapause.

Quiescence is a spontaneous response to unpredictable unfavourable environmental conditions, and normal activity and development resumes when favourable conditions return. Quiescence can be obligate or facultative. Obligate quiescence occurs when the environmental cue affects a specific receptive stage of the life cycle, whereas facultative quiescence is not stage-specific.

Diapause is a cessation of development, which is only reversible when specific requirements have been satisfied, even if favourable conditions return. Diapause is a strategy to overcome cyclic long-term adverse circumstances, such as seasonal conditions and/or the absence of the host. The incidence of diapause varies greatly between species of *Meloidogyne* and between populations of the same species (see section 9.3.5). De Guiran and Villemin (1980) found that the percentage of unhatched J2 that enter diapause varies from less than 10% for the predominantly tropical *M. arenaria* to 94% for the temperate *M. naasi*.

As with quiescence, diapause can be obligate or facultative. Obligate diapause is initiated by endogenous factors, and normal development and activity can resume after exogenous stimuli have been experienced for a required period of time. Temperature is the most important environmental cue for the termination of obligate diapause, with a fixed period of exposure to low temperatures relieving the arrested development. Facultative diapause is initiated by exogenous, rather than endogenous, stimuli and is terminated by endogenous factors after a set period of time.

It is difficult to separate types of dormancy on the basis of metabolic activity, and the cause of the arrest in development is a more relevant criterion (Evans and Perry, 1976). Evans (1987) further distinguished between dormancy affecting ontogenetic development and that affecting somatic development. There are several factors that are involved in the survival of unfavourable conditions by *Meloidogyne*, and these will be discussed in this chapter in the context of quiescence and diapause.

9.3 Embryonation and the Egg Mass Environment

Embryonation within the egg may have begun in the maternal reproductive tract but is not completed until after deposition in the egg mass. It continues with the experiences of the egg mass environment, which may be either transitory or lengthy, depending on circumstances, until environmental signals promote eclosion (hatching). The egg mass and components of the eggshell are important for the survival of the developing embryo and the fully formed juvenile stages within the egg. Timing of the hatching response is also central to optimizing the changes of successful host location by the pre-parasitic J2. The details of the hatching response and the mechanisms of eclosion by the J2 of *Meloidogyne* and the influence of the host plant are detailed by Curtis *et al.*, Chapter 6, this volume. In the following sections aspects of development and hatching in relation to survival are discussed.

9.3.1 The egg mass

Development of the embryo proceeds to the first-stage juvenile (J1), which moults in the egg to the J2, and all *Meloidogyne* spp. start their mobile life when the J2 successfully hatch and emerge from the gelatinous matrix into which they were deposited as early-stage embryos. Within the gelatinous matrix, the egg may already have been influenced by the life experiences of the female, in terms of the maturation status of its host plant and the experience of seasonal changes in its environment.

De Guiran (1980) reported that high percentages of unhatched J2 of *Meloidogyne* in diapause were associated, to some degree, with plant resistance. Huang and Pereira (1994) also found that J2 from egg masses on hosts that were less susceptible to *M. javanica* showed a marked delay in hatch. These authors speculated that plants close to the end of their growing season may produce some substance affecting the feeding female and indirectly resulting in delayed hatch of J2. If true, this illustrates the influence of the plant on development and survival of *Meloidogyne* and reflects a change of priority during the plant growing season from hatch from egg masses on young plants to survival after host senescence (see Curtis *et al.*, Chapter 6, this volume).

The egg mass forms the first line of defence against the hostile elements in the form of the soil, its predators and parasites, its atmosphere, often low in oxygen, and the desiccating effects of low soil moisture. Thus, the egg mass provides several benefits that enhance survival. When first produced, the egg mass contains only eggs with early embryos developing into J1, but as further eggs are laid into the egg mass the first eggs contain J1, which immediately complete their first moult into J2 ready to hatch when conditions permit. Older egg masses may comprise a mixture of eggs at various stages of embryo development and differentiation, together with J1 and both unhatched and hatching J2.

Several investigations have explored the egg mass as a biological entity (Wallace, 1966; Bird and Soeffky, 1972; de Guiran, 1980). It clearly provides the first and major barrier to desiccation of the contents. Until the end of the J1 stage, the eggshell protects the embryo from water loss. Embryos and J1 survive drying conditions more effectively than unhatched J2 because there is a change in the eggshell membrane after the first moult in the egg. Immediately prior to hatch of the J2, enzyme activity erodes layers of the eggshell, resulting in a change in permeability and a loss of desiccation protection (Wallace, 1968a; Perry *et al.*, 1992; Curtis *et al.*, Chapter 6, this volume). The pharyngeal glands may be the source of enzymes involved in hatching, but the enzymes may be present in the egg fluid surrounding the J2 and are kept inactive either by separation from their substrate by the eggshell lipid membrane or by an inhibitor (Perry, 1989). In this scenario, a change in the permeability of the membrane would precede the activity of enzymes responsible for eroding other layers of the eggshell.

9.3.2 The effect of soil moisture

In pioneering research that is still relevant today, Wallace (1966) recognized the essential role played by soil moisture in survival and hatch, and investigated hatch in a population of *M. javanica* that gave up to 95% emergence of J2 from fresh egg masses. The fresh, translucent egg masses had a matrix with an agar-like consistency. With age, the outer layers appear tanned to form a brown skin, which gives increased rigidity but appears not to form a barrier to the emergence of J2. Water availability to eggs in the soil can be affected either by osmotic forces or by capillary forces producing a suction pressure (matric potential). Results with electrolytes and non-electrolytes exerting a range of osmotic potential were compared with situations where suction potential withdraws water from soil pores. Whereas low levels of osmotic potential had no effect on the percentage of J2 hatching from eggs freed from their egg masses, some electrolytes could decrease the percentage hatch. Wallace used the non-electrolyte glycerol to expose egg masses to a range of osmotic pressures. He concluded that eggs containing J2 and ready to hatch have eggshells readily permeable to water. In dry soil, loss of water forces the J2 into a quiescent state, readily reversed by rehydration, whereas embryos that have experienced osmotic pressure high enough to affect their development may enter a more prolonged dormancy. This may contribute to longer-term survival.

In a further study on *M. javanica*, Wallace (1968a) showed that egg masses do not resist water loss and shrink considerably, even at low soil suction potentials; however, they do provide a barrier to water loss from the eggs they surround. Embryo development continued in eggs within egg masses at 98% relative humidity (RH) over 5 days, whereas embryos in eggs freed earlier from their eggs mass were immediately affected by this treatment. Hatched J2 also showed little resistance to desiccation at 98% RH.

In a dry soil, a dehydrating egg mass may also provide a small mechanical pressure on the eggs, which inhibits hatch (Wallace, 1968a; Bird and Soeffky, 1972). An extra protective layer, which appears as an extracuticular subcrystalline layer in *M. charis* (Demeure and Freckman, 1981), may function to slow the rate of water loss of unhatched J2 and help protect against desiccation.

9.3.3 The effect of soil aeration

Wallace (1968b) explored how hatch of the J2 was affected by soil aeration. J2 in egg masses need some oxygen to hatch, even though soil atmospheres containing as little as 5% oxygen only slightly delayed the hatching period rather than decreasing the cumulative number hatched. Increasing the length of time egg masses were kept in water without oxygen decreased cumulative hatch over 8 days to about 30%, compared with a hatch of 80% with no oxygen deprivation. Embryos were the most sensitive stage; as little as 1 day without oxygen prevented their further development in these experiments. Some J2 that had already developed could still hatch after 3 days of anaerobiosis, whereas hatched J2 remained unaffected for up to 4 days, but were irreversibly inhibited after 8 days of anoxia. In practical terms, this means that while a completely flooded soil will allow little diffusion of oxygen, in many situations, once soil pores begin draining, soil may rapidly become adequately aerated, allowing resumption of embryo development and hatch of J2.

Using glass beads of different diameters to model soils with different pore size, there was evidence of how egg masses contribute to survival under soil conditions. In saturated soil, hatch is inhibited by lack of oxygen. As soil pores begin to drain, the entry of air provides oxygen for development and hatching, but as the pores empty of water, shrinkage of the egg mass applies pressure to the eggs within, and hatching is progressively curtailed, conserving the population of J2 within during conditions that are lethal to hatched J2. Soils with larger pores will allow earlier and greater hatch than soils with smaller pores, indicating why *Meloidogyne* is often favoured by coarse soils well watered by rainfall or irrigation.

Some of Wallace's results were later refined in work on soil moisture effects (Baxter and Blake, 1969a), by using glass sinters to apply suction to single eggs of *M. javanica*. Some discrepancies, which could have been due to artefacts, were observed, but Baxter and Blake (1969a) concluded that J2 of *Meloidogyne* are well adapted to hatch at the same moisture-level ranges over which their host plants grow. Further studies (Baxter and Blake, 1969b) on oxygen relations in eggs and J2 of the same population of *M. javanica* confirmed that embryo development and J2 emergence were maximized at 15% oxygen and were not restricted by the presence of the egg mass.

9.3.4 Other roles for the egg mass

Evidence for cellulolytic, proteolytic and pectinolytic enzymes was the interpretation placed on the observation by Orion and Frank (1990) that the egg mass passes through disrupted plant tissue before reaching the root surface. Lectin binding by the mucopolysaccharides in the gel matrix has indicated a role in deterring egg parasites, while in egg masses of *M. hapla* there is evidence of colonization by bacteria of kinds different from those found elsewhere in the rhizosphere (Papert and Kok, 2000). Orion *et al.* (2001) demonstrated that the gelatinous matrix of *M. javanica* served to protect the enclosed eggs from invasion of some microorganisms. However, Stirling (1991) considered that, although the dry egg matrix may deter some microorgansisms, the hydrated matrix may provide a food source for some parasites.

Many authors have noted among the 'tropical' species of *Meloidogyne* that, whereas the fresh egg mass has a matrix that is colourless, sticky and jelly-like when first extruded by the young female, with age the egg mass changes its appearance, the surface hardening and assuming a brown colour; some authors considered this to be indicative of tanning. Differences were also observed in egg masses of *M. incognita* produced in dry and subsaturated soils (de Guiran, 1980) (Fig. 9.1; Plate 16). Observations that development in and hatching from the 'old' egg masses differ from those of the young colourless ones has prompted a number of investigations. Ishibashi and Kunii (1963) noted that hardened, dark-coloured egg masses were most often seen approaching harvest time in the autumn season. The females that produced them had often suspended egg production but could produce more eggs when placed in physiological saline, suggesting that the hardened egg mass had inhibited further egg laying while on the host

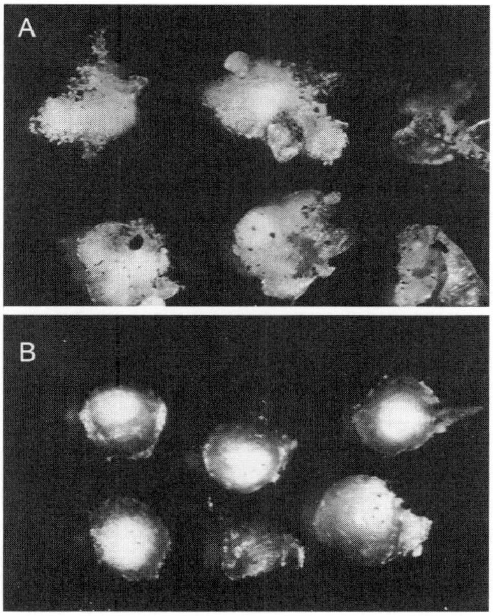

Fig. 9.1. Differences in the egg masses of *Meloidogyne incognita* produced in (A) dry or (B) subsaturated soils. In the dry soil, the gelatinous matrix was colourless, soft and expanded. In subsaturated soils, the matrix rapidly turned hard and brown and contracted. (From de Guiran, 1980.)

plant. Nutrient deficiency in plants hosting a *M. incognita* infection did not seem to induce brown egg masses (Ishibashi *et al.*, 1964), but to investigate the subject further, the proteins of several *Meloidogyne* samples were examined by gel electrophoresis. Differences in lactate dehydrogenase isozymes were recognized between those females producing egg masses that were pale-coloured and soft, compared with those that were brown and harder. Furthermore, J2 in eggs from old brown egg masses could not be induced to hatch in response to host root exudates. The role of host exudates (= diffusates) in the hatching of some species of *Meloidogyne* has been reviewed by Perry and Wesemael (2008), and is discussed by Curtis *et al.*, Chapter 6, this volume.

9.3.5 The egg mass and dormancy

De Guiran and colleagues studied how water relations in soils affected the development of eggs and juveniles in egg masses of *M. incognita* from a population from Côte d'Ivoire, especially focusing on those egg masses that had survived adverse conditions, yet the J2 were unable to hatch shortly after return to favourable conditions. De Guiran and Demeure (1978) subjected egg masses to 1 week in dry soil and found that on subsequent exposure to optimum conditions for 20 days J2 hatched quickly; by contrast, if the egg masses had been exposed for 1 week in poorly aerated soil, there was a subsequent delay in hatch when exposed to optimal conditions. Hatch was low from egg masses subjected to waterlogged conditions in either clay or sandy soils, but their response depended on the stage of development; hatched J2 recovered quickly, but earlier stages, whose development was delayed, subsequently hatched less, and some became arrested at an earlier (J1) stage and were postulated to be 'in diapause'.

Subsequently, de Guiran (1979a) sought to explore this state of arrest or dormancy. J2 in egg masses withstood dry soil at wilting point for up to 6 weeks and recovered within a day. In naturally infested soil, J2 could be recovered for up to 12 weeks, although viable eggs could be recovered for periods of up to 22 weeks. These results demonstrated that J2 within egg masses withstood short periods of desiccation and were able to become infective individuals in favourable conditions. In saturated soils, cumulative numbers of J2 hatching from egg masses after 8 days comprised only 10% of numbers hatched after 12 h, and no hatch took place after 2 months, although viable eggs were present at 12 weeks and embryos were longer lived than J1 stages. Nematode antagonists were considered responsible for the loss of more J2 from aerated soils than from saturated soils.

To make more explicit some aspects of the dormancy, which was proposed 'to be diapause', de Guiran (1979b) carried out various treatments in water and in soil. The results obtained for an extended hatch under optimum conditions provided the most persuasive evidence for delayed-hatching abilities. Following an initially large hatch, the rate of emergence fell near to zero after 20 days, until between 45 and 70 days a second (smaller) peak of hatching was evident. This was considered as typical of the pattern of daily emergence during this prolonged hatch, although attempts to repeat the results failed.

Despite the inability of any imposed treatment to stimulate the resumption of development of eggs containing embryos or unhatched juveniles, they were still considered to be 'in diapause'; however, it is arguable that, if still alive, they more properly meet the requirements of an obligate quiescence.

To elucidate factors that could be inducing 'diapause', de Guiran (1980) raised further *M. incognita* females on tomato. Three weeks after the females started egg laying, small egg masses contained only eggs with embryos and unhatched J2, but by 10 weeks egg masses contained about one-third embryos and two-thirds empty shells, with only a few unhatched J2 remaining. To apply a purely physical stress to egg masses and their contents, a number were selected and chilled at 4°C for up to 16 weeks. Results showed the proportion of eggs from which J2 had hatched (i.e. empty shells) progressively decreased at the expense of embryos ('in diapause'), unhatched J2 and dead eggs, in approximately equal proportions. When similar batches of egg masses as used in the 4°C observations were placed in anoxic conditions, all development was blocked at the J1 stage, as Wallace (1968a) had also observed when using these conditions with *M. javanica*. A 'split root' technique was used to allow the egg masses produced by females on roots to be observed in both moist (subsaturated) soil and in very dry conditions. Eggs masses produced in dry soil were translucent with a soft, loose texture and lacking a defined form; 14% of eggs in these egg masses contained J2 considered to be 'in diapause'. By contrast, in moist soil the egg masses were subspherical, brown in colour with a firm, contracted texture, and 37% of their unhatched J2 were 'in diapause'. Nutritive stress was applied to tomato plants at 20 days post-inoculation by either a withdrawal of mineral nutrients or by defoliation, and counts of eggs in egg masses placed 39% of eggs in 'diapause'.

This 'diapause' state was noted, therefore, as existing even in the first eggs deposited into the egg mass in the absence of any age-related stress, but with increasing age of females the numbers 'in diapause' increased; there was also an increase with other stressing conditions such as degree of water saturation of soil, defoliation and nutrition. Contrary to the proposal of Ishibashi and Kunii (1963) that browning of egg masses induces a diapause in the enclosed eggs, de Guiran (1980) concluded that the exposure to moist soil alone was sufficient to explain both the browning of egg masses and the increased numbers of eggs 'in diapause', and both conditions were considered to have been independent of the physiological state of the female. Any treatment or artificial means to remove the 'diapause' remains to be found.

Extending this line of investigation to the other major 'thermophil' (see section 9.4) species of *Meloidogyne* and their hosts, de Guiran and Villemin (1980) showed that fecundity of females was not correlated with host infection, as revealed by root-galling indices, and that developmental arrest of embryos was not only present in other species but also varied in degree between host plants and also between different geographical isolates. The embryonic 'diapause' was examined in eggs within egg masses produced in either complete or nutrient-deficient regimes in monoxenic culture on excised host roots. Fewer eggs per female were produced by potassium-deficient females and the percentage of eggs 'in diapause' decreased with age of the female, but the percentage 'in diapause' was unchanged with age on the complete medium. The authors speculated that the 'diapause' has its origin in a quiescence that gradually lost its (rapidly) reversible characteristics and permitted recovery when favourable conditions had become stabilized.

Thus, it appears that this phenomenon in the thermophil species of *Meloidogyne*, discussed by the authors of these studies, bears all the characteristics of a quiescence, albeit one from which recovery, although prolonged, affords little penalty. To conform to the presently understood characteristics of a diapause, some specific trigger to induce many individuals to resume development within a short space of time would be necessary. A critique of the conclusions of these authors is provided in section 9.4.7.

9.4 Temperature Effects on Development of Eggs and Infective Stages

Temperature is the factor most easily recognized as important in determining not only the rate of development of nematodes, but also affecting their distribution, especially in relation to the ability to survive the effects of extremes of heat and cold. It is the third of the triumvirate of physical factors influencing the suitability for life of any habitat, the others being water and oxygen, to which must be added the vital biological component of food.

Within the genus *Meloidogyne* there are two groups, thermophils and cryophils, based on their ability to survive lipid-phase transitions that occur at 10°C (Van Gundy, 1985). *M. chitwoodi, M. hapla* and probably *M. naasi* are cryophils and can survive in soil at temperatures down to and below 0°C, whereas *M. javanica, M. arenaria* and probably *M. exigua* are thermophils and do not survive in soils at temperatures below 10°C.

9.4.1 Temperature as an isolated factor

It is not always easy to compare results on temperature tolerance of species of *Meloidogyne* from different workers because experimental conditions vary and often only generalizations can be made. The earliest studies on how *Meloidogyne* responds to temperature were by Tyler (1933), working with the root-knot nematode known then as '*Heterodera marioni*'. She used a population of unknown provenance (but thought by Thomason and Lear (1961) to be *M. hapla* because Tyler reported her population as having a maximum temperature for reproduction of 31.5°C). An early report on the 'thermal death times' of *M. incognita* compared it with *Panagrellus redivivus*, where the characteristics of both nematode species converged on a single lethal value (exposure to a temperature of 46°C for 1 h), although while exposure for 4 h at the lower temperature of 43°C was also lethal for *M. incognita, P. redivivus* took 6 h to succumb (Santmeyer, 1955).

One of the first comparative studies of temperature relationships with life stages and their survival among species of *Meloidogyne* was made by Bergeson (1959). He used a relatively simple but apparently effective bioassay for testing viability following temperature treatments: gall production on the roots of a common host plant, tomato. Eggs and J2 of *M. incognita* '*acrita*' were compared with those from *M. hapla* and *M. javanica*, using exposures to eight temperatures ranging from 0 to 38°C.[1] Some J2 of *M. incognita* survived 0°C for 5 days and 5°C for 12 days, but at 10°C about 50% of treated J2 produced galls after 6 months (and 10% at 12 months), indicating their resilience to prolonged cool temperatures in moist soil. By contrast, at all higher temperatures (15–38°C) fewer than 10% of J2 reproduced after 3 months (cf. nearly 90% at 10°C).

Egg survival was assessed similarly in terms of numbers of galls produced on hosts after treatments and showed 100% survival after 3 months at 10°C (50% at 5°C). By 6 months, more than 70% were alive (0% at 5°C) and almost 20% were alive at 12 months. Once again, no eggs survived beyond 5 months at temperatures above 10°C. Attempts at preconditioning J2 at or below 10°C for several weeks gave no evidence of adaptation. J2 of *M. hapla* survived at 0°C for up to 16 days (3 days for *M. incognita*), and almost no mortality of *M. hapla* occurred at 5°C over 4 weeks (7 days for *M. incognita*).

Time of survival of eggs of *M. incognita* (50 days at 0°C) was compared with two populations of *M. hapla* ('San Francisco' and 'Davis'), both of which survived more than 90 days. In comparison, few eggs of *M. javanica* survived the 50 days' exposure to freezing temperatures. A concluding experiment showed that early emergence of J2 from egg masses over a 10–15-day period at 5, 10 and 20–30°C was not sustained, whereas at 15°C a slower but persistent emergence rate continued undiminished up to the conclusion of observations at 22 days. For this population of *M. incognita*, the optimum survival temperature for J2 was 10°C, but it was unable to survive damage by freezing temperatures for more than a few days. However, some eggs tolerated freezing for a few months.

Another study used several isolates of four species of *Meloidogyne* sourced from different locations in California (Thomason and Lear, 1961). They were assayed for tolerance to high temperatures for up to 35 days, again by the production of galls containing egg masses (Gall Index Value;[2] here abbreviated to GIV) on tomato or the weed, *Sesbania exultata*, which grew at high temperatures. Most populations reproduced best at 25 or 30°C, but at 35°C, in both isolates of *M. hapla* and in *M. arenaria*, galling and egg production were much reduced. *Meloidogyne javanica* was substantially unaffected at this temperature. When both *M. incognita* isolates were compared with the three *M. javanica* isolates at temperatures from 15 to 35°C, one *M. javanica* isolate failed to reproduce

[1] To facilitate comparisons with all work mentioned at this point and subsequently, the results reported here have been converted to the nearest temperature on the Celsius scale at 5°C intervals; thus 3°F = 0°C, and 40°F = 4.4°C but is given here as 5.0°C.

[2] This GIV scale refers to Sasser's 0–4 scale (0 = 0% root galling, and 4 = 100% root galling); see Fig. 2 in Barker (1985).

at the lowest temperature and all other populations produced fewer egg masses than at 20–30 °C. A similar effect was seen at 35 °C, where all but one *M. incognita* isolate produced very few egg masses. Using the heat-tolerant weed *S. exultata*, each of the species reproduced at 32.6 °C but only *M. javanica* produced egg masses at 36.6 °C, although immature eggs were unable to complete their development at this temperature.

A population of *M. javanica* from Southern Africa was compared with two populations from eastern North America by Daulton and Nusbaum (1961). Their responses were compared with a population of *M. hapla*, also from eastern North America, using the root-galling bioassay method, this time employing a scale of 0–100%, covering 'eight infection classes'. By testing the tolerance of eggs in egg masses to temperature extremes in both wet and dry soils, it was evident that survival in moist soil was almost always less than that in dry soil. Galling by *M. hapla* was greater than that for the three *M. javanica* populations at 2 °C, whereas at 33 °C the *M. hapla* population survived least well and the African population of *M. javanica* was almost unchanged over 10 days. The North American populations were similar to the Southern African populations for 6 days, but then declined rapidly until 10 days. A repeat of these treatments at −2 °C again found galling by all *M. javanica* isolates decreased rapidly to 0% after 12 days in moist soil, whereas *M. hapla* responded differently, having a GIV of 18% in moist soil at 10 days but only 8% at 12 days in dry soil. By contrast, at 36 °C all populations quickly expired (within 12 h) in dry soil, but in moist soil, despite an initial rapid decline over the first 4 days, all four populations still had survivors after 24 days. Thus 'moist cold' −2 °C) was more damaging than 'dry cold', whereas 'moist warmth' (36 °C) favoured survival over 'dry warmth'. J2 survivors of the temperature extremes produced a greater number of galls, and the egg masses yielded progeny tolerant of the temperatures survived by their parents. However, egg masses surviving the 'cold moist' exposure showed no greater survival abilities than controls when exposed to the 'warm moist' conditions.

Information on lethal temperatures is important in assessing the possibility of control of plant-parasitic nematodes using soil solarization (Gaur and Perry, 1991). To determine lethal temperature–time combinations, Wang and McSorley (2008) incubated eggs and J2 of *M. incognita* in test tubes containing a sand–peat mix in a water bath at eight temperatures of 1 °C intervals between 38 and 42 °C. Complete suppression of hatch required 164.5, 32.9, 19.7 and 13.1 h at 38, 39, 40, 41 and 42 °C, respectively, whereas killing of hatched J2 took 47.9, 46.2, 17.5 and 13.8 h at 39, 40, 41 and 42 °C, respectively. The authors concluded that the effect of temperature on nematode mortality is not determined by heat units. Field trials in Florida with soil solarization demonstrated that temperatures exceeded the lethal temperatures in the top 15 cm of soil but nematodes were able to survive in lower depths. Stapleton and Duncan (1998) and Ploeg and Stapleton (2001) considered that temperature and time treatments using soil solarization to control *M. incognita* and *M. javanica* could be reduced in soils amended with broccoli residues. The use of soil solarization as a strategy to control *Meloidogyne* spp. is discussed in detail by Coyne *et al.* (Chapter 19, this volume).

9.4.2 Low temperature survival

For some species of *Meloidogyne*, overwinter survival may involve withstanding temperatures below freezing. In *in vitro* studies, Sayre (1964) found that both *M. hapla* and *M. incognita* survived supercooling for short periods, but that ice crystal formation was not invariably fatal. A physiological salt solution assisted more *M. hapla* to survive −7 °C than *M. incognita*. When egg masses were frozen to −30 °C, some J2 of both species subsequently hatched. Sayre (1964) tested whether a period of storage at different temperatures prior to freezing allowed some adaptation before freezing for 30 min at −30 °C. He found that 60% of *M. hapla* survived to hatch but only 30% of *M.incognita* survived, which may reflect the greater tolerance to chilling of *M. hapla*, or perhaps that it is freeze resistant and remains undamaged.

The tolerance of *M. hapla* and *M. incognita* to chilling and freezing temperatures was tested by Vrain *et al.* (1978). When, after a progressive temperature decrease over several weeks, *M. incognita* egg masses experienced 14 days of exposure to freezing, few embryonating eggs survived and many showed abnormalities. By contrast, of those eggs containing J1 and J2 stages, numbers surviving to enter the roots of a bioassay host were about half the numbers in untreated controls. *M. hapla* exposed to −4 °C for 10 days similarly produced eggs with abnormalities, but about 10% of embryonating eggs survived. Similar exposures were arranged wherein eggs were subjected to either saturated or subsaturated soil conditions with chilling or freezing. *M. incognita* eggs contained infective J2 after 2 days exposure at 0 °C in subsaturated soil; *M. hapla* tolerated 2 weeks and temperatures down to −8 °C for 6 days with similar levels of infectivity. Few eggs produced infective J2 after exposures in saturated soil but J2 of both species tolerated short periods of freezing, even in saturated soils.

Lyons *et al.* (1975) compared the biophysical and biological correlates of the chilling-resistant species *M. hapla* with the chilling-sensitive species *M. javanica* using the nematode's oxygen uptake over a temperature range from 5 to 30 °C as an indicator of respiration rates. When an Arrhenius plot of log rate of oxygen consumption against 1/T (reciprocal of the absolute temperature) was made for *M. hapla*, a continuous and linear respiration rate was observed over the whole temperature range. By contrast, the Arrhenius plot for *M. javanica* was discontinuous and non-linear, with the temperature of change at between 12 and 15 °C.

Polar lipids are characteristic components of membranes whose integrity is essential for mitochondrial function. Extracts of polar phospholipids from species of *Meloidogyne* were probed using spin-labelling compounds, and the resulting Arrhenius plots for *M. javanica* again showed a discontinuity characteristic of a phase change at 10 °C, whereas the plot for *M. hapla* was again continuous. It is interesting that, of the two functions investigated – respiration rate and phase changes in polar lipids – it is the latter that matches most closely the temperature below which *M. javanica*, at best, suffers some chilling injury and, at worst, is lethal to the nematode.

9.4.3 The influence of soil type and moisture content on temperature effects

One difficulty in assessing temperature effects on survival of species of *Meloidogyne* is that *in vitro* experiments on effects of temperature as a single factor are difficult to relate to the *in vivo* situation, where seasonal effects, diurnal rhythms of temperature changes, soil type and moisture are some of the many factors that impact survival in the field. In addition, optimum conditions for activity and normal life-cycle processes may not be ideal for long-term survival. For example, the optimum temperature for growth and reproduction of *M. javanica* is 25–30 °C, whereas the optimum temperature for survival of eggs and J2 in the soil is 10–15 °C (Van Gundy, 1985). Analysis of the effects of a combination of some or all of the components of the agroecosystem is difficult, but some workers have attempted such studies in order to improve information on the field condition.

Daulton and Nusbaum (1961) conducted a field test of overwintering survival of egg masses placed in a clay loam soil and kept weed free, from August to April (temperature range 0–31 °C). Periodic sampling on five occasions enabled survival to be estimated using the GIV method. At the first sampling interval (98 days), all *M. javanica* populations had much reduced GIV, unlike *M. hapla*, which had levels similar to controls (GIV at day 0). This trend was continued, and at the final sampling (250 days), there was negligible survival of any *M. javanica* population, while the GIV for *M. hapla* had reduced by only half. The positive feature of these experiments includes the use of egg masses rather than free eggs or J2 as the inoculum; however, the assay of galling was never verified by the more certain (but more time consuming) method of assuring that each gall had yielded a reproducing female.

Variability between *M. javanica* isolates appears to confirm that the difference between them and between other species seems to be relatively small. To understand better the way that species of *Meloidogyne* respond to various climatic conditions ranging from interior valleys to the coastal plains of southern California, Thomason (1962) tested reproduction of *M. hapla*, *M. javanica* and two populations of *M. incognita* on a range of commonly grown cereals and Sudan Grass. Soil temperatures were below 15 °C from mid-December to mid-March, reaching a minimum of 11.1 °C in mid-February. The number of J2 of *M. incognita* free in the soil was low in March but rose eightfold to a peak in May, reflecting a higher development rate of embryos in overwintering egg masses. A laboratory hatching test for *M. javanica* showed an initially rapid hatch (at 4 days), with similar numbers emerging at 12, 15, and 18 °C, but with a peak at 30 °C. Over a longer period (to 21 days), more J2 emerged at all temperatures and hatching peaked at 33 °C. Clearly, if the opportunity for invasion by *M. javanica* were to be restricted by changing soil conditions in spring, an ability to hatch rapidly and to seek and penetrate the host in the shortest time could be a behaviour rewarded by successful reproduction.

In addressing the question of how long nematodes retain viability in storage, and how mobility relates to infectivity, Thomason *et al.* (1964) subjected J2 from a population of *M. javanica* to 3, 15 and 27 °C for up to 10 days. They found that movement decreased rapidly at 3 °C (by over 85% over 8 days) compared with higher temperatures. Only at 27 °C after 16 days was mobility decreased by 50%; at 15 °C, it decreased by only around 20%. Infectivity rates at these temperatures were always lower than mobility, falling to values under 10% after 16 days at 3 °C and 27 °C, but were relatively conserved (55%) at 15 °C. When fresh J2 were harvested daily from egg masses on roots, those harvested at days 3 and 9 were placed at the three storage temperatures and again tested up to 16 days for motility and infectivity; those collected at 9 days generally had greater mobility and infectivity.

In a study of the comparative development of populations of *M. incognita* and *M. hapla* in the laboratory and the field, Vrain *et al.* (1978) showed that both species invaded and developed in the roots of clover at 12, 16 and 20 °C but not at 8 °C. After invasion, *M. hapla* developed at all temperatures more quickly than *M. incognita*, first as sexually undifferentiated J2, subsequently as female-developing J2, and at 20 °C only as J3/J4 females. *M. incognita* failed to complete further development at 12 and 16 °C. These times were reflected in accumulation of heat units (measured in degree-hours) above a minimum temperature for development, which was recorded as 8.8 °C (*M. hapla*) and 10.08 °C (*M. incognita*). Two sites in North Carolina were compared in field studies; one site (Clayton) was sandy loam (sand 82%, clay 4%) while the other (Fletcher) had a fine sandy loam (40% sand, 18% clay). White clover seedlings were infected with 1000 J2 of either species in October (soil temperature 18–20 °C at 10 cm depth) and November (soil temperature 12 °C at 10 cm depth). Plants inoculated in October at Clayton had only 5.6% of *M. hapla* as adult females by early December, compared with 42% for *M. incognita*, which had produced some empty egg masses. Reproduction of both species began in early March. In the heavier soil at Fletcher, slower development resulted in 1.4% females of *M. hapla* and 12% *M. incognita* by December, and only in April was reproduction evident. There was low invasion of plants infected in November at Clayton but no females of *M. incognita* developed. J2 of *M. hapla* resumed development in February, and by March 79% became females, which reproduced in April. By comparison, *M. incognita* in plants at Fletcher infected in November failed to survive the winter. Comparative values for basal development temperatures for the two species were surprisingly close: *M. hapla* 9.4 °C and *M. incognita* 9–10 °C.

Survival of eggs in egg masses and hatched J2 of *M. javanica* in a naturally infested soil that had been stored at 24 °C for periods up to 6 months was greatest in the driest soil conditions (Towson and Apt, 1983). Desiccated J2 in a coiled state were most frequently found in the driest storage conditions, and, in the absence of rainfall, these are likely to contribute to the long-term survival of this species. Winter survival of *M. incognita* and *M. arenaria* was studied at nine fallow sites in Texas (Jeger and Starr, 1985a), each of which were classed as loamy sands (containing more than 85% sand), which had previously grown cotton or peanut. At most sites, populations in April were on average only about 9% of those from the previous November, but ranged from 0.4 to 33.0%. This variation could not be easily explained by rainfall or temperature history, but at one site in one winter more survived below 20 cm depth than at the surface. Through the winter the majority of the population existed as eggs but had decreased to an average of 24.0%. By contrast, numbers of J2 increased until mid-January due to hatch of J2, but then declined at rates similar to eggs.

Survival in Texas was less than in California or North Carolina (Jeger and Starr, 1985a), and numbers surviving from larger initial populations were not always as large as from small ones. Hatch of J2 was likely to be the reason for the continuing J2 population found throughout the winter, but there is little evidence for any hatch which is delayed pending an environmental signal. It may be that eggs of both *M. incognita* and *M. arenaria* in Texas can survive the low temperature better than the J2 stage and are principally responsible for survival when the soil is subject to winter rainfall. The results from this study were incorporated into a model (Jeger and Starr, 1985b) that aimed to predict initial populations of *Meloidogyne* available to attack crops planted in spring.

Subsequently, a closer examination was made of the temperature and moisture content (as matrix potential) of soils during winter and the effects that these have on longevity of eggs of *M. incognita* (Starr, 1993). Egg masses produced on glasshouse-grown tomato or cotton were placed in field soil (91% sand) at high levels of saturation and levels around field capacity to simulate a series of winter conditions with declining temperatures over 10–12 weeks. In wet soils, the number of eggs and J2 in browning egg masses from tomato declined faster than in dry soil; by 6 weeks eggs and J2 had experienced three successive temperature decreases to 10 °C. Over the following 4 weeks, with temperatures declining to 5 °C, numbers of viable eggs per egg mass had recovered from a low at 6 weeks of 260 eggs per egg mass to around 420. In egg masses from cotton, a similar population decline during 6 weeks of falling temperatures was seen, but by 12 weeks numbers of eggs per egg mass appeared to have stabilized at around 150–200. Hatching of J2 from eggs exposed to these conditions was found to be greatest at 5 weeks in soil at field capacity, declining quickly thereafter to zero at 12 weeks.

In *in vitro* experiments, Vrain *et al.* (1978) investigated the effects of overwintering temperatures from 10 to 20 °C in soil on the rate of development of eggs in egg masses of *M. incognita* and *M. hapla*. A large proportion of eggs developed abnormally, especially in *M. incognita* (20–30% abnormal or dead at 10–12 °C), perhaps because of chilling injuries due to a lack of adaptation to lower temperatures. *M. hapla* proved more robust, with fewer eggs developing abnormally and with the same proportions abnormal or dead at low and higher temperatures. J2 of *M. hapla* were ready to hatch in 2.3 days at 20 °C (3.4 days at 16 °C), but at 10 and 12 °C development of *M. hapla* was at the early J2 stage by 20.0 days and 8.5 days, respectively. By contrast, *M. incognita* was unable to complete development at 10–12 °C, and at 16 and 20 °C took 5.3 and 6.3 days, respectively. The authors speculated that some eggs may have remained viable while development was suppressed. This may be indicative of a temperature-induced quiescence. These results gave a threshold egg development temperature of 8.26 °C for *M. incognita* and 6.74 °C for *M. hapla*, which may be compared with minimum temperatures for development of J2 stages of 10.08 °C and 8.78 °C, respectively.

Goodell and Ferris (1989) explored normal ranges of temperature and soil moisture to observe the effects on hatch and survival in soil of J2 of *M. incognita*. At the lower temperatures (9–12 °C), replicating late-autumn temperatures, hatch was observed as the environment accumulated 20 degree-days above the temperature minimum of 10 °C. This was contrary to previous indications of reduced motility below 18 °C in work by Roberts *et al.* (1981). Nevertheless, some hatch was possible at 12 °C and a minimum amount occurred at 10 °C. Decreasing percentage hatch was associated with declining moisture content in the sandy loam soil. While suction pressures extending below field capacity can delay eclosion, they will not inhibit development; thus, unhatched J2 will accumulate and only hatch when favourable conditions pertain.

In the absence of a host, populations of *Meloidogyne* in the field decline at different rates, according to temperature and moisture levels. Goodell and Ferris (1989) reported an average rate of decline of 0.2% per degree-days above the developmental minimum of 10 °C. The value of energy reserves may be greater and longer lasting at temperatures below the 'activity threshold' if no great demand is placed on them by inactive juveniles. The practical value of 'wet fallow' as a means of population depletion in the absence of host roots is evident from this work.

9.4.4 A case study investigating factors affecting infectivity of *Meloidogyne javanica* J2

In an approach that integrated almost all the factors that affect survival in the soil, Van Gundy *et al.* (1967) analysed the components contributing to the concept of infectivity, described as the ability successfully to complete the free-living soil phase of the *M. javanica* life history by infecting a host. They assessed: (i) food reserves by quantifying the rate of disappearance of the lipid globules and protein granules stored mainly in the intestine; (ii) motility, using a sand column migration assay; and (iii) physiological age of the J2, by measuring changing respiration rates with time and in conjunction with histochemical tests for esterase and acid phosphatase activity. The experiments were made at 27 °C, the optimum temperature for 'ageing' of J2 in *M. javanica* as revealed by previous work (Thomason *et al.*, 1964).

Van Gundy *et al.* (1967) showed that infectivity began to decline in many J2 within hours of the start of the experiments. Initially (over 4 days) this decline was faster in soils than in water. By 16 days, infectivity of J2 stored in soil fell to 2% (compared with 7% in water), but after 32 days infectivity of both groups was less than 2%; by 64 days infectivity in soil remained at about 1%, whereas no J2 in water was infective. The decline of infectivity in water was closely reflected in the decline of motility (0% after 32 days), and a linear loss of food reserves to zero by 25 days. Lipid loss from J2 kept agitated in a Warburg respirometer fell to near 30% of initial values, while the QO_2 (quotient of oxygen consumption) declined over this period from 6.75 to less than 1.0 μl of O_2 per mg per hour.

Using a range of storage temperatures from 5 to 35 °C, Van Gundy *et al.* (1967) found that J2 of *M. javanica* responded poorly to the lowest and highest temperatures in motility and/or infectivity assays, while food reserves were preserved best at 5 °C but were exhausted at 35 °C. Of the intermediate temperatures (15, 25 and 30 °C), the response at 15 °C showed motility, infectivity and body contents sustained at moderate levels, and gave a remarkable infectivity rate above 20% for more than 40 days. Oxygen regimes *in vitro* ranging from supra-aerobic to anaerobic were tested for up to 24 days at the temperature optimum (27 °C) because lipids require the use of oxygen in their metabolism. In oxygen concentrations of 10%, motility, infectivity and food reserves were not reduced by as much as they were in air over this period, and were even less reduced in 3% oxygen, which equated to survival for up to 60 days. Anaerobiosis spared almost all the food reserves and inhibited motility, although allowing 10% infectivity (one-third of control values) on return to aerobic conditions. Supra-aerobic oxygen levels resulted in total consumption of body contents, with accompanying total loss of motility and infectivity.

Food reserves remained above 30% in wetter soils but were below 5% in drier soils. In either moist or dry soils, food reserves were utilized over 10 days at a faster rate in the presence of roots than when roots were absent, but J2 recovered from roots after 7 days had food reserves at initial levels. It was concluded that much energy was needed to find an invasion site in a root but, having reached a suitable root, invasion itself consumed little energy.

These detailed studies allow conditions for optimum survival of J2 of *M. javanica* to be specified as cool, moist soils with low oxygen and an absence of host roots. Some microniches may exist in many soils, in which some J2 become quiescent and from which they emerge when the presence of roots suitable for invasion is detected.

9.4.5 Overwintering of adult stages

Meloidogyne species attacking annual crops must rely on survival of progeny free in the soil through unfavourable winter or summer conditions, whereas those on perennial crops may have a proportion of the population already infecting hosts and be potentially able, like the hosts, to tolerate suboptimal conditions in dormancy. The extent to which this was possible was explored by Melakeberhan *et al.* (1989) in grapes growing in the central valley of California that were infested with *M. incognita*. By sampling all stages of nematodes from the roots and soil in the vineyard from the onset of dormancy in autumn to bud-break the following spring, it was found that mature females were always present in roots, with numbers declining following the onset of host dormancy at leaf fall. At bud-break, many surviving females were able to increase their egg production.

9.4.6 Diapause in *Meloidogyne naasi*

The requirement for a period of chilling illustrated by *M. naasi* is the only known example in the genus where a period of low temperature exposure is integral to the life cycle for a major proportion of the eggs. Its role is to deny hatch of J2 from eggs experiencing an unseasonably early exposure to hatching temperatures in the environment; most J2 will not emerge into the soil before spring-like conditions have taken hold and host roots are available for invasion. The cold requirement was found to enhance subsequent hatch at 20 °C by Watson and Lownsbery (1970) and Franklin *et al.* (1971). The reason for this was not clear; one proposal was the need to break down hatch inhibitors in the egg, but, as J2 mechanically liberated from the eggshell were infective, it seemed that a quiescence, rather than diapause, was involved (Gooris and d'Herde, 1977). After this phenomenon was further investigated in a population from Wales by Ogunfowora and Evans (1971), and proposed to be a diapause affecting stages in the egg, it was demonstrated that populations from elsewhere behaved similarly (Antoniou and Evans, 1987). Little further work has been done since then, other than a report by Al Zubaidy and Evans (1997) that chilling was also effective in eggs held under anaerobic conditions in the laboratory, reproducing the waterlogged soil in which the nematodes often spend the winter. Photoperiod of host plants (light for 8, 16 or 24 h per day) had no effect on the extent of diapause. The period of chilling was a necessary but not a sufficient requirement for hatch, as conditions unsuitable for emergence could prevail in the soil following the end of this obligate quiescence. Thus after diapause is completed, the nematode remains quiescent until favourable oxygen, moisture and temperatures are experienced, allowing eclosion.

These studies demonstrate that diapause can end but can be followed by a period of quiescence. Diapause in *M. naasi* and other parasitic nematodes has been reviewed by Sommerville and Davey (2002), and they propose mechanisms that may underlie the control of nematode development.

9.4.7 A critique of de Guiran's use of 'diapause' as an explanation of late-emerging J2

De Guiran and colleagues interpreted J2 remaining unhatched after the 3-week-long 'initial hatch' as being in a state of dormancy 'like a diapause'. This seems inappropriate, as the essence of a diapause (see section 9.2) is a dormancy that requires a specific signal *before* members of the population are released into the next phase of development – in this case, eclosion. Those eggs which de Guiran and others have described as being in a state 'like a diapause', appear to be in a state of incomplete development, reaching different stages of embryogenesis at a slow rate, 'little by little over time', and largely unaffected by further external signals, such as chilling or freezing shocks, or changes in O_2 or CO_2 concentrations.

It is important to note that eggs in this condition and responding in this way have been recorded in several species, specifically in other thermophils (de Guiran and Demeure, 1978), and also in different generations in a single host or through the cropping season (Wesemael *et al.*, 2006), where weather conditions or the age of the host have been invoked in explanation. Their importance is to provide a second or third chance for host invasion, leading to enhanced population survival. Page (1984) reported the absence of any eggs in a 'diapause' (*sensu* de Guiran) in the *M. acronea* population she studied, noting that they had been invaded by fungi and destroyed. J2 of *M. naasi* remaining unhatched after their chill/warm exposure remained viable (Ogunfowora and Evans, 1971).

Since diapause is an inappropriate term to apply to these slowly developing, late-maturing eggs, we propose the term *tardicultus* (literally, 'late-developing') be used as descriptive of their condition. Further work on this tardicult state is warranted to quantify its importance in population survival in the field. It may prove to be a particularly important feature of most species in the genus.

9.5 The Effect of Osmotic Stress on Infective Stages in Soil

Under normal environmental conditions, the majority of plant-parasitic nematodes show limited ability for osmotic and ionic regulation (Wright and Newall, 1980; Wright, 1998). Little is known about the ability of *Meloidogyne* to survive osmotic and ionic extremes. Prot (1978) found that *M. javanica* J2 migrated away from areas of high salt concentration. Reversat (1981) found that quiescence was induced in J2 of *M. javanica* by placing them in 0.3 M NaCl; food reserves continued to be utilized but the rate was significantly slower compared with worms not subjected to the osmotic and ionic stress. *Meloidogyne hapla* was intolerant of 0.8 M NaCl (Viglierchio *et al.*, 1969), and infectivity, development and hatch of *M. incognita* were reduced by elevated NaCl and $CaCl_2$ concentrations (Edongali and Ferris, 1981; Edongali *et al.*, 1982). However, Robinson (2002) pointed out the high permeability of nematodes to ions and the deleterious effects that unbalanced salt solutions have on many physiological processes, and he agreed with Wright (1998), who criticized the use of single salt solutions in experiments evaluating the effects of osmotic pressure. Thus, it is difficult to interpret experiments using single salt solutions rather than 'balanced' salt solutions, and it is also difficult to relate data from *in vitro* experiments to the field situation. Tenuta and Ferris (2004) attempted to relate sensitivity of different nematode groups to ions and osmotic stress generated by nitrogenous solutions that were common stressors in agricultural systems, and found that *M. javanica* was more tolerant of stress than omnivorous and predacious nematodes.

It is not clear whether J2 of *Meloidogyne* are liable to a loss of essential inorganic mineral salts in a soil with an excess of water or, conversely, whether a dry soil allows the retention of salts while imposing greater stress due to potential loss of water to the more concentrated soil solution. Such a loss would be partially compensated for by the yield of metabolic water in conversion of fatty acids to carbon dioxide and water by their complete oxidation, which, in the case of each 18-carbon palmytic acid molecule consumed, yields 23 water molecules. In many situations this may be a critically important product of the food reserves.

When investigating the effect of soil salinity and an infection of *M. javanica* on growth of tomatoes, Maggenti and Hardin (1973) found that plant height and number of galls per root system were reduced as the salinity increased. Their data indicated that about one-third of J2 in the experimental inoculum were able to osmoregulate, thereby adjusting to the salinity and surviving in the soil long enough to infect the plants and form galls.

Robinson and Carter (1986) found that J2 of *M. incognita* remained mobile and retained volume at 100 m osmol/kg (−2.5 bar) polyethylene glycol solution supplemented with balanced salts. When nematodes were exposed to the respiratory inhibitor sodium cyanide to detect any respiration-dependent processes that regulate volume, these authors found lethal effects of solutions of osmotic pressure that were innocuous in aerobic conditions. This indicates that respiration is essential to survive changes in water potential and control nematode volume.

9.6 Survival Mechanisms Deployed: Life History Strategies in *Meloidogyne* Species

The structures and behaviours outlined in the foregoing analysis of survival mechanisms possessed by those few intensively studied species together form an integrated and coherent strategy by which the nematode can continue to reproduce and adapt in its own ever-changing environment. As sophisticated parasites of few or many host plants, each species already carries those adaptations common to the genus, but others have evolved to be deployed in a species-specific **life history strategy (LHS)**.

One basis for analysing and comparing the factors affecting individual species is to refer to their geographical dispersion, particularly relating to climatic temperatures. This approach is not new, having been used in the pioneering studies of Tyler (1933), when she recognized that the energy needed for nematode development was the product of time and temperature (above the minimum temperature needed for development), and was measured in **degree-days**. In its modern guise it has become 'thermal time', championed by Trudgill and colleagues (Trudgill *et al.*, 2005), of which the following is a brief summary.

Poikilothermic organisms (whose body temperature is governed by their environment) function within the range of temperatures to which they have adapted. The temperature below which development is not possible is termed the **base temperature (T_b)**, above which development will occur and can be expressed in the sum of heat units or the **heat sum (S)**, measured in degree-days above the base temperature, and sometimes referred to as the thermal constant or physiological time. Development rates produce a linear plot as temperature rises, until an upper limit is reached, termed the **optimum temperature** (**T_o**). Beyond this temperature, rates of development again decline, until they cease at the **maximum temperature** (**T_m**) at which the organism can function, giving a plot that has been described as a wigwam shape. It has been shown for many organisms, including species of *Meloidogyne*, that the linear rate of development can be back-projected to zero, where it intersects at the basal temperature.

Thermal time studies can offer explanations of how T_b and S influence competitive abilities between species, or how intraspecific 'thermotypes' (populations showing different T_b and/or T_o values) can be found from different places within the species range. They may also inform the building of population models and the use of solarization as a control measure for nematodes in the field.

The values for T_b have been measured for several of the most common thermophil species of *Meloidogyne*, and, together with records of geographical distribution (Whitehead, 1969) and climatic measures (temperature means and ranges, and rainfall data; Sasser et al., 1980, 1983), allow the responses of different species to be understood in terms of their LHS.

One useful approach to categorizing the LHS among species has been to analyse which characteristics have been selected to maximize fitness in the species of interest. Traits favouring a fast rate of development (**r**), together with short generation times, may be contrasted with those favouring a slower rate of development but allowing a population to grow to a large size, close to the environment's carrying capacity (**K**), and indicate the type of resource allocation that evolves with different lifestyles. Environmental sex determination (Trudgill, 1972) is another r-selected feature that is brought into play whenever environmental stress is critical. The r-selected strategy achieves fitness when a small parent maximizes reproduction whenever and wherever conditions are favourable in a variable, even ephemeral, environment, and is contrasted with a K-selected strategy, where a parent may grow to a large size before producing a few well-adapted progeny in a habitat that is frequently stable and more predictable. In practice, these strategies are recognized as being part of a spectrum of characteristics, which tend towards either **r** – or **K** – selection and in which co-related factors are recognized (Pianka, 1970).

These concepts have been incorporated into nematology with soil fauna analyses, e.g. the maturity index of Bongers (1990), where the 'colonizer–persister' (c–p) scale rating of a taxon reflects much of its r–K characteristics. As a way of life, parasitism gives many examples where r-selected traits like prodigious egg production appear to predominate, but this is not always the case (Esch *et al.*, 1977). In some instances, multiple solutions to the problems posed by the free-living phase are available to parasites as a form of 'bet-hedging'. In *Meloidogyne*, all species share common features in their biology which facilitate rapid development and reproduction: the development to J1 completed in the egg, from which the J2 emerges to invade and establish in the host, before omitting feeding by the J3 and J4, after which the mature female (often parthenogenetic) rapidly begins egg laying (Trudgill and Perry, 1994). Most species also produce a proportion of tardicult eggs, whose late development can leave a small residual population of J2 to emerge when all others have hatched.

Among the closely related thermophil species there are populations and thermotypes that differ in their observed T_b and S values, and these species are now briefly reviewed to explore how survival mechanisms are integrated in LHS.

9.6.1 *Meloidogyne javanica*

Meloidogyne javanica comprised 31% of the samples taken from many countries (Sasser *et al.*, 1983) and was the species apparently best adapted for climates with distinct wet and dry seasons, and least represented where there were no dry months. *Meloidogyne javanica* is present when the average temperature is at least 3 °C for the coldest month, and up to maximum of 36 °C. It is found where rainfall is well distributed though the year but can tolerate 4 or 5 months that lack rainfall. It has a preference for soils low in silt or clay.

Hatch is rapid, with 45–50% of the J2 emerging after 6 days in suitable conditions, and mobility in soil is greatest at 25 °C (80%) but lower at 30 °C (20%) (Bird and Wallace, 1965). Similar results were found by Thomason (1962) in moist, well-aerated soil, where, at temperatures of 24–35 °C, J2 emerged rapidly and available plant roots were invaded within 4 days; Arens *et al.* (1981) found that the J2 became swollen as they established in the root, producing large galls. This rapid invasion may be crucial in marginal environments with short windows of opportunity for host infection. Should soil conditions for the J2 become suboptimal due to the soil slowly drying out over 5 days, the juvenile may employ the tactic of entering a quiescence in a desiccated state, potentially allowing survival for several weeks or months (Towson and Apt, 1983). Thereafter, host suitability and durability in the environment may allow successful reproduction by mitotic parthenogenesis, so that at 20 days after invasion 73% of females had an average of 86 eggs per female, increasing quickly to 270 eggs by 25 days (Arens *et al.*, 1981).

The response of a Tanzanian population of *M. javanica* in terms of thermal time for a life cycle has been analysed by Madulu and Trudgill (1994), who found a T_b value of 12.9 °C and an S of about 350 degree-days. The species has an effective temperature range starting at ~15 °C and increasing to a T_m of ~35 °C and a T_o of ~30 °C. When thermal time requirement for embryogenesis of the Tanzanian population was compared with a population from Crete, the T_b and S values for the two populations were almost identical (Tzortzakakis and Trudgill, 2005). Davila and Dickson (2004) reported a Florida population with T_b = 10.2 °C (and for development from J2 to females with eggs only, S = 379 degree-days). Davila *et al.* (2005) confirmed a T_b = 10.2 °C for the Florida population under black polythene mulch, but development (also from J2 to females with eggs only) was decreased to S =320 degree-days.

This combination of attributes allows *M. javanica* to thrive on almost any host it encounters, accommodating intermittently favourable periods to complete its development in minimum time and produce several hundreds of progeny. Immediate and rapid egg embryonation means that a succession of J2 will be ready to hatch during favourable windows of opportunity for invasion, offering the chance of persistence in different environments at variable densities. It has adaptations for survival through periods of unpredictable duration, not least among which are the tardicult eggs, allowing it to remain in marginal habitats over long periods or rapidly to re-colonize after surviving several poor seasons if the chance arises. It shows strong r-selected characteristics.

9.6.2 *Meloidogyne arenaria*

Meloidogyne arenaria comprised 8% of the populations examined by Sasser *et al.* (1983) and were 68% of populations received from climates where there were no dry months in the year. The similarities between *M. arenaria* and *M. incognita* have been noted when comparisons are made between species. This is not surprising, given that the differences are small and environmental effects can disguise differences that might be more obvious in other conditions. They produce large and often compound galls on their hosts.

The study by Thomason and Lear (1961) is revealing as to the maximum soil temperatures that were tolerated by Californian populations of *M. arenaria*. *Meloidogyne arenaria thamesi* produced egg masses at a soil temperature of 35 °C, whereas '*M. arenaria arenaria*' reproduced on the heat-tolerant host *Sesbania* at a soil temperature of 32.6 °C but failed at 36.5 °C, unlike *M. javanica*. The authors noted that these temperature responses placed *M. arenaria* between *M. javanica* and *M. incognita*.

Meloidogyne arenaria had smaller galls than *M. javanica* (Arens *et al.*, 1981), and invasion rates and the rate of development of J2 inside the root were much slower, but fecundity rates were similar until 35 days, when *M. javanica* had more eggs in egg masses and a greater percentage of females with egg masses. The thermal time for embryonation in a Californian population of *M. arenaria* was measured at T_b = 10.2 and S = 176 degree-days (Ferris *et al.*, 1978). When developing a simulation model of a Californian vineyard population (egg–J2), the T_b was estimated at 10.11 °C for egg development. Hatching tests demonstrated that 24% of the J2 in egg masses remained unhatched, and a 'cold shock' had no effect on the rate of hatch or final hatch after a further 16 weeks. Davila and Dickson (2004) reported a Florida population with a T_b = 8.8 °C (and for development from J2 to mature females with eggs only, S = 386 degree-days). Davila *et al.* (2005) confirmed the T_b = 8.8 °C for this population under polythene mulch (and for development from J2 to females with eggs only, S = 343 degree-days).

De Guiran and Villemin (1980) noted three French populations had <10% of the eggs in egg masses in the state of arrest they called 'diapause', but this increased in eggs from different hosts, depending on the population. In comparing *M. arenaria* with *M. javanica* and *M. incognita*, Arens *et al.* (1981) found a slow initial rate of three J2 invading per root system of tobacco over the first 2 days, increasing to 27 J2 by 4 days, which was a median number between the 128 *M. javanica* and eight *M. incognita* invading during the same time. Also, the galls of *M. arenaria* were intermediate in size and the females produced eggs as rapidly as *M. javanica* within 20 days, but by 35 days the numbers were less. Arens *et al.* (1981) rated *M. arenaria* as 'moderately aggressive' in invasion rate and gall induction. It may be speculated that it shows moderately r-selected features, with a preference for warm temperatures and possibly perennial hosts.

9.6.3 *Meloidogyne incognita*

Meloidogyne incognita comprised 51% of the populations examined by Sasser *et al.* (1983). It was reported frequently (46% of samples) from habitats with no dry months in the year, with fewer (8%) from climates with 4 or 5 dry months each year and, together with *M. javanica*, its occurrence had a positive correlation with high soil pH. It was often found where *M. javanica* was also present but with characteristics suggesting it is better suited to less extreme conditions.

The optimum warm-month temperature for *M. incognita* was 27 °C, but among several studies that were made on Californian isolates, Bergeson (1959) found embryonation was halted by 0–4 °C for longer than 12 days, but J2 survived 10 °C well, with up to 20% of a population alive after 12 months. Populations reported by Thomason and Lear (1961) preferred moist sandy loam, in which J2 easily find host roots whose internal environment is more secure for overwintering than soil (Thomason, 1962). The Californian isolate studied by Ploeg and Maris (1999) indicated a T_b of 10.1 °C and an S of 400 degree-days, qualities that help it to persist in the moist, often lowland tropics, but equally useful to tolerate periods of lower temperatures in a seasonal environment. For many populations, life cycles between 20 and 30 °C will be longer than those of *M. javanica* (Trudgill *et al.*, 2005)

Dao (1971) recognized thermotypes within *M. incognita* in differences between The Netherlands population in a temperate climate and that from the seasonally wet and dry tropics of Venezuela. Vrain *et al.* (1978) showed an isolate from North Carolina had a T_b = 10.08 °C and S = 354 degree-days. Hatch was possible between 5 and 35 °C for 15–20 days but at 15 °C hatch could continue for 47 days. At 20–30 °C the rate of infection was relatively slow compared with *M. javanica* (Arens *et al.*, 1981). Davila *et al.* (2005) reported a T_b = 9.3 °C for a Florida population under a black polythene mulch and S = 307 degree-days (for development from J2 to females with eggs only).

The behaviour observed by Arens *et al.* (1981) showed slow invasion rates at 26 °C, and evidence for mass invasion is frequently reported. Differences in invasion and egg production on hosts of different suitability were measured by Anwar *et al.* (1994). The adaptability of this species ensures that some reproduction is usually possible, even on poor hosts, whereas the high invasion rates on favourable hosts produce overcrowding and a higher incidence of sex-reversed males; such density-dependent competition may be the selection pressure for forming races. These indications suggest that *M. incognita* populations generally may be best adapted to slower rates of reproduction and may be more K-selected in a more predictable environment than is exploited by the warmer-adapted *M. javanica* and *M. arenaria*.

9.6.4 *Meloidogyne hapla*

Long known as the 'northern' root-knot nematode, *M. hapla* comprised only 8% of the populations received by Sasser *et al.* (1983). It has a significant presence in warmer latitudes where rainfall is well distributed throughout the year, as well as in places having seasonally cold winters and warm, wet summers, and it may be limited to sites with average annual temperatures of 24–27 °C. It is also present at elevations adjacent to humid lowland tropical locations (Whitehead, 1969).

The ability of *M. hapla* to survive low temperature, both as eggs and as J2, was noted by Bergeson (1959), using two Californian populations, and was confirmed in both the laboratory and the field by Daulton and Nusbaum (1961) with a North Carolinian isolate that was more susceptible to soil temperatures of 36–40 °C in dry soil. Eggs developed slowly at 10–12 °C but more rapidly at 16–20 °C. A number were retained in 'suspended development', with development resuming later.

Several important features of *M. hapla* were revealed by Bird and Wallace (1965) in comparison with that of *M. javanica* under identical conditions. The hatch of J2 of *M. hapla* at 20 °C (80%) was greater than that of *M. javanica* (50%), but at 25 °C the percentages were reversed. Invasion rate of *M. hapla* was maximal in the optimum temperature range of 15–20 °C, compared with *M. javanica* at 20–25 °C. Over 28 days, growth rate of *M. hapla* biomass at 20–25 °C was double that of *M. javanica* at the same temperature (or at 25–30 °C), and *M. hapla* produced 39 eggs per egg mass over this time compared with 28 by *M. javanica.*

Thermal time details were measured by Vrain *et al.* (1978) at a range of temperatures, revealing a T_b = 8.8 °C and S = 350 degree-days. Embryos were sensitive to freezing temperatures, but 90% of the unhatched J2 survived −4 °C for 10 days, and egg development had a threshold temperature of 6.74 °C. More J2 of *M. hapla* than *M. incognita* migrated towards host roots when their hosts experienced a 12h/12h light/dark photoperiod over 7 days (Prot and Van Gundy, 1981). In Finland, a *M. hapla* population was estimated to have a T_b of 8.25 °C and S = 553 degree-days minimum to complete the life cycle from J2 to J2 (Lahtinen *et al.*, 1988). Populations from Sweden and Finland gave similar values.

The freezing tolerance mechanism documented to exist in this species has been mentioned previously, but whether low temperature treatments prior to freezing could increase tolerance was the question Forge and MacGuidwin (1990) sought to answer. Evidence for induced tolerance was found but could be partially lost during later exposure to higher temperatures.

The behaviour of *M. hapla* in relation to other species of *Meloidogyne* is in accord with its observed geographical distribution (Madulu and Trudgill, 1994; Ploeg and Maris, 1999). It is clear that *M. hapla* at its optimum temperature behaves in a different way to *M. javanica.* By allocating more of its resources to growing a larger body and limiting early egg production so that it invests in an ability eventually to produce a larger number of eggs per female, it is possible to speculate that *M. hapla* shows many aspects of a K-selected LHS, thus efficiently exploiting its environment.

9.7 Conclusions and Future Directions

The categories of dormancy (see section 9.2) outlined by Evans and Perry (1976) and modified by Evans (1987) have been widely, if not always critically, applied and have been found sufficiently robust for use in other fields (Alexseev *et al.*, 2007). It is some time since diapause was explored by plant nematologists and it awaits the ministrations of molecular biologists. In spite of the descriptions of several species of *Meloidogyne* that reproduce in seasonal environments with cool or cold winters, none has been found sharing requirements similar to those reported by Ogunfowora and Evans (1971) and Antoniou and Evans (1987) for *M. naasi.* However, the temperature requirements of the majority of the 97 species of *Meloidogyne* (as of June 2009) have yet to be appropriately examined. In a review on diapause in parasitic nematodes (Sommerville and Davey, 2002), *M. naasi* together with *Globodera rostochiensis* and *Heterodera avenae* were recognized as best exemplifying the features of dormancy to be found in plant-parasitic nematodes.

Since the study of *Caenorhabditis elegans* mutants have become the method of choice in explorations of dauer-stage differences (either lacking the ability to enter the dauer stage or being unable to avoid entering it), much progress has been made; the *daf*-gene pathways have been followed successfully and revealed the interactions by which one or more of three dauer-forming pheromones ('Daumones') is sensed by the J1 stage of the worm. Its reception of the signal leading to 'the dauer diapause' in the J2 involves a network of further signal and receptor molecules coordinating the entry into this 'active' quiescence in the third juvenile stage, which, on the receipt of appropriate environmental signals, then resumes development, leading to maturity and reproduction (Fielenbach and Antebi, 2008) These pathways are seen to be so interactive that they are involved in all aspects of the life of the worm, controlling not only development but also longevity, growth and fecundity. They are highly conserved and their homologues in more elaborated forms are present in higher animal forms, including mammals.

With the use of modern biotechnological tools, the equivalent key pathways and gene products may now be sought in plant-parasitic nematodes. Sommerville and Davey recognized as recently as 2002 that 'there were a few straws in the wind' offered for parasitologists by work on *C. elegans*. In a few short years these straws have become so numerous as to afford a veritable cornucopia of new knowledge of how multicellular animals work. This is likely to mean that the *daf*-gene homologues are at work in the plant-parasitic nematodes, no doubt controlling diapause mechanisms as well as being responsible for the tardicult state of late-developing embryos. Both of these features can be seen as important measures and mechanisms in a life history to extend the chances for one generation bridging the often hazardous free-living phase, until their progeny find a host for creating their own new generation. Having a complete sequence of the genomes of *M. hapla* and *M. incognita* (see Abad and Opperman, Chapter 16, this volume) now opens the way for rapid progress in our understanding of the pathways leading to diapause in *Meloidogyne*.

9.8 References

Alexseev, V.R., DeStasio, B.T. and Gilbert, J.J. (2007) *Diapause in Aquatic Invertebrates. Theory and Human Use*. Springer, Dordtrecht, The Netherlands.

Al Zubaidy, M.A.J. and Evans, A.A.F. (1997) Effects of an aerobic chilling and day length of maternal host plant on diapause in eggs of *Meloidogyne naasi. Journal of Nematology* 29, 567.

Antoniou, M. and Evans, A.A.F. (1987) Diapause in *Meloidogyne naasi* eggs. The effect of constant temperature on subsequent hatch. *Nematologica* 33, 186–198.

Anwar, S.A., Trudgill, D.L. and Phillips, M.S. (1994). The contribution of variation in invasion and development rates of *Meloidogyne incognita* to host status differences. *Nematologica* 40, 579–586.

Arens, M.L., Rich, J.R. and Dickson, D.W. (1981) Comparative studies on root invasion, root galling, and fecundity of three *Meloidogyne* spp. on a susceptible tobacco cultivar. *Journal of Nematology* 13, 201–205.

Barker, K.R. (1985) Nematode extraction and bioassays. In: Barker, K.R. Carter, C.C. and Sasser, J.R. (eds) *An Advanced Treatise on* Meloidogyne. *Volume II: Methodology.* North Carolina State University Graphics, Raleigh, North Carolina, pp. 19–35.

Baxter, R.I. and Blake, C.D. (1969a) Some effects of suction on the hatching of eggs of *Meloidogyne javanica. Annals of Applied Biology* 63, 183–190.

Baxter, R.I. and Blake, C.D. (1969b) Oxygen and the hatch of eggs and migration of larvae of *Meloidogyne javanica. Annals of Applied Biology* 63, 191–203.

Bergeson, G.R. (1959) The influence of temperature on the survival of some species of *Meloidogyne. Nematologica* 4, 344–354.

Bird, A.F. and Soeffky, A. (1972). Changes in the ultrastructure of the gelatinous matrix of *Meloidogyne javanica* during dehydration. *Journal of Nematology* 4, 166–169.

Bird, A.F. and Wallace, H.R. (1965) The influence of temperature on *Meloidogyne hapla* and *M. javanica. Nematologica* 11, 581–589.

Bongers, T. (1990) The maturity index: an ecological measure of environmental disturbance based on nematode species composition. *Oecologia* 83, 14–19.

Dao, D.F. (1971) Climatic influence on distribution pattern of plant parasitic and soil inhabiting nematodes. *Mededelingen Landbouwhogeschool, Wageningen* No. 70-2.

Daulton, R.A.C. and Nusbaum, C.J. (1961) The effect of soil temperature on the survival of the root-knot nematodes *Meloidogyne javanica* and *M. hapla. Nematologica* 6, 280–294.

Davila, M. and Dickson, D.W. (2004) Base temperature and heat unit requirements for development of *Meloidogyne arenaria* and *Meloidogyne javanica. Journal of Nematology* 36, 314.

Davila, M., Allen, L.H. and Dickson, D.W. (2005) Influence of soil temperatures under polythene mulch and bare soil on root-knot nematode egg laying. *Journal of Nematology* 37, 364–365.

de Guiran, G. (1979a) Survie des nématodes dans les sols seds et saturés d'eau: oeufs et larvae de *Meloidogyne incognita. Revue de Nématologie* 2, 65–77.

de Guiran, G. (1979b) A necessary diapause in root-knot nematodes. Observations on its distribution and inheritance. *Revue de Nématologie* 2, 223–231.

de Guiran, G. (1980) Facteurs induisant chez *Meloidogyne incognita* un blocage du développement des œufs considéré comme une diapause. *Revue de Nématologie* 3, 61–69.

de Guiran, G. and Demeure, Y. (1978) Influence du potentiel hydrique des sols sur les masses d'œufs de *Meloidogyne incognita* (Nematoda :Meloidogynidae). *Revue de Nématologie* 1, 119–134.

de Guiran, G. and Villemin, M.-A. (1980) Spécificité de la diapause embryonnaire des œufs de *Meloidogyne* (Nematoda). *Revue de Nématologie* 3, 115–121.

Demeure, Y. and Freckman, D.W. (1981) Recent advances in the study of anhydrobiotic nematodes. In: Zuckerman, B.M. and Rohde, R.F. (eds) *Plant Parasitic Nematodes*, Volume 3. Academic Press, London, pp. 205–226.

Edongali, E.A. and Ferris, H. (1981) Effect of salinity and temperature on reproduction and egg hatching of *Meloidogyne incognita* on tomato. *Nematologica Mediterranea* 9, 123–132.

Edongali, E.A., Duncan, L. and Ferris, H. (1982) Influence of salt concentration on infectivity and development of *Meloidogyne incognita* on tomato. *Revue de Nématologie* 5, 111–117.

Esch, G.W., Hazen, T.C. and Aho, J.M. (1977) Parasitism and r- and K-selection. In: Esch, G.W. (ed.) *Regulation and Parasite Populations*. Academic Press, New York, pp. 9–62.

Evans, A.A.F. (1987) Diapause in nematodes as a survival strategy. In: Veech, J.A. and Dickson, D.W. (eds) *Vistas on Nematology*. Society of Nematologists, Hyattsville, Maryland, pp. 180–187.

Evans, A.A.F. and Perry, R.N. (1976) Survival strategies in nematodes. In: Croll, N.A. (ed.) *The Organisation of Nematodes*. Academic Press, London, pp. 383–424.

Ferris, H., Du Vernay, S. and Small, R.H. (1978) Development of a soil temperature data base on *Meloidogyne arenaria* for a simulation model. *Journal of Nematology* 10, 39–42.

Fielenbach, N. and Antebi, A. (2008) *C. elegans* dauer formation and the molecular basis of plasticity. *Gene and Development* 22, 2149–2165.

Forge, T.A. and MacGuidwin, A.E. (1990) Cold hardening of *Meloidogyne hapla* second-stage juveniles. *Journal of Nematology* 22, 101–105.

Franklin, M.T., Clark, S.A. and Course, J.A. (1971) Population changes and development of *Meloidogyne naasi* in the field. *Nematologica* 17, 575–590.

Gaur, H.S. and Perry, R.N. (1991) The use of soil solarization for control of plant parasitic nematodes. *Nematological Abstracts* 60, 153–167.

Goodell, P.B. and Ferris, H. (1989) Influence of environmental factors on the hatch and survival of *Meloidogyne incognita. Journal of Nematology* 21, 328–334.

Gooris, J. and d'Herde, C.J. (1977) *Study on the Biology of* Meloidogyne naasi *Franklin 1965*. State Agricultural Research Centre, Gent, Belgium.

Huang, S.P. and Pereira, A.C. (1994) Influence of inoculum density, host, and low-temperature period on delayed hatch of *Meloidogyne javanica* eggs. *Journal of Nematology* 26, 72–75.

Ishibashi, N. and Kunii, Y. (1963) Studies on the egg production of the root-knot nematode, *Meloidogyne incognita* var. *acrita* Chitwood. 1. The reduction of the egg productivity of female adults accompanied by coloration and hardening of the gelatinous substance of the egg masses. *Japanese Journal of Zoology* 72, 217–223.

Ishibashi, N. Kegasawa, K. and Kunii, Y. (1964) Studies on the egg production of the root-knot nematode, *Meloidogyne incognita* Chitwood. 2. Effects of the growth condition of the host plant on the propagation of the nematode. *Japanese Journal of Applied Entomology and Zoology* 8, 245–249.

Jeger, M.J. and Starr, J.L. (1985a) Dynamics of the winter survival of eggs and juveniles of *Meloidogyne incognita* and *M. arenaria*. *Journal of Nematology* 17, 252–256.

Jeger, M.J. and Starr, J.L. (1985b) A theoretical model of the winter survival dynamics of *Meloidogyne* spp. eggs and juveniles. *Journal of Nematology* 17, 257–260.

Kutywayo, V. and Been, T.H. (2006) Host status of six major weeds to *Meloidogyne chitwoodi* and *Pratylenchus penetrans*, including a preliminary field survey concerning other weeds. *Nematology* 8, 647–657.

Lahtinen, A.E., Trudgill, D.L. and Tiilikkala, K. (1988) Threshold temperature and minimum thermal time requirements for the complete life cycle of *Meloidogyne hapla* from Northern Europe. *Nematologica* 34, 443–451.

Lyons, J.M., Keith, A.D. and Thomason, I.J. (1975) Temperature-induced phase transitions in nematode lipids and their influence on respiration. *Journal of Nematology* 7, 98–104.

Madulu, J. and Trudgill, D.L. (1994) Influence of temperature on *Meloidogyne javanica. Nematologica* 40, 230–243.

Maggenti, A.R. and Hardin, A. (1973) The effects of soil salinity and *Meloidogyne javanica* on tomato. *Journal of Nematology* 5, 231–233.

Melakeberhan, H., Ferris, H., McKenry, M.V. and Gaspard, J.T. (1989) Overwintering stages of *Meloidogyne incognita* in *Vitis vinifera*. *Journal of Nematology* 21, 92–98.

Ogunfowora, A.O. and Evans, A.A.F. (1971) Factors affecting the hatch of eggs of *Meloidogyne naasi*, an example of diapause in a second-stage larva. *Nematologica* 23, 137–146.

Orion, D. and Frank, A. (1990) An electron microscopy study of cell wall lysis by *Meloidogyne javanica* matrix. *Revue de Nématologie* 13, 105–107.

Orion, D., Kritzman, G., Meyer, S.L.F., Erbe, E.F. and Chitwood, D.J. (2001) A role of the gelatinous matrix in the resistance of root-knot nematode (*Meloidogyne* spp.) eggs to microorganisms. *Journal of Nematology* 33, 203–207.

Page, S.L.J. (1984) Effects of physical properties of two tropical cotton soils on the permanent wilting points and relative humidity, in relation to survival and distribution of *Meloidogyne acronea*. *Revue de Nématologie*, 7, 227–232.

Papert, A. and Kok, C.J. (2000) Size and community metabolic profile of the bacterial population of the *Meloidogyne hapla* egg masses. *Nematology* 2, 581–584.

Perry, R.N. (1989) Dormancy and hatching of nematode eggs. *Parasitology Today* 5, 377–383.

Perry, R.N. and Wesemael, W. (2008) Host plant effects on hatching of root-knot nematodes. *Russian Journal of Nematology* 16, 1–5.

Perry, R.N., Knox, D.P. and Beane, J. (1992) Enzymes released during hatching of *Globodera rostochiensis* and *Meloidogyne incognita*. *Fundamental and Applied Nematology* 15, 283–288.

Pianka, E.R. (1970) On 'r' and 'K' selection. *American Naturalist* 104, 592–597.

Ploeg, A.T. and Maris, P.C. (1999) Effects of temperature on the duration of the life cycle of a *Meloidogyne incognita* population. *Nematology* 1, 389–393.

Ploeg, A.T. and Stapleton, J.J. (2001) Glasshouse studies on the effects of time, temperature and amendment of soil with broccoli plant residues on the infestation of melon plants by *Meloidogyne incognita* and *M. javanica*. *Nematology* 3, 855–861.

Prot, J.C. (1978) Vertical migration of four natural populations of *Meloidogyne*. *Revue de Nématologie* 1, 109–112.

Prot, J.C. and Van Gundy, S.D. (1981) Influence of photoperiod and temperature on migration of *Meloidogyne* juveniles. *Journal of Nematology* 13, 217–220.

Reversat, G. (1981) Consumption of food reserves by starved second-stage juveniles of *Meloidogyne javanica* under conditions inducing osmobiosis. *Nematologica* 27, 207–214.

Roberts, P.A., Van Gundy, S.D. and McKinney, H.E. (1981) Effects of soil temperature and planting date of wheat on *Meloidogyne incognita* reproduction. *Journal of Nematology* 13, 338–345.

Robinson, A.F. (2002) Soil and plant interactions' impact on plant-parasitic nematode host finding and recognition. In: Lewis, E.E., Campbell, J.F. and Sukhdeo, M.V.K. (eds) *The Behavioural Ecology of Parasites*. CAB International, Wallingford, UK, pp. 89–110.

Robinson, A.F. and Carter, W.W. (1986) Effects of cyanide ion and hypoxia on the volumes of second-stage juveniles of *Meloidogyne incognita* in polyethylene glycol solutions. *Journal of Nematology* 18, 563–570.

Santmeyer, P.H. (1955) A comparison of the thermal death time of two dissimilar species of nematodes: *Panagrellus redivivus* (Linn. 1767) Goodey 1945, and *Meloidogyne incognita* var. *acrita* Chitwood, 1949. *Proceedings of the Helminthological Society of Washington* 22, 22–25.

Sasser, J.N., Taylor, A.L. and Nelson, L.A. (1980) Ecological factors influencing survival and pathogenicity of *Meloidogyne* species. *Journal of Nematology* 12, 237.

Sasser, J.N., Eisenbach, J.N. and Carter, C.C. (1983) The international *Meloidogyne* project – its goals and accomplishments. *Annual Review of Phytopathology* 21, 271–288.

Sayre, R.M. (1964) Cold-hardiness of nematodes. I. Effects of rapid freezing on the eggs and larvae of *Meloidogyne incognita* and *M. hapla*. *Nematologica* 10, 168–179.

Sommerville, R.I. and Davey, K.G. (2002) Diapause in parasitic nematodes: a review. *Canadian Journal of Zoology* 80, 1817–1840.

Stapleton, J.J. and Duncan, R.A. (1998) Soil disinfestations with cruciferous amendments and sublethal heating: effects on *Meloidogyne incognita*, *Sclerotium rolfsii* and *Pythium ultimum*. *Plant Pathology* 47, 737–742.

Starr, J.L. (1993) Recovery and longevity of egg masses of *Meloidogyne incognita* during simulated winter survival. *Journal of Nematology* 25, 244–248.

Stirling, G.R. (1991) *Biological Control of Plant-Parasitic Nematodes*. CAB International, Wallingford, UK.

Tenuta, M. and Ferris, H. (2004) Sensitivity of nematode life-history groups to ions and osmotic stress tensions of nitrogenous solutions. *Journal of Nematology* 36, 85–94.

Thomason, I.J. (1962) Reaction of cereals and sudan grass to *Meloidogyne* spp. and the relation of soil temperature to *Meloidogyne javanica* population. *Phytopathology* 52, 787–791.

Thomason, I.J and Lear, B. (1961) Rate of reproduction of *Meloidogyne* spp. as influenced by soil temperature. *Phytopathology* 51, 520–524.

Thomason, I.J., Van Gundy, S.D. and Kirkpatrick, J.D. (1964) Motility and infectivity of *Meloidogyne javanica* as affected by storage time and temperature in water. *Phytopathology* 54, 192–195.

Timper, P., Davis, R.F. and Tillman, P.G. (2006) Reproduction of *Meloidogyne incognita* on winter clover crops used in cotton production. *Journal of Nematology* 38, 83–89.

Towson, A.J. and Apt, W. (1983) Effects of soil water potential on survival of *Meloidogyne javanica* in fallow soil. *Journal of Nematology* 15, 110–115.

Trudgill, D.L. (1972) Influence of feeding duration on moulting and sex determination of *Meloidogyne incognita. Nematologica* 18, 476–481.

Trudgill, D.L. and Perry, J.N. (1994) Thermal time and ecological strategies – a unifying hypothesis. *Annals of Applied Biology* 125, 521–532.

Trudgill, D.L., Honek, A., Li, D. and Van Straalen, N.M. (2005) Thermal time – concepts and utility. *Annals of Applied Biology* 146, 1–14.

Tyler, J. (1933) Development of the root-knot nematode as affected by temperature. *Hilgardia* 7, 374–415.

Tzortzakakis, E.A. and Trudgill, D.L. (2005) A comparative study of the thermal time requirements for embryogenesis in *Meloidogyne javanica* and *M. incognita. Nematology* 7, 313–315.

Van Gundy, S.D. (1985) Ecology of *Meloidogyne* spp. – emphasis on environmental factors affecting survival and pathogenicity. In: Sasser, J.N. and Carter, C.C. (eds) *An Advanced Treatise on* Meloidogyne. North Carolina State University Graphics, Raleigh, North Carolina, pp. 177–182.

Van Gundy, S.D., Bird, A.F. and Wallace, H.R. (1967) Aging and starvation in larvae of *Meloidogyne javanica* and *Tylenchulus semipenetrans. Phytopathology* 57, 559–571.

Viglierchio, D.R., Croll, N.A. and Gortz, J.H. (1969) The physiological response of nematodes to osmotic stress and an osmotic treatment for separating nematodes. *Nematologica* 15, 15–21.

Vrain, T.C., Barker, K.R. and Holtzman, G.I. (1978) Influence of low temperature on rate of development of *Meloidogyne incognita* and *M. hapla* larvae. *Journal of Nematology* 10, 166–171.

Wallace, H.R. (1966) Factors influencing the infectivity of plant-parasitic nematodes. *Proceedings of the Royal Society of London, Series B* 164, 592–614.

Wallace, H.R. (1968a) The influence of soil moisture on the survival and hatch of *Meloidogyne javanica. Nematologica* 14, 231–242.

Wallace (1968b) The influence of aeration on the survival and hatch of *Meloidogyne javanica. Nematologica* 14, 223–230.

Wang, K.-H. and McSorley, R. (2008) Exposure times to lethal temperatures for *Meloidogyne incognita* suppression and its implications for soil solarization. *Journal of Nematology* 40, 7–12.

Watson, T.R. and Lownsbery, B.R. (1970) Factors influencing the hatching of *Meloidogyne naasi*, and a comparison with *M. hapla. Phytopathology* 60, 457–460.

Wesemael, W.M.L., Perry, R.N. and Moens, M. (2006) The influence of root diffusate and host age on hatching of the root-knot nematodes, *Meloidogyne chitwoodi* and *M. fallax. Nematology* 8, 895–902.

Whitehead, A.G. (1969) The distribution of root-knot nematodes (*Meloidogyne* spp.) in tropical Africa. *Nematologica* 15, 315–333.

Wright, D.J. (1998) Respiratory physiology, nitrogen excretion and osmotic and ionic regulation. In: Perry, R.N. and Wright, D.J. (eds) *The Physiology and Biochemistry of Free-living and Plant-parasitic Nematodes*. CAB International, Wallingford, UK, pp. 103–131.

Wright, D.J. and Newall, D.R. (1980) Osmotic and ionic regulation in nematodes. In: Zuckerman, B.M. (ed.) *Nematodes as Biological Models*, Vol. 2. Academic Press, London, pp. 143–164.

10 Interactions with Other Pathogens

Rosa H. Manzanilla-López[1] and James L. Starr[2]
[1]*Rothamsted Research, Harpenden, Hertfordshire, UK;* [2]*Texas A&M University, College Station, Texas, USA*

10.1 Introduction	**223**
10.2 Interactions with Microbial Pathogens	**226**
10.3 Interactions with Other Plant-parasitic Nematodes	**230**
10.4 Basis for Interactions	**238**
10.5 Conclusions and Future Directions	**239**
10.6 References	**240**

10.1 Introduction

The environments inhabited by plant-parasitic nematodes are complex ecosystems, plant-parasitic nematodes being only one of the many different biotic components of these systems. It would be unrealistic, therefore, to view nematode–plant interactions as if they occurred in isolation from the other components of the environment. None the less, this is what many nematologists do far too often. The germ theory of disease and the need to validate disease based on the aetiology of a single pathogen has caused many scientists to overlook the complexities of disease with multiple pathogens. Given the complexity of the typical environment in which nematodes parasitize plants, we should not be surprised that nematode parasitism and pathogenesis of plants are affected, often greatly, by other organisms in the environment, and conversely. In this chapter we examine root-knot nematode interactions with other pathogens, especially those with other nematode species, fungi and bacteria.

Atkinson (1892), with his observation that the incidence and severity of Fusarium wilt of cotton was greater in fields also infested with root-knot nematodes, was among the first to comment on plant disease complexes involving nematodes. Since those early observations, there have been many reports published on disease complexes, and several excellent reviews are available (see Powell, 1971; Wallace, 1978; Taylor, 1979; Webster, 1985; Mai and Abawi, 1987).

Associations of *Meloidogyne* species with other plant-parasitic nematodes in naturally infested fields are common. For example, *Helicotylenchus multicinctus*, *Pratylenchus coffeae* and *Radopholus similis* are frequently found in banana roots in association with *Meloidogyne* spp. in all *Musa*-producing areas (Moens *et al.*, 2006). In North Carolina, *M. incognita*, *M. hapla* and *Pratylenchus brachyurus* occur commonly in mixed populations, the first species being injurious to tobacco and the second to peanut (Johnson and Nusbaum, 1970). In South Carolina, 25% of the samples collected from cotton contained *Hoplolaimus columbus*. Of these, 48% also contained *Scutellonema brachyurus* and 7% *Meloidogyne* spp. (Kraus-Schmidt and Lewis, 1981). *M. incognita* has been commonly found co-infecting roots of soybean with *Heterodera glycines* (Ross, 1964) and *M. javanica* (McGawley and Winchell, 1987). *Ditylenchus dipsaci* and *M. hapla* on lucerne (*Medicago sativa* L.) can co-occur in the same field, both being pathogenic to lucerne and both involved in disease complexes (Hawn, 1963; Hunt *et al.*, 1971; Griffin, 1980). *Meloidogyne incognita* and *M. javanica* have similar ecological requirements and thrive together in many different parts of the world (Khan and Haider, 1991). *Meloidogyne hapla* and *Heterodera schachtii* commonly occur in the intermountain region of the western USA, occasionally cohabiting the same soil and, although not usually parasitizing the same host, both may occur together in tomato fields in northern Utah (Griffin and Waite, 1982). In India, both *Meloidogyne* species and the reniform nematode have been observed frequently to be associated with black gram (*Vigna mungo*) (Mishra and Gaur, 1981).

Although different *Meloidogyne* species are common in complex communities, they may occupy distinct niches in the community. As an example, *Globodera pallida*, *Meloidogyne* sp. and *Nacobbus aberrans* have been reported to occur together in northern Argentina in native Andean potatoes (*Solanum tuberosum* subsp. *andigenum*) yet infecting different organs of the potato plant, with *G. pallida* and *N. aberrans* in roots and *Meloidogyne* sp. in stolons (Doucet *et al.*, 2007). The environment may exert an overwhelming influence on potential interactions. In the case of *M. incognita* and *Rotylenchulus reniformis*, both of which reproduce well on cotton, they are rarely found infesting the same field at high population densities. *Meloidogyne incognita* is generally the dominant species in more coarsely textured, sandy soils, whereas *R. reniformis* is generally dominant in the more finely textured, silty soils (Robinson *et al.*, 1987; Starr *et al.*, 1993).

Observations that one nematode species increases while another decreases do not necessarily indicate antagonism, interaction or competition between species but may result from the natural life cycle of the nematode(s) and changes in niche dimensions. It has also been argued that the term 'competition' should preferably be used in relation to a shortage of substrate and niche overlap (Norton, 1989). Most population increments or decrements through interactions have been reported as competition *sensu lato* in *Meloidogyne* interaction studies. Competition for food and space can also result in the production of more males, a strategy related to adverse environmental conditions (Khan *et al.*, 1985, 1986a).

Much of the early work on interactions of *Meloidogyne* species with other nematodes and microbes focused on four main topics: (i) identification of the other microorganisms or nematodes involved; (ii) whether the interaction was of a synergistic or additive nature; (iii) whether infection of a plant by one nematode species greatly affects the interaction of that plant with a different nematode species; and (iv) did nematode parasitism overcome or 'break' plant resistance to the particular fungal or bacterial pathogen involved in the disease complex? Much less effort has been expended in unravelling the basis for these many interactions or their effects on disease epidemics in agroecosystems (and hence on nematode and disease management). Rather than summarize all of the vast volume of reports describing various interactions involving *Meloidogyne* species, we will review those contributions that enhance our understanding of these complexes.

The term interaction has been used quantitatively and qualitatively to describe interrelationships between two or more factors involved in plant diseases. Perhaps it would be useful here to consider the definition of an interaction, especially from a statistical perspective. Statistically, an interaction is a term in the statistical model referring to the fact that the effect of two or more variables is not additive. Such a term reflects the fact that the effect of one variable depends on the values of one or more other variables (Kuehl, 1994). This has implications in experimental design as it can be misleading to vary only one factor at a time. Such experimental considerations were often ignored in early studies of disease complexes and the interaction of multiple nematodes and other pathogens with the host. The experimental design frequently used in early studies of nematode complexes with fungal or bacterial plant pathogens was simply that of a non-inoculated control, plants inoculated with a single, typically high, nematode density, plants inoculated with a single density of the fungal or bacterial pathogen, and then plants inoculated with both the nematode and the other pathogen at the same inoculum density used in the single pathogen treatments. If the effects of the combined inoculation was greater than the expected additive effects (Fig. 10.1A), then the system was judged to be a synergistic

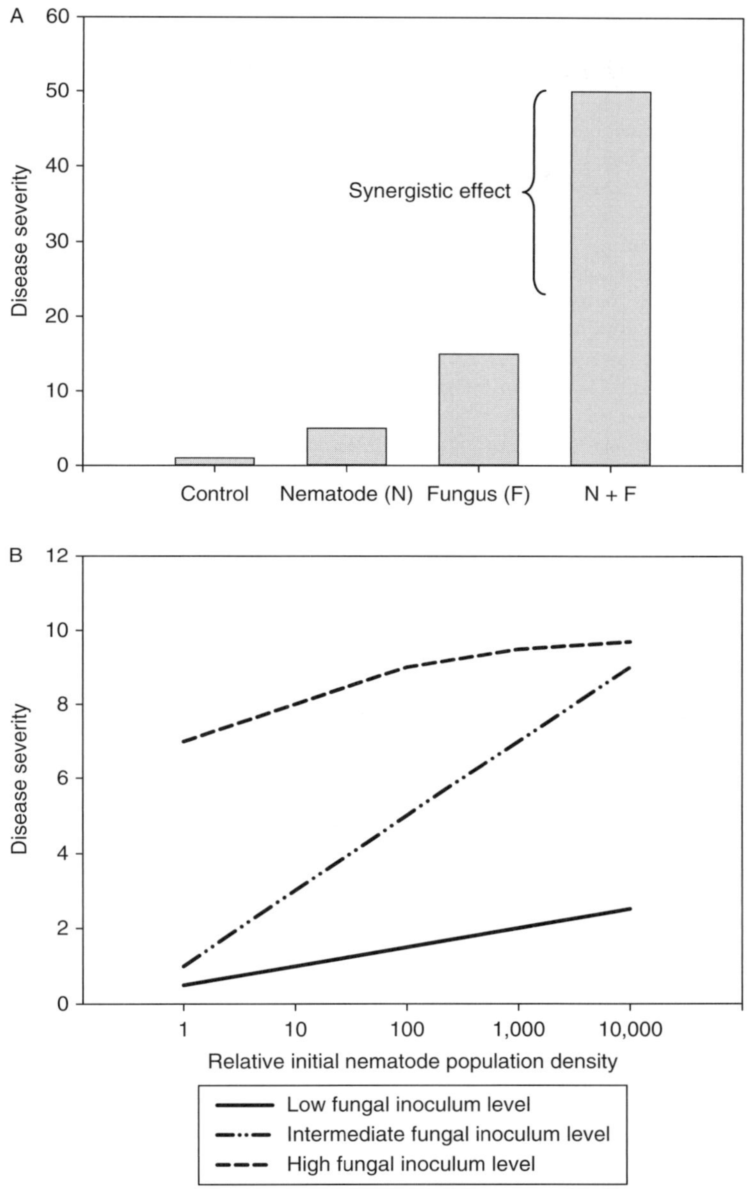

Fig. 10.1. A; expected relationship for a synergistic interaction between *Meloidogyne* species and another plant pathogen, typically a fungus, where the overall disease severity in the presence of both pathogens is greater than a simple additive effect. B; a more complex interaction between a fungal and nematode pathogen, where a strong synergistic effect is observed only with the intermediate concentration of the fungal inoculum.

interaction. A more informative method would be to use a factorial experimental design, testing multiple densities of each pathogen and then employing regression analysis to determine if one pathogen affected the plant's response to the other pathogen (Fig. 10.1B). Using such a factorial experimental design facilitates examination of the effects of treatments on both the intercept and rate parameters of the resulting models, thereby allowing a more complete description of any possible interaction. As detection of 'statistical interaction' by ANOVA and subsequent inference of 'biological interaction' refer to two different things (Khan and Dasgupta, 1993), care must be taken when using this analysis for assessing joint responses to associations of two or more types of organism.

10.2 Interactions with Microbial Pathogens

10.2.1 Vascular wilt pathogens

Among the most common disease complexes in which the root-knot nematodes are key components are those involving the vascular wilt pathogen, *Fusarium oxysporum*, and the large number of crop species susceptible to this pathogen. The bacterial vascular pathogen, *Ralstonia* (formerly *Pseudomonas*) *solanacearum* (Johnson and Powell, 1969), on several solanaceous crops, has been studied somewhat less frequently. It is interesting that there are no convincing reports of interactions of *Meloidogyne* spp. with Verticillium wilt diseases, even though interactions among *Pratylenchus* spp. and Verticillium wilt are well documented (see Powell, 1971). For example, in a field infested with both *M. incognita* and *Verticillium dahliae*, reduction of the nematode population density with soil fumigation did not reduce the incidence of wilt disease relative to plots not fumigated (McClellan *et al.*, 1955). The lack of interaction may be due in part to the fact that Verticillium wilts are favoured by finer-textured soils, which are less conducive for root-knot nematodes than the more coarsely textured soils favoured by *F. oxysporum* and *Meloidogyne* spp.

The influence of infection by *Meloidogyne* spp. on Fusarium wilt (Plate 30) has been most often studied on tomato and cotton, although the effects are probably similar on nearly every plant susceptible to these pathogens. Further, although this disease complex was first noted in field situations, most research has been based on studies conducted in small pots in more controlled environments. In the field, wilt symptoms are more severe, develop more rapidly and at greater frequency when the plants are also infected by the nematode than in the absence of nematode infection. This is true for both wilt-susceptible and wilt-resistant plant genotypes. In many cases the presence of both pathogens results in much greater plant mortality than when only one is present. Relatively few studies have examined the effects of such complexes on epidemiological parameters of disease caused by either pathogen. Roberts *et al.* (1985) noted that in a cotton field infested with *M. incognita* and *F. o.* f. sp. *vasinfectum*, the slope of the regression model relating seed cotton yield to pre-plant nematode densities was greater than in a field infested only with *M. incognita*. Similarly, Starr *et al.* (1989) reported from microplot experiments that the presence of *F. o.* f. sp. *vasinfectum* had little effect on the threshold parameter (T) of the Seinhorst model relating cotton yield to initial densities of *M. incognita*, but that the minimum yield parameter (m) was lower in the presence of the wilt pathogen. Starr *et al.* (1989) also reported that increased plant mortality was one of the most important consequences of the interaction of these two pathogens, and that interaction was most dramatic at intermediate population densities of both pathogens. At very high or very low densities of either pathogen there was no apparent interaction.

One of the most common observations from pot studies of the interaction between *Meloidogyne* spp. and *F. oxysporum* has been that the effects of the nematode on wilt disease are most pronounced if the nematode is inoculated on to the host *c.* 4 weeks prior to inoculation with the wilt pathogen (Porter and Powell, 1967). This common phenomenon has been used to argue that the basis of the interaction must be physiological rather than due to a simple response to wounds made by the invading nematodes. It is well known that physiological activity in galls and nematode-induced giant cells is substantially different from that of non-infected host roots, yet it has been difficult to link such physiological and biochemical changes in the host tissues to mechanism(s) of increased susceptibility to *F. oxysporum*.

Many plant pathologists ascribe to the hypothesis that wounding of plant roots during nematode parasitism is the major factor contributing to the plant's increased susceptibility to other pathogens (Lucas *et al.*, 1955; Stewart and Schindler, 1956; Agrios, 1997). We are unaware of any report that has documented infection of *Meloidogyne*-induced galls by *F. oxysporum* via any detectible wound made by the nematode. Several studies have noted that the giant cells appear to be highly susceptible to fungal infection, as they degenerate rapidly when infected by the fungus (Meléndez and Powell, 1967; Fattah and Webster, 1983). Sidhu and Webster (1977) suggested that the effects of the nematode on susceptibility to *Fusarium* were due to a translocatable factor of unknown identity or origin. They observed that, when adventitious roots were induced from several locations on a tomato stem, and when the nematode and fungus were inoculated on to physically separate parts of the root system, there was still an increase in wilt severity due to infection by the nematode. While their research provided strong mitigating evidence against the role of nematode-induced wounds as the principal cause of the interaction, their 'translocatable factor' was highly speculative. In a similar experiment using a split-root system on tobacco, nematodes increased wilt severity only when they were inoculated on to the same portion of the root system as *F. o.* fp. *nicotianae* (Moorman *et al.*, 1980).

Breaking of disease resistance by plant-parasitic nematodes is an expression commonly used to denote the predisposition of plants that are resistant to a given pathogen. The gene(s) for resistance to a pathogen is rendered ineffective through physiological alterations in the plant caused by the nematode, even though the gene(s) remain operational (Khan, 1993). With regard to the concept of nematodes 'breaking' genetic resistance to Fusarium wilt, we think that this is an unfortunate and misleading terminology, as it implies that the nematode infection acts in some direct manner on the resistance mechanism. Resistance to Fusarium wilt in cotton is a multigenic and quantitative trait (Hillocks, 1992). The mechanism(s) of resistance are poorly understood and we are unaware of any study that has specifically examined the effects of nematode parasitism on any specific component of resistance to the fungal pathogen. None the less, it is well established that cotton genotypes that have high levels of resistance to Fusarium wilt in the absence of nematodes are much more susceptible to wilt when nematodes are present. As a consequence, many cotton breeding programmes have used a nematode-infested wilt nursery to screen cotton genotypes for wilt resistance. This practice has resulted in the development of Fusarium wilt-resistant cotton genotypes that also suppresses wilt incidence in the presence of the nematode.

One unfortunate consequence of the recognition of the importance of *Meloidogyne* infection on the strength of resistance to Fusarium wilt was the assumption by some cotton breeders that, if they had resistance to wilt in the presence of the nematode, then they also had resistance to the nematodes. In a study that compared 20 cotton varieties that varied from poor to excellent in reported resistance to the nematode–wilt disease complex (Starr and Martyn, 1991), all cultivars with some reported resistance to the complex exhibited lower levels of vascular browning and plant mortality than susceptible genotypes, even in the presence of the nematode. However, none of the wilt-resistant genotypes supported less nematode reproduction than either the wilt-complex-susceptible genotypes or the nematode-susceptible control cultivar. Cotton cultivars reported to be highly resistant to the wilt disease complex had greater incidence of vascular browning and plant mortality in the presence of *M. incognita* than when the nematodes were absent. These genotypes still expressed useful levels of resistance to wilt disease in the presence of the nematode (Table 10.1). Thus, wilt resistance was less effective when the plants were also infected by the nematodes, but it was not broken in the sense that there was no longer any effective resistance to the wilt pathogen. However, none of the cultivars with good to excellent resistance to the complex exhibited any resistance to nematode reproduction.

Evidence that *Meloidogyne* spp. can affect resistance to Fusarium wilt in tomato is contradictory, some reports indicating an increase in the wilt symptoms of resistant tomato genotypes, while others showed no effect of the nematodes on wilt symptoms. It is likely that variation in experimental methodology is responsible for some of the confusion. Inoculum levels certainly affect the degree to which wilt disease is influenced by the nematodes (Sidhu and Webster,

Table 10.1. Effect of *Meloidogyne incognita* and *Fusarium oxysporum* f. sp. *vasinfectum* on vascular browning and mortality of cotton in cultivars that differ in resistance to the wilt complex. (Data from Starr and Martyn, 1991.)

		Vascular browning (%)		Mortality (%)		Nematodes/ 500 cm^3 soil
Resistance class	N	F	F + N	F	F + N	F + N
Resistant control	1	20	31	10	1	10
Excellent	4	28	82	9	22	1700
Good	4	35	67	8	20	1440
Moderate	6	50	74	8	35	1330
Poor	2	55	82	38	79	1630
Susceptible control	1	49	83	15	73	1340

The resistant and susceptible controls refer to host status relative to *M. incognita*, whereas excellent, good, moderate and poor refer to reported resistance to the nematode–wilt disease complex. N = number of cultivars; F = Fusarium; N = nematodes.

1981). In one report indicating no effect of the nematodes on wilt resistance, plants were inoculated with few juveniles (Jones *et al.*, 1976), whereas in another, with poorly quantified but apparently very high levels of inoculum, *M. incognita* and *M. hapla* increased wilt severity in tomato genotypes with moderate and high levels of wilt resistance (Jenkins and Coursen, 1957). Sidhu and Webster (1974, 1981) presented strong evidence that *M. incognita* does increase wilt severity in tomato genotypes having a single gene for resistance to Fusarium wilt. In a later study, Abawi and Barker (1984) were unable to show any effect of the nematodes on Fusarium wilt of wilt-resistant tomato genotypes, but did observe an increase in non-specific 'wilt' symptoms due to root damage from nematode parasitism and root necrosis typically caused by secondary pathogens. In contrast to cotton, most of the resistance to Fusarium wilt in the tomato genotypes tested in these studies is inherited as a single dominant gene.

In most crops, it would appear that moderate to high population densities of *Meloidogyne* spp. reduce the effectiveness of Fusarium wilt resistance that is inherited as a polygenic, quantitative trait. The available data are less clear when resistance is inherited as a monogenic trait. In summer squash (*Curcurbita pepo* var. *melopepo*) high population densities of *M. incognita* resulted in increased wilt severity in a cultivar with moderate resistance to wilt but not in a cultivar with high resistance to wilt (Caperton *et al.*, 1986). Similarly, in chickpea (*Cicer arietinum*) varieties with partial resistance to Fusarium wilt, plants also infected with *M. artiellia* showed greater wilt than when not infected by the nematodes (Castillo *et al.*, 2003). In chickpea with high levels of resistance to Fusarium wilt, co-infection by *M. artiellia* overcame the plant's resistance to wilt in two varieties but not in a third, highly resistant, variety (Castillo *et al.*, 2003). The complete resistance phenotype of selected chickpea genotypes to *F. o.* f. sp. *ciceris* pathogenic races *Foc*-0, *Foc*-1A and *Foc*-2 was not modified by co-infection with *M. artiellia* (Navas-Cortés *et al.*, 2008). In cowpea (*Vigna unguiculata*) with a single dominant gene for resistance to Fusarium wilt, *M. incognita* had no effect on wilt disease (Roberts *et al.*, 1995). Collectively, these observations suggest that the ability of *Meloidogyne* spp. to overcome resistance to Fusarium wilt is conditioned by a single, major-effect resistance gene to the fungus, or at least there is a very high level of resistance, which is variable and likely to be based upon the exact nature of that resistance.

10.2.2 Root-rot pathogens

Disease complexes involving *Meloidogyne* spp. and various root- and stem-rot pathogens may be among the most important disease complexes in terms of total economic impact. Nearly every soil environment is infested with one or more root-rot pathogens, whereas a much lower percentage of agricultural soils is infested with vascular wilt pathogens. Thus, root- and stem-rot disease complexes are likely to occur with a greater frequency than vascular wilts. There are numerous reports on many crop species where root-rots exhibit a synergistic, rather than an additive, increase in the presence of root-knot nematodes. As is the case with Fusarium wilt diseases, the effects of the nematodes on the fungal disease are most pronounced if infection by the nematodes precedes that by the fungal pathogen by 3–4 weeks (Powell *et al.*, 1971; Starr and Mai, 1976).

Among the best studied of the root-rot pathogens that are affected by root-knot nematodes are the seedling disease pathogens of cotton and black shank of tobacco caused by *Phytophthora parasitica* (Powell and Nusbaum, 1960). Cotton seedling disease is caused by any of several fungal pathogens. The most common of these pathogens include *Pythium* spp., *Fusarium* spp., *Rhizoctonia solani* (Brodie and Cooper, 1964) and *Thielaviopsis basicola* (Fichtner *et al.*, 2005). In general, moderate to high population densities of *M. incognita* increase both the incidence and severity of cotton seedling disease caused by any of these fungal pathogens.

In addition to disease complexes involving known root-rot pathogens, there is evidence that fungi with no recognized or only weak pathogenic abilities can cause substantial root necrosis when plants are heavily infected with different *Meloidogyne* spp. In a study on tobacco, Powell *et al.* (1971) demonstrated that species of *Trichoderma* and *Penicillium* that were not recognized as pathogens caused substantial root disease when the plants were also infected with *M. incognita*. Mayol and Bergeson (1970), in a study on tomato infected with *M. incognita*, reported that galls on plants grown in sterile soil remained white and 'healthy', whereas galls from plants grown in sterilized soil that also received a treatment of a small amount of water extract from non-sterile field soil (and presumably the microflora of the field soil) were necrotic. Although they reported that the common pathogenic fungi *Fusarium* spp. and *R. solani*, in addition to *Trichoderma* spp., were isolated from some necrotic galls, they were unable to detect any fungi specifically associated with most of the necrotic gall tissue. Because bacteria were frequently isolated from the galls, Mayol and Bergeson (1970) suggested that much of the necrosis observed in field soils was due to these unidentified bacteria. Based on their own observations, both Powell *et al.* (1971) and Mayol and Bergeson (1970) suggested that many soil-borne fungi and bacteria not commonly recognized as plant pathogens were important in the root necrosis typically associated with severe infection by *Meloidogyne* spp. In contrast to their observations, root necrosis of celery associated with infection by *M. hapla* in organic soils was found to be caused only by *Pythium polymorphon* (Starr and Mai, 1976; Plate 31). None of the more than 40 isolates of unidentified bacteria that were obtained from necrotic gall tissue, nor any of the other fungi (including numerous isolates of *Fusarium* spp.), caused any root necrosis when inoculated on to celery roots previously infected by the nematode. Collectively, these observations suggest that, whereas many soil-inhabiting fungi can cause necrosis of *Meloidogyne*-induced galls, there are many other fungi, including many fusaria, which are not pathogenic, even on severely galled root tissue. Although disease complexes are known to be caused by several bacterial pathogens, including the vascular wilt pathogens *R. solanacearum* (Lucas *et al.*, 1955) and *Pseudomonas caryophylli* (Stewart and Schindler, 1956), the crown gall pathogen *Agrobacterium tumefaciens* (Griffin *et al.*, 1968), and *Pseudomonas marginata*, which causes scab of gladiolus corms (El-Goorani *et al.*, 1974), it is surprising that no bacteria have been proven to be involved in the root necrosis commonly associated with high levels of infection by root-knot nematodes.

In some disease complexes, it is not clear if there is truly an interaction or if the effects are only additive. There are several reports of a reduced incidence of southern blight of groundnut caused by *Sclerotium rolfsii* when *M. arenaria* was controlled either by use of nematicides (Rodríguez-Kábana *et al.*, 1982) or by rotation with non-hosts for the nematode (Rodríguez-Kábana *et al.*, 1991). However, when more controlled experiments were conducted in microplots with a factorial experimental design utilizing multiple densities of each pathogen, no statistical interaction was observed (Starr *et al.*, 1996). None of the nematode inoculation densities affected the amount of disease caused by the fungus. *Sclerotium rolfsii* is a necrotrophic pathogen that typically attacks plant stems above the soil surface, rather than operating as a root pathogen. This difference, in contrast to the mode of attack by *Pythium* and *Rhizoctonia* spp., two genera which are often involved in disease interactions with root-knot species, may play a role in the fact that the association of *S. rolfsii* and *Meloidogyne* spp. is more additive than synergistic.

10.2.3 More recently described disease complexes

Although the preceding sections describe some of the most intensively studied disease complexes, new disease complexes and interactions involving nematodes and microorganisms continue to be reported. These include *M. incognita* and *Macrophomina phaseolina* on *Coleus forksohlii* (Poornima and Subramanian, 2006), *M. incognita* and *Phytophthora* in betel vine (Sitaramaiah and Parvathi, 1994; Jonathan *et al.*, 2006), wilt disease expression in the presence of *M. incognita* and *Fusarium moniliforme* on grapevine (Senthilkumar and Rajendran, 2003), and decline of guava trees caused by the association of *Fusarium solani*, *Pythium aphanidermatum*, *V. dahliae*, *Trichothecium roseum* and *Trichoderma* sp. with *M. incognita*, *M. javanica* and *M. arenaria* (Avelar-Mejía *et al.*, 2001). In Mexico, Central and South America the 'corchosis del cafeto' on coffee trees has been related to interactions between *Meloidogyne* spp., *Pratylenchus*, *Fusarium* and *Trichoderma* (Téliz-Ortíz *et al.*, 1993). Recent studies have shown that more than one *Meloidogyne* species can occur simultaneously in all major coffee-growing countries and potentially be involved in disease complexes (Carneiro *et al.*, 2000; Campos and Villain, 2005; see Moens *et al.*, Chapter 1 and Hunt and Handoo, Chapter 3, this volume). Regardless of the importance of some crops, a more thorough understanding of the interactions between *Meloidogyne* spp. and other soil-borne pathogens, including plant-parasitic nematodes, is needed, especially from tropical regions where multiple infections by nematodes are common on a single crop. Sugarcane, for example, can be damaged by 20 different nematode species (Luc *et al.*, 2005).

10.3 Interactions with Other Plant-parasitic Nematodes

Interactions among plant-parasitic nematodes are complex and *Meloidogyne*–nematode interactions are no exception. It is important to note that *Meloidogyne* and species of other plant-parasitic nematodes do occur together, but that their interaction/relationships have not been so intensively studied as the other *Meloidogyne*–microbe interactions discussed above. There are few studies involving *Meloidogyne* and more than two nematode species, and less information is available for even more complex relationships, such as those that can be expected when, apart from *Meloidogyne* and other plant-parasitic nematodes, pathogenic soil-borne microorganisms are also involved.

Much of the available information on *Meloidogyne* spp. and other nematodes has been generated from glasshouse and laboratory observations under controlled experimental conditions that usually include one species of nematode on one plant species, an uncommon situation in agricultural soils, where multi-species and polyphagous nematode communities occur (Johnson and Nusbaum, 1970; Norton, 1989). Only few studies of *Meloidogyne* interactions with other nematodes are supported with observations from field studies.

In glasshouse experiments, host response to plant-parasitic nematodes (such as *Meloidogyne* spp.) may be altered by the presence of another nematode in different ways, depending on the experimental conditions (e.g. temperature, nematode species, host, etc.). One nematode may enhance or delay development and reproduction of another nematode, but the effect may be reversed with the same two species of nematodes on a different host plant. Some nematode species act independently, and penetration by one species may not be affected by another species, although subsequent development may be affected (Norton, 1989). Interactions may also result in mutual benefit, when physiological changes may enhance nutrition or reduce resistance of the host to the parasite. Potential versus effective or null interactions depend on host suitability (i.e. resistance/susceptibility) and environment interactions. Studies of nematodes in cohabitance (i.e. nematodes living in the same root, rhizosphere or soil) that include *Meloidogyne* sp. have been reviewed elsewhere (Eisenback and Griffin, 1987; Eisenback, 1993), with several types of interactions being identified, including antagonistic interactions resulting from spatial competition, physical alteration and destruction of feeding sites, and a decrease in host suitability mediated by physiological change.

Interactions of *Meloidogyne* spp. with other nematodes have been examined through different levels of complexity and experimental conditions: intraspecific interactions, including single species of *Meloidogyne*; interspecific populations of two or more species (with at least one of them being a *Meloidogyne* spp.); and, occasionally, whole communities (e.g. *Meloidogyne* and all other nematode populations occurring in specific agroecosystems or pathosystems). Relationships of plant-parasitic nematodes in the rhizosphere are regulated by environmental factors and host genetics. Under field conditions, many opportunities for competition, synergism and other interactions occur between plant-parasitic nematodes, but these have been rarely explored under such natural conditions, *Meloidogyne* spp. interactions being no exception. Field mixed populations of different or the same species are not uncommon, but dominance of one species may depend on different factors, such as coincident infections (spatial and temporal), different temperature and inoculum levels, and whether host response might be more favourable to one of the species involved (Kinloch and Allen, 1972). To separate the effects of factors such as soil, host and competition, among others that influence community composition and structure, is one of the most difficult problems when working with nematode communities (Norton, 1989). Because of the effect of these interactions on crop health and yield, an approach that separates nematode–nematode interactions into ecological and etiological types can be used to facilitate their study (Eisenback and Griffin, 1987).

Meloidogyne–nematode interactions that occur or may occur in a given crop (or cultivar) and at a specific location have been studied to assess host status, crop damage and yield loss to potential pests, as well as their resistance or susceptibility to other soil-borne pathogens such as *Phytophthora* spp. (Powell and Nusbaum, 1960; Sitaramaiah and Parthavi, 1994; Jonathan *et al.*, 2006). Hence, real or hypothetical '*Meloidogyne*–nematode partners' have been used to investigate a number of selected, specific interactions (see Table 10.2). Some of these studies have even highlighted required changes in pest management programmes where interactions could increase crop losses, such as the combined occurrence of *H. columbus*, *S. brachyurus* and *M. incognita* in South Carolina (Kraus-Schmidt and Lewis, 1981).

10.3.1 Interactions and parasitic habits

When a plant supports a polyspecific nematode community, niche differentiation can become evident based on differences in feeding habits, the tissue being parasitized, host physiological alterations and succession of nematode species through the crop cycle (e.g. annual, semi-perennial, perennial). In general, the same ecological niche cannot be occupied indefinitely by two closely related species as they will compete until eventually one dominates (Eisenback and Griffin, 1987). In natural habitats the same species may be present in different proportions in different environments and at different times (Norton, 1989), thus allowing or avoiding direct interaction with other species. The host root system not only attracts *Meloidogyne* second-stage juveniles (J2), but also ectoparasitic and other endoparasitic (e.g. migratory and sedentary) species. Although ectoparasites and sedentary endoparasites occupy different niches, they may parasitize the same root with or without any interaction. The ectoparasites *Tylenchorhynchus agri* and *Paratrichodorus minor* reduced *M. naasi* infection by inhibiting root growth of creeping bent grass, thus effectively decreasing the availability of feeding sites for the latter (Sikora *et al.*, 1979). Reproduction of the sedentary endoparasite *Meloidogyne* can be suppressed by an ectoparasite by direct or indirect competition for feeding sites and by damage to the root system, but the opposite effect can also occur, as in the inhibitory effect of *Meloidogyne* on the development of *Pratylenchus penetrans*, which was considered to be caused by more than competition for feeding sites, since *Meloidogyne* galls were smaller and fewer in the presence of *Pratylenchus* (Estores and Chen, 1970). Sedentary endoparasitic nematodes can suppress ectoparasites through physiological rather than mechanical effects, and also through environmental and edaphic factors, as in the interaction between *M. hapla* and *Xiphinema americanum* (Norton, 1969; Eisenback and Griffin, 1987).

Population dynamics and pathogenicity of plant-parasitic nematodes in combination may differ from those of the monospecific populations usually studied experimentally under glasshouse conditions (Kraus-Schmidt and Lewis, 1981), particularly in terms of abundance, host invasion timing and dominance through crop development. Organisms with a short life and rapid

Table 10.2. List of plant-parasitic nematode species that have been studied in association with *Meloidogyne* spp. (Modified from Eisenback, 1993.)

Meloidogyne species	Other nematode species	Host	Reference
M. arenaria	*Pratylenchus brachyurus*	Cotton	Gay and Bird (1973)
M. graminicola	*Heterodera oryzicola*	Rice	Rao and Prasad (1981)
M. hapla	*Ditylenchus dipsaci*	Lucerne	Griffin, (1980); Hunt *et al.* (1971); Hawn (1963)
M. hapla	*Ditylenchus dipsaci*	Sugarbeet	Griffin (1985)
M. hapla	*Heterodera schachtii*	Tomato	Griffin and Waite (1982)
M. hapla	*H. schachtii*	Sugarbeet	Jatala and Jensen (1976)
M. hapla	*Macroposthonia xenoplax*	Concord grapes	Santo and Bolander (1977)
M. hapla	*Meloidogyne javanica*	Tomato	Kinloch and Allen (1972)
M. hapla	*Pratylenchus brachyurus*	Tobacco	Johnson and Nusbaum (1970)
M hapla	*Xiphinema americanum*	Lucerne	Norton (1969)
M. incognita	*Aphelenchus avenae*	Tomato	Ishibashi and Choi (1991)
M. incognita	*Belonolaimus longicaudatus*	Cotton	Yang *et al.* (1976)
M. incognita	*Criconemella ornata*	Aubergine	Misra and Das (1979)
M. incognita	*Helicotylenchus multicinctus*	Banana	Moens *et al.* (2006)
M. incognita	*Heterodera cajani*	Cowpea	Sharma and Sethi (1978)
M. incognita	*Heterodera glycines*	Soybean	Ross (1964); Niblack *et al.* (1986)
M. incognita	*H. schachtii*	Sugarbeet	Inserra *et al.* (1984)
M. incognita	*Heterodera zeae*	Maize	Kaul and Sethi (1982)
M. incognita	*Hoplolaimus columbus*	Cotton	Bird *et al.* (1974); Kraus-Schmidt and Lewis (1981)
M. incognita	*Hoplolaimus columbus*	Soybean	Guy and Lewis (1987a,b)
M. incognita	*Hoplolaimus galeatus*	Cotton	Yang *et al.* (1976)
M. incognita	*Hoplolaimus indicus*	Aubergine	Misra and Das (1979)
M. incognita	*Meloidogyne arenaria*	Soybean	Ibrahim and Lewis (1986)
M. incognita	*Meloidogyne javanica*	Soybean	McGawley and Winchell (1987); Khan and Haider (1991)
M. incognita	*Nacobbus aberrans*	Tomato	López-Portillo *et al.* (1984)
M. incognita	*Nacobbus aberrans*	Sugarbeet	Inserra *et al.* (1984)
M. incognita	*Pratylenchus brachyurus*	Cotton	Gay and Bird (1973)
M. incognita	*Pratylenchus brachyurus*	Tobacco	Johnson and Nusbaum (1970)
M. incognita	*Pratylenchus brachyurus*	Soybean	Herman *et al.* (1988)
M. incognita	*Pratylenchus coffeae*	Banana	Moens *et al.* (2006)
M. incognita	*Pratylenchus penetrans*	Lucerne	Chapman and Turner (1975)
M. incognita	*Pratylenchus penetrans*	Red ceover	Amosu and Taylor (1975); Chapman and Turner (1975)

M. incognita (= *M. incognita acrita*)	*Pratylenchus penetrans*	Tomato	Estores and Chen (1970)
M. incognita	*Pratylenchus vulnus*	Grape	Chitamber and Raski (1984)
M. incognita	*Radopholus similis*	Banana	Moens *et al.* (2006)
M. incognita	*Radopholus similis*	Black pepper	Sheela and Venkitesan (1981)
M. incognita	*Rotylenchulus reniformis*	Black gram	Mishra and Gaur (1981)
M. incognita	*Rotylenchulus reniformis*	Cowpea	Taha and Kassab (1980); Khan and Husain (1989)
M. incognita	*Rotylenchulus reniformis*	Grapevine	Ras and Seshadri (1981)
M. incognita	*Rotylenchulus reniformis*	Aubergine	Khan *et al.* (1986b)
M. incognita	*Rotylenchulus reniformis*	Pigeonpea	Pathak *et al.* (1985)
M. incognita	*Rotylenchulus reniformis*	Soybean	Singh (1976)
M. incognita	*Rotylenchulus reniformis*	Sweet potato	Thomas and Clarke (1980, 1981, 1983a,b)
M. incognita	*Rotylenchulus reniformis*	Tomato	Khan *et al.* (1986a)
M. incognita	*Scutellonema brachyurus*	Cotton	Kraus-Schmidt and Lewis (1981)
M. incognita	*Tylenchorhynchus agri*	Red clover	Amosu and Taylor (1975)
M. incognita	*Tylenchorhynchus brassicae*	Tomato	Khan et al. (1986b)
M. incognita	*Tylenchorhynchus nudus*	Aubergine	Misra and Das (1979)
M. incognita	*Tylenchorhynchus vulgaris*	Pearl millet	Vaishnav and Sethi (1978)
M. incognita	*Tylenchorhynchus vulgaris*	Maize	Kaul and Sethi (1982)
M. javanica	*Scutellonema cavenessi*	Cowpea	Diop *et al.* (2002)
M. javanica	*Scutellonema cavenessi*	Tomato	Diop *et al.* (2002)
M. javanica	*M. incognita*	Tomato	Khan and Haider (1991)
M. javanica	*Hemicycliophora arenaria*	Tomato	Van Gundy and Kirkpatrick (1975)
M. javanica	*Pratylenchus sefaensis*	Cowpea	Egunjobi *et al.* (1986)
M. javanica	*Rotylenchulus reniformis*	Cowpea	Taha and Kassab (1979)
M. naasi	*Paratrichodorus minor*	Bent grass	Sikora *et al.* (1979)
M. naasi	*Tylenchorhynchus agri*	Bent grass	Sikora *et al.* (1979)
Meloidogyne sp.	*Globodera pallida*	Potato	Doucet *et al.* (2007)
Meloidogyne sp.	*Nacobbus aberrans*	Potato	Doucet *et al.* (2007)

reproduction rate are known as r-strategists and are usually opportunists. Trudgill and Phillips (1997) considered *Meloidogyne* spp. as r-strategists, but the life history strategies of *M. incognita* and *M. hapla*, for example, show features more associated with K-strategists (see Evans and Perry, Chapter 9, this volume). Those with a long life, but relatively low reproductive rate, are K-strategists and are more abundant in stable ecosystems. The r-selected nematode species thrive in the continuously cultivated agroecosystems, while K-selected species usually dominate in more stable natural environments (Norton, 1989).

Some migratory endoparasites (e.g. *Pratylenchus* spp., *Radopholus* spp., *Hoplolaimus* spp., etc.) invade and move faster through root tissue than do *Meloidogyne* species, thereby inhibiting penetration and disturbing feeding sites of sedentary endoparasitic nematodes and resulting in suppression or dominance over *Meloidogyne* (Eisenback and Griffin, 1987). However, suppression and dominance can be exerted by *Meloidogyne* on other endoparasites, although this effect may not be immediately apparent.

Not many experiments that have been conducted under glasshouse conditions can be followed for periods of time longer than just the average duration of the whole cycle of both host and parasite, especially in the case of non-perennial or semi-perennial crops. Under field conditions (e.g. experimental plots and microplots) and whenever it is possible to repeat experiments for more than one season, other features of the interaction can be shown. For example, studies with *H. glycines* and *M. incognita* over three seasons revealed that cyst nematode populations in microplots were largely unaffected by low *M. incognita* populations, but high initial *M. incognita* populations curtailed cyst nematode reproduction during the latter part of the season. However, cyst nematode populations were greater in plots containing both nematodes than in plots containing only *H. glycines* (Ross, 1964).

10.3.2 Sequential infections

Reproduction and dominance can be affected by different factors (and also by their combination), including sequence of inoculation (or natural succession under field conditions), parasitic habit, species combination, initial inoculum quantities, host cultivar, penetration rate and other environmental conditions. Root invasion by migratory endoparasites can be faster than that by root-knot nematodes, but earlier invasion does not always result in long-term dominance. *Pratylenchus penetrans* invaded roots more rapidly than J2 of *M. incognita*, but roots began to respond to the presence of *M. incognita* J2 even before they were invaded. The invasion of red clover roots and lucerne by J2 of *M. incognita* was reduced when the ratio of entrant *P. penetrans* to *M. incognita* inoculum was 2–3:1 and there were 150–200 nematodes per root. Invasion of roots by adults of *P. penetrans* was not reduced in the reciprocal combinations 72 h after inoculation, and neither *M. incognita* nor *P. penetrans* affected root penetration by the other (Turner and Chapman, 1972; Chapman and Turner, 1975). *Rotylenchulus reniformis* penetration and multiplication were unaffected by a few *M. incognita* (10 J2/pot) in simultaneous inoculations of both nematodes, while higher *M. incognita* inoculum levels (100 J2/pot) affected multiplication of both species, although penetration was unaffected (Khan *et al.*, 1985).

Attractiveness and suitability of roots for nematode feeding and reproduction are also affected by prior nematode activity and may alter the ability of another species to penetrate. *Meloidogyne incognita* migrated to non-infected roots rather than to *M. incognita*-infected or *H. columbus*-infected roots of cotton cultivars Davis (susceptible to *M. incognita* and *H. columbus*) and Centennial (resistant to *M. incognita* and tolerant to *H. columbus*). *Hoplolaimus columbus* showed a similar preference pattern. It appeared that *M. incognita*-infected roots were more attractive than *H. columbus*-infected roots to both nematodes. *Meloidogyne incognita* stimulated reproduction of *H. columbus* on cotton to the extent that the latter may become the dominant species in cotton fields where mixed populations occur (Guy and Lewis, 1987a).

The parasitic habit of interacting nematode species and infection sequence may adversely affect *Meloidogyne*, as shown in studies involving cyst and root-knot nematodes. Treatments in which *H. schachtii* preceded *M. hapla* on sugarbeet (*Beta vulgaris*) (Jatala and Jensen, 1976) resulted in less gall formation, and there were no differences in total cyst formation when plants were inoculated alone, with both nematodes simultaneously or with *H. schachtii* preceding *M. hapla*. Increases in cyst formation occurred when inoculation with *M. hapla* preceded *H. schachtii* inoculation by 10 days. The optimum number of *M. hapla* as a predisposing factor for *H. schachtii* was 250–500 J2/pot (Jatala and Jensen, 1976).

Meloidogyne incognita did not seem to interfere with penetration of *Heterodera zeae* in simultaneous inoculations, although *M. incognita* was adversely affected by *H. zeae* alone or in combination with *Tylenchorhynchus vulgaris*, the latter species increasing penetration of both endoparasites unless *T. vulgaris* was already established. Prior establishment of any of the three species, either singly or in combination, reduced the invasiveness of both of the other species (Kaul and Sethi, 1982).

In multi-species inoculations on a host, the sequence in which the species are inoculated can affect the population growth of a subsequent species, as well as dominance. Prior invasion of *M. incognita* suppressed *P. brachyurus* populations on tomato while it had no effect when the host was lucerne or tobacco. *Meloidogyne incognita* populations on cotton were generally inhibited by *P. brachyurus* (Gay and Bird, 1973). In banana, the interaction between *R. similis*, *H. multicinctus*, *M. incognita* and *P. coffeae* was investigated in a concomitancy experiment in pots. *Meloidogyne incognita* was the only previously inoculated species that suppressed *R. similis* populations. Both species have a similar habitat, which allows interspecific competition to occur, with *M. incognita* showing a slight tendency to decrease when *R. similis* numbers increased. This is probably due to destruction of galled tissues and feeding sites of *M. incognita* by the migratory habit of *R. similis*. Stabilization, or even slight decrease, in final *H. multicinctus* and *M. incognita* numbers in fresh roots was related to intraspecific competition at penetration. The frequency and population levels of *M. incognita* in commercial plantations are very low as a consequence of an interspecific competition with *R. similis* (Moens *et al.*, 2006). In cotton, populations of *H. columbus* increased in simultaneous inoculations with either *M. incognita* or *S. brachyurus* both at 60 and 90 days after inoculation. Most J2 of *M. incognita* did not develop in the presence of *H. columbus*. The rate of *Meloidogyne* penetration and reproduction was also affected by the presence of concomitant *H. columbus* and *S. brachyurus*, although *S. brachyurus* did not suppress penetration of *M. incognita* (Kraus-Schmidt and Lewis, 1981).

Simultaneous inoculations of lucerne with *M. hapla* J2 and *D. dipsaci* at 10, 20, 24 and 28 °C did not depress penetration of either nematode on root-knot-resistant and susceptible cultivars of lucerne. Inoculations with *D. dipsaci* at 2, 4, 6 and 8 weeks before inoculation with *M. hapla* at the different temperatures did not influence the resistance or susceptibility of cvs Nev syn xx (resistant to both nematode species), Lahontan (resistant to *M. hapla* and susceptible to *D. dipsaci*) or Ranger (susceptible to both species) (Griffin, 1980). There was a direct correlation between galling of inoculated seedlings of resistant progeny and temperature, inoculated 8-week-old cuttings of resistant plants being galled only at 32 °C (Griffin and Hunt, 1972).

10.3.3 Additive interaction

The combined effect of *H. glycines* and *M. incognita* on host yield was among early examples of an additive effect. Three years' plant growth and yield data showed that an interaction occurred between *H. glycines* and *M. incognita* on soybean and was related to changes in nematode population. In the first year, *M. incognita* reduced yield more in the presence of *H. glycines* than when either nematode was present alone. The combined effect of these species on soybean yield was considered to be more than additive. In the two subsequent crop seasons, plant weight reductions were equal when lower population levels of *M. incognita*, alone or in combination with *H. glycines*, occurred. This interaction response was described as additive, but the response to a high level of *M. incognita* and *H. glycines* was not additive, presumably because of the higher root-knot population (Ross, 1964).

10.3.4 Competition

The interaction of *H. schachtii*, *M. hapla* and *N. aberrans* fits the definition of competition for resources. Combined inoculations of *H. schachtii*, *M. hapla* and *N. aberrans* suppressed the growth of sugarbeet, and, over time, *Nacobbus* was antagonistic to the reproduction of *H. schachtii* and

M. hapla, particularly to the latter. Simultaneous root infection resulted in different spatial distributions of the adult females of the three species, with *M. hapla* females located in the external layers of the galls, females and cysts of *H. schachtii* more peripherally at the base of *N. aberrans*-induced galls and syncytia of *N. aberrans* in the central part of the root (Inserra *et al.*, 1984). Simultaneous infection by *M. incognita* and *N. aberrans* on tomato resulted in the formation of feeding sites for both species, located in either the same or separate vascular bundles. It was also noted that *N. aberrans* affected the development of the giant cells induced by *M. incognita* (López-Portillo *et al.*, 1984).

Although *H. glycines* and *M. incognita* are endoparasites, infection sites of the two nematodes and their effect on the host differ; J2 of *Meloidogyne* invade the region just behind the root tip, while J2 of *H. glycines* invade more mature tissues (Ross, 1964). The different physiological alterations caused by the two nematodes are manifest by an increase in percentage of protein in soybeans harvested from plants infected with *M. incognita* compared with a decrease in plants infected with *H. glycines*. Inhibition of nitrogen-nodule formation on soybean roots infected with *H. glycines* and nitrogen deficiency may be involved in reduction of *Meloidogyne* populations in plants that are also infected with *H. glycines* (Ross, 1964).

Weak pathogens, such as *Tylenchorhynchus brassicae*, did not inhibit the rate of population increase of *M. incognita* when their own density was high (Khan *et al.*, 1986b). However, migratory endoparasitic nematodes, including some species that can be considered as stronger pathogens, can or cannot exert dominance over *Meloidogyne*. Simultaneous inoculations of *M. incognita* and *R. reniformis* mutually suppressed the rate of population increase of each other, the effect being more marked for *M. incognita*, allowing *R. reniformis* to survive, adapt and compete better than the former in a community system (Khan *et al.*, 1986b). *Hoplolaimus columbus* may inhibit *M. incognita* on cotton from becoming the predominant species (Bird *et al.*, 1974). Experimental evidence has also indicated that reproduction of *H. columbus* was enhanced in the presence of *M. incognita* (Kraus-Schmidt and Lewis, 1981). The final root population of *M. incognita* or *H. columbus* declined linearly with increasing initial population density of the other species, the suppressive effect of *H. columbus* on *M. incognita* being attributed to retardation, cessation of maturation and suppression of penetration, as well as minor changes in host cells that may interfere with *M. incognita* feeding and reproduction through physiological alteration of the plant. Conversely, *M. incognita* may render the root more attractive to *H. columbus* by providing either penetration sites due to increased root mass or more attractive chemical cues than in roots not infected with root-knot nematodes. The interaction between the two species may not be strongly competitive based on a strict definition of competition (Kraus-Schmidt and Lewis, 1981).

In cotton, the preferred penetration sites for *R. reniformis* are the fully differentiated young roots, not the root tip, but, according to Khan *et al.* (1985), the penetration zone for both *M. incognita* and *R. reniformis* in tomato is the root tip, and this may affect or even reduce penetration when cohabitance occurs, nematodes effectively competing for penetration sites and hence establishment, thereby affecting the future rate of population increase of each species. *Hoplolaimus columbus* is also a migratory endoparasite, but feeds almost exclusively in the cortex. Interaction with *Meloidogyne* occurs probably via physiological changes in the roots and translocatable factors. Although *H. columbus* may penetrate the galled tissue, it stays within the swollen cortex of the gall, but aborted and necrotic *M. incognita* giant cells have been attributed to a physiological effect of *H. columbus* on the host (Kraus-Schmidt and Lewis, 1981).

10.3.5 Interactions between *Meloidogyne* species

Dominance of *M. javanica* over *M. hapla* has been reported (Johnson and Nusbaum, 1970; Kinloch and Allen, 1972). The ratio of mature females of *M. hapla* and *M. javanica* in mixed-species infections demonstrated the dominance of the latter species in experiments involving divided tomato root systems. Depressive effects of *M. incognita* upon *M. hapla* in plants of tobacco 'NC95' and 'NC2512' (which are resistant to *M. incognita* but susceptible to *M. hapla*) was related to rapid necrosis of root tips caused by invasion of J2 of *M. incognita* (Johnson and Nusbaum, 1970). Reduction of infection sites available for *M. hapla* has also been attributed to a hypersensitive reaction, but root tip necrosis did not interfere with infection by *P. brachyurus* populations (Johnson and Nusbaum, 1970; Gay and Bird, 1973).

According to Khan and Haider (1991), mutually inhibitory interactions occurred between *M. javanica* and *M. incognita*, but these negative interactions did not impact on plant growth. Mutual suppression effects were smaller for *M. javanica* with respect to root galls, egg mass production, total number of females, total population and reproduction than for *M. incognita*, but variations occurred among the host races of *M. incognita*, with R2 appearing to be more competitive than the other races (Khan and Haider, 1991), but it is difficult to establish if differences observed in race interactions studies are due to a more general (species) or race-specific effect when only one population of a race is included. To include multiple populations of each race would help to establish that interaction differences can be attributed confidently to a specific race and not to a more general 'species' effect.

Different *Meloidogyne* species often have different basal temperature or thermal requirements for physiological processes such as embryogenesis, host penetration, reproduction and generation time. Interactions between *M. incognita* and *M. hapla* are greatly influenced by temperature, the latter not being able to invade roots at 35 °C (Johnson and Nusbaum, 1970; Khan and Haider, 1991). *Meloidogyne incognita* dominance over *M. javanica* and *M. hapla* occurs at high temperatures (25–32 °C). Kinloch and Allen (1972) reported that populations of *M. hapla* and *M. javanica* matured slowly at 15 °C, with *M. hapla* maturing a few days sooner than the latter. Development of *M. hapla* was most rapid at 25 °C, with *M. javanica* maturing before *M. hapla* at 30 °C. A soil temperature of 20 °C was equally suitable for the invasion and development of *M. hapla* and *M. javanica*, although the latter species predominated in a mixed-species infection at this temperature (Kinloch and Allen, 1972).

In mixed infections of *M. javanica* and *M. hapla*, maximum invasion of tomato roots by *M. javanica* occurred 2 days before maximum invasion by *M. hapla*. An intermediate condition prevailed at higher inoculum levels, with nematodes being equally distributed between terminal and non-terminal galls (Kinloch and Allen, 1972). Terminal galls, more prevalent in infections by *M. hapla*, may result from a cessation of the mitotic activity of the root meristem, thus resulting in a limited number of nematodes per gall, changes in gall size and reduced penetration at increased inoculum densities. Different plant growth regulators associated with *M. hapla* have been suggested as the explanation for suppression of root tip meristems by this species (Kinloch and Allen, 1972).

10.3.6 Effect on host

Estores and Chen (1970) found that *M. incognita acrita* (= *M. incognita*) alone caused more severe stunting on tomato than when combined with *P. penetrans*. Cotton seedlings inoculated with *S. brachyurus*, *H. columbus* and *M. incognita* singly and in all possible combinations showed that treatments containing *M. incognita* had lower shoot weights. Root weights were often suppressed in treatments with *M. incognita* at P_i (initial population as nematodes/15 cm pot) 4500 but not at P_i 1500, and combined species suppressed shoot weight more than any species alone under glasshouse conditions (Kraus-Schmidt and Lewis, 1981). Single or combined inoculations of *M. incognita*, *M. hapla* and *P. brachyurus* depressed tobacco plant height in most treatments of *Meloidogyne*-susceptible varieties ('Hicks' and 'NC2336'), but not resistant varieties ('NC95' and 'NC2512'). The mechanisms by which *M. incognita* spp. suppressed reproduction of *P. brachyurus* in 'Hicks' and 'NC2336' were attributed to an indirect expression of host response, and not only to individual nematode species but also to the combination of species (Johnson and Nusbaum, 1970). Yield reduction due to the combined effect of *H. glycines* and *M. incognita* on soybean varied, depending on nematode population levels, from slightly less to slightly more than the additive effects (Ross, 1964).

Growth reduction of tomato plants in response to combined inoculations of *M. incognita*, *R. reniformis* and *T. brassicae* was much less than the total sum of reductions caused by the same inoculum levels of nematodes when inoculated separately as single species (Khan *et al.*, 1986b). A linear relationship was produced between the initial inoculum and plant dry weight in *M. incognita*, *R. reniformis* and *T. brassicae* single-species inoculations, but reductions in plant weight in combined inoculations were relatively lower than the total sum of reductions caused by the same inoculum levels in single-species inoculations. Mutual inhibitory interactive effects that caused decline in their rate of population increase resulted in reduction of their damaging potentials (Khan *et al.*, 1986a,b). Relationships between different population densities of *M. incognita* and *R. reniformis* singly or in combination in pot experiments showed that both species caused significant growth reduction on black gram (*V. mungo*) at the level of one infective juvenile per cm^3 of soil. The extent of growth reduction was relatively less in concomitant inoculations than the effect of single-species inoculations. Fresh shoot weight was more sensitive than shoot length to the presence of nematodes. Pod formation and the number of rhizobial nodules were reduced (Mishra and Gaur, 1981).

A direct correlation between plant age and resistance and susceptibility of lucerne to *M. hapla* was reported by Griffin and Hunt (1972). Plant age, soil temperature and use of single species versus combined inocula of *M. hapla* and *D. dipsaci* affected the galling (root-knot indices) on 'Vernal 298' lucerne (a selection resistant to *M. hapla* and susceptible to *D. dipsaci*). The highest root-knot index was observed at 28 °C in plants inoculated as seedlings with a combination of *M. hapla* and *D. dipsaci* (Griffin, 1980).

Occasionally, the effect of nematode interactions can result in host growth stimulation. For example, *M. incognita* and *M. hapla* increased the fresh root weight of susceptible (but not resistant) tobacco (Johnson and Nusbaum, 1970). Fresh shoot weights of plants inoculated alone or simultaneously with *M. hapla* or *H. schachtii* were greater than weights of non-inoculated plants and plants inoculated sequentially with both nematodes, regardless of order (Jatala and Jensen, 1976). *Rotylenchulus reniformis* feeds in the endodermis and pericycle region, the generating centre for lateral roots. Therefore, parasitism in this region might hamper the regeneration of lateral roots, but regeneration and proliferation of lateral roots in response to infection by *M. incognita* may increase the water and nutrient uptake by plant roots (Khan *et al.*, 1985).

10.4 Basis for Interactions

There have been numerous studies that have attempted to determine why plants infected by *Meloidogyne* spp. are more susceptible to many other pathogens. Initially it was assumed that root wounding played a major role in these disease complexes. In complexes involving bacteria, root wounding probably contributes, as the effect of the nematodes declines with increasing time between inoculation with the nematodes and the bacteria (Lucas *et al.*, 1955; Stewart and Schindler, 1956). However, there are no reports that clearly demonstrate an increased plant susceptibility to fungi due to wounding similar to that induced when plants are infected by *Meloidogyne* spp. Indeed, other nematode species that cause more physical wounding are not involved in disease complexes of the same nature as those involving *Meloidogyne* spp. (Starr and Mai, 1976). That disease complexes are more pronounced when nematode infection precedes inoculation with the fungal pathogen by 3–4 weeks is taken as evidence against wounds per se having a major role in increasing susceptibility to other pathogens. However, in at least one study, *R. solani* was observed to invade the galls following the path in the gall made when the egg matrix was secreted by the mature *Meloidogyne* female (Golden and Van Gundy, 1975).

In the case of cotton seedling disease, delayed development of cotton plants infected with *M. incognita* may contribute to increased seedling disease. As cotton plants grow beyond the seedling stage, their innate resistance to seedling diseases increases (Brodie and Cooper, 1964). Delaying cotton development by nematode infection or by removing the cotyledons lengthens the time during which the plants are susceptible to seedling disease pathogens, thus increasing the incidence and severity of seedling disease (Brodie and Cooper, 1964).

Nematode-induced giant cells degenerate rapidly when infected by various root pathogenic fungi (Meléndez and Powell, 1967; Starr and Aist, 1977). Since giant cells and associated gall tissues contain higher concentrations of amino acids (Owens and Specht, 1966; Sidhu and Webster, 1977), it has been suggested that this tissue provides an enriched nutrient source for the fungi that colonize the galls (Sidhu and Webster, 1977). *In vitro* growth rates of *P. polymorphon* on media prepared from root-gall tissues and uninfected roots were similar, but the fungus produced more oospores on the media prepared from gall tissues compared with media from non-galled roots (Starr and Aist, 1977). In complexes with fungal pathogens, the galls appear to be the initiation site of fungal penetration, because root necrosis is first observed associated with the galls (Golden and Van Gundy, 1975; Starr and Mai, 1976). This suggests that the galls are more susceptible than adjacent non-galled tissues and/or that root-rot fungi are specifically attracted to the gall tissues. In the case of *R. solani* and *M. javanica* on okra, the fungus appears to be attracted to the galls, where it forms sclerotia on the gall surface, but actual penetration of the host tissues is delayed until the galls are 4–5 weeks old (Golden and Van Gundy, 1975). This delayed penetration corresponded in time to when there was an increase in nitrogenous compounds in root leachates, which favours pathogenic activity of *R. solani* (Van Gundy *et al.*, 1977).

More recently, it has been demonstrated that *Meloidogyne* spp. induce profound changes in plant gene expression, both in the giant cells specifically (Wang *et al.*, 2003), and globally in the roots (Schaff *et al.*, 2007). Expression of some genes is increased, in some cases by more than 50-fold (Wang *et al.*, 2003), whereas expression of other genes is suppressed. In one study, a greater portion of the genes whose expression was altered in response to infection by *M. incognita* were downregulated rather than upregulated (Schaff *et al.*, 2007). Many of the host genes with altered patterns of expression due to infection by root-knot nematodes are genes related to plant defence pathways (Bird and Wilson, 1994; see Abad *et al.*, Chapter 7, this volume). Given the profound effects *Meloidogyne* spp. have on gene expression both in giant cells and more globally, and the resulting physiological and biochemical changes in affected tissues, perhaps we should not be surprised that such tissues have altered responses to challenge by other pathogens.

10.5 Conclusions and Future Directions

Most research on root-knot nematodes has focused on the four *Meloidogyne* species (*M. arenaria*, *M. hapla*, *M. javanica* and *M. incognita*) widely recognized as the most common and economically important species. Few studies are available on those species characterized by more restricted host ranges, fewer generations per season, or less extensive gall formation. In the case of interactions with microbial pathogens, there is little understanding of how the nematodes increase plant susceptibility to these various pathogens. It will probably require extensive studies of nematode effects on host gene expression to develop testable hypotheses that may eventually shed light on these complex biological systems. Although all *Meloidogyne* species induce apparently similar giant cells within host roots, the galling response varies greatly among hosts and nematode species. This suggests that, whereas many plant responses to root-knot nematodes are quite general, some *Meloidogyne* species/host species interactions are quite specific. Thus, while there are many common characteristics of the interaction of *Meloidogyne* species with other pathogens, we should not be surprised to find that some interactions have unique features.

With respect to the interactions of *Meloidogyne* species with other plant-parasitic nematodes, there are still few long-term studies, particularly under field conditions, and even fewer interactions have been epidemiologically characterized. Reviews of methodologies and of the study of interactions by authors such as Sikora and Carter (1987) and Eisenback and Griffin (1987) should be considered when planning methodologies. Norton (1989) has questioned whether nematode–nematode 'interactions' that operate indirectly, such as by causing physiological change in the host, are really host–parasite relationships that result in nematode succession by a decrease or increase of the population of other species. One potentially useful line of investigation would be the effect of introducing a low population density of a *Meloidogyne* species (a few egg masses or a small volume of infested soil) into an established nematode community in typical cropping systems that have one or more susceptible hosts. We know little about the ability of the *Meloidogyne* species to become established and spread from a point source of introduction. There has been much recent speculation in the southern USA about the reniform nematode *R. reniformis* displacing *M. incognita* in fields planted regularly to cotton.

Unfortunately, there are few data to support this speculation. Further, given the general tendency for *M. incognita* to be better adapted to soils with high sand content and for the reniform nematode to be better adapted to soils with greater silt content (Robinson *et al.*, 1987; Starr *et al.*, 1993), how would soil types affect the ability of either species to become dominant? Similarly, these agroecosystems often contain other crops, such as soybean, that are hosts to both nematode species, as well as crop species that may be a non-host for one but a moderate to good host to the other nematode. These alternative crops in the system will affect long-term shifts in the densities of both species. Both nematode species have many weed hosts that would similarly affect population dynamics. There are many opportunities for important research on the behaviour of *Meloidogyne* species in natural and agroecosystems.

Studies on interactions should include both a thorough statistical background and its translation into the 'how to' of the biological basis of interactions. Molecular approaches and understanding of the host–parasite relationship are advancing rapidly, and the information from the genome of *Meloidogyne* (see Abad and Opperman, Chapter 16, this volume), along with the genomes of other nematodes, microbes and hosts, will aid greatly our study of these biological systems. This will facilitate the gathering of new information concerning *Meloidogyne* interactions with other pathogens through physiological processes and metabolic pathways (Williamson and Gleason, 2003; Bird, 2004).

10.6 References

Abawi, G.S. and Barker, K.R. (1984) Effects of cultivar, soil temperature, and population levels of *Meloidogyne incognita* on root-necrosis and Fusarium wilt of tomato. *Phytopathology* 74, 433–438.

Agrios, G.N. (1997) *Plant Pathology*, 4th edn. Academic Press, New York.

Amosu, J.O. and Taylor, D.P. (1975) Interaction of *Meloidogyne hapla*, *Pratylenchus penetrans* and *Tylenchorhynchus agri* on Kenland clover, *Trifolium pratense. Indian Journal of Nematology* 4, 124–131.

Atkinson, G.F. (1892) Some diseases of cotton. *Alabama Polytechnical Institute and Agricultural Experiment Station Bulletin* No. 41, 61–65

Avelar-Mejía, J.J., Téliz-Ortíz, D. and Zavaleta-Mejía, E. (2001) Pathogens associated to 'guava decline'. *Mexican Journal of Phytopathology* 19, 223–229.

Bird, D.M. (2004) Signalling between nematodes and plants. *Current Opinion in Plant Biology* 7, 372–376.

Bird, D.M. and Wilson, M.A. (1994) DNA sequence and expression analysis of root-knot nematode-elicited giant cell transcripts. *Molecular Plant–Microbe Interactions* 7, 419–424.

Bird, G.W., Brooks, O.L. and Perry, C.E. (1974) Dynamics of concomitant field populations of *Hoplolaimus columbus* and *Meloidogyne incognita. Journal of Nematology* 6, 190–191.

Brodie, B.B. and Cooper, W.E. (1964) Relation of parasitic nematodes to post-emergence damping-off of cotton. *Phytopathology* 54, 1023–1027.

Campos, V.P. and Villain, L. (2005). Nematode parasites of coffee and cocoa. In: Luc, M., Sikora, R.A. and Bridge, J. (eds) *Plant Parasitic Nematodes in Subtropical and Tropical Agriculture*, 2nd edn. CAB International, Wallingford, UK, pp. 529–579.

Caperton, C.M., Martyn, R.D. and Starr, J.L. (1986) Effects of *Fusarium* inoculum density and root-knot nematodes on wilt resistance in summer squash. *Plant Disease* 70, 207–209.

Carneiro, R.M.D.G., Almeida, M.R. and Quénéhervé, P. (2000) Enzyme phenotypes of *Meloidogyne* spp. populations. *Nematology* 2, 645–654.

Castillo, P., Navas-Cortes, J.A, Gomar-Tinco, D., Di Vito, M. and Jiménez-Díaz, R.M. (2003) Interactions between *Meloidogyne artiellia*, the cereal and legume root-knot nematode, and *Fusarium oxysporum* f. sp. *ciceris* race 5 in chickpea. *Phytopathology* 93, 1513–1523.

Chapman, R.A. and Turner, D.R. (1975) Effect of *Meloidogyne incognita* on reproduction of *Pratylenchus penetrans* in red clover and alfalfa. *Journal of Nematology* 7, 6–10.

Chitamber, J.J. and Raski, D.J. (1984) Reactions of grape rootstocks to *Pratylenchus vulnus* and *Meloidogyne* spp. *Journal of Nematology* 16, 166–170.

Diop, M.T., Ndiaye, S., Mounport, D. and Thierry, M. (2002) Développement des populations de *Meloidogyne javanica* et de *Scutellonema cavenessi* dans les systèmes de culture maraîchère au Sènègal. *Nematology* 2, 535–540.

Doucet, M.E., Lax, P., Lorenzo, E., Gallardo, C. and de L'Argentier, M. (2007) Histological observations of sedentary plant-parasitic nematodes in two varieties of Andean potato (*Solanum tuberosum* L. subsp. *andigenum*) from northern Argentina. *Nematropica* 37, 121–125.

Egunjobi, O.A., Akonde, P.T. and Caveness, F.E. (1986) Interaction between *Pratylenchus sefaensis*, *Meloidogyne javanica* and *Rotylenchulus reniformis* in sole and mixed crops of maize and cowpea. *Revue de Nématologie* 9, 61–70.

Eisenback, J.D. (1993) Interactions between nematodes in cohabitance. In: Khan, W.M. (ed.) *Nematode Interactions.* Chapman and Hall, London, pp. 134–174.

Eisenback, J.D. and Griffin, G.D. (1987) Interactions with other nematodes. In: Veech, J.A. and Dickson, D.W. (eds) *Vistas on Nematology: a Commemoration of the Twenty-fifth Anniversary of the Society of Nematologists.* Society of Nematologists, Hyattsville, Maryland, pp. 313–320.

El-Goorani, M.A., Abo-El-Dahab, M.K. and Mehiar, F.F. (1974) Interaction between root-knot nematodes and *Pseudomonas marginata* on gladiolus corms. *Phytopathology* 64, 271–272.

Estores, R.A. and Chen, T.A. (1970) Interaction of *Pratylenchus penetrans* and *Meloidogyne incognita acrita* as cohabitants on tomatoes. *Journal of Nematology* 4, 170–174.

Fattah, F.A. and Webster, J.M. (1983) Ultrastructural changes caused by *Fusarium oxysporum* f. sp. *lycopersici* in *Meloidogyne javanica* induced giant cells in Fusarium resistant and susceptible tomato cultivars. *Journal of Nematology* 15, 128–135.

Fichtner, S.M., Isakeit, T., Wheeler, T.A., Kaufman, H.W. and Gannaway, J.R. (2005) Evaluation of several approaches to management of *Meloidogyne incognita* and cotton seedling disease complexes in the high plains of Texas. *Journal of Nematology* 37, 66–70.

Gay, C.M. and Bird, G.W. (1973) Influence of concomitant *Pratylenchus brachyurus* and *Meloidogyne* spp. on root penetration and population dynamics. *Journal of Nematology* 5, 212–217.

Golden, J.K. and Van Gundy, S.D. (1975) A disease complex of okra and tomato involving the nematode, *Meloidogyne incognita*, and the soil-inhabiting fungus *Rhizoctonia solani*. *Phytopathology* 65, 265–273.

Griffin, G.D. (1980) Interrelationship of *Meloidogyne hapla* and *Ditylenchus dipsaci* on resistant and susceptible alfalfa. *Journal of Nematology* 12, 287–293.

Griffin, G.D. (1985) The interrelationship of *Heterodera schachtii* and *Ditylenchus dipsaci* on sugarbeet. *Journal of Nematology* 15, 426–432.

Griffin, G.D. and Hunt, O.J. (1972) Effect of plant age on resistance of alfalfa to *Meloidogyne hapla*. *Journal of Nematology* 4, 87–90.

Griffin, G.D. and Waite, W.W. (1982) Pathological interaction of a combination of *Heterodera schachtii* and *Meloidogyne hapla* on tomato. *Journal of Nematology* 14, 182–187.

Griffin, G.D., Anderson, J.L. and Jorgenson, E.C. (1968) Interaction of *Meloidogyne hapla* and *Agrobacterium tumefaciens* in relation to raspberry cultivars. *Plant Disease Reporter* 52, 492–493.

Guy, D.W. Jr and Lewis, S.A. (1987a) Selective migration and root penetration by *Meloidogyne incognita* and *Hoplolaimus columbus* on soybean roots in vitro. *Journal of Nematology* 19, 390–392.

Guy, D.W. Jr and Lewis, S.A. (1987b) Interaction between *Meloidogyne incognita* and *Hoplolaimus columbus* on Davis soybean. *Journal of Nematology* 19, 346–351.

Hawn, E.J. (1963) Transmission of bacterial wilt of alfalfa by *Ditylenchus dipsaci* (Kühn). *Nematologica* 9, 65–68.

Herman, M., Hussey, R.S. and Boerma, H.R. (1988) Interactions between *Meloidogyne incognita* and *Pratylenchus brachyurus* on soybean. *Journal of Nematology* 20, 79–84.

Hillocks, R.J. (1992) Fusarium wilt. In: Hillocks, R.J. (ed.) *Cotton Diseases.* CAB International, Wallingford, UK, pp. 127–160.

Hunt, P.J., Griffin, G.D., Murray, J.J., Pedersen, M.W. and Peaden, R.N. (1971) The effects of root-knot nematodes on bacterial wilt in alfalfa. *Phytopathology* 61, 250–259.

Ibrahim, I.K.A. and Lewis, S.A. (1986) Interrelationships of *Meloidogyne arenaria* and *Meloidogyne incognita* on tolerant soybean. *Journal of Nematology* 18, 106–111.

Inserra, R.N., Griffin, G.D., Vovlas, N., Anderson, J.L. and Kerr, D. (1984) Relationship between *Heterodera schachtii*, *Meloidogyne hapla* and *Nacobbus* on sugarbeet. *Journal of Nematology* 16, 135–140.

Ishibashi, N. and Choi, D.R. (1991) Biological control of soil pests by mixed application of entomopathogenic and fungivorous nematodes. *Journal of Nematology* 23, 175–181.

Jatala, P. and Jensen, H.J. (1976) Parasitism of *Beta vulgaris* by *Meloidogyne hapla* and *Heterodera schachtii* alone and in combination. *Journal of Nematology* 8, 200–205.

Jenkins, W.R. and Coursen, B.W. (1957) The effect of root-knot nematodes, *Meloidogyne incognita acrita* and *M. hapla*, on Fusarium wilt of tomato. *Plant Disease Reporter* 41, 182–186.

Johnson, A.W. and Nusbaum, C.J. (1970) Interactions between *Meloidogyne incognita*, *M. hapla* and *Pratylenchus brachyurus* in tobacco. *Journal of Nematology* 4, 334–340.

Johnson, H.A. and Powell, N.T. (1969) Influence of root knot nematodes on bacterial wilt development in flue-cured tomato. *Phytopathology* 59, 486–491.

Jonathan, E.J., Umamaheswari, R. and Bommaraju, P. (2006) Bioefficacy of native plant growth promoting rhizobacteria against *Meloidogyne incognita* and *Phytophthora capsici* disease complex in betelvine. *Indian Journal of Nematology* 36, 234–240.

Jones, J.P., Overman, A.J. and Crill, P. (1976) Failure of root-knot nematode to affect Fusarium wilt resistance of tomato. *Phytopathology* 66, 1339–1341.

Kaul, R.K. and Sethi, C.L. (1982) Effect of simultaneous inoculations and prior establishment of *Heterodera zeae* and *M. incognita* and *Tylenchorhynchus vulgaris* singly and in combination on penetration of *H. zeae* and *M. incognita* into maize roots. *Indian Journal of Nematology* 10, 11 (Abstr.).

Khan, M.W. (1993) Mechanisms of interactions between nematodes and other plant pathogens. In: Khan, M.W. (ed.) *Nematode Interactions*. Chapman and Hall, London, pp. 55–78.

Khan, M.W. and Dasgupta, M.K. (1993) The concept of interaction. In: Khan, M.W. (ed.) *Nematode Interactions*. Chapman and Hall, London, pp. 42–54.

Khan, T.A. and Haider, S.R. (1991) Interaction of *Meloidogyne javanica* with different races of *Meloidogyne incognita*. *Journal of Nematology* 23, 298–305.

Khan, T.A. and Husain, S.I. (1989) Relative resistance of six cowpea cultivars as affected by the concomitance of two nematodes and a fungus. *Nematologia Mediterranea* 17, 39–41.

Khan, R.M., Khan, M.W. and Khan, A.M. (1985) Cohabitation of *Meloidogyne incognita* and *Rotylenchulus reniformis* in tomato roots and effect on multiplication and plant growth. *Nematologia Mediterranea* 13, 153–159.

Khan, R.M., Khan, M.W. and Khan, A.M. (1986a) Interactions of *Meloidogyne incognita*, *Rotylenchulus reniformis* and *Tylenchorhynchus brassicae* as cohabitants on eggplant. *Nematologia Mediterranea* 14, 201–206.

Khan, R.M., Khan, A.M. and Khan, M.W. (1986b) Interaction between *Meloidogyne incognita*, *Rotylenchulus reniformis* and *Tylenchorhynchus brassicae* on tomato. *Revue de Nématologie* 9, 245–250.

Kinloch, R.A. and Allen, M.W. (1972) Interaction of *Meloidogyne hapla* and *M. javanica* infecting tomato. *Journal of Nematology* 4, 7–16.

Kraus-Schmidt, H. and Lewis, S.A. (1981) Dynamics and concomitant populations of *Hoplolaimus columbus*, *Scutellonema brachyurum*, and *Meloidogyne incognita* on cotton. *Journal of Nematology* 13, 41–48

Kuehl, R.D. (1994) Statistical principles of research design and analysis. Duxbury Press, Belmont, California.

López-Portillo, A.C., Sosa-Moss, C. and Cardenas-Soriano, E. (1984) Histology of tomato roots attacked by *Meloidogyne incognita* race 1 and *Nacobbus aberrans*, separately and interaction. In: *Proceedings of the First International Congress of Nematology, Guelph, Ontario Canada, August 5–10, 1984*, p. 49 (Abstr.).

Luc, M., Sikora, R.A. and Bridge, J. (2005) *Plant Parasitic Nematodes in Subtropical and Tropical Agriculture*, 2nd edn. CAB International, Wallingford, UK.

Lucas, G.B., Sasser, J.N. and Kelman, A. (1955) The relationship of root-knot nematodes to Granville wilt resistance in tobacco. *Phytopathology* 45, 537–540.

Mai, W.F. and Abawi, G.S. (1987) Interactions among root-knot nematodes and Fusarium wilt fungi on host plants. *Annual Review of Phytopathology* 25, 317–338.

Mayol, P.S. and Bergeson, G.B. (1970) The role of secondary invaders in *Meloidogyne incognita* infection. *Journal of Nematology* 2, 80–83.

McClellan, W.D., Wilhelm, S. and George, A. (1955) Incidence of Verticillium wilt in cotton not affected by root-knot nematodes. *Plant Disease Reporter* 39, 226–227.

McGawley, E.C. and Winchell, K.L. (1987) Greenhouse reproduction of single and combined *Meloidogyne incognita* and *M. javanica* populations on soybean. *Journal of Nematology* 19, 542.

Meléndez, P.L. and Powell, N.T. (1967) Histological aspects of the Fusarium wilt–root-knot complex in flue-cured tobacco. *Phytopathology* 57, 286–292.

Mishra, S.D. and Gaur, H.S. (1981) Effect of individual and concomitant inoculation with *Meloidogyne incognita* and *Rotylenchulus reniformis* on the growth of black gram (*Vigna mungo*). *Indian Journal of Nematology* 11, 25–28.

Misra, C. and Das, S.N. (1979) Interaction of some plant parasitic nematodes on the root-knot development in brinjal. *Indian Journal of Nematology* 7, 46–53.

Moens, T., Araya, M., Swennen, R. and De Waele, D. (2006) Reproduction and pathogenicity of *Helicotylenchus multicinctus*, *Meloidogyne incognita* and *Pratylenchus coffeae*, and their interactions with *Radopholus similis* on *Musa*. *Nematology* 8, 45–48.

Moorman, G.W., Huang, J.S. and Powell, N.T. (1980) Localized influence of *Meloidogyne incognita* on Fusarium wilt resistance of flue-cured tobacco. *Phytopathology* 70, 969–970.

Navas-Cortés, J.A, Landa, B.B., Rodríguez-López, J., Jiménez-Díaz, R.M. and Castillo, P. (2008) Infection by *Meloidogyne artiellia* does not break down resistance to races 0, 1a and 2 of *Fusarium oxysporum* f. sp. *ciceris* in chickpea genotypes. *Phytopathology* 98, 709–718.

Niblack, T.L., Hussey, R.S. and Boerma, H.R. (1986) Effects of *Heterodera glycines* and *Meloidogyne incognita* on early growth of soybean. *Journal of Nematology* 18, 444–450.

Norton, D.C. (1969) *Meloidogyne hapla* as a factor in alfalfa decline in Iowa. *Phytopathology* 59, 1824–1828.

Norton, D.C. (1989) Abiotic factors and plant-parasitic nematode communities. *Journal of Nematology* 21, 299–307.

Owens, R.G. and Specht, H.N. (1966) Biochemical alterations induced in host tissues by root-knot nematodes. *Contributions from Boyce Thompson Institute* 23, 181–198.

Pathak, K.N., Nath, R.P. and Haider, M.G. (1985) Effect of initial inoculum level of *Meloidogyne incognita* and *Rotylenchulus reniformis* on pigeonpea and their interrelationship. *Indian Journal of Nematology* 15, 177–179.

Poornima, M.S. and Subramanian, S. (2006) Nematode–fungal disease complex involving *Meloidogyne incognita* and *Macrophomina phaseolina* on *Coleus forskohlii* Briq. *Indian Journal of Nematology* 36, 181–184.

Porter, D.M. and Powell, N.T. (1967) Influence of certain *Meloidogyne* species on Fusarium wilt development in flue-cured tobacco. *Phytopathology* 57, 282–285.

Powell, N.T. (1971) Interaction between nematodes and fungi in disease complexes. *Annual Review of Phytopathology* 9, 253–274.

Powell, N.T. and Nusbaum, C.J. (1960) The black shank-root knot complex in flue cured tobacco. *Phytopathology* 50, 899–906.

Powell, N.T., Melendez, P.L. and Batten, C.K. (1971) Disease complexes in tobacco involving *Meloidogyne incognita* and certain soil-borne fungi. *Phytopathology* 61, 1332–1337.

Rao, Y.S. and Prasad, J.S. (1981) Interaction of *Heterodera oryzicola* and *Meloidogyne graminicola* in artificial inoculation in rice. *Indian Journal of Nematology* 11, 104 (Abstr.).

Ras, K.T. and Seshadri, A.R. (1981) Studies on interactions between *Meloidogyne incognita* and *Rotylenchulus reniformis* on grapevine seedlings. *Indian Journal of Nematology* 11, 101–102.

Roberts, P.A., Smith, S.N. and Matthews, W.C. (1985) Quantitative aspects of the interaction of *Meloidogyne incognita* with Fusarium wilt on Acala cotton in field plots. *Proceedings Beltwide Cotton Conferences*, National Cotton Council of America. Memphis, Tennessee, p. 21 (Abstr.).

Roberts, P.A., Frate, C.A., Matthews, W.C. and Osterli, P.P. (1995) Interactions of virulent *Meloidogyne incognita* and Fusarium wilt on resistant cowpea genotypes. *Phytopathology* 85, 1288–1295.

Robinson, A.F., Heald, C.M, Flannagan, S., Thames, W.H. and Amador, J. (1987) Geographic distribution of *Rotylenchulus reniformis*, *Meloidogyne incognita*, and *Tylenchulus semipenetrans* in the Lower Rio Grande Valley as related to soil texture and land use. *Annals of Applied Nematology* 1, 20–25.

Rodríguez-Kábana, R., Shelby, R.A., King, P.S. and Pope, M.H. (1982) Application time and effectiveness of four systemic nematicides against *Meloidogyne arenaria* on Florunner peanut. *Nematropica* 12, 85–96.

Rodríguez-Kábana, R., Robertson, D.G., Wells, L., Weaver, C.F. and King, P.S. (1991) Cotton as a rotation crop for the management of *Meloidogyne arenaria* and *Sclerotium rolfsii* in peanut. *Supplement to Journal of Nematology* 23, 652–657.

Ross, J.P. (1964) Interaction of *Heterodera glycines* and *Meloidogyne incognita* on soybean. *Phytopathology* 54, 304–307.

Santo, G.S. and Bolander, W.J. (1977) Separate and concomitant effects of *Macroposthonia xenoplax* and *Meloidogyne hapla* on concord grapes. *Journal of Nematology* 4, 282–283.

Schaff, J.E., Nielsen, D.M., Smith, C.P., Scholl, E.H. and Bird, D.M. (2007) Comprehensive transcriptome profiling in tomato reveals a role for glycosyltransferase in Mi-mediated nematode resistance. *Plant Physiology* 144, 1079–1092.

Senthilkumar, T. and Rajendran, G. (2003) Studies on wilt disease expression in the presence of *Meloidogyne incognita* and *Fusarium moniliforme* in grapevine. *Nematologia Mediterranea* 31, 187–189.

Sharma, N.K. and Sethi, C.L. (1978) Interaction between *Meloidogyne incognita* and *Heterodera cajani* on cowpea. *Indian Journal of Nematology* 6, 1–12.

Sheela, M.S. and Venkitesan, T.S. (1981). Interrelationships of infectivity between the burrowing and root-knot nematode in black pepper, *Piper nigrum*. *Indian Journal of Nematology* 11, 105 (Abstr.).

Sidhu, G. and Webster, J.M. (1974) Genetics of resistance in the tomato to root-knot nematode–wilt-fungus complex. *The Journal of Heredity* 65, 153–156.

Sidhu, G. and Webster, J.M. (1977) Predisposition of tomato the wilt fungus (*Fusarium oxysporum lycopersici*) by the root-knot nematode (*Meloidogyne incognita*). *Nematologica* 23, 436–442.

Sidhu, G. and Webster, J.M. (1981) Influence of population levels of root-knot nematode on Fusarium wilt severity of tomato. *Phytoprotection* 62, 61–66.

Sikora, R.A. and Carter, W.W. (1987) Nematode interactions with fungal and bacterial plant pathogens – factor fantasy. In: Veech, J.A. and Dickson, D.W. (eds) *Vistas on Nematology: a Commemoration of the Twenty-fifth Anniversary of the Society of Nematologists*. Society of Nematologists, Hyattsville, Maryland, pp. 307–312.

Sikora, R.A., Malek, R.B., Taylor, D.P. and Edwards, D.I. (1979) Reduction of *Meloidogyne naasi* infection of creeping bentgrass by *Tylenchorhynchus agri* and *Paratrichodorus minor*. *Nematologica* 25, 179–183.

Singh, N.D. (1976) Interaction of *Meloidogyne incognita* and *Rotylenchulus reniformis* on soybean. *Nematropica* 6, 76–81.

Sitaramaiah, K. and Parvathi D.G. (1994) Nematodes, soil microorganisms in relation to *Phytophthora* wilt disease in betelvine. *Indian Journal of Nematology* 24, 16–25.

Starr, J.L. and Aist, J.R. (1977) Early development of *Pythium polymorphon* on celery roots infected by *Meloidogyne hapla*. *Phytopathology* 67, 497–501.

Starr, J.L. and Mai, W.F. (1976) Effect of soil microflora on the interaction of three plant parasitic nematodes with celery. *Phytopathology* 66, 1224–1228.

Starr, J.L. and Martyn, R.D. (1991) Reaction of cotton cultivars to Fusarium wilt and root-knot nematodes. *Nematropica* 21, 51–58.

Starr, J.L., Jeger, M.J., Martyn, R.D. and Schilling, K. (1989) Effects of *Meloidogyne incognita* and *Fusarium oxysporum f. sp. vasinfectum* on plant mortality and yield of cotton. *Phytopathology* 79, 640–646.

Starr, J.L., Heald, C.M., Robinson, A.F., Smith, R.G. and Krausz, J.P. (1993) *Meloidogyne incognita* and *Rotylenchulus reniformis* and associated soil textures from some cotton production areas of Texas. *Supplement to Journal of Nematology* 25, 895–899.

Starr, J.L., Shim, M.-Y., Lee, T.A. Jr and Simpson, C. (1996) Additive effects of *Meloidogyne arenaria* and *Sclerotium rolfsii* on peanut. *Journal Nematology* 28, 99–106.

Stewart, R.N. and Schindler, A.F. (1956) The effect of some ectoparasitic and endoparasitic nematodes on expression of bacterial wilt of carnations. *Phytopathology* 46, 219–222.

Taha, A.H.Y. and Kassab, A.S. (1979). The histological reactions of *Vigna sinensis* to separate and concomitant parasitism by *Meloidogyne javanica* and *Rotylenchulus reniformis*. *Journal of Nematology* 11, 117–123.

Taha, A.H.Y. and Kasab, A.S. (1980) Interrelationships between *Meloidogyne incognita*, *Rotylenchulus reniformis*, and *Rhizobium* sp. on *Vigna sinensis*. *Journal of Nematology* 12, 57–62.

Taylor, C.E. (1979). *Meloidogyne* interrelationships with micro-organisms. In: Lamberti, F. and Taylor, C.E. (eds) *Root-knot Nematodes* (Meloidogyne *species*). Academic Press, London and New York, pp. 375–398.

Téliz-Ortíz, D., Castillo-Ponce, G. and Nieto-Angel, D. (1993) La corchosis del cafeto en México. *Revista Mexicana de Fitopatología* 11, 5–12.

Thomas, R.J. and Clark, C.A. (1980) Interactions between *Meloidogyne incognita* and *Rotylenchulus reniformis* on sweet potato. *Journal of Nematology* 12, 239 (Abstr.).

Thomas, R.J. and Clark, C.A. (1981) *Meloidogyne incognita* and *Rotylenchulus reniformis* interactions in a sweet potato field. *Phytopathology* 71, 908 (Abstr.).

Thomas, R.J. and Clark, C.A. (1983a) Effects of concomitant development on reproduction of *Meloidogyne incognita* and *Rotylenchulus reniformis*. *Journal of Nematology* 15, 215–221.

Thomas, R.J. and Clark, C.A. (1983b) Population dynamics of *Meloidogyne incognita* and *Rotylenchulus reniformis* alone and in combination, and their effects on sweet potato. *Journal of Nematology* 15, 204–211.

Trudgill, D.L. and Phillips, M.S. (1997) Nematode population dynamics, threshold levels and estimation of crop losses. In: Maqbool, M.A. and Kerry, B.R. (eds) *Plant Nematode Problems and their Control in the Near East Region.* FAO Plant Production and Protection Paper 144, Rome, pp. 45–58.

Turner, D. and Chapman, R.A. (1972) Infection seedlings of alfalfa and red clover by concomitant populations of *Meloidogyne incognita* and *Pratylenchus penetrans. Journal of Nematology* 4, 280–286.

Vaishnav, M.U. and Sethi, C.L. (1978) Pathogenicity of *Meloidogyne incognita* and *Tylenchorhynchus vulgaris* on bajra and their interrelationship. *Indian Journal of Nematology* 8, 1–8.

Van Gundy, S.D. and Kirkpatrick, J.D. (1975) Nematode–nematode interactions on tomato. *Journal of Nematology* 7, 330 (Abstr.).

Van Gundy, S.D., Kirkpatrick, J.D. and Golden, J. (1977) The nature and role of metabolic leakage from root-knot nematode galls and infection by *Rhizoctonia solani. Journal of Nematology* 9, 113–121.

Wallace, H.R. (1978) Diagnosis of diseases of complex etiology. *Annual Review of Phytopathology* 16, 379–402.

Wang, Z., Potter, R.H. and Jones, M.G.K. (2003) Differential display analysis of gene expression in the cytoplasm of giant cells induced in tomato roots by *Meloidogyne javanica. Molecular Plant Pathology* 4, 361–371.

Webster, J.M. (1985) Interaction of *Meloidogyne* with fungi on plants. In: Sasser, J.N and Carter, C.C. (eds) *An Advanced Treatise on* Meloidogyne. *Vol. I. Biology and Control.* North Carolina University Graphics, Raleigh, North Carolina, pp. 183–192.

Williamson, V.M. and Gleason, C.A. (2003) Plant nematode interactions. *Current Opinion in Plant Biology* 6, 327–333.

Yang, H., Powell, N.T. and Barker, K.R. (1976) Interaction of concomitant species of nematodes and *Fusarium oxysporum* f. sp. *vasinfectum* on cotton. *Journal of Nematology* 8, 74–80.

11 Population Dynamics and Damage Levels

Nicola Greco and Mauro Di Vito
Consiglio Nazionale delle Ricerche, Istituto per la Protezione delle Piante, Bari, Italy

11.1	Introduction	246
11.2	Patterns of Population Dynamics	246
11.3	Factors Affecting Population Dynamics	248
11.4	Modelling Population Dynamics	251
11.5	Damage Levels	253
11.6	Pattern of Nematode Damage to Crop Plants	254
11.7	Factors Affecting Nematode Damage	256
11.8	Modelling Damage Levels	259
11.9	Implementing Experiments to Assess Nematode Dynamics and Crop Damage	260
11.10	Yield Loss Assessment	268
11.11	Importance of Information on Nematode Damage Levels and Dynamics in Management Strategies	268
11.12	Conclusions and Future Directions	269
11.13	Acknowledgements	269
11.14	References	269

11.1 Introduction

Nematodes, like all living organisms, need to feed, reproduce and survive during unsuitable environmental conditions. Reproduction, survival and damage to host crops will vary and will be greater under environmental conditions that are optimal for the nematodes. As environmental conditions usually vary from basal to optimal and to extremely unsuitable for nematode development and survival, nematode populations in soil and living plant organs will change accordingly. The variation in the numbers of nematodes over time is defined as population dynamics. Understanding, and possibly modelling, population dynamics as affected by different factors is key information for predicting the yield loss the nematode may cause to crop plants and for implementing the most appropriate management tactics. As nematode populations will increase in the presence of a host plant and survive in its absence according to specific life cycle strategies, information on the rates of both increase and survival of the nematode is necessary for reasonably long-term predictions.

11.2 Patterns of Population Dynamics

A measure of the reproduction of a nematode species is given by the reproduction rate, which is defined as the ratio between the density of the

 ©CAB International 2009. *Root-knot Nematodes* (eds R.N. Perry, M. Moens and J.L. Starr)

nematode population in the soil at the end of the crop cycle – the final population density (P_f), and that present at sowing or planting, usually called the initial population density (P_i). However, for practical purposes with *Meloidogyne* assessments, such as for screening test plants, instead of a precise measure of the nematode reproduction, a reproduction index or egg mass index, or both, of the infected roots are used. The reproduction index (RI) usually refers to the nematode population on the root system and may be defined as the total number of eggs and infective juvenile stages produced on the roots of the test plant, and is expressed as a percentage of that on the roots of a reference plant, usually a fully susceptible cultivar of the same plant species (Fassuliotis, 1979). An alternative definition of RI used by many nematologists is RI = P_f / P_i (Cook and Noel, 2002). The egg mass index (EI) is, instead, based on the number of egg masses per root system and can be rated on a 0–5 scale, similar to that for the root gall index (Taylor and Sasser, 1978; see section 11.7.3). Theoretically, if there is sufficient food such that there is no competition among individuals for resources, the reproduction rate of the nematode would be constant and the relationship between P_f and P_i a straight line. In fact, with increasing initial nematode population density, food and space available to each individual nematode of the population is reduced. Also, with increasing P_i the root system becomes increasingly damaged, and this results in further reduction of food supply and, therefore, of the nematode reproduction rate (Fig. 11.1). Thus, the relation between P_f and P_i is better described by a logistic curve. With amphimictic nematode species at very low P_i, mating may be difficult, thus resulting in a reproduction rate smaller than at a larger P_i, at which the probability of mating is increased. In these cases, P_f/P_i relationships would be better described by a sinusoidal curve (Oostenbrink, 1966).

The initial nematode population density at which the reproduction rate is 1 (P_f/P_i = 1) is called the equilibrium density (E) or maintenance density, and the line in a graph representing a range of P_i at which P_f remains at the same level as P_i is referred to as the equilibrium density or maintenance line (Fig. 11.2). At very large initial population densities, because the roots of the host crop can be severely damaged and there is substantial competition among individuals for limited resources, the reproduction rate of the nematode may be <1, and the final nematode density may be smaller than the initial population density.

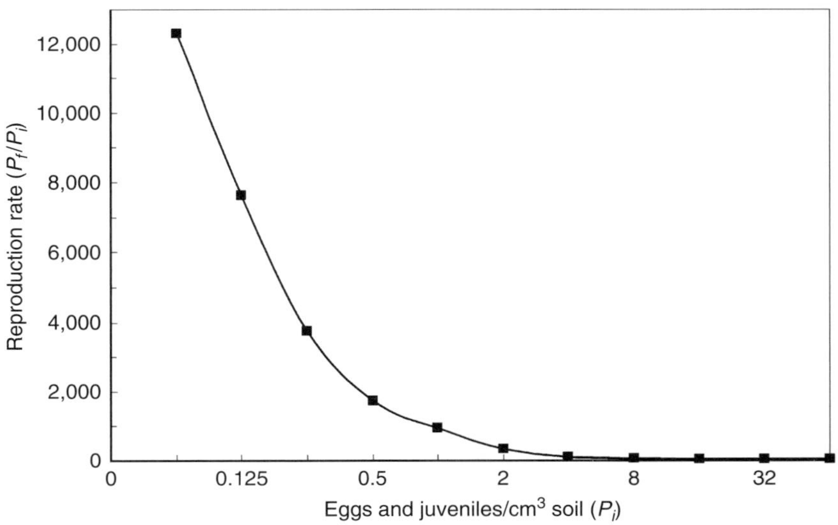

Fig. 11.1. Relationship between initial population densities (P_i) and reproduction rate (P_f/P_i) of *Meloidogyne incognita* on aubergine in microplots at Castellaneta (Apulia region, Italy). Note the decrease of the reproduction rate of the nematode with the increase of its P_i. (Adapted from Di Vito *et al.*, 1986.)

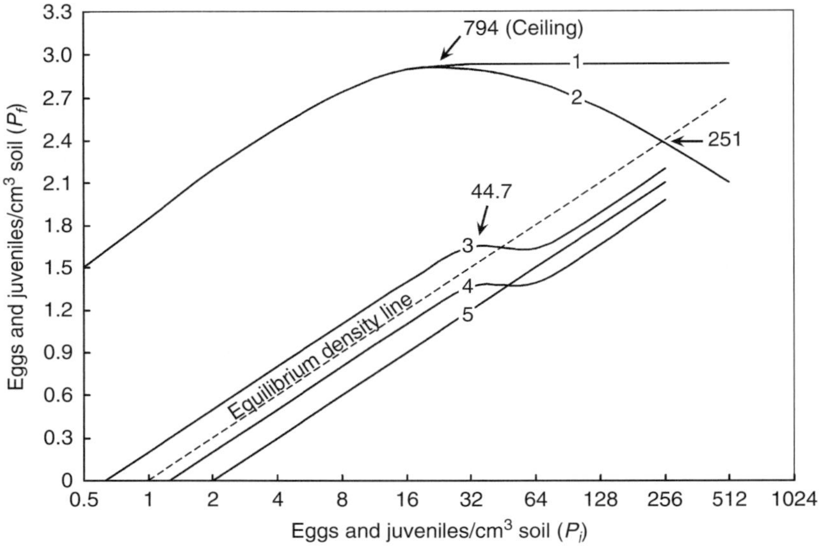

Fig. 11.2. Relationship between initial (P_i) and final (P_f) population densities of *Meloidogyne* spp. Line 1 represents the shape of the curve according to the first addendum of equation (6) of the Seinhorst (1970) model. Line 2 is typical of susceptible plants and shows: maximum P_f (ceiling level) of 794 eggs/cm³ soil at P_i = 16 eggs/cm³ soil; and equilibrium density (E) of 251 eggs/cm³ soil. Line 3 is typical of a moderately resistant plants and shows a maximum P_f of 44.7 eggs/cm³ soil at P_i = 32 eggs/cm³. Line 4 is typical of a very resistant plant and line 5 of a non-host plant. The dashed line is the equilibrium density line (where $P_f = P_i$).

Moreover, as soon as the roots begin to rot or die, the nematode population declines naturally, and will continue to do so in the absence of the host plant. Therefore, for a given nematode species/host crop combination, the nematode population will not increase indefinitely with increasing initial population densities, but will reach a maximum that cannot be surpassed. This maximum value of the final nematode density is called the 'ceiling level of the nematode population' (Fig. 11.2) (Oostenbrink, 1966), and will depend upon the intrinsic reproduction ability of the nematode species, the suitability and size of the root system of the host crop, the ability of the host crop to tolerate nematode parasitism, and the environmental conditions.

11.3 Factors Affecting Population Dynamics

11.3.1 The nematode species

The development of a nematode population depends very much on the number of generations it may complete and on its survival strategy (see Evans and Perry, Chapter 9, this volume). The tropical *Meloidogyne* species hatch and develop most rapidly in the temperature range 25–30 °C and develop more slowly below 20 °C. Further, they have rather short life cycles, which allows them to complete several generations during the growing season on a good host with favourable temperatures and moisture (Van Gundy, 1985). A few species have optimum temperatures in the range 20–25 °C, and have fewer generations per year. Moreover, the cool period required by *M. naasi*, for second-stage juveniles (J2) to hatch may limit both the number of generations per year (to as few as one) and the decline of the nematode population, which will start only after the cool period. In the Mediterranean area, *M. artiellia* completes only one generation per growing season (autumn–mid-spring), but J2 do not usually hatch during the dry season in late spring–autumn. However, if there are rains and some J2 hatch, they may survive any subsequent dry period in a state of anhydrobiosis (Di Vito and Greco, 1988a). The number of eggs per egg mass and the hatch strategy of the *Meloidogyne* species (see Curtis *et al.*, Chapter 6, this volume) also play important roles in nematode population dynamics. J2 of *M. naasi* hatch within a couple weeks (Gooris and d'Herde, 1977), while hatching of other species may last for several (4–8) weeks, and a small proportion of J2 may not hatch during the first host crop season and may be in the physiological state of diapause (de Guiran and Demeure, 1978; de Guiran 1980; de Guiran and Villemin, 1980; see Evans and Perry, Chapter 9, this volume).

11.3.2 Crop and cropping system

The crop is the only source of food for the nematode. Crop plants with large root systems or those that can replace dead or damaged roots with new adventitious roots will favour nematode reproduction and thus support a higher final nematode population density than plants lacking this ability. The response of the host crop to the nematode is also an important factor to consider. At large nematode densities, roots of intolerant crop plants are severely damaged, whereas those of tolerant plants will suffer less.

Sugarbeet is very intolerant to *M. naasi* and infected roots die; however, most J2 of *M. naasi* hatch over a short period of time and new roots may be produced when most of the hatched J2 have already died, thus giving little or no chance for the nematode to reproduce. Usually, infective J2 penetrate roots of resistant plants but, depending on the degree of resistance, none or only a few may reach the adult stage and produce eggs. Therefore, the rate of reproduction of the nematode will be close to or less than one and the final nematode density in the soil slightly more, or perhaps even less, than that at sowing or transplanting.

As nematode populations densities are related to the number of generations the nematode completes per growing cycle of the crop, the length of the growing cycle is important. Short-cycle crops will allow the nematode to complete fewer generations, perhaps only one, compared with long-cycle crops, and thus limit nematode populations. Several crop plants can be either sown or transplanted. In pots, transplanted tomatoes more than doubled reproduction rates and final soil population densities of *M. incognita* compared with sown tomatoes of the same cultivar and growing-cycle length (Ekanayake and Di Vito, 1984). Similarly, perennial crops such as trees allow root-knot nematodes to reproduce continuously for several consecutive years, so ceilings of the nematode population will be reached only after a few years. Thereafter, nematode population declines and increases will alternate due to rotations of seasons, and severe root damage and flushes of new root growth (Fig. 11.3) (Souza *et al.*, 2008).

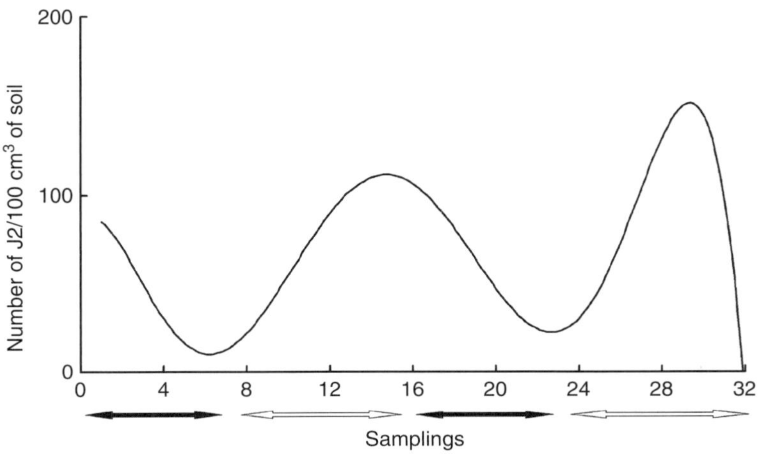

Fig. 11.3. Soil fluctuation of second-stage juveniles (J2) of *Meloidogyne exigua* in the rhizosphere of a coffee plantation in south-east Brazil during a 2-year period, at 40 cm distance from the trunk and at a depth of 0–25 cm. Samples were collected every 3 weeks. Rainy and dry seasons are marked by black and white arrows, respectively. (From Souza *et al.*, 2008.)

An interesting example of the relationship between a host and parasitism by root-knot nematodes is the difference between upland and lowland rice. Several root-knot nematode species are parasitic on rice, but most will infect and reproduce in roots only in upland rice. *M. graminicola*, however, reproduces on and damages rice under lowland conditions, but root infection will occur only before rice is flooded (Bridge *et al.*, 2005).

Cropping sequences greatly affect the dynamics of the nematode populations, with continuous growing of a host crop giving rise to larger nematode densities than a crop rotation in which the host crop is grown only once in 3 or 4 years. In soil left fallow or planted with non-host crops, the populations of *Meloidogyne* spp. in warm and dry areas can decline greatly and to nearly undetectable levels after a year. Johnson *et al.* (2000), in a winter wheat/cotton rotation in a field infested with *M. incognita*, observed marked decline of the nematode population following cultivation of wheat, and an increase following cotton. Similar results were obtained with a winter wheat/groundnut rotation in soil infested with *M. arenaria*, with the nematode population declining during wheat cultivation and increasing during groundnut cultivation. The nematode decline was greater when non-host plants were grown for two consecutive years. Different non-host plants will affect nematode decline differently and some may also produce root exudates possessing nematicidal activity (Ferraz and Grassi de Freitas, 2004). Moreover, as root-knot nematodes have large host ranges that also include weeds and grasses, the adoption or not of weeding practices will also affect nematode dynamics. Other crop maintenance practices that improve crop performance and favour nematode movement and infection, such as fertilization and irrigation, will also favour nematode reproduction and increase their density in the soil.

11.3.3 The season

Each nematode species has an optimal temperature requirement for hatch, infection and development. Therefore, whenever a host plant is cultivated during a season conducive to nematode development, a great increase of the nematode population can be expected. In summer crops in Italy, such as tomato, pepper, aubergine, tobacco, etc., population densities of *M. incognita* reach high levels by harvest (end of summer–early autumn), but they decline by as much as 87% 1 month later and by 94% after 6 months (Fig. 11.4) (Di Vito *et al.*, 1985a). In Belgium,

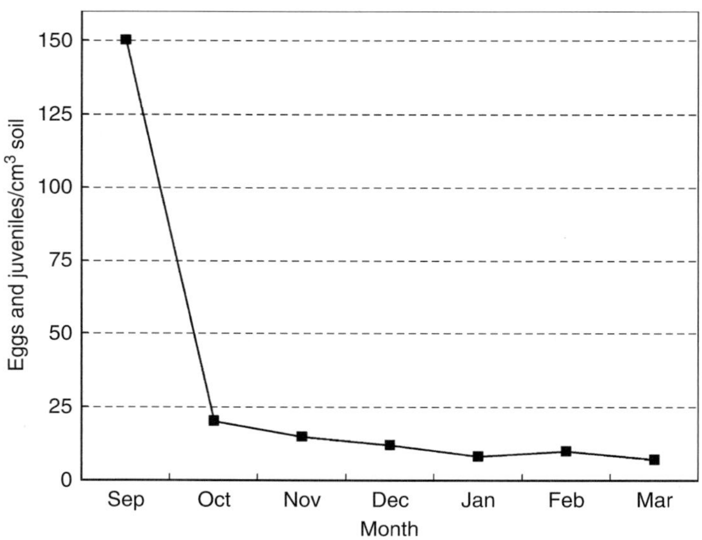

Fig. 11.4. Soil population dynamics of race 1 of *Meloidogyne incognita* in microplots at Castellaneta (Apulia region, Italy) after harvest of sweet pepper. (Adapted from Di Vito *et al.*, 1985a.)

M. chitwoodi peaked in the summer after cultivation of a host crop and declined after its harvest throughout the autumn–winter (Wesemael and Moens, 2008). In temperate climates, planting host crops early in spring or late in autumn, when soil temperature has dropped below 18 °C, will allow the crops to escape infection by most of the warm-temperature root-knot species. Potato is susceptible to tropical root-knot species, but under Mediterranean conditions this crop is planted from late summer to midwinter. In crops planted at the end of summer, potato roots can be invaded and damaged by *M. incognita*, but subsequently nematode development will cease, because of the drop in temperature, and the nematode soil population at harvest can be smaller than that at planting. In potato planted in midwinter, root invasion by the nematode will occur at a later growth stage of the crop, and few juveniles will complete their life cycle by harvest, thus leading to only a slight population increase (Hlaoua and Horrigue Raouani, 2007; Russo *et al.*, 2007).

In vineyards in California, the numbers of eggs and juveniles of most common nematodes, including root-knot nematodes, remained fairly constant during winter. The following spring, with the rise in temperature, J2 hatch and therefore the numbers of eggs in the soil declined, while there was an increase in the numbers of infective juveniles (Ferris and McKenry, 1974). Thereafter, J2 in the soil also declined because they penetrated into the roots of the vines. Numbers of eggs and J2 in the soil peaked at the beginning of spring and at the end of summer, respectively.

In tropical regions, where temperatures do not vary greatly between seasons, root-knot nematodes would reproduce continuously in the presence of a suitable host and favourable soil moisture content. However, population fluctuation can be observed because of alternate rainy and dry seasons (Fig. 11.3) (Souza *et al.*, 2008).

11.3.4 The soil

The soil represents the environment in which the infective J2 move in search of host roots or in which hatched and unhatched J2 survive in the absence of a host. As sufficient soil aeration and adequate moisture content are necessary for nematode hatching, movement and infection, sandy or well-structured and drained soils, combined with appropriate irrigation regimes or sufficient rainfall, favour nematode reproduction (Wallace, 1966, 1968, 1971). Root-knot nematodes are typically found in sandy or sandy loam soils. However, *M. artiellia* infects and reproduces well even in soil containing up to 30–35% clay under Mediterranean conditions. Other factors, such as soil pH and salinity, may affect build-up of nematode population densities but, generally, if they allow satisfactory plant growth they will not suppress populations of *Meloidogyne* significantly.

11.4 Modelling Population Dynamics

Probably the best model describing the relationship between initial (P_i) nematode density at planting of a host crop and final (P_f) density at harvest is that of Equation 11.6, proposed by Seinhorst (1966, 1967a,b, 1970, 1986a). The model is based on the competition model of Nicholson (1935) and on a logistic curve. To understand Seinhorst's model, some basic definitions and concepts must be considered. The first is the maximum reproduction rate, *a*, of the nematode. As the reproduction rate of a nematode decreases with increase of the population density of the nematode, *a* is defined as the reproduction rate occurring at a very low nematode density (that at $P_i \rightarrow 0$) on a given host crop. Thus, the final nematode population (P_f) on a given crop inoculated at planting with a low population density (P_i) would be:

$$P_f = aP_i \quad (11.1)$$

Assuming that the value of *a* has been estimated, Equation 11.1 would apply only to very low values of P_i, with the food requirement of every individual nematode completely fulfilled (i.e. no competition), and with all individual nematodes constituting P_i equally infective. In fact, not all of the P_i but only a proportion (x) of it will affect the crop. This proportion will vary between 1 (x = 1), if all individuals of the P_i could infect the plant, and 0 (x = 0) when no specimens could infect the plant. Moreover, of the proportion (x)

not all individuals may reproduce. Also, as the initial nematode population (P_i) increases, the plant will be damaged more and more, and therefore the size of the food resource (y) available to the nematode will be reduced. The food size (y) will also vary between 1 (plant undamaged, as when P_i is very low and ≤ damage threshold level) and 0 (= plant and roots dead because of a very large P_i). Therefore, P_f would actually be given by:

$$P_f = aP_i xy \tag{11.2}$$

Also, it must be considered that for a given nematode/crop combination, the nematode's final population density (P_f) cannot surpass a certain limit (ceiling), which depends upon the maximum population that the roots of the host plant would sustain assuming that no damage occurred. Also, there will be an initial nematode level (P_i) at which the final nematode population (P_f) will remain unchanged (the equilibrium density or maintenance density) (E) (Fig. 11.2). Therefore, at the end of a growing cycle, the maximum density that the final nematode population could reach is the ceiling. However, if the proportion x of P_i that infects the host plant is less than one, at the end of the crop cycle in the soil there will be a residual portion of the inoculum at planting (P_i) of:

$$(1-x)P_i \tag{11.3}$$

If, of the proportion x only an amount xy will reproduce, because of limited food availability, the proportion

$$(x-xy)\,P_i \tag{11.4}$$

will not be affected by the host plant and will (theoretically) behave as in the absence of a host. Therefore, a proportion s (= that remaining in the soil in the absence of a host)

$$(x-xy)P_i = sx(1-y)P_i \tag{11.5}$$

will also be found in the soil at the end of the crop cycle. The value of s will also vary between 1 (no nematode decline in the absence of the host) and 0 (complete mortality of the nematode population in the absence of a host). For most species of *Meloidogyne*, s is expected to be close to 0.

Based on these assumptions, to describe the relationship between the final nematode population at harvest of the crop (P_f) and that at planting (P_i), Seinhorst (1966, 1967a,b, 1970, 1986a) derived the equation:

$$P_f = xy\frac{aP_iE}{(a-1)P_i + E} + (1-x)P_i + sx(1-y)P_i \tag{11.6}$$

In this equation, the first addendum represents the amount of P_f deriving from true reproduction of the nematode on the host crop plant, and has the shape of the line 1 in Fig. 11.2.

Except for the value of P_i, which is known *a priori*, the other values (a, x, y, s, E) must be estimated from properly designed experiments. As root-knot nematodes feed on roots, the proportion of the roots available to the nematode should be considered in Equation 11.6. However, quantification of roots in soil is difficult. As root size is related (although not necessarily directly proportional) to the size of the above-ground plant parts and/or yield of the host crop, as an approximation, y in Equation 11.6 is usually considered the same as y in the damage model (see Equation 11.8, section 11.8). Seinhorst (1970, 1986a) also derived another equation in which the first addendum

$$\text{the first addendum } xya\frac{1-q^{P_i}}{-{}^e\log q} \tag{11.7}$$

is slightly different. Here q is a constant <1 (but very close to 1) and can be considered as the proportion of the total food available to the nematode that is not exploited at $P_i = 1$ (similar to q in Nicholson's (1935) competition model). The shape of the curves according to the two models is only slightly different (Seinhorst, 1967a). According to Seinhorst (1966), the model of Equation 11.6 is more appropriate for migratory nematodes, and having Equation 11.7 as first addendum is more appropriate for sedentary nematodes like *Meloidogyne* spp. (Seinhorst, 1967a).

As most root-knot nematodes have several generations per year, maximum reproduction rates as large as ×12,000 have been observed (Di Vito *et al.*, 1986). With a few exceptions, as most J2 of root-knot nematodes will hatch during the first crop cycle, the value of x can reasonably be assumed to be 1 or close to it and that of s nearly

0, as no eggs or J2 would survive in fallow soil. However, the situation is more complex with J2 that need to be stimulated by root diffusates (exudates) to hatch (see Curtis *et al.*, Chapter 6, this volume). As hatching factors may diffuse outside the rhizosphere, a proportion of the J2 in the soil volume larger than that explored by the roots will hatch, and therefore the proportion s and the value of Equation 11.5 (the third addendum of equation 11.6) could be smaller than expected.

The model can be useful to predict the nematode population increase following the cultivation of a host crop. However, it requires the estimation of a number of parameters. Also, many factors other than the host plant affect nematode reproduction and these vary from year to year. Therefore, to obtain data on nematode dynamics of predictive value, it is necessary that properly designed experiments are conducted and repeated in conditions as close as possible to those of a given area. Moreover, when different cropping systems, including the use of non-host plants in a rotation, are adopted, it is necessary that the dynamics of the nematodes be investigated throughout the rotation cycle to provide data of practical relevance. For species that complete several generations per growing season, at values of P_i smaller than those at which the nematode ceiling is reached, P_f would be very large and approach the nematode ceiling, making it difficult to fit a curve for the whole growing season according to the above model. Therefore, the model would need to estimate the nematode population at the end of each generation and the P_f of the first generation be considered as a P_i for the second generation, and so on. Also, it must be realized that not all nematodes of a P_i complete their life cycles at the same time; instead, overlapping of generations is the norm. All this makes long-term prediction of the nematode population increase problematic.

The model proposed by Seinhorst is based on the logistic curve of nematode increase and is referred to as a critical point model, as it makes predictions of the nematode change at a given time (usually the end of the crop cycle) based on initial conditions.

Nowadays simulation models are used in forecasts. They would be very attractive also in nematology because, at least theoretically, they would allow simulation of nematode increase in any conditions at any time, and they can be conveniently installed on a computer. However, every simulation model requires a huge amount of information. For example, it would require exact information on the change of every nematode P_i at every temperature and when exposed to a continuum of soil types and lengths of the plant growth cycle, and to different plant crops and cultivars. In other words, the model would require that the effect on nematode dynamics be assessed for all possible combinations among factors known to affect nematode changes. Although each combination could be investigated only within given and accepted ranges and for given degrees of approximation, this would still require a huge amount of work.

11.5 Damage Levels

As they derive nutrients from tissues of plant organs, nematodes attacking plants are described as parasitic. All plant-parasitic nematodes, including root-knot nematodes, are obligate parasites. The reaction of plants to a parasitic nematode differs greatly according to plant species and cultivar. Root-knot and many other parasitic nematodes are pathogens and cause disease that is expressed at histological, morphological, physiological and molecular levels, and which results in reduced growth, yield, lifespan and resistance to environmental stresses of affected plants. Typical symptoms of nematode-incited disease include stunted growth, wilting, leaf discoloration (mostly yellowing) and deformation of plant organs. The degree of the pathogenicity of a nematode also depends upon its aggressiveness and the reaction to it of a plant species or cultivar. Crop damage from nematode-incited disease may consist of reduced quantity and/or quality of the yield. For instance, produce from nematode-infested soil may have a larger unmarketable proportion because of small size or deformation. Also, sugar content of sugarbeet tap roots from fields infested with *M. incognita* (Di Vito and Lamberti, 1977) can be reduced, as is protein content of legumes in soil infested with *M. artiellia* (Di Vito and Greco, 1988b). Plant species or cultivars that support nematode reproduction are referred to as susceptible. Among them, some may suffer greatly from nematode parasitism and are considered intolerant, while others will suffer much less and produce satisfactory yield, and are referred to as tolerant. Also, other plant species or cultivars may be infected by the nematode but then activate a defence mechanism that limits the nematode's development and reproduction; these are referred to as resistant (see Williamson and Roberts, Chapter 13, this volume). Different plant species or cultivars may express different degrees of resistance, susceptibility and tolerance.

The evaluation of crop response to root-knot nematodes is based on the severity of the symptoms caused by the nematode on the roots (gall index) and the reproduction of the nematode on that species. Several scales are used to rate root gall index (GI); they are based on the number of galls, as suggested by Taylor and Sasser (1978), or also on the size of the galls and appearance of the root system (Di Vito *et al.*, 1979). Taylor and Sasser (1978) rated root gall index on a 0 to 5 scale, in which 0 = no gall on the root, 1 = 1–2 galls, 2 = 3–10, 3 = 11–30, 4 = 31–100, and 5 = more than 100 galls per root. The same scale is also used to rate egg mass index of the roots (EI). Based on root gall and egg mass indices or reproduction rate, Canto-Sáenz (1985) classified plant response to root-knot nematodes as susceptible (good nematode reproduction and severe galling), tolerant (good nematode reproduction and little galling), resistant (no or poor nematode reproduction and little or no galling) and hypersusceptible (poor nematode reproduction and severe galling). Examples of hypersusceptible plants are sugarbeet infected by *M. naasi* (Gooris and d'Herde, 1977), maize infected by *M. artiellia* (Di Vito *et al.*, 1985b) and some cowpea cultivars infected by *M. incognita* (Olowe, 2007).

Information on the extent of damage a nematode may cause, especially the yield loss, is basic to implementation of the most appropriate control strategies. For example, to include a nematode in the list of quarantine organisms, the pathogenic potential of the nematode in a country must be demonstrated. Also, national policy makers will consider the impact a nematode may have on given crops and areas for their decisions. At farm level, information on the potential of the nematode populations to cause yield loss is a prerequisite to deciding if and when to apply a treatment, and choosing the most appropriate control strategy.

11.6 Pattern of Nematode Damage to Crop Plants

Contrary to most pests and diseases that attack aerial plant parts, soil-borne nematodes have rather long life cycles and move actively only over short distances. Therefore, it is reasonable to assume that the damage a nematode may cause to a crop plant depends on its soil population density. A number of investigations have demonstrated that, in annual crops, the extent of damage is related to the nematode density at planting. When a wide range of population densities at planting (P_i) are considered, it is observed that at low population levels no yield loss occurs. On the contrary, some small increases in yield and/or plant size have been reported following the production of auxin-like compounds by the plants after nematode infection (see Abad *et al.*, Chapter 7, this volume). Crop yield thus remains more or less unaffected up to a certain nematode population density, above which yield loss does occur. The nematode population density above which yield loss starts to occur is defined as the nematode damage threshold level or tolerance limit of the crop, and is indicated by the term T. At nematode population levels above the tolerance limit, the yield reduction becomes obvious, and up to a certain P_i it appears to be inversely related to the $\log_{10}$ of P_i. With further increase of P_i, the yield declines towards a lower limit, identified as the minimum yield (m). The minimum yield (m) can be as low as 0 in very susceptible plants at rather large P_i, but generally is a proportion greater than 0 of the yield occurring at a $P_i \leq T$. The values of both the tolerance limit (T) and minimum yield (m) are important parameters characterizing the response of a crop plant to a nematode (Figs 11.5 and 11.6). At very large P_i, plant growth may be completely arrested (Fig. 11.7). Seinhorst (1981) reported three mechanisms of growth reduction, with the most

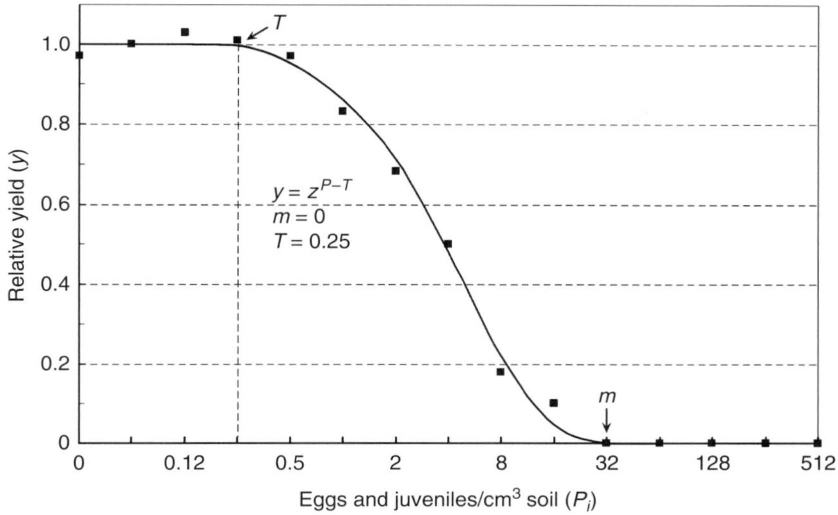

Fig. 11.5. Relationship between initial population densities of *Meloidogyne incognita* (P_i) and relative yield (y) of common bean grown in pots in a glasshouse. (Adapted from Di Vito *et al.*, 2004a.)

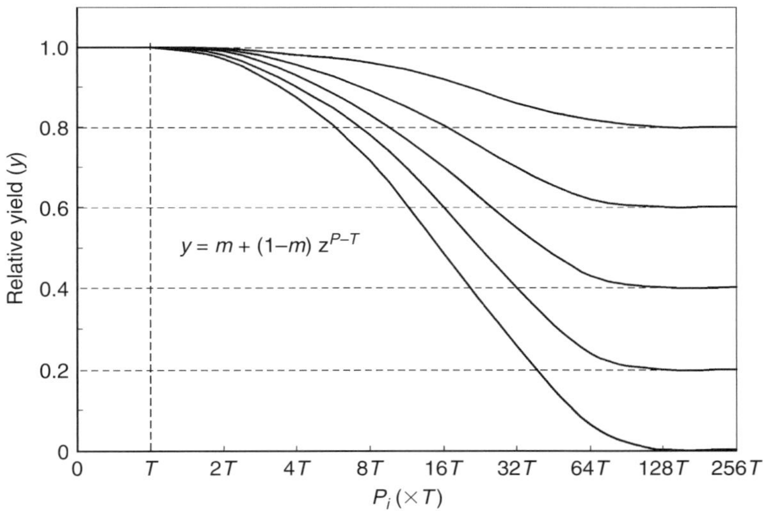

Fig. 11.6. Relationship between initial population densities (P_i expressed as $\times T$) of a nematode and relative yields (y). Different curves show the same T but different values of m (minimum yield; 0, 0.2, 0.4, 0.6, 0.8).

obvious being the first, in which P_i is larger than T, and there is simple growth reduction with no other evident effect; the other two mechanisms involve concomitant changes in plant water consumption and other parameters. With nematodes affecting planting materials such as seeds, bulbs or tubers, the extent to which a crop can be damaged also depends on the degree of nematode infection and, although tolerance thresholds have not been estimated for such crops, for practical purposes, only planting material free of nematodes must be used, i.e. a threshold value of 0.

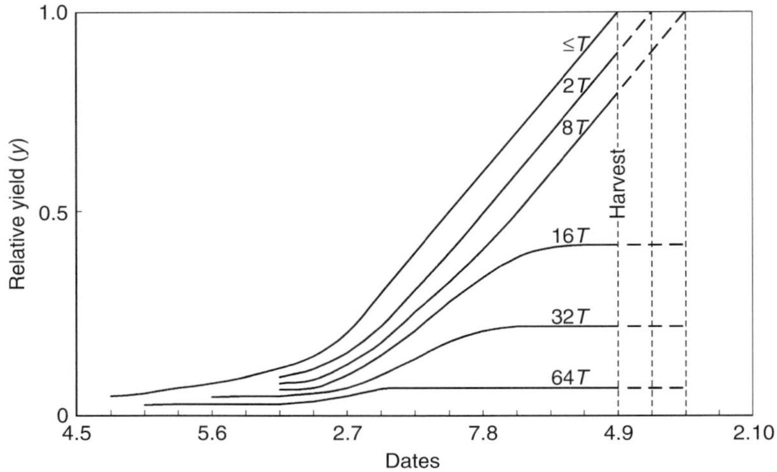

Fig. 11.7. Growth curves of a susceptible crop grown in soil infested with increasing initial population densities (P_i) of a nematode. Crops infected with P_i of 2*T*–8*T* show only reduced growth rates and would reach the same size as crops in soil at P_i of 0 ≤*T*, provided they are allowed to grow for a longer time. Crops in soil at P_i ≥16*T* show arrested growth.

11.7 Factors Affecting Nematode Damage

In addition to general plant sensitivity (= tolerance) to a given nematode species, several factors may affect the extent of damage a nematode may cause. In general, all factors affecting nematode reproduction will, in turn, affect the extent of the nematode damage to a host plant.

11.7.1 Nematode species and population level

Under given environmental conditions, different nematode species are known to cause different degrees of damage. In North Carolina in the USA, *M. arenaria* caused greater yield loss and galling to groundnut than did *M. hapla* at similar P_i (Koenning and Barker, 1992). Similarly, species of *Meloidogyne* differ in their aggressiveness on tobacco, with *M. hapla* again being the least aggressive species (Barker *et al.*, 1976, 1981). Although an estimate of the yield lost annually has never been determined with accuracy, on a worldwide basis root-knot nematodes are recognized as the most damaging nematodes of many crop plants (Sasser and Freckman, 1987). Most of the species cause measurable yield loss under field conditions, while others, such as *M. chitwoodi* on potato, mostly affect the yield (tubers) during storage, as they may continue to develop in the infected tubers even after harvest. Generally, the tolerance limit of most crop plants to root-knot nematodes is less than one egg per cm^3 soil, and complete crop failure of chickpea may occur at population densities of *M. artiellia* as low as two eggs/cm^3 soil (Table 11.1) (Di Vito and Greco, 1988b).

11.7.2 Soil and environmental conditions

As root-knot nematodes prefer sandy soils, the damage will be more severe as soil texture approaches these conditions (Wallace, 1989; Barker and Weeks, 1991). Root-knot nematodes are usually suppressed in clayey soils, with few exceptions, among which is *M. artiellia*. In the Mediterranean area, this species develops on and severely damages host crops in sandy soil as well as in those containing 30–35% clay, the constitution of most of the soils used for growing cereals, crucifers and legumes. In the same conditions, the other warm-season species would not cause noticeable damage on either annual or perennial crops, even if irrigated.

Table 11.1. Tolerance limit (*T*) and minimum yield (*m*) of some crop plant/*Meloidogyne* spp. combinations, under different environmental conditions and with different types of inoculum.

Meloidogyne spp.	Crop plant	Type of experiment	Type of inoculum	Growth variable	Tolerance limit (eggs/cm^3 soil)	Minimum yield	Reference
M. arenaria	White mulberry	Pots	Chopped roots	Top weight	1.38	0	Castillo *et al.* (2000)
M. artiellia	Chickpea (winter)	Microplots	Chopped roots	Yield	0.14	0.07	Di Vito and Greco (1988b)
M. artiellia	Chickpea (spring)	Microplots	Chopped roots	Yield	0.01	0.18	Di Vito and Greco (1988b)
M. artiellia	Wheat	Microplots	Chopped roots	Yield	0.43	0.1	Di Vito and Greco (1988c)
M. chitwoodi	Sugarbeet	Pots	Eggs	Roots	2.8	0.35	Griffin *et al.* (1982)
M. exigua	Coffee	Pots	Chopped roots	Top weight	1.2	0.5	Di Vito *et al.* (2000)
M. hapla	Lucerne	Pots	Eggs	Top weight	0.12	0.16	Inserra *et al.* (1983)
M. hapla	Sugarbeet	Pots	Eggs	Roots	0.6	0.4	Griffin *et al.* (1982)
M. javanica	Banana	Pots	Chopped roots	Top weight	0.15	0.1	Vovlas *et al.* (1993)
M. javanica	Coffee	Pots	Chopped roots	Top weight	1.34	0.5	Vovlas and Di Vito (1991)
M. javanica	Common bean	Pots	Eggs	Top weight	0.6	0	Di Vito *et al.* (2007)
M. javanica	Olive	Pots	Chopped roots	Top height	0.4	0.5	Sasanelli *et al.* (2002)
M. javanica	Peach	Pots	Chopped roots	Top weight	0.85	0.29	Di Vito *et al.* (2005)
M. javanica	Groundnut	Pots	Chopped roots	Top weight	1.8	0.3	Di Vito *et al.* (1999)
M. javanica	Pepper	Pots	Chopped roots	Top weight	0.36	0	Mekete *et al.* (2003)
M. javanica	Potato	Pots	Chopped roots	Top weight	0.5	0.6	Vovlas *et al.* (2005)
M. javanica	Rice	Pots	Chopped roots	Top weight	0.26	0	Di Vito *et al.* (1996a)
M. javanica	Sunflower	Microplots	Chopped roots	Yield	0.74	0	Di Vito *et al.* (1996b)
M. javanica	Tomato	Pots	Chopped roots	Top weight	0.28	0	Mekete *et al.* (2003)
M. incognita	Artichoke	Pots	Chopped roots	Top weight	1.1	0	Di Vito and Zaccheo (1991)
M. incognita	Cabbage	Pots	Chopped roots	Top weight	0.5	0.05	Sasanelli *et al.* (1992)
M. incognita	Cassava	Pots	Chopped roots	Top weight	1	0.79	Crozzoli and Parra (1999)
M. incognita	Coffee	Pots	Chopped roots	Top weight	2.09	0.4	Vovlas and Di Vito (1991)
M. incognita	Common bean	Pots	Chopped roots	Top weight	0.25	0	Di Vito *et al.* (2004a)

(continued)

Table 11.1. Continued

Meloidogyne spp.	Crop plant	Type of experiment	Type of inoculum	Growth variable	Tolerance limit (eggs/cm^3 soil)	Minimum yield	Reference
M. incognita	Corn	Pots	Eggs	Top weight	10	0.1	Di Vito *et al.* (1980)
M. incognita	Cowpea	Pots	Eggs	Top weight	0.03	0.28	Crozzoli *et al.* (1997)
M. incognita	Aubergine	Microplots	Chopped roots	Yield	0.05	0.05	Di Vito *et al.* (1986)
M. incognita	Grape	Pots	Chopped roots	Top weight	0.78	0.55	Sasanelli *et al.* (2006)
M. incognita	Guava	Pots	Chopped roots	Top weight	0.05	0.71	Casassa *et al.* (1998)
M. incognita	Kenaf	Microplots	Chopped roots	Top weight	0.13	0.05	Di Vito *et al.* (1997)
M. incognita	Melon	Microplots	Eggs	Yield	0.19	0	Di Vito *et al.* (1983)
M. incognita	Papaya	Pots	Eggs	Top weight	0.16	0.77	Bustillo *et al.* (2000)
M. incognita	Parsley	Pots	Chopped roots	Top weight	0.17	0.5	Aguirre *et al.* (2003)
M. incognita	Pepper	Microplots	Chopped roots	Yield	0.3	0.16	Di Vito *et al.* (1992)
M. incognita	Potato	Microplots	Chopped roots	Yield	1.2	0.2	Russo *et al.* (2007)
M. incognita	Rice	Pots	Chopped roots	Top weight	4	0	Greco *et al.* (2000)
M. incognita	Spinach	Pots	Chopped roots	Top weight	0.25	0	Di Vito *et al.* (2004b)
M. incognita	Sugarbeet	Microplots	Eggs	Yield	1.1	0.1	Di Vito *et al.* (1981)
M. incognita	Sunflower	Pots	Chopped roots	Top weight	1.85	0.25	Sasanelli and Di Vito (1992)
M. incognita	Tobacco	Microplots	Eggs	Yield	2	0	Di Vito *et al.* (1983)
M. incognita	Tomato	Microplots	Chopped roots	Yield	0.55	0	Di Vito *et al.* (1991)

11.7.3 Crop and cropping system

Many species of *Meloidogyne* have wide host ranges, and the extent of damage a susceptible crop plant may suffer is much affected by the crop plant species (Table 11.1) and crop sequence, the damage being less in long rotations with non-host crops than in short rotations or with no rotation at all, as occurs in glasshouses. Also, the period of the year in which the crop is established and the planting material used may greatly affect the extent of damage by root-knot nematodes. For example, in the Mediterranean area, planting potato in February–March results in less yield loss than in potato planted in late summer (Russo *et al.*, 2007). Several vegetables that can be cultivated year round, such as umbelliferous, cruciferous and salad crops, can be severely damaged in summer but are less damaged if planted in late summer or early spring, and will suffer no damage if planted in mid-autumn. Roberts (1987) investigated the effect of different sowing dates on the percentage of marketable yield of carrots in soil infested with *M. incognita* in California, and obtained 50% of marketable yield with a mid-October sowing and about 90% of marketable yield with sowing in mid-November or early December. In India, postponing the sowing of chickpea from mid-October to mid-November reduced damage by root-knot nematodes (Gaur *et al.*, 1979). In Belgium, carrot tap-root damage by *M. chitwoodi* increased from 10% when harvested 100 days after sowing to 70% when harvested 40 days later (Wesemael and Moens, 2008). Moreover, if planted during periods conducive to nematode infection, sown crops will suffer much more damage than those that have been transplanted (Ekanayake and Di Vito, 1984). Also, plants will suffer less damage if transplanted with the entire (undamaged) root system, such as those produced in polypot trays in modern agriculture rather than those produced in seed beds and transplanted with at least partly damaged root systems. The transplanting shock of the latter method, coupled with early nematode infection, can be very deleterious to the crop.

Whether a plant is annual or perennial also affects the damage it may suffer. However, this is not due to intrinsic properties of the plant itself, but is mainly due to the degree of infestation of the field into which an annual crop is sown or, if the crop is transplanted, and in the case of trees, the use of seedlings much smaller than tree nursery stocks which therefore suffer more damage at early growth stages. However, in the following years, even trees grown from good nursery stock may also be severely damaged because of the increased nematode population.

In modern agriculture, nematode management is a routine practice. The effects of application of nematicides and the use of other means of control on the nematode, and then on the damage they may cause to plants, depends on their mode of action. For example, fumigant nematicides, and to some extent also soil solarization, kill a large proportion of the nematodes before sowing or planting. Therefore, if the nematode populations before treatments are considered in the damage model it will appear as if the tolerance limit (T) of the host plant has increased. If, instead, the effect of the treatments is nematostatic, such as a delay in hatching or a delay of plant invasion by some means, as with non-fumigant nematicides, the main effect on the damage curve will be an increase of the minimum yield (m). However, the same control system, depending on its rate of application (or the rate at which it is acquired by the nematode), may act in both these modes, and therefore an increase of both T and m may occur.

The degree of resistance or tolerance of crop plants also affects the damage they may suffer. Resistant plants are still invaded by sedentary nematodes and, even if nematode feeding and development is prevented by a hypersensitive resistance reaction (see Williamson and Roberts, Chapter 13, this volume), these plants nevertheless suffer some damage. Comparing the responses of susceptible and resistant cultivars of the same plant species to a range of nematode P_i levels will show little difference in the tolerance limit (T), but usually a much larger value of the minimum yield (m) in the resistant cultivars (Ekanayake and Di Vito, 1984; Di Vito *et al.*, 1991, 1992; Zhou and Starr, 2003).

11.8 Modelling Damage Levels

Modelling the damage caused to a crop plant by a range of initial soil population densities of a nematode species is necessary for the estimation of the damage the nematode may cause in a given area and to predict yield loss in a field, provided the infestation level is known. Also, damage functions are basic to the estimation of economic damage thresholds, which are a prerequisite for making decisions on the management of the nematodes. As mentioned before (section 11.6), the yield of a crop plant in an infested soil may be unaffected up to a certain (rather low) nematode infestation level or completely lost at large nematode infestation levels. Therefore, the model describing the yield of a crop plant as affected by the nematode, to be predictive, must fit a wide range of initial population densities of the nematode.

Seinhorst (1965, 1972, 1979, 1986b, 1998) found that the model

$$y = m + (1 - m)z^{P-T} \qquad (11.8)$$

or, for $m = 0$,

$$y = z^{P-T} \qquad (11.9)$$

adequately fits such relationships (Figs 11.5 and 11.6). In this model P (= P_i) is the nematode soil population at sowing or planting (expressed in eggs and/or J2 per cm^3 or g of soil); T is the tolerance limit of the crop to the nematode (the nematode density (P_i) up to which no yield loss occurs); y is the relative yield (the yield at a given P_i divided by the yield at $P_i \leq T$, with $y = 1$ at $P_i \leq T$); m is the minimum yield, the value of y at very large P_i, with $m = 0$ when all the yield is lost; z is a constant ≤ 1 with z^{-T} = approximately 1.05 or $z^T = 0.95$. Seinhorst (1998), based on the results of 36 experiments, 31 of which were published, established Equation 11.8 as the general relationship between nematode density and plant weight for all combinations of plant species and nematode species in the range of nematode soil population densities from 0 to about $100T$. The average value of z^T was 0.95, and the value of z varied according to the nematode/host crop combination – it is smaller at larger values of T.

The model proposed by Seinhorst fits well to data for annual plants and especially with nematodes that develop only one generation per growing season. With nematodes developing several generations per crop cycle, such as the majority of *Meloidogyne* spp., the effects of the succeeding generations may be small at low initial nematode densities and low reproduction rates, while at larger nematode densities and with large nematode reproduction rates the effect of a second and later generations can be considerable (Seinhorst, 1995). With perennials, the same model is appropriate only during the first growing season; in subsequent seasons it will be dependent on the nematode populations that occur at the beginning of each growing season. In the following years, the plant may even suffer less damage. Also, if the effect on yield is considered, and yield is represented by fruits, it must be kept in mind that the root damage suffered by a plant in a given year will usually affect the yield of the succeeding year(s). All this makes it difficult to fit a model to the damage suffered by perennials.

In addition to the Seinhorst models, other models have been useful in relating P_i to crop yield. Most modern computer graphics programs allow one to fit a curve that relates P_i to yield or crop growth. In a comparison of groundnut cultivars resistant or susceptible to *M. arenaria*, Starr *et al.* (2002) reported that this relationship best fits a negative exponential model for the susceptible cultivars. Further, they observed that there was no detectable effect of increasing P_i on yield of the resistant cultivar. Barker *et al.* (1976, 1981) have used linear and quadratic models to estimate the relationship of P_i to yield of tomato and tobacco for several species of *Meloidogyne.*

11.9 Implementing Experiments to Assess Nematode Dynamics and Crop Damage

To investigate the dynamics of root-knot nematodes and the damage they may cause to host crops, properly designed experiments are necessary. To obtain predictive data, environmental and crop conditions and the type of inoculum must be the same as, or at least as close as possible to, those under field conditions. Moreover, as reproduction rates and the extent of damage caused by nematodes varies with the inoculum level, a range of nematode levels, from the non-damaging to highly damaging, must be investigated. As suggested by Seinhorst (1965, 1986a), a range of inoculum densities that follow a geometric series is the most suitable, as the $\log_{10}$ of the inoculum density can then be plotted against the crop yield, thus making the fitting of a model easier. In several experiments, including those of the authors, a range of inoculum densities of 0, 0.125, 0.25, 0.5, 1, … to 512 eggs and/or J2/cm^3 soil was adopted to obtain a satisfactory picture of the relationship between nematode densities at planting and nematode densities at harvest or yield of the host crop, and to fit the described model to the experimental data. Each of the inoculum levels tested should be replicated five-fold, as a minimum, or better eight- to tenfold. As Seinhorst's model of nematode dynamics also requires information on the decline of the nematode population in the absence of a host crop, an additional P_i of about 30–60 × T and left fallow is necessary.

In experiments to derive growth curves of the crop plant, as affected by a range of P_i, measurements of variables to describe the growth of the crop must be recorded periodically, usually weekly or fortnightly, throughout the growing season. The best variable to record is the plant weight (fresh or dried). As such measurements involve destructive samplings, this means that the number of replicates of each treatment has to be multiplied by the number of times the measurements have to be recorded. This approach is possible in glasshouse pot experiments, but an experiment would become extremely large and nearly impossible under microplot and field-plot conditions. In the glasshouse, an acceptable compromise is the use of a single set of replicated P_i. Periodically, the weight of the plants in each pot can be obtained by weighing each pot at planting and at suitable time intervals. However, the pots may not have the same soil moisture content and, therefore, each weighing must be accompanied by a determination of the soil moisture content at each observation. In order to allow for this frequent determination of soil moisture content, the experiment must be performed in relatively large pots. The difference between the pot weight at a given observation date and that of the same pot at planting will give a good estimate of the size of the plant of each replicated P_i throughout the growing season. As the height of a plant or the length of all branches of a plant is usually directly proportional to plant weight, monitoring these growth components instead of the plant weight will also provide an estimate of the plant growth throughout the entire growing season.

11.9.1 Preparation and type of inoculum

The nematode population used should be as aggressive as the average nematode population occurring in a given area. Therefore, preliminary investigations are necessary to ascertain not only the prevalent nematode species but also its aggressiveness towards the most common cultivar of the target host plant. Once such a population is identified, it must be increased to obtain the required amount of nematodes.

Some questions must then be answered. Among them are: how to increase the population? What kind of inoculum must be used? The answers can be given only if the aim of the investigation is clear. Usually, the nematode population is increased on a plant species that is susceptible or, better, tolerant to the target nematode species, easy to grow and having a reasonably large root system to allow production of a large number of eggs. In many studies on pathogenicity, the nematode is reared in steam-sterilized sandy soil in pots or trays in which one or more seedlings of a host plant are transplanted and maintained in a glasshouse at 25–28 °C for the warm-climate species of root-knot nematodes, or at 20–22 °C for species preferring lower temperatures, such as *M. artiellia* and *M. chitwoodi*. A week after transplanting, when new roots have already developed, each plant is inoculated with a water suspension of eggs and J2 of the nematodes. This can be obtained by shaking infected roots for 3–4 min in a 0.5–1% sodium hypochlorite (NaOCl) solution, as suggested by Hussey and Barker (1973). Plants are then left to grow for about 45–50 days, to allow the nematode to develop a second generation and plenty of well-developed egg masses on the roots. Longer growing periods may not be suitable as roots may rot and the nematode population begins its natural decline. Plants are then uprooted and the roots washed free of adhering soil. A suitable inoculum must then be prepared.

If J2 are required, these can be obtained by setting up a system for hatching using eggs collected from egg masses, and collecting emerging J2 every day or every other day. This method is time consuming but efficient. Infective J2 may also be obtained by incubating infected roots in a beaker or similar container containing just enough water to cover the roots, into which air is continuously bubbled. J2 can be separated from eggs by pouring the suspension on to a Baermann's funnel, or one of its modifications, and collecting motile J2 at regular intervals. However, with this method it is difficult to obtain a large number of infective J2, so it is only suitable for producing inoculum for tests in small pots (10–15 cm diameter) in glasshouses. Lambert *et al.* (1992) have reported on a system using tomato plants grown in hydroponic culture for collecting large numbers of infective J2. When using J2 for inoculum, it is important to coordinate inoculation with plant growth, because if the roots of the host plant are not sufficiently developed when inoculated, the J2 that do not enter the roots quickly may not survive.

In most published experiments on the damage potential and reproduction ability of *Meloidogyne* spp., the inoculum was made of a mixture of eggs and J2 prepared by the less laborious method of shaking infected roots in 0.5–1% NaOCl, as noted above. This method allows the preparation of the large amounts of inoculum required for large microplots, containing 30–40 dm^3 soil, even providing inoculum levels as high as 128–512 eggs and J2 per cm^3 soil. Also, the NaOCl acts as a disinfectant for the many microorganisms carried by egg masses and roots. However, egg masses provide the eggs with mechanical defence, as during the soil mixing, and antibiotic substances that prevent infection by several antagonistic microorganisms occurring in the soil. Without the protection of the egg mass, a large proportion of the eggs may be killed during soil mixing and after inoculation. The authors used this type of inoculum in microplots planted to melon and obtained almost no damage at nematode population levels as large as 256 eggs/cm^3 soil. They then shifted to the use of infected roots with very satisfactory results (Di Vito *et al.*, 1985a, 1986).

When used as inoculum, infected roots are cut into small pieces (0.5–1 cm long), put in a bucket containing tap water and mixed (mixing in water is more efficient). The roots are then collected and excess water eliminated from them by putting them between sheets of laboratory paper. Roots are then weighed and subsamples (five or six, each of 3–6 g) are used to estimate the density of nematodes by the NaOCl method, which allows estimation of the total number of eggs and J2 available on and in the roots. As the inoculum density per g of roots is usually very large, the roots are then thoroughly mixed into a known amount of sterilized sand and the resulting mixture used as inoculum for the soil of each replicated pot or microplot. For this mixing, the authors used a concrete mixer rotating at low speed for at least 3–5 min during each mixing.

The use of infected roots instead of free eggs has proved useful in both glasshouse pot and microplot experiments. However, if small pots such as 12–15 cm diameter are used, it may not be possible to prepare a very low inoculum level, as even a single egg mass would contain more eggs than are required to inoculate a single pot. In such a case, some pots may receive more eggs than planned and others may not receive any.

Another aspect of inocula to consider is the physiological state of the nematode population, which can be greatly affected by soil type, temperature, moisture content and aeration, and the growth stage of the host plants on which the nematode has been reared (de Guiran and Demeure, 1978; de Guiran, 1980; de Guiran and Villemin, 1980; Van Gundy, 1985; and Curtis *et al.*, Chapter 6, this volume). Generally, eggs of nematode populations produced in glasshouses hatch immediately. Therefore, when they are used as inoculum, a large proportion of the emerging J2 may not survive long enough to infect newly formed roots, thus resulting in much less root damage than expected, especially under field conditions. Under field conditions, daily temperature variation and the decreasing average soil temperature at the end of the crop season (early or late autumn) and the senescence of the host may affect hatch and survival (Wesemael *et al.*, 2006; Wesemael and Moens, 2008), which declines rather rapidly immediately after harvest, but much less so during winter (Fig. 11.4). Therefore, the aggressiveness of the nematode population in the soil by the start of the crop cycle the next spring may differ from that of populations reared in glasshouses under constant temperature. The authors failed to obtain significant damage to durum wheat and chickpea with a population of *M. artiellia* produced on chickpea in the glasshouse, while severe damage was observed, even at low nematode densities, using a population that had been reproduced on wheat in microplots in the previous season, and from which different nematode inoculum levels were

1

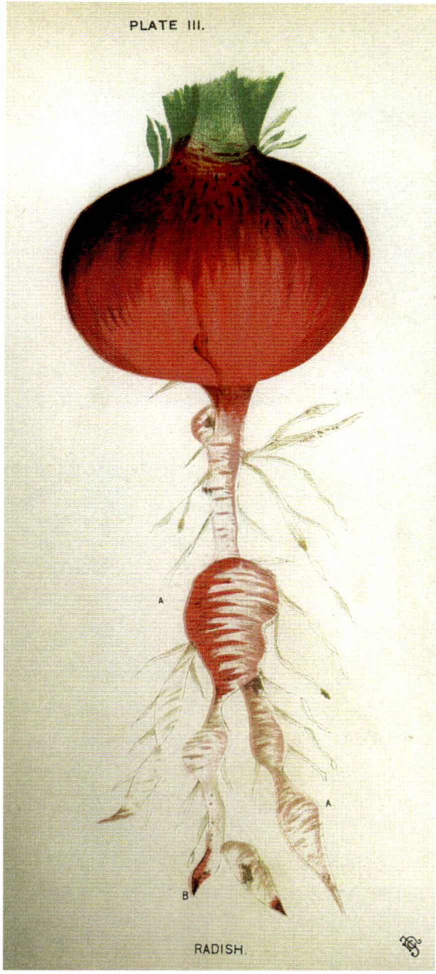

2

3

Plate 1. Root galling on radish caused by *Anguillula* (= *Meloidogyne*) *'arenaria'*. (After Neal (1889) courtesy USDA; see Chapter 3.)

Plate 2. Galls on tomato roots infected by *Meloidogyne incognita*. (Courtesy P. Abad, INRA, France.)

Plate 3. Melon roots showing characteristic large galls due to infection by *Meloidogyne incognita*. (Courtesy J.L. Starr, Texas A&M University, USA.)

4

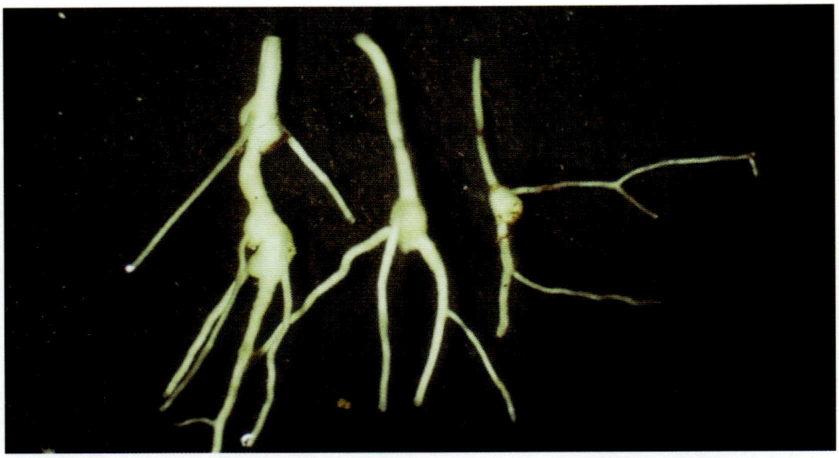

5

6

Plate 4. *Meloidogyne hapla* galls, which are generally smaller than those caused by *M. incognita*, but have several adventitious roots emerging from most galls. (Courtesy E.C. Bernard, University of Tennessee, USA.)
Plate 5. *Meloidogyne marylandi* on St. Augustine grass; note the limited galling and protruding egg masses. (Courtesy T.R. Faske, Tarleton State University, USA.)
Plate 6. Galls on clover caused by *Meloidogyne trifoliophilia*. Galls tend to be elongated in shape and most egg masses are completely embedded within the gall tissue. (Courtesy C.F. Mercer, AgResearch Grasslands, New Zealand.)

7

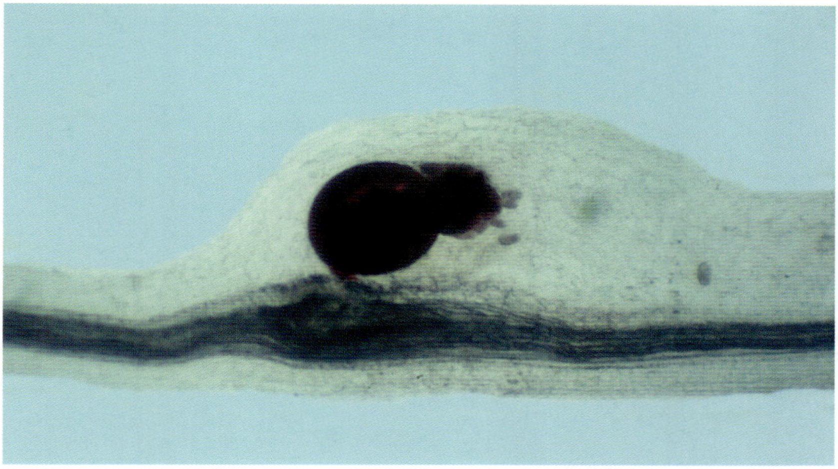

8

9

Plate 7. *Meloidogyne trifoliophilia* female on clover with eggs deposited completely within the gall. (Courtesy E.C. Bernard, University of Tennessee, USA.)

Plate 8. Pecan roots infected by *Meloidogyne partityla* with mature females exposed on the root surface. (Courtesy J.L. Starr, Texas A&M University, USA.)

Plate 9. Tomato root piece with galls and females of *Meloidogyne chitwoodi*. (Courtesy W.M.L. Wesemael. ILVO, Belgium.)

10

11

12

Plate 10. Bean (*Phaseolus vulgaris*) root piece with galls and females of *Meloidogyne chitwoodi* stained with acid fuchsin. (Courtesy W.M.L. Wesemael. ILVO, Belgium.)

Plate 11. Galls on onion roots caused by *Meloidogyne fallax*. (Courtesy G. Korthals, PPO-AGV, WUR, The Netherlands.)

Plate 12. Galls on lettuce caused by *Meloidogyne minor*. (Courtesy G. Korthals, PPO-AGV, WUR, The Netherlands.)

13

14

15

Plate 13. *Meloidogyne minor* on potato. A, healthy crop; B, infested patches; C, damaged root system with 'stubby' root form; D, close up of damaged roots. (Courtesy S.J. Turner, Agri-Food and Biosciences Institute, UK.)
Plate 14. Damaged patches of creeping bentgrass caused by *Meloidogyne minor* on golf course turfgrass. (Courtesy C. Fleming, Agri-Food and Biosciences Institute, UK.)
Plate 15. Damaged patches of creeping bentgrass caused by *Meloidogyne minor* on golf course turfgrass. (Courtesy C. Fleming, Agri-Food and Biosciences Institute, UK.)

16

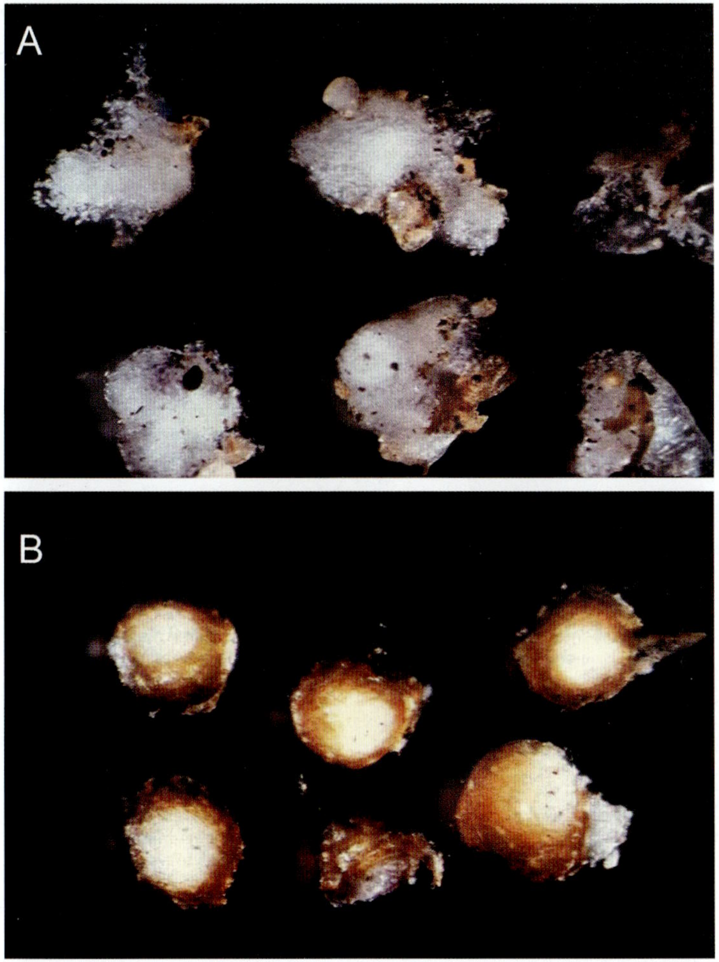

17

18

Plate 16. Differences in the egg masses of *Meloidogyne incognita* produced in (A) dry or (B) subsaturated soils. In the dry soil the gelatinous matrix was colourless, soft and expanded. In subsaturated soils the matrix rapidly turned hard and brown and contracted. (From de Guiran, 1980; see Chapter 9.)

Plate 17. Symptoms of *Meloidogyne izalcoensis* on roots of *Coffea arabica* 'Catuaí Vermelho' stained with Phoxine B. (Courtesy R.M.D.G. Carneiro, EMBRAPA, Brazil.)

Plate 18. Tubers of yakon (*Polymia sonchifolia*) infected (right) or non-infected (left) by *Meloidogyne ethiopica*. (Courtesy R.M.D.G. Carneiro, EMBRAPA, Brazil.)

19

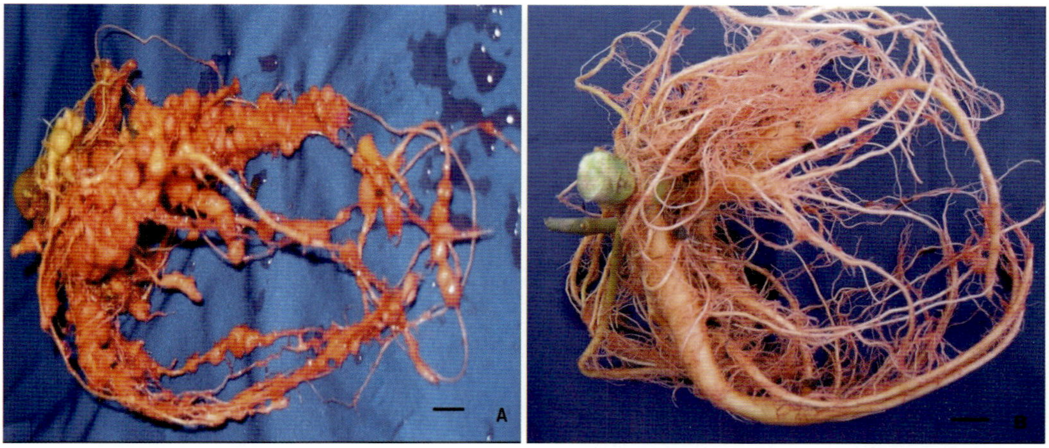

20

21

Plate 19. Roots of Brazilian ginseng (*Pfaffia glomerata*) inoculated with *Meloidogyne incognita*. A, acession Farmacotécnica (susceptible); B, acession UVF (resistant). Scale bar = 2 cm. (Courtesy R.M.D.G. Carneiro, EMBRAPA, Brazil.)

Plate 20. Symptoms of *Meloidogyne enterolobii* (= *M. mayaguensis*) on guava tree 'Paluma' in Pernambuco State, Brazil. (Courtesy R.M.D.G. Carneiro, EMBRAPA, Brazil.)

Plate 21. Damage by *Meloidogyne* spp. on carrot. (Courtesy D. Coyne, IITA, Uganda.)

22

23

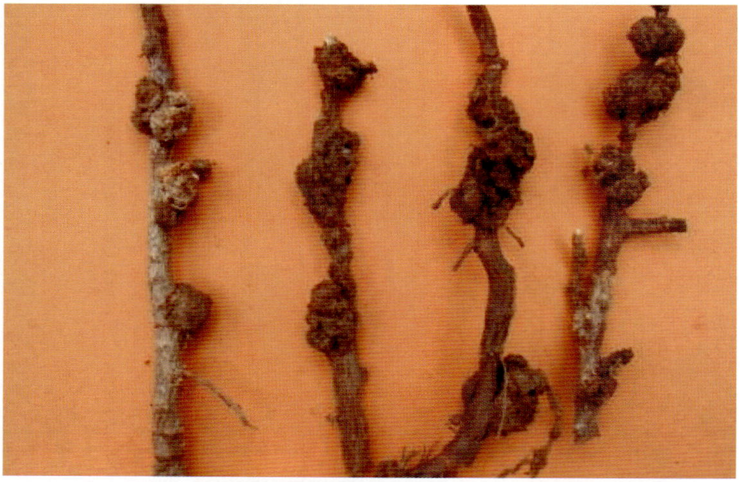

24

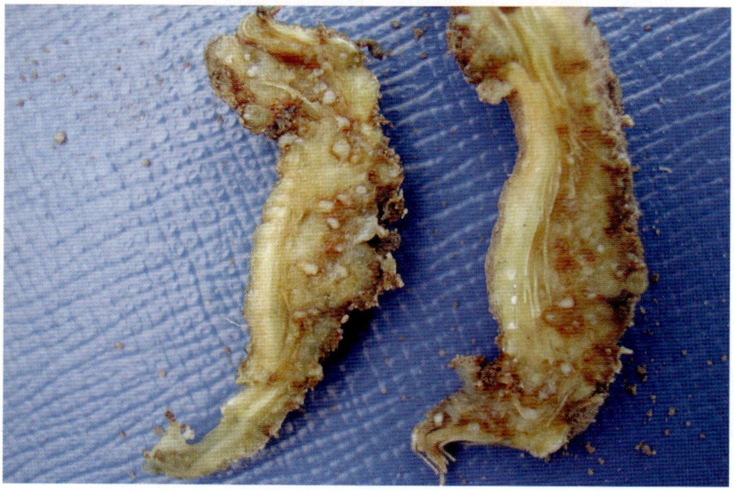

Plate 22. Damage by *Meloidogyne chitwoodi* on carrots. (Courtesy W.M.L. Wesemael, ILVO, Belgium.)
Plate 23. Cassava roots deformed with cankerous galls caused by infection with *Meloidogyne* spp. (Courtesy D. Coyne, IITA, Uganda.)
Plate 24. Sweet potato, extensively infected with *Meloidogyne* spp. (Courtesy D. Coyne, IITA, Uganda.)

25

26

27

Plate 25. Potato tubers showing damage caused by *Meloidogyne chitwoodi.* (Courtesy W.M.L. Wesemael, ILVO, Belgium.)
Plate 26. Potato tuber sliced open to show females and egg masses of *Meloidogyne chitwoodi* inside the potato. (Courtesy W.M.L. Wesemael, ILVO, Belgium.)
Plate 27. Galls on black salsify caused by *Meloidogyne chitwoodi.* (Courtesy W.M.L. Wesemael, ILVO, Belgium.)

28

29

30

Plate 28. Poor growth of *Meloidogyne incognita*-susceptible pepper (foreground) relative to the growth of nematode-resistant pepper (background) in a *Meloidogyne*-infested field. (Courtesy J.A. Thies, USDA, USA; see Chapter 14.)

Plate 29. Comparison of growth of *Meloidogyne*-resistant peanut (left) and susceptible peanut cultivar (right) in *Meloidogyne*-infested soil. (Courtesy J.L. Starr, Texas A&M University, USA; see Chapter 14.)

Plate 30. Symptoms of the *Meloidogyne*/Fusarium wilt complex on cotton. (Courtesy J.L. Starr, Texas A&M University, USA; see Chapter 10.)

31

32

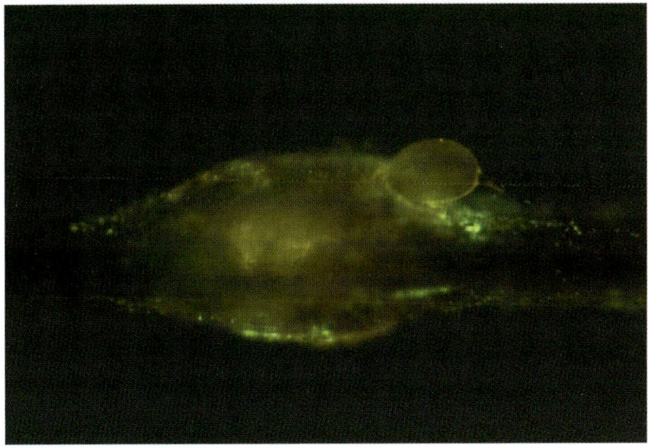

33

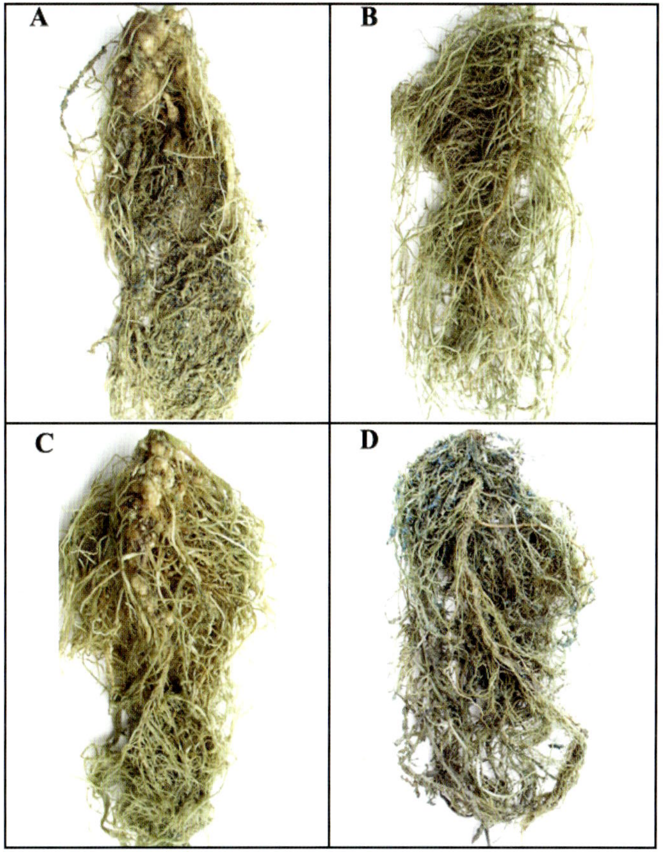

Plate 31. Interaction of a *Pythium* species with *Meloidogyne hapla* on celery. Two necrotic root systems on the left were from soil infested with both pathogens, whereas the root systems on the right were from soil infested only with *M. hapla*. (Courtesy J.L. Starr, Texas A&M University, USA; see Chapter 10.)

Plate 32. Epifluorescence image of a nematode gall plus attached female of *Meloidogyne incognita* on *Arabidopsis thaliana* showing strong GFP fluorescence of *Rhizobium etli* G12 within the galled tissue. (Courtesy J. Hallmann, Julius Kühn-Institut, Germany; see Chapter 17.)

Plate 33. Root-galling symptoms and egg mass production of *Meloidogyne incognita* (A-C) and *M. javanica* (D) 60 days after inoculation of Lima bean recombinant inbred lines from the cross of susceptible Henderson Bush × resistant L-136 possessing different combinations of resistance genes; A, susceptible to both primary root galling and reproduction; B, resistant to both primary root galling and reproduction; C, susceptible to primary root galling and resistant to reproduction; D, resistant to primary root galling and susceptible to reproduction. Egg masses are stained blue with erioglaucine. (From Roberts *et al.*, 2008; see Chapter 13.)

34

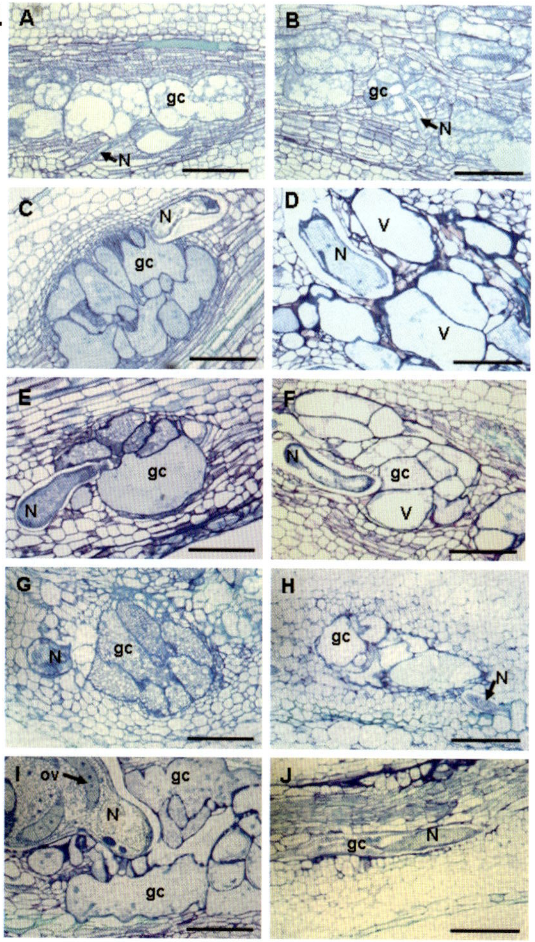

35

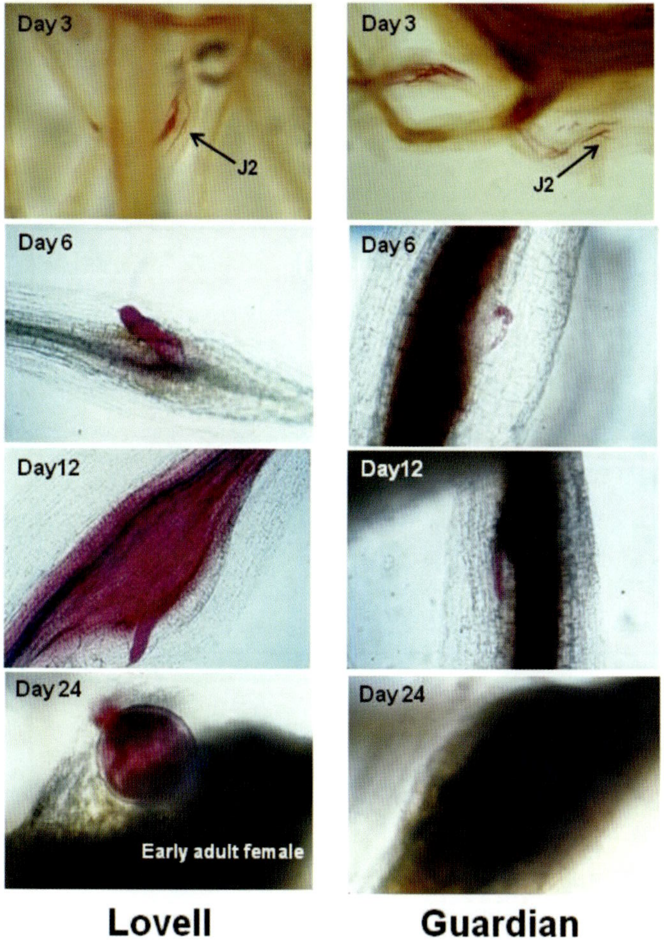

Plate 34. Longitudinal sections of *Meloidogyne incognita* feeding sites in inoculated cowpea roots. Sections were stained with toluidine blue O. Panels A, C, E, G and I are null-*Rk* (susceptible) root sections at 5, 9, 14, 19 and 21 days post-inoculation (dpi), respectively. Panels B, D, F, H and J are CB46 (resistant) root sections at 5, 9, 14, 19 and 21 dpi, respectively. gc = giant cell, N = nematode, ov = ovaries, and V = vacuole. Bar = 200 μm. (From Das *et al.*, 2008; see Chapter 13.)

Plate 35. Development of *Meloidogyne incognita* in root systems of Lovell (susceptible) and Guardian (resistant) peach stocks after 24 days (note that second-stage juveniles failed to mature and reproduce in Guardian roots). (Courtesy A.P. Nyczepir, USDA, USA; see Chapter 18.)

36

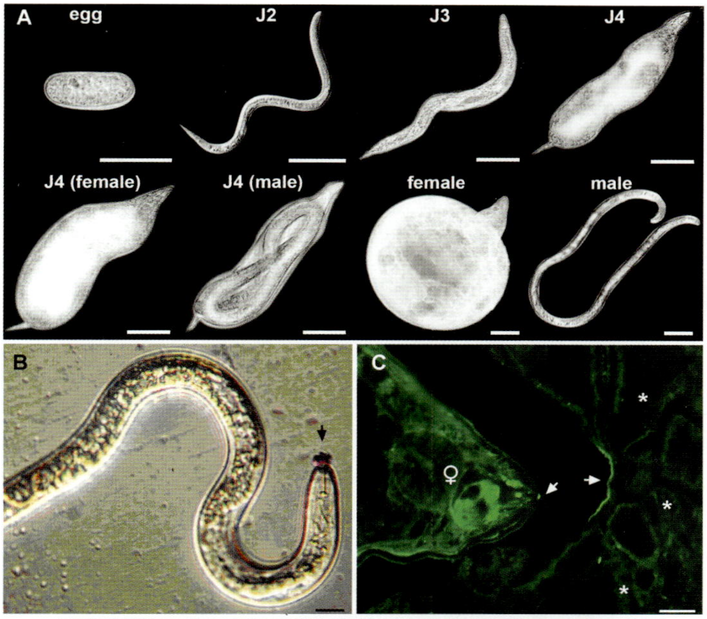

37

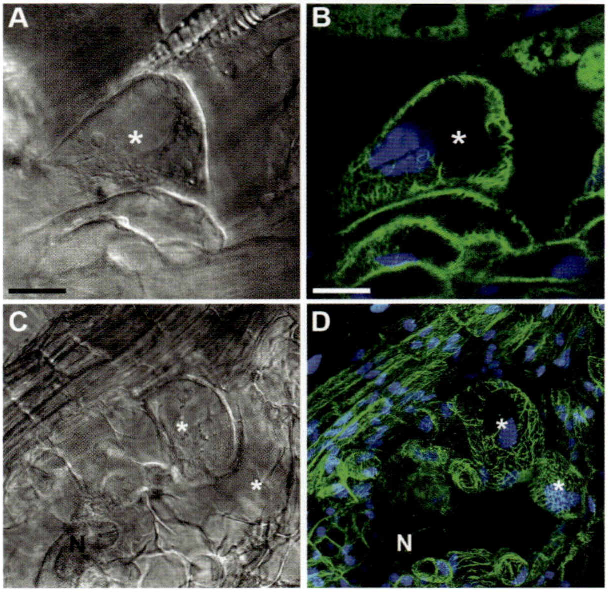

Plate 36. Developmental stages and stylet secretions of *Meloidogyne incognita*. A, developmental stages, from eggs to adult nematodes; B, proteins secreted *via* the stylet by second-stage juveniles are visualized with Coomassie staining (arrow); C, fluorescence immuno-labelling of a calreticulin (arrows) secreted during parasitism. The calreticulin accumulates at the stylet tip of the female and along the cell wall of adjacent giant cells. Asterisks, giant cell; ♀, female. Scale bars = 40 μm (A), 10 μm (B and C). (A, courtesy P. Abad, INRA, France; B and C from Caillaud *et al*., 2008a; see Chapter 7.)

Plate 37. Giant cells induced by *Meloidogyne incognita* in *Arabidopsis thaliana*. A-B, formation of vascular binucleate cell, first sign of giant cell development; A, differential interference contrast (DIC) image; B, *in vivo* confocal microscopy of galls. *Arabidopsis* root cells co-expressed two marker proteins, microtubule-binding domain-Green Fluorescent Protein (MBD-GFP, green) and nuclear Histone 2B-Yellow Fluorescent Protein (H2B:YFP, blue); C-D, multinucleate giant cells with bundles of cortical microtubules; C, DIC image; D: *In vivo* confocal microscopy of galls co-expressing MBD-GFP (green) and H2B:YFP (blue). Sections through a gall at 10 days post infection stained with toluidine blue. *, giant cell; N, nematode; nu, nuclei. Scale bars = 20 μm. (Courtesy P. Abad, INRA, France; see Chapter 7.)

38

39

40

Plate 38. Fodder radish and mustard cultivars used for biofumigation just before incorporation into the soil. (Courtesy Julius Kühn-Institut, Braunschweig, Germany; see Chapter 17.)

Plate 39. Incorporation of fodder radish at flowering using a chopper (front) in combination with a cultivator and rototiller, followed by a plain roll to reduce evaporation. (Courtesy Julius Kühn-Institut, Braunschweig, Germany; see Chapter 17.)

Plate 40. Application of 1,3-dichloropropene using cultipactor to seal soil surface prior to establishing peach orchard in *Meloidogyne*-infested site. (Courtesy A.P. Nyczepir, USDA, USA; see Chapter 18.)

41

42

43

Plate 41. Incorporation of *Brassica* species (i.e. green manure) as a preplant nematode control strategy. (Courtesy A.P. Nyczepir, USDA, USA; see Chapter 18.)

Plate 42. Enhanced growth and lint production in cotton following preseason application of 56 l/ha 1,3-dicholoropropene for suppression of *Meloidogyne incognita* (left) compared with untreated cotton (right). (Courtesy S.H. Thomas; see Chapter 18.)

Plate 43. Enhanced growth and stand density in pepper (*Capsicum annuum*) following preseason application of 56 l/ha 1,3-dicholoropropene for suppression of *Meloidogyne incognita* (left) compared to untreated pepper (right). (Courtesy S.H. Thomas; see Chapter 18.)

44

45

Plate 44. Pepper (*Capsicum annuum*) crop demonstrating successful suppression of the *Meloidogyne incognita*/yellow nutsedge (*Cyperus esculentus*)/purple nutsedge (*C. rotundus*) pest complex following a 3-year rotation with *M. incognita*-resistant alfalfa (*Medicago sativa*) (left panel) compared with pepper following a standard 3-year rotation with cotton and preseason treatment with 56 l/ha 1,3-dichloropropene (right panel). Pepper following alfalfa received no preseason nematicide. (Courtesy S.H. Thomas; see Chapter 18.)
Plate 45. Cotton trial using seed treated with Avicta® (left of red line, showing enhanced growth and stand) compared with non-treated seed (right of red line) in a field infested with *Meloidogyne incognita*. (Courtesy Syngenta, Switzerland.)

prepared by mixing infested and non-infested soil in appropriate proportions (Di Vito and Greco, 1988a,b). Therefore, to obtain damage data as close as possible to those occurring under field conditions, we suggest that the nematode population be reared in conditions similar to those of the growing season preceding that of the test crop, as can be obtained in large pots or microplots outdoors. As most crops are planted in spring, the nematode should be produced during the previous summer and left overwinter for use the following spring. Before use, the roots must be finely chopped, thoroughly mixed into the same potting soil, the nematode population density estimated and then proper amounts used as inoculum by mixing different proportions of infested and non-infested soil.

In assessing the nematode population of the infested soil before planting or sowing and after harvest, the extraction method used must be able to extract eggs. If only J2 and not eggs are extracted, the tolerance limit would be overestimated and the reproduction of the nematode underestimated.

11.9.2 Glasshouse experiments

Investigations in glasshouses are useful whenever data on the pathogenicity of a nematode population or comparison among different nematode populations and/or plant species/cultivars are required. As both resistant and susceptible cultivars may show the same or very close tolerance limit (T) but different minimum yield (m) (Ekanayake and Di Vito, 1984; Di Vito *et al.*, 1991, 1992, 1999), the use of several inoculum levels would be useful to differentiate different nematode populations or tested plants on the base of the growth reduction occurring at P_i larger than T. However, even if the growth/yield of a crop plant is investigated over a range of nematode P_i, because glasshouse conditions may differ greatly from those in the field, the data obtained may not be very useful for prediction under field conditions. To minimize differences, large pots (containing 2–10 dm^3 soil, depending on the crop plant) that allow normal plant growth must be used. Also, the experiment should be discontinued before the plants become pot-bound.

11.9.3 Field experiments

Without doubt, field plots best represent the environmental conditions of the area under study, and would produce yield data that are comparable to those in farmers' fields and that are of high predictive value. The main limitation of field plots is that it may be difficult to select a field with a range of P_i at sowing or planting that varies from very small (non-infested to $\leq T$) to very large (about $60 \times T$). Preparing plots with a range of P_i obtained by artificial inoculation would require a huge amount of work and therefore it is not suggested. This limit can be partly overcome by managing the plots differently in the preceding year(s), such as planting different crops (very susceptible, resistant or partially resistant) in the different replicated plots, treating with different rates of a nematicide and/or different agronomic practices. To obtain plots with almost no nematodes, use of a high rate of a nematicide very effective against nematodes and having little or no effect on other soil-borne microorganisms is very helpful. However, these management practices will also affect other biotic and abiotic soil characteristics, thus resulting in plots that differ in characteristics additional to that of their nematode soil population densities (Kinlock, 1982; McSorley and Parrado, 1986). The size of the plots will vary according to the aim of the experiment. If the experiment is to last only one growing season and is designed to investigate the relationships between a range of P_i and yield of a host crop, plots of 4–10 m^2 would be sufficient, while they must be much larger if the dynamics of the nematodes in soil under different crop rotation regimes is to be investigated throughout the rotation term (Fig. 11.8).

11.9.4 Microplots

The use of microplots is a compromise between pots and field plots that minimizes the limitations of working in pots. The reduced volume of soil allows the preparation of a wide range of P_i, but this requires organizational skill. A microplot must have a minimum size sufficient to allow at least a single crop plant to grow as easily as in the field throughout the growing season. For annual plants in a number of experiments, tubes or tiles

Fig. 11.8. Field plots of a rotation experiment to investigate the dynamics of *Meloidogyne artiellia*.

about 50 cm deep and 30 cm in diameter or with a square section of 30 cm × 30 cm have been used (Fig. 11.9). Tubes or tiles are sunk into the soil to within 5–10 cm of the top edge and filled with infested soil up to ground level. Such microplots may contain 30–40 dm^3 soil and are suitable for many crop plants, including wheat, pea, broad bean, potato, sugarbeet, tomato, chickpea, pepper, aubergine and carrot. Microplots can be arranged in rows, but maintaining an adequate distance between microplots within and between rows. With tree crops that will be allowed to grow for several years, microplots must be of at least 1 m^2 section and about 1 m depth. More information on the use of microplots for nematological studies is given by Barker (1985).

Ideally, the soil used to fill the microplots should be the same as that in the study area in terms of texture and biotic and abiotic characteristics, but plant-parasitic nematodes and other soil-borne pests and diseases should be absent or present at non-damaging levels. This is best achieved by collecting the soil from a field in which crop plants susceptible to the target *Meloidogyne* species, and to pests and diseases that might damage the crop plant under investigation, had not been cultivated for at least the past 5 years. To kill other nematodes, the soil can be treated with a chemical having only nematicidal activity. If the only possible method of obtaining soil free of major pathogens is steaming or treating with a biocide, these treatments must be applied several months before using the soil.

11.9.5 Maintenance of experiments

During the experiments, crop plants must be fertilized, irrigated and maintained free of weeds and major pests and diseases, to allow plants to express their maximum potential growth and yield, and the nematode to infect and reproduce at its best. Stresses must be avoided as they would be added to those caused by the nematode, and it is impossible to partition their effects. As root-knot nematodes live in the soil and attack roots and other below-ground plant parts, care must be taken to maintain suitable soil moisture content and aeration. In pot and microplot experiments, irrigation is very important. This must be regulated carefully to avoid overwetting of the soil and to ensure that eggs and J2 of the nematodes are not flushed out of the pots and microplots or into deeper parts of the soil profile not explored by roots. The use of automated irrigation dispensers may not be the best solution as they would dispense the same

Fig. 11.9. Different materials used to prepare microplots for the study of the relationships between a range of population densities of a nematode and yield of a host-plant crop and the population of the nematode at harvest. Concrete tiles (top) and rigid (below right) and thin (below left) plastic tube, buried into the soil. Note different growth of pepper (top), as affected by different population densities of *Meloidogyne incognita*.

amount of water to all pots or microplots; water consumption will differ according to plant size at the different P_i and pots will require different amounts of water. Also, the required amount of water must be dispensed slowly. Simulating rainfall would avoid water runoff and is the most appropriate way of adding water to pots and microplots. Soil compaction must be avoided and therefore periodic shallow soil cultivation may be necessary. In rainfed crops, irrigation must be given only in dry periods and, if necessary, should simulate an average growing season.

Even if destructive sampling has not been undertaken, growth or yield components of the host crop must be determined at the end of the experiment and the final nematode population density (P_f) in each replicate estimated. However, some investigations may also require monitoring of the nematode dynamics periodically after planting. This is the case in experiments on nematode control with treatments applied at planting or on the established crop, and also when monitoring the nematode populations throughout the growing cycle of a host crop, in different seasons or on investigations on nematode decline.

With pots, each plant is carefully uprooted and roots and top parts separated. The roots are gently washed free of adhering soil, weighed, cut in 0.5–1-cm-long pieces and eggs and juveniles in them extracted by shaking in 1% NaOCl solution for 4 min (Hussey and Barker, 1973) and counted. Whenever the roots contain large galls, many egg masses may be embedded in the roots of some plant species. Satisfactory extraction of eggs from such roots can be achieved by macerating a 5–10 g subsample in a blender at a speed and time to be previously assessed according to the type of blender. Then the eggs can be extracted from the water/root suspension by the centrifugal method of Coolen (1979). Also, the soil of each pot is thoroughly mixed and a subsample used to extract and estimate eggs and J2. The sum of the total eggs in the soil and that in the roots of a plot will be considered as the final (P_f) population of the nematode in that pot.

In microplot experiments, the plant is cut at ground level and the largest possible proportion of the roots collected and processed as described above. A soil sample is then collected after thorough mixing of the soil of each microplot, or by taking multiple cores with a soil sampler. For microplots of 30 × 30 cm, the authors collect 20 soil cores per microplot, using an auger 30 cm long and 1.5–2.5 cm in diameter, giving a total of about 1.5 dm^3 soil. Eggs and J2 are extracted from a subsample.

In field plot experiments, the sampling strategy will depend upon the plot size. In plots of about 10–20 m^2, collecting 40–50 soil cores per plot (1.5 cm diameter and 30 cm deep) should be satisfactory. However, as nematode population densities would vary greatly in different plots, it may be necessary to adapt the sampling to the expected P_f to minimize sampling error (see Duncan and Phillips, Chapter 12, this volume). In long-term experiments, as with crop rotations, it is suggested that when several months elapse between the harvest of the preceding crop and the planting of the succeeding crop, sampling be done both at harvest and at planting, in order to assess the soil population increase in the presence of the host plant and the decrease in the presence of different non-host crops and during the non-cultivated period. When investigating the dynamics of the nematodes in perennial crops, soil and/or root samples must be collected periodically (every 15–30 days) for several years, always at a given depth and distance from the trunk, and possibly in the same orientation.

In all situations, the method used to process soil samples must also extract eggs and therefore may require some modification depending, for example, on the soil texture, organic matter content, roots of the host plant and characteristics of the blender used to macerate the roots. The authors use the method of Coolen (1979) with a slight modification (Di Vito *et al.*, 1986) consisting of the addition of 150 ml of commercial bleach (6% NaOCl solution) and a few drops of silicon antifoam in the mixing cylinder containing 2.5 l of the water–soil suspension. The NaOCl dissolves the matrix of egg masses and the silicon antifoam avoids the formation of foam when NaOCl is added. However, other methods or their modifications are used in the USA (Byrd *et al.*, 1972, 1976) and elsewhere (Demeure and Netscher, 1973) with success.

11.9.6 Fitting the models to data

Table 11.2 shows data on the weight of top parts of common bean plants as affected by a range of P_i of *M. incognita* in a glasshouse. There are no obvious growth changes at P_i 0–0.5 egg/cm3 soil, and a growth reduction occurring at $P_i \geq 1$ egg/cm3 soil. Therefore, the average yield at 0–0.5 egg/cm^3 soil (23.1 g) is considered as the yield at $P \leq T$ (= (22.3 + 23.8 + 23.6 + 23.3 + 22.5)/5), and is used to calculate the relative yield at different P_i. This is done by dividing the yield at each P_i with the average yield at $P_i \leq T$ (23.1 g). The calculated relative yields (y) (Table 11.2) are plotted on the y-axis, and the P_i values are plotted on a logarithmic scale on the X-axis (Fig. 11.5). Also, the data suggest a tolerance limit (T) of common bean to *M. incognita* of 0.25 egg/cm^3 soil and a minimum relative yield (m) of 0 occurring at P_i of 32 eggs/cm^3 soil. However, to estimate T and m with accuracy, the damage model must be fitted to the data. With m=0, the model represented by Equation 11.8 becomes:

$$y = 0 + (1-0)z^{P-0.25}$$

and

$$y = z^{P-0.25}$$

Considering that

$$z^{-T} = 1.05,$$

in our example

$$z^{-0.25} = 1.05$$

The value of z can be derived by solving this equation with a logarithmic calculation. This gives:

$$-0.25\log z = \log 1.05$$

and

$$\log z = \log 1.05/-0.25$$

Then

$$z = \text{antilog } 0.21189299/-0.25 = 0.8227$$

Table 11.2. Values of the relative yield (y) calculated according to the observed yield data (3) and the model $y = z^{P-T}$ (4), from an experiment to relate initial population densities (P_i) of *Meloidogyne incognita* (1) and top plant weight (2) of common bean. (Adapted from Di Vito *et al.*, 2004a.)

P_i (eggs/cm³ soil) (1)	Top weight of plants (g) (2)	Relative yield (top weight) (3)	y according to $y = 0.8227^{P-T}$ (4)
0	22.3	0.97	1
0.0625	23.8	1.03	1
0.125	23.6	1.02	1
0.25	23.3	1	1
0.5	22.5	0.97	0.95
1	19.2	0.83	0.86
2	15.7	0.68	0.71
4	11.5	0.5	0.48
8	4.1	0.18	0.22
16	2.3	0.10	0.046
32	0	0	0
64	0	0	0
128	0	0	0
256	0	0	0
512	0	0	0

The points of the curves according to the model will be $y = 1$ up to $P_i = 0.25$ egg/cm³ soil and those obtained by solving the equation $y = z^{P-T}$ by logarithmic calculation at all remaining P_i (0.5–512 eggs/cm³ soil) (Table 11.2). The values of y derived with the model at the respective P_i are now plotted on the same graph. Connecting the points given by the coordinates (P_i;y) will originate the curve representing the y/P_i relationship according to the model. Now the fit of the curve to the data must be evaluated. This can be done visually or, better, with statistical calculations. Should the model not fit well to the data, another trial must be done by estimating new values of T and m. Visually, this can be done by drawing one or more curves parallel to the original one and selecting the one that best fits the observed values of y. Then new T and m are estimated, and new points of the curves according to the model are calculated. The drawing of the curve according to the model can also be done using graphic computer programs. Alternatively, the computer program Seinfit (Viaene *et al.*, 1997) can be used to fit the Seinhorst model to experimental data.

A simple, naïve, but practical way to fit the damage model to the data is as follows. A number of curves having the same T, but different values of m, are drawn (Fig. 11.6) and then printed or photocopied on to a transparent sheet. This sheet can be used as an overlayer to the graph representing the observed values of y to select the curve that best fit to the data. In doing so, it is necessary that the scales of the x- and y-axes of the graphs be the same. This procedure usually provides good estimations of T and m.

As already stated, fitting the model represented by Equation 11.6 (for population dynamics) to P_f and P_i data is rather difficult for root-knot nematodes. With short-term experiments, the nematode may have developed only one generation and the fitting of the model to the data is possible. Therefore, the values of P_i and P_f are plotted on graph paper with logarithmic scales on both axes. If the points show a pattern similar to that of lines 1–3 in Fig. 11.2, then a way to fit the model to the data is as follows. A curve according to the first addendum of Equation 11.6 can be drawn by selecting certain values for the parameters a, y, x and E, and solving the equation for different P_i and corresponding y values in the damage model. It must be considered that changing the values of the parameters mentioned will not cause a change in the shape of the curve, but only in its shift horizontally or perpendicularly to the axis. The observed values of P_f at different P_i are plotted, and a curve derived by hand or by using a graphic computer program. The curve can now be printed or photocopied on to a transparent sheet, and this can be used to fit the experimental data, as explained for the damage model. Then the values of a, x and E are estimated graphically.

11.10 Yield Loss Assessment

Sasser and Freckman (1987) provided tables reporting the yield losses caused by nematodes, including root-knot nematodes, to a number of crops of worldwide importance. The information given by these authors has been very valuable in making many scientists aware of the impact of nematodes on agriculture, and has enabled them to set priorities for investigations on nematodes in their home countries. The information given in these tables consists of estimated losses from a survey among nematologists from all over the world, rather than of scientific data from properly designed experiments. These data are not precise estimates of actual losses.

Models describing the relationship between a range of nematode *Pi* and yield of a host crop are a basic requirement for assessing, with acceptable accuracy and precision, the yield loss a nematode may be causing in a given area: information required by those who will make decisions on management strategies. Such information would also be valuable to farmers, especially those using precision agriculture systems (Evans and Barker, 2004), to decide if, where, how and when a host crop can be planted without incurring yield loss of economic importance (Ferris and Noling, 1987). Accurate estimates of yield losses also require that proper sampling be done (Ferris, 1981; Ferris and Noling, 1987; Evans and Barker, 2004). To be representative of the study area (farm, state, country), sampling must be based on a sound statistical scheme (Ferris, 1984; Ferris and Noling, 1987; see Duncan and Phillips, Chapter 12, this volume), must consider only fields to be planted to the target crop plant, and must be undertaken during the period of the year for which the relationship between P_i and yield of the host plant has been established. This is generally just before sowing or planting of annual plants. However, the distribution of root-knot nematodes in the soil profile may vary depending on the previous crop and date of sampling (Wesemael and Moens, 2008). Also, species of *Meloidogyne* are known to move within the soil profile when attracted by roots of hosts (Prot and Netscher, 1977), and this may cause underestimation of the actual population density. Once the nematode population density (P_i) of a field is known, the use of models as represented in Figs 11.5 and 11.6, but derived for the specific nematode/crop plant combination, will allow the yield loss in a given field to be estimated. Averaging the yield losses of different sampled fields will give an estimation of the yield and its value lost annually in an area. In addition to the work of Seinhorst, the problems of yield loss assessment and prediction of nematode dynamics have been discussed by various other authors (Barker and Olthof, 1976; Barker *et al.*, 1985; Teng, 1985; Ferris and Noling, 1987; McSorley and Duncan, 2004; Schomaker and Been, 2006).

11.11 Importance of Information on Nematode Damage Levels and Dynamics in Management Strategies

Data from experiments assessing the relationship between P_i and yield of a host crop and nematode density at harvest (P_f), and on the dynamics of the nematode population in the absence and presence of a host, are basic to analysis of the economics of possible management tactics (Ferris and Noling, 1987). In nematode management, the entire crop sequence of a rotation must be considered rather than a single crop cycle. Predictions can then be made with acceptable accuracy of the dynamics of target nematodes considered throughout the term of all possible rotation options for a given field or area. However, information is also necessary on the relationship between the application rates of any control treatments and the mortality of the target nematode. Based on this information, the economic threshold can be calculated. This is defined as the nematode P_i at which the value of the predicted yield increase following a treatment is equal to the predicted cost of that treatment (Ferris, 1978). From an economic point of view, the optimizing economic threshold rather than the actual economic threshold must be considered. This is defined as the P_i at which the difference between the value of the expected yield increase and the cost of the management tactics is maximum (Ferris, 1978; Ferris and Greco, 1992). As prices of the produce generated by the crop final yield and prices of the management options may vary from year to year, and from one area or country to another, the economic and maximizing economic thresholds, may vary accordingly. An example of how to use information on damage and economic thresholds, and on nematode dynamics, to select the most convenient management strategy, is given by Ferris and Greco (1992).

The information on the relationship between nematode population density and yield of a host crop should be considered whenever, in the assessment of the efficacy of a control measure, the weight of the plants or of any part of the plant is used as an evaluation criterion. This is important when selecting infested fields for control experiments. It is not sufficient that the field be infested by the nematode; the level of infestation (P_i) must also have potential to cause yield losses of about 30–50% in order to discriminate between treatments.

11.12 Conclusions and Future Directions

Despite the general agreement on its importance in nematode management, information on relationships between nematode P_i and yield of host crops, and between P_f and reduction of nematode P_i as affected by different rates of a control treatment, is poor or lacking, even for the major root-knot nematode/crop combinations in most countries. Of the available data, many refer to tests using glasshouse pots (usually rather small) (Table 11.1), and/or narrow ranges of nematode densities. Therefore, it is to be hoped that, depending on countries needs, information will be gathered on both damage levels and nematode dynamics, and also on the overall yield losses that root-knot nematodes are causing, in order to set priorities in both research and management decision making at different levels. Despite the fact that the damage model was developed for nematodes having one generation per growing season, a re-examination of data from the literature (Seinhorst, 1998) demonstrated that, with a few exceptions, it adequately describes the relationship between population densities of root-knot nematodes and their host crops. Also, there is evidence that when environmental conditions are similar, information from one study area can also be of use in other areas. For example, curves relating soil population densities of potato cyst nematode to yield of susceptible potato cultivar in three different areas in Italy (Greco *et al.*, 1982), during two different growing seasons in Chile (Greco and Moreno, 1992) and in Venezuela (Jiménez *et al.*, 2000), using the same methodology but with different cultivars, were very similar. The tolerance limits (T) were in the range of 1.3–2.1 nematode eggs/g soil, and minimum yields (m) varied from 0.03 to 0.3. Such variations are not great from the point of view of management decisions. However, users must carefully consider the available information before using it for prediction and management decisions in their areas of interest.

Nematodes usually occur in complex communities. Ideally, each nematode should be investigated in all possible communities. In reality this would be impossible. However, in many cases, when present at damaging levels, root-knot nematodes are the most prevalent nematodes, so information on their dynamics and the extent of their damage to the host crop, determined as explained, would still be useful.

11.13 Acknowledgements

We wish to thank Dr Ken Evans for valuable suggestions and editing the manuscript, and Dr Alberto Troccoli for help with the figures.

11.14 References

Aguirre, Y., Crozzoli, R. and Greco, N. (2003) Effect of the root-knot nematode *Meloidogyne incognita* on parsley. *Russian Journal of Nematology* 11, 27–31.

Barker, K.R. (1985) The application of microplot techniques in nematological research. In: Barker, K.R., Carter, C.C. and Sasser, J.N. (eds) *An Advanced Treatise on Meloidogyne. Volume II – Methodology.* North Carolina State University Graphics, Raleigh, North Carolina, USA pp. 127–134.

Barker, K.R. and Olthof, T.H.A. (1976) Relationship between nematode population densities and crop responses. *Annual Review of Phytopathology* 14, 327–353.

Barker, K.R and Weeks, W.W. (1991) Relationships between soil and levels of *Meloidogyne incognita* and tobacco yield and quality. *Journal of Nematology* 23, 82–90.

Barker, K.R., Shoemaker, P.B. and Nelson, L.A. (1976) Relationships of initial population densities of *Meloidogyne incognita* and *M. hapla* to yield of tomato. *Journal of Nematology* 8, 232–239.

Barker, K.R., Todd, F.A., Shane, W.W. and Nelson, L.A. (1981) Interrelationships of *Meloidogyne* species with flue-cured tobacco. *Journal of Nematology* 13, 67–79.

Barker, K.R., Schmitt, D.P. and Imbriani, J.L. (1985) Nematode population dynamics with emphasis on determining damage potential to crops. In: Barker, K.R., Carter, CC. and Sasser, J.N. (eds) *An Advanced Treatise on Meloidogyne. Volume II – Methodology.* North Carolina State University Graphics, Raleigh, North Carolina, pp. 135–148.

Bridge, J., Plowright, R.A. and Peng, D. (2005) Nematode parasites of rice. In: Luc, M., Sikora, R.A. and Bridge, J. (eds) *Plant Parasitic Nematodes in Subtropical and Tropical Agriculture*, 2nd edn. CAB International, Wallingford, UK, pp. 87–130.

Bustillo, Y.Y.C., Crozzoli, R., Greco, N. and Lamberti, F. (2000) Efecto del nematodo agallador *Meloidogyne incognita* sobre el crescimiento de la lechosa (*Carica papaya*) en vivero. *Nematologia Mediterranea* 28, 163–170.

Byrd, D.W. Jr, Ferris, H. and Nusbaum, C.J. (1972) A method for estimating numbers of eggs of *Meloidogyne* spp. in soil. *Journal of Nematology* 4, 266–269.

Byrd, D.W. Jr, Barker, K.R., Ferris, H., Nusbaum, C.J., Grifin, W.E., Small, R.H. and Stone, C.A. (1976) Two semi-automatic elutriators for extracting nematodes and certain fungi from soil. *Journal of Nematology* 8, 206–212.

Canto-Sáenz, M. (1985) The nature of resistance to *Meloidogyne incognita* (Kofoid & White) Chitwood, 1949. In: Sasser, J.N. and Carter, C.C. (eds) *An Advanced Treatise on* Meloidogyne. *Volume I – Biology and Control.* North Carolina State University Graphics, Raleigh, North Carolina, pp. 225–231.

Casassa, A.M., Crozzoli, R., Matheus, J., Bravo, V. and Marin, M. (1998) Efecto del nematodo agallador *Meloidogyne incognita* sobre el crecimiento del guayabo (*Psidium* spp.) en vivero. *Nematologia Mediterranea* 26, 237–242.

Castillo, P., Di Vito, M., Vovlas, N. and Jimenez-Diaz, R.M. (2000) Host–parasite relationship in root-knot disease of white mulberry. *Plant Disease* 85, 277–281.

Cook, R. and Noel, G.R. (2002) Cyst nematodes: *Globodera* and *Heterodera* species. In: Starr, J.L., Cook, R. and Bridge, J. (eds) *Plant Resistance to Parasitic Nematodes.* CAB International, Wallingford, UK, pp. 71–105.

Coolen, W.A. (1979) Methods for the extraction of *Meloidogyne* spp. and other nematodes from roots and soil. In: Lamberti, F. and Taylor, C.E. (eds) *Root-knot Nematodes* (Meloidogyne *Species) – Systematics, Biology and Control.* Academic Press, London, pp. 317–329.

Crozzoli, R. and Parra, N. (1999) Patogenicidad del nematodo agallador, *Meloidogyne incognita*, en yuca (*Manihot esculentum*). *Nematologia Mediterranea* 27, 95–100.

Crozzoli, R., Greco, N., Suarez, A.C. and Rivas, D. (1997) Patogenicidad del nematodo agallador, *Meloidogyne incognita*, en cultivares de *Phaseolus vulgaris* y *Vigna unguiculata. Nematropica* 27, 61–67.

de Guiran, G. (1980) Facteurs induisant chez *Meloidogyne incognita* un blocage du développement des oeufs considèré comme une diapause. *Revue de Nématologie* 3, 61–69.

de Guiran, G. and Demeure, Y. (1978) Influence du potentiel hydrique des sols sur le masses d'œufs de *Meloidogyne incognita* (Nematoda: Meloidogynidae). *Revue de Nématologie* 1, 119–134.

de Guiran, G. and Villemin, M.-A. (1980) Spécificité de la diapause embryonnaire des œufs de Meloidogyne (Nematoda). *Revue de Nématologie* 3, 115–121.

Demeure, Y. and Netscher, C. (1973) Méthode d'estimation des populations de *Meloidogyne* dans le sol. *Cahier ORSTOM, Serie Biologie*, No. 21, 85–90.

Di Vito, M. and Greco, N. (1988a) Investigation on the biology of *Meloidogyne artiellia. Revue de Nématologie* 11, 223–227.

Di Vito, M. and Greco, N. (1988b) The relationship between initial population densities of *Meloidogyne artiellia* and yield of winter and spring chickpea. *Nematologia Mediterranea* 16, 163–166.

Di Vito, M. and Greco, N. (1988c) Effect of population densities of *Meloidogyne artiellia* and yield of wheat. *Nematologia Mediterranea* 16, 167–169.

Di Vito, M. and Lamberti, F. (1977) Prove di lotta chimica contro i nematodi galligeni su barbabietola da zucchero. *Nematologia Mediterranea* 5, 31–38.

Di Vito, M. and Zaccheo, G. (1991) Population density of root-knot nematode, *Meloidogyne incognita* and growth of artichoke (*Cynara scolimus*). *Advances in Horticultural Science* 5, 81–82.
Di Vito, M., Lamberti, F. and Carella, A. (1979) La resistenza del pomodoro nei confronti dei nematodi galligeni: prospettive e possibilità. *Rivista di Agronomia* 13, 313–322.
Di Vito, M., Vovlas, N. and Inserra, R.N. (1980) Influence of *Meloidogyne incognita* on growth of corn in pots. *Plant Disease* 64, 1025–1026.
Di Vito, M., Greco, N. and Carella, A. (1981) Relationship between population densities of *Meloidogyne incognita* and yield of sugarbeet and tomato. *Nematologia Mediterranea* 9, 99–103.
Di Vito, M., Greco, N. and Carella, A. (1983) The effect of population densities of *Meloidogyne incognita* on the yield of cantaloupe and tobacco. *Nematologia Mediterranea* 11, 169–174.
Di Vito, M., Greco, N. and Carella, A. (1985a) Population densities of *Meloidogyne incognita* and yield of *Capsicum annuum*. *Journal of Nematology* 17, 45–49.
Di Vito, M., Greco, N. and Zaccheo, G. (1985b) On the host range of *Meloidogyne artiellia*. *Nematologia Mediterranea* 13, 207–212.
Di Vito, M., Greco, N. and Carella, A. (1986) The effect of *Meloidogyne incognita* and importance of the inoculum on yield of eggplant. *Journal of Nematology* 18, 487–490.
Di Vito, M., Cianciotta, V. and Zaccheo, G. (1991) The effect of population densities of *Meloidogyne incognita* on yield of susceptible and resistant tomato. *Nematologia Mediterranea* 19, 265–268.
Di Vito, M., Cianciotta, V. and Zaccheo, G. (1992) Yield of susceptible and resistant pepper in microplots infested with *Meloidogyne incognita*. *Nematropica* 22, 1–6.
Di Vito, M., Vovlas, N., Lamberti, F., Zaccheo, G. and Catalano, F. (1996a) Pathogenicity of *Meloidogyne javanica* on Asian and African rice. *Nematologia Mediterranea* 24, 95–99.
Di Vito, M., Zaccheo, G., Della Gatta, C. and Catalano, F. (1996b) Relationship between initial population densities of *Meloidogyne javanica* and yield of sunflower in microplots. *Nematologia Mediterranea* 24, 109–112.
Di Vito, M., Piscioneri, I., Pace, S., Zaccheo, G. and Catalano, F. (1997) Pathogenicity of *Meloidogyne incognita* on kenaf. *Nematologia Mediterranea* 25, 165–168.
Di Vito, M., Vovlas, N., Lamberti, F., Zaccheo, G. and Catalano, F. (1999) Difference in growth reduction potential of three Indian populations of *Meloidogyne javanica* on peanut. *International Journal of Nematology* 9, 168–173.
Di Vito, M., Crozzoli, R. and Vovlas, N. (2000) Pathogenicity of *Meloidogyne esigua* on coffee (*Coffea arabica* L.) in pots. *Nematropica* 30, 55–61.
Di Vito, M., Parisi, B. and Catalano, F. (2004a) Effect of population densities of *Meloidogyne incognita* on common bean. *Nematologia Mediterranea* 32, 81–85.
Di Vito, M., Vovlas, N. and Castillo, P. (2004b) Host–parasite relationship of *Meloidogyne incognita* on spinach. *Plant Pathology* 53, 508–514.
Di Vito, M., Simeone, A.M. and Catalano, F. (2005) Effect of the root-knot nematode, *Meloidogyne javanica* on the growth of peach (*Prunus persica*) rootstock in pots. *Nematologia Mediterranea* 33, 87–90.
Di Vito, M., Parisi, B. and Catalano, F. (2007) Pathogenicity of *Meloidogyne javanica* on common bean (*Phaseolus vulgaris* L.) in pots. *Nematropica* 37, 339–344.
Ekanayake, H.M.R.K. and Di Vito, M. (1984) Effect of population densities of *Meloidogyne incognita* on growth of susceptible and resistant tomato plants. *Nematologia Mediterranea* 12, 11–16.
Evans, K. and Barker, A.D.P. (2004) Economies in the management from precision agriculture – limitations and possibilities. In: Cook, R. and Hunt, D.J. (eds) *Nematology Monographs and Perspectives 2 – Proceedings of the Fourth International Congress of Nematology*. Brill, Leiden, The Netherlands, pp. 23–32.
Fassuliotis, G. (1979) Plant breeding for root-knot nematode resistance. In: Lamberti, F. and Taylor, C.E. (eds) *Root-knot Nematodes* (Meloidogyne *Species*) *– Systematics, Biology and Control*. Academic Press, London, pp. 425–453.
Ferraz, S. and Grassi de Freitas, L. (2004) Use of antagonistic plants and natural products. In: Chen, Z.X., Chen, S.Y and Dickson, D.W. (eds) *Nematology Advances and Perspectives – Volume 2, Nematode Management and Utilization*. CAB International, Wallingford, UK, pp. 931–977.
Ferris, H. (1978) Nematode economic thresholds: derivation, requirements and theoretical consideration. *Journal of Nematology* 10, 341–350.
Ferris, H. (1981) Dynamic action thresholds for diseases induced by nematodes. *Annual Review of Phytopathology* 19, 427–436.
Ferris, H. (1984) Probability range in damage prediction as related to sampling decision. *Journal of Nematology* 16, 246–251.

Ferris, H. and Greco, N. (1992) Management strategies for *Heterodera goettingiana* in a vegetable cropping system in Italy. *Fundamental and Applied Nematology* 15, 25–33.

Ferris, H. and McKenry, M.V. (1974) Seasonal fluctuation in the spatial distribution of nematode populations in a California vineyard. *Journal of Nematology* 6, 203–210.

Ferris, H. and Noling, J.W. (1987) Analysis and prediction as a basis for management decision. In: Brown, R.H. and Kerry, B.R. (eds) *Principles and Practice of Nematode Control in Crops.* Academic Press, Sydney, Australia, pp. 49–85.

Gaur, H.S., Mishra, S.D. and Sud, U.C. (1979) Effect of date of sowing on the relation between the population density of the root-knot nematode, *Meloidogyne incognita* and the growth of three varieties of chickpea, *Cicer arietinum. Indian Journal of Nematology* 9, 152–159.

Gooris, J. and d'Herde, C.J. (1977) *Study on biology of* Meloidogyne naasi *Franklin 1965.* State Nematology and Entomology Research Station, Merelbeke, Belgium.

Greco, N. and Moreno, I.L. (1992) Influence of *Globodera rostochiensis* on yield of summer, winter and spring sown potato in Chile. *Nematropica* 22, 165–173.

Greco, N., Di Vito, M., Brandonisio, A., Giordano, I. and De Marinis, G. (1982) The effect of *Globodera pallida* and *G. rostochiensis* on potato yield. *Nematologica* 28, 379–386.

Greco, N., Crozzoli, R., Lamberti, F. and Brandonisio, A. (2000) Pathogenic potential of the root-knot nematode, *Meloidogyne incognita*, on rice in Venezuela. *Nematologia Mediterranea* 28, 47–55.

Griffin, G.D., Inserra, R.N. and Di Vito, M. (1982) Comparative relationship between *Meloidogyne chitwoodi* and *M. hapla* population densities and growth of sugarbeet seedlings. *Journal of Nematology* 14, 409–411.

Hlaoua, W. and Horrigue Raouani, N. (2007) Effet de *Meloidogyne incognita* sur la culture de pomme de terre. *Nematologia Mediterranea* 35, 213–220.

Hussey, R.S. and Barker, K.R. (1973) A comparison of methods of collecting inocula of *Meloidogyne* spp., including a new technique. *Plant Disease Reporter* 57, 1025–1028.

Inserra, R.N., O'Bannon, J.H., Di Vito, M. and Ferris, H. (1983) Response of two alfalfa cultivars to *Meloidogyne hapla. Journal of Nematology* 15, 644–646.

Jiménez, N., Crozzoli, R. and Greco, N. (2000) Effect of *Globodera rostochiensis* on the yield of potato in Venezuela. *Nematologia Mediterranea* 28, 295–299.

Johnson, A.W., Dowler, C.C. and Handoo, Z.A. (2000) Population dynamics of *Meloidogyne incognita, M. arenaria*, and other nematodes and crop yields in rotations of cotton, peanut, and wheat under minimum tillage. *Journal of Nematology* 32, 52–61.

Kinlock, R.A. (1982) The relationship between soil populations of *Meloidogyne incognita* and yield reduction of soybean in the coastal plain. *Journal of Nematology* 14, 162–167.

Koenning, S.R. and Barker, K.R. (1992) Relative damage functions and reproductive potentials of *Meloidogyne arenaria* and *M. hapla* on peanut. *Journal of Nematology* 24, 187–192.

Lambert, K.N., Tedford, E.C., Caswell, E.P. and Williamson, V.M. (1992) A system for continuous production of root-knot nematode juveniles in hydroponic culture. *Phytopathology* 82, 512–515.

McSorley, R. and Duncan, L.W. (2004) Population dynamics. In: Chen, Z.X., Chen, S.Y and Dickson, D.W. (eds) *Nematology Advances and Perspectives – Volume 2, Nematode Management and Utilization.* CAB International, Wallingford, UK, pp. 469–492.

McSorley, R. and Parrado, J.L. (1986) Relationship between height of kenaf and root galling by *Meloidogyne incognita. Nematotropica* 16, 205–211.

Mekete, T., Mandefro, W. and Greco, N. (2003) Relationship between initial population densities of *Meloidogyne javanica* and damage to pepper and tomato in Ethiopia. *Nematologia Mediterranea* 31, 169–171.

Nicholson, A.J. (1935) The balance of animal populations. *Journal of Animal Ecology* 2, 132–178.

Olowe, T. (2007) Host reaction of cowpea cultivars to root-knot nematode, *Meloidogyne incognita. Nematologia Mediterranea* 35, 177–182.

Oostenbrink, M. (1966) Major characteristics of the relation between nematodes and plants. *Mededeelingen van de Landbouwhogeschool te Wageningen* 66(4), 1–46.

Prot, J.C. and Netscher, C. (1977) Influence of movement of juveniles on direction of fields infested with *Meloidogyne*. In: Lamberti, F. and Taylor, C. E. (eds) *Root-knot Nematodes* (Meloidogyne *Species*) – *Systematics, Biology and Control.* Academic Press, London, pp. 193–203.

Roberts, P.A. (1987) The influence of planting date of carrot on *Meloidogyne incognita* reproduction and injury to roots. *Nematologica* 33, 335–342.

Russo, G., Greco, N., D'Errico, F.P. and Brandonisio, A. (2007) Impact of the root-knot nematode, *Meloidogyne incognita*, on potato during two different growing seasons. *Nematologia Mediterranea* 35, 29–34.

Sasanelli, N. and Di Vito, M. (1992) The effect of *Meloidogyne incognita* on growth of sunflower in pots. *Nematologia Mediterranea* 20, 9–12.

Sasanelli, N., Di Vito, M. and Zaccheo, G. (1992) Population densities of *Meloidogyne incognita* and growth of cabbage in pots. *Nematologia Mediterranea* 20, 21–23.

Sasanelli, N., D'Addabbo, T. and Lemos, R.M. (2002) Influence of *Meloidogyne javanica* on growth of olive cuttings in pots. *Nematropica* 32, 59–63.

Sasanelli, N., D'Addabbo, T. and Lišková, M. (2006) Influence of the root-knot nematode *Meloidogyne incognita* r.1 on growth of grape. *Helminthologia* 43, 168–170.

Sasser, J.N. and Freckman, D.W. (1987) A world perspective on nematology: the role of the Society. In: Veech, J.A. and Dickson, D.W. (eds) *Vistas on Nematology: a Commemoration of the Twenty-fifth anniversary of the Society of Nematology.* Society of Nematologists, Hyattsville, Maryland, pp. 7–14.

Schomaker, C.H. and Been, T.H. (2006) Plant growth and population dynamics. In: Perry, R.N. and Moens, M. (eds) *Plant Nematology.* CAB International, Wallingford, UK, pp. 275–301.

Seinhorst, J.W. (1965) The relationship between nematode density and damage to plants. *Nematologica* 11, 137–154.

Seinhorst, J.W. (1966) The relationship between population increase and population density in plant parasitic nematodes – I. Introduction and migratory nematodes. *Nematologica* 12, 157–169.

Seinhorst, J.W. (1967a) The relationship between population increase and population density in plant parasitic nematodes – II. Sedentary nematodes. *Nematologica* 13, 157–171.

Seinhorst, J.W. (1967b) The relationship between population increase and population density in plant parasitic nematodes – V. Influence of damage to the host on multiplication. *Nematologica* 13, 481–492.

Seinhorst, J.W. (1970) Dynamics of populations of plant parasitic nematodes. *Annual Review of Phytopathology* 8, 131–156.

Seinhorst, J.W. (1972) The relationship between yield and square root of nematode density. *Nematologica* 18, 585–590.

Seinhorst, J.W. (1979) Nematodes and growth of plants: formulation of the nematode–plant system. In: Lamberti, F. and Taylor, C.E. (eds) *Root-knot Nematodes* (Meloidogyne *Species*) – *Systematics, Biology and Control.* Academic Press, London, pp. 231–256.

Seinhorst, J.W. (1981) Water consumption of plant attacked by nematodes and mechanisms of growth reduction. *Nematologica* 27, 34–51.

Seinhorst, J.W. (1986a) The development of individuals and populations of cyst nematodes on plants. In: Lamberti, F. and Taylor, C.E. (eds) *Cysts Nematodes.* Academic Press, London, pp. 101–117.

Seinhorst, J.W. (1986b) Effects of nematode attack on the growth and yield of crop plants. In: Lamberti, F. and Taylor, C.E. (eds) *Cysts Nematodes.* Academic Press, London, pp. 191–209.

Seinhorst, J.W. (1995) The reduction of the growth and weight of plants by a second and later generations of the nematode. *Nematologica* 41, 592–602.

Seinhorst, J.W. (1998) The common relation between population density and plant weight in pot and microplot experiments with various nematode plant combinations. *Fundamental and Applied Nematology* 21, 459–468.

Souza, R.M., Volpato, A.R. and Viana, A.P. (2008) Epidemiology of *Meloidogyne exigua* in upland coffee plantation in Brazil. *Nematologia Mediterranea* 36, 13–17.

Starr, J.L., Morgan, E. and Simpson, C.E. (2002) Management of the peanut root-knot nematode, *Meloidogyne arenaria*, with host resistance. Online. *Plant Health Progress.* doi:10.1094/PHP-2002-1121-01-HM.

Taylor, A.L. and Sasser, J.N. (1978) *Biology, Identification and Control of Root-knot Nematodes* (Meloidogyne *species*). North Carolina State University Graphics, Raleigh, North Carolina.

Teng, P.S. (1985) Crop loss assessment methods: current situation and needs. In: Barker, K.R, Carter, C.C. and Sasser, J.N. (eds) *An Advance Treatise on* Meloidogyne. *Volume II – Methodology.* North Carolina State University Graphics, Raleigh, North Carolina, pp. 149–158.

Van Gundy, S.D. (1985) Ecology of *Meloidogyne* spp. – emphasis on environmental factors affecting survival and pathogenicity. In: Sasser, J.N and Carter, C.C. (eds) *An Advanced Treatise on* Meloidogyne – *Volume I. Biology and Control.* North Carolina State University Graphics, Raleigh, North Carolina, pp. 177–182.

Viaene, N.M., Simoens, P. and Abawi, G.S. (1997) Seinfit, a computer program for the estimation of the Seinhorst equation. *Journal of Nematology* 29, 474–477.

Vovlas, N. and Di Vito, M. (1991) Effect of root-knot nematodes *Meloidogyne incognita* and *M. javanica* on the growth of coffee (*Coffea arabica* L.) in pots. *Nematologia Mediterranea* 19, 253–258.

Vovlas, N., Di Vito, M. and Grammatikaki, G. (1993) Growth response of in vitro produced banana plantlets to *Meloidogyne javanica* in pots. *Nematropica* 23, 203–208.

Vovlas, N., Mifsud, D., Landa, B.B. and Castillo, P. (2005) Pathogenicity of the root-knot nematode *Meloidogyne javanica* on potato. *Plant Pathology* 54, 657–664.

Wallace, H.R. (1966) The influence of moisture stress on the development, hatch and survival of eggs of *Meloidogyne javanica. Nematologica* 12, 57–69.

Wallace, H.R. (1968) The influence of soil moisture on survival and hatch of *Meloidogyne javanica. Nematologica* 14, 231–242.

Wallace, H.R. (1971) Abiotic influences in the soil environmental. In: Zuckerman, B.M., Mai, W.F. and Rohde, R.A. (eds) *Plant Parasitic Nematodes.* Academic Press, London, pp. 257–280.

Wallace, H.R. (1989) Environmental and plant health, a nematological perception. *Annual Review of Phytopathology* 27, 59–75.

Wesemael, W.M.L. and Moens, M. (2008) Quality damage on carrots (*Daucus carota* L.) caused by the root-knot nematode *Meloidogyne chitwoodi. Nematology* 10, 261–270.

Wesemael, W.M.L., Perry, R.N. and Moens, M. (2006) The influence of root diffusate and host age on hatching of the root-knot nematodes, *Meloiodogyne chitwoodi* and *M. fallax. Nematology* 8, 893–899.

Zhou, E. and Starr, J.L. (2003) A comparison of the damage functions, root galling and reproduction of *Meloidogyne incognita* on resistant and susceptible cotton cultivars. *Journal of Cotton Science* 7, 224–230.

12 Sampling Root-knot Nematodes

Larry W. Duncan[1] and Mark S. Phillips[2]
[1]University of Florida, Institute of Food and Agricultural Sciences (IFAS), Lake Alfred, Florida, USA; [2]Scottish Crop Research Institute, Invergowrie, Dundee, UK

12.1	Introduction	275
12.2	Nematode Spatial Patterns	276
12.3	Characterizing Sample Accuracy and Reliability	278
12.4	Sample Processing	281
12.5	Extracting Nematodes from Soil	281
12.6	Extracting Nematodes from Plant Material	282
12.7	Root Gall Indices	282
12.8	Other Plant Symptoms	283
12.9	Research to Optimize Sampling Programmes for Root-knot Nematodes	283
12.10	Examples of Results from Sampling Programmes	285
12.11	Conclusions and Future Directions	293
12.12	References	295

12.1 Introduction

Nematode population levels are estimated from samples of soil or plant tissue and the results are used for various purposes. They may be used in an advisory capacity to predict the need to impose management tactics to suppress populations to desired levels, or in research to study the responses of nematode populations to management tactics and other anthropogenic factors, or to derive relationships between population size and environmental factors. For most of these purposes an ideal nematode sample would reveal the spatial pattern of population densities within the sampled area. Such information would permit growers to select planting material best suited for various locations or to adopt variable rate methods for application of nematicides at levels needed to suppress local populations to non-damaging levels (Ferris, 1978; Wheeler and Bronson, 2004; Monfort *et al.*, 2007). It would facilitate research by permitting comparisons between spatial patterns of nematodes and those of the myriad physical and biotic forces to which they respond (Perry, 1998; Spiridinov *et al.*, 2007). However, nematode samples are expensive to collect and process: to estimate the population density of *Pratylenchus penetrans* in an area of 0.01 ha within 20% of the true mean with 95% confidence required approximately 7 h of labour (Proctor and Marks, 1974). Therefore, rather than estimating population densities at many discrete locations in a field, soil sampled at those locations (referred to hereafter as **cores**, after the method commonly used to collect them) is combined in a composite or bulk **sample**, which is mixed and then usually subsampled. Currently, growers can only approximate precision agriculture methods by delineating a sampling area into smaller units based on affordability and factors such as soil texture, cropping history or past crop performance, which are known to affect nematode population size (Barker and Campbell, 1981; Monfort *et al.*, 2007). Often as few as one to several bulk soil samples are taken in each delineated unit in a field. After mixing the soil, nematodes are commonly extracted from an additional subsample(s), which may comprise only 50–500 cm^3 of soil taken from areas as large as 2–5 ha (Barker and Imbriani, 1984). Population estimates from such small samples are clearly subject to substantial error. Thus, sampling intensity is usually dictated by the time and money growers and researchers are willing to invest rather than by what is required to achieve a specified level of accuracy (Been and Schomaker, 2000). The need for mechanized sampling devices has long been recognized as a major impediment to improving sampling effectiveness (Barker and Imbriani, 1984; Ferris *et al.*, 1990; Duncan, 1991; Been and Schomaker, 2000). Nevertheless, to make rational management decisions based on information from soil or plant samples requires some understanding of their reliability – the probability of achieving a specified degree of accuracy. Sample methods are devised and characterized in a variety of ways, most of which require knowledge of nematode spatial patterns in nature and how those patterns affect the frequency distributions of the sample data.

12.2 Nematode Spatial Patterns

The spatial patterns of nematodes in nature are patchy or aggregated at almost any resolution at which they have been studied (Goodell and Ferris, 1980, 1981; Duncan *et al.*, 1994a, 1995; Been and Schomaker, 1998). Samples from discrete locations within field plots as small as 1 m^2 or as large as a field will almost always result in counts with the variance larger than the mean (Wheeler *et al.*, 1992; Siddiqui and Shaukat, 2002). As populations grow they become relatively less aggregated (Taylor *et al.*, 1979; Noe and Barker, 1985). The degree of patchiness also depends on the spatial patterns of edaphic factors that affect nematodes (and their competitors and natural enemies), cultural practices that move soil and dictate the location and species of plant roots, and the biology of the nematode species itself. For example, reproduction of *Meloidogyne incognita* on cotton was much greater in coarse-textured soil than in fine-textured soil, whereas that for *Rotylenchulus reniformis* was highest in soils with moderate levels of clay and silt (Koenning *et al.*, 1996). The generally wide host range of most *Meloidogyne* species means that populations have greater opportunity to survive on weed hosts when non-host crops are grown, providing them with more and larger foci to initiate population growth on a subsequent host crop compared with species with very restricted host range, such as *Globodera pallida* (Been and Schomaker, 2006). The migratory behaviour, longevity, fecundity and oviposition behaviours of nematodes are characteristics that affect their spatial patterns in a field. Most species of *Meloidogyne* migrate readily as second-stage juveniles (J2), become sedentary after infection of the host, develop in 3–4 weeks, produce several hundreds of eggs in egg masses on roots, under optimum conditions, populations grow rapidly (Prot and Netscher, 1978; Prot and Van Gundy, 1981; Anwar and McKenry 2007). Even though distributions of eggs of *Meloidogyne* spp. are extremely patchy on a small scale compared with those of migratory nematodes, the combined traits of relatively many foci, ability to migrate readily and rapid population growth suggest that populations of most of the widely occurring species of *Meloidogyne* tend to be less aggregated at the field level than some of the other economically important nematode species (Goodell and Ferris, 1980; Noe and Campbell, 1985; Been and Schomaker, 2006; Herve *et al.*, 2005). By comparison, nematodes, such as *Tylenchulus semipenetrans*, that also oviposit large numbers of eggs in masses on roots, but which migrate less readily, may show greater aggregation across a field (Davis, 1984; Duncan *et al.*, 1995) than do most species of root-knot nematodes.

Nematode spatial patterns dictate the sampling patterns most likely to achieve accurate population estimates. Nematodes are rarely randomly distributed across a field, so random sampling is less likely than systematic sampling to detect and to describe accurately the average population density (Goodell and Ferris, 1981; Duncan *et al.*, 1994a,b; Been and Schomaker, 2000, 2006). Random sampling could easily fail to encounter a small focus of nematodes in a field that is otherwise non-infested, whereas systematic sampling on a grid (including a regular zigzag pattern) with dimensions smaller than the focus should encounter nematodes and will also give appropriate weight to the larger, non-infested area. Similarly, in orchards, single core samples from more trees were shown to provide more accurate population estimates than multiple core samples taken from fewer trees (McSorley and Parrado, 1982a). For the same reason, collecting more small cores provides a more accurate mean estimate than an equivalent amount of soil collected as fewer large cores (Goodell and Ferris, 1981; McSorley and Parrado, 1982a, 1983; Been and Schomaker, 2006). Prot and Ferris (1992) demonstrated that for large-scale surveys a ten-core sample reliably detected most nematode species in two Californian fields as large as 7 ha, and that detection probability increased with distance between cores because microhabitats that influence the nematode spatial patterns were sampled more adequately. Grid dimensions can also be optimized with respect to spatial patterns. In row crops, nematodes are moved in the direction of the row by cultivation and they migrate along adjacent root systems, so densities are more highly concentrated within than across rows (Noe and Cambell, 1985; Been and Schomaker, 1998). Therefore, samples from grids with more points across rows than within rows are likely to estimate an average population density more accurately (Been and Schomaker, 2000). Similarly, the variability of population densities between plots tends to be lower in experimental plots with dimensions longer across than within rows (Noe and Campbell, 1985). The same principle applies when separate samples are collected from different strata in fields that comprise different soil conditions or previous or present host material. Stratification is intended to reduce the overall variability of the sampled populations by separately sampling areas likely to have lower or higher nematode numbers (Goodell and Ferris, 1981).

The depth to which samples are taken is another form of stratification, because nematodes have vertical spatial patterns that usually extend below the depth of a soil core. In a Californian vineyard, numerous species of *Meloidogyne* occurred at 120 cm and were detected at 330 cm depth (Ferris and McKenry, 1974). Therefore, the depth of sampling will affect the density of nematodes recovered per sample or per volume of soil. Nematode patterns are closely related to the vertical pattern of host root abundance (Ingham *et al.*, 1985; Duncan, 1986; Rodriguez-Kabana and Robertson, 1987; Verschoor *et al.*, 2001). Host root density is greatest in the surface soil horizons, so sampling to depths of 30 cm or less is usually adequate to reduce sample variability by recovering nematodes in the zone of their greatest density. However, in deep-rooted crops such as grape, samples to a depth of 60 cm were required to obtain the highest density and least variable estimates of *Meloidogyne* spp. (Ferris and McKenry, 1974). The highest density of *M. incognita* on banana was at 0–10 cm depth in the first-year crop but at 0–20 cm in the ratoon crop (Jonathan and Rajendran, 2003). The density of nematodes near the soil surface can provide a reasonable estimate of the total population to greater depths. Numbers of *T. semipenetrans* to a depth of 30 cm explained 73% of the variability in samples from 0 to 60 cm (Duncan 1986). *Meloidogyne hapla* on carrot, onion and weed fallow had similar patterns of abundance to a depth of 40 cm, with two-thirds of the population at 0–20 cm depth (Bélair 1998). Half of the numbers of *M. chitwoodi* to a depth of 70 cm were recovered in samples to either 20 or 34 cm in two fields with different cropping regimes (Wesemael and Moens, 2008).

Crop damage functions and nematode population models (see Greco and Di Vito, Chapter 11, this volume) that are functions of initial nematode density are non-linear because root damage and yield loss per nematode decreases with increasing population density (Seinhorst, 1965, 1998). These functions are usually developed in glasshouse, microplot or small-plot field trials using artificially uniform populations over a range of densities. As nematodes are not distributed uniformly within fields, the accuracy with which the mean nematode density in a field is predictive of yield loss and population change depends on the population spatial pattern. With increased aggregation, predictions from a mean tend to overestimate crop loss due to the increasing frequency of locations with no or fewer than the mean number of nematodes, and because damage per nematode is inflated for the fewer locations contributing high counts to the mean (Seinhorst, 1973; Perry, 1983; Noe and Barker, 1985). Damage functions can also underestimate damage that occurs in a few clusters in a field for which the mean density of the entire field is below a damaging level. Approaches to resolving the problem of prediction from advisory samples have been addressed theoretically, but are not widely applied (Seinhorst, 1973; Perry, 1983; Noe and Barker, 1985; Hughes, 1990, 1999). Noe and Barker (1985) fitted the negative binomial model to frequency distributions of the counts of *M. incognita* from many quadrats in each of ten tobacco fields, and showed that the mean nematode density for the field was directly related to the aggregation parameter (k) of the negative binomial distribution. Using a published damage function, yield loss predictions from field means were compared with those from summing the effects of densities at discreet areas, as estimated from the associated probabilities from the negative binomial, to show that overestimation of yield loss in a field was inversely related to nematode mean density. Noe and Barker (1985) suggested deriving these relationships for specific crop–nematode systems in order to adjust estimates of yield loss from the field mean density. Others have suggested approximating this approach empirically by stratifying fields to estimate yields in discreet areas, although this approach is often cost-prohibitive using current sampling and processing methods (Perry, 1983).

12.3 Characterizing Sample Accuracy and Reliability

Sample accuracy and reliability can be characterized using the mean–variance relationship for normally distributed data (Ferris *et al.*, 1990; Buntin, 2000). The confidence interval (CI) for a population of known mean (μ) and standard deviation (σ) is described by:

$$CI = \mu \pm z_{\alpha/2}\, \sigma / \sqrt{n} \qquad (12.1)$$

where z is the standard normal variate (1.96 at $P = 0.05$), α is the probability and n is the number of samples. Accuracy can be specified by equating the half-length of CI to a specified proportion (D) of the mean density

$$D\,\mu = z_{\alpha/2}\, \sigma / \sqrt{n} \qquad (12.2)$$

where b is an aggregation parameter (see Equation 12.7). Equation 12.2 can be rearranged to solve for z by estimating μ and σ from a preliminary sample mean ($\bar{x}$) and standard deviation (s)

$$z_{\alpha/2} = \sqrt{n}\, D\, \bar{x}/s. \qquad (12.3)$$

Equation 12.3 can be used to estimate the reliability of a population estimate obtained from a predetermined number of samples at a specified accuracy level (D $\bar{x}$). For example, if we choose to collect three samples from a population with mean and standard deviation equal to 1000 and 353, respectively, and we wish to know the frequency with which our population estimate will be no greater or smaller than 40% of the true mean, then $z = (\sqrt{3} \times 0.4 \times 1000) / 353 = 1.96$, and from a table of probabilities associated with standard normal variates, $P = 0.05$. We expect 95% of our population estimates from such a sample plan to be within 40% of the true mean.

Alternatively, rather than predefining n, we can select the value of z needed to achieve the desired precision and estimate an appropriate sample size by rearranging Equation 12.2 as:

$$n = z_{\alpha/2}^{2}\, s^2 / D^2\, \bar{x}^2 \qquad (12.4)$$

For a population at a given place and time, the parameter estimates used in Equations 12.3 and 12.4 can be derived several ways. Often, several bulk samples comprising a predetermined number of cores are collected, each across the entire sample area. Extraction of each bulk sample, or subsample(s) of each bulk sample, provides an estimate of the mean number of nematodes in the individual cores that comprise each sample. From the Central Limit Theorem, as the number of cores in a bulk sample increases, the distribution of sample estimates will approach normality, regardless of the spatial pattern in the field (Duncan and McSorley, 1987). Thus, Equations 12.3 and 12.4 are appropriate for characterizing data from bulk samples. However, samples comprising single cores or multiple cores taken at many discrete locations across the sample area (as in grid sampling) will yield population estimates that reflect the population spatial pattern, and the data will often be distributed according to the negative binomial model. The model can be fitted to such data to estimate the parameter k describing the degree of aggregation (McSorley and Parrado, 1982b). The relationship between the sample variance and k is approximated by:

$$s^2 \approx \bar{x} + \bar{x}^2 / k \qquad (12.5)$$

and has been used to estimate the variance or standard deviation in Equations 12.1–12.4 as in:

$$z_{\alpha/2} = D\sqrt{n} / \sqrt{(1/\bar{x} + 1/k)} \qquad (12.6)$$

Alternatively, algorithms can be written to sample from data sets of counts from single cores or samples from discrete areas within the sample area to develop simulated bulk samples with means that approach normality (Goodell and Ferris, 1981; McSorley and Parrado, 1983; Perry, 1983; Duncan *et al.*, 1993). Such an approach is especially useful to estimate optimum sampling as a function of numbers of cores per sample and samples per field, taking into account the time and cost to collect cores and to process samples (Goodell and Ferris, 1981).

As populations increase or decline, the variance changes as a function of the mean, a relationship that is well described by Taylor's Power Law (TPL) (Taylor, 1961)

$$\sigma^2 = a\,\mu^b \qquad (12.7)$$

where a is a scaling factor and b describes the degree of aggregation (b<1 approaches uniformity, b = 1 is random and b>1 is increasingly aggregated). Of the two parameters, b tends to be species-specific, whereas a is affected by the methodology – habitat, sample size, sample area size, collection and processing methodology (Ferris *et al.*, 1990; McSorley and Dickson, 1991; Duncan *et al.*, 1993, 2001). Taylor's law shows clearly that estimating the accuracy and reliability of different-sized samples based on sampling one field at one time applies only to populations of similar size. However, data from populations sampled in the same manner at different locations and/or times can be used to estimate parameters of TPL by regressing the log-transformed variances on the log-transformed means from each sampling event (i.e. $\log s^2 = \log a + b \log \bar{x}$) (Fig. 12.1). This approach permits the variance to be defined by the mean in order to optimize sampling across a range of population densities as in:

$$z_{\alpha/2} = \sqrt{n\,D^2 / a\,\bar{x}^{b-2}} \qquad (12.8)$$

Use of Equations 12.1–12.8 to investigate sample accuracy and reliability reveals some general principles of sampling. For a given sample size, accuracy and reliability increase dramatically with population density, especially at the lower density ranges (Fig. 12.2). Whereas σ^2 increases faster than μ when the spatial pattern is aggregated, σ increases more slowly than μ, resulting in decreasing coefficients of variation as populations grow. Therefore, the most accurate samples are obtained from locations and at times when population size is greatest (Singh and Gaur, 1997; Souza *et al.*, 2007). Similarly, for sampling methods that can accurately estimate population size at sub-damaging levels, the accuracy of predictions from damage functions will increase at

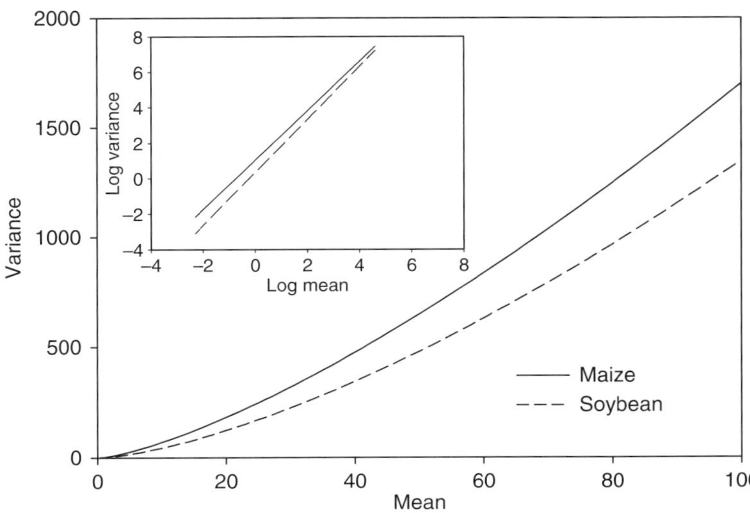

Fig. 12.1. The relationship between the variance and the mean population density for *Meloidogyne incognita* on maize and soybean as given by Taylor's Power Law ($\sigma^2 = a\,\mu^b$). Inset shows data transformed to permit parameter estimation through linear regression according to $\log s^2 = \log a + b \log \bar{x}$. Parameter values from McSorley and Dickson (1991).

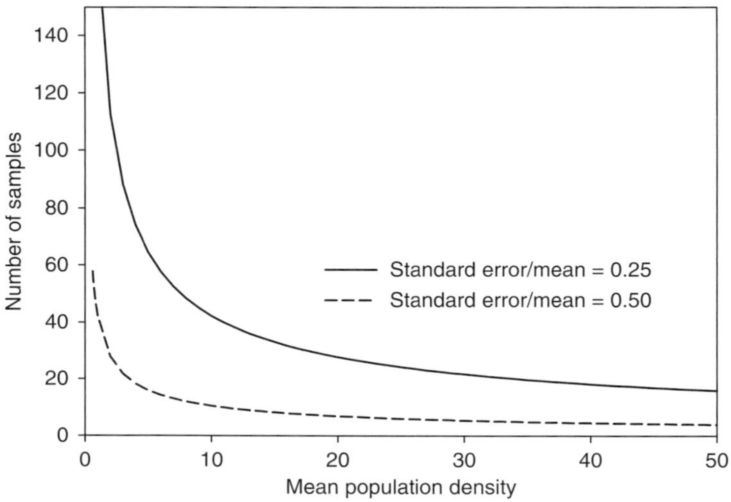

Fig. 12.2. Numbers of 12-core samples per 3 × 3 m area needed to estimate the population density of *Meloidogyne incognita* on soybean with estimated (from Equation 12.8, rearranged to solve for *n*) accuracy levels of 0.25 or 0.5 as a function of mean population density. Parameter values from McSorley and Dickson (1991).

higher, more-damaging densities, whereas less accurate estimates for smaller populations are of no economic consequence (Ferris *et al.*, 1990). The relationship between sample size and reliability (α) is not linear (Duncan *et al.*, 1996, 2001). It can be useful to characterize the reliability of samples of different size, rather than merely derive a sample size to achieve a specified accuracy and reliability (Neher and Campbell, 1996). For example, from Equations 12.3 and 12.4 it would require three bulk samples to estimate the mean with a 95% confidence interval half-length of 40% for a population with μ = 1000 and σ = 580. However, the likelihood of achieving the same accuracy with just two samples is almost the same (94%), but at a saving of one-third of the expense for sample acquisition and processing. Finally, whether for research or advisory purposes, nematode sampling plans rarely achieve confidence interval half-lengths of less than 40–50% with reasonable reliability (McSorley and Dickson, 1991). The utility for growers of population estimates with large error terms depends primarily on the size of the action threshold density. For species of *Meloidogyne* on various crops, typical tolerance limits range from just a few to more than 100 eggs and J2 per 100 cm^3 soil, with action thresholds being higher (Ehwaeti *et al.*, 1998; Asuaje *et al.*, 2004; Russo *et al.*, 2007). For other nematode species, such as *T. semipenetrans*, the action thresholds are as high as several thousand eggs and J2 per 100 cm^3 soil (Sorribas *et al.*, 2008). In Florida, most samples from citrus orchards reveal no or low numbers of *T. semipenetrans* (Duncan, 1999). Confidence interval half-lengths as high as 100% or more at low population density still result in small population estimates, which can be distinguished from the high action threshold for *T. semipenetrans*. Greater accuracy is required as the action threshold decreases, or as the frequency of damaging population levels increases, and the profitability of sampling becomes increasingly problematic.

Regulatory situations in which no nematodes can be tolerated represent the extreme case of high sampling costs associated with low threshold levels. Although applied sampling programmes do not include enough samples to allow mapping of nematode spatial patterns in a field, the most precise sampling programmes developed to date are based on such information (Been and Schomaker, 2000). The programmes are designed to detect the potato cyst nematode *G. pallida*, a regulated species for seed potato production. Fields with a wide range of different densities of potato cyst nematodes were sampled on grid patterns to identify discrete patches of nematodes. These patches were further sampled at a finer grid resolution, which permitted accurate modelling of the dimensions of nematode patches as a function of the mean number of nematodes at the central point in the patch (Schomaker and Been, 1999). With this information, it is possible to optimize sampling programmes that address the challenging task of identifying fields with only a single patch of nematodes at densities below which symptoms will occur in the potato crop. A key aspect of the optimization is the ability to define the grid dimensions in the direction of rows and across rows to minimize both the labour needed to collect samples and the likelihood of not adequately sampling a patch due to an overly coarse grid. Sample plans derived by these methods were estimated to detect very small foci of cyst nematodes (two eggs/g soil) with 90% reliability, compared with just 12% using conventional sampling plans. Achieving such accuracy and reliability is expensive, requiring more than 300 cores and 6–7 bulk samples per ha. However, by substantially increasing the reliability of detecting small foci of cyst nematodes, potato growers in The Netherlands can confidently restrict soil fumigation to just infested parts of fields or grow less-profitable resistant varieties only when needed. The widespread adoption of the new sampling method resulted in an 80% reduction of soil fumigation between 1989 and 1994. A somewhat analogous situation is that in which plant propagation material must be certified as free of certain regulated nematodes. Assuming that infested material is properly identified if sampled, the probability of detection can be derived from the binomial distribution and depends on the size of the plant population, the percentage of plants infested, and the number of plants sampled. Tables have been published to determine, for example, that in a lot of 10,000 plants with either 1 or 5% infestation rates, sampling 500 or 100 plants, respectively, should detect the infestation 99% of the time (McSorley and Littell, 1993).

12.4 Sample Processing

Root-knot nematode infestation levels in fields can be estimated directly by enumerating some life stage of the nematode, or indirectly by measuring plant symptoms or damage.

When a soil/plant sample is taken, decisions have to be made about what is to be examined. It could be eggs and/or J2 in the soil, life stages within roots, root galls, females or number of egg masses. These are obtained variously, either directly from a field sample or from a bioassay of the soil sample. Different methods differ in the efficiency with which they separate nematodes from soil or plant tissue, and factors such as sample size and soil type have major effects on extraction efficiency (Campos and Campos, 2005; Qiu *et al.*, 2006). A commonly employed method (Baermann incubation) recovered only 30% as many plant-parasitic nematodes, but more than four times more omnivorous and predatory nematodes, than did centrifugal flotation (McSorley and Frederick 2004). The efficiency with which centrifugal flotation recovered *M. incognita* varied from 19 to 65%, depending on soil type (Hernandez and Lopez, 1989). Female and juvenile *M. incognita* were more efficiently separated from root tissue by a method that employed grinding rather than stirring, whereas the opposite occurred for eggs (Stetina *et al.*, 1997). Knowledge of extraction efficiency permits the comparison of results from different laboratories or methods, and protocols for efficiency estimation have been proposed (Ferris, 1987). Although detection thresholds and sample variability are inversely related to extraction efficiency, depending on the objectives, use of methods with lower efficiency may be adequate and justified for reasons of cost or convenience.

12.5 Extracting Nematodes from Soil

There are a number of methods for extracting eggs and juveniles. These are mainly some form of Baermann trays/funnels, decanting and sieving, elutriation (semi-automated) and centrifugal flotation. Baermann techniques are good for the recovery of clean, live samples; sieves are not needed and the necessary equipment is relatively inexpensive to construct. However, they rely on the nematodes being active and recovery percentages can be low. Recovery of *M. incognita* and other nematodes was accelerated by ten- to 100-fold by covering dishes to reduce evaporation and the resulting temperature gradient across the soil layer (Robinson and Heald, 1989).

Barker *et al.* (1969) found that Baermann methods worked very well for the extraction of *Meloidogyne* in the summer but not in the winter. Baermann trays or dishes are preferable to funnels. Sieving techniques or bucket sieving methods (Cobb, 1918) are relatively rudimentary but useful for the rapid extraction of active and inactive nematodes. Only two buckets and a set of sieves are required.

Elutriation techniques extract nematodes by using a vertical current of water, the rate of which is adapted to the particular size of the nematode. The efficiency of extraction is little different from that of the sieving techniques but does produce a cleaner sample. Various designs have been published, including those of Seinhorst (1956) and Oostenbrink (1960). Trudgill *et al.* (1973) and Winfield *et al.* (1987) produced fluidizing columns which are simple and adaptable. Semi-automated versions have also been produced (Byrd *et al.*, 1976).

Probably the most efficient extraction methods are flotation techniques. These rely on solutions of particular specific gravity such that the target organism(s) to be extracted float or are suspended in the solution. These methods can be useful for fast- or slow-moving nematodes as well as dead ones. The solutions may be made of sugars (sucrose), $MgSO_4$ or $ZnSO_4$, as well as of the rather more expensive manufactured compounds – Percol, Ficoll and Ludox (Viglierchio and Yamashita, 1983). Centrifugal flotation involves the mixing of water and soil (stones removed) followed by centrifugation, which pellets soil and nematodes. The supernatant with organic matter can then be removed. The pellet is resuspended in the suspending solution and re-centrifuged. The supernatant then contains the nematodes (Caveness and Jenson, 1955), which must be washed to minimize osmotic pressure and subsequent damage. This method was improved by Jenkins (1964) to handle larger volumes of soil up to 500 cm^3. Byrd *et al.* (1972) describe a method for the extraction of eggs by first extracting egg masses by elutriation. The egg masses are then dissolved in a 0.525% NaOCl solution to release the eggs, which are subsequently washed and stained by adding acid fuchsin solution and lactic acid. As well as extracting directly from soil, these techniques can be used to clean up samples which have been extracted by other means.

12.6 Extracting Nematodes from Plant Material

If it is felt that counting the number of egg masses is appropriate, the whole or parts of root systems can be sampled, washed free of soil and stained in a solution of Phloxine B (Hartman and Sasser, 1985) or an alternative dye (Thies *et al.*, 2002). The roots can be treated with a NaOCl solution in order to degrade the egg masses and release the eggs (Hussey and Barker, 1973). Thus, an egg count rather than an egg mass count can be obtained.

Various versions of the Baermann technique can be used to extract active nematodes from plant tissue, although it is worthwhile not to submerge the plant material or to let it dry out. The plant material will usually be chopped into small segments. The use of trays or dishes is preferable to funnels, as this maintains better oxygenation (Rodriguez-Kabana and Pope, 1981). Better still in this respect are mistifier techniques, whereby a mist is sprayed over the sample, usually intermittently.

Plant material can also be chopped up and placed in a domestic food blender. The time required to macerate the tissue depends on the type of plant material. The main aim is to blend for the minimum time needed to induce the nematodes to escape the tissue without harm to themselves. Comparing techniques, it has been observed that a brief exposure to 0.5% NaOCl (for 10 s) can enhance the efficiency of extraction (Stetina *et al.*, 1997). Enzymatic predigestion of tissue prior to mechanical maceration is seldom used but can increase extraction efficiency substantially (Kaplan and Davis, 1990; Julio *et al.*, 2003). Samples produced in these ways can be cleaned up using Baermann techniques if the nematodes are active, or by centrifugal flotation.

12.7 Root Gall Indices

Extraction of nematodes from soil samples does not always reveal their presence, especially at low densities. Olowe (2004) found in a survey of species of *Meloidogyne* in Nigeria that many negative samples proved positive when a bioassay was conducted using a susceptible host grown in the apparently *Meloidogyne*-free soil. Barker (1985) also comments on bioassays as being the most sensitive in terms of low population densities. Typically the roots are recovered and the degree of galling assessed. This sampling of roots may be from a bioassay (Gugino *et al.*, 2008) or from hosts in the field directly (Bélair and Boivin, 1988). Various scales have been suggested, such as that of Bridge and Page (1980), which uses a 0–10 point scale, while others have proposed narrower scales. For example, Kinloch *et al.* (1987) used a 0–4 scale, where 0 = no galling, 0.2 = < 5%, 1 = 5–25%, 2 = 26–50%, 3 = 51–75%, and 4 = > 75% of the root surface galled. Barker (1985) summarizes these and suggests that a 0–6 scale is adequate in evaluating nematicide trials. Gall indices that are assessed directly in the field at harvest have the advantages of being relatively inexpensive and of measuring the effects of populations at their highest densities, which reduces sample variability. Bélair and Boivin (1988) related end-of-season gall indices of *M. hapla* to economic damage in the following carrot crop, thus providing a simple method by which growers can assess the need for nematode management in fields or even in just parts of fields. Mapping of maximum gall indices per ha on a large scale (350 ha) is being used to predict damage, monitor treatment efficacy and adjust management tactics in Morocco (Mateille *et al.*, 2005). A similar approach, using gall indices on lettuce seedlings to bioassay pre-plant soil samples for *M. hapla*, is currently used by New York vegetable growers (Gugino *et al.*, 2008). Although the assay requires 6 weeks for appropriate gall development, it does not require production of a susceptible host in the previous season, as do end-of-season field assays. Gall index bioassays of soil samples were an effective means of identifying edaphic factors associated with soils conducive to damage by *Meloidogyne* spp. in Mexico (Guzman-Plazola *et al.*, 2006).

12.8 Other Plant Symptoms

Other symptoms that can be observed above ground include stunting, chlorosis, wilting and leaf curling, but, being non-specific, they are not useful for quantification of nematode populations. However, above-ground symptoms can be used to prioritize sampling efforts and reduce the cost of an effective sampling programme. A two-step process was developed to predict crop loss in tomato in Australia, in which previous disease symptoms, combined with other risk factors such as soil texture and regional temperature, are used to estimate a numerical 'hazard index', which is used to determine whether to enumerate nematodes through soil sampling (Stirling *et al.*, 2004). Symptoms may occur on other plant parts, such as tubers. Coyne *et al.* (2006) were able to assess the distribution of *Meloidogyne* spp. on yam in West Africa by sampling yams from markets, and were able to determine in which countries and on which species of yam there were most infections by simple observation of the tubers. In Europe, where *M. chitwoodi* is a quarantine pest, more careful inspection is required to detect these nematodes and prevent their spread if found. Studies of the distribution of the nematodes within potato tubers and of the best way of extracting them revealed that zonal centrifugation yielded two to three times more eggs and juveniles of *M. chitwoodi* than elutriation and conventional centrifugation (Viaene *et al.*, 2007).

12.9 Research to Optimize Sampling Programmes for Root-knot Nematodes

Field sampling for agricultural prediction, survey sampling for ecological studies and agricultural resource allocation, and small-plot sampling for experimental purposes, have all been studied to optimize the cost of measuring population density of root-knot nematodes. Barker and Imbriani (1984) reviewed advisory sampling programmes, and noted that most recommended 20–50 cores collected systematically rather than randomly in areas of 2 ha or less in order to achieve population estimates of 30–50% of the mean in row crops, and slightly better in solidly planted perennials (Goodell and Ferris, 1981). Their overview remains largely current because sampling and extraction methods have changed little in the subsequent quarter-century. For example, the systematic collection of 50 cores in pineapple fields, recommended for management decisions, was shown to estimate *M. javanica* in two quarter-hectare fields with a standard error to mean ratio of *c.* 30% (Stirling and Kopittke, 2000). McSorely and Parrado (1982b) reported similar levels of precision (standard error to mean ratio = 25%, or 95% confidence interval of *c.* 50%) for *M. incognita* estimated from 22 cores per quarter- to half-hectare areas of fallow vegetable fields. Gall index bioassays of *M. incognita* in vegetable fields (2–6 ha), required three to four 20-core composite samples to achieve 25% standard error to mean ratios, whereas use of 20 assays of individual cores resulted in ratios of 50% (i.e. confidence interval half-lengths = 100% of the mean) (McSorley and Parrado, 1983). Bélair and Boivin (1988) recommended a sequential sampling method with an error level of 0.05 to determine whether plant gall intensity on end-of-season carrots represented an economic threshold for the succeeding crop. The method requires comparing the cumulative gall index of a minimum of ten and maximum of 72 plants with a range of upper and lower index limits. At the point that the cumulative index exceeds (management required) or falls below (management unnecessary) the uncertainty range, sampling stops. Samples of 72 plants that remain within the uncertainty range are considered above the economic threshold.

The ability to identify and quantify DNA of soil organisms and the deregistration of increasing numbers of soil fumigants are two phenomena that may provide the impetus to develop more efficient methods of soil sampling for advisory purposes. DNA analyses have the advantage of potentially identifying any life stage of *Meloidogyne* to the species level (see Blok and Powers, Chapter 4, this volume). Tests can be designed to identify and quantify a suite of pest organisms from a single DNA sample. An Australian testing service processes up to 2500, 500-g soil samples annually, using real-time PCR to identify and quantify key lesion and cyst nematode pests of cereals, as well as key fungal disease organisms (Ophel-Keller *et al.*, 2008). Detection limits of these methods are lower than threshold levels, and population estimates for key nematodes have been validated against conventional methods and by using samples with augmented nematodes. Samples comprise 45 cores from 15 locations in fields previously stratified on the basis of historical yield maps, elevation, soil conductivity surveys, and even satellite reflectance bioimagery. The resulting information is currently the basis for nematode and disease management decisions on up to 100,000 ha of cereal. Similar tests have been developed and validated in a variety of soil types to identify and quantify species of *Meloidogyne* and other disease organisms in tomato, pineapple and other horticultural crops (Stirling *et al.*, 2004). With the decreasing availability and increased cost of soil fumigants and nematicides, soil sampling is likely to become increasingly important as a basis for more complex integrated pest management (IPM) approaches that involve greater use of rotations, resistance and biological control tactics (Stirling and Pattison, 2008). The use of partially mechanized DNA-based analyses that can accurately identify and quantify multiple pest and disease organisms will become increasingly more cost effective than current systems requiring greater time, labour and biological expertise (Stirling and Pattison, 2008).

A review of nematode survey data showed that reported species richness is highly correlated with sampling intensity, yet little work has been done to optimize sampling methods for regional nematode surveys (Boag and Yeates, 1998). Neher and Campbell (1996) investigated sample requirements necessary to distinguish regional differences in several ecological indices (maturity and diversity) used to describe nematode communities. Samples consisted of 20 cores (2.5 cm diameter × 20 cm depth) taken at equidistant intervals on a 90 m diagonal transect per field. They estimated that an optimum plan to achieve minimum correlations between samples in a field of 0.6, and a 90% probability of detecting a 10% difference in indices over time, requires two samples each from between 50 and 100 fields. The probability of detecting plant-parasitic nematodes in 10-core samples from lucerne fields between 2.5 and 7 ha was as high as 62–93% for species with relatively low average density (<30 per 100 g soil), and species such as *M. arenaria* with densities between 80 and 2000 per 100 g soil were always detected (Prot and Ferris, 1992). Sample pattern had little effect on species detection in these fields, although variability was lower in systematic than random samples.

Despite the widespread use of small-plot experiments in nematology, few studies report variability of counts among plots to optimize replication. For example, six to eight plots (2 rows × 2 m × 10–12 cores per plot) were needed to estimate most mean densities (10–100 J2 per 100 cm^3 soil) of *M. hapla* in potato fields with a coefficient of variation of 0.5 (Wheeler *et al.*, 1994, 2000).

12.10 Examples of Results from Sampling Programmes

12.10.1 Surveys

Sampling patterns, numbers, timing and the material sampled all depend on the objectives of the work. Survey work often only requires sampling at its most basic level, the objective usually being to investigate what nematodes are present and where. Sampling of soil and/or roots will often be carried out once. Such surveys may also be done in order to develop or verify molecular diagnostic tools. Hallmann *et al.* (2007), investigating the occurrence and importance of plant-parasitic nematodes (including *Meloidogyne*) in Germany, sampled 246 fields, taking 1 ha as a minimum area for a sample. Fields were selected mainly on the basis that they had known or suspected nematode problems, although some sites with no nematode history were used. To sample so many sites at a specific time of year was impractical, and samples were taken from each site over the course of a year. Each site was sampled in a zigzag manner, with 50 cores making up some 6 l of soil, which was mixed and subsampled (8% or 250 cm^3) for nematode extraction by centrifugation (Hooper *et al.*, 2005), followed by microscopic identification. However, there are difficulties in using this identification approach (see Blok and Powers, Chapter 4, this volume), even for experts, because of the high similarity between some species, as well as the fact that some *Meloidogyne* species have been poorly described (Eisenback, 1985). Olowe (2004) took a similar strategy in a survey of *Meloidogyne* spp. on cowpea in Nigeria but sampled both soil and roots from 248 farms (0.5–1.0 ha) with a spade to a depth of 10–15 cm. Again the samples were taken in a systematic zigzag path. The 'cores' from a farm were mixed and a 1.5 kg subsample taken. Rather than extract nematodes from the soil, galled roots were recovered by hand and females (a minimum of 30) recovered to be examined for perineal patterns. Samples without galled roots were bioassayed by planting a susceptible cowpea and looking for galling after 60 days. Using this approach, all fields sampled were found to harbour root-knot nematodes. The most commonly occurring species (*M. incognita*, *M. javanica*, *M. arenaria)* were identified and shown to vary with geographic location. The resulting information is being used to develop individual control strategies in different parts of Nigeria. These two examples illustrate similarities (sample pattern, sample area, composite sampling) and variability (core, sample and subsample sizes) in sampling methods commonly employed by different laboratories. They also illustrate the desirability of employing multiple detection methods, such as the use of a bioassay in conjunction with direct extraction of nematodes.

More intensive approaches can be taken, such as that of Fourie *et al.* (2001), who studied a range of nematode species (including *Meloidogyne* spp.) on soybean in South Africa. They sampled 17 locations containing replicated field trials of 30 different soybean genotypes each. Samples were taken by harvesting the roots and soil from 12 plants in each plot during flowering. A range of methods was then employed to examine the samples and for species identification and quantification. Nematodes from soil and roots were extracted by the sugar centrifugal–flotation method (Caveness and Jenson, 1955) or the maceration, sugar centrifugal–flotation method (Coolen and d'Herde, 1972), respectively. Nematode numbers were counted and some were fixed for species identification. A bioassay on tomato was also done to provide *Meloidogyne* specimens for species identity. Such a study was able to give more quantitative data about relative species abundance, rather than merely a qualitative assessment.

12.10.2 Field experimentation

When the objectives include determining the biology of the nematode/host interaction and ultimately investigating management strategies, sampling is required to evaluate experimental treatments (nematicide use, rotations, etc.). It is then most likely that sampling for nematodes will occur at least twice, pre- and post-planting or treatment, to estimate effectively the responses to treatments and sometimes to measure density-dependent population change. With root-knot nematodes, where there can often be more than one generation in a growing season, more frequent sampling may be undertaken.

In order to compare galling and yield on a number of soybean cultivars with a view to developing management strategies, experiments were conducted in a field naturally infested with *M. arenaria* (Kinloch *et al.*, 1987). Nematode population density in the soil was determined immediately before planting and at harvest. Seven soil cores 2.5 cm diameter and 20 cm deep were taken from a 15-cm-wide band along the centre row of each plot. The cores were mixed and the nematodes extracted from a 100 cm^3 sample (15% of the total) by centrifugal flotation (Jenkins, 1964). Similar methods were used to study pre-planting and end-of season population levels of *M. incognita* during 8 years of various soybean–maize rotations (Kinloch *et al.*, 1985; Kinloch, 1986). These studies showed that in most situations there was a significant correlation between galling and yield, and between pre-plant densities, crop yield and final nematode densities.

A study of winter cover crops by McSorley and Gallaher (1992) compared nematode population growth during the summer on forage crops of corn, sorghum, soybean, cowpea and velvet bean at several sites in north Florida. They had seven experiments at seven different sites, which differed in the cover crops maintained on them during the winter of 1990/1 (wheat, rye, lupin, clover, crimson clover, hairy vetch and fallow (no crops or weeds)). Individual plots comprised four rows, 3.0 m long and 76 cm apart. Soil samples for nematode analysis were collected by removing and compositing six cores 2.5 cm deep × 20 cm long (*c.* 600 cm^3). Sampling took place at planting and post harvest. Nematodes were extracted from 100 cm^3 (17% of the total) subsamples with a modified sieving and centrifugation procedure. Of the nematodes observed in this study, *M. incognita* was the most suitable target for management by crop selection (velvet bean and resistant cowpea) and population growth on all crops was inversely density-dependent. McSorley and Gallaher (1993) extended this work, investigating maize and sorghum as cover crops using similar sampling methods They went on to pool their data from these field trials and were able to demonstrate that the relationship between $\log_n (P_f/P_i)$ and $\log_n (P_i)$, where P_f is the final population density and P_i is the initial population density, adequately estimated parameters in the population dynamics model of Seinhorst (1966).

Johnson and Campbell (1980) evaluated 31 crop-rotation systems during 5 years to determine the effects of cropping sequences, and of the nematicide fensulfothion, on nematode population densities of tomatoes seeded in the field, where the tomato plants were then harvested to be transplanted elsewhere. The soil was naturally infested with *M. incognita*, *M. javanica*, *Macroposthonia ornata*, *Paratrichodorus minor* and *Pratylenchus* spp. Soil samples (800 cm^3) were taken in February, June and October each year, beginning in June 1972 and continuing until June 1976. Soil samples consisted of a composite of 20 cores (2 cm diameter × 20 cm deep) collected randomly throughout the plot from the root zone of plants in each replicate. A 150 cm^3 (*c.* 19%) aliquot for each replicate was processed by a centrifugal–flotation method. Extracted nematodes were placed in calibrated dishes for identification and counting. In addition to the soil sampling, plants were selected randomly from each plot and examined for root galls. Roots were washed in tap water and indexed on a scale of 1–5. Immediately after the tomato plants were harvested, summer cover crop treatments were imposed on each plot. Sampling methods in this study were sufficient to demonstrate the inadequacy of the nematicide to control the nematodes, and demonstrated the need for extensive rotation between crops of transplant tomato to guarantee plants that meet certification requirements.

A common approach to investigating the biology of the nematode/host interactions is to examine the relationship by fitting models, usually between yield and initial and/or final nematode population density (P_i and or P_f). In order to determine the economic threshold level for oriental melon production, Kim and Ferris (2002) examined the relationship of the yield of oriental melon (*Cucumis melo* L. cv. Geumssaragi-euncheon) grafted on to cv. Shintozoa (*Cucurbita maxima* × *Cucurbita moschata*) to population densities of *M. arenaria*. In order to obtain a range of P_i values of *M. arenaria*, plots were established by either no cultivation or continuous cultivation of oriental melon during the summer, and by addition of non-infested soil in autumn the year before the main trial. Pre-plant P_i levels were determined 4 months and 1 month before planting. Post-plant population densities were measured 2 and 4 months after planting. Composite samples consisted of 14 soil cores per plot (3 m of bed length). J2 of *M. arenaria* were extracted from 300 cm^3 (20%) of each sample by sugar-flotation–sieving (Southey, 1986). Due to the low egg densities before planting, these were not assessed. Advisory threshold levels were estimated from the study because the resulting data were adequate to explain 42–45% of the variability of fruit yield either regressed against log-transformed initial numbers of nematodes in each plot, or fitted to Seinhorst's (1965) damage function as the model of the relationship between relative yield and the initial nematode population levels (Fig. 12.3). The data were also robust enough to reveal a strong relationship between population growth rate and initial densities (Fig. 12.4). By comparison, Ploeg and Phillips (2001) also examined the relationship between pre-planting densities of *M. incognita* and yield of melons (*Cucumis melo*) cv. Durango using both pot tests and a field trial. The field trial area was divided into four blocks of three 18.3-m-long beds. Markers were placed at 61 cm intervals in the centre of each bed and a composite soil sample consisting of three cores was collected from around each marker for nematode extraction. Nematodes were extracted from 100 g (10–20%) soil subsamples in a modified Baerman funnel technique (Rodriguez-Kabana and Pope, 1981). Melon growth in pots was well described by the Seinhorst (1965) damage function. However, despite a relatively intensive level of sampling in the field trial, just 26% of the variation was accounted for by the Seinhorst (1965) yield loss model (Fig. 12.5) and only 3% of the total variation in final population densities was accounted for when attempting to fit the Seinhorst (1967) population model to the data (Fig. 12.6).

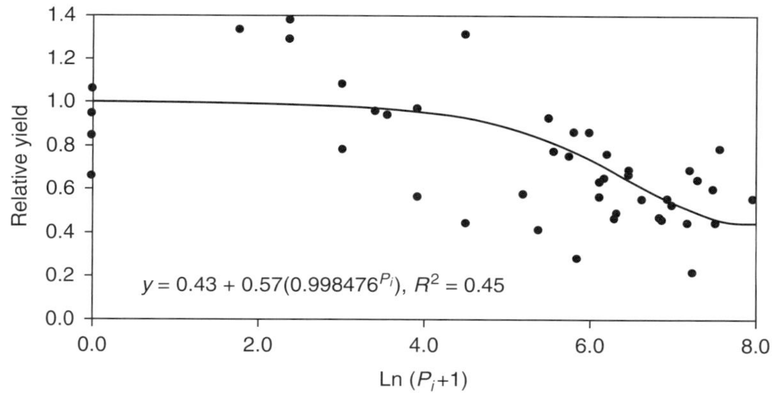

Fig. 12.3. The relationship between relative yield of oriental melon and initial nematode population density (P_i). (From Kim and Ferris, 2002.)

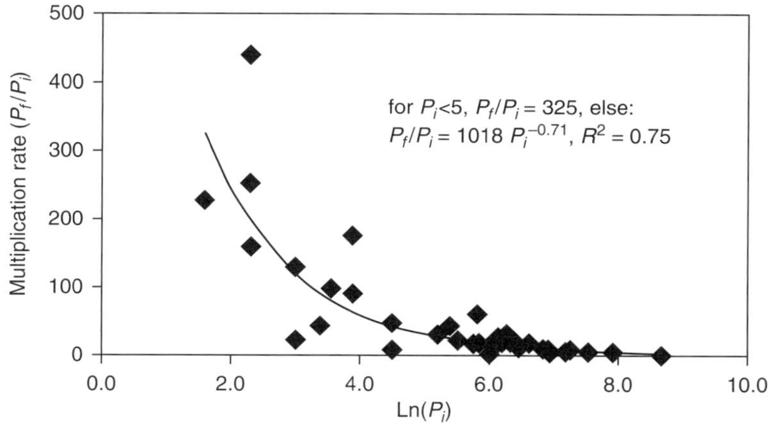

Fig. 12.4. The relationship between nematode multiplication rate and initial population density (P_i). (From Kim and Ferris, 2002.)

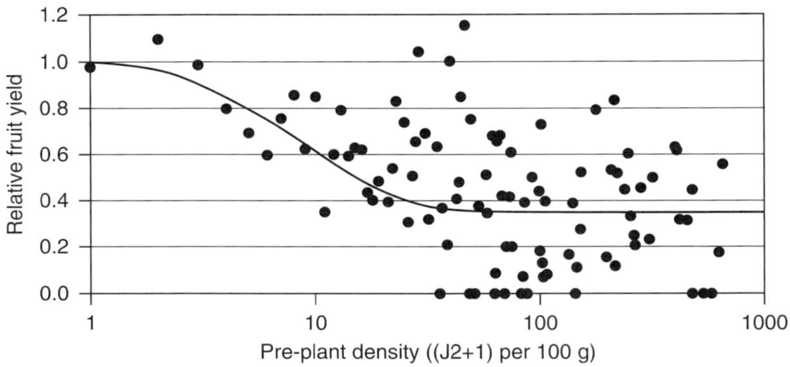

Fig. 12.5. The relationship between relative fruit yield of melon and initial (P_i) population density (second-stage juveniles/100 g soil). (From Ploeg and Phillips, 2001.)

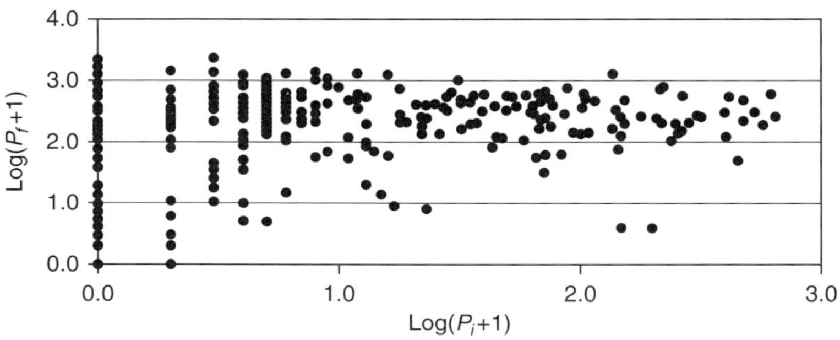

Fig. 12.6. The relationship between final (P_f) and initial (P_i) population density (second-stage juveniles/100 g soil). (From Ploeg and Phillips, 2001.)

Setting out to determine the effect of different crop rotations on *M. incognita* and *P. penetrans*, Kratochvil *et al.* (2004) conducted randomized trials in Maryland. To increase sample size and improve sampling technique to account for the spatial variability likely to be encountered, they split their plots into three subplots for sampling purposes, and took 20 soil cores from each. A composite sample of 500 cm^3 was then produced. Sampling of soil and roots amounted to a pre-planting, during plant growth and post-harvest sample. Half of each sample was used for extraction of nematodes using a modified Baermann method. Just before harvest, root samples from five plants chosen at random from each subplot were assessed for root galling. Although the sampling was concentrated between the spring and autumn, it was possible to obtain some idea of the population density fluctuations under the different management treatments applied (Fig. 12.7).

Sampling more frequently can give a more complete picture of seasonal population change. Johnson *et al.* (1998) were interested in looking at rotations made up of combinations of cotton and soybean with triticale with and without nematicide treatment (fenamiphos). Field plots were established in an area naturally infested with *M. incognita* race 3, *P. brachyurus* and *Helicotylenchus dihystera*. The soil was sampled monthly, with 20 cores taken from the rows of each subplot. Nematodes were extracted from a 150 cm^3 subsample with centrifugal flotation (Fig. 12.8). Twenty plants were also dug from each plot and rated for root galling by *M. incognita* on a scale of 1 to 5 (Barker *et al.*, 1986). Similarly, Thomas and Clark (1983) conducted experiments in 1979 and 1980, sampling every 30 days during the growing

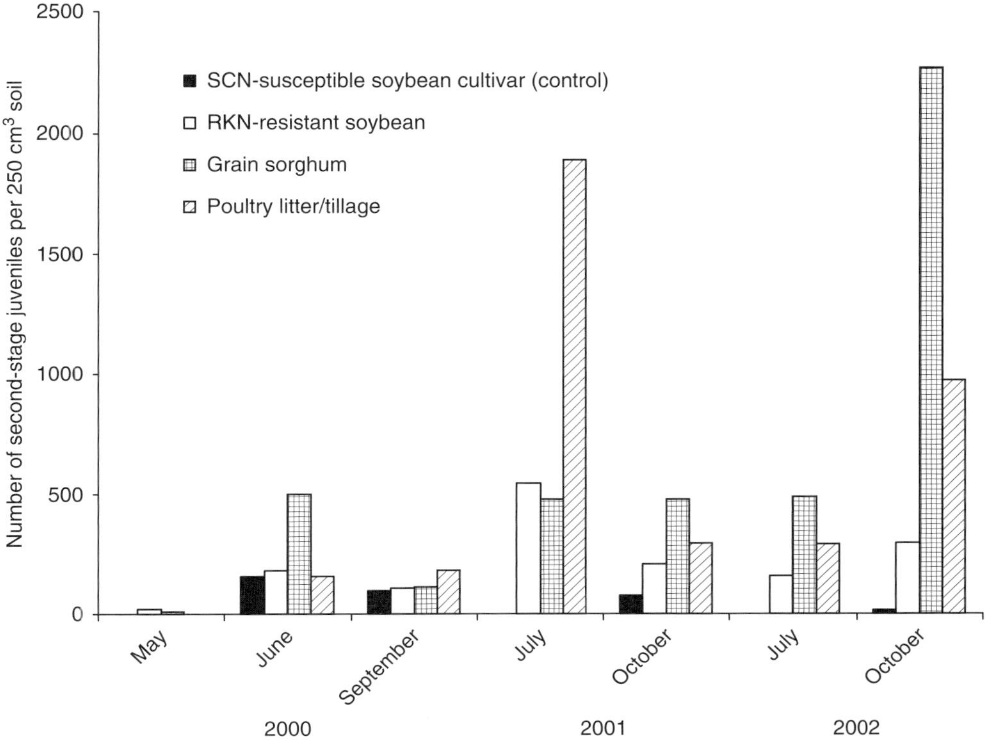

Fig. 12.7. Numbers of second-stage juveniles of root-knot nematodes per 250 cm^3 soil for seven sampling dates and four crop (or tillage) treatments. SCN = soybean cyst nematode; RKN = root knot nematode. (Data from Table 2 in Kratochvil *et al.*, 2004.)

season, to determine if an interaction exists between *R. reniformis* and *M. incognita* and, if so, the effects of this on yield and quality of sweet potatoes. Soil samples were taken approximately every 30 days using a 2 cm diameter soil probe, with five probes per plot to a depth of 15–20 cm, yielding approximately 250 cm^3. Unlike most other studies, the juveniles of *M. incognita* and the juveniles + young adults of *R. reniformis* were extracted from entire samples in a semi-automatic elutriator (Jenkins, 1964). In field plots with a high natural population density of *R. reniformis*, artificial infestations with high levels of *M. incognita* in both fumigated and non-fumigated treatments inhibited *R. reniformis*, while the final *M. incognita* juvenile population density was not affected. The results indicate that a competitive interaction exists, with each species capable of inhibiting the other and becoming the dominant population.

Even more intensive sampling over time was done by Pinkerton *et al.* (1991), who set out to determine the relationship between degree-day accumulation and population dynamics of *M. chitwoodi* in soil and potato tubers. Soil was sampled monthly during the overwintering period from harvest through to planting each year, and every 14 days during the growing season. Root samples from two plants selected from each plot were collected and examined for egg masses on each sampling date, until egg masses were detected (Dickson and Struble, 1965). J2 were extracted from 250 cm^3 soil samples (17% of the total) by wet sieving–centrifugation, and counted, thus allowing very detailed analysis of the relationship between temperature and seasonal population dynamics. This kind of study can have practical implications in terms of management, i.e. time of harvest, but could also be valuable in considering the effects of global warming on the potential distribution *M. chitwoodi* (Fig. 12.9). Samples of field soil were also collected at planting and harvest each season, and bioassayed with

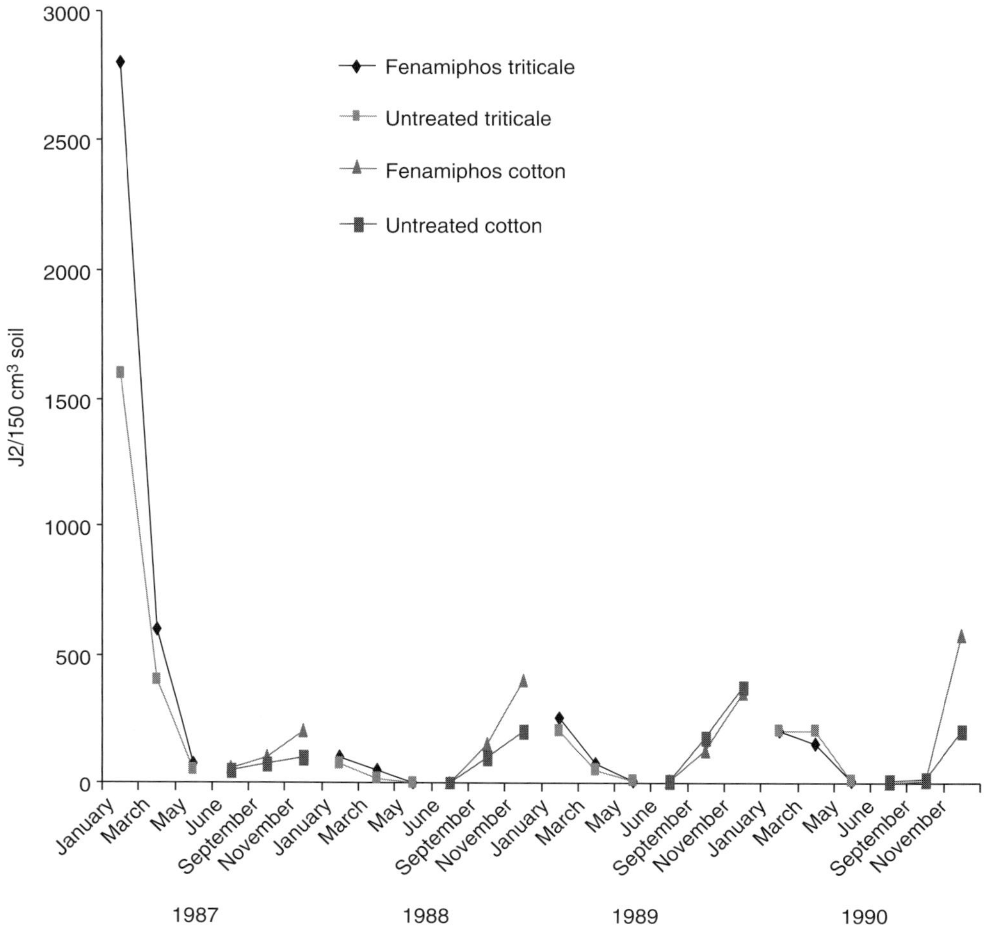

Fig. 12.8. The fluctuations of second-stage juveniles of root-knot nematodes per 150 cm^3 soil both temporally and with crop rotation, with or without nematicide. (Modified from Fig. 1 in Johnson *et al.*, 1998.)

a 3-week-old Columbia tomato plant to determine nematode infectivity. Plants were harvested after 21 days, the roots were stained (Byrd *et al.*, 1983) and the nematodes in the roots were counted. Five tubers from each plot on each sampling date were carefully scrubbed with a nylon scouring pad to remove the epidermis. A series of slices (0.75–1.5 mm thick) was cut tangentially from the tuber surface through the vascular ring. Approximately 5 g tissue per tuber was collected from four surfaces of each tuber. The slices were soaked in 1.5% NaOC1 for 5 min, rinsed in tap water for at least 30 min, and stained. By combining the bioassay with the soil sampling, Pinkerton *et al.* (1991) were able to provide a detailed etiological record of the temporal development of all stages of nematode (J2, swollen J2 through to fourth-stage juveniles, females, and females with egg masses) in the plant parts and soil.

Sampling for nematodes in relation to biological control organisms, such as *Pasteuria penetrans* (see Hallmann *et al.*, Chapter 17, this volume), presents the same questions, as it is usual that the sampling is for the nematodes rather than the endospores. Chen *et al.* (1994) studied a mixed population of *M. incognita* and *M. javanica* in a tobacco field where the severity of root-knot in tobacco had decreased over time. *Pasteuria*

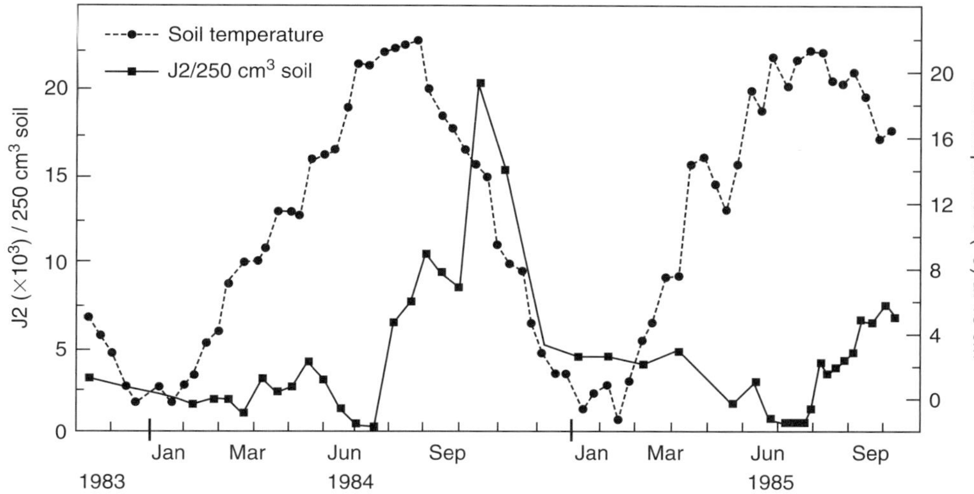

Fig. 12.9. The seasonal population dynamics of *Meloidogyne chitwoodi* and soil temperature 15 cm deep in an eastern Washington potato field. (From Fig.1 in Pinkerton *et al.*, 1991.)

penetrans was observed attached to J2 of *Meloidogyne* extracted from the soil. The site had been planted to tobacco continuously for 6 years prior to experimentation, with factorial design including different management strategies. Soil and root samples were taken with a bucket auger (10 cm diameter) from each plot in the root rhizosphere, to 25 cm depth. Soil was sampled four times through the growing season. The soil was mixed, and a 100 cm^3 (<2% of the total) subsample was taken from each sample to extract nematodes by a centrifugal-flotation technique. The numbers of J2 of *Meloidogyne* per 100 cm^3 soil were determined and the numbers of endospores attached to the cuticles of J2 were counted from 20 J2 per sample with an inverted light microscope.

Cetintas and Dickson (2005) also studied *P. penetrans*, but focused on its vertical distribution in field soil at a groundnut field in Florida that had been used to investigate the persistence and suppressiveness of *P. penetrans* to *M. arenaria*. Three different crop regimes were used. The soil was sampled annually for 4 years at five depths, down to 75 cm. Five cores per depth were taken using a bucket auger (10 cm diameter). One litre of soil at each depth was combined and mixed, and nematodes were extracted by centrifugal-flotation. The number of *P. penetrans* endospores on the first 20 J2 of *Meloidogyne* spp. observed was counted (Fig. 12.10).

Wesemael and Moens (2008) wanted to examine the importance of vertical distribution of *M. chitwoodi* in different rotations of field-grown vegetables. They chose two naturally infested fields and selected a 20 × 2 m plot in each. The plots were sampled every 3 or 4 weeks for 2 years. Fifteen soil cores 2.5 cm diameter × 70 cm were taken; each was divided into 10 cm segments and the corresponding segments were pooled. They found that the vertical distribution was consistent within each field irrespective of crop, suggesting that the main factors involved in any differences were soil type.

Investigating the vertical distribution of *M. arenaria* in a groundnut field, Rodriguez-Kabana and Robertson (1987) sampled soil over a 15-month period. They used a 5 cm diameter, hydraulically operated cylindrical probe to obtain cores 1 m long. They then cut this into 20 cm sections to produce a model of the depth distribution of the nematode (Fig. 12.11).

The results from soil sampling are usually obtained by extracting nematodes from the soil in some way and counting them, and sometimes by using the soil to conduct a bioassay. For example, a simple, visual soil bioassay was developed to assess *M. hapla* soil-infestation levels in order to facilitate the implementation of management practices (Gugino *et al.*, 2008). This involved the evaluation of the galling of lettuce as a bioassay plant in pots representing

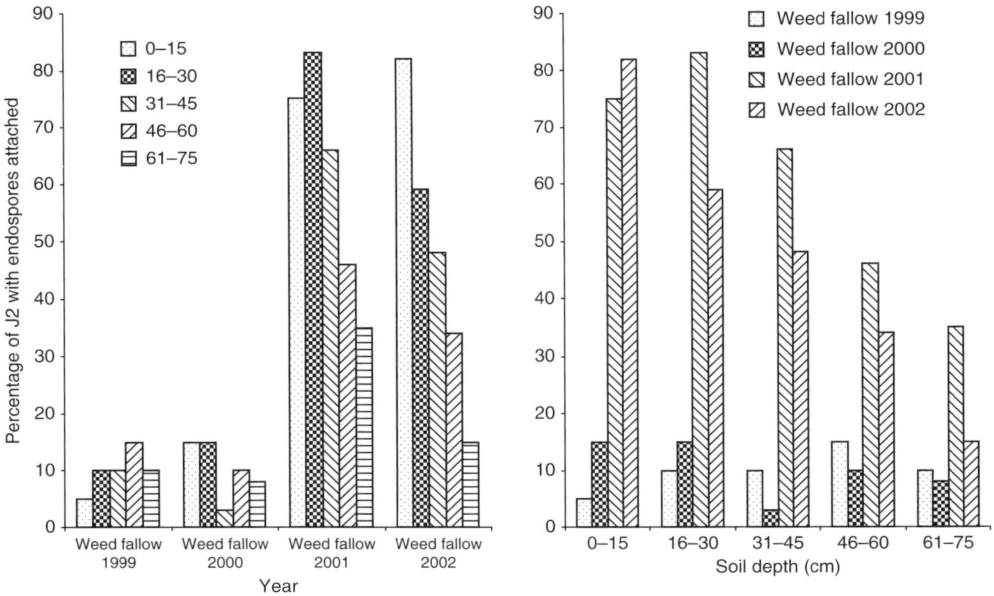

Fig. 12.10. The percentages of second-stage juveniles (J2) of *Meloidogyne* spp. with *Pasteuria penetrans* endospores attached and average numbers of endospores attached per J2 (as determined by soil extraction) from soil collected at five soil depths during groundnut harvest for the years 1999 to 2002 following 9 years of weed fallow. (Data from Table 2 in Cetintas and Dickson, 2005.)

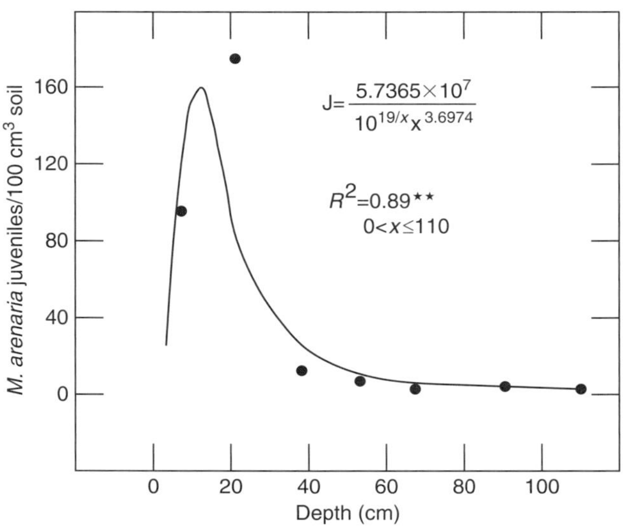

Fig. 12.11. The numbers of juveniles of *Meloidogyne arenaria* in a groundnut field as a function of soil depth. J = number of juveniles; x = soil depth. (From Fig. 2 in Rodriguez-Kabana and Robertson, 1987.)

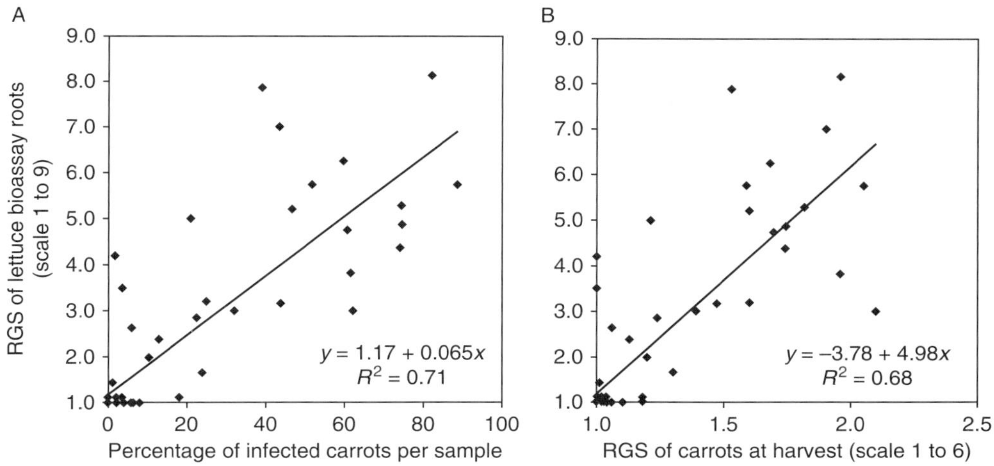

Fig. 12.12. The relationship between the percentage of infected carrot roots (A) and root-galling severity (RGS) (B) of mature carrots collected from 33 commercial fields over 3 years compared with the RGS ratings obtained from the glasshouse soil bioassay with lettuce conducted with corresponding composite soil samples. (From Fig. 2 in Gugino *et al.*, 2008.)

four composite soil samples per field. This proved an effective way of relating nematode infestation to disease levels on carrot (Fig. 12.12).

12.11 Conclusions and Future Directions

The familiar disease triangle concept – that crop loss occurs only when a host and pathogen coexist in an appropriate environment – reflects an aspect central to some current efforts to extend the utility of sampling nematode populations. When Australian pineapple growers lost access to ethylene dibromide, they relied increasingly on sampling programmes to make informed decisions on the need for and efficacy of new nematode management tactics. Stirling and Pattison (2008) point out that 'as with many monitoring programmes for crop pests', growers eventually submitted fewer samples each year as they began to recognize the edaphic conditions conducive to damage by nematodes. In some soils, the numbers of root-knot nematodes may be periodically above or below threshold densities, whereas in others they may never merit management. Elucidating soil factors that favour population growth by different nematode species should permit the adoption of site-specific management of nematodes using existing GPS (global positioning system)-based, variable-rate technologies that are currently uneconomic if nematode populations must be assessed directly (Wheeler *et al.*, 1999; Wrather *et al.*, 2002; Avendaño *et al.*, 2004). Indeed, technologies that integrate GPS to monitor variables, such as crop yield, plant height, canopy density, aerial reflectance and bulk soil electrical conductivity, continuously in real time are increasingly exploited to find relationships between soils, nematodes and crop damage (Ortiz *et al.*, 2007). These technologies have dramatically reduced the cost of collecting large data sets to develop soil and yield maps that can also be used to select experimental sites with the appropriate levels of variation in which to study nematode spatial patterns. For example, bulk soil electrical conductivity (EC_a) was inversely related to clay content and population density of *M. incognita* in Louisiana cotton fields (Wolcott *et al.*, 2004). In fields with multiple nematode pests, the spatial patterns of *M. incognita* that aggregate in coarse-textured soil could be distinguished from those of *R. reniformis*, found mainly in finer-textured soil, as estimated by EC_a (Overstreet *et al.*, 2009). These relationships permitted delineation of nematode management zones in fields based on the mean

EC_a value of the various zones (Fig. 12.13). Nematicide application was shown to increase yield more in management zones with low EC_a that contain large numbers of *M. incognita* than in zones with greater clay content and few root-knot nematodes (Overstreet *et al.*, 2005). Some growers are now attempting to adopt this technology to identify parts of fields, without the need for nematode sampling, in which management of root-knot nematodes is not profitable (C. Overstreet, 2009, personal communication). Engineering advances that improve the efficiency and precision of measuring edaphic and plant variables are increasingly likely to leverage research and extension efforts to estimate nematode spatial patterns in fields.

Grower impetus to quantify root-knot nematodes in conducive habitats will increase with advances in the variety and quality of information available from individual soil samples, and in response to the increased cost or reduced availability of chemical controls. The theoretical basis exists to relate sampling intensity to control costs and loss-prediction reliability, but the usefulness of such approaches is constrained by the high cost of sampling (Ferris, 1984; Ferris *et al.*, 1990). The ability of molecular methods, particularly real-time PCR, to identify and quantify root-knot nematodes in samples of nematodes extracted from soil with precision as good as or better than conventional methods is well established (Qiu *et al.*, 2006; Adam *et al.*, 2007; Berry *et al.*, 2008; Toyota *et al.*, 2008). This approach largely obviates the need for taxonomic expertise in diagnostic laboratories, and has tremendous potential for increased throughput via mechanization. Further cost reduction can be achieved by technical improvements in methodology, such as the design of lower-cost probes and the development of multiplex systems to quantify multiple nematode species in addition to other types of soil-borne pathogens (Ophel-Keller *et al.*, 2008). Given the promise shown by these new technologies to detect and infer nematode spatial patterns in order to predict the need for resistant planting material, or to guide the precision application of other control measures, it would seem likely that the development of mechanized sampling and extraction methods will receive renewed attention.

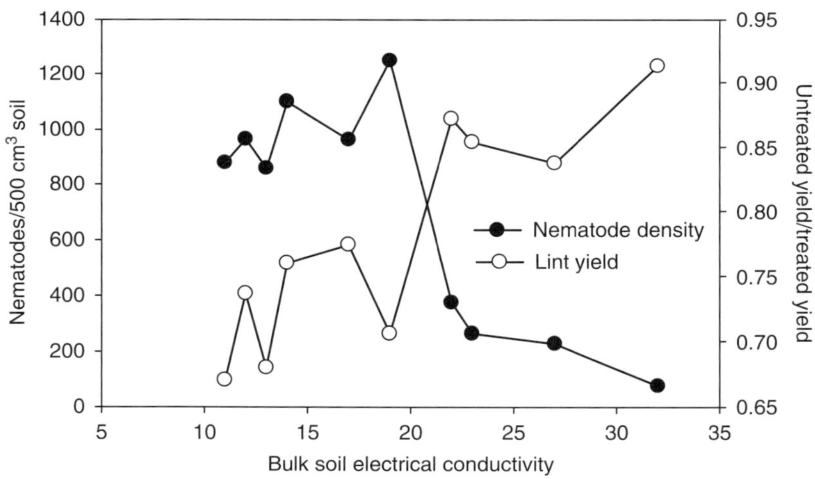

Fig. 12.13. The relationship between mean population density of *Meloidogyne incognita*, cotton yield response to treatment with Telone and mean bulk soil electrical conductivity in a 32-ha field in north-eastern Louisiana infested with root-knot nematode. (Modified from Overstreet *et al.*, 2005.)

12.12 References

Adam, M.A.M., Phillips, M.S. and Blok, V.C. (2007) Molecular diagnostic key for identification of single juveniles of seven common and economically important species of root-knot nematode (*Meloidogyne* spp.). *Plant Pathology* 56, 190–197.

Anwar, S.A. and McKenry, M.V. (2007) Variability in reproduction of four populations of *Meloidogyne incognita* on six cultivars of cotton. *Journal of Nematology* 39, 105–110.

Asuaje, L., Jimenez, M.A., Jimenez-Perez, N. and Crozzoli, R. (2004) Effect of the root-knot nematode, *Meloidogyne incognita* on growth of three lettuce cultivars. *Fitopatologia Venezolana* 17, 2–5.

Avendaño, F., Pierce, F.J., Schabenberger, O. and Melakeberhan, H. (2004) The spatial distribution of soybean cyst nematode in relation to soil texture and soil map unit. *Agronomy Journal* 96, 181–194.

Barker, K.R. (1985) Nematode extraction and bioassays. In: Barker, K.R., Carter, C.C. and Sasser, J.N. (eds) *An Advanced Treatise on* Meloidogyne. *Volume II: Methodology.* North Carolina State University Graphics, Raleigh, North Carolina, *pp.* 19–35.

Barker, K.R. and Campbell, C.L. (1981) Sampling nematode populations. In: Zuckerman B.M. and Rohde R.A. (eds) *Plant Parasitic Nematodes*, Vol. III. Academic Press, New York, pp. 451–474.

Barker, K.R. and Imbriani, J.L. (1984) Nematode advisory programs – status and prospects. *Plant Disease* 68, 735–741.

Barker, K.R., Nusbaum, C.J. and Nelson, L.A. (1969) Seasonal population dynamics of selected plant-parasitic nematodes as measured by 3 extraction procedures. *Journal of Nematology* 1, 232–239.

Barker, K.R., Townshend, J.L., Bird, G.W., Thomason, I.J. and Dickson, D.W. (1986) Determining nematode population responses to control agents. In: Hickey, K.D. (ed.) *Methods of Evaluating Pesticides for Control of Plant Pathogens.* APS Press, St Paul, Minnesota, pp. 282–296.

Been, T.H. and Schomaker, C.H. (1998) Quantitative studies on the management of potato cyst nematodes (*Globodera* spp.) in the Netherlands. PhD thesis, Agricultural University, Wageningen, The Netherlands.

Been, T.H. and Schomaker, C.H. (2000) Development and evaluation of sampling methods for fields with infestation foci of potato cyst nematodes (*Globodera rostochiensis* and *G. pallida). Phytopathology* 90, 647–656.

Been, T.H. and Schomaker, C.H. (2006) Distribution patterns and sampling. In: Perry R.N. and Moens M. (eds) *Plant Nematology.* CAB International, Wallingford, UK, pp. 302–326.

Bélair, G. (1998) Seasonal and vertical distribution of *Meloidogyne hapla* in organic soil. *Phytoprotection* 79, 1–8.

Bélair, G. and Boivin, G. (1988) Spatial pattern and sequential plan for *Meloidogyne hapla* in muck-grown carrots. *Phytopathology* 78, 604–607.

Berry, S.D., Fargette, M., Spaull, V.W., Morand, S. and Cadet, P. (2008) Detection and quantification of root-knot nematode (*Meloidogyne javanica*), lesion nematode (*Pratylenchus zeae*) and dagger nematode (*Xiphinema elongatum*) parasites of sugarcane using real-time PCR. *Molecular and Cellular Probes* 22, 168–176.

Boag, B. and Yeates, G. (1998) Soil nematode biodiversity in terrestrial ecosystems. *Biodiversity and Conservation* 7, 617–630.

Bridge, J. and Page, S.L.J. (1980) Estimation of root-knot nematode infestation levels on roots using a rating chart. *Tropical Pest Management* 26, 296–298.

Buntin, G.D. (2000) Developing a primary sampling program. In: Pedigo, L.P. and Buntin, G.D. (eds) *Handbook of Sampling Methods for Arthropods in Agriculture.* CRC Press, Boca Raton, Florida, pp. 99–118.

Byrd, D.W., Nusbaum, C.J. and Ferris, H. (1972) Method for estimating numbers of eggs of *Meloidogyne* spp in soil. *Journal of Nematology* 4, 266–269.

Byrd, D.W., Barker, K.R., Ferris, H., Nusbaum, C.J., Griffin, W.E., Small, R.H. and Stone, C.A. (1976) Two semiautomatic elutriators for extracting nematodes and certain fungi from soil. *Journal of Nematology* 8, 206–212.

Byrd, D.W., Kirkpatrick, T.L. and Barker, K.R. (1983) An improved technique for clearing and staining plant tissues for detection of nematodes. *Journal of Nematology* 15, 142–143.

Campos, H.D. and Campos, V.P. (2005) Studies on inoculum, inoculation and extraction of root-knot nematodes, *Meloidogyne javanica. Nematologia Brasileira* 29, 75–82.

Caveness, F.E. and Jenson, H.J. (1955) Modification of the centrifugal-flotation technique for the isolation and concentration of nematodes and their eggs from soil and plant tissue. *Proceedings of the Helminthological Society of Washington* 22, 87–89.

Cetintas, R. and Dickson, D.W. (2005) Distribution and downward movement of *Pasteuria penetrans* in field soil. *Journal of Nematology* 37, 155–160.

Chen, S.Y., Dickson, D.W. and Whitty, E.B. (1994) Response of *Meloidogyne* spp. to *Pasteuria penetrans*, fungi, and cultural practices in tobacco. *Journal of Nematology* 26, 620–625.

Cobb, N.A. (1918) Estimating the nema population of the soil. *Agricultural Technology Circular 1*. Bureau of Plant Industry, United States Department of Agriculture, Washington, DC.

Coolen, W.A. and d'Herde, C.J. (1972) *A Method for the Quantitative Extraction of Nematodes from Plant Tissue*. Ghent State Agricultural Research Centre, Ministry of Agriculture, Belgium.

Coyne, D.L., Tchabi, A., Baimey, H., Labuschagne, N. and Rotifa, I. (2006) Distribution and prevalence of nematodes (*Scutellonema bradys* and *Meloidogyne* spp.) on marketed yam (*Dioscorea* spp.) in West Africa. *Field Crops Research* 96, 142–150.

Davis, R.M. (1984) Distribution of *Tylenchulus semipenetrans* in a Texas grapefruit orchard. *Journal of Nematology* 16, 313–317.

Dickson, D.W. and Struble, F.B. (1965) A sieving-staining technique for extraction of egg masses of *Meloidogyne incognita* from soil. *Phytopathology* 55, 497.

Duncan, L.W. (1986) The spatial distribution of citrus feeder roots and of the citrus nematode, *Tylenchulus semipenetrans*. *Revue de Nématologie* 9, 233–240.

Duncan, L.W. (1991) Current options for nematode management. *Annual Review of Phytopathology* 29, 469–490.

Duncan, L.W. (1999) Nematode diseases of citrus. In: Timmer, L.W. and Duncan, L.W. (eds) *Citrus Health Management*. APS Press, St Paul, Minnesota, pp. 136–148.

Duncan, L.W. and McSorley, R. (1987) Modeling nematode populations. In: Veech, J.A. and Dickson, D.W. (eds) *Vistas on Nematology*. Society of Nematologists, Hyattsville, Maryland, pp. 377–389.

Duncan, L.W., Graham, J.H. and Timmer, L.W. (1993) Seasonal patterns associated with *Tylenchulus semipenetrans* and *Phytophthora parasitica* in the citrus rhizosphere. *Phytopathology* 83, 573–581.

Duncan, L.W., El-Morshedy, M.M. and McSorley, R. (1994a) Sampling citrus fibrous roots and *Tylenchulus semipenetrans*. *Journal of Nematology* 26, 442–451.

Duncan, L.W., Toole, J.D., Inserra, R.N., Castle, W.S. and Obannon, J.H. (1994b) Comparing 2 sampling methods to detect *Radopholus citrophilus* in Florida citrus. *Soil and Crop Science Society of Florida Proceedings* 53, 42–45.

Duncan, L.W., Mashela, P., Ferguson, J., Graham, J.H., Abou-Setta, M.M. and El-Morshedy, M.M. (1995) Estimating crop loss in orchards with patches of mature citrus trees infected by *Tylenchulus semipenetrans*. *Nematropica* 25, 43–51.

Duncan, L.W., Graham, J.H. and Terranova, A.C. (1996) Estimating sample size and persistence of entomogenous nematodes in sandy soils and their efficacy against the larvae of *Diaprepes abbreviatus* in Florida. *Journal of Nematology* 28, 56–67.

Duncan, L.W., McCoy, C.W., Stansley, P.A., Graham, J.H. and Mizell, R.F. (2001) Estimating the relative abundance of adult citrus root weevils (Coleoptera : Curculionidae) with modified Tedders traps. *Environmental Entomology* 30, 939–946.

Ehwaeti, M.E., Phillips, M.S. and Trudgill, D.L. (1998) Dynamics of damage to tomato by *Meloidogyne incognita*. *Fundamental and Applied Nematology* 5, 627–635.

Eisenback, J.D. (1985) Diagnostic characters useful in the identification of the four most common species of root-knot nematodes (*Meloidogyne* spp.). In: Sasser, J.N. and Carter, C.C. (eds) *An Advanced Treatise on* Meloidogyne. *Vol I. Biology and Control*. North Carolina State University Graphics, Raleigh, North Carolina, pp. 95–112.

Ferris, H. (1978) Nematode economic thresholds: derivation, requirements, and theoretical considerations. *Journal of Nematology* 10, 341–350.

Ferris, H. (1984) Probability range in damage predictions as related to sampling decisions. *Journal of Nematology* 16, 246–251.

Ferris, H. (1987) Extraction efficiencies and population estimation. In: Veech, J.A. and Dickson, D.W. (eds) *Vistas on Nematology*. Society of Nematologists, Hyattsville, Maryland, pp. 377–389.

Ferris, H. and McKenry, M.V. (1974) Seasonal fluctuations in spatial distribution of nematode populations in a California vineyard. *Journal of Nematology* 6, 203–210.

Ferris, H., Mullens, T.A. and Food, K.E. (1990) Stability and characteristics of spatial description parameters for nematode populations. *Journal of Nematology* 22, 427–439.

Fourie, H., Mc Donald, A.H. and Loots, G.C. (2001) Plant-parasitic nematodes in field crops in South Africa. 6. Soybean. *Nematology* 3, 447–454.

Goodell, P.B. and Ferris, H. (1980) Plant parasitic nematode distributions in an alfalfa field. *Journal of Nematology* 12, 136–141.

Goodell, P.B. and Ferris, H. (1981) Sample optimization for five plant-parasitic nematodes in an alfalfa field. *Journal of Nematology* 13, 304–313.

Gugino, B.K., Ludwig, J.W. and Abawi, G.S. (2008) An on-farm bioassay for assessing *Meloidogyne hapla* infestations as a decision management tool. *Crop Protection* 27, 785–791.

Guzman-Plazola, R.A., de Dios Jaraba Navas, J., Caswell-Chen, E., Zavaleta-Mejia, E., and Cid-del Prado-Vera, I. (2006) Spatial distribution of *Meloidogyne* species and races in the tomato (*Lycopersicon esculentum* Mill.) producing region of Morelos, Mexico. *Nematropica* 36, 215–229.

Hallmann, J., Frankenberg, A., Paffrath, A. and Schmidt, H.S. (2007) Occurrence and importance of plant-parasitic nematodes in organic farming in Germany. *Nematology* 9, 869–879.

Hartman, K. and Sasser, J. (1985) Identification of *Meloidogyne* species on the basis of differential host test and perineal-pattern morphology. In: Barker, K.R., Carter, C.C. and Sasser, J.N. (eds) *An Advanced Treatise on* Meloidogyne. *Volume II: Methodology*. North Carolina State University Graphics, Raleigh, North Carolina, pp. 69–77.

Hernandez, O. and Lopez, R. (1989) Extraction efficiency of *Meloidogyne incognita* (Nemata:Heteroderidae) in two tropical soils. *Agronomia Costarricense* 13, 169–174.

Herve, G., Bertrand, B., Villain, L., Licardie, D. and Cilas, C. (2005) Distribution analysis of *Meloidogyne* spp. and *Pratylenchus coffeae sensu lato* in coffee plots in Costa Rica and Guatemala. *Plant Pathology* 54, 471–475.

Hooper, D.J., Hallman, J. and Subbotin, S.A. (2005) Methods for extraction, processing and detection of plant and soil nematodes. In: Luc, M., Sikora, R.A. and Bridge, J. (eds) *Plant Parasitic Nematodes in Subtropical and Tropical Agriculture*. CAB International, Wallingford, UK, pp. 53–86.

Hughes, G. (1990) Characterizing crop responses to patchy pathogen attack. *Plant Pathology* 39, 2–4.

Hughes, G. (1999) Sampling for decision making in crop loss assessment and pest management introduction. *Phytopathology* 89, 1080–1083.

Hussey, R.S. and Barker, K.R. (1973) Comparison of methods of collecting inocula of *Meloidogyne* spp, including a new technique. *Plant Disease Reporter* 57, 1025–1028.

Ingham, R.E., Anderson, R.V., Gould, W.D. and Coleman, D.C. (1985) Vertical distribution of nematodes in a short grass prairie. *Pedobiologia* 28, 155–160.

Jenkins, W.R. (1964) A rapid centrifugal flotation technique for separating nematodes from soil. *Plant Disease Reporter* 48, 692.

Johnson, A.W. and Campbell, G.M. (1980) Managing nematode population-densities on tomato transplants using crop-rotation and a nematicide. *Journal of Nematology* 12, 6–19.

Johnson, A.W., Dowler, C.C., Baker, S.H. and Handoo, Z.A. (1998) Crop yields and nematode population densities in triticale–cotton and triticale–soybean rotations. *Journal of Nematology* 30, 353–361.

Jonathan, E.I. and Rajendran, G. (2003) Spatial distribution of root-knot nematode, *Meloidogyne incognita* in banana, *Musa* sp. *Indian Journal of Nematology* 33, 47–51.

Julio, V.A., Freitas, L.G., Coelho, J.L.C., Muscardi, D.C. and Ferraz, S. (2003) Extraction of *Meloidogyne javanica* females from tomato roots through maceration with fungal enzymes. *Nematologia Brasileira* 27, 75–80.

Kaplan, D.T. and Davis, E.L. (1990) Improved nematode extraction from carrot disk culture. *Journal of Nematology* 22, 399–406.

Kim, D. and Ferris, H. (2002) Relationship between crop losses and initial population densities of *Meloidogyne arenaria* in winter-grown oriental melon in Korea. *Journal of Nematology* 34, 43–49.

Kinloch, R.A. (1986) Soybean and maize cropping models for the management of *Meloidogyne incognita* in the coastal plain. *Journal of Nematology* 18, 451–457.

Kinloch, R.A., Hiebsch, C.K. and Peacock, H.A. (1985) Comparative root-knot galling yield responses of soybean cultivars to *Meloidogyne incognita. Plant Disease* 69, 334–336.

Kinloch, R.A., Hiebsch, C.K. and Peacock, H.A. (1987) Galling and yields of soybean cultivars grown in *Meloidogyne arenaria*-infested soil. *Journal of Nematology* 9, 233–239.

Koenning, S.R., Walters, S.A. and Barker, K.R. (1996) Impact of soil texture on the reproductive and damage potentials of *Rotylenchulus reniformis* and *Meloidogyne incognita* on cotton. *Journal of Nematology,* 28, 527–536.

Kratochvil, R.J., Sardanelli, S., Everts, K. and Gallagher, E. (2004) Evaluation of crop rotation and other cultural practices for management of root-knot and lesion nematodes. *Agronomy Journal* 96, 1419–1428.

Mateille, T., Schwey, D. and Amazouz, S. (2005) Gall index mapping: a tool for root-knot nematode (*Meloidogyne* spp.) infestation analysis in vegetable crops in Morocco. *Phytoma* 584, 40–43.

McSorley, R. and Dickson, D.W. (1991) Determining consistency of spatial dispersion of nematodes in small plots. *Journal of Nematology* 23, 65–72.

McSorley, R. and Frederick, J.J. (2004) Effect of extraction efficiency on perceived composition of the soil nematode community. *Applied Soil Ecology* 27, 55–63.

McSorley, R. and Gallaher, R.N. (1992) Comparison of nematode population-densities on 6 summer crops at 7 sites in North Florida. *Journal of Nematology* 24, 699–706.

McSorley, R. and Gallaher, R.N. (1993) Population-dynamics of plant-parasitic nematodes on cover crops of corn and sorghum. *Journal of Nematology* 25, 446–453.

McSorley, R. and Littell, R.C. (1993) Probability of detecting nematode infestations in quarantine samples. *Nematropica* 23, 177–181.

McSorley, R. and Parrado, J.L. (1982a) Plans for the collection of nematode soil samples from fruit groves. *Nematropica* 12, 257–267.

McSorley, R. and Parrado, J.L. (1982b) Estimating relative error in nematode numbers from single soil samples composed or multiple cores. *Journal of Nematology* 14, 522–529.

McSorley, R. and Parrado, J.L. (1983) A bioassay sampling plan for *Meloidogyne incognita. Plant Disease* 67, 182–184.

Monfort, W.S., Kirkpatrick, T.L., Rothrock, C.S. and Mauromoustakos, A. (2007) Potential for site-specific management of *Meloidogyne incognita* in cotton using soil textural zones. *Journal of Nematology* 39, 1–8.

Neher, D.A. and Campbell, C.L. (1996) Sampling for regional monitoring of nematode communities in agricultural soils. *Journal of Nematology* 28, 196–208.

Noe, J.P. and Barker, K.R. (1985) Overestimation of yield loss of tobacco caused by the aggregated spatial pattern of *Meloidogyne incognita. Journal of Nematology* 17, 245–251.

Noe, J.P. and Campbell, C.L. (1985) Spatial pattern-analysis of plant-parasitic nematodes. *Journal of Nematology* 17, 86–93.

Olowe, T. (2004) Occurrence and distribution of root-knot nematodes, *Meloidogyne* spp., in cowpea growing areas of Nigeria. *Nematology* 6, 811–817.

Oostenbrink, M. (1960) Estimating nematode populations by some selected methods. In: Sasser, J.N. and Jenkins, W.R. (eds) *Nematology.* The University of North Carolina Press, Chapel Hill, North Carolina, pp. 85–202.

Ophel-Keller, K., McKay, A., Hartley, D., Herdina and Curran, J. (2008) Development of a routine DNA-based testing service for soilborne diseases in Australia. *Australasian Plant Pathology* 37, 243–253.

Ortiz, B., Sullivan, D., Perry, C., Vellidis, G., Seymour, L. and Rucker, K. (2007) Delineation of management zones for site-specific management of parasitic nematodes using geostatistical analysis of measure field characteristics. *Precision Agriculture '07, Papers presented at the 6th European Conference of Precision Agriculture.* Skiathos, Greece, pp. 615–623.

Overstreet, C., Wolcott, M.C., Burris, D., Cook, R., Sullivan, D., Padgett, G.B. and Goodson, R. (2005) Telone application against root-knot nematode in Tensas Parish, Louisiana during 2004. *Beltwide Cotton Conferences Proceedings*, pp. 171–176.

Overstreet, C., Wolcott, M.C., Burris, G. and Burns, D. (2009) Management zones for cotton nematodes. *Beltwide Cotton Conferences Proceedings*, pp. 167–176.

Perry, J.N. (1983) Effects of spatial heterogeneity on Jones model for cyst nematode population dynamics and crop root damage. *Journal of Applied Ecology* 20, 849–856.

Perry, J.N. (1998) Measures of spatial pattern for counts. *Ecology* 79, 1008–1017.

Pinkerton, J., Santo, G. and Mojtahedi, H. (1991) Population dynamics of *Meloidogyne chitwoodi* on Russet Burbank potatoes in relation to degree-day accumulation. *Journal of Nematology* 23, 283–290.

Ploeg, A.T. and Phillips, M.S. (2001) Damage to melon (*Cucumis melo* L.) cv. Durango by *Meloidogyne incognita* in Southern California. *Nematology* 3, 151–157.

Proctor, J.R. and Marks, C.F. (1974) The determination of normalizing transformations for nematode count data from soil samples and efficient sampling schemes. *Nematologica* 20, 395–406.

Prot, J.C. and Ferris, H. (1992) Sampling approaches for extensive surveys in nematology. *Supplemental to Journal of Nematology* 24, 757–764.

Prot, J.C. and Netscher, C. (1978) Improved detection of low population densities of *Meloidogyne*. *Nematologica* 24, 129–132.

Prot, J.C. and Van Gundy, S.D. (1981) Effect of soil texture and the clay component on migration of *Meloidogyne incognita* 2nd-stage juveniles. *Journal of Nematology* 13, 213–217.

Qiu, J.J., Westerdahl, B.B., Anderson, B. and Williamson, V.M. (2006) Sensitive PCR detection of *Meloidogyne arenaria, M. incognita,* and *M. javanica* extracted from soil. *Journal of Nematology* 38, 434–441.

Robinson, A.F. and Heald, C.M. (1989) Accelerated movement of nematodes from soil in Baermann funnels with temperature gradients. *Journal of Nematology* 21, 370–378.

Rodriguez-Kabana, R. and Pope, M.H. (1981) A simple incubation method for the extraction of nematodes from soil. *Nematropica* 11, 175–186.

Rodriguez-Kabana, R. and Robertson, D.G. (1987) Vertical-distribution of *Meloidogyne arenaria* juvenile populations in a peanut field. *Nematropica* 17, 199–208.

Russo, G., Greco, N.D., Errico, F.P. and Brandonisio, A. (2007) Impact of the root-knot nematode, *Meloidogyne incognita* on potato during two different growing seasons. *Nematologia Mediterranea* 35, 29–34.

Schomaker, C.H. and Been, T.H. (1999) A model for infestation foci of potato cyst nematodes *Globodera rostochiensis* and *G. pallida. Phytopathology* 89, 583–590.

Seinhorst, J.W. (1956) The quantitative extraction of nematodes in soil. *Nematologica* 1, 249–267.

Seinhorst, J.W. (1965) The relation between nematode density and damage to plants. *Nematologica* 11, 137–154.

Seinhorst, J.W. (1966) Relationships between population increase and population density in plant parasitic nematodes. I. Introduction and migratory nematodes. *Nematologica* 12, 157–169.

Seinhorst, J.W. (1967) The relationship between population increase and population density in plant parasitic nematodes. II. Sedentary nematodes. *Nematologica* 13, 157–171.

Seinhorst, J.W. (1973) Relation between nematode distribution in a field and loss in yield at different average nematode densities. *Nematologica* 19, 421–427.

Seinhorst, J.W. (1998) The common relation between population density and plant weight in pot and microplot experiments with various nematode plant combinations. *Fundamental and Applied Nematology* 21, 459–468.

Siddiqui, I.A. and Shaukat, S.S. (2002) Spatial pattern analysis of root rot–root knot disease complex in an infested eggplant field. *Nematologia Mediterranea* 30, 131–135.

Singh, J. and Gaur, H.S. (1997) Seasonal variation in the horizontal spatial pattern of the root-knot nematode, *Meloidogyne incognita* and application of the negative binomial and Taylor's power law models in developing sampling schemes. *Indian Journal of Nematology* 26, 226–236.

Sorribas, F.J., Verdejo-Lucas, S., Pastur, J., Ornat, C., Pons, J. and Valero, J. (2008) Population densities of *Tylenchulus semipenetrans* related to physicochemical properties of soil and yield of Clementine mandarin in Spain. *Plant Disease* 92, 445–450.

Southey, J.F. (1986) *Laboratory Methods for Work with Plant and Soil Nematodes.* Ministry of Agriculture, Fisheries and Food, Her Majesty's Stationary Office, London.

Souza, R.M., Volpato, A.R. and Viana, A.P. (2007). Field assessment of different sampling strategies for coffee plantations parasitized by *Meloidogyne exigua. Nematropica* 37, 345–355.

Spiridonov, S.E., Moens, M. and Wilson, M.J. (2007) Fine scale spatial distributions of two entomopathogenic nematodes in a grassland soil. *Applied Soil Ecology* 37, 192–201.

Stetina, S.R., McGawley, E.C. and Russin, J.S. (1997) Extraction of root-associated *Meloidogyne incognita* and *Rotylenchulus reniformis. Journal of Nematology* 29, 209–215.

Stirling, G.R. and Kopittke R. (2000) Sampling procedures and damage thresholds for root-knot nematode (*Meloidogyne javanica*) on pineapple. *Australian Journal of Experimental Agriculture* 7, 1003–1010.

Stirling, G.R. and Pattison, A.B. (2008) Beyond chemical dependency for managing plant-parasitic nematodes: examples from the banana, pineapple and vegetable industries of tropical and subtropical Australia. *Australasian Plant Pathology* 37, 254–267.

Stirling, G.R., Griffin, D., Ophel-Keller, K., McKay, A., Hartley, D., Curran, J., Stirling, A.M., Monsour, C., Winch, J. and Hardie, B. (2004) Combining an initial risk assessment process with DNA assays to improve prediction of soilborne diseases caused by root-knot nematode (*Meloidogyne* spp.) and *Fusarium oxysporum* f. sp. *lycopersici* in the Queensland tomato industry. *Australasian Plant Pathology* 33, 285–293.

Taylor, L.R. (1961) Aggregation, variance and the mean. *Nature* 189, 732–735.

Taylor, L.R., Woiwod, I.P. and Perry, J.N. (1979) Negative binomial as a dynamic ecological model for aggregation, and the density dependence of K. *Journal of Animal Ecology* 48, 289–304.

Thies, J.A., Merrill, S.B. and Corley, E.L. (2002) Red food coloring stain: new, safer procedures for staining nematodes in roots and egg masses on root surfaces. *Journal of Nematology* 34, 179–181.

Thomas, R. and Clark, C. (1983) Population dynamics of *Meloidogyne incognita* and *Rotylenchulus reniformis* alone and in combination and their effects on sweet potato. *Journal of Nematology* 15, 204–211.

Toyota, K., Shirakashi, T., Sato, E., Wada, S. and Min, Y.Y. (2008) Development of a real-time PCR method for potato-cyst nematode *Globodera rostochiensis* and root-knot nematode *Meloidogyne incognita. Soil Science and Plant Nutrition* 54, 72–76.

Trudgill, D., Evans, K. and Faulkner, G. (1973) A fluidising column for extracting nematodes from soil. *Nematologica* 18, 469–475.

Verschoor, B.C., de Goede, R.G.M., de Hoop, J.W. and de Vries, F.W. (2001) Seasonal dynamics and vertical distribution of plant-feeding nematode communities in grasslands. *Pedobiologia* 45, 213–233.

Viaene, N., Mahieu, T. and de la Pena, E. (2007) Distribution of *Meloidogyne chitwoodi* in potato tubers and comparison of extraction methods. *Nematology* 9, 143–150.

Viglierchio, D.R. and Yamashita, T.T. (1983) On the methodology of nematode extraction from field samples – density flotation techniques. *Journal of Nematology* 15, 444–449.

Wesemael, W.M.L. and Moens, M. (2008) Vertical distribution of the plant-parasitic nematode, *Meloidogyne chitwoodi,* under field crops. *European Journal of Plant Pathology* 120, 249–257.

Wheeler, T.A. and Bronson, K.F. (2004) Precision agriculture: tools of the trade. In: Cook, R. and Hunt, D. J. (eds) *Proceedings of the Fourth International Congress of Nematology. Nematology Monographs and Perspectives*, Vol. 2. Brill, Leiden, The Netherlands, pp. 5–12.

Wheeler, T.A., Madden, L.V., Rowe, R.C. and Riedel, R.M. (1992) Modeling of yield loss in potato early dying caused by *Pratylenchus penetrans* and *Verticillium dahliae. Journal of Nematology* 24, 99–102.

Wheeler, T.A., Madden, L.V., Riedel, R.M. and Rowe, R.C. (1994) Distribution and yield-loss relations of *Verticillium dahliae*, *Pratylenchus penetrans*, *P. scribneri*, *P. crenatus*, and *Meloidogyne hapla* in commercial potato fields. *Phytopathology* 84, 843–852.

Wheeler, T.A., Kaufman, H.W., Baugh, B., Kidd, P., Schuster, G. and Sider, K. (1999) Comparison of variable and single-rate applications of aldicarb on cotton yield in fields infested with *Meloidogyne incognita. Journal of Nematology* 31, 700–708.

Wheeler, T.A., Madden, L.V., Rowe, R.C. and Riedel, R.M. (2000) Effects of quadrat size and time of year for sampling of *Verticillium dahliae* and lesion nematodes in potato fields. *Plant Disease* 84, 961–966.

Winfield, A.L., Enfield, M.A. and Foreman, J.H. (1987) A column elutriator for extracting cyst nematodes and other small invertebrates from soil samples. *Annals of Applied Biology* 111, 223–231.

Wolcott, M.C., Overstreet, C., Padgett, B. and Burris, E. (2004) Using soil electrical conductivity to denote potential nematode management zones. *Beltwide Cotton Conferences Proceedings*, pp. 349–353.

Wrather, J.A., Stevens, W.E., Kirkpatrick, T.L. and Kitchen, N.R. (2002) Effects of site-specific application of aldicarb on cotton in a *Meloidogyne incognita*-infested field. *Journal of Nematology* 34, 115–119.

13 Mechanisms and Genetics of Resistance

Valerie M. Williamson[1] and Philip A. Roberts[2]
[1]University of California, Davis, California, USA; [2]University of California, Riverside, California, USA

13.1 Introduction 301
13.2 Sources and Inheritance of Root-knot Nematode Resistance 302
13.3 Mechanisms of Resistance to Pathogens in Plants 307
13.4 Structure and Function of the Nematode Resistance Gene *Mi-1* 308
13.5 What is Known about Other Nematode R-Genes 310
13.6 Nematode Virulence and Durability of Resistance 311
13.7 Management of Resistance and Virulence in the Field 315
13.8 Conclusions and Future Directions 317
13.9 References 319

13.1 Introduction

Root-knot nematodes (*Meloidogyne* spp.) of economic importance in agriculture have broad, often global, distributions and typically have very wide host ranges (Sasser, 1977). For these species, especially *M. incognita*, *M. javanica*, *M. arenaria*, *M. hapla*, *M. fallax*, *M. chitwoodi* and *M. enterolobii* (=*M. mayaguensis*), the identification of effective host plant resistance that can be introgressed into crop cultivars and rootstocks by plant breeding constitutes an important component of nematode pest management programmes. By common definition, resistant plants support low or no nematode reproduction (Cook and Evans, 1987; Roberts, 2002, 2004). Resistance within a plant species is often due to specific genes that can segregate within the species. By contrast, for non-host species the nematode cannot reproduce on that species or group of plants due to a broader absence of host traits required for parasitism. To reproduce, the infective second-stage juveniles (J2) must be attracted to host roots, penetrate the epidermis and migrate through the root cortex to establish a feeding site in the vascular parenchyma that provides sufficient nutrition for development and egg production (see Abad *et al.*, Chapter 7, this volume). Resistance genes, in response to nematode infection, block or suppress one or more of these critical steps in nematode parasitism. An additional feature of root-knot nematode resistance is the effect on the development of root galls ('root knots') typically associated with compatible nematode–plant interactions on susceptible host plants. In many incompatible interactions on resistant host plants, root galling surrounding the infection site is reduced or lacking, depending on the resistance mechanism. However, root galling and nematode reproduction are not always coupled in root-knot nematode–host plant interactions, and genes that mediate reduced root galling but do not affect nematode reproduction have been identified (Garcia *et al.*, 1996; Roberts *et al.*, 2008).

While the subject of host tolerance (defined as hosts that display little or no plant damage or crop loss in response to nematode infection) is not covered in any detail in this chapter, it is worth noting that, in most cases, nematode-resistant plants grown in infested soil are much more tolerant of nematode infection than comparable susceptible genotypes of the same crop plant (Roberts, 1982; Trudgill, 1991). Most root-knot nematode resistance mechanisms are post-infection based; that is, the J2 penetrate the roots of resistant plants and then migrate through root tissues towards the stele, where they attempt to establish a feeding site in a similar manner to that in susceptible plants. Therefore, ability to tolerate this initial infection process is vital to the practical success of genetic resistance in nematode management for farmers. There are examples of intolerant resistant crops; for example, resistant sweet potato cultivars are intolerant to *M. incognita*, as the strong hypersensitive reaction damages root growth and function (Roberts and Scheuerman, 1984). Also, in some compatible *Meloidogyne*–host plant interactions, susceptible crops are tolerant; for example, some maize cultivars tolerate *Meloidogyne* spp. infection due to rapid compensatory regeneration of roots to maintain adequate root function (Roberts, 1992). Fortunately, most *Meloidogyne* resistance responses do not disrupt root growth, so the genetic potential of resistance can be exploited through plant breeding for crop protection.

Root-knot nematode resistance traits comprise genotypic and phenotypic features of utility in guiding resistance breeding programmes. Resistance can be dominant, recessive or additive in expression, and can be conferred by single major genes or by combinations of two or more genes or quantitative trait loci (QTLs). For example, there is evidence of strong epistatic gene interactions in some sources of root-knot nematode resistance, such as for *M. incognita* resistance in cotton, in which transgressive segregation for extreme resistance phenotypes is found in progenies from resistant × susceptible parent crosses (Shepherd, 1974; Wang *et al.*, 2008). The resistance phenotype can be characterized as strong or partial, depending on the extent to which nematode reproduction and root galling are suppressed. *Meloidogyne* resistance genes differ in mechanisms of action, strength and durability, and also in their practical utility in plant breeding and nematode management. Resistance genes are often specific in their elicitation and expression of the resistance response at several levels. Several distinct genes for resistance can occur in a plant species. Their specific character can include differential responses to different species of *Meloidogyne*, and to different populations of the same species based on presence or absence of interacting avirulence factors in the nematode. Some resistance genes also function less well at high soil temperatures. In the following sections we draw on examples of the better-understood *Meloidogyne* resistance genes to illustrate the primary features and characteristics of these complex interactions. In particular, the first-cloned of these genes, *Mi-1* from tomato, is used for in-depth examination of the structure and function of a root-knot nematode resistance gene.

13.2 Sources and Inheritance of Root-knot Nematode Resistance

Several reviews have described the availability and use of genetic resources for host plant resistance to nematodes, including *Meloidogyne* spp. (Cook and Evans, 1987; Hussey and Janssen, 2002; Roberts, 1982, 1992; Roberts *et al.*, 1998; Starr *et al.*, 2002; Starr and Roberts, 2004). In Table 13.1, the better-characterized root-knot nematode resistance traits are summarized, with information on gene designation and number of resistance genes identified, the mode of inheritance, and indications of gene sensitivity to temperature and presence of virulence in the target nematode species, where known. This listing, although not comprehensive, indicates a broad availability of effective resistance genes in a wide range of host plant taxa. Further, it shows that in some host plants, e.g. tomato, pepper and some agronomic crops, such as grain legumes and cotton, multiple resistance genes have been identified, often with unique specificities for particular species of *Meloidogyne*, for matching avirulence at the population level or for temperature sensitivity. While these genes reveal a valuable resource,

Table 13.1. A summary of genes for resistance to root-knot nematodes (*Meloidogyne* spp.), indicating the plant host, the numbers and inheritance mode of the genes, their unique specificities (for matching avirulence at population level, V; or temperature sensitivity, T) and key references.

Host plant	Resistance gene(s) or source	*Meloidogyne* species	Inheritance/expression	Virulence/ Temperature	Reference
Lucerne		*M. hapla*; *M. incognita*			Potenza *et al.*, 2001
Carrot	*Mj-1*	*M. incognita; M. javanica*	Dominant/additive	T	Boiteux *et al.*, 2000; Simon *et al.*, 2000
Clover	*Trifolium repens* *TRKR (Trifolium semipilosum)*	*M. trifoliophila*	Polygenic/ recessive Single dominant		Barrett *et al.*, 2002; Mercer *et al.*, 2004
Coffee	*Mex-1*	*M. exigua*	Dominant		Anthony *et al.*, 2005
Common bean	*Me1*; *Me2*; *me3*	*M. hapla*; *M. incognita*; *M. javanica*	Dominant (1,2) Recessive (3)	V; T	Omwega and Roberts, 1992; Chen and Roberts, 2003a
Cotton	*rkn1*; *RKN2*	*M. incognita*	Recessive/additive (1) Dominant/additive (2)	V	Wang *et al.*, 2006a, b; Wang *et al.*, 2008
Cowpea	*Rk*; Rk^2; *rk3*	*M. arenaria*; *M. hapla*; *M. incognita*; *M. javanica*	Dominant (1,2) Recessive (3)	V	Fery and Dukes, 1980; Roberts *et al.*, 1996; Ehlers *et al.*, 2000
Grape	*N*; *Mur1*	*M. arenaria*; *M. incognita*; *M. javanica*	Single dominant	V	Cousins *et al.*, 2003
Lima bean	*mir-1* *Mig-1* *Mjg-1*	*M. incognita – reproduction* *M. incognita – galling* *M. javanica – galling*	Recessive Dominant Dominant		Roberts *et al.*, 2008
Groundnut	*Arachis* spp. hybrids	*M. arenaria*; *M. javanica*	1 Dominant and 1 Recessive		Choi *et al.*, 1999; Church *et al.*, 2005
Pepper	*Me1; Me3; Me4; Me7;* *Mech1*; *Mech2*	*M. arenaria*; *M. incognita*; *M. javanica* *M. chitwoodi;*	Single genes, mix of dominant and recessive	T	Djian-Caporalino *et al.*, 2007
Potato	*Rmc1 (Solanum bulbocastanum)* *MfaXIIspl (Solanum sparsipilum)*	*M. chitwoodi*; *M. hapla*; *M. fallax*; *M. incognita*	Dominant Dominant, additive	V	Brown *et al.*, 1996; Janssen *et al.*, 1997; Kouassi *et al.*, 2006

(continued)

Table 13.1. Continued.

Host plant	Resistance gene(s) or source	*Meloidogyne* species	Inheritance/expression	Virulence/ Temperature	Reference
Prunus	*Ma*; R_{MiaNem}	*M. arenaria*; *M. incognita*; *M. javanica*	Dominant	T	Dirlewanger *et al.*, 2004
Soybean	*2 QTLs*	*M. javanica*;	Additive		Tamulonis *et al.*, 1997a
	Rmi1	*M. incognita*;	Single additive		Li *et al.*, 2001
	2 QTLs	*M. arenaria*	1 additive, 1 dominant		Tamulonis *et al.*, 1997b
Sugarbeet	*Beta vulgaris* ssp., maritime source	*M. arenaria*; *M. chitwoodi*; *M. fallax*; *M. hapla*; *M. incognita*; *M. javanica*	Single locus		Yu *et al.*, 1999, 2001
Sweet potato		*M. incognita*; *M. javanica*; *M. arenaria*	Additive	V	Jones and Dukes, 1980
Tobacco	*Rk*	*M. incognita*	Single/dominant	V	Yi *et al.*, 1998
Tomato	*Mi-1– Mi-9*	*M. arenaria*; *M. incognita*; *M. javanica*	Single genes/dominant	T, V	Yaghoobi *et al.*, 1995; Veremis and Roberts, 1996; Ammiraju *et al.*, 2003
Wheat	*Triticum tauschii* (syn. *Aegilops squarrosa*)	*M. incognita*; *M. javanica*; *M. chitwoodi*	Single dominant		Kaloshian *et al.*, 1990

many of them are present so far only in wild species, and not in cultivars. For example, even though at least nine resistance genes have been identified in wild tomato relatives, only *Mi-1* is currently available in cultivated tomato.

The inheritance of root-knot nematode resistance genes does not conform to a single pattern; some are expressed in dominant or recessive fashion and others are additive, showing allelic dosage effects on resistance phenotype. The advent of molecular marker techniques and other genomic resources for genetic and physical mapping, including detection of QTLs, has broadened our understanding of these resistance trait determinants and has helped to define their genomic arrangements and inheritance behaviour. Historically, the single major gene resistance traits have been the focus of most breeding efforts due to their relative ease of manipulation in plant progeny development and selection. Molecular markers have facilitated resolution of resistance gene organization and action (Collard *et al.*, 2005), including genomic arrangements, novel phenotypes that result from gene combinations, and the use of co-dominant markers to reveal effects of homo- and heterozygosity.

Genomic arrangements of root-knot nematode resistance genes in host plants show a range of organizational patterns. The organization of genes *Mi-1* and *Mi-9* and their numerous homologues clustered with other pathogen resistance genes on chromosome 6 in tomato *Lycopersicon esculentum* and its wild relative species is described later in this chapter (section 13.4). Also within tomato and the wild relative species *Solanum peruvianum* and *S. arcanum* are several other genes (*Mi-2* to *Mi-8*) that assort independently from *Mi-1*, indicating unlinked genomic locations. For example, gene *Mi-3*, which resists *M. incognita* and *M. javanica* populations virulent to gene *Mi-1*, is located on tomato chromosome 12, whereas other genes such as *Mi-2* are unlinked to both the chromosome 6 and chromosome 12 nematode resistance gene regions (Veremis and Roberts, 1996). A group of six root-knot nematode genes in pepper (*Me4*, resistant to *M. arenaria*; *Mech1* and *Mech2*, each resistant to *M. chitwoodi*; and *Me1*, *Me3* and *Me7*, each broadly resistant to *M. arenaria*, *M. incognita* and *M. javanica*) are clustered on pepper chromosome P9 (Djian-Caporalino *et al.*, 2007). Interestingly, this region of P9 corresponds to the chromosome 12 region of tomato where gene *Mi-3* resides and is also syntenous with chromosome XII of potato, which harbours a QTL for resistance to *M. fallax* in *Solanum sparsipilum*, indicating that these genes are arranged in conserved orthologous genomic regions in the Solanaceae (Kouassi *et al.*, 2006; Djian-Caporalino *et al.*, 2007). By contrast, a series of three specific *M. incognita* and *M. javanica* galling and reproduction resistance genes in Lima bean recombine fully and are independently located in the *Phaseolus lunatus* genome (Roberts *et al.*, 2008).

The clustering of nematode resistance genes is seen to some extent in other host plants. For example, in cowpea (*Vigna unguiculata*), the *Rk* gene maps to linkage group 1 (Ouédraogo *et al.*, 2002), in a location which also contains a second gene, *Rk*^{2}. Gene *Rk*^{2} confers broader and stronger resistance than *Rk* and is either an allele of *Rk* or a tightly linked tandem locus (Roberts *et al.*, 1996). This *Rk* region also contains a closely linked additional resistance gene that shows a low recombination frequency with *Rk* and confers partial resistance to *M. incognita* reproduction and strong resistance to *M. incognita* galling (P.A. Roberts, 2008, unpublished data). Cowpea also possesses another gene, *rk3*, which is recessive, assorts independently from *Rk*, and acts with *Rk* in an additive manner to produce a stronger and broader resistance phenotype (Ehlers *et al.*, 2000). In cotton (*Gossypium hirsutum*), a genomic region harbouring several pathogen resistance genes occurs on chromosome 11. Two genes for resistance to *M. incognita*, a gene for resistance to the reniform nematode, *Rotylenchulus reniformis*, and genes for resistance to both Verticillium wilt and Fusarium wilt reside in this chromosome 11 region (Starr *et al.*, 2007), making it a focus for marker development and cotton breeding. Genes for *M. incognita* resistance are also present in other parts of the cotton genome, such as on chromosome 14, and play a role in polygenic resistance expression (Ynturi *et al.*, 2006).

In upland cotton (*G. hirsutum*) and its pima cotton relative (*G. barbadense*), an important relationship between genes contributing to resistance is emerging from recent molecular marker mapping studies. The first nematode resistance gene mapped in cotton was *rkn1*, located on chromosome 11 (Wang *et al.*, 2006a). This gene confers resistance to *M. incognita*, and the resistance is expressed in a recessive or additive manner, depending on the genotypic background in *G. hirsutum* crosses (Wang *et al.*, 2006b). Subsequent analysis of *rkn1* effects on resistance phenotype in interspecific crosses of resistant *G. hirsutum* × susceptible *G. barbadense* revealed a strong epistatic gene interaction between *rkn1* and another gene, *RKN2*, contributed by *G. barbadense* (Wang *et al.*, 2008). In F_{2-3} and backcross progenies from this interspecific cross, transgressive segregants were recovered that had resistant and susceptible phenotypes beyond the range of the parents. The ultraresistant progenies had genotypes with at least one allele of both *rkn1* and *RKN2*. Interpretations of this gene interaction were possible through identification of co-dominant SSR (simple sequence repeat) markers for each gene. Based only on evaluations of resistance phenotype for level of nematode reproduction and root galling, *rkn1* appears to have recessive inheritance in intraspecific *G. hirsutum* crosses, with heterozygous plants, as in the F_1 generation, being susceptible. However, in the interspecific cross with a highly susceptible parent, *rkn1* appeared to be dominantly expressed, with the F_1 and other heterozygotes highly resistant due to transgressive effects of the gene combination with *RKN2* (Wang *et al.*, 2008). Gene *RKN2* could not be detected by phenotype, having no measurable effect on nematode reproduction and galling in the absence of *rkn1*.

The phenomenon of epistatic gene interactions and transgressive segregation is quite common for diverse traits in cotton, and may be more widespread and important than previously recognized for optimal function of root-knot nematode resistance genes. Effects of 'genotype background' have been recognized to modify levels of nematode resistance in other crops. For example, different levels of *Mi-1*-mediated resistance have been reported in different tomato cultivars (Jacquet *et al.*, 2005; López-Pérez *et al.*, 2006), suggesting the presence of modifying genes that have epistatic interactions with the primary resistance determinants. The application of molecular marker analyses to resistance gene inheritance and expression should lead to better understanding of these genetic processes. From a practical standpoint, cases such as the *M. incognita* resistance in cotton, where progenies with heightened resistance levels were identified from crosses of resistant × susceptible parents, create opportunities for breeding improved crop cultivars with novel resistance genotypes.

An important consideration regarding root-knot nematode resistance is the relationship between suppression of the root-galling reaction and effects on nematode reproduction. Galling typically accompanies root-knot nematode infection in host plants, and results from expansion of the cortical tissue surrounding the giant cells used for nematode feeding in the infection site (Williamson and Hussey, 1996). The extent of galling can be considerable in both glasshouse and field-grown host root systems, and may contribute to diversion of plant photosynthate to roots, to the detriment of shoot growth and yield. The function of the cell division and expansion in the root cortex to create galls, and whether it serves to benefit the host or the nematode, is not known. Galling is not necessary for nematode reproduction, because in some compatible interactions there is little or no galling produced around the egg-laying female root-knot nematodes.

In many well-known major resistance gene reactions, such as *Mi-1* in tomato and *Rk* in cowpea, galling is greatly reduced, with only slight residual swelling around the feeding site early in infection (Das *et al.*, 2008), and this is true for many other root-knot resistance gene reactions. However, in some root-knot nematode resistance responses, suppression of reproduction and suppression of galling are separable phenotypes, as, for example, in Lima bean (*P. lunatus*) in response to *M. incognita* and *M. javanica* (Garcia *et al.*, 1996; Helms *et al.*, 2004; Roberts *et al.*, 2008). A donor breeding line 'L-136' with resistance to both nematodes was used to develop a Lima bean cultivar carrying the comprehensive resistance profile. Genetic analysis of the resistance revealed that the donor carried three resistance genes, each with a unique resistance phenotype. One gene (*mir-1*) controlled resistance to *M. incognita* reproduction, a second gene (*Mig-1*) controlled *M. incognita* primary root galling, and a third gene (*Mjg-1*) controlled resistance to *M. javanica*-induced primary root galling. None of the genes suppressed *M. javanica* reproduction. The phenotypic reactions of the roots are shown in Plate 33. In susceptible plants, Lima bean responds to infection by producing large, coalesced galls on the upper portion of the primary root, a response that seems to be at least partially systemic, because in plants resistant to *M. incognita* reproduction, few nematodes were found developing in this galled tissue. It appears that the few nematodes feeding on secondary and, later, lateral roots stimulate this massive galling response through systemic action in the absence of resistance genes. In the presence of the galling resistance gene (*Mig-1*), with or without the *mir-1* gene for reproduction resistance, this galling reaction is almost entirely lacking, although a few small galls around infection sites are present on lateral roots. A similar galling resistance gene (*Mjg-1*) that affects only *M. javanica* was also discovered. In the presence of these galling resistance genes, each species is able to reproduce freely. In the presence of the reproduction resistance gene, *M. incognita* reproduction is strongly suppressed, even though the primary root galling is still stimulated. Combinations of any two or all three of these genes showed limited interaction effects (Roberts *et al.*, 2008).

The differential actions of these resistance genes reinforce the notion that the galling response in roots is not required for successful parasitism, and puts into question the function of root-galling stimulation. These genes are unlinked in the Lima bean genome, based on their independent assortment in recombinant inbred line populations (Roberts *et al.*, 2008). As such, they required a pyramiding approach for breeding the full complement of resistance, which involved co-screening breeding progenies for the three traits (Helms *et al.*, 2004). This differential plant response to nematode root galling and reproduction observed in Lima bean is not common among root-knot nematode host plants, although it has also been observed in two other leguminous plants, common bean (*Phaseolus vulgaris*) (Fassuliotis *et al.*, 1970) and soybean (*Glycine max*) (Harris *et al.*, 2003), as well as in cotton (Shepherd, 1979). The differential resistance reactions in these other hosts were not followed by a formal genetic analysis of the trait determinants involved. A clearer definition of independent resistance gene effects on nematode reproduction and induced root galling, as described for Lima bean, may provide useful genotypes for molecular analysis of resistance pathways.

13.3 Mechanisms of Resistance to Pathogens in Plants

Plants in agricultural settings and in their native environments are attacked by a wide variety of pathogens and pests, and have developed diverse strategies to defend themselves (reviewed in Glazebrook, 2005; Chisholm *et al.*, 2006; Hückelhoven, 2007). These defences include physical barriers, repellents, toxins and other preformed defences. Plants also have innate immunity in each cell against broad groups of pathogens and produce systemic signals that emanate from infection sites (Jones and Dangl, 2006). Complex defence responses are triggered upon perception by the host's basal immune receptors of the presence of common microbial products (pathogen-associated molecular patterns, or PAMPs). Pathogens have evolved strategies, often mediated by effector molecules, to render the host's defences ineffective. As the next line of defence, plants have a repertoire of resistance genes (R-genes) that mediate recognition of specific pathogens, including viruses, bacteria, fungi and nematodes (Martin *et al.*, 2003; Belkhadir *et al.*, 2004). Upon recognition of the presence of a specific pathogen-produced effector, the product of what is sometimes called an avirulence (*Avr*) gene, the R-protein initiates an array of active defences, typically including a rapid, localized cell death, often referred to as the hypersensitive response (HR), at the site of infection. Because the resistance response requires the presence of matched gene products from the host and pathogen, it has been referred to as gene-for-gene resistance (Flor, 1971; Dangl and McDowell, 2006).

More than 40 R-genes from various plant species have been cloned. The large majority of these genes encode proteins having a domain structure including a central conserved region with a nucleotide-binding (NB) site and C-terminal leucine-rich repeat (LRR) domain (Belkhadir *et al.*, 2004). These R-proteins are generally constitutively expressed and act as surveillance molecules that, upon recognition of the presence of pathogen *Avr* gene products, undergo conformational changes that lead to signalling of defence responses. *Avr* genes have been cloned from bacteria, viruses, oomycetes and fungi (Belkhadir *et al.*, 2004). *Avr* gene products are diverse, although there is increasing evidence that many of these products have roles in pathogenicity or in stopping the basal defence response of the host (Chisholm *et al.*, 2006). In some cases, direct interaction between the *Avr* gene product and the R-protein has been demonstrated (Jia *et al.*, 2000; Dodds *et al.*, 2006). However, for several examples of resistance against bacterial pathogens, the interaction is indirect, in that Avr proteins modify host products and those modifications are recognized by the R-gene products 'guarding' the host products (Dangl and Jones, 2001).

Activation of R-genes triggers a large set of responses, including an oxidative burst and major changes in gene expression. In general, these changes are similar to those induced in the basal defence response, but occur more rapidly and are of greater magnitude (Glazebrook, 2005). R-gene-mediated defences generally are regulated by a salicylic acid-dependent defence pathway (Glazebrook, 2005). While the HR is widely seen associated with R-gene-mediated resistance, it is not always present, and in some cases it has been demonstrated that the HR is not required for resistance. Salicylic acid-mediated defences are effective against biotrophic pathogens. This pathway competes with the jasmonic acid and ethylene-mediated defence responses that are triggered by necrotrophic pathogens and wounding caused by attack of chewing insects (Bostock, 2005; Glazebrook, 2005). The components of the salicylic acid/R-gene response that are effective in protecting the host are poorly understood and are likely to depend on the pathogen.

13.4 Structure and Function of the Nematode Resistance Gene *Mi-1*

The tomato gene *Mi-1* is currently the best-characterized nematode resistance gene and it serves as a useful basis for comparison with others. *Mi-1* confers effective resistance against three root-knot nematode species: *M. incognita*, *M. javanica* and *M. arenaria* (Williamson, 1998). *Mi-1*-mediated nematode resistance was originally discovered in the wild species *Lycopersicon peruvianum* (synonymous with *S. peruvianum*), and introgression began with embryo rescue of a single plant from a cross of resistant *L. peruvianum* with cultivated tomato (*L. esculentum*, synonymous with *Solanum lycopersicum*) (Smith, 1944). Genetically linked molecular markers, first the isozyme acid phosphatase and later PCR markers, were used as aids in introgression (Williamson, 1998). *Mi-1* confers effective field resistance against prevalent root-knot nematodes in California, and is widely deployed; currently the majority of the processing tomato acreage in California is planted with cultivars that carry this gene. However, there are limitations to the efficacy of *Mi-1*. The gene is ineffective at high soil temperatures (> 28°C). *Mi-1* is not effective against the species *M. hapla* or *M. enterolobii*, both of which can be problematic on tomato (Brito *et al.*, 2004; Liu and Williamson, 2006). In addition, some isolates of *M. incognita*, *M. javanica* and *M. arenaria* virulent on *Mi-1* tomato have been identified (reviewed in Jacquet *et al.*, 2005; Williamson and Kumar, 2006).

Resistance mediated by *Mi-1* is characterized by rapid, localized host cell death. The earliest visible indications of this HR can be seen about 12 h after inoculation of roots with J2. Microscopic observation indicates that the J2 do not elicit an extensive HR while penetrating the host root tip or while migrating through the root tissue, but do so while attempting to establish a feeding site (Paulson and Webster, 1972; Ho *et al.*, 1992). This timing and localization suggests that cell penetration by the nematode's stylet or other events in initiation of the feeding site may elicit the host defence response.

Mi-1 was the first root-knot nematode resistance gene to be cloned (Milligan *et al.*, 1998; Vos *et al.*, 1998). Detailed genetic mapping using molecular markers, and based on recombination in progeny from controlled genetic crosses of *L. peruvianum* plants with and without resistance, resulted in localization of the gene to a small region of the short arm of chromosome 6 (Kaloshian *et al.*, 1998). DNA sequence analysis of this region of the genome identified three closely related candidate genes: *Mi-1.1*, *Mi-1.2* and *Mi-1.3*. The functional resistance gene was identified to be *Mi-1.2* by transformation of susceptible tomato followed by assays for resistance to nematodes (Milligan *et al.*, 1998). Additional assays of these transgenic plants determined that *Mi-1.2* was also responsible for resistance to potato aphids and whiteflies (Rossi *et al.*, 1998; Vos *et al.*, 1998; Nombela *et al.*, 2003). So far, *Mi-1* remains the only known R-gene that confers resistance against such a diverse spectrum of pests.

Molecular analysis revealed that the gene *Mi-1.2* encodes a large (1257 amino acid) protein (Mi-1.2) that belongs to the NB-LRR class of plant defence proteins discussed in section 13.3. *Mi-1.2* is a member of a small gene family with seven highly homologous copies clustered together on the short arm of chromosome 6 in resistant tomato (Seah *et al.*, 2007). The same number of closely related homologues is present in the corresponding region of susceptible tomato, but analysis of flanking sequences indicates that some of the sequences between the genes are inverted, possibly accounting for the severe recombination repression observed near the *Mi-1* gene (Seah *et al.*, 2004). Two homologues from each source are pseudogenes, and one from each source encodes a truncated product. DNA sequence identity between the homologues, excluding the pseudogenes, ranges from 92.9 to 96.7%. All of the *Mi-1* family members in both susceptible and resistant tomato that appear to be intact genes are transcribed (Seah *et al.*, 2007). However, so far a function has only been shown for *Mi-1.2*. The highest similarity of *Mi-1.2* to a gene whose product has known function is that to *Rpi-blb2* (82% identity), which is located on chromosome 6 in the corresponding genomic position to *Mi-1* in the wild potato *Solanum bulbocastanum* and confers broad resistance against the oomycete *Phytophthora infestans* (van der Vossen *et al.*, 2005). Genetic studies have localized *Mi-9*, another root-knot nematode resistance gene (which confers resistance to the same spectrum of nematodes as *Mi-1*, but is capable of functioning at higher temperatures), to the same region of the genome as *Mi-1* in *S. arcanum* accession LA2157 (Ammiraju *et al.*, 2003). Molecular studies using RNA interference (RNAi) to silence genes in the *Mi-1* family indicate that *Mi-9* is a homologue of *Mi-1* (Jablonska *et al.*, 2007). Interestingly, genes for resistance to the fungus *Oidium neolycopersici*, and to alfalfa mosaic virus, Gemini viruses and specific bacterial pathogens, also map to this region of the chromosome in other *Solanum* species (Zamir *et al.*, 1994; Thoquet *et al.*, 1996; Gebhardt and Valkonen, 2001; Parella *et al.*, 2004; Bai *et al.*, 2005). This suggests that the *Mi-1* region may be a hot spot for evolution of diverse resistance specificities.

To investigate the function of specific regions of the protein Mi-1, hybrid constructs were produced *in vitro* of the functional copy of the gene *Mi-1.2* and of a closely related, expressed homologue *Mi-1.1* (Hwang *et al.*, 2000). Some of these constructs did not confer nematode resistance when expressed in tomato roots and had no other observed phenotype, while other constructs were lethal when expressed in plant tissue, suggesting that they produced a massive HR and corresponded to constitutively active versions of *Mi-1*. These findings led to the development of a model for the regulation of signalling, in which intramolecular interactions hold the protein in an inactive conformation in the absence of nematodes (Hwang and Williamson, 2003). According to the model, detection of the presence of the nematode results in conformational changes in the protein that direct signals, leading to an HR and other manifestations of the defence response. *Mi-1* is constitutively present and surveys the cell to detect effector molecules that indicate the presence of an attacking nematode (Fig. 13.1). Consistent with this, *Mi-1* transcript has been found to be constitutively and ubiquitously present (Martinez de Ilarduya and Kaloshian, 2001; Goggin *et al.*, 2004). *Mi-1* has a domain that can bind and hydrolyse ATP (Tameling *et al.*, 2002). Binding and hydrolysis of nucleotide triphosphates is required for the conformational changes that occur in *Mi-1* and other R-proteins (reviewed in van Ooijen *et al.*, 2007).

Little is known about how the presence of nematodes is recognized by *Mi-1*. While *Mi-1* mediates resistance against only certain species of nematodes and aphids, two very different organisms, there are isolates of each species that can circumvent the resistance (Kaloshian *et al.*, 1996; Goggin *et al.*, 2000). It may be that *Mi-1* is capable of recognizing different avirulence products or effectors from each group of organisms. It is also possible that *Mi-1* recognizes modifications to a plant product altered by both pests. Mutations in an unlinked locus called *Rme1* eliminate *Mi-1*-mediated resistance to nematodes, aphids and whiteflies, but not resistance mediated by other pathogen R-genes, indicating that the *Rme1* product is required specifically for *Mi-1*-mediated resistance (Martinez de Ilarduya *et al.*, 2001). Additional experiments indicate that *Rme1* acts at the same step of the recognition/signalling pathway as *Mi-1*, or possibly upstream (Martinez de Ilarduya *et al.*, 2001). Thus, *Rme1* may encode the plant gene product that is guarded by *Mi-1*, or may in some other way be involved in nematode and aphid recognition or early defence activation.

Studies using chemicals, mutants and gene silencing have provided information on the defence signalling that is initiated by *Mi-1.2*, but the story is not complete. Application of cytokinin to resistant roots results in loss of resistance (Dropkin *et al.*, 1969). The presence of salicylic acid is required for

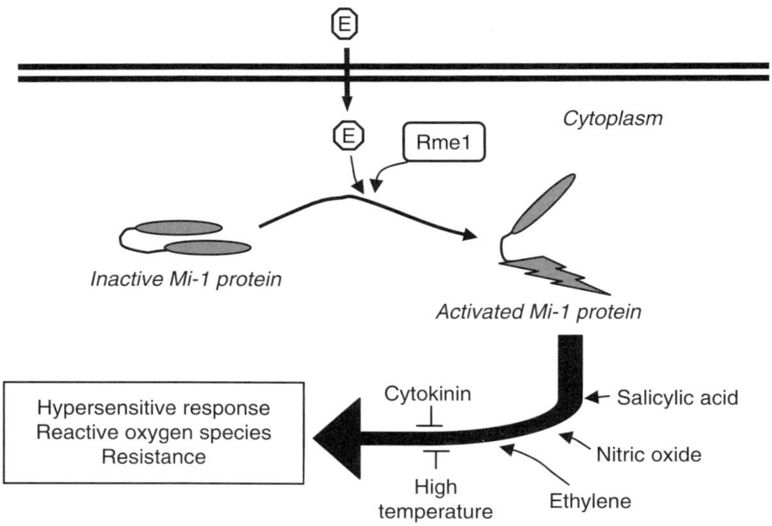

Fig. 13.1. General model for the role of the protein Mi-1 in resistance to root-knot nematodes. The presence of an effector molecule (E) from the nematode in the cytoplasm of a plant cell is recognized by Mi-1, resulting in activation of this protein. Additional proteins such as Rme1 may be involved in the recognition/activation, which initiates molecular signalling, leading to a response that includes resistance to the nematode. Some of the agents or conditions that activate (arrow) or inhibit (T-bar) the signalling pathway are indicated.

both nematode and aphid resistance (Branch *et al.*, 2004; Li *et al.*, 2006). Plant genes *Hsp90* and *Sgt1*, which are required for resistance mediated by other R-genes, are also required for *Mi-1* (Bhattarai *et al.*, 2007). However, the genes *EDS1* and *Rar1*, which are required for R-gene-mediated defence against some pathogens, are not required for *Mi-1*-mediated resistance (Hu *et al.*, 2005; Bhattarai *et al.*, 2007). A glycosyltransferase that is increased in expression after infection of resistant tomato has been shown to have a role in *Mi*-mediated resistance (Schaff *et al.*, 2007). Microscopic and biochemical studies have shown that, as for other examples of R-gene-mediated resistance, an enhanced production of reactive oxygen species (ROS) with altered timing is associated with nematode infection of resistant compared with susceptible tomato roots (Zacheo *et al.*, 1993; Molinari, 2001; Melillo *et al.*, 2006; see Moens *et al.*, Chapter 1, this volume). The HR associated with the resistance response is a logical candidate for the aspect of the response that confers the resistance phenotype, as this biotrophic pathogen feeds only on the cytoplasm of living plant cells. However, some experiments suggest that the hypersensitive response is not required for *Mi-1*-mediated resistance (Sawhney and Webster, 1979). Thus, additional factors associated with the activation of *Mi-1* are likely to contribute to the resistance response. Variation has been noted in the level of resistance in diverse tomato genotypes with *Mi-1* (Jacquet *et al.*, 2005; López-Pérez *et al.*, 2006). Genetic dissection of this variation may lead to identification of additional genes and responses that contribute to the resistance phenotype. Function of *Mi-1* and, interestingly, of several other R genes is lost at high temperature (Ammiraju *et al.*, 2003). However, the basis of this temperature sensitivity is not known. Characterization of the gene *Mi-9*, which appears to be an allele of *Mi-1* that is not temperature sensitive (Jablonska *et al.*, 2007), may provide clues.

13.5 What is Known About Other Nematode R-Genes

Histological and physiological studies of resistance mediated by several other root-knot nematode R-genes indicate that response to infection is similar to the *Mi-1*-mediated HR in tomato characterized by rapid, localized host cell death. For example, some combinations of the *Me* genes in pepper challenged with *M. incognita*, *M. javanica* or *M. arenaria* (Pegard *et al.*, 2005) and the *Mex-1* gene in coffee infected by *M. exigua* (Anthony *et al.*, 2005) confer strong HR-type responses to infection. In coffee, both penetration and development of J2 of *M. exigua* were decreased in plants with *Mex-1*, and the earliest observations, made at 4–6 days post-inoculation (dpi), revealed altered organelle structure and dark-stained cytoplasm in root cells around the nematode head rather than the normal giant cell development seen in susceptible roots (Anthony *et al.*, 2005). How soon after infection the HR response occurred was not reported. In pepper, the timing, location and specific aspects of the resistance response appear to depend on the plant genotype, specific R-gene and *Meloidogyne* species combination, indicating the likely presence of diverse mechanisms of resistance.

Pegard *et al.* (2005) summarized their own and other reported observations on responses of the R-genes Me_1, Me_3, Me_5 and Me_7 to three root-knot species. In ten incompatible R-gene–nematode species combinations, these responses ranged from immediate or rapid necrosis in the root epidermis or cortex within 1–2 dpi to formation of small giant cells and necrosis appearing by 5 dpi. The rapid HR response mediated by gene Me_7 in response to *M. arenaria* limited migration of J2 in roots and was associated with the accumulation of phenolic compounds, especially chlorogenic acid, at the infection sites. This effect on J2 migration in resistant pepper roots contrasts with *Mi-1*-mediated resistance response in tomato, where an extensive HR is not seen during penetration of or migration through the root tissue, but occurs when the J2 attempts to establish a feeding site (Paulson and Webster, 1972; Ho *et al.*, 1992). An early HR, similar to that seen for the *Mi-1* response in tomato, was also reported for the resistance reaction in soybean to *M. incognita* (Kaplan *et al.*, 1979), in which a visible response was apparent by 2 dpi during giant cell initiation.

In contrast to the above responses, which are characterized by a rapid HR, studies of the gene *Rk*-mediated-resistance response in cowpea to *M. incognita* revealed a delayed response and no associated HR reaction (Das *et al.*, 2008). The resistance conferred by *Rk* in cowpea does not decrease the rate of root penetration or J2 migration to the stele, where the J2 establish apparently functional feeding sites, each comprising several giant cells. The giant cells support nematode development to the third- and fourth-stage juveniles, similar to giant cells in susceptible cowpea roots, up to 9 dpi. By 14 dpi, an increased vacuolation of giant cells in resistant roots is visible, and by 19–21 dpi, the cells become fully vacuolated and thin-walled, and collapse. These incompatible reactions are associated with deterioration and death of the immature nematodes. Even though a few juveniles reach the young female adult stage, rarely do any produce eggs and egg masses (Fig. 13.2; Plate 34) (Das *et al.*, 2008). Profiles of ROS modulation in resistant cowpea roots following infection with *M. incognita* confirmed the lack of a classical HR reaction (Das *et al.*, 2008). Both susceptible and resistant infected roots, and also mechanically wounded positive-control roots, show an early oxidative burst (OB) during 1–3 dpi, representing a basal defence response in both compatible and incompatible interactions. However, the second OB typical of the biphasic profile associated with a pathogen-elicited HR does not occur, and the OB profile does not differ between infected resistant and susceptible roots, indicating an independence from *Rk*-mediated resistance (Das *et al.*, 2008).

In contrast to both *Mi-1* and *Rk* resistance gene responses to infection, lucerne (*Medicago sativa*) resistant to *M. incognita* blocks J2 development in the absence of an HR (Potenza *et al.*, 1996). In resistant roots, initial root penetration and migration of J2 to the root apex occurs up to 3 dpi, but the J2 remain clumped at the root apex rather than migrating into the vascular cylinder, as do the J2 in susceptible roots, and by 7 dpi, no J2 were observed in resistant roots (Potenza *et al.*, 1996).

In summary, these comparative observations between root-knot nematode resistance reactions indicate that diverse resistance mechanisms and processes have evolved in host plants, including among different genes in the same plant species and among different major genes across host plant taxa.

13.6 Nematode Virulence and Durability of Resistance

Differences within nematode species in their ability to reproduce on hosts carrying specific

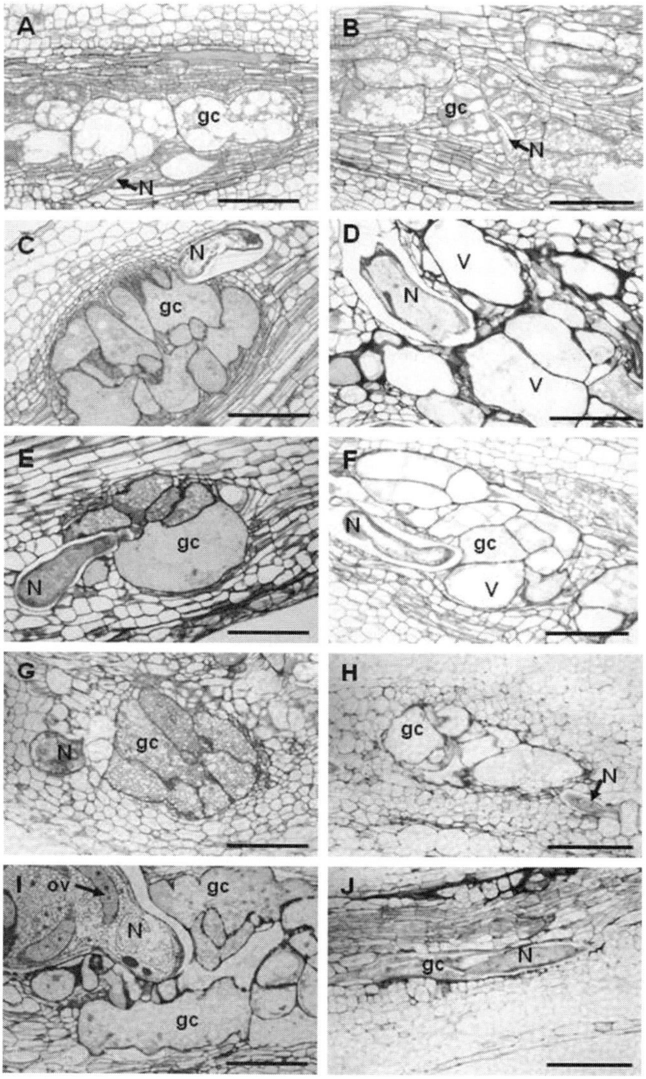

Fig. 13.2. Longitudinal sections of *Meloidogyne incognita* feeding sites in inoculated cowpea roots. Sections were stained with toluidine blue O. Panels A, C, E, G and I are null-*Rk* (susceptible) root sections at 5, 9, 14, 19 and 21 days post-inoculation (dpi), respectively. Panels B, D, F, H and J are CB46 (resistant) root sections at 5, 9, 14, 19 and 21 dpi, respectively. gc = giant cell, N = nematode, ov = ovaries and V = vacuole. Scale bar = 200 μm. (From Das *et al.*, 2008.)

nematode resistance genes have been found in many nematode–host interactions, including in important *Meloidogyne* species–host crop combinations (Roberts, 1995; Castagnone-Sereno, 2006). Avirulent nematode strains are defined here as those against which the resistance gene is effective and the nematodes do not reproduce, while virulent or 'resistance breaking' strains are those able to reproduce on resistant plants. Acquisition of virulence could be due to loss of, or modification to, the nematode gene product recognized by the plant resistance gene or to a gain of ability to circumvent resistance, for example by producing antioxidants or altering the hormone balance to compromise the defence response. The durability of resistance depends on the natural occurrence of virulent forms of the target nematode species and on whether the virulent nematode forms arise and can be selected by exposure to resistant host plants. Root-knot nematodes differ in reproduction modes (see Chitwood and Perry, Chapter 8, this volume), and these differences may also affect their potential to develop virulence.

The species *M. incognita*, *M. javanica* and *M. arenaria*, generally considered among the world's most damaging plant-parasitic nematodes, reproduce exclusively by mitotic parthenogenesis (Triantaphyllou, 1985; Castagnone-Sereno, 2006). This asexual mode of reproduction, without sexual recombination, is expected to give rise to clonal progeny. Yet these species are widely distributed, polyphagous and amenable to selection and adaptive variation (Trudgill and Blok, 2001). How these species are able to adapt and evolve genetically is an important question for both evolutionary studies and practical management. One clue may be in the variable chromosomal complement of isolates of a given species. For example, the chromosome number of *M. javanica* can range from 42 to 48 and that of *M. incognita* from 32 to 38 (Triantaphyllou, 1985). The variable chromosome number suggests that these species are aneuploid, with different copy numbers of some chromosomes. Changes in ploidy of a specific chromosome could lead to phenotypic differences, especially if the individual is heterozygous for the relevant alleles. Gene conversion, transposable elements or other specialized mechanisms of genetic change could also have a role in generating genetic variability. So far, the mechanisms for host range adaptability and virulence in those species have not been determined.

There have been several reports of *M. incognita* and *M. javanica* populations that can infect tomato plants with *Mi-1* (Riggs and Winstead, 1959; Bost and Triantaphyllou, 1987; Kaloshian *et al.*, 1996; Eddaoudi *et al.*, 1997). Virulent populations include both field isolates and populations selected on *Mi*-containing tomato in the glasshouse. Two groups investigating the development of virulence in the glasshouse to plants with *Mi* found a progressive increase in virulence, and concluded that several genes may be involved in the development of this virulence (Bost and Triantaphyllou, 1987; Jarquin-Barberena *et al.*, 1991). Some field isolates that do not have a history of cropping with tomato nevertheless are able to infect tomato with *Mi*, while other isolates do not develop virulence even after repeated exposure to resistant host plants (Roberts, 1995). In some cases, there appears to be an adverse cost of fitness to *Mi-1*-virulent lines, as seen by a lower infection capacity or fecundity on susceptible tomato (Roberts, 1995; Castagnone-Sereno *et al.*, 2007). In one case, laboratory-selected *Mi*-virulent *M. incognita* isolates were found to have lost the ability to reproduce on pepper plants (Castagnone-Sereno *et al.*, 1994). A field population of *Mi*-virulent *M. incognita* was also found to be unable to reproduce on pepper (Tzortzakakis and Blok, 2007). Thus, it may be that gain of virulence is associated with other alterations in host range in at least some cases.

Comparison of protein profiles of *Mi*-virulent and avirulent nematode pairs by 2-D gels revealed that these strains are very similar (Castagnone-Sereno *et al.*, 1995). However, differential expression analyses with transcripts from closely related strains of nematodes that differ in virulence have revealed loss of expression of specific transcripts. A gene called *map-1* was reported to be missing from virulent compared with avirulent strains of *M. incognita* (Semblat *et al.*, 2001). The gene product was found to be localized to the amphids. Similar genes (*mjap-1* and *mjap-2*) have been identified in *M. javanica*, but these genes are expressed in both virulent and avirulent lines and are localized to the subventral pharyngeal glands of J2 (Adam *et al.*, 2009). In some of the paired virulent and avirulent strains, expression of multiple genes is reduced in the virulent lines (Neveu *et al.*, 2003). However, direct evidence for a role of any of these genes in *Mi-1*-mediated resistance is lacking so far. In another case, a strain of *M. javanica* that was able to reproduce well on tomato carrying *Mi-1* was obtained after glasshouse selection from an avirulent *M. javanica* strain (Gleason *et al.*, 2008). Here, full virulence had developed suddenly, after only a few generations on a resistant plant, and its development was a low-frequency event. Differential expression analysis indicated that the strains were very similar, but identified a gene *Cg-1* that was transcribed in the avirulent *M. javanica* but was absent from its virulent derivative. Further analysis showed that the virulent strain carried a deletion in the genomic region corresponding to *Cg-1*. In addition, silencing *Cg-1* by soaking avirulent nematodes in double-stranded RNA corresponding to its sequence resulted in a gain of virulence, strongly suggesting that *Cg-1* is required for the *Mi-1*-mediated recognition of the nematode by its host. However, *Cg-1* does not encode a known gene product, and it is possible that it is not translated into protein but acts at the RNA level (Gleason *et al.*, 2008). Additional experiments will be needed to determine the role of *Cg-1* in *Mi-1*-mediated resistance and may lead to the development of diagnostic indicators of virulence or predictors for the potential of a root-knot nematode population to develop virulence.

Virulence selection studies with *M. incognita* and gene *Rk* in cowpea also indicated that genetic adaptability is present in these asexual species (Petrillo and Roberts, 2005a,b). Isofemale lines from single egg masses were developed from both *Rk*-virulent and *Rk*-avirulent parent isolates of *M. incognita* and cultured on cowpea plants with and without gene *Rk* for up to 27 generations (Petrillo and Roberts, 2005a). The avirulent parent isolate initially produced ~7% virulent isofemale lines, but high levels of virulence were selected within eight generations on resistant plants. Such differential lineage selection probably explains the rapid virulence selection that can occur in fields planted with resistant cowpea. Virulent progeny were selected from single egg masses of avirulent females, indicating that there is some mechanism by which these mitotically reproducing nematodes can produce progeny with different phenotypes. The complex, possibly multiple-mechanism nature of the changes in virulence to *Rk* was indicated by the variation between isofemale lines in their virulence over 27 generations. At least five distinct profiles with stable or variable (a)virulence were discerned (Petrillo and Roberts, 2005a). That is, while some lines remained avirulent on susceptible plants and became extinct on resistant plants, others remained virulent when cultured on resistant and susceptible plants; still others became virulent on resistant plants, became avirulent on susceptible plants or became virulent when cultured on susceptible plants. In some, but not all, cases there was a reduction in egg mass production and fecundity on susceptible cowpea associated with virulence to *Rk*, indicating a trade-off between reproductive fitness and virulence (Petrillo and Roberts, 2005b).

The multiple-generation behaviour of *Rk*-avirulent and virulent *M. incognita* lineages is explained in part by the relative fitness of the lines. In studying life history traits of the isofemale lines, Petrillo and Roberts (2005b) found that hatch and root penetration were not correlated with virulence status, whereas reproductive potential measured by egg mass production and fecundity (eggs per egg mass) on susceptible plants was associated with virulence. Some virulent lines showed significantly lower reproductive potential than avirulent lines, indicating a trade-off between reproductive fitness and virulence. These virulent lines would be expected to decline on susceptible hosts and, in populations comprising both virulent and avirulent lineages, to represent a diminishing proportion of the population as a whole, due to being outcompeted by avirulent forms. Interestingly, not all virulent lines had lower relative fitness, as some were able to reproduce effectively for many generations on susceptible cowpea and were characterized by stable virulence (Petrillo and Roberts, 2005b). These findings were the first to demonstrate fitness costs associated with root-knot nematode virulence, even though this was suspected to occur since it was established as an important feature of stabilizing selection in animals (Caswell and Roberts, 1987). However, in earlier studies on tomato *Mi-1* virulence in *M. incognita* (Castagnone-Sereno *et al.*, 1994) and *M. javanica* (Tzortzakakis and Gowen, 1996) no differences in reproductive fitness between virulent and avirulent lines were found, indicating stable virulence (Castagnone-Sereno *et al.*, 1993), and there were few and differing reports of presence and absence of fitness costs associated with virulence in the amphimictic cyst nematodes (Turner, 1990; Lasserre *et al.*, 1996). The trade-off between reproductive fitness and virulence in *M. incognita* in the cowpea *Rk*-gene study was supported by another study of the tomato *Mi-1* virulence in *M. incognita*, in which lower reproductive fitness, based on a combination of hatch of J2 and egg mass production, was found in virulent compared with avirulent near-isogenic nematode lines (Castagnone-Sereno *et al.*, 2007). These variations in fitness profiles further support the existence of multiple mechanisms of virulence in *M. incognita*, and no doubt related species, that contribute to adaptation and maintenance of virulence under stabilizing selection.

While *M. incognita*, *M. javanica* and *M. arenaria* reproduce by obligate mitotic parthenogenesis, other nematode species can reproduce sexually. Some isolates of *M. hapla* reproduce by mitotic parthenogenesis (cytological race B), but most isolates (cytological race A) undergo meiosis and reproduce by facultative meiotic parthenogenesis (Triantaphyllou, 1985; Liu and Williamson, 2006). In this form of reproduction, sexual crosses occur when males are present, but in the absence of males the haploid sister nuclei of a single meiosis fuse together to restore diploidy (Triantaphyllou, 1985; Liu *et al.*, 2007). Other important root-knot nematode species, including *M. chitwoodi* and *M. fallax*, are also facultative parthenogens. These species tend to have a narrower host range than the asexual species, but more variation within isolates. For example, considerable variation in virulence was found between and within isolates of *M. chitwoodi*, *M. fallax* and *M. hapla* on *Solanum* spp. (van der Beek *et al.*, 1998). *Meloidogyne hapla*-resistant lucerne is susceptible to a Californian isolate of *M. hapla* (Griffin and McKenry, 1989). *Meloidogyne* new *M. chitwoodi* pathotype has been identified that infects potato with the $R_{mc1(blb)}$ resistance gene (Mojtahedi *et al.*, 2007). Isolates of *M. hapla* differ in virulence on the common bean cultivar NemaSnap, which carries a single, dominant gene for *M. hapla* resistance (Chen and Roberts, 2003a). It is possible that resistance is less durable in these species due to within-population variability and recombination of traits by sexual reproduction.

Although attempts to identify traits responsible for increased host range (virulence factors) are handicapped by lack of genetic possibility in species that reproduce mitotically, species capable of sexual reproduction have the potential for genetic analysis to identify such traits. For example, genetic crosses between *M. hapla* strains show segregation consistent with the interpretation that virulence on common bean is a gene-for-gene system (Chen and Roberts, 2003a,b). In this case, there is genetic evidence that avirulence in the nematode segregates as a single, dominant locus (Chen and Roberts, 2003b). Genetic crosses have also been carried out between inbred lines of *M. hapla* that differ in pathogenicity on various hosts and in molecular markers (Liu and Williamson, 2006; Liu *et al.*, 2007). By monitoring progeny with molecular markers, it has been possible to establish 183 F_2 lines as glasshouse cultures. Characterization of polymorphic DNA markers in these lines has resulted in the production of a genetic map (Liu *et al.*, 2007; Opperman *et al.*, 2008). Because the parental lines differ in pathogenicity and attraction to specific hosts, it should be possible to map these traits on to the genetic map. In addition, a 10× draft of the genome sequence of *M. hapla* has been completed (Opperman *et al.*, 2008), and integration of the genetic and physical maps will eventually provide the resources to clone genetic factors contributing to pathogenicity, attraction, host range and other important traits (see Abad and Oppermann, Chapter 16, this volume). Functional analysis by RNAi may aid to confirm identity of genes responsible for virulence acquisition.

13.7 Management of Resistance and Virulence in the Field

With increased environmental concerns and restrictions on nematicides, the importance of host resistance for nematode management has grown substantially in the past few years. A major challenge in the use of host resistance, especially of single resistance genes, is that the root-knot and other nematodes can undergo genetic alterations allowing them to reproduce on the resistant host, or existing virulent subpopulations can be selected, as described in the previous section. Due to the limited availability of root-knot nematode resistance genes, it is important to manage existing resources to extend their field utility as long as possible (Roberts, 1995). Virulence among root-knot nematodes exposed to resistant crop cultivars or rootstocks is not universal, but it is widespread. The virulence to tomato *Mi-1* and cowpea *Rk* genes discussed in detail above has occurred in infested fields planted with resistant cultivars of these crops. In the case of *Mi-1*, reports from many countries indicate that virulence is widely distributed geographically and may be quite common. In California, resistant processing tomatoes carrying *Mi-1* have been grown intensively on infested fields for more than 20 years, and the recent identification of several virulent populations of *M. incognita* and *M. javanica* in this primary tomato-growing region stresses the immediacy of the problem (Kaloshian *et al.*, 1996; Williamson and Kumar, 2006; Williamson, unpublished data). Processing-tomato growers have come to rely on host resistance for nematode management, instead of soil fumigation with nematicides, and concern about resistance-breaking populations has increased. The extensive use of cowpea cultivars with gene *Rk* has resulted in several fields in California with virulent *M. incognita* populations (Petrillo *et al.*, 2006). Examples from other crops and *Meloidogyne* species illustrate the breadth of problems associated with virulence selection. Virulence in *M. chitwoodi* to the resistance gene $R_{mc1(blb)}$ in potato threatens to undermine the use of resistance in the US Pacific Northwest (Mojtahedi *et al.*, 2007). Resistant lucerne shows susceptibility to some populations of *M. hapla* (Griffin and McKenry, 1989), and *Meloidogyne* populations virulent on resistant grapevines have been reported in California (Anwar and McKenry, 2002). However, in some cases, such as the resistance to *Meloidogyne* spp. in the *Prunus* rootstock Nemaguard, virulent populations have not been found during more than 50 years of use in commercial orchards. This strong durability may reflect unique features of the resistance mechanism and its expression in a perennial root system background, or the inability of nematodes to become virulent on *Prunus* with this trait.

For annual crops, alternating resistant and susceptible cultivars, and using resistant crops in rotation with other host crops, including those with resistance, are approaches to managing virulence selection (Roberts, 1995; Petrillo *et al.*, 2006). Monocultures provide strong selection for acquisition of virulence, and rotation-based approaches relieve the selection pressure in the cropping system. Two considerations are important to the success of this integrated management approach. When resistance is first implemented, rotation of resistant and susceptible host crops, and non-hosts, if available, can reduce the probability that virulent forms will be selected. In addition, when virulent populations occur, rotation will be necessary, and, in populations in which the virulent forms have lower reproductive or survival fitness, reproduction on susceptible host plants in a rotation may result in a decline in the overall level of virulence in the field. This potential is illustrated by the cowpea *Rk*-virulence in *M. incognita* populations. Field populations of *M. incognita* may comprise a mixture of virulent and avirulent lineages, in which the virulent forms were shown to have reduced reproductive fitness (Petrillo and Roberts, 2005a,b). When these populations were maintained on susceptible tomato host plants for multiple generations, the virulence levels declined significantly with time (Petrillo *et al.*, 2006). This decline is described by an exponential decay (Fig. 13.3). While this study was based on glasshouse-grown cultures and has not been demonstrated in the field, the dynamics of the virulent population indicate that rotations with susceptible host crops should reduce virulence and the damage potential of field populations.

Another approach to managing nematode virulence is to identify new genes for resistance that are effective against root-knot nematode populations virulent on the currently used resistance gene(s) and to introduce these genes into the same crop. This approach, sometimes portrayed as an 'arms-race', with plant breeders trying to keep one step ahead of the pathogen, depends on the availability of additional effective resistance genes. Several resistance genes have been identified in wild relatives of tomato in addition to *Mi-1*, including *Mi-2*–*Mi-9*, some of which, such as *Mi-3*, are effective against *Mi-1*-virulent root-knot nematode populations (Roberts, 1995; Yaghoobi *et al.*, 1995; Veremis and Roberts, 1996; Williamson, 1998). The *Mi-1*-virulent populations are unable to reproduce on tomatoes carrying these other genes, and, more broadly, they do not show virulence to resistance genes in other crops, including in related species such as pepper (Castagnone-Sereno *et al.*, 1996). However, these novel *Mi* genes have proved difficult to introgress from the wild *S. peruvianum* donor into cultivated *S. lycopersicum* and, as yet, are unavailable for tomato growers. In cowpea, gene *Rk*^{2} is being bred into cultivars to manage *Rk*-virulent populations, and blackeye-bean cultivars of cowpeas with *Rk*^{2} are nearing release. The *Rk*^{2} gene has a broader and stronger resistance expression than *Rk*, is effective against virulent populations of *M. incognita* and *M. javanica*, and is either an allele of *Rk* or a tightly linked tandem locus (Roberts *et al.*, 1996). Effective use of this strategy will require integration of the resistance into rotations, because field microplot studies showed that,

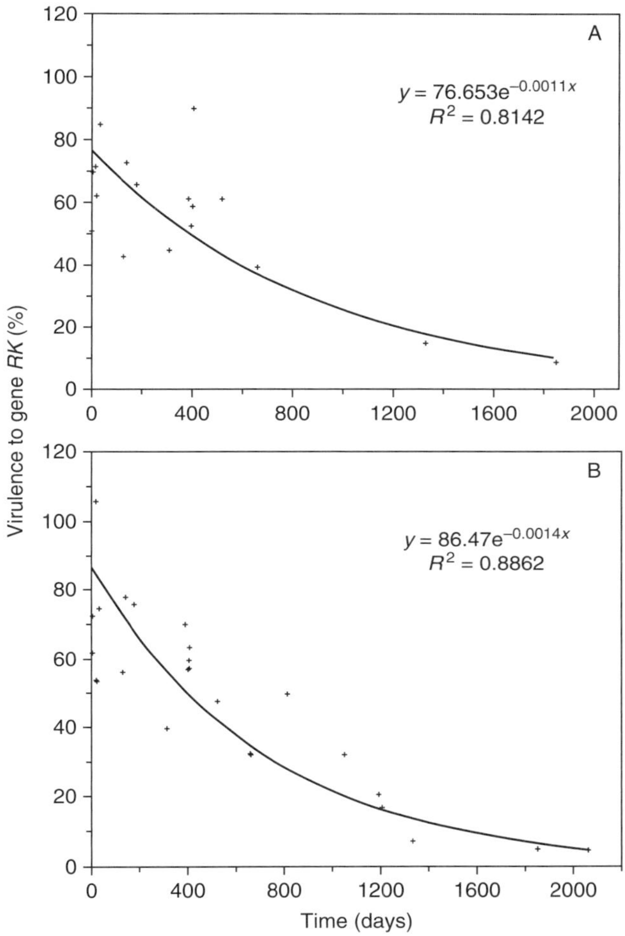

Fig. 13.3. The effect of continuous culturing on a susceptible host (tomato) for several years on the virulence to resistance gene *Rk* in cowpea of a *Meloidogyne incognita* population isolated from a cowpea field showing breakdown of *Rk*-resistance. Percent virulence was determined at various time points on cowpea genotypes CB5 (A) and CB46 (B) carrying gene *Rk*, based on egg mass production on the resistant genotypes as a proportion of that on susceptible cowpea CB3. The *x*-axis indicates the length of time the *Rk*-virulent population was cultured on susceptible tomato. (From Petrillo *et al.*, 2006.)

to some extent, *Rk*-virulent isolates of *M. incognita* can be cross-selected for virulence to Rk^2 and vice versa (Petrillo *et al.*, 2006). In this case, the (a)virulence-matching specificity of the genes must overlap to some extent. For perennial tree and vine crops, virulence management is less amenable to rotation of hosts or resistance genes because of long-term planting horizons and costs of re-planting. However, pyramiding or stacking of resistance genes may increase the durability of resistance in such crops.

13.8 Conclusions and Future Directions

This chapter describes the valuable genetic resource of natural host plant resistance traits effective against the most important species of root-knot nematodes. An increased knowledge and understanding of the resistance genes governing these traits is critical to optimizing their use for root-knot nematode management. Unlike the use of nematicides, which in general are broad spectrum and are not dependent on biological specificity of action among nematodes, the effective use of host resistance requires knowledge of the biological specificity inherent in the resistance traits and their target root-knot nematode populations.

The genetic basis of the resistance and the current understanding of resistance mechanisms elicited upon host infection represent a broad range of gene actions and interactions that do not conform to a single recognized pattern. The inheritance of resistance includes single and multiple genes, which may operate by dominant, recessive, additive or epistatic interacting mechanisms. The rapid advances in molecular marker technologies are providing important tools for better understanding of the genetic organization, function and interaction attributes of root-knot nematode and other resistance genes, and of how the additive effects of gene alleles operate. In turn, marker-based approaches to resistance gene analysis are providing important plant-breeding tools through application of easy-to-use markers for marker-assisted selection (see Starr and Mercer, Chapter 14, this volume).

Some root-knot nematode resistance genes, including *Mi-1* in tomato, show an early classical HR-associated cell death around the feeding site coupled with a significant oxidative burst. Other genes operate without an HR-based process and act either earlier in the infection process, as in lucerne, or later, as in *Rk* gene action in cowpea. Regardless of the mechanism, the net result of these resistance responses is suppression of nematode reproduction and, in most cases, tolerance in the infected plant. Gene *Mi-1* in tomato was the first root-knot nematode resistance gene to be cloned and, along with the wealth of resources on disease-resistance genes in model systems, has provided an important model system for dissecting the mechanisms of resistance at the cellular and molecular levels. In addition, the recent characterization of a *Mi-1* matching avirulence factor (*Cg-1*) that is deleted in a virulent isolate of *M. javanica* opens the possibility for defining the nature of the gene-for-gene-level interactions between host resistance and nematode avirulence.

In this chapter we have also provided a descriptive analysis of other host resistance–nematode (a)virulence interactions, including those for the obligate parthenogen *M. incognita* and gene *Rk* in cowpea, in which isofemale line analyses have proved informative, and for the sexually reproducing *M. hapla* and resistance in common bean, in which a Mendelian approach to genetic analysis is possible. Studies of these interactions provide understanding of virulence development and stability in nematodes challenged with resistance gene selection. The biological processes governing the interactions have important consequences for the durability of resistance in root-knot nematode management programmes, and optimization of resistance durability will require additional knowledge of the specific mechanisms involved. In current work on *M. hapla*, opportunities are presented for development of genetic and physical mapping of avirulence and other host determinant traits, and for cloning and characterization of the genes conditioning these traits, particularly with the recent whole genome sequencing of *M. hapla* and *M. incognita* (see Abad and Opperman, Chapter 16, this volume).

A continuing challenge for the use of resistance is that most root-knot nematode resistance genes are effective only against one or a few of the target *Meloidogyne* species, and in some cases effective against only some populations of a species. In a few crops, resistance genes can be pyramided or stacked to provide a broader platform of resistance phenotypes within a crop cultivar, as described for resistance in Lima bean, but in many cases this is not currently possible because the genes are unavailable in the crop species. Therefore, continued research and development efforts are needed to exploit transgenic forms of resistance, either by using cloned natural resistance genes between crop plants, such as the use of *Mi-1* from tomato in other solanaceous crops (Goggin *et al.*, 2006), or by engineering novel resistance, such as demonstrated recently by transforming host plants with gene-silencing capability specific to conserved *Meloidogyne* genes (see Atkinson *et al.*, Chapter 15, this volume). In the latter case, development of resistance effective against four common *Meloidogyne* spp. was achieved (Huang *et al.*, 2006). It may well be possible, and perhaps necessary, in the future to pyramid combinations of novel transgenic resistance and natural resistance genes to develop broad-based and durable forms of root-knot nematode resistance.

13.9 References

Adam, M.A.M., Phillips, M.A., Tzortzakakis, E.A. and Blok, V.C. (2009) Characterisation of *mjap* genes encoding novel secreted proteins from the root-knot nematode, *Meloidogyne javanica. Nematology* 11, 255–267.

Ammiraju, J.S.S., Veremis, J.C., Huang, X., Roberts, P.A. and Kaloshian, I. (2003) The heat-stable, root-knot nematode resistance gene *Mi-9* from *Lycopersicon peruvianum* is localized on the short arm of chromosome 6. *Theoretical and Applied Genetics* 106, 478–484.

Anthony, F., Topart, P., Martinez, A., Silva, M. and Nicole, M. (2005) Hypersensitive-like reaction conferred by the *Mex-1* resistance gene against *Meloidogyne exigua* in coffee. *Plant Pathology* 54, 476–482.

Anwar, S.A. and McKenry, M.V. (2002) Developmental response of a resistance-breaking population of *Meloidogyne arenaria* on *Vitis* spp. *Journal of Nematology* 34, 28–33.

Bai, Y., van der Hulst, R., Bonnema, G., Marcel, T.C., Meijer-Dekens, F., Niks, R.E. and Lindhout, P. (2005) Tomato defense to *Oidium neolycopersici*: dominant *Ol* genes confer isolate-dependent resistance via a different mechanism than recessive *ol-2. Molecular Plant–Microbe Interactions* 18, 354–362.

Barrett, B.B., Mercer, C.F., Moore, K.K. and Woodfield, D.R. (2002) Microsatellite markers for a nematode resistance locus in *Trifolium. Proceedings of the 12th Australasian Plant Breeding Conference*, pp. 367–371.

Belkhadir, Y., Subramaniam, R. and Dangl, J.L. (2004) Plant disease resistance protein signaling: NBS-LRR proteins and their partners. *Current Opinion in Plant Biology* 7, 391–399.

Bhattarai, K.K., Li, Q., Liu, Y., Dinesh-Kumar, S.P. and Kaloshian, I. (2007) The *Mi-1* mediated pest resistance requires *Hsp90* and *Sgt1. Plant Physiology* 144, 312–323.

Boiteux, L.S., Belter, J.G., Roberts, P.A. and Simon, P.W. (2000) RAPD linkage map of the genomic region encompassing the root-knot nematode (*Meloidogyne javanica*) resistance locus in carrot. *Theoretical and Applied Genetics* 100, 439–446.

Bost, S.C. and Triantaphyllou, A. (1982) Genetic basis of the epidemiological effects of resistance to *Meloidogyne incognita* in tomato cultivar Small Fry. *Journal of Nematology* 14, 540–544.

Bostock, R.M. (2005) Signal crosstalk and induced resistance: straddling the line between cost and benefit. *Annual Review of Phytopathology* 43, 545–580.

Branch, C., Hwang, C.F., Navarre, D.A. and Williamson, V.M. (2004) Salicylic acid is part of the *Mi-1*-mediated defense response to root-knot nematode in tomato. *Molecular Plant–Microbe Interactions* 17, 351–356.

Brito, J., Powers, T.O., Mullin, P.G., Inserra, R.N. and Dickson, D.W. (2004) Morphological and molecular characterization of *Meloidogyne mayaguensis* isolates from Florida. *Journal of Nematology* 36, 232–240.

Brown, C.R., Yang, C.-P., Mojtahedi, H. and Santo, G.S. (1996) RFLP analysis of resistance to Columbia root-knot nematode derived from *Solanum bulbocastanum* in a BC2 population. *Theoretical and Applied Genetics* 92, 572–576.

Castagnone-Sereno, P. (2006) Genetic variability and adaptive evolution in parthenogenetic root-knot nematodes. *Heredity* 96, 282–289.

Castagnone-Sereno, P., Bongiovanni, M. and Dalmasso, A. (1993) Stable virulence against the tomato resistance *Mi* gene in the parthenogenetic root-knot nematode *Meloidogyne incognita. Genetics* 83, 803–805.

Castagnone-Sereno, P., Bongiovanni, M. and Dalmasso, A. (1994) Reproduction of virulent isolates of *Meloidogyne incognita* on susceptible and *Mi*-resistant tomato. *Journal of Nematology* 26, 324–328.

Castagnone-Sereno, P., Rosso, M.N., Bongiovanni, M. and Dalmasso, A. (1995) Electrophoretic analysis of near-isogenic avirulent and virulent lineages of the parthenogenetic root-knot nematode *Meloidogyne incognita. Physiological and Molecular Plant Pathology* 47, 293–302.

Castagnone-Sereno, P., Bongiovanni, M., Palloix, A. and Dalmasso, A. (1996) Selection for *Meloidogyne incognita* virulence against resistance genes from tomato and pepper and specificity of the virulence/resistance determinants. *European Journal of Plant Pathology* 102, 585–590.

Castagnone-Sereno, P., Bongiovanni, M. and Wajnberg, E. (2007) Selection and parasite evolution: a reproductive fitness cost associated with virulence in the parthenogenetic nematode *Meloidogyne incognita. Evolutionary Ecology* 21, 259–270.

Caswell, E.P. and Roberts, P.A. (1987) Nematode population genetics. In: Veech, J.A. and Dickson, D.W. (eds) *Vistas on Nematology*. Society of Nematologists, Hyattsville, Maryland, pp. 390–397.

Chen, P. and Roberts, P.A. (2003a) Virulence in *Meloidogyne hapla* differentiated by resistance in common bean (*Phaseolus vulgaris*). *Nematology* 5, 39–47.
Chen, P. and Roberts, P.A. (2003b) Genetic analysis of (a)virulence in *Meloidogyne hapla* to resistance in bean (*Phaseolus vulgaris*). *Nematology* 5, 687–697.
Chisholm, S.T., Coaker, G., Day, B. and Staskawicz, B.J. (2006) Host–microbe interactions: shaping the evolution of the plant immune response. *Cell* 124, 803–814.
Choi, K., Burow, M.D., Church, G.T., Burow, G., Paterson, A.H., Simpson, C.E. and Starr, J.L. (1999) Genetics and mechanism of resistance to *Meloidogyne arenaria* in peanut germplasm. *Journal of Nematology* 31, 283–290.
Church, G.T., Starr, J.L. and Simpson, C.E. (2005) A recessive gene for resistance to *Meloidogyne arenaria* in interspecific *Arachis* spp. hybrids. *Journal of Nematology* 37, 178–184.
Collard, B.C.Y., Jahufer, M.Z.Z., Brouwer, J.B. and Pang, E.C.K. (2005) An introduction to markers, quantitative trait loci (QTL) mapping and marker-assisted selection for crop improvement: the basic concepts. *Euphytica* 142, 169–196.
Cook, R. and Evans, K. (1987) Resistance and tolerance. In: Brown, R.H. and Kerry, B.R. (eds) *Principles and Practice of Nematode Control in Crops*. Academic Press, New York, pp. 179–231.
Cousins, P., Lauver, M. and Boyden, L. (2003) Genetic analysis of root-knot nematode resistance derived from *Vitis mustangensis. Acta Horticulturae* 603, 149–155.
Dangl, J.L. and Jones, J.D.G. (2001) Plant pathogens and integrated defense responses to infection. *Nature* 411, 826–833.
Dangl, J.L. and McDowell, J.M. (2006) Two modes of pathogen recognition by plants. *Proceedings of the National Academy of Sciences of the USA* 103, 8575–8576.
Das, S., DeMason, D.A., Ehlers, J.D., Close, T.J. and Roberts, P.A. (2008) Histological characterization of root-knot nematode resistance in cowpea and its relation to reactive oxygen species modulation. *Journal of Experimental Botany* 59, 1305–1313.
Dirlewanger, E., Cosson, P., Howad, W., Capdeville, G., Bosselut, N., Claverie, M., Voisin, R., Poizat, C., Lafargue, B., Baron, O., Laigret, F., Kleinhentz, M., Arús, P. and Esmenjaud, D. (2004) Microsatellite genetic linkage maps of myrobalan plum and an almond-peach hybrid – location of root-knot nematode resistance genes. *Theoretical and Applied Genetics* 109, 827–838.
Djian-Caporalino, C., Fazari, A., Arguel, M.J., Vernie, T., VandeCasteele, C., Faure, I., Brunoud, G., Pijarowski, L., Palloix, A., Lefebvre, V. and Abad, P. (2007) Root-knot nematode (*Meloidogyne* spp.) *Me* resistance genes in pepper (*Capsicum annuum* L.) are clustered on the P9 chromosome. *Theoretical and Applied Genetics* 114, 473–486.
Dodds, P.N., Lawrence, G.J., Catanzariti, A-M., The, T., Wang, C-I. A., Ayliffe, M.A., Kobe, B. and Ellis, J.E. (2006) Direct protein interaction underlies gene-for-gene specificity and coevolution of the flax resistance genes and flax rust avirulence genes. *Proceedings of the National Academy of Sciences of the USA* 103, 8888–8893.
Dropkin, V.H., Helgeson, J.P. and Upper, C.D. (1969) The hypersensitivity reaction of tomatoes resistant to *Meloidogyne incognita*: reversal by cytokinins. *Journal of Nematology* 1, 55–61.
Eddaoudi, M., Ammati, M. and Rammah, A. (1997) Identification of the resistance breaking populations of *Meloidogyne* on tomatoes in Morocco and their effect on new sources of resistance. *Fundamental and Applied Nematology* 20, 285–289.
Ehlers, J.D., Matthews, W.C., Hall, A.E. and Roberts, P.A. (2000) Inheritance of a broad-based form of root-knot nematode resistance in cowpea. *Crop Science* 40, 611–618.
Fassuliotis, G., Deakin, J.R. and Hoffman, J.C. (1970) Root-knot nematode resistance in snap beans: breeding and nature of resistance. *Journal for the American Society of Horticultural Science* 95, 640–645.
Fery, R.L. and Dukes, P.D. (1980) Inheritance of root-knot nematode resistance in cowpea [*Vigna unguiculata* (L.) Walp.]. *Journal for the American Society of Horticultural Science* 105, 671–674.
Flor, H.H. (1971) Current status of the gene-for-gene concept. *Annual Review of Phytopathology* 9, 275–296.
Garcia, G.M., Stalker, H.T., Shroeder, E. and Kochert, G. (1996) Identification of RAPD, SCAR, and AFLP markers tightly linked to nematode resistance genes introgressed from *Arachis hypogaea. Genome* 39, 836–845.
Gebhardt, C. and Valkonen, J.P.T. (2001) Organization of genes controlling disease resistance in the potato genome. *Annual Review of Phytopathology* 39, 79–102.
Glazebrook, J. (2005) Contrasting mechanisms of defense against biotrophic and necrotrophic pathogens. *Annual Review of Phytopathology* 43, 205–227.

Gleason, C.A., Liu, Q.L. and Williamson, V.M. (2008) Silencing a candidate nematode effector gene corresponding to the tomato resistance gene *Mi-1* leads to acquisition of virulence. *Molecular Plant Microbe Interactions* 21, 576–585.

Goggin, F., Williamson, V.M. and Ullman, D. (2000) Variability in the response of *Macrosiphum euphorbiae* Thomas and *Myzus persicae* Shull (Hemiptera: Aphididae) to the tomato resistance gene, *Mi. Environmental Entomology* 30, 101–106.

Goggin, F.L., Shah, G., Williamson, V.M. and Ullman, D.E. (2004) Developmental regulation of *Mi*-mediated aphid resistance is independent of *Mi-1.2* transcript levels. *Molecular Plant–Microbe Interactions* 17, 532–536.

Goggin, F.L., Jia, L., Shah, G., Hebert, S., Williamson, V.M. and Ullman, D.E. (2006) Heterologous expression of the *Mi-1.2* gene from tomato confers resistance against nematodes but not aphids in eggplant. *Molecular Plant–Microbe Interactions* 19, 383–388.

Griffin, G.D. and McKenry, M.V. 1989. Susceptibility of Nevada synthetic XX germplasm to a California race of *Meloidogyne hapla. Journal of Nematology* 21, 292–293.

Harris, D.K., Boerma, H.R., Hussey, R.S. and Finnerty, S.L. (2003) Additional sources of soybean germplasm resistant to two species of root-knot nematode. *Crop Science* 43, 1848–1851.

Helms, D.M., Matthews, W.C., Temple, S.R. and Roberts, P.A. (2004) Registration of 'Cariblanco N' Lima bean. *Crop Science* 44, 352–353.

Ho, J.-Y., Weide, R., Ma, H.M., Wordragen, M.F., Lambert, K.N., Koornneef, M., Zabel, P. and Williamson, V.M. (1992) The root-knot nematode resistance gene (*Mi*) in tomato: construction of a molecular linkage map and identification of dominant cDNA markers in resistant genotypes. *The Plant Journal* 2, 971–982.

Hu, G., deHart, A.K.A., Li, Y., Ustach, C., Handley, V., Navarre, R., Hwang, C.F., Aegerter, B.J., Williamson, V.M. and Baker, B. (2005) *EDS1* in tomato is required for resistance mediated by TIR-class R genes and the receptor-like R gene *Ve. The Plant Journal* 42, 376–391.

Huang, G., Allen, R., Davis, E.L., Baum, T.J. and Hussey, R.S. (2006) Engineering broad root-knot resistance in transgenic plants by RNAi silencing of a conserved and essential root-knot nematode parasitism gene. *Proceedings of the National Academy of Sciences of the USA* 103, 14302–14306.

Hückelhoven, R. (2007) Cell wall-associated mechanisms of disease resistance and susceptibility. *Annual Review of Phytopathology* 45, 101–127.

Hussey, R.S. and Janssen, G.J.W. (2002) Root-knot nematodes: *Meloidogyne* species. In: Starr, J.L., Cook, R. and Bridge, J. (eds) *Plant Resistance to Parasitic Nematodes.* CAB Interntional, Wallingford, UK, pp. 43–70.

Hwang, C.F. and Williamson, V.M. (2003) Leucine-rich repeat-mediated intramolecular interactions in nematode recognition and cell death signaling by the tomato resistance protein Mi. *The Plant Journal* 34, 585–593.

Hwang, C., Bhakta, A., Truesdell, G., Pudlo, W. and Williamson, V.M. (2000) Evidence for a role of the N terminus and leucine-rich repeat region of the *Mi* gene product in regulation of localized cell death. *Plant Cell* 12, 1319–1329.

Jablonska, B., Ammiraju, J.S.S., Bhattarai, K.K., Mantelin, S., Martinez de Ilarduya, O., Roberts, P.A. and Kaloshian, I. (2007) The *Mi-9* gene from *Solanum arcanum* conferring heat-stable resistance to root-knot nematodes is a homologue of *Mi-1. Plant Physiology* 143, 1044–1054.

Jacquet, M., Bongiovanni, M., Martinez, M., Verschave, P., Wajnberg, E. and Castagnone-Sereno, P. (2005) Variation in resistance to the root-knot nematode *Meloidogyne incognita* in tomato genotypes bearing the *Mi* gene. *Plant Pathology* 54, 93–99.

Janssen, G.J.W., Van Norel, A., Janssen, R. and Hoogendoorn, J. (1997) Dominant and additive resistance to the root-knot nematodes *Meloidogyne chitwoodi* and *M. fallax* in Central American *Solanum* species. *Theoretical and Applied Genetics* 94, 692–700.

Jarquin-Barberena, H., Dalmasso, A., de Guiran, G. and Cardin, M. (1991) Acquired virulence in the plant parasitic nematode *Meloidogyne incognita.* I. Biological analysis of the phenomenon. *Revue de Nématologie* 14, 261–275.

Jia, Y., McAdams, S.A., Bryan, G.T., Hershey, H.P. and Valent, B. (2000) Direct interaction of resistance gene and avirulence gene products confers rice blast resistance. *EMBO Journal* 19, 4004–4014.

Jones, A. and Dukes, P.D. (1980) Heritabilities of sweet potato resistance to root-knot caused by *Meloidogyne incognita* and *M. javanica. Journal of the American Society for Horticultural Science* 105, 154–156.

Jones, J.D.G. and Dangl, J.L. (2006) The plant immune system. *Nature* 444, 323–329.

Kaloshian, I., Roberts, P.A., Waines, J.G. and Thomason, I.J. (1990) Inheritance of resistance to root-knot nematodes in *Aegilops squarrosa* L. *Journal of Heredity* 81, 170–172.

Kaloshian, I., Williamson, V.M., Miyao, G., Lawn, D.A. and Westerdahl, B.B. (1996) 'Resistance-breaking' nematodes in California tomatoes. *California Agriculture* 50, 18–19.

Kaloshian, I., Yaghoobi, J., Liharska, T., Hontelez, J., Hanson, D., Hogan, P., Jesse, T., Wijbrandi, J., Simons, G., Vos, P., Zabel, P. and Williamson, V.M. (1998) Genetic and physical localization of the root-knot nematode resistance locus *Mi* in tomato. *Molecular Genetics and Genomics* 257, 376–385.

Kaplan, D.T., Thomason, I.J. and Van Gundy, S.D. (1979) Histological study of compatible and incompatible interaction of soybeans and *Meloidogyne incognita. Journal of Nematology* 11, 338–343.

Kouassi, A.B., Kerlan, M.-C., Caromel, B., Dantec, J.-P., Fouville, D., Manzanares-Dauleux, M., Ellisseche, D. and Mugniery, D. (2006) A major gene mapped on chromosome XII is the main factor of a quantitatively inherited resistance to *Meloidogyne fallax* in *Solanum sparsipilum. Theoretical and Applied Genetics* 112, 699–707.

Lasserre, F., Gigault, F., Gauthier, J.P., Henry, J.P., Sandmeier, M. and Rivoal, R. (1996) Genetic variation in natural populations of the cereal cyst nematode (*Heterodera avenae* Woll.) submitted to resistant and susceptible cultivars of cereals. *Theoretical and Applied Genetics* 93, 1–8.

Li, Q., Xie, Q.G., Smith-Becker, J., Navarre, D.A. and Kaloshian, I. (2006) Mi-1-mediated aphid resistance involves salicylic acid and mitogen-activated protein kinase signaling cascades. *Molecular Plant–Microbe Interactions* 19, 655–664.

Li, Z., Jakkula, L., Hussey, R.S., Tamulonis, J.P. and Booerma, H.R. (2001) SSR mapping and confirmation of the QTL from PI96354 conditioning soybean resistance to southern root-knot nematode. *Theoretical and Applied Genetics* 103, 1167–1173.

Liu, Q.L. and Williamson, V.M. (2006) Host-specific pathogenicity and genome differences between inbred strains of *Meloidogyne hapla. Journal of Nematology* 38, 158–164.

Liu, Q.L., Thomas, V.P. and Williamson, V.M. (2007) Meiotic parthenogenesis in a root-knot nematode results in rapid genomic homozygosity. *Genetics* 176, 1483–1490.

López-Pérez, J.A., Le Strange, M., Kaloshian, I. and Ploeg, A.T. (2006) Differential response of *Mi* gene-resistant tomato rootstocks to root-knot nematodes (*Meloidogyne incognita*). *Crop Protection* 25, 382–388.

Martin, G.B., Bogdanove, A.J. and Sessa, G. (2003) Understanding the functions of plant disease resistance genes. *Annual Review of Plant Biology* 54, 23–61.

Martinez de Ilarduya, O. and Kaloshian, I. (2001) *Mi-1.2* transcripts accumulate ubiquitously in root-knot nematode resistant *Lycopersicon esculentum. Journal of Nematology* 33, 116–120.

Martinez de Ilarduya, O., Moore, A.E. and Kaloshian, I. (2001) The tomato *Rme1* locus is required for *Mi-1*-mediated resistance to root-knot nematodes and the potato aphid. *The Plant Journal* 27, 417–425.

Melillo, M.T., Leonetti, P., Bongiovanni, M., Castagnone-Sereno, P. and Bleve-Zacheo, T. (2006) Modulation of reactive oxygen species activity and H_2O_2 accumulation during compatible and incompatible tomato–root-knot nematode interactions. *New Phytologist* 170, 501–512.

Mercer, C.F., Hussain, S.W. and Moore, K.K. (2004) Resistance reactions to *Meloidogyne trifoliophila* in *Trifolium repens* and *T. semipilosum. Journal of Nematology* 36, 499–504.

Milligan, S., Bodeau, J., Yaghoobi, J., Kaloshian, I., Zabel, P. and Williamson, V.M. (1998) The root knot nematode resistance gene *Mi* from tomato is a member of the leucine zipper, nucleotide binding, leucine-rich repeat family of plant genes. *Plant Cell* 10, 1307–1319.

Mojtahedi, H., Brown, C.R., Riga, E. and Zhang, L.H. (2007) A new pathotype of *Meloidogyne chitwoodi* Race 1 from Washington State. *Plant Disease* 91, 1051.

Molinari, S. (2001) Inhibition of H_2O_2-degrading enzymes in the response of *Mi*-bearing tomato to root-knot nematodes and salicylic acid treatment. *Nematologica Mediterranea* 29, 235–239.

Neveu, C., Jaubert, S., Abad, P. and Castagnone-Sereno, P. (2003) A set of genes differentially expressed between avirulent and virulent *Meloidogyne incognita* near-isogenic lines encode secreted proteins. *Molecular Plant–Microbe Interactions* 16, 1077–1084.

Nombela, G., Williamson, V.M. and Muniz, M. (2003) The root-knot nematode resistance gene *Mi-1.2* of tomato is responsible for resistance against the whitefly *Bemisia tabaci. Molecular Plant–Microbe Interactions* 16, 645–649.

Omwega, C.O. and Roberts, P.A. (1992) Inheritance of resistance to *Meloidogyne* spp. in common bean and the genetic basis of its sensitivity to temperature. *Theoretical and Applied Genetics* 83, 720–726.

Opperman, C.H., Bird, D.M., Williamson, V.M., Rokhsar, D.S., Burke, M., Cohn, J., Cromer, J., Diener, S., Gajan, J., Graham, S., Houfek, T.D., Liu, Q., Mitros, T., Schaff, J., Schaffer, R., Scholl, E., Sosinski,

B.R., Thomas, V.P. and Windham, E. (2008) Sequence and genetic map of *Meloidogyne hapla*: a compact nematode genome for plant parasitism. *Proceedings of the National Academy of Sciences of the USA* 105, 14802–14807.

Ouédraogo, J.T., Gowda, B.S., Jean, M., Close, T.J., Ehlers, J.D., Hall, A.E., Gillaspie, A.G., Roberts, P.A., Ismail, A.M., Bruening, G., Gepts, P., Timko, M.P. and Belzile F.J. (2002) An improved genetic linkage map for cowpea (*Vigna unguiculata* L.) combining AFLP, RFLP, RAPD, biochemical markers and biological resistance traits. *Genome* 45, 175–188.

Parella, G., Moretti, A., Gognalons, P., Lesage, M.-L., Marchoux, G., Gebre-Selassie, K. and Caranta, C. (2004) The *Am* gene controlling resistance to *Alfalfa mosaic virus* in tomato is located in the cluster of dominant resistance genes on chromosome 6. *Phytopathology* 94, 345–350.

Paulson, R.E. and Webster, J.M. (1972) Ultrastructure of the hypersensitive reaction in roots of tomato, *Lycopersicon esculentum* L., to infection by the root-knot nematode, *Meloidogyne incognita. Physiological Plant Pathology* 2, 227–234.

Pegard, A., Brizzard, G., Fazari, A., Soucaze, O., Abad, P. and Djian-Caporalino, C. (2005) Histological characterization of resistance to different root-knot nematode species related to phenolics accumulation in *Capsicum annuum. Phytopathology* 95, 158–165.

Petrillo, M.D. and Roberts, P.A. (2005a). Isofemale line analysis of *Meloidogyne incognita* virulence to cowpea resistance gene *Rk. Journal of Nematology* 37, 448–456.

Petrillo, M.D. and Roberts, P.A. (2005b) Fitness of virulent *Meloidogyne incognita* isolates on susceptible and resistant cowpea. *Journal of Nematology* 37, 457–466.

Petrillo, M.D., Matthews, W.C. and Roberts, P.A. (2006) Dynamics of *Meloidogyne incognita* virulence to resistance genes *Rk* and Rk^2 in cowpea. *Journal of Nematology* 38, 90–96.

Potenza, C., Thomas, S.H., Higgins, E.A. and Sengupta-Gopalan, C. (1996) Early root response to *Meloidogyne incognita* in resistant and susceptible alfalfa cultivars. *Journal of Nematology* 28, 475–484.

Potenza, C., Thomas, S.H. and Sengupta-Gopalan, C. (2001) Genes induced during early response to *Meloidogyne incognita* in roots of resistant and susceptible alfalfa cultivars. *Plant Science* 161, 289–299.

Riggs, R.D. and Winstead, N.N. (1959) Studies on resistance in tomato to root-knot nematodes and on occurrence of pathogenic biotypes. *Phytopathology* 49, 716–724.

Roberts, P.A. (1982) Plant resistance in nematode pest management. *Journal of Nematology* 14, 24–32.

Roberts, P.A. (1992) Current status of the availability, development, and use of host plant resistance to nematodes. *Journal of Nematology* 24, 213–227.

Roberts, P.A. (1995) Conceptual and practical aspects of variability in root-knot nematodes related to host plant resistance. *Annual Review of Phytopathology* 33, 199–221.

Roberts, P.A. (2002) Concepts and consequences of resistance. In: Starr, J.L., Cook, R. and Bridge, J. (eds) *Plant Resistance to Parasitic Nematodes.* CAB International, Wallingford, UK, pp. 23–41.

Roberts, P.A. (2004) Nematodes and host resistance. In: Goodman, R.M. (ed.) *Encyclopedia of Plant and Crop Science.* Marcel Dekker, New York, pp. 805–808.

Roberts, P.A. and Scheuerman, R.W. (1984) Field evaluation of sweet potato clones and cultivars for reaction to root-knot and stubby root nematodes in California. *Hortscience* 19, 270–273.

Roberts, P.A., Matthews, W.C. and Ehlers, J.D. (1996) New resistance to virulent root-knot nematodes linked to the *Rk* locus of cowpea. *Crop Science* 36, 889–894.

Roberts, P.A., Matthews, W.C. and Veremis, J.C. (1998) Genetic mechanisms of host-plant resistance to nematodes. Plant–nematode interactions. In: Barker, K.R., Pederson, G.A. and Windham, G.L. (eds) *Agronomy Monograph.* ASA, CSSA and SSSA, Madison, Wisconsin, pp. 209–238.

Roberts, P.A., Matthews, W.C., Ehlers, J.D. and Helms, D. (2008) Genetic determinants of differential resistance to root-knot nematode reproduction and galling in lima bean. *Crop Science* 48, 553–561.

Rossi, M., Goggin, F., Milligan, S.B., Kaloshian, I., Ullman, D. and Williamson, V.M. (1998) The nematode resistance gene *Mi* of tomato confers resistance against the potato aphid. *Proceedings of the National Academy of Sciences of the USA* 95, 9750–9754.

Sasser, J.N. (1977) Worldwide dissemination and importance of the root-knot nematodes, *Meloidogyne* spp. *Journal of Nematology* 9, 26–29.

Sawhney, R. and Webster, J.M. (1979) The influence of some metabolic inhibitors on the response of susceptible/resistant cultivars of tomato to *Meloidogyne incognita. Nematologica* 25, 86–93.

Schaff, J.E., Nielsen, D.M., Smith, C.P., Scholl, E.H. and Bird, D.M. (2007) Comprehensive transcriptome profiling in tomato reveals a role for glycosyltransferase in *Mi*-mediated nematode resistance. *Plant Physiology* 144, 1079–1092.

Seah, S., Yaghoobi, J., Rossi, M., Gleason, C.A. and Williamson, V.M. (2004) The nematode-resistance gene, *Mi-1*, is associated with an inverted chromosomal segment in susceptible compared to resistant tomato. *Theoretical and Applied Genetics* 108, 1635–1642.

Seah, S., Telleen, A.C. and Williamson, V.M. (2007) Introgressed and endogenous *Mi-1* gene clusters in tomato differ by complex rearrangements in flanking sequences and show sequence exchange and diversifying selection among homologues. *Theoretical and Applied Genetics* 114, 1289–1302.

Semblat, J.P., Rosso, M.N., Hussey, R.S., Abad, P. and Castagnone-Sereno, P. (2001) Molecular cloning of a cDNA encoding an amphid-secreted putative avirulence protein from the root-knot nematode *Meloidogyne incognita. Molecular Plant–Microbe Interactions* 14, 72–79.

Shepherd, R.L. (1974) Transgressive segregation for root-knot nematode resistance in cotton. *Crop Science* 14, 872–875.

Shepherd, R.L. (1979) A quantitative technique for evaluating cotton for root-knot nematode resistance. *Phytopathology* 69, 427–430.

Simon, P.W., Matthews, W.C. and Roberts, P.A. (2000) Evidence for a simply inherited dominant resistance to *Meloidogyne javanica* in carrot. *Theoretical and Applied Genetics* 100, 735–742.

Smith, P.G. (1944) Embryo culture of a tomato species hybrid. *Proceedings of the American Society of Horticultural Science* 44, 413–416.

Starr, J.L. and Roberts, P.A. (2004) Resistance to plant-parasitic nematodes. In: Chen, Z.X., Chen, S.Y. and Dickson, D.W. (eds) *Nematology, Advances and Perspectives. Vol. 2. Nematode Management and Utilization.* CAB International, Wallingford, UK, pp. 879–907.

Starr, J.L., Bridge, J. and Cook, R. (2002) Resistance to plant-parasitic nematodes: history, current use and future potential. In: Starr, J.L., Cook, R. and Bridge, J. (eds) *Plant Resistance to Parasitic Nematodes.* CAB International, Wallingford, UK, pp. 1–22.

Starr, J.L., Koenning, S.R., Kirkpatrick, T.L., Robinson, A.F., Roberts, P.A. and Nichols, R.L. (2007) The future of nematode management in cotton. *Journal of Nematology* 39, 283–294.

Tameling, W.I.L., Elzinga, S.D.J., Darmin, P.S. *et al.* (2002) The tomato R gene products I-2 and Mi-1 are functional ATP binding proteins with ATPase activity. *Plant Cell* 14, 2929–2939.

Tamulonis, J.P., Luzzi, B.M. and Hussey, R.S. (1997a) DNA marker associated with resistance to Javanese root-knot nematode in soybean. *Crop Science* 37, 783–788.

Tamulonis, J.P., Luzzi, B.M., Hussey, R.S., Parrott, W.A. and Booerma, H.R. (1997b) DNA marker analysis of loci conferring resistance to peanut root-knot nematode in soybean. *Theoretical and Applied Genetics* 95, 664–670.

Thoquet, P., Oliver, J., Sperisen, C., Rogowsky, P., Laterrot, H. and Grimsley, N. (1996) Quantitative trait loci determining resistance to bacterial wilt in tomato cultivar Hawaii. *Molecular Plant–Microbe Interactions* 9, 826–836.

Triantaphyllou, A.C. (1985) Cytogenetics, cytotaxonomy and phylogeny of root-knot nematodes. In: Sasser, J. and Carter, C. (eds) *An Advanced Treatise on* Meloidogyne, Vol. I. North Carolina State University Graphics, Raleigh, North Carolina, pp. 113–126.

Trudgill, D.L. (1991) Resistance to and tolerance of plant parasitic nematodes in plants. *Annual Review of Phytopathology* 29,167–193.

Trudgill, D.L. and Blok, V.C. (2001) Apomictic, polyphagous root-knot nematodes: exceptionally successful and damaging biotrophic root pathogens. *Annual Review of Phytopathology* 39, 53–77.

Turner, S.J. (1990) The identification and fitness of virulent potato cyst-nematode populations (*Globodera pallida*) selected on resistant *Solanum vernei* hybrids for up to eleven generations. *Annals of Applied Biology* 117, 385–397.

Tzortzakakis, E.A. and Blok, V.C. (2007) Differentiation in two populations of *Meloidogyne incognita* from Greece in relation to reproduction on resistant tomato and pepper. *Journal of Plant Diseases and Protection* 114, 276–277.

Tzortzakakis, E.A. and Gowen, S.R. (1996) Occurrence of a resistance breaking pathotype of *Meloidogyne javanica* on tomatoes in Crete, Greece. *Fundamental and Applied Nematology* 19, 283–288.

van der Beek, J.G., Poleij, L.M., Janssen, R. and Janssen, G.J.W. (1998) Variation in virulence within *Meloidognye chitwoodi, M. fallax* and *M. hapla* on *Solanum* spp. *Phytopathology* 88, 658–665.

van der Vossen, E.A.G., Gros, J., Sikkema, A., Muskens, M., Wouters, D., Wolters, P., Pereira, A. and Allefs, S. (2005) The *Rpi-blb2* gene from *Solanum bulbocastanum* is an *Mi-1* gene homolog conferring broad-spectrum late blight resistance in potato. *The Plant Journal* 44, 208–222.

van Ooijen, G., van den Burg, H.A., Cornelissen, B.J.C. and Takken, F.L.W. (2007) Structure and function of resistance proteins in solanaceous plants. *Annual Review of Phytopathology* 45, 43–72.

Veremis, J.C. and Roberts, P.A. (1996) Relationships between *Meloidogyne incognita* resistance genes in *Lycopersicon peruvianum* differentiated by heat sensitivity and nematode virulence. *Theoretical and Applied Genetics* 93, 950–959.

Vos, P., Simons, G., Jesse, T., Wijbrandi, J., Heinen, L., Hogers, R., Frijters, A., Groenendijk, J., Diergaarde, P., Reijans, M., Fierens-Onstenk, J., de Both, M., Peleman, J., Liharska, T., Hontelez, J. and Zabeau, M. (1998) The tomato *Mi-1* gene confers resistance to both root-knot nematodes and potato aphids. *Nature Biotechnology* 16, 1365–1369.

Wang, C., Matthews, W.C. and Roberts, P.A. (2006a) Phenotypic expression of *rkn1*-mediated *Meloidogyne incognita* resistance in *Gossypium hirsutum* populations. *Journal of Nematology* 38, 250–257.

Wang, C., Ulloa, M. and Roberts, P.A. (2006b) Identification and mapping of microsatellite markers linked to a root-knot nematode resistance gene (*rkn1*) in Acala NemX cotton (*Gossypium hirsutum* L.). *Theoretical and Applied Genetics* 112, 770–777.

Wang, C., Ulloa, M. and Roberts, P.A. (2008) A transgressive segregation factor (*RKN2*) in *Gossypium barbadense* for nematode resistance clusters with gene *rkn1* in *G. hirsutum*. *Molecular Genetics and Genomics* 279, 41–52.

Williamson, V.M. (1998) Root-knot nematode resistance genes in tomato and their potential for future use. *Annual Review of Phytopathology* 36, 277–293.

Williamson, V.M. and Hussey, R.S. (1996) Nematode pathogenesis and resistance in plants. *Plant Cell* 8, 1735–1745.

Williamson, V.M. and Kumar, A. (2006) Nematode resistance in plants: the battle underground. *Trends in Genetics* 22, 396–403.

Yaghoobi, J., Kaloshian, I., Wen, Y. and Williamson, V.M. (1995) Mapping a new nematode resistance locus in *Lycopersicon peruvianum*. *Theoretical and Applied Genetics* 91, 457– 464.

Yi, Y.-H., Rufty, R.C., Wernsman, E.A. and Conkling, M.C. (1998) Mapping the root-knot nematode resistance gene (*Rk*) in tobacco with RAPD markers. *Plant Disease* 82, 1319–1322.

Ynturi, P., Jenkins, J.N., McCarty, J.C. Jr, Gutierrez, O.A. and Saha, S. (2006) Association of root-knot nematode resistance genes with simple sequence repeat markers on two chromosomes in cotton. *Crop Science* 46, 2670–2674.

Yu, M.H., Heijbroek, W. and Pakish, L.M. (1999) The sea beet source of resistance to multiple species of root-knot nematode. *Euphytica* 108, 151–155.

Yu, M.H., Pakish, L.M. and Zhou, H. (2001) An isozyme marker for resistance to root-knot nematode in sugarbeet. *Crop Science* 41, 1051–1053.

Zacheo, G., Orlando, C., and Bleve-Zacheo, T. (1993) Characterization of anionic peroxidases in tomato isolines infected by *Meloidogyne incognita*. *Journal of Nematology* 25, 249–256.

Zamir, D., Ekstein-Michelson, I., Zakay, Y., Navot, N., Zeidan, M., Sarfatti, M., Eshed, Y., Harel, E., Pleban, T., van-Oss, H., Kedar, N., Rabinowitch, H.D. and Czosnek, H. (1994) Mapping and introgression of a *tomato yellow leaf curl virus* tolerance gene, *TY-1*. *Theoretical and Applied Genetics* 88, 141–146.

14 Development of Resistant Varieties

James L. Starr[1] and Chris F. Mercer[2]
[1]*Texas A&M University, College Station, Texas, USA;* [2]*AgResearch Grasslands, Palmerston North, New Zealand*

14.1 Introduction – the Plus Side of Resistance	**326**
14.2 Introduction – a Look at the Other Side	**326**
14.3 Successful Use of Resistance – Room for Wider Deployment	**327**
14.4 Planning a Resistance-breeding Programme	**328**
14.5 Screening Methods, Including Marker-assisted Selection	**331**
14.6 Quality of Candidate Resistant Material	**333**
14.7 Engineered Resistance	**334**
14.8 Conclusions and Future Directions	**335**
14.9 References	**335**

14.1 Introduction – the Plus Side of Resistance

Many consider resistance, when available, to be the best option for nematode management because of its cost effectiveness, because resistance is typically compatible with other management tactics, and because resistance is environmentally benign (Starr *et al.*, 2002). Resistance to nematodes is most often defined based on the effects on nematode reproduction, with resistant plants supporting lower levels of reproduction than reproduction on susceptible plant genotypes (Roberts, 2002; Cook and Starr, 2006). Over time, this suppression of nematode reproduction leads to lower nematode population densities, with obvious long-term benefits in an agricultural system that encompasses crops susceptible to the nematode species of concern (Plates 19, 28, 29, 35). Thus, resistance can be a powerful tool for the management of root-knot nematodes. Host resistance is generally viewed as a management tool that can be used without an increase in production costs because seeds of the resistant varieties are typically priced the same as susceptible varieties (although this might not be the case for resistance based on a transgene). The widespread distribution of several *Meloidogyne* species and their general aggressiveness in terms of crop damage further justifies the use of resistance as a management tactic.

14.2 Introduction – a Look at the Other Side

Even when resistant varieties are available, host resistance has characteristics that can limit its usefulness. For example, resistance is typically species specific and often race specific. Even though some sources of resistance to *Meloidogyne* spp., such as the *Mi* gene in tomato (Williamson, 1998) and resistance to *M. arenaria* and *M. javanica* in groundnut (Simpson *et al.*, 2003), are effective against more than one species, most resistance is effective against only a single *Meloidogyne* species.

 ©CAB International 2009. *Root-knot Nematodes* (eds R.N. Perry, M. Moens and J.L. Starr)

Additionally, a root-knot-resistant variety is likely to be susceptible to most other nematode genera. Lack of durability due to the increased frequency of virulent nematode populations is another potential limitation. Luckily, the frequency of nematode populations with virulence on specific resistance genes has not yet been a major problem with *Meloidogyne* spp., as it has been with *Heterodera* and *Globodera* spp. (Cook and Noel, 2002). None the less, one should expect that with widespread and frequent deployment of any one resistance gene that there will be an increased probability of the development of nematode populations virulent on that gene. In at least one case, five successive crops of tomato with the *Mi* gene were sufficient for the development of a virulent population of *M. incognita* (Noling, 2000).

The occurrence of virulent nematode populations need not render that particular resistance gene completely ineffective. With care, the distribution of a virulent population can be restricted to a few fields, and the rate of spread of the virulent population limited. Several gene-deployment strategies can also be used to prolong the usefulness of the resistance gene. Such strategies would include only occasional use of resistance in a cropping system that includes rotation to both non-host crops and occasional use of susceptible varieties. Virulent individuals will have a selective advantage over avirulent individuals when the resistant variety is grown, but not when a susceptible variety is grown. Therefore, the occasional use of susceptible crops in the cropping system will slow the rate at which the virulent individuals increase as a proportion of the total population in the field.

Another major limitation with resistance can be yield potential. Unfortunately, nematologists often place greatest emphasis on the effects of host resistance on nematode population densities rather than on crop yield potential. By contrast, it is yield and profitability that are of primary concern for the grower. Resistance suppresses nematode reproduction rates, so most resistant plants have higher yield potentials in fields infested with the target *Meloidogyne* species than do susceptible varieties in infested fields. As an example, the resistant Acala NemX cotton had lint yields that were twofold greater than a susceptible variety in cotton fields with moderate to high initial densities of *M. incognita* (Ogallo *et al.*, 1997). However, the higher yield potential of resistant crops in nematode-infested fields is not sufficient to guarantee success of a resistant variety. Nematode-resistant varieties, especially if the source of resistance was an unadapted accession of the crop or a related species, usually will not have the yield potential of the best-yielding varieties currently available. This 'yield drag', which is often associated with newly developed sources of resistance, is a serious impediment to widespread adoption of a resistant variety. In most developed countries, new crop varieties are adopted by growers only after extensive yield testing by agronomists in fields not infested with damaging population densities of nematodes. Unless a resistant variety performs well under these conditions it gets little attention from the agronomists or from potential producers. In a worst-case scenario, if the resulting market for the resistant variety is viewed as limited only to those fields that are infested with the *Meloidogyne* species of concern, then large-scale seed producers are less likely to invest resources in the production and marketing of a resistant variety. The end result is limited (or no) availability of a useful tool for nematode management. The importance of developing resistant varieties that are competitive with the best varieties being grown in terms of yield and other agronomically or horticulturally important traits cannot be overemphasized.

14.3 Successful Use of Resistance – Room for Wider Deployment

There are numerous examples of where crop varieties bred specifically for resistance to one or more *Meloidogyne* spp. have been widely adopted by growers, and have made important contributions to nematode management. The *Mi* gene, which confers resistance to three species (*M. arenaria*, *M. incognita* and *M. javanica*; see Williamson and Roberts, Chapter 13, this volume) has been introgressed into numerous tomato varieties. Its use was shown to return an additional profit of US$5100/ha (€8000/ha) over the use of methyl bromide in the production of tomatoes in plastic houses that were infested with *M. incognita* (Sorribas *et al.*, 2005). Resistance to *M. incognita* has been widely used in tobacco and peach, although in the case of peach the use of the resistant rootstock Nemaguard was limited by its susceptibility to the peach tree short life syndrome (Nyczepir and Halbrendt, 1993; see Nyczepir and Thomas, Chapter 18, this volume). More recently developed peach varieties with resistance to *M. incognita* and *M. javanica*, such as Nemared and Okinawa, are less susceptible to the peach tree short life syndrome. The use of resistant rootstocks for annual crops is being investigated and used in some special cases (Cohen *et al.*, 2007); susceptible cucurbit (Siguenza *et al.*, 2005) and pepper scions (Oka *et al.*, 2004) grafted on to resistant rootstocks show promise for nematode management in some intensive production systems. In lucerne, resistance to *M. hapla* has been important for improved productivity (Cook and Yeates, 1993).

In other cases, introgression of resistance into high-yielding varieties has been relatively recent, and thus the resistance is still limited to a few varieties and has not yet had a major impact. Resistance to *M. arenaria* and *M. javanica* in groundnut (Simpson and Starr, 2001; Simpson *et al.*, 2003) and resistance to *M. arenaria*, *M. incognita* (Plate 28) and *M. javanica* in pepper (Thies and Fery, 2001; Thies *et al.*, 2003) are examples of more recently developed resistance that have not yet had a major impact on nematode management in these crops. Unfortunately, there are also cases where resistance, although known to exist for many years, has not been incorporated into crop varieties that are widely grown. In the case of cotton, excellent sources of resistance were described in the 1970s (Shepherd, 1974), but to date only a limited number of resistant varieties are available to growers.

There are numerous reasons why greater emphasis has not been given to the development of crop varieties with resistance to root-knot nematodes. Historically, and continuing with some crops, there has been a lack of appreciation for the total economic losses attributable to nematodes. As an example, in 1989 a horticulturalist in Botswana who was working with an international bean and cowpea improvement project was complaining about the difficulty of collecting descriptive data on a collection of cowpea accessions because local fields were infested with a *Meloidogyne* sp. that was causing severe damage. When asked why, if the nematodes were causing such damage, the project did not include screening for nematode resistance, the horticulturalist replied that nematodes were not sufficiently important to warrant the effort! (Starr, personal observation.)

The ready availability of highly effective nematicides in the 1960s and 1970s also limited the interest in developing genetic resistance (which is typically a long-term effort). With some crops, identification of useful sources of resistance has been difficult, although successful identification of resistant accessions is likely to be correlated with the effort given to such searches. It should be noted that in the volumes on nematodes affecting subtropical and tropical (Luc *et al.*, 2005), and temperate (Evans *et al.*, 1993) crops, potentially useful sources of resistance were identified for nearly every crop discussed.

It may require a major effort over a period of years for the community of scientists, crop consultants and growers of a particular crop to appreciate fully the importance of nematode parasitism to crop productivity. In the USA, in an effort to sell more nematicides, a major agrochemical company sponsored a multi-year, multistate effort to document the importance of nematodes as pathogens of cotton. This extensive survey, when coupled with a reduced importance of insect pests because of successful boll weevil eradication programmes, and the development of transgenic cottons that are resistant to lepidopteran pests, has resulted in a much greater endeavour to develop nematode-resistant cotton varieties (Starr *et al.*, 2007).

14.4 Planning a Resistance-breeding Programme

When is a resistance-breeding programme appropriate? Development of a nematode-resistant variety is typically a long-term effort, and thus costly, so initiation of a breeding effort should begin with careful planning. An example of how long may be needed is the case of resistance to *M. arenaria* in groundnut. The initial interspecific crosses were made in the 1970s, resistance in the germplasm collection was identified in 1987 (Nelson *et al.*, 1989), and the first variety released only in 2001 (Simpson and Starr, 2001). Because of the time required to develop a commercially viable resistant variety, one needs to be sure that the problem being tackled is sufficient to justify this effort. Thus breeding for resistance to *Meloidogyne* spp. may not be practical for some crops that are grown on a small area, or where the yield losses due to nematode parasitism are relatively low compared with other more conspicuous pests. Roberts (1990) emphasized the importance of defining the problem from social, economic or environmental points of view before starting a breeding programme. Fassuliotis (1979, 1985) listed seven areas of responsibility for a nematologist in a breeding programme, and these are still valid; they are detailed below.

14.4.1 Identification of the root-knot nematode species present

Identification of the species present in the area in which the new variety is to be grown is important, as it allows the searching of records for existing resistant germplasm. Identification of all root-knot nematode species present in the area allows surveys of distribution and importance, so that resistance-breeding priorities can be established.

14.4.2 Establishing pure cultures

Pure culture of the target nematode for the resistance-breeding programme is preferred over the mixed-inoculum approach. One step (one nematode species) at a time is the method favoured by nematologists. By screening plants against a mixed population of root-knot nematodes, valuable germplasm with resistance to only one species may be discarded. Pure cultures are best established from a single egg mass, with concurrent identification of the corresponding female. However, starting a pure species culture with more than one identified root-knot nematode egg mass would maintain more of the genetic variability encountered in field populations (Hartman and Sasser, 1985). If one is working with a crop–*Meloidogyne* species system for which virulence to a specific resistance gene is known but that resistance gene is still of value because the virulence is of limited distribution, then one must screen segregating plant populations with a confirmed avirulent isolate of the nematode species. If one uses by mistake a virulent race of the nematode then all plants will be susceptible and still valuable resistance will be discarded. If, however, no virulence is known, then using cultures that were established by using several egg masses, each collected from separate fields, will help ensure the broadest possible diversity within the nematode inoculum. Increased diversity within the nematode population used for screening plants in the breeding programme will increase the probability that the resulting resistant variety will be applicable over the widest possible geographic region (Hussey and Boerman, 1981)

Another question that is frequently overlooked is: How long should a given nematode isolate be maintained in culture? Despite attempts to begin with one or more populations derived from several individual isolates, with time in culture there will be some genetic drift, and one should expect a reduction in diversity of the population. Indeed, at least one report documents how maintenance of *M. hapla* on different susceptible tomato hosts has caused a change in virulence towards resistance in *Solanum* spp. (van der Beek and Poleij, 2008). Therefore, stock cultures should be renewed periodically. This will require several experiments with new cultures to ensure that their behaviour is similar to the older cultures that were used previously in the breeding programme.

14.4.3 Nematode variability

Variability within a nematode species, based on virulence to specific resistance genes, will contribute to the breakdown of resistance, as virulent subpopulations (pathotypes or races) overcome the resistance and eventually dominate over the avirulent subpopulations that reproduce poorly or not at all on the resistant variety. As soon as resistant material is available, the target area of land should be surveyed for pathotypes to give an early indication of the presence of virulence that will limit the durability of the resistant material.

14.4.4 Screening methods

The choice of screening methods is important to the programme as success depends on the ability to screen large numbers of genotypes with optimal chances for infection. Glasshouse or field techniques have merits and limitations, depending on the budget, crop species or stage in the breeding programme (Kinloch, 1990). Glasshouse (or growth chamber) methods allow inoculation with the same number of eggs or juveniles per plant, with some control over environmental conditions, but at greater cost for use and maintenance of such facilities. The number of individuals that can be screened for resistance in a growth chamber in any one test is likely to be greatly reduced from the numbers that can be screened in glasshouse or field tests, which is a serious limitation to a breeding programme. Field methods have the advantage of subjecting plants to other biotic and abiotic factors, thus allowing selection for yield and other important agronomic or horticultural traits in addition to resistance. However, in the field plants are subject to more variable inoculum levels and possibly a mixture of *Meloidogyne* spp. The variation in nematode population densities increases the probability of susceptible plants being incorrectly identified as resistant. Details of several effective screening techniques have been covered by Fassuliotis (1985), Kinloch (1990) and Hussey and Janssen (2002).

14.4.5 Sources of resistance

There are two general sources of resistance: that found within the crop species and that found in closely related species (the secondary gene pool). The transfer of resistance is 'greatly simplified' (Fassuliotis, 1979) if the source of resistance can be found within varieties adapted to the locality where root-knot nematodes are presenting a problem. Simple pair-wise or polycrosses using such sources of resistance, followed by screening and possibly further crosses, will rapidly develop a new plant variety. However, many crops have a narrow genetic base, and a more intensive breeding effort may be needed to incorporate nematode resistance. Induced mutants, as from irradiation, and plant regenerates (somoclonal variants) may facilitate improved resistance within adapted germplasm. The secondary gene pool has played an important part in contributing genes for resistance in crop species. However, the first hybrids contain many undesirable traits, which must be eliminated by several rounds of backcrossing and screening.

Success of the effort to develop a resistant variety is likely to be highly dependent on the level of support from a plant breeder. In addition to the significance of the problem, the breeder will want to know how good is the source of resistance that will be used in the breeding programme, the heritability of resistance (if known), and impact of resistance on yield potential both in nematode-infested fields and non-infested fields. Other important considerations for the plant breeder will be whether the source of resistance is a currently grown variety (which is the most desirable), an old variety, an unimproved wild accession of the crop species, or a related wild species (which is the least desirable). The more distant genetically the source of resistance is from a modern variety, the greater the effort that will be required to introgress the resistance into such a plant type. The breeder will be helpful in determining what susceptible varieties or elite breeding lines are likely to be suitable parents into which the resistance can be introgressed. Unfortunately, the best-yielding variety available may not be the best choice as a parent in a breeding programme. Plant breeders, especially those working to develop inbred lines used for hybrid crops, have long recognized that different plant genotypes may differ in combining ability. General combining ability is a measure of how well a particular genotype combines with several other lines to produce superior progeny for the trait(s) of interest (Poehlman and Sleper, 1995). Plants may also differ in specific combining ability, which is the ability of that genotype to produce superior progeny when crossed with a single genotype of interest (Poehlman and Sleper, 1995). A knowledgeable plant breeder will be very helpful in making the important decisions on which crop genotypes should be used as parents in the breeding programme.

Resistance-breeding methodology deals with both binomial (qualitative) and continuous (quantitative) distribution in terms of trait inheritance. The mode of sexual reproduction – self- or cross-pollination – is also a significant determinant of breeding strategy. The simplest breeding programme to improve resistance is that presented by a self-pollinated crop with a single-gene resistance. Selection of one or more resistant parents to be crossed with susceptible parents is followed by selection of resistant individuals from among the progeny (usually the F_2 generation). The resistant individuals are selfed and subsequent generations screened for resistance, yield and other important traits. Both backcross and pedigree breeding strategies (Poehlman and Sleper, 1995) can be used to develop varieties with a resistant phenotype. The use of the backcross method limits the potential for genetic gain with respect to yield potential, so it is used primarily when the source of resistance is a poorly adapted genotype. If the source of resistance is a well-adapted genotype with high yield potential, then a pedigree strategy will be preferred. These breeding strategies have been used successfully for nematode resistance in a number of crops, including cotton (*M. incognita*), groundnut (*M. arenaria*), pepper (*M. incognita*), tobacco (*M. incognita*) and tomato (*M. arenaria*, *M. incognita* and *M. javanica*).

14.4.6 Mass selection

With outcrossing species, decisions on crossing procedures are usually made on a population basis, e.g. in mass selection the purpose is to increase the proportion of superior genotypes in a population. Here, the efficiency of the selection depends mainly on gene frequency and heritability. However, in mass selection, there is no control over the general combining ability of the plants that are intercrossed.

14.4.7 Recurrent selection

Recurrent selection is a breeding strategy in which superior plants are selected and propagated, all possible intercrosses are made, and the resulting intercross population serves as source material for future cycles of selection and crossing. In recurrent selection, the combining ability of the potential parental plants is determined by evaluation of the progeny. Parents that produce superior progeny are said to have good combining ability. These are intercrossed to produce the next population.

The advantage of recurrent selection is that it allows greater opportunity for recombination than does selection within selfed or inbred lines. If care is taken to keep the rate of inbreeding at a low level, then it is possible to maintain high genetic variability, and hence provide for effective selection over a long period.

The recurrent selection approach for polygenetically controlled resistance has been used successfully to improve resistance to *M. trifoliophila* in white clover (Mercer *et al.*, 2008). This legume is a significant contributor to New Zealand's grazed pasture systems, but its potential yield is rarely reached due to several constraints, including nematodes. In a recurrent selection programme that relied on screening in a glasshouse environment, large numbers of seeds were sown individually in pots and the soil infested with eggs of *M. trifoliophila*. After 4 weeks, soil was washed from the roots and galls per root system were counted. Individuals with the fewest galls, and a few with high counts, were then crossed and the progeny screened for resistance in a like manner. Over several generations, resistant selections exhibited reduced mean gall numbers as a proportion of the mean for susceptible material – in the fifth generation the resistant material had 34% of the number of galls on the susceptible selections. Resistant and susceptible material was crossed with susceptible varieties, and in the following two seasons, selections were made within the new outcrossed lines without further crosses with varieties. It is risky to compare the results between generations (seasons) due to environmental factors, so resistant and susceptible material from several generations was compared in one season. Over seven cycles of selection, the number of galls per g root dry weight decreased by 2.3% per cycle of selection, and there was no response to selection for susceptibility (Fig. 14.1) (Mercer *et al.*, 2008).

The relatively slow progress in improving the resistance of white clover to *M. trifoliophila* indicated that several recessive genes were conferring the resistance. The recessive nature was tested by screening the progeny of a diallel cross of resistant and susceptible genotypes. Numbers of galls on progeny of the crosses between resistant and susceptible material were higher than the average of the parents (Fig. 14.2), indicating that most of the genes for the trait are recessive in nature.

In this example, there were no highly resistant genotypes identified among the parents or early progenies. Resistance could only be improved by repeatedly recombining genes and selecting the most resistant progeny for further crosses (recombinations).

14.5 Screening Methods, Including Marker-assisted Selection

The plant breeder will assume that the nematologist can accurately identify resistant individuals from segregating populations, but will want to know how many plants the nematologist can

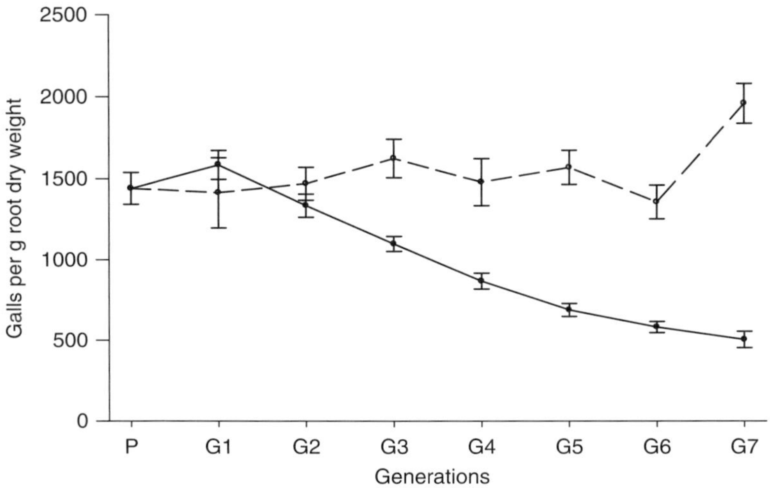

Fig. 14.1. Means of samples from each generation of the white clover recurrent selection programme improving resistance to *Meloidogyne trifoliophila* show the rate of separation of resistant (solid trace) from susceptible (dotted trace) lines. Bars indicate standard errors. (From Mercer *et al.*, 2008.)

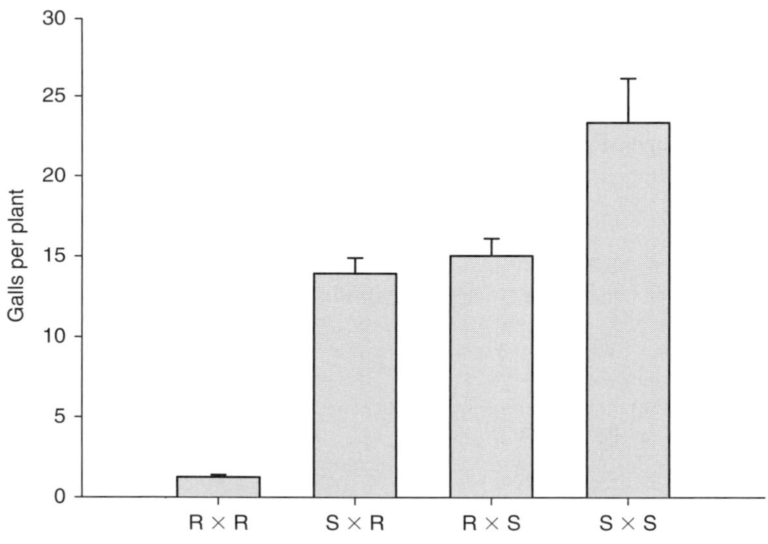

Fig. 14.2. Numbers of *Meloidogyne trifoliophila* galls per plant from progeny of crosses of four combinations of resistant (R) or susceptible (S) parents. Error bars indicate standard errors (5%). (From Mercer *et al.*, 2008.)

screen for the resistant phenotype in one growing season or one year. The expected answer to this question will be in the hundreds or even thousands. The more plants that can be evaluated in a given period of time, the more rapidly will the programme advance, and with a greater probability of timely success. Fortunately, root-galling indices often can be used for rapid identification of resistant individuals (Hussey and Janssen, 2002). Measures of nematode reproduction, such as numbers of egg masses per root system or eggs per g roots, are more precise estimates of resistance but are much more time consuming and costly. These more precise measures are justified in the later stages of a breeding programme, when greater accuracy is required than in the initial screening of large numbers of genotypes. Hussey and Janssen (2002) outlined several useful protocols for screening large numbers of plants.

Modern tools of molecular biology allow the use of molecular markers to identify resistant individuals from within segregating populations (Francia *et al.*, 2005). Co-dominant markers that allow one to distinguish an individual that is homozygous for resistance from a heterozygous individual are preferred to dominant markers that only identify the resistance phenotype. The ideal marker is one derived from the sequence of the resistance gene's open reading frame. The more common situation is the identification of a unique DNA sequence that is genetically linked to the resistance locus. If the marker is mapped to ≤5 cM from the resistance locus, then the efficiency of the identification of the resistance phenotype will be sufficient for use in most breeding programmes. Marker-assisted-selection techniques have the advantage of being relatively immune to environmental fluctuations and the problems encountered with regard to highly variable phenotypes, as is often the case when measuring nematode reproduction. Further, tissue samples for DNA extraction can be obtained from seedlings without sacrificing the entire plant. Breeders and nematologists using marker-assisted selection will still be required to confirm the resistance phenotype via a direct measurement of resistance at one or more times during development of the resistant variety. High-efficiency and high-throughput, DNA-based, marker-assisted selection protocols are becoming increasing common (Claverie *et al.*, 2004; Wang *et al.*, 2006; Chu *et al.*, 2007) and are likely to play an ever-increasing role in breeding for resistance to *Meloidogyne* spp. Hussey and Janssen (2002) listed seven crops for which molecular marker technologies are being used for breeding for resistance, or for which markers tightly linked to root-knot nematode resistance have been identified. The crops were tomato, soybean, groundnut, tobacco, potato, *Prunus* spp. and wheat. The list may now be extended to include cotton (Niu *et al.*, 2007), coffee (Dinitz *et al.*, 2005), carrot (Boiteux *et al.*, 2004), Kenya white clover (Barrett, *et al.*, 2005) and pepper (Dijian-Caporalino *et al.*, 2001).

14.6 Quality of Candidate Resistant Material

One important consideration for the breeding programme is the quality of the available resistance phenotypes. Fig. 14.3 illustrates a continuum of host responses, from highly susceptible to highly resistant (= immunity, where there is no detectable nematode reproduction on that plant genotype). One would like to have a resistance phenotype that is near immunity and highly heritable, i.e. governed by one or a few genes. However, any genotype that supports a level of resistance that is to the left of the susceptible standard on the continuum is, by definition, resistant relative to that susceptible standard and of some potential value. Most nematologists would be likely to adopt the division of this continuum into the categories of highly resistant, resistant, moderately resistant and susceptible. This arbitrary division of the plant response continuum is highly skewed and implies that low levels of resistance (that in the middle to right side of the scale)

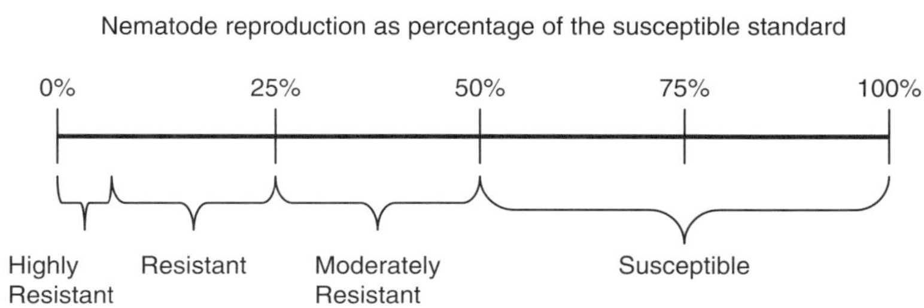

Fig. 14.3. A diagrammatic representation of the continuum of susceptibility and resistance to nematode reproduction within a crop germplasm pool.

are of little value, such as that which may suppress nematode reproduction by 25–50%. However, this may not always be the case. Reports in the literature suggest that when resistant and susceptible crop varieties are compared based on the Seinhorst model ($y = m + (1 - m)z^{P-T}$) (see Greco and Di Vito, Chapter 11, this volume), the minimum yield parameter (m) is often greater for the resistant genotype than for the susceptible genotype (DiVito *et al.*, 1992; Zhou and Starr, 2003). Thus, yield loss at any nematode density above the threshold density is less for the resistant variety than for a susceptible variety. Therefore, even low levels of resistance would be beneficial, although admittedly less than ideal, with respect to managing the nematode population density and protecting yield potential. Low levels of resistance would probably be a very useful component of a more complex management system, which might also include crop rotation, some use of nematicides, and possibly even biological control. Moderate levels of resistance in cotton have been shown to have a substantial impact on *M. incognita* population densities (Koenning *et al.*, 2001). In that report, glasshouse tests to compare nematode reproduction on several cotton varieties showed that the resistant cotton variety Auburn 634 RNR had a reproduction index of *c.* 0.2; two varieties with more moderate levels of resistance had reproduction indices of *c.* 2.0; and two susceptible varieties had indices of nearly 50. In multiple-location field tests, nematode population densities at crop harvest for the resistant cotton varieties were 25% of those for the susceptible varieties. Such reductions in nematode population densities will be beneficial to cropping systems that include susceptible crops.

14.7 Engineered Resistance

Naturally occurring resistance is conditioned by a variety of genes and involves a large number of complex plant responses (Fuller *et al.*, 2008). Introgression of such complex traits is often difficult using traditional plant breeding systems, and effective resistance against many *Meloidogyne* spp. has yet to be discovered. Thus, there is great interest in the use of genetic engineering to develop effective resistance in modern crop cultivars. Although, to our knowledge, no crop variety with resistance to *Meloidogyne* or other nematode species has yet been developed via genetic transformation, the promise of this approach remains high (Thomas and Cottage, 2006; see Atkinson *et al.*, Chapter 15, this volume). Current areas of research into the root-knot nematode/host plant interaction are being conducted both on what the nematodes secrete into feeding sites in the host plant and on how the plant responds (see Abad *et al.*, Chapter 7, this volume). Techniques of molecular biology are adding to the list of nematode-secreted compounds; Gheysen and Jones (2006) have summarized findings on chorismate mutase, CLAVATA3-like peptides, ubiquitin, cytokinins and nodulation factors. Huang *et al.* (2006a) reported evidence that at least one protein produced by *M. incognita* that appears to be injected into the host is similar to a plant transcription factor. Gheysen and Jones (2006) pointed out that there have been numerous suggestions as to how these molecules might affect plant metabolism, but the mechanisms nematodes use to parasitize plants still remain largely unknown. Several tools have been used to study the plant's response to nematodes successfully establishing feeding sites (Bird *et al.*, 2009). Microarrays allow discovery of genes up- or downregulated upon infection, but it is likely that many of these genes do not play an essential role in nematode parasitism. Knocking out a gene or overexpressing it has taken researchers closer to understanding the function of a gene, but this has only been done in a few cases (see Atkinson *et al.*, Chapter 15, this volume). Comparisons with other better-known systems in plant roots can be helpful in studying feeding site development, e.g. root-knot nematodes and rhizobium bacteria produce similar structures in roots, and also the transcription factors *PHAN* and *KNOX* are induced in both nematode feeding sites and in rhizobium nodules. More detail must be elucidated on the interactions between plant-parasitic nematodes and their plant hosts before a unifying model of root-knot nematode parasitism can be proposed (Gheysen and Jones, 2006).

Genetically engineered resistance has been demonstrated experimentally using antisense to disrupt genes that are specifically upregulated in the developing giant cells (Opperman and Conkling, 1996). More recently, transformation of *Arabidopsis* to produce small RNA molecules that silenced a required nematode parasitism gene via RNA interference rendered the transformed plants resistant to the four major root-knot nematodes species (Huang *et al.*, 2006b). It is likely that, in the relatively near future, genetic transformations will provide useful sources of resistance for improvement of many crops. In most instances, it is likely that only a small number of varieties or breeding lines (possibly only one) of a crop species will be directly transformed, and then the resistance trait will be moved into other varieties by traditional plant breeding techniques. At that point, further development of the resistance will be very similar, if not identical, to that of any other source of resistance. Marker-assisted selection will probably be used in these systems because the sequence of the 'resistance' gene will be known, allowing for the development of high-throughput protocols for marker-assisted selection. It is hypothesized that such engineered resistance mechanisms will be highly durable because of their unique designs and/or because they are 'downstream' from initial recognition events in the interaction between the nematode and the host (see Atkinson *et al.*, Chapter 15, this volume).

14.8 Conclusions and Future Directions

Genetic resistance, from a variety of sources and in a range of effectiveness, has much greater potential for improving management of *Meloidogyne* spp. than is currently being achieved. Unfortunately, development of resistant crop varieties is a long-term and costly endeavour. Greatest successes will come from those efforts that will place as much emphasis on crop yield as on nematode resistance. Even low levels of resistance will be used if the crop variety has superior yield potential. The active and close involvement of a plant breeder in such efforts is also essential. As nematologists, we need to move beyond the mere identification and characterization of resistance, and focus on introgression of the resistance into modern, competitive crop varieties. It is unlikely that this will occur without us actively seeking collaboration with plant breeders.

14.9 References

Barrett, B., Mercer, C. and Woodfield, D.R. (2005) Genetic mapping of a root-knot nematode resistance locus in *Trifolium. Euphytica* 143, 85–92.

Bird, D.McK., Opperman, C.H. and Williamson, V.M. (2009) Plant infection by root-knot nematode. In: Berg, R.H. and Taylor, C.G. (eds) *Cell Biology of Plant Nematode Parasitism.* Springer, Berlin and Heidelberg, pp. 1–13.

Boiteux, L.S., Hyman, J.R., Bach, I.C., Fonseca, M.E.N., Matthews, W.C., Roberts, P.A. and Simon, P.W. (2004) Employment of flanking co-dominant STS markers to estimate allelic substitution effects of a nematode resistance locus in carrot. *Euphytica* 136, 37–44.

Chu, Y., Holbrook, C.C., Timper, P. and Ozais-Akins, P. (2007) Development of a PCR-based molecular marker for to select for nematode resistance in peanut. *Crop Science* 47, 841–847.

Claverie, M.B., Voisin, N., Esmenjaud, R., Chalhoub, D., Dirlewanger, B., Klienhentz, E. and Laigret, M. (2004) High resolution map of the *Ma* gene for resistance to root-knot nematodes in Myrobalan plum (*Prunus cerasifera*). *International Society for Horticultural Sciences* 663, 69–74.

Cohen, R., Burger, Y., Horev, C., Koren, A. and Edelstein, M. (2007) Introducing grafted cucurbits to modern agriculture: the Israeli experience. *Plant Disease* 91, 916–923.

Cook, R. and Noel, G.R. (2002) Cyst nematodes: *Globodera* and *Heterodera* species. Starr, J.L., Cook, R. and Bridge, J. (eds) *Plant Resistance to Parasitic Nematodes.* CAB International, Wallingford, UK, pp. 71–105.

Cook, R. and Starr, J.L. (2006) Resistant cultivars. In: Perry, R.N. and Moens, M. (eds) *Plant Nematology.* CAB International, Wallingford, UK, pp. 370–391.

Cook, R. and Yeates, G.W. (1993) Nematode pests of grassland and forage crops. In: Evans, K., Trudgill, D.L. and Webster J.M. (eds) *Plant Parasitic Nematodes in Temperate Agriculture.* CAB International, Wallingford, UK, pp. 305–350.

Diniz, L.E., Sakiyama, N.S., Lashermes, P., Caixeta, E.T., Oliveira, A.C., Zambolim, E.M., Lureuro, M.F., Pereira, A.A. and Zambolium, L. (2005) Analysis of AFLP markers associated to the *Mex-1* resistance locus in Icatu progenies. *Crop Breeding and Applied Biotechnology* 5, 387–393.

Di Vito, M., Cianciotta, V. and Zaccheo, G. (1992) Yield of susceptible and resistant peppers in microplots infested with *Meloidogyne incognita. Nematropica* 22, 1–6.

Djian-Caporalino, C., Pijarowski, L., Fazari, A., Samson, M., Gaveau, L., O'Byrne, C., Lefebvre, V., Caranta, C., Palloix, A. and Abad, P. (2001) High-resolution genetic mapping of pepper (*Capsicum annuum* L.) resistance loci *Me-3* and *Me-4* conferring heat-stable resistance to root-knot nematodes (*Meloidogyne* spp.). *Theoretical and Applied Genetics* 103, 592–600.

Evans, K., Trudgill, D.L. and Webester, J.M. (eds) (1993) *Plant Parasitic Nematodes in Temperate Agriculture.* CAB International, Wallingford, UK.

Fassuliotis, G. (1979) Plant breeding for root-knot nematode resistance. In: Lamberti, F. and Taylor, C.E. (eds) *Root-knot Nematodes (*Meloidogyne *spp) – Systematics, Biology and Control.* Academic Press, London, pp. 425–453.

Fassuliotis, G. (1985) The role of the nematologists in the development of resistant cultivars. In: Sasser, J.N. and Carter, C.C. (eds) *An Advanced Treatise on* Meloidogyne. *Vol I. Biology and Control.* North Carolina State University Graphics, Raleigh, North Carolina, pp. 233–240.

Francia, E., Tacconi, G., Crosatti, C., Barabaschi, D., Dall'Aglio, E. and Vale, G. (2005) Marker assisted selection in crop plants. *Plant Cell, Tissue and Organ Culture* 82, 317–342.

Fuller, V.L., Lilley, C. J. and Urwin, P.E. (2008) Nematode resistance. *New Phytologist* 180, 27–44.

Gheysen, G. and Jones, J.T. (2006) Molecular aspects of plant–nematode interactions. In: Perry, R.N. and Moens, M. (eds) *Plant Nematology.* CAB International, Wallingford, UK, pp. 234–254.

Hartman, K.M. and Sasser, J.N. (1985) Identification of *Meloidogyne* species on the basis of differential host test and perineal-pattern morphology. In: Barker, K.R., Carter, C.C. and Sasser, J.N. (eds) *An Advanced Treatise on* Meloidogyne. *Vol II. Methodology.* North Carolina State University Graphics, Raleigh, North Carolina, pp. 69–77.

Huang, G., Dong, R., Allen, R., Davis, E.L., Baum, T.J. and Hussey, R.S. (2006a) A root-knot nematode secretory peptide functions as a ligand for a plant transcription factor. *Molecular Plant–Microbe Interactions* 19, 463–470.

Huang, G.Z., Allen, R., Davis, E.L., Baum, T.J. and Hussey, R.S. (2006b) Engineering broad root-knot resistance in transgenic plants by RNAi silencing of a conserved and essential root-knot nematode parasitism gene. *Proceedings of the National Academy of Sciences of the USA* 103, 14302–14306.

Hussey, R.S. and Boerman, H.R. (1981) A greenhouse screening procedure for root-knot nematode resistance in soybean. *Crop Science* 21, 794–796.

Hussey, R.S. and Janssen, G.J.W. (2002) Root-knot nematodes: *Meloidogyne* species. In: Starr, J.L., Cook, R. and Bridge, J. (eds) *Plant Resistance to Parasitic Nematodes.* CAB International, Wallingford, UK, pp. 43–70.

Kinloch, R.A. (1990) Screening for resistance to root-knot nematodes. In: Starr, J.L. (ed.) *Methods for Evaluating Plant Species for Resistance to Plant-parasitic Nematodes.* Society of Nematologists, Hyattsville, Maryland, pp. 16–23.

Koenning, S.R., Barker, K.R. and Bowman, D.T. (2001) Resistance as a tactic for management of *Meloidogyne incognita* on cotton in North Carolina. *Journal of Nematology* 33,126–131.

Luc, M., Sikora, R.A. and Bridge, J. (eds) (2005) *Plant Parasitic Nematodes in Subtropical and Tropical Agriculture*, 2nd edn. CAB International, Wallingford, UK.

Mercer, C.F., Bell, N.L. and Yeates, G.W. (2008) Plant-parasitic nematodes on pasture in New Zealand. *Australasian Plant Pathology* 37, 279–288.

Nelson, S.C., Simpson, C.E. and Starr, J.L. (1989) Resistance to *Meloidogyne arenaria* in *Arachis* spp. germplasm. *Journal of Nematology* 21, 654–660.

Noling, J. (2000) Effects of continuous culture of a resistant tomato cultivar on *Meloidogyne incognita* soil population density and pathogenicity. *Journal of Nematology* 32, 452 (abstract).

Nui, C., Hinchliffe, D.J., Cantrell, R.G., Wang, C.L., Roberts, P.A. and Zhang, J.F. (2007) Identification of molecular markers associated with root-knot nematode resistance in upland cotton. *Crop Science* 47, 951–960.

Nyczepir, A.P. and Halbrendt, J.M. (1993) Nematode pests of deciduous fruit and nut trees. In: Evans, K., Trudgill, D.L. and Webster, J.M. (eds) *Plant Parasitic Nematodes in Temperate Agriculture*. CAB International, Wallingford, UK, pp. 381–425.

Ogallo, J.L., Goodell, P.B., Eckert, J. and Roberts, P.A. (1997) Evaluation of NemX, a new cultivar of cotton with high resistance to *Meloidogyne incognita. Journal of Nematology* 29, 531–537.

Oka, Y., Offenbach, R. and Pivonia, S. (2004) Pepper rootstocks graft compatibility and response to *Meloidogyne javanica* and *M. incognita. Journal of Nematology* 36, 137–147.

Opperman, C.H. and Conkling, M.A. (1996) Root-knot nematode induces *TobRB7* expression and antisense transgenic resistance strategies. *Advances in Molecular Gentics Plant–Microbe Interactions* 4, 521–526.

Poehlman, J.C. and Sleper, D.A. (1995) *Breeding Field Crops*, 4th edn. Iowa State University Press, Ames, Iowa.

Roberts, P.A. (1990) Resistance to nematodes: definitions, concepts, and consequences. In: Starr, J.L. (ed.) *Methods for Evaluating Plants Species for Resistance to Plant-parasitic Nematodes*. Society of Nematologists, Hyattsville, Maryland, pp. 1–15.

Roberts, P.A. (2002) Concepts and consequences of resistance. In: Starr, J.L., Cook, R. and Bridge, J. (eds) *Plant Resistance to Parasitic Nematodes*. CAB International, Wallingford, UK, pp. 23–41.

Shepherd, R.L. (1974) Registration of Auburn 623 RNR cotton germplasm. *Crop Science* 14, 911.

Siguenza, C., Schochou, M., Turini, T. and Ploeg, A. (2005) Use of *Cucumis metaliferus* as a rootstock for melon to manage *Meloidogyne incognita. Journal of Nematology* 37, 256–280.

Simpson, C.E. and Starr, J.L. (2001) Registration of 'COAN' peanut. *Crop Science* 41, 918.

Simpson, C.E., Starr, J.L., Church, G.T., Burow, M.D. and Paterson, A.H. (2003) Registration of 'NemaTAM' peanut. *Crop Science* 43, 1561.

Sorribas, F.J., Ornat, C., Verdejo-Lucas, S., Galeano, M. and Valero, J. (2005) Effectiveness and profitability of the *Mi*-resistance gene in tomato over three consecutive growing seasons. *European Journal of Plant Pathology* 111, 29–38.

Starr, J.L., Bridge, J. and Cook, R. (2002) Resistance to plant parasitic nematodes: history, current use future potential. In: Starr, J.L., Cook, R. and Bridge, J. (eds) *Plant Resistance to Parasitic Nematodes*. CAB International, Wallingford, UK, pp. 1–22.

Starr, J.L., Koenning, S.R., Kirkpatrick, T.L., Robinson, A.F., Roberts, P.A. and Nichols, R.L. (2007) The future of nematode management in cotton. *Journal of Nematology* 39, 283–294.

Thies, J.A. and Fery, R.L. (2001) Characterization of *Capsicum chinense* cultigens for resistance to *Meloidogyne arenaria*, *M. hapla*, and *M. javanica. Plant Disease* 85, 267–270.

Thies, J.A., Fery, R.L., Mueller, J.D., Miller, G. and Varne, J. (2003) Response of bell pepper varieties near-isogenic for the *N* gene to *Meloidogyne incognita* in field trials. *HortScience* 38, 1394–1396.

Thomas, C. and Cottage, A. (2006) Genetic engineering for resistance. In: Perry, R.N. and Moens, M. (eds) *Plant Nematology*. CAB International, Wallingford, UK, pp. 255–272.

van der Beek, J.G. and Poleij, L.M. (2008) Altered reproduction response of *Meloidogyne hapla* on *Solanum* genotypes caused by propagation on different tomato cultivars. *Nematology* 10, 777–779.

Wang, C., Ulloa, M. and Roberts, P.A. (2006) Identification and mapping of microsatellite markers linked to a root-knot nematode resistance gene (*rkn1*) in Acala NemX cotton (*Gossypium hirsutum* L.). *Theoretical and Applied Genetics* 112, 770–777.

Williamson, V.M. (1998) Root-knot nematode resistance genes in tomato and their potential for future use. *Annual Review of Phytopathology* 36, 277–293.

Zhou, E. and Starr, J.L. (2003) A comparison of the damage functions, root galling, and reproduction of *Meloidogyne incognita* on resistant and susceptible cotton cultivars. *Journal of Cotton Science* 7, 224–230.

15 Plant Biotechnology and Control

Howard J. Atkinson[1], Peter E. Urwin[1] and Richard S. Hussey[2]
[1]University of Leeds, Leeds, West Yorkshire, UK; [2]University of Georgia, Athens, Georgia, USA

15.1	Introduction	338
15.2	Proteinase Inhibitors	339
15.3	Cry Proteins of *Bacillus thuringiensis* as Biopesticides	340
15.4	*In planta* RNAi to Target Plant-parasitic Nematodes	342
15.5	Repellents	345
15.6	The *Mi-1*-mediated Resistance Response	346
15.7	Efficacy and Durability	347
15.8	Promoters for Transgenic Control of *Meloidogyne*	349
15.9	Biosafety	349
15.10	Developing World Needs	352
15.11	Conclusions and Future Directions	353
15.12	References	356

15.1 Introduction

Development of resistant plants that suppress nematode growth and reproduction is the most cost-effective and environmentally sustainable strategy for reducing *Meloidogyne* damage to food and fibre crops (Hussey and Janssen, 2002). However, use of host plant resistance for control of *Meloidogyne* is limited for several reasons, the most fundamental of which is that only a small number of plant species are resistant to this nematode and there are many crops for which appropriate resistance loci have not been identified (Roberts, 2002; Williamson and Kumar, 2006; see Williamson and Roberts, Chapter 13, and Starr and Mercer, Chapter 14, this volume). As with other plant resistance genes, the function of available root-knot resistance genes involves recognition of specific biotypes of *Meloidogyne* spp., making the resistant crops vulnerable to selection for virulent field populations (Roberts, 1995). Consequently, there is a need for new approaches to developing root-knot-resistant plants, and bioengineering provides a strategy to design effective and durable forms of *Meloidogyne*-resistant crops.

Meloidogyne spp. have evolved highly specialized and complex feeding relationships with their hosts (see Abad *et al.*, Chapter 7, this volume). Aspects of such information underpin the rapidly developing field of plant biotechnology for control of this nematode. The substantial advances in the application of biotechnology for control of *Meloidogyne* that have been made in the last few years range from approaches to reduce root invasion to disrupting parasitic success of the nematode. RNA interference has now been targeted successfully at the molecular signals secreted from this nematode that modify host plant root cells. Other approaches rely on disrupting the utilization of dietary protein or delivering a bionematicide to the gut of the feeding *Meloidogyne*. The disruption of parasitic success has been achieved,

 ©CAB International 2009. *Root-knot Nematodes* (eds R.N. Perry, M. Moens and J.L. Starr)

and useful and durable control under field conditions should be demonstrated soon. Biosafety issues must be addressed before deployment of novel resistance is considered. There are distinct prospects and need for utilization in both developed-world and developing-world agriculture. This field will continue to expand rapidly and remain one of the most interesting and challenging areas within plant pathology.

15.2 Proteinase Inhibitors

Proteinase inhibitors (PIs) are widely expressed in plants, where they are often induced by wounding and herbivory. The cowpea trypsin inhibitor (CpTI), a serine PI, was the first PI to be studied for delivering an anti-nematode defence. CpTI expressed in transgenic potato influenced the sexual fate of newly established *Globodera pallida*. As a result, the population was biased toward a predominance of the much smaller and less damaging males, but no reduction in the fecundity of established females of *G. pallida* was observed (Hepher and Atkinson, 1992). Conversely, CpTI did reduce the fecundity of females of *M. incognita* without influencing their sexual fate (Hepher and Atkinson, 1992; Urwin *et al.*, 1998). Transgenic expression of other serine PIs has focused on providing resistance to cyst nematodes. The sweet potato serine PI, sporamin, when expressed in sugarbeet hairy roots, inhibited growth and development of female beet cyst nematodes, *Heterodera schachtii* (Cai *et al.*, 2003). The severity of the effect was clearly correlated with the level of trypsin-inhibitory activity detected in the transformed root lines. Transgenic wheat that expressed the potato serine PI (PIN2) also showed a positive correlation with plant growth and yield under challenge with the cereal cyst nematode *Heterodera avenae* (Vishnudasan *et al.*, 2005).

Protein engineering, utilizing crystallographic data, was used to increase the efficacy of the rice cystatin (Oc-I) to produce an improved cystatin PI with increased inhibitory activity. The level of resistance against cyst nematodes conferred by the expression of the engineered variant (Oc-IΔD86) in root cultures was higher than that of the unaltered molecule (Urwin *et al.*, 1995). The same cystatin was shown to be effective against *M. incognita* following challenge of *Arabidopsis* with this species. The plants were also resistant to *H. schachtii*. This was the first demonstration of a transgenic technology working against two major groups of economically important nematodes. The uptake of the cystatin was correlated with loss of nematode proteinase activity (Urwin *et al.*, 1997). This cystatin-based technology has been shown to be effective against other species. The *Arabidopsis* plants showing resistance to *M. incognita* and *H. schachtii* also suppressed reproduction of the reniform nematode *Rotylenchulus reniformis*, with higher levels of PI expression again correlated with reduced reproductive success (Urwin *et al.*, 2000). This seems to indicate that similar cysteine proteinases are involved in digestion in *Meloidogyne* and other species. Rice cystatins expressed at a low level in lucerne under the control of a wound-inducible promoter conferred some resistance to the root-lesion nematode *Pratylenchus penetrans* (Samac and Smigocki, 2003). Cystatins have also been used to protect rice. Plants of this crop expressing chicken egg-white cystatin displayed 55% resistance to *M. incognita* (Vain *et al.*, 1998) and 91 ± 7% when a maize cystatin was expressed under control of the root- specific *Arabidopsis* promoter, Tub-1 (J. Green, Leeds, UK, 2008, personal communication). Banana can also act as a host for *Meloidogyne*; therefore, having resistance that is effective against this species and the other damaging nematode pests of the crop is of particular value. Cavendish dessert bananas that express a cystatin show resistance to the migratory endoparasite *Radopholus similis* (Atkinson *et al.*, 2004a) and also show a higher level of resistance (83 ± 4%) to *M. incognita* (J. Green, Leeds, UK, 2008, personal communication) when the same cystatin is expressed under control of the Tub-1 promoter (Green *et al.*, 2002; Lilley *et al.*, 2004)

Additive resistance has also been demonstrated by expressing two proteinase inhibitors in the same transgenic plant. The two proteinase inhibitors were expressed as a translational fusion, held together with a linker refractory to proteolytic cleavage. The work provided a platform for delivering a transgenic resistance with improved efficacy and durability (Urwin *et al.*, 1998). The development of PI-mediated nematode resistance culminated in successful field trials of transgenic potatoes expressing a cystatin. The best transgenic lines of the fully susceptible potato cv. Désirée were shown to have commercially useful resistance (Urwin *et al.*, 2001). Potato plants in which the expression of the cystatin is limited mainly to the roots, and in particular to the unique feeding structure induced by the nematode within the plant root, were shown to have similar resistance levels to those achieved with constitutive expression for both *G. pallida* and *M. incognita* (Lilley *et al.*, 2004). In the UK, full resistance was observed in the field by stacking natural partial and transgenic resistance. Two potato cultivars, Sante and Maria Huanca, which exhibit partial natural resistance to *G. pallida*, were transformed with Oc-IΔD86 expressed constitutively. In both cases, the natural partial resistance was enhanced to full resistance, i.e. there were fewer eggs in the soil postharvest than pre-planting (Urwin *et al.*, 2003). This was the first demonstration of full resistance to a nematode using a combination of an R-gene and a transgene.

Characterization of further native plant cystatins has revealed that those from maize and sunflower have better binding affinities than that of the engineered Oc-I variant. The design of proteinase-inhibitor-based biotechnology has been directed by information about the target nematode proteinases. cDNAs (complementary DNAs) encoding serine and cysteine digestive proteinases have been cloned and their developmental expression profiles determined. The biochemical activity has been localized to the intestine of all feeding stages (Lilley *et al.*, 1996, 1997; Urwin *et al.*, 1997, 2002; Neveu *et al.*, 2003; Shingles *et al.*, 2007).

15.3 Cry Proteins of *Bacillus thuringiensis* as Biopesticides

15.3.1 Cry proteins

The spore-forming bacterium *Bacillus thuringiensis* (*Bt*) is a soil microorganism capable of killing insects. The δ-endotoxins it produces have been the centre of particular attention. The δ-endotoxins are encoded by the *cry* group of genes and are produced in the sporulation phase of *B. thuringiensis*. In addition, this bacterium also produces β-exotoxins that have effects against nematodes (Devidas and Rehberger, 1992). The β-exotoxins are encoded by the gene *vip*3a in vegetative stages and released from the bacterium. The δ-endotoxin Cry proteins are the basis of the 32.1 million ha of plant biotechnology lines planted in 2006 for insect resistance either alone or stacked with herbicide tolerance (http://www.isaaa.org/resources/publications/briefs/35/executivesummary/default.html).

A basic model for the function of a Cry protein against an insect is that the ingested endospore crystals are digested in its gut. The Cry protein is activated by proteinase activity, binds to a membrane receptor and induces the formation of a pore in the midgut, with lethal consequences (de Maagd *et al.*, 2001). In the current nomenclature, the term *cry* indicates one of many members of the gene superfamily, the following number indicates a distinct subgroup (e.g. 1–22; Crickmore *et al.*, 1998), and subsequent letters and numbers provide further subdivisions of sequences.

The tertiary structures of six different three-domain Cry proteins from within the main Cry lineage have been determined by X-ray crystallography (Bravo *et al.*, 2007). The exposed regions in domains II and III are involved in receptor binding (Bravo *et al.*, 2007), with domain III sharing structural similarity with other carbohydrate-binding proteins (de Maagd *et al.*, 2003). These similarities suggest that carbohydrate moieties could have an important role in the mode of action of Cry proteins (Bravo *et al.*, 2007). An insect midgut proteinase activity may be important in processing the ingested Cry protein to an active form, with the degree or rate of activation being important (Crickmore, 2005). This, and receptor binding, help to define specificity, but a subsequent ability to oligomerize or insert into a membrane are also important events (Crickmore, 2005).

15.3.2 Activity of Cry proteins against nematodes

Borgonie *et al.* (1996) first reported a Cry protein inducing a pathology to bacterial-feeding nematodes that resembled the more rapid disease progress reported previously for insects. Activity of a Cry protein against both *Caenorhabditis* and *Pratylenchus* was also claimed in a patent granted in 1995 (Sick *et al.*, 1994). It is now termed Cry5Aa1 (Crickmore *et al.*, 1998). A number of subsequent patents claim efficacy of different Cry proteins, including Cry21Aa1 (Payne *et al.*, 1996;

Crickmore *et al.*, 1998), Cry6, Cry13 (see Kotze *et al.*, 2005). Detailed results are also available for Cry14A plus Cry21A (Wei *et al.*, 2003). The Cry 5, 13, 14 and 21 proteins cluster in different secondary ranks within the main Cry lineage, based on amino acid sequence identity.

Cry5B, Cry6A, Cry14A and Cry21A proteins are toxic to species of the bacterial-feeding genera *Caenorhabditis*, *Acrobeloides*, *Panagrellus* and *Disolabrellus* after ingestion of *Escherichia coli* expressing the corresponding *cry* genes (Wei *et al.*, 2003). There may be differential toxicity of these Cry proteins among nematodes. All but Cry6A also strongly inhibited development of pre-infective first-stage larvae of the rat intestinal parasite *Nippostrongylus brasiliensis* when they ingested *E. coli* clones transgenic for these Cry proteins. A further two *Bt* strains, expressing Cry5A plus Cry5b in one case and Cry13 in the second, were toxic to both larvae and adults of other animal-parasitic nematodes: *Trichostrongylus colubriformis* and *Ostertagia circumcincta* (Kotze *et al.*, 2005). Subsequent work discounted Cry5a as an effective toxin of these nematodes (Lenane *et al.*, 2008), which is consistent with the results of Wei *et al.* (2003).

The tertiary structures for Cry proteins that are toxic to nematodes are not yet described and none of these genes cluster close to those for which tertiary structure has been defined (Crickmore *et al.*, 1998). However, Cry5, 13, 14 and 21 have sufficient amino acid sequence similarity to those already studied (Crickmore *et al.*, 1998) to suggest they too are three-domain proteins.

15.3.3 Activity of Cry6A against *Meloidogyne incognita*

Cry6A is as yet the only nematicial Cry protein studied for its efficacy against a feeding plant-parasitic nematode. It is one of three known outlying Cry lineages (Crickmore *et al.*, 1998), and more needs to be known about its structure. It is toxic to *M. incognita* when expressed in hairy roots of tomato at 0.2–0.4% tsp (total soluble protein). It reduced egg mass number to 50–60% and egg production per root system to 24–36% of controls. Altering gene sequence from microbial to plant-favoured codon usage enhanced expression in the roots. The effect is probably dose responsive and may be enhanced by higher expression levels *in planta* (Li *et al.*, 2007). Whole-plant transformation and challenge of an efficient host under field conditions is necessary before the resistance levels that Cry6A can deliver against *M. incognita* can be determined reliably.

15.3.4 Resistance to Cry proteins in nematodes

Resistance to Cry proteins in insects involves many mechanisms (Gill and Ellar, 2002; Soberón *et al.*, 2007). Resistance of *Caenorhabditis elegans* to Cry5B involves two mechanisms. Four mutants lack the glycosphingolipid chains that are a receptor for this Cry protein. The glycolipids have a core tetrasaccharide, GalNAcb1-4GlcNAcb1-3Manb1-4Glc, also found in insects. Their absence from vertebrates is a possible key to the lack of toxicity of Cry proteins to such animals (Griffitts *et al.*, 2005). Moderate resistance of a fifth *C. elegans* mutant to Cry5B involves a different mechanism. It encodes for a GDP (guanosine diphosphate)-mannose-4,6-dehydratase that is involved in synthesis of GDP-fucose. The mutant has little fucosylated protein and the resistance this provides has a fitness cost (Barrows *et al.*, 2007).

The long evolutionary history between nematodes in soil and *B. thuringiensis* provides a basis for the evolution of resistance to the δ-endotoxins of *B. thuringiensis*. Such a possibility seems less likely for *Meloidogyne* than for nematodes that ingest soil bacteria. A second mechanism of potential resistance could be pertinent to *Meloidogyne*. Two mitogen-activated protein kinases that may upregulate an efflux transporter are induced in *C. elegans* on exposure to Cry5B (Huffman *et al.*, 2004). This response may protect against either cytotoxic cations in the diet or the consequences of pore formation by a Cry protein (Huffman *et al.*, 2004). A key issue is whether or not a diet-related detoxification mechanism functions in *Meloidogyne*. It may predispose the nematode to Cry-protein resistance if the novel protein is ingested. However, forward genetics has not readily isolated mutants of *C. elegans* that are resistant to Cry6A (Li *et al.*, 2007), so this protein may provide durable resistance to *Meloidogyne*.

15.4 *In planta* RNAi to Target Plant-parasitic Nematodes

Application of RNA interference (RNAi) for gene silencing in plant-parasitic nematodes has culminated in the demonstration that plants expressing double-stranded RNA (dsRNA) targeting a nematode gene display resistance to infection (reviewed by Gheysen and Vanholme, 2007; Lilley *et al.*, 2007).

Gene silencing triggered by dsRNA was first demonstrated for *C. elegans* (Fire *et al.*, 1998), and the underlying mechanism of RNAi has subsequently been studied in depth for this free-living nematode. A similar phenomenon had previously been described for plants as post-transcriptional gene silencing (Jorgensen *et al.*, 1996; Waterhouse *et al.*, 1998). The detailed molecular mechanisms and various component proteins involved in RNAi are still being elucidated for *C. elegans* (e.g. Pak and Fire, 2007; Sijen *et al.*, 2007; Winston *et al.*, 2007). A number of reviews provide a more complete description of the earlier work (e.g. Grishok, 2005; May and Plasterk, 2005; Joyce *et al.*, 2006).

RNAi is widely employed as a tool for analysis of plant gene function. Both sense and antisense cDNA sequences of the target gene, separated by a spacer region or intron, are cloned into a binary vector under the control of a plant promoter. The transcribed RNA then forms into a self-complementary hairpin structure with either the spacer region forming a loop or the intron sequence being removed by splicing. More recently there has been interest in using RNAi to engineer novel plant traits (Kusaba, 2004; Mansoor *et al.*, 2006), with a number of potential commercial applications already described (e.g. Byzova *et al.*, 2004; Ogita *et al.*, 2004; Davuluri *et al.*, 2005). RNAi has also been used in plants to confer resistance to the bacterial pathogen *Agrobacterium tumefaciens* (Escobar *et al.*, 2001), to viruses (Waterhouse *et al.*, 1998; Pooggin *et al.*, 2003) and to phytophagous insects (Baum *et al.*, 2007; Mao *et al.*, 2007). The utility of RNAi for functional analysis of plant-parasitic nematode genes was first reported in 2002. Genes encoding a cysteine proteinase, a C-type lectin and a major sperm protein of cyst nematodes were targeted by soaking infective second-stage juveniles (J2) in a solution of dsRNA with the addition of the neurochemical octopamine to stimulate ingestion (Urwin *et al.*, 2002). A number of genes, expressed in a range of different tissues and cell types, have now been successfully targeted for silencing in both cyst and root-knot nematodes (reviewed by Fleming *et al.*, 2007; Lilley *et al.*, 2007). The possibility of engineering nematode resistance by the *in planta* production of dsRNA to target essential nematode genes has been recognized since the first demonstration of RNAi in plant-parasitic nematodes (Urwin *et al.*, 2002; Atkinson *et al.*, 2003). The feeding nematodes could ingest either dsRNA or primary short interfering RNA (siRNA) continually from the plant cell cytoplasm, allowing genes expressed at all parasitic stages of development to be targeted.

Uptake of dsRNA from the gut is a proven route to systemic RNAi in *C. elegans*. The systemic nature of RNAi in plant-parasitic nematodes following ingestion of dsRNA suggests that they share similar uptake and dispersal pathways with *C. elegans*. Alternative routes to dsRNA uptake may also exist for plant-parasitic nematodes. RNAi of a chitin synthase gene expressed in the eggs of *M. artiellia* was achieved by soaking intact eggs contained within their gelatinous matrix in a solution containing dsRNA (Fanelli *et al.*, 2005). The enzyme plays a role in the synthesis of the chitinous layer in the eggshell. Depletion of its transcript by RNAi led to a reduction in stainable chitin in eggshells and a delay in hatching of J2 from treated eggs. The results imply that the eggs of this nematode, and possibly others, are permeable to dsRNA. Neuronally expressed genes of *C. elegans* can be refractory to RNAi (Kamath *et al.*, 2000; Timmons *et al.*, 2001), although RNAi effects in these cells can be enhanced by using a mutant strain defective in the RdRP (RNA-directed RNA polymerase) *rrf-3* gene (Simmer *et al.*, 2003). A study by Kimber *et al.* (2002) describes the silencing of five FMRFamide-like (*flp*) neuropeptide genes of *G. pallida*, each with a unique neuronal expression pattern. All five genes were readily susceptible to RNAi, as evidenced by an absence of transcript in treated worms and abnormal behavioural phenotypes.

The genes targeted by RNAi to date are expressed in a range of different tissues and cell types. The ingested dsRNA can silence genes in the intestine (Urwin *et al.*, 2002; Shingles *et al.*, 2007) and also in the female reproductive system (Lilley *et al.*, 2005), sperm (Urwin *et al.*, 2002; Steeves *et al.*, 2006) and both subventral and dorsal pharyngeal glands (Chen *et al.*, 2005; Rosso *et al.*, 2005; Huang *et al.*, 2006a; Bakhetia *et al.*, 2007). The pharyngeal gland cells express the parasitism proteins that are encoded by parasitism genes (Davis *et al.*, 2004; Mitchum, *et al.*, 2007). Since the secreted parasitism proteins mediate the dynamic interaction of *Meloidogyne* with its host plants during migration through roots, establishment of giant cells and feeding, they are viable RNAi targets for directly disrupting nematode parasitism of plants.

Meloidogyne spp. express a large array of parasitism genes in their pharyngeal glands, including 20 genes in the two subventral glands and 28 genes in the single dorsal gland, with their expression patterns varying throughout the nematode's life cycle (Huang *et al.*, 2003, 2004). While some of the parasitism proteins have similarities to known proteins, e.g. pectate lyases, cellulose binding proteins, chorismate mutases and endoglucanases, the majority of the parasitism genes encode novel proteins of unknown function in the parasitic process.

The functions of the novel secreted parasitism proteins coded by the *Meloidogyne* parasitism genes are beginning to be revealed. Functional analysis of a parasitism gene designated as *16D10*, which is expressed in the two subventral pharyngeal gland cells, has revealed that the secreted parasitism protein appears to mediate an early signalling event in the *Meloidogyne*–host interaction (Huang *et al.*, 2003, 2006b). This parasitism gene encodes a small, novel secreted peptide of 13 amino acids, including a 30 amino acid N-terminal hydrophobic signal peptide, and is conserved in *Meloidogyne* spp. When the 16D10 peptide is overexpressed in *Arabidopsis*, root growth is significantly accelerated, giving rise to a much-enlarged root system without affecting shoot growth (Huang *et al.*, 2006b). The 16D10 peptide was shown to bind directly to the SAW domain of two *Arabidopsis* SCARECROW-like (SCL) transcription factors, AtSCL6 and AtSCL21, in a yeast two-hybrid screen for 16D10-interacting proteins (Huang *et al.*, 2006b). SCL transcription factors are members of the GRAS protein family, which plays important roles in plant development and signalling (Bolle, 2004). These data suggest that the 16D10 peptide functions as a signalling peptide and specifically induces root growth by directly interacting with a host intracellular SCL transcription regulator. Furthermore, since the conserved *Meloidogyne* secreted signalling peptide is strongly expressed in the subventral pharyngeal gland cells of J2 at the time when the giant cells are being developed, this peptide is speculated to have a role in the reprogramming of gene expression required for giant cell formation.

The first study demonstrating silencing of *Meloidogyne* genes by RNAi delivered from host tobacco plants was reported by Yadav *et al.* (2006). Nematode splicing factor and integrase genes were targeted for their presumed essential role in basic cellular processes. Plants expressing hairpin constructs for each of the two sequences displayed >95% resistance to *M. incognita.* The few nematodes that formed galls appeared developmentally compromised and lacked detectable transcript for the targeted genes (Yadav *et al.*, 2006). No evidence was presented for the presence of either dsRNA or siRNAs in the transgenic plants, so the route by which silencing occurred cannot be deduced.

Another example in which resistance to *Meloidogyne* has recently been demonstrated involves *Arabidopsis* that has been engineered to produce dsRNA molecules complementary to the *16D10* parasitism gene. The *16D10* dsRNAs were processed into approximately 21-bp siRNA for ingestion by feeding root-knot nematodes, with subsequent RNAi silencing of *16D10* in the subventral glands (Huang *et al.*, 2006a). When inoculated with each of the four major *Meloidogyne* species (*M. incognita*, *M. javanica*, *M. arenaria* and *M. hapla*), the *16D10* dsRNA *Arabidopsis* transgenic lines showed a significant reduction (63–90%) in the number of galls, as well as an overall decrease in gall size, compared with control vector-transformed lines (Fig. 15.1; Huang *et al.*, 2006a). This silencing of *16D10* by RNAi demonstrates that parasitism gene *16D10* is essential for successful *Meloidogyne* infection of plants and, more importantly, that *in planta* delivery of RNAi of *16D10* can provide broad-spectrum host resistance to *Meloidogyne* spp.

Evidence supporting the RNAi silencing of *16D10* was obtained by crossing a transgenic *Arabidopsis* line overexpressing *16D10* with a *16D10* dsRNA transgenic line to generate F_1 hybrid lines simultaneously expressing both transgenes. *16D10* mRNA was not detected in the F_1 hybrid line, but a higher level of *16D10* siRNA was present in the hybrid line when compared

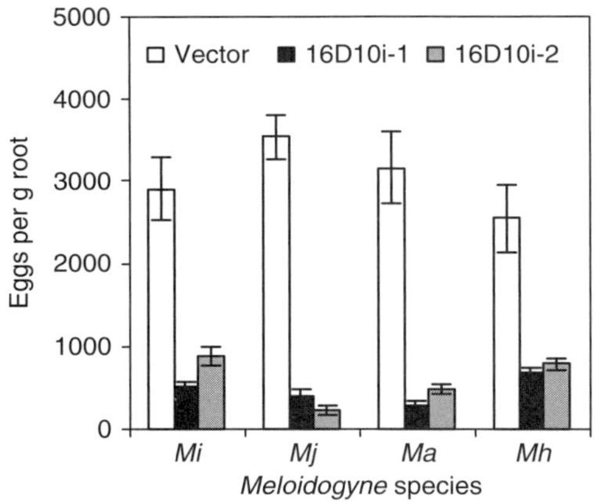

Fig. 15.1. Reproduction (eggs per g root) of four *Meloidogyne* species (*Mi*: *M. incognita*; *Mj*: *M. javanica*; *Ma*: *M. arenaria*; *Mh*: *M. hapla*) on transgenic *Arabidopsis thaliana* expressing *16D10* dsRNA is significantly decreased compared with control plants. Each bar value represents the mean ± SD of n = 24–48 (Student's t-test, $P < 0.001$ versus control). (From Huang *et al.*, 2006a. Copyright 2006 National Academy of Sciences, USA.)

with the paternal *16D10* dsRNA transgenic line (Huang *et al.*, 2006a). The increased levels of siRNAs, called secondary siRNAs, may be generated by the antisense primary siRNAs serving as primers in RNA-dependent RNA polymerase-mediated synthesis of new dsRNA on the target mRNA (Nishikura, 2001; Baulcombe, 2007). Moreover, the RNAi silencing of *16D10* expression in the F_1 hybrid line restored the *16D10*-stimulated root growth phenotype of the maternal *16D10*-overexpressing transgenic line to a wild-type normal root growth phenotype. This *in planta* silencing of overexpressed *16D10* confirms that host-generated *16D10* dsRNA can trigger RNAi of *16D10* to subsequently impede the function of the *Meloidogyne* 16D10 parasitism signalling peptide in plants. Furthermore, the amplification of the siRNA population by simultaneously expressing the target gene and its complementary dsRNA in the same transgenic plant might lead to a higher level of nematode resistance than expressing only the dsRNA transgene.

An important aspect of silencing parasitism gene *16D10* by expressing *16D10* dsRNA in transgenic *Arabidopsis* is the development of transgenic plants that are highly resistant to multiple *Meloidogyne* species. Because no known natural resistance gene has this wide effective range of root-knot resistance, bioengineering crops expressing dsRNA that silence target parasitism genes directly to disrupt the parasitic process represents a viable and flexible means to develop novel, durable root-knot-resistant crops with unprecedented broad-spectrum resistance to *Meloidogyne* spp. Indeed, *in planta* RNAi silencing of *16D10* in *Meloidogyne* spp. could provide a strategy for developing root-knot-resistant crops for which natural resistance genes are not known. In addition, this type of resistance has the potential to be more durable than natural resistance genes. The specificity of RNAi-mediated resistance is based on RNA hybridization rather than on the receptor-ligand binding interactions characteristic of traditional plant resistance genes, which should make the RNAi-mediated resistance highly durable (Escobar *et al.*, 2001).

There are several other advantages in using RNAi silencing of parasitism genes for developing transgenic resistant crops. Targeting a parasitism gene through *in planta* RNAi guarantees that the dsRNA (siRNA) uptake is very closely timed to the expression of the actual target gene, since parasitism genes by definition are expressed during nematode feeding. Targeting genes not tied to parasitism may not be as effective, as their expression may not be timed with the nematode's feeding activity. Furthermore, silencing an essential parasitism gene ensures that the RNAi silencing will impair a process that is vital for the nematode to be able to infect plants. The targeting of parasitism genes that are *Meloidogyne* specific should also minimize the possibility of the siRNAs inducing 'off-target' gene silencing effects (Jackson *et al.*, 2003), which should facilitate regulatory approval of the transgenic plants. Homologues of nematode housekeeping genes are found in many organisms, and silencing these genes could become a potential issue in getting regulatory approval. Last, targeting parasitism genes should be durable because a mutation in the parasitism gene in response to RNAi silencing would knock out the function of a nematode gene that is essential for parasitism, inhibiting nematode infection and the potential for the development of new virulent biotypes.

The evidence for successful *in planta* delivery of RNAi to *Meloidogyne* spp. continues to grow. More recently, partial silencing of a putative transcription factor of *M. javanica* (*MjTis11*) was achieved through the expression of dsRNA hairpin constructs in tobacco (Fairbairn *et al.*, 2007). In this case, downregulation of the *MjTis11* transcript did not significantly affect either nematode development or fecundity.

There is currently no clear evidence to suggest whether it is the longer dsRNA or the siRNAs that act as the trigger for silencing of nematode genes. Huang *et al.* (2006a) demonstrated the presence of both full-length *16D10* dsRNA and processed siRNAs in transgenic plants, whereas Steeves *et al.* (2006) detected siRNAs corresponding to the *Heterodera glycines MSP* sequence but failed to detect the full-length dsRNA. Transformation of hairpin constructs into *Arabidopsis* mutants defective in Dicer functions may help to elucidate the relative importance of the two RNA forms in conferring nematode resistance in plants. Successful RNAi-mediated gene silencing targeted from transgenic plants towards phytophagous insects has been reported (Baum *et al.*, 2007; Mao *et al.*, 2007). Mao *et al.* (2007), using triple mutants lacking three of the four Dicer-like nucleases of *Arabidopsis thaliana* (*dcl2*; *dcl3*; *dcl4*), demonstrated that plants accumulating higher levels of intact dsRNA induced more profound silencing of the targeted insect gene, even when siRNAs were largely absent from the leaves. If this proves also to be the case for plant-parasitic nematodes, the efficacy of the approach as a control strategy may depend upon the stabilization of the dsRNAs.

15.5 Repellents

Chemoreception is important for J2 of *Meloidogyne* when locating roots to invade (see Curtis *et al.*, Chapter 6, this volume). Disrupting this process is a proven target. Low levels of aldicarb impair orientation without inducing either paralysis or death of plant-parasitic nematodes (Trett and Perry, 1985). Such effects relate to uptake of certain compounds, including some dyes, by open-ended, chemoreceptive neurons in the amphids of both *C. elegans* (Hedgecock *et al.*, 1985) and cyst nematodes (Winter *et al.*, 2002). This provided a lead for work to identify peptides that show this ability. The initial effort focused on biopanning a phage display library for those phages able to bind to acetylcholinesterase. This enzyme is specifically inhibited by aldicarb and functions at cholinergic neurons (Winter *et al.*, 2002). The consensus sequence for the selected phages informed synthesis of a peptide. It disrupted chemoreception of cyst nematodes in the same timescale as dye-filling of certain of their chemoreceptive neurons. A site of action at chemoreceptive cell bodies and interneurons close to the nerve ring of cyst nematodes was strongly indicated by the uptake of fluorescently tagged peptide with a concurrent loss of chemoreception (Dong Wang, Leeds, UK, 2008, personal communication). BY2 tobacco cells were transformed to express the repellent with a signal for plant cell secretion. Soaking J2 of cyst nematodes in the supernatant fluid of the BY2 cells suppressed their ability to invade roots. The same effect can be achieved using the hydroponics solution in which the roots of similarly transformed transgenic potato plants have grown. These latter plants also suppress multiplication of both *Globodera* (Liu *et al.*, 2005) and *Meloidogyne* (Dong Wang, Leeds, UK, 2008, personal communication). A second repellent was isolated, which has a different neuronal target. It binds to nematode membranes and can be displaced from them by levamisole, which is a nicotine-like drug. Levamisole binds to some nicotinic acetylcholine receptors (nAChR) of nematodes, of which there are substantially more in *C. elegans* than in mammals (Brown *et al.*, 2006). Levamisole is a safe drug, widely used as a veterinary anthelmintic for gastrointestinal and pulmonary nematode infections in cattle and sheep (http://www.vmd.gov.uk/espcsite/documents/191903.doc). It has also been used as an adjuvant of fluorouracil in colon cancer therapy (Belle, 1972). To date, most attention on nACh receptors has concentrated on nematode muscle and so it is uncertain which member(s) of the large family occur at the synapses of nematode chemoreceptive neurons. Results to date suggest the second repellent is highly effective against *G. pallida* and *Meloidogyne*, particularly when controlled by a specific promoter that provides expression at the root tips where *Meloidogyne* invades. The levels of repellent detectable in hydroponics are low ($< 1\ \mu g/ml$) and it is not stable enough to persist in soil water. Therefore, the effect is probably due to suppressing invasion at the rhizoplane, or at least very close to the root surface. Further work is required if this technology, which is targeted at pre- rather than post-establishment, is to become a powerful component for stacked resistance against *Meloidogyne*.

15.6 The *Mi-1*-mediated Resistance Response

Natural resistance in tomato to root-knot nematodes is mediated by a rapid hypersensitive response. It is characterized by the generation of reactive oxygen species in plant root cells associated with the J2, which precludes the establishment of J2, and they either leave the root or die (Williamson, 1998). This response is mediated by a specific resistance gene (R-gene) in the plant. The simplistic view is that R-proteins are direct receptors for pathogen avirulence (Avr) proteins that trigger a defence response. This elicitor/receptor model may be true for some systems, but for many others there is no evidence of a direct R/Avr interaction, suggesting that a more complex system operates. The 'guard hypothesis' was originally proposed to rationalize the observed molecular interactions that occur during resistance to bacterial speck disease in tomato caused by *Pseudomonas syringae* (van der Biezen and Jones, 1998). It provides a premise for the action of disease effectors and the R-protein complex beyond simple R/Avr interactions. The hypothesis suggests that the R-gene product acts to guard the cellular components, which include virulence targets. The Avr proteins interact with and modify cellular proteins; the R-gene protein then perceives the altered status of the virulence target and induces a defence response. (for an R-gene review see Jones and Dangl, 2006).

The tomato *Mi-1.2*-based resistance to *Meloidogyne* spp. is the best-characterized example of natural nematode resistance. The same gene also offers protection against the potato aphid, *Macrosiphum euphorbiae* (Rossi *et al.*, 1998), and sweet potato whitefly, *Bemisia tabaci* (Nombela *et al.*, 2003). The product of *Mi-1.2* has been mechanistically well characterized (see Williamson and Kumar, 2006). The signal cascade that mediates the resistance has also been investigated. The recessive mutation, termed *rme1* (resistance to *Meloidogyne* spp.), is located in a single locus distinct from *Mi-1.2*. This mutation completely and specifically abolishes *Mi-1.2*-mediated resistance. The gene product of *RME1* remains unidentified and it may be a protein kinase acting either upstream of *Mi-1* or at the same early stage of the transduction pathway (Martinez de Ilarduya and Kaloshian, 2001; Martinez de Ilarduya *et al.*, 2004). Work that compared susceptible and *Mi-1*-based resistant varieties has suggested a role for a glycosyltransferase in the *Mi-1* pathway (Schaff *et al.*, 2007). At least one mitogen-activated protein-kinase (MAPK) and the cytosolic heat-shock protein Hsp90 have also been shown to be involved in the *Mi-1*-based signal cascade (Bhattarai *et al.*, 2007).

One of the aims of identifying and cloning natural resistance genes is to make them available for transfer to susceptible but agronomically important crops. Attempts to transfer *Mi*-mediated root-knot nematode resistance from tomato to tobacco have been unsuccessful (Williamson, 1998). The guard hypothesis provides a rationale for the lack of transgenic R-gene function in heterologous systems. The virulence targets that are guarded by the R-gene may be too divergent to be recognized by an R-gene ectopically expressed in a transgenic system. Results that are more promising were obtained when transgenic *Mi-1.2* was expressed in aubergine, which is taxonomically related to tomato (Goggin *et al.*, 2006). Transgenic lines sustained significantly lower levels of *Meloidogyne* reproduction and numbers of egg masses, but aphid performance was not compromised. In order for R-genes to produce a resistance response following transfer to a susceptible plant, it is a prerequisite that downstream components of the response cascade also be present. Failure of R-genes to produce resistance in heterologous species may be due to restricted taxonomic functionality (Tai *et al.*, 1999). The choice of promoters must also be considered. Under the expression of its native promoter, *Mi-1.2* produced a large range in levels of resistance in a susceptible tomato background, and showed a significant silencing effect in T_2 and T_3 generations (Goggin *et al.*, 2004). Additionally, there is an apparent dosage effect of R-gene copy number on the level of resistance. Jacquet *et al.* (2005) report that tomato lines homozygous at the *Mi* locus have significantly greater nematode resistance, while Chen *et al.* (2006) noted a decrease in resistance correlated with an increase in copy number of the transgene. Further work is required before natural resistance genes can become effective components of plant biotechnology programmes.

15.7 Efficacy and Durability

15.7.1 Efficacy

Many *Meloidogyne* spp. have wide host ranges, with a variation in the rate of multiplication that different hosts support. High host efficiency supporting a rapid population increase results in a higher chance of crop loss. For example, over 88 days, damage occurred with a starting density of 20 viable eggs/g soil for a good host (tomato) but not for intermediate or poor hosts. It is typical of *Meloidogyne* that, even if pre-planting population density is low, crop damage will occur if subsequent multiplication is high. Therefore, in effect, an aim for plant biotechnology is to render good hosts much less efficient so that damaging population densities do not accrue on roots.

Tomato can be taken as a good host, and three *Meloidogyne* generations are assumed to be typical over the growth of an annual crop. The relationship between plant weight and initial population density has been modelled using standard models (Seinhorst, 1967; Seinhorst and den Ouden, 1971), and also the relationship between initial and final egg densities (Ehwaeti *et al.*, 2000). This latter model can be applied to determine the levels of suppression of multiplication required by an effective transgenic resistance (Fig. 15.2). This suggests that 80% control per generation is sufficient to suppress multiplication over many generations, providing that the host is vigorous and has an efficiency that is no higher than that of tomato (Ehwaeti *et al.*, 2000). However, full control requires no more eggs post-harvest than pre-plant, and this requires a somewhat higher level of resistance (Fig. 15.3).

The level of resistance may not be constant throughout the growing season. For instance, protein synthesis declines generally when plants senesce, and roots may show an earlier decline in growth. Superimposed on this general pattern is the activity of the particular promoter driving the transgenic resistance gene. This unwanted effect will be either eliminated or reduced if the defence is expressed within the giant cells by promoters that are upregulated in these cells, such as RPL16A, ARSK1 and TUB-1 (Green *et al.*, 2002; Lilley *et al.*, 2004).

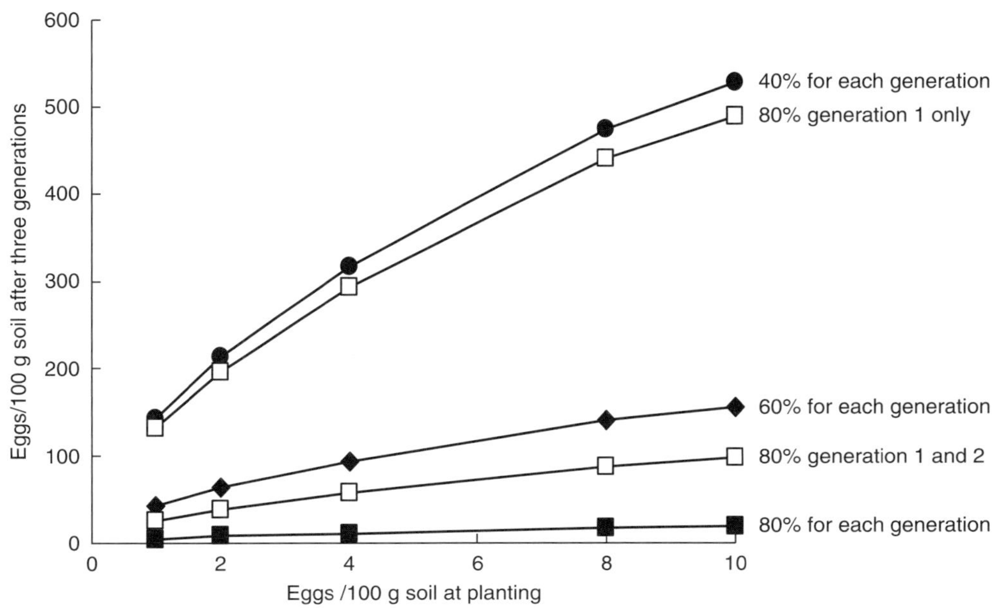

Fig. 15.2. A model of the relationship between initial and final population density of *Meloidogyne incognita* over three generations on tomato as a good host when all generations have an imposed transgenic resistance of 80, 60 or 40% mortality, or 80% mortality for either the first generation or the first two generations only. Based on model data of Ehwaeti *et al.* (2000) and restricted to final population densities that do not suppress yield.

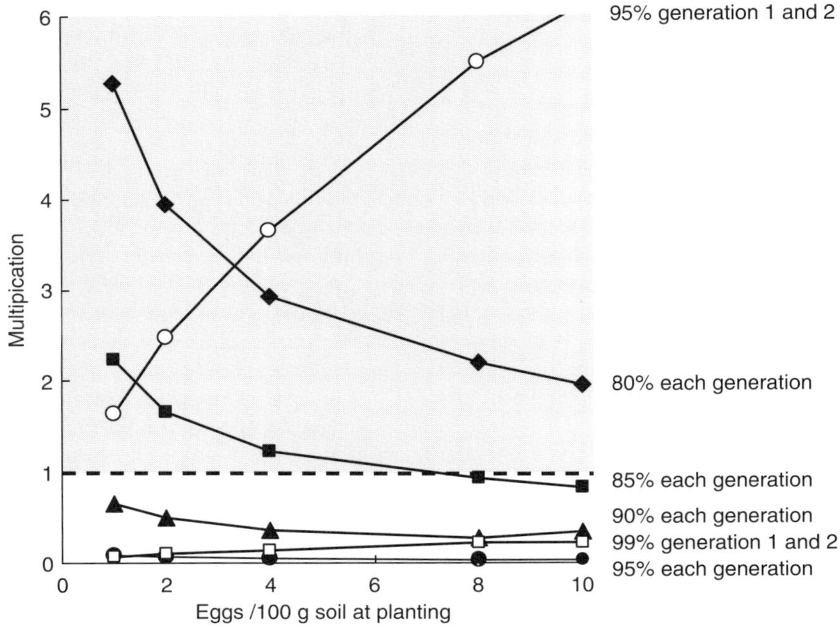

Fig. 15.3. Model of the relationship between multiplication and initial population density of *Meloidogyne incognita* after three generations on tomato as a good host when all generations have an imposed transgenic resistance of 80%, 85% 90% or 95% mortality imposed by transgenic resistance, or 90% or 95% mortality for the first two generations. Full control requires multiplication to <1×. Values above that are within the grey shaded area.

15.7.2 Durability

The section above suggests that a single transgene can prevent damage to a good host if 80% resistance is maintained for each generation of *Meloidogyne*. This value must be raised to 90% if a second susceptible crop is to be grown shortly after the first is harvested. The weakness of this approach is that partial resistance may occur. Even a loss of resistance from 80 to 60% causes an eight times increase in population density over three generations, which may not result in crop damage, but could compromise a follow-on crop. There is, therefore, a value in combinatorial approaches. Two defences of 80% each acting independently would be expected to provide 96% resistance and full virulence against one line of defence, or partial breakdown of both defences from 80 to 60% would still provide at least 80% resistance overall.

The value of stacked defences is evident if comparison is made with the deployment of *cry* genes against insects. Growers of transgenic, insect-resistant cotton in the USA plant non-transgenic cotton refuges to hinder selection of homozygous resistant insects (Bates *et al.*, 2005). This reduces the risk of resistance-breaking insects, but imposes a yield penalty. As a refuge strategy, it is inappropriate for *Meloidogyne* spp. as they lack the mobility it requires. A more recent approach is the stacking of dissimilar *cry* genes to widen the spectrum of protection. The advantages are a better life expectancy of the cultivar, better control of the whole insect pest complex and the need for smaller refuges. Australia now requires stacked *cry* genes in all planted transgenic, insect-resistant cotton, and the Environmental Protection Agency (EPA) has sanctioned that approach for the USA (Bates *et al.*, 2005). Combinatorial transgenic resistance would enhance the durability of any resistance deployed against nematodes. For insects it is considered superior to strategies that use resistance singly, but sequentially, at the field level (Zhao *et al.*, 2003).

The choice of targets for RNAi-mediated *Meloidogyne* resistance could have an impact on the durability of the resistance. For example, parasitism gene *16D10* encodes a 13 amino acid signalling peptide that is identical in the four common *Meloidogyne* spp., and is essential for parasitism. Therefore, any mutation in this gene to circumvent the RNAi resistance would more than likely alter the function of the peptide and the ability of the nematode to infect plants. However, there is the theoretical possibility that the mRNA sequence that encodes the 16D10 protein may vary without changing the amino acid sequence at all. If this were to happen, RNAi-based resistance strategies may fail due to the sequence-specific requirement of the technology.

15.8 Promoters for Transgenic Control of *Meloidogyne*

CaMV35S has been widely used in many biotechnological applications, and is often the first choice to demonstrate the proof of principle of new biotechnologies. Promoters that are spatially regulated in their pattern of expression can be used subsequently to restrict the expression profile of the transgenic technology. Promoters of TUB-1, a β-tubulin gene of *A. thaliana*, of RPL16, which encodes an *Arabidopsis* ribosomal protein L16, and of ARSK-1, a probable serine/threonine kinase, all direct expression of sufficient cystatin to provide partial resistance to *G. pallida* in the field and to *M. incognita* in containment. The ARSK promoter lines provide more resistance to *G. pallida* than *M. incognita*, whereas the other promoters were associated with less resistance against the cyst nematodes. All three promoters were active in the giant cells induced by *M. incognita*, but only ARSK-1 was also active in the syncytium of the cyst nematode (Lilley *et al.*, 2004). Molecular engineering can lead to promoters that are extremely specific. Deletion of the 5′ flanking region of a root-preferential promoter *Tob*RB7 resulted in a 300 bp promoter fragment just upstream of the coding region, which remained active within the giant cells induced by *M. incognita* and silenced in root meristems (Opperman *et al.*, 1994). While such promoters may have specificity, the strength of expression must also be considered. When Fairbairn *et al.* (2007) targeted a transcription factor of *M. javanica* in an *in planta* RNAi biotechnological strategy (see section 15.4), they tested the CaMV35S promoter and the Δ0.3*Tob*RB7 promoter. None of the lines containing the Δ0.3*Tob*RB7 promoter, in sharp contrast to those harbouring CaMV35S, showed any signs of silencing the targeted gene. Reporter plants with GUS (β-glucuronidase) under the control of Δ0.3*Tob*RB7 were subsequently analysed and only a small percentage of galls showed GUS activity, and those that did revealed only weak activity (Fairbairn *et al.*, 2007).

15.9 Biosafety

15.9.1 Food

Risks are the product of hazard and exposure. In the case of a novel protein developed for control of *Meloidogyne*, the hazard is its mammalian toxicity or allergenicity, whereas exposure is related to the quantity of the protein that could be ingested.

The EPA is responsible for food and environmental food safety in the USA. Its approach is directed at the novel ingredient in food and it requires data of several types to be obtained. Its approach provides one basis for assessing the food safety of any novel protein used to control *Meloidogyne* in the future. As one aspect, it seeks to define if the new protein behaves like any other in the diet, and is not structurally related to a known food allergen or toxic protein (Mendelsohn *et al.*, 2003). Current approaches are probably more reliable for assessment of toxicity than allergenicity. The latter is clearly important, but evaluation is not yet fully developed on a sound scientific basis. Some current assessments, such as the predictive value of short-peptide matches or predictions based on animal models, have uncertain value (Goodman *et al.*, 2008).

In vitro digestibility assays determine whether the protein, like many in the diet, is unstable in the presence of digestive fluids. This has often involved use of simulated gastric fluid. The cystatin OcIΔD86 was subject to this approach and lost >95% of its ability to inhibit papain after 15 s in simulated gastric fluid. It is therefore not stable to the effects of the first part of the digestion process (Atkinson *et al.*, 2004b). This favours its lack of toxicity and makes it less likely that the whole protein is an allergen. A second useful assay is the heat stability of the protein. OcIΔD86 lost 50% of its activity in fewer than 10 min of boiling (Atkinson *et al.*, 2004b). The EPA also sees value in delivering a single, high dose of the novel protein to mice (3.28–5.0 g/kg) orally to evaluate toxic effects. The weakness of this experimental approach is the high need for isolated protein. This cannot be overcome by delivering high doses within the food derived from the transgenic plant unless enrichment is used to ensure an adequate diet of the rodent (http://royalsociety.org/displaypagedoc.asp?id=6170). A different approach is the only one attempted to date for anti-nematode proteins. Groups of rats were given 0.1–10 mg cystatin/kg body weight by oral gavage daily for 28 days, and a detailed analysis of organ change and blood chemistry conducted. The results suggest that the no-effect level for OcIΔD86 is >10 mg/kg/day. This provides a range of dietary exposure of at least 200, which was increased to 2000-fold when a promoter to favour expression of OcIΔD86 in roots was used rather than potato tubers (Atkinson *et al.*, 2004b).

Food safety will be less of an issue with RNAi-derived nematode resistance, as the transgenic plants do not express new proteins. Another approach for assessing food safety was devised by The Organisation for Economic Cooperation and Development (OECD). It is termed substantial equivalence (Kok and Kuiper, 2003). It was introduced with the aim of establishing a scientifically sound, globally accepted approach to safety, based on the whole food rather than the individual novel ingredient. It is a comparative approach, and difficulties arise in defining the control plants if the non-transgenic parent is not available. It is meant only as a basis for identifying effects deserving close scrutiny. It does take into account indirect effects of transformation, which are not evaluated by studying the isolated novel protein. There is a considerable range of opinion about the scientific basis of this approach. One concern is any bias associated with changes in food composition that are associated with transgenic crops, without parallel effort to scrutinize any such changes in conventional crops (arising from factors such as cultivar and environmental stress).

15.9.2 Environment

To date, any environmental impact of transgenic plants with resistance to *Meloidogyne* spp. has not been reported. However, some principles can be established from work with such plants for control of other nematodes. A key issue is that it is necessary for risk assessment to determine both the hazard and the exposure, and not just one or the other. An example where this was not done, according to Johnson *et al.* (2006), was the report of the hazard of Bt expressed in maize pollen to monarch butterfly (Losey *et al.*, 1999) without appropriate consideration of exposure. Hopefully future work with *Meloidogyne* spp. will avoid this error. A second key issue is that any assessment must be based on a case-by-case basis, considering the crop, its geographical location in the world and the transgene product.

Crops can be placed into different risk categories based on the likelihood of gene transfer from them to crops or wild plants of other species (Stewart *et al.*, 2003). At the safest extreme are plants that are fully sterile, such as banana, followed by those for which there is no molecular evidence of introgression to wild plants. The next category is for those plants for which introgression to wild plants is rare, and consequently the transgenic trait defines the risk. The highest-risk crops are those for which introgression occurs to wild relatives that are known agricultural weeds. This occurs for sorghum grown within cross-pollination distance of Johnston's grass (Stewart *et al.*, 2003; Schmid and Bothma, 2006).

Any introgression is most likely to be stably inherited if the novel trait imposes an ecological advantage on the recipient plant. Any transfer of *Meloidogyne* resistance from a crop would have ecological consequences if the recipient species is susceptible to the nematode or other organism to which resistance is also gained. It is valuable to know the influence of *Meloidogyne* on the vigour or seed production of any wild plant that might benefit from such stable introgression. Such an event may eventually influence its prevalence in the habitat. *Meloidogyne* are known to affect plant ecology in at least one natural community (Brinkman *et al.*, 2005), so this issue is worthy of further consideration, particularly where this nematode occurs outside fields. This issue is also pertinent to natural resistance. For example, gene flow occurs within Peruvian fields from potato to its wild relatives that are susceptible to at least *Globodera*, and so any natural resistance gene introduced to potato could flow to any wild relative able to express it stably and become resistant. This illustrates the point that where the crop is grown also needs to be considered in relation to introgression to wild relatives. Potato has *c.* 130 wild relatives in Peru (Celis *et al.*, 2004) but very few in many countries where it is grown. The situation is more complex when gene flow may occur. It depends on pollen viability, pollen completion, the extent of overlapping flowering time, flowering dynamics and a range of landscape, canopy and meteorological variables (Kuparinen *et al.*, 2007).

A distinct issue is the chance of lateral movement of the gene of interest to adjacent fields of the same or a cross-fertile crop species. This raises difficulties for the coexistence of transgenic and conventional crops. In that case, the isolation distance is an important issue, as is the tolerance level of rare cross-pollination events. This will differ according to the extent of outcrossing that is likely and for wind- and insect-pollinated crops. Any analysis of exposure of the recipient crop should also take account of local conditions and the biology of the pollinator.

Another of the risk assessments is at the level of the gene product. Both dsRNA and a protein could, in theory, have targets in other organisms. The more *Meloidogyne*-specific the defence, the less likely are such concerns. The only detailed consideration for non-target organisms to date is for cystatin-expressing potato. These plants have efficacy against *Meloidogyne*, and so many of the conclusions reached are valid for this nematode as well as for *Globodera* in UK field trials. The mode of action of any novel protein will be known. Therefore, it is possible to devise assays based on the known mode of action to evaluate which non-target invertebrates associated with the crop may be vulnerable. Cystatins inhibit cysteine proteinases, which are important digestive enzymes of many invertebrates, but not of mammals. Simple histochemical assays can determine which non-target invertebrate crop associates have digestive cysteine proteinases. The impact on each can then be studied. This was done for aphids, leafhoppers, mites and collembola for cystatin-expressing potato in UK field trials; it was followed by monitoring for any changes in their abundance under field conditions. No such effects were detected, even for potatoes expressing the cystatin constitutively (Cowgill *et al.*, 2002a; Keizebrink and Atkinson, 2004). There was also no adverse effect on the hymenopteran parasitoids that interact with the potato aphids (Cowgill *et al.*, 2004). It is valuable to compare effects with those imposed by conventional control, particularly when that involves nematicide use to control *Meloidogyne*. For instance, cystatin-expressing plants do not harm earthworm numbers, in contrast to the nematicide aldicarb (Keizebrink and Atkinson, 2004).

The soil microbial community requires careful study for any novel protein targets directed at *Meloidogyne*, particularly as an impact on them has been reported for plants expressing a Cry protein (Donegan *et al.*, 1995). Appropriate approaches include phospholipid fatty acid profiles, and the use of BIOLOG plates to determine the range of substrates used by the microbial community. In the case of potato plants expressing a cystatin, 3 years of data at the same site indicated that soil microbes were not adversely affected by the transgenic lines used (Cowgill *et al.*, 2002b; Celis *et al.*, 2004; Kiezebrink and Atkinson, 2004). The changes that occurred reflected both spatial and temporal variations in soil abiotic conditions. The reality is that most transgenic plants have only minor effects on the soil system compared with differences between cultivars, or those associated with weather or season (Lilley *et al.*, 2006). Tissue-specific expression also reduces the exposure of non-target organisms. Appropriate promoters are available for targeted expression of bionematicides to the feeding cells of *Meloidogyne* (Lilley *et al.*, 2004). Their use should be limited to enhancing the biosafety of inherently benign novel proteins.

One way forward after the extent of impacts on a soil organism has been defined is to apply principal component analysis (Celis *et al.*, 2004). The aim is to define an acceptable impact based on it being no more than results from common agricultural practices – such as the current free choice of crop species or variety by the grower (Fig. 15.4). It is necessary to ensure that any changes in soil do not have consequences for follow-on crops. Currently, population size and activity measurement for soil bacteria, fungi and fauna may detect interesting effects (Lilley *et al.*, 2006), but robust methodology is not yet in place

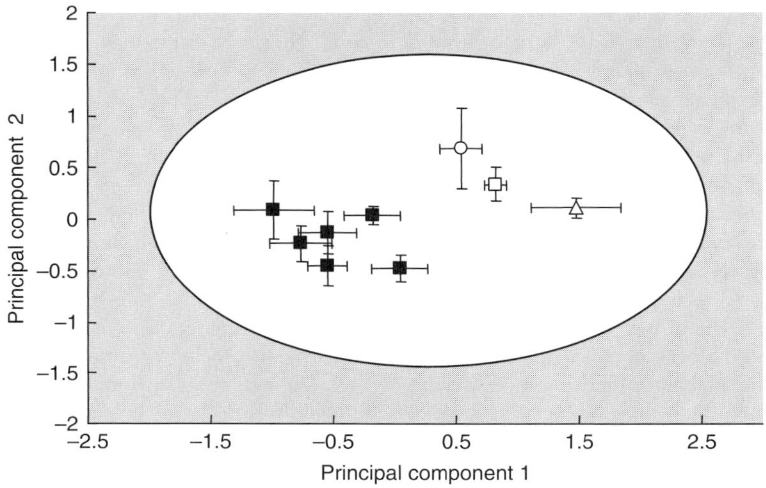

Fig. 15.4. Community-level physiological profiles for microorganisms in soil supporting a range of crops. It is proposed that any transgenic crop used for *Meloidogyne* control imposes acceptable change to the microorganisms if this remains within limits set by changes (unshaded area) imposed by conventional crops freely chosen by the grower. Data in this example are varieties of South American potato (closed squares), potato bred in Europe (cv. Désirée, open squares), broad bean (open circles) and lupin (open triangles). Illustrative data are based on Celis *et al.* (2004).

to evaluate any long-term impact of such plans on soil health.

Unfortunately, environmental risk assessment research does not seem to assist decision makers, partly because it rarely defines problems in relation to policy objectives (Raybould, 2007). Whatever methodology prevails in the future, it would seem reasonable to suggest that the level of scrutiny should fall progressively once initial studies identify no concerns, and again if subsequent secondary and tertiary sites also suggest that the plant biotechnology under scrutiny is benign. It should end if widespread uptake does not result in detection of either substantial incidents or progressive adverse effects.

15.10 Developing World Needs

15.10.1 The need for biotechnology to control *Meloidogyne* in the developing world

Global food insecurity is predicted to decline from 930 million people in 1970 to an estimated 576 million by 2015, but 33% of sub-Saharan Africans are undernourished and that region will remain food insecure (ftp://ftp.fao.org/docrep/fao/008/a0200e/a0200e.pdf). Predicted improvements may prove optimistic if world food prices increase or climate change effects on crop yields are larger than previously predicted. Kofi Annan, while Director General of the UN, called for a uniquely African Green Revolution to address that challenge (http://www.un.org/News/Press/docs/2004/sgsm9405.doc.htm). He also favoured considering the potential of biotechnology, and this has been advocated by many scientists, including 25 Nobel laureates (http://www.agbioworld.org/declaration/nobelwinners.html). Plant biotechnology is required to address the key agricultural challenges of countries left behind by the green revolution (http://hdr.undp.org/reports/global/2001/en/). Improved *Meloidogyne* control is one need, given that it causes nematode damage globally, with common association with subsistence farming, often on orphan crops and those for which R-genes are not well developed.

Rice provides a good example of the need for biotechnology to improve control of *Meloidogyne* in the developing world. It also demonstrates that the issue is not limited to orphan crops. In Asia, the annual growth rate of yields was about 2.5%

for a decade from 1965, *c.* 3.3% in the following 10 years and only about 1.6% in the following decade (http://www.fao.org/docrep/V6017T/V6017T03.htm). The yields from the rice–wheat cropping system are now static (Ladha *et al.*, 2003). This is a concern because a further increase in the Asian cereal harvest is needed, from 545 million t to about 700 million t by 2025 (http://www.fao.org/docrep/008/y5682e/y5682e0c.htm#bm12). This requires increasing yields as no additional, suitable land is available. If biotechnology was fully deployed, control of *Meloidogyne* and other nematodes could be provided in rice while other traits are still under development.

Much of Indian agriculture is rainfed, and about 30% of its rice is lost to the three most damaging nematode species, i.e. *M. graminicola*, *Pratylenchus indicus* and *Heterodera oryzicola* (Prasad *et al.*, 1987). Rice attacked by *Meloidogyne* often shows limited galling or other obvious symptoms, and its presence and the damage it causes is underestimated. The limited genetic pool of resistance available, the concurrent need to control other nematodes and the limitations of chemical treatment in low-input systems all combine to ensure a timely and essential opportunity for plant biotechnology. *Meloidogyne* control has added importance because the nematode damages other crops grown in rotation. Early-maturing cereals now allow three crops per year (e.g. wheat, rice and vegetables). This has brought great benefits to the Ghangetic plain but intensified the pest status of *Meloidogyne*. Options discussed earlier are needed and should be deployed soon for rice in India and elsewhere.

15.10.2 Appropriate technology

The criticisms raised of biotechnology include promotion of inappropriate traits, crops and cultivars, a dependence on additional inputs, overdependence on companies and a lack of focus on the needs of small farmers (http://www.ids.ac.uk/idS/KNOTS/PDFs/Briefing10.pdf). Therefore, there is a need to establish the value of *Meloidogyne* control. This can succeed if benefits are demonstrated under local conditions. It requires an environment in which progress can be made, ensuring the importance of countries like China and India in the process. A national commitment to agricultural development must extend beyond the scientific base, to government and other stakeholders. This requires a scientific infrastructure able to be involved in product development and evaluation, plus biosafety scrutiny. A media willing to accept that plant biotechnology has a role in development is essential while adopting a pro-consumer stance that scrutinizes all aspects. An objective and informed press is an important alternative to non-governmental organizations (NGOs) influencing opinion with unsubstantiated and misleading information, as apparently has occurred in Zambia (Bodulovic, 2005).

Donation of genes and other technology is the best route forward for most developing countries for the foreseeable future, given the high cost of new trait discovery. There is also background intellectual priority to consider, which can be a complex issue (Beyer *et al.*, 2002). Therefore, an intellectual property (IP) audit must be conducted before deploying any *Meloidogyne* resistance, to assess which background patent assignees should be requested to provide a freedom to operate on the chosen crop. Forceful points to include are humanitarian aims, and several not-for-profit organizations now exist to promote this possibility.

15.11 Conclusions and Future Directions

15.11.1 Proteinase inhibitors

To date, the best-characterized approach for control is delivery of proteinase inhibitors from the plant to disrupt feeding of *Meloidogyne*. The efficacy of the defence has been established against a number of plant-parasitic nematode species. This is an important feature, as controlling just one nematode on one or more crops may often be insufficient to justify the investment levels needed to bring the novel resistance to the market. Cystatins have been established to provide effective nematode resistance in a number of host plant species; the molecular targets have been well characterized and the mode of action of the technology has been described. The effects on the environment and non-target organisms have been shown to be benign. Allergenicity and toxicological studies have shown that this approach does not pose a risk to human health. While the technology provides high levels of control, industry has not commercialized the approach. An important factor here is that it probably judges that transgenic resistance currently lacks consumer acceptance in food for major markets.

15.11.2 Cry proteins

Work on the tertiary structure of Cry6 protein is needed because it is an outlier of the main *cry* lineage (Crickmore *et al.*, 1998; Griffitts and Aroian, 2005). Description of its intestinal epithelium receptor(s) in *Meloidogyne* and *C. elegans* would also be of value. The toxicity of the main *cry* lineages providing Cry 5, 13, 14A and 21A to *Meloidogyne* should also be determined, as this would allow much knowledge to be drawn on from the study of insects. Mutants of *C. elegans* have been identified that are resistant to Cry5B toxin, but they remain susceptible to the Cry6A toxin, suggesting that the toxin-binding domains show different receptor specificity (Marroquin *et al.*, 2000). Therefore, there would be value in defining the diversity of the receptors that Cry6 recognizes in *Meloidogyne*. Screening for novel anti-nematode Cry proteins and receptor specificities would help the development of novel and durable resistance to *Meloidogyne*. Such work would be wise to anticipate resistance to Cry proteins and help underpin an appropriate protein-engineering response.

More work is needed to determine the efficacy achieved against *Meloidogyne* spp. and if one or a combination of Cry proteins can control a range of nematode species. A key advantage of the approach is its proven record for successful commercialization for the control of insects (Li *et al.*, 2007).

15.11.3 RNAi

Effective control of *Meloidogyne* been reported for the emerging technology of RNAi, but the literature is less developed for the *in planta* application of this technology to cyst nematodes. As the technology develops, field trials will be conducted to assess its potential and to determine if the technology can protect against several parasitic species of nematode that attack the same host. Targeting just *Meloidogyne* spp. may be commercially viable because of its high economic importance (Huang *et al.*, 2006a). Key features that can be advocated are lack of impact on non-target species and no transgenic protein presence. This is an important advantage for both food and environmental safety.

15.11.4 Commercial prospects of deployment of transgenic resistance to *Meloidogyne*

Advances in plant biotechnology have allowed several approaches for control of *Meloidogyne* to be demonstrated. A high efficacy must be shown at the prototype level before substantial investment is made by agribusiness to develop commercial products. One approach is to place transgenic resistance in a background of partial natural resistance (Urwin *et al.*, 2003). Efficacy can also be improved by stacking biotechnological defences together in the same plant for combinatorial effects. Stacking supports the durability required to ensure that investment in new cultivar development can be recovered.

If attitudes common in Europe prevail elsewhere, then commercialization may be very slow. The precautionary principle that is embedded in some European polices is one source of delay to progress, to the extent that it does not consider transgenic plants and non-transgenic plants offering the same trait and environmental risk similarly (Morris, 2007). The precautionary approach supports public policy action when there is a potentially serious or irreversible threat to health or the environment. It has been applied to issues as diverse as food safety and climate change (Harremoës, 2002; Immordino, 2003). The weakness is that it is not clear when sufficient knowledge has accumulated for a transgene to render this approach redundant. This seems a matter more of politics than scientific accuracy (Morris, 2007). One example of the precautionary principle's application internationally is to trans-boundary movement of any living, modified organism resulting from modern biotechnology that may have an adverse effect on the conservation and sustainable use of biological diversity (The Cartagena Protocol 2000; http://www.biodiv.org/doc/legal/cartagena-protocol-en.pdf). Currently, the protocol is a disincentive for the uptake of benefits such as plant biotechnology-based resistance to *Meloidogyne* in developing nations. It does not balance concerns against benefits for future food security (De Grief, 2004).

One consequence of the limited uptake of transgenic plants in Europe is that some of its plant biotechnology businesses have relocated to the USA, so reducing European influence on the directions of the industry. Companies have also contributed to slow market uptake by needing to recover their investment with short-term benefits. Pressure groups and public activists have influenced public opinion by arguing a lack of benefits that the consumer appreciates (Tait and Chataway, 2007). As a result, products of clear agricultural utility, such as Cry-expressing potato for insect control, are not on the market. The transgenic crop base was just 11 crops in 2007, with most of the area of 114.3 million ha dominated by soybean, cotton, maize and canola (James, 2007). The narrow crop base may impede deployment of transgenic vegetables and other crops for *Meloidogyne* control.

A reduction in pesticide use on cotton has been one outcome of the uptake of Cry technology in many parts of the world, such as China (Huang *et al.*, 2005). Policy makers and government bodies should have done more to develop policies that rewarded companies for developing products with public benefits, such as the social benefits reported in China. For instance, *Meloidogyne*-resistant strawberries to replace the former use of methyl bromide could be used to demonstrate both consumer and environmental benefits. The absence of such progressive policies contributed to emphasis on labelling in Europe, rather than the safety-of-ingredients approach that prevails in the USA. Many EU citizens consider that they can afford any premium needed to avoid transgenic crop products. Any cost imposed by the lack of ability to choose them in the marketplace is also borne by about 78 million people in Europe who are at risk of poverty (http://www.poverty.org.uk/summary/eapn.shtml).

15.11.5 Prospects of uptake in support of food security

There is a pro-poor need to modify the attitudes of industry, scientists, policy makers and governments if *Meloidogyne*-resistant transgenic plants are to be widely deployed. The EU's unwillingness to accept transgenic crops even in the light of favourable evidence may be having an adverse impact on development in Africa (Bodulovic, 2005), and presumably elsewhere. This is unfortunate, as the contrast in needs is stark. Africa and Europe have 10 and 15% of the world's population and an estimated 1.1 and 29.5% of global wealth distribution, respectively (http://www.wider.unu.edu/research/2006-2007/2006-2007-1/wider-wdwh-launch-5-12-2006/wider-wdhw-report-5-12-2006.pdf). Those in the developed world need to give more weight to the life or death concerns of the hungry than to the less pressing concerns of the well fed (http://www.nuffieldbioethics.org/go/textonly/ourwork/gmcrops/introduction.html). Forgoing possible benefits for unlikely risks invokes the fallacy of thinking that doing nothing is itself without risk to the poor (2003 follow-up to Nuffield bioethics report, URL as above).

15.11.6 Rate of uptake possible

Unfortunately, even if their deployment was certain, the development of substantially improved crops takes longer than is ideal for countries keen to adopt plant biotechnology. Scientific progress is necessarily limited by the duration of extensive field trials to establish benefits. Collection of biosafety-related information to assess risks is also required for transgenic crops. The most favourable approach to advance plant biotechnology is for countries to have a national biotechnology strategy and well-founded national research laboratories (Eicher *et al.*, 2006), populated with trained plant biotechnologists whom they can retain. This provides a basis for progress and equitable interaction with international laboratories. The national capacity must also extend to biosafety plus regulatory expertise and effective IP management (Raney, 2006). Scientific independence is necessary for developing nations to determine when plant biotechnology can underpin food security and development.

15.12 References

Atkinson, H.J., Urwin, P.E. and McPherson, M.J. (2003) Engineering plants for nematode resistance. *Annual Review of Phytopathology* 41, 615–639.

Atkinson, H.J., Grimwood, S., Johnston, K. and Green, J. (2004a) Prototype demonstration of transgenic resistance to the nematode *Radopholus similis* conferred on banana by a cystatin. *Transgenic Research* 13, 135–142.

Atkinson, H.J., Johnston, K.A. and Robbins, M. (2004b) Prima facie evidence that a phytocystatin for transgenic plant resistance to nematodes is not a toxic risk in the human diet. *Journal of Nutrition* 134, 431–434.

Bakhetia, M., Urwin, P.E. and Atkinson, H.J. (2007) qPCR analysis and RNAi define pharyngeal gland cell-expressed genes of *Heterodera glycines* required for initial interactions with the host. *Molecular Plant–Microbe Interactions* 20, 306–312.

Barrows, B.D., Haslam, S.M., Bischof, L.J., Morris, H.R., Dell, A. and Aroian, R.V. (2007) Resistance to *Bacillus thuringiensis* toxin in *Caenorhabditis elegans* from loss of fucose. *Journal of Biological Chemistry* 282, 3302–3311.

Bates, S.L., Zhao, J-Z., Roush, R.T. and Shelton, A.M. (2005) Insect resistance management in GM crops: past, present and future. *Nature Biotechnology* 23, 57–62.

Baulcombe, D.C. (2007) Amplified silencing. *Science* 315, 199–200.

Baum, J.A., Bogaert, T., Clinton, W., Heck, G.R., Feldmann, P., Ilagan, O., Johnson, S., Plaetinck, G., Munyikwa, T., Pleau, M., Vaughn, T. and Roberts, J. (2007) Control of coleopteran insect pests through RNA interference. *Nature Biotechnology* 25, 1322–1326.

Belle, H. (1972) Kinetics and inhibition of rat and avian alkaline phosphatases. *General Pharmacology: the Vascular System* 7, 53–58.

Beyer, P., Al-Babili, S., Ye, X., Lucca, P., Schaub, P., Welsch, R. and Potrykus, I. (2002) Golden rice: introducing the β-carotene biosynthesis pathway into rice endosperm by genetic engineering to defeat vitamin A deficiency. *Journal of Nutrition* 132, 506S–510S.

Bhattarai, K.K., Li, Q., Liu, Y., Dinesh-Kumar, S.P. and Kaloshian, I. (2007) The *Mi-1*-mediated pest resistance requires *Hsp90* and *Sgt1*. *Plant Physiology* 144, 312–323.

Bodulovic, G. (2005) Is the European attitude to GM products suffocating African development? *Functional Plant Biology* 32, 1069–1075.

Bolle, C. (2004) The role of GRAS proteins in plant signal transduction and development. *Planta* 218, 683–692.

Borgonie, G., Claeys, M., Leyns, F., Arnaut, G., De Waele, T. and Coomans, D.A. (1996) Effect of nematicidal *Bacillus thuringiensis* strains on free-living nematodes. 1. Light microscopic observations, species and biological stage specificity and identification of resistant mutants of *Caenorhabditis elegans*. *Fundamental and Applied Nematology* 19, 391–398.

Bravo, A., Gill, S.S. and Soberón, M. (2007) Mode of action of *Bacillus thuringiensis* Cry and Cyt toxins and their potential for insect control. *Toxicon* 49, 423–435.

Brinkman, E.P., Duyts, H. and Van der Putten, W.H. (2005) Consequences of variation in species diversity in a community of root-feeding herbivores for nematode dynamics and host plant biomass. *Oikos* 110, 417–427.

Brown, L.A., Jones, A.K., Buckingham, S.D., Mee, C.J. and Sattelle, D.B. (2006) Contributions from *Caenorhabditis elegans* functional genetics to antiparasitic drug target identification and validation: nicotinic acetylcholine receptors, a case study. *International Journal for Parasitology* 36, 617–624.

Byzova, M., Verduyn, C., De Brouwer, D. and De Block, M. (2004) Transforming petals into sepaloid organs in *Arabidopsis* and oilseed rape: implementation of the hairpin RNA-mediated gene silencing technology in an organ-specific manner. *Planta* 218, 379–387.

Cai, D., Thurau, T., Tian, Y., Lange, T., Yeh, K-W. and Jung, C. (2003) Sporamin-mediated resistance to beet cyst nematodes (*Heterodera schachtii* Schm.) is dependent on trypsin inhibitory activity in sugar beet (*Beta vulgaris* L.) hairy roots. *Plant Molecular Biology* 51, 839–849.

Celis, C., Scurrah, M., Cowgill, S.E., Chumbiauca, S., Franco, J., Main, G., Keizebrink, D.T., Green, J. and Atkinson, H.J. (2004) Environmental biosafety and transgenic potato in a centre of diversity for this crop. *Nature* 432, 222–225.

Chen, Q., Rehman, S., Smant, G. and Jones, J.T. (2005) Functional analysis of pathogenicity proteins of the potato cyst nematode *Globodera rostochiensis* using RNAi. *Molecular Plant–Microbe Interactions* 18, 621–625.

Chen, R.G., Zhang, L.Y., Zhang, J.H., Zhang, W., Wang, X., Ouyang, B., Li, H.X. and Ye, Z.B. (2006) Functional characterization of *Mi*, a root-knot nematode resistance gene from tomato (*Lycopersicon esculentum* L.). *Journal of Integrative Plant Biology* 48, 1458–1465.

Cowgill, S.E., Wright, C. and Atkinson, H.J. (2002a) Transgenic potatoes with enhanced levels of nematode resistance do not have altered susceptibility to nontarget aphids. *Molecular Ecology* 11, 821–827.

Cowgill, S.E., Bardgett, R.D., Kiezebrink, D.T. and Atkinson, H.J. (2002b) The effect of transgenic nematode resistance on non-target organisms in the potato rhizosphere. *Journal of Applied Ecology* 39, 915–923.

Cowgill, S.E., Danks, C. and Atkinson, H.J. (2004) Multitrophic interactions involving genetically modified potatoes, nontarget aphids, natural enemies and hyperparasitoids. *Molecular Ecology* 13, 639–647.

Crickmore, N. (2005) Using worms to better understand how *Bacillus thuringiensis* kills insects. *Trends in Microbiology* 13, 347–350.

Crickmore, N., Zeigler, D.R., Feitelson, J., Schnepf, E., van Rie, J., Lereclus, D., Baum, J. and Dean, D.H. (1998) Revision of the nomenclature for the *Bacillus thuringiensis* pesticidal crystal proteins. *Microbiology and Molecular Biology Reviews* 62, 807–813.

Davis, E.L., Hussey, R.S. and Baum, T.J. (2004) Getting to the roots of parasitism by nematodes. *Trends in Parasitology* 20, 134–141.

Davuluri, G.R., van Tuinen, A., Fraser, P.D., Manfredonia, A., Newman, R., Burgess, D., Brummell, D.A., King, S.R., Palys, J., Uhlig, J., Bramley, P.M., Pennings, H.M.J. and Bowler, C. (2005) Fruit-specific RNAi-mediated suppression of *DET1* enhances carotenoid and flavonoid content in tomatoes. *Nature Biotechnology* 23, 890–895.

De Grief, W. (2004) The Cartagena Protocol and the future of agbiotech. *Nature Biotechnology* 22, 811–812.

de Maagd, R.A., Bravo, A. and Crickmore, N. (2001) How *Bacillus thuringiensis* has evolved specific toxins to colonize the insect world. *Trends in Genetics* 17,193–99.

de Maagd, R.A., Bravo, A., Berry, C., Crickmore, N. and Schnepf, H.E. (2003) Structure, diversity and evolution of protein toxins from spore-forming entomopathogenic bacteria. *Annual Review of Genetics* 37, 409–433.

Devidas, P. and Rehberger, L.A. (1992) The effects of exotoxin Thuringiensin from *Bacillus thuringiensis* on *Meloidogyne incognita* and *Caenorhabditis elegans*. *Plant and Soil* 145, 115–120.

Donegan, K.K., Palm, C.J., Fieland, V.J., Porteous, L.A., Ganio, L.M., Schaller, D.L., Bucao, L.Q. and Seidler, R.J. (1995) Changes in levels, species and DNA fingerprints of soil-microorganisms associated with cotton expressing the *Bacillus thuringiensis* var kurstaki endotoxin. *Applied Soil Ecology* 2, 111–124.

Ehwaeti, M.E., Elliot, M.J., McNicol, J.M., Phillips, M.S. and Trudgill, D.L. (2000) Modelling nematode population growth and damage, *Crop Protection* 19, 739–745.

Eicher, C.K., Maredia, K. and Sithole-Naing, I. (2006) Crop biotechnology and the African farmer. *Food Policy* 31, 504–527.

Escobar, M.A., Civerolo, E.L., Summerfelt, K.R. and Dandekar, A.M. (2001) RNAi-mediated oncogene silencing confers resistance to crown gall tumorigenesis. *Proceedings of the National Academy of Sciences of the USA* 98, 13437–13442.

Fairbairn, D.J., Cavallaro, A.S., Bernard, M., Mahalinga-Iyer, J., Graham, M.W. and Botella, J.R. (2007) Host-delivered RNAi: an effective strategy to silence genes in plant parasitic nematodes. *Planta* 226, 1525–1533.

Fanelli, E., Di Vito, M., Jones, J.T. and De Giorgi, C. (2005) Analysis of chitin synthase function in a plant parasitic nematode, *Meloidogyne artiellia*, using RNAi. *Gene* 349, 87–95.

Fire, A., Xu, S., Montgomery, M.K., Kostas, S.A.., Driver, S.E. and Mello, C.C. (1998) Potent and specific genetic interference by double-stranded RNA in *Caenorhabditis elegans*. *Nature* 391, 806–811.

Fleming, C.C., McKinney, S., McMaster, S., Johnston, M.J.G., Donnelly, P., Kimber, M.J. and Maule, A.G. (2007) Getting to the root of neuronal signaling in plant-parasitic nematodes using RNA interference. *Nematology* 9, 301–315.

Gheysen, G. and Vanholme, B. (2007) RNAi from plants to nematodes. *Trends in Biotechnology* 25, 89–92.

Gill, M. and Ellar, D. (2002) Transgenic *Drosophila* reveals a functional *in vivo* receptor for the *Bacillus thuringiensis* toxin Cry1Ac1. *Insect Molecular Biology* 11, 619–625.

Goggin, F.L., Shah, G., Williamson, V.M., and Ullman, D.E. (2004) Instability of *Mi*-mediated nematode resistance in transgenic tomato plants. *Molecular Breeding* 13, 357–364.

Goggin, F.L., Jia, L.L., Shah, G., Hebert, S., Williamson, V.M. and Ullman, D.E. (2006) Heterologous expression of the *Mi-1.2* gene from tomato confers resistance against nematodes but not aphids in eggplant. *Molecular Plant–Microbe Interactions* 19, 383–388.

Goodman, R.E., Veiths, S., Sampson, H.A., Hill, D., Ebisawa, M., Taylor, S.L. and van Ree, R. (2008) Allergenicity assessment of genetically modified crops – what makes sense? *Nature Biotechnology* 26, 73–81.

Green, J., Vain, P., Fearnehough, M.T., Worland, B., Snape, J.W. and Atkinson, H.J. (2002) Analysis of the expression patterns of the *Arabidopsis thaliana* tubulin-1 and *Zea mays* ubiquitin-1 promoters in rice plants in association with nematode infection. *Physiological and Molecular Plant Pathology* 60, 197–205.

Griffitts, J.S. and Aroian, R.V. (2005) Many roads to resistance: how invertebrates adapt to Bt toxins. *Bioessays* 27, 614–624.

Griffitts, J.S., Halsam, S.M., Yang, T., Garczynski, S.F., Mulloy, B., Morris, H., Cremer, P.S., Dell, A., Adang, M.J. and Aroian, R.V. (2005) Glycolipids as receptors for *Bacillus thuringiensis* crystal toxin, *Science* 307, 922–925.

Grishok, A. (2005) RNAi mechanisms in *Caenorhabditis elegans. FEBS Letters* 579, 5932–5939.

Harremoës, P., Gee, D., MacGarvin, M., Stirling, A., Keys, J., Wynne, B. and Guedes Vaz, S. (eds) (2002) Introduction. In: *The Precautionary Principle in the 20th Century; Late Lessons from Early Warnings.* Earthscan, London.

Hedgecock, E.M., Culotti, J.G., Thomson, J.N. and Perkins, L.A. (1985) Axonal guidance mutants of *Caenorhabditis elegans* identified by filling sensory neurons with fluorescein dyes. *Developmental Biology* 111, 158–170.

Hepher, A. and Atkinson, H.J. (1992) Nematode control with protease inhibitors. European Patent Publication Number 0 502 730 A1.

Huang, G.Z., Gao, B., Maier, T., Allen, R., Davis, E.L., Baum, T.J. and Hussey, R.S. (2003) A profile of putative parasitism genes expressed in the esophageal gland cells of the root-knot nematode *Meloidogyne incognita. Molecular Plant–Microbe Interactions* 16, 376–381.

Huang, G., Dong, R., Maier, T., Allen, R., Davis, E.L., Baum, T.J. and Hussey, R.S. (2004) Use of solid-phase subtractive hybridization for the identification of parasitism gene candidates from the root-knot nematode *Meloidogyne incognita. Molecular Plant Pathology* 5, 217–222.

Huang, G., Allen, R., Davis, E.L., Baum, T.J. and Hussey, R.S. (2006a) Engineering broad root-knot resistance in transgenic plants by RNAi silencing of a conserved and essential root-knot nematode parasitism gene. *Proceedings of the National Academy of Sciences of the USA* 103, 14302–14306.

Huang, G.Z., Dong, R., Allen, R., Davis, E.L., Baum, T.J. and Hussey, R.S. (2006b). A root-knot nematode secretory peptide functions as a ligand for a plant transcription factor. *Molecular Plant–Microbe Interactions* 19, 463–470.

Huang, J., Hu, R., Rozelle, S. and Pray, C. (2005) Insect-resistant GM rice in farmers fields: assessing productivity and health effects in China. *Science* 308, 688–690.

Huffman, D.L., Abrami, L., Sasik, L., Corbeil, J., van der Goot, F.G. and Aroian, R.V. (2004) Mitogen-activated protein kinase pathways defend against bacterial pore-forming toxins. *Proceedings of the National Academy Sciences of the USA* 101, 10995–11000.

Hussey, R.S. and Janssen, G.J.W. (2002) Root-knot nematodes: *Meloidogyne* species. In: Starr, J., Cook, R. and Bridge, J. (eds) *Plant Resistance to Parasitic Nematodes.* CAB International, Wallingford, UK, pp. 43–70.

Immordino, G. (2003) Looking for a guide to protect the environment: the development of the precautionary principle. *Journal of Economic Surveys* 17, 629–643.

Jackson, A.L., Bartz, S.R., Schelter, J., Kobayashi, S.V., Burchard, J., Mao, M., Li, B., Cavet, G. and Linsley P.S. (2003) Expression profiling reveals off-target gene regulation by RNAi. *Nature Biotechnology* 21, 635–637.

Jacquet, M., Bongiovanni, M., Martinez, M., Verschave, P., Wajnberg, E. and Castagnone-Sereno, P. (2005) Variation in resistance to the root-knot nematode *Meloidogyne incognita* in tomato genotypes bearing the *Mi* gene. *Plant Pathology* 54, 93–99.

James, C. (2007) ISAAA Brief 37-2007: Executive Summary. Global Status of Commercialized Biotech/GM Crops: 2007 http://www.isaaa.org/resources/publications/briefs/37/executivesummary/default.html.

Johnson, K.L., Raybould, A.F., Hudson, M.D. and Poppy, G.M. (2006) How does scientific risk assessment of GM crops fit within the wider risk analysis? *Trends in Plant Science* 12, 1–5.

Jones, J.D.G. and Dangl, J.L. (2006) The plant immune system, *Nature* 444, 323–329.

Jorgensen, R.A., Cluster, P.D., English, J., Que, Q. and Napoli, C.A. (1996) Chalcone synthase cosuppression phenotypes in petunia flowers: comparison of sense vs antisense constructs and single-copy vs complex T-DNA sequences. *Plant Molecular Biology* 31, 957–973.

Joyce, P.I., Gallagher, J.M. and Kuwabara, P.E. (2006) Manipulating and enhancing the RNAi response. *Journal of RNAi and Gene Silencing* 2, 118–125.

Kamath, R.S., Martinez-Campos, M., Zipperlen, P., Fraser, A.G. and Ahringer, J. (2000) Effectiveness of specific RNA-mediated interference through ingested double-stranded RNA in *Caenorhabditis elegans. Genome Biology* 2(1), Research0002.

Keizebrink, D.T. and Atkinson, H.J. (2004) Determining risks to soil organisms associated with a genetically modified (GM) crop expressing a biopesticide in its roots. Final report, DEFRA project code RG0122 http://www.defra.gov.uk/science/project_data/DocumentLibrary/RG0122/RG0122_6115_FRP.pdf.

Kimber, M.J., Fleming, C.C., Prior, A., Jones, J.T., Halton, D.W. and Maule, A.G. (2002) Localisation of *Globodera pallida* FMRFamide-related peptide encoding genes using *in situ* hybridisation. *International Journal for Parasitolology* 32, 1095–1105.

Kok, E.J. and Kuiper, H.A (2003) Comparative safety assessment for biotech crops. *Trends in Biotechnology* 21, 439–444.

Kotze, A.C., O'Grady, J., Gough, J.M., Pearson, R., Bagnall, N.H., Kemp, D.H. and Akhurst, R.J. (2005) Toxicity of *Bacillus thuringiensis* to parasitic and free-living life-stages of nematode parasites of livestock. *International Journal for Parasitology* 35, 1013–1022.

Kuparinen, A., Schurr, F., Tackenberg, O. and O'Hara, R.B. (2007) Air-mediated pollen flow from genetically modified to conventional crops. *Ecological Applications* 17, 431–440.

Kusaba, M. (2004) RNA interference in crop plants. *Current Opinions in Biotechnology* 15, 139–143.

Ladha, J.K., Dawe, D., Pathak, H., Padre, A.T., Yadav, R.L., Singh, B., Singh, Y., Singh, Y., Singh, P., Kundu, A.L., Sakal, R., Ram, N., Regmi, A.P., Gami, S.K., Bhandari, A.L., Amin, R., Yadav, C.R., Bhattarai, E.M., Das, S., Aggarwal, H.P, Gupta, R.K. and Hobbs, P.R. (2003) How extensive are yield declines in long-term rice–wheat experiments in Asia? *Field Crops Research* 81, 159–180.

Lenane, I.J., Bagnall, N.H., Josh, P.F., Pearson, R.D., Akhurst, R.J. and Kotze, A.C. (2008) A pair of adjacent genes, cry5Ad and orf2-5Ad, encode the typical N- and C-terminal regions of a Cry5A*d*-endotoxin as two separate proteins in *Bacillus thuringiensis* strain L366. *FEMS Microbiology Letters* 278, 115–120.

Li, X.Q., Wei, J.Z., Tan, A. and Aroian, R.V. (2007) Resistance to root-knot nematode in tomato roots expressing a nematicidal *Bacillus thuringiensis* crystal protein. *Plant Biotechnology Journal* 5, 455–464.

Lilley, A.K., Bailey, M.J., Cartwright, C., Turner, S.L. and Hirsch, P.R. (2006) Life in earth: the impact of GM plants on soil ecology? *Trends in Biotechnology* 24, 9–14.

Lilley, C.J., Urwin, P.E., McPherson, M.J. and Atkinson, H.J. (1996) Characterisation of intestinally active proteases of cyst-nematodes. *Parasitology* 113, 415–424.

Lilley, C.J., Urwin, P.E., Atkinson, H.J. and McPherson, M.J. (1997) Characterisation of cDNAs encoding serine proteases from the soybean cyst nematode *Heterodera glycines. Molecular and Biochemical Parasitology* 89, 195–207.

Lilley, C.J., Urwin, P.E., Johnston, K.A. and Atkinson, H.J. (2004) Preferential expression of a plant cystatin at nematode feeding sites confers resistance to *Meloidogyne incognita* and *Globodera pallida. Plant Biotechnology Journal* 2, 3–12.

Lilley, C.J., Atkinson, H.J. and Urwin, P.E. (2005) Molecular aspects of cyst nematodes. *Molecular Plant Patholology* 6, 577–588.

Lilley, C.J., Bakhetia, M., Charlton, W.L. and Urwin, P.E. (2007) Recent progress in the development of RNA interference for plant parasitic nematodes. *Molecular Plant Pathology* 8, 701–711.

Liu, B., Hibbard, J.K., Urwin, P.E. and Atkinson, H.J. (2005) The production of synthetic chemodisruptive peptides *in planta* disrupts the establishment of cyst nematodes *Plant Biotechnology Journal* 3, 487–496.

Losey, J.E., Rayor, L.S. and Carter, M.E. (1999) Transgenic pollen harms monarch larvae *Nature* 399, 214.

Mansoor, S., Amin, I., Hussain, M., Zafar, Y. and Briddon, R.W. (2006) Engineering novel traits in plants through RNA interference. *Trends in Plant Science* 11, 559–565.

Mao, Y-B., Cai, W-J., Wang, J-W., Hong, G-J., Tao, X-Y., Wang, L-J., Huang, Y-P. and Chen, X-Y. (2007) Silencing a cotton bollworm P450 monooxygenase gene by plant-mediated RNAi impairs larval tolerance of gossypol. *Nature Biotechnology* 25, 1307–1313.

Marroquin, L.D., Elyassnia, D., Griffitts, J.S., Feitelson, J.S. and Aroian, R.V. (2000) *Bacillus thuringiensis* (Bt) toxin susceptibility and isolation of resistance mutants in the nematode *Caenorhabditis elegans. Genetics* 155, 1693–1699.

Martinez de Ilarduya, O. and Kaloshian, I. (2001) *Mi-1.2* transcripts accumulate ubiquitously in resistant *Lycopersicon esculentum. Journal of Nematology* 33, 116–120.

Martinez de Ilarduya, O., Moore, A.E. and Kaloshian, I. (2004) The tomato *Rme1* locus is required for *Mi-1*-mediated resistance to root-knot nematodes and the potato aphid. *Plant Journal* 27, 417–425.

May, R.C. and Plasterk, R.H.A. (2005) RNA interference spreading in *C. elegans. Methods in Enzymology* 392, 308–315.

Mendelsohn, M., Kough, J., Vaituzis, Z. and Matthews, K. (2003) Are *Bt* crops safe? *Nature Biotechnology* 21, 1003–1009.

Mitchum, M., Hussey, R.S., Davis, E.L. and Baum, T.J. (2007) Application of biotechnology to understand pathogenesis in nematode plant pathogens. In: Punja, Z.K., DeBoer, S. and Sanfacon, H. (eds) *Biotechnology and Plant Disease Management.* CAB International, Wallingford, UK, pp. 58–86.

Morris, S.H. (2007). EU biotech crop regulations and environmental risk: a case of the emperor's new clothes? *Trends in Biotechnology* 25, 2–6.

Neveu, C., Abad, P. and Castagnone-Sereno, P. (2003) Molecular cloning and characterization of an intestinal cathepsin L protease from the plant-parasitic nematode *Meloidogyne incognita. Physiological and Molecular Plant Pathology* 63, 159–165.

Nishikura, K. (2001) A short primer on RNAi: RNA-directed RNA polymerase acts as a key catalyst. *Cell* 107, 415–418.

Nombela, G., Williamson, V.M. and Muniz, M. (2003) The root-knot nematode resistance gene *Mi-1.2* of tomato is responsible for resistance against the whitefly *Bemisia tabaci. Molecular Plant–Microbe Interactions* 16, 645–649.

Ogita, S., Uefuji, H., Morimoto, M. and Sano, H. (2004) Application of RNAi to confirm theobromine as the major intermediate for caffeine biosynthesis in coffee plants with potential for construction of decaffeinated varieties. *Plant Molecular Biology* 54, 931–941.

Opperman, C.H., Taylor, C.G. and Conkling, M.A. (1994) Root-knot nematode-directed expression of a plant root-specific gene. *Science* 263, 221–223.

Pak, J. and Fire, A. (2007) Distinct populations of primary and secondary effectors during RNAi in *C. elegans. Science* 315, 241–244.

Payne, J., Narva, K.E. and Fu, J. (1996) *Bacillus thuringiensis* genes encoding nematode-active toxins. US patent 5,589,382.

Pooggin, M., Shivaprasad, P.V., Veluthambi, K. and Mohn, T. (2003) RNAi targeting of DNA virus in plants. *Nature Biotechnology* 21, 131–132.

Prasad, J.S., Panwar, M.S. and Rao, Y.S. (1987) Nematode problems of rice in India. *Tropical Pest Management* 33, 127–136.

Raney, T. (2006) Economic impact of transgenic crops in developing countries. *Current Opinion in Biotechnology* 17, 1–5.

Raybould, A. (2007) Ecological versus ecotoxicological methods for assessing the environmental risks of transgenic crops. *Plant Science* 173, 589–602.

Roberts, P.A. (1995) Conceptual and practical aspects of variability in root-knot nematodes related to host plant resistance. *Annual Review of Phytopathology* 33, 199–221.

Roberts, P.A. (2002) Concepts and consequences of resistance. In: Starr, J., Cook, R. and Bridge, J. (ed.) *Plant Resistance to Parasitic Nematodes.* CAB International, Wallingford, UK, pp. 23–41.

Rossi, M., Goggin, F.L., Milligan, S.B., Kaloshian, I., Ullman, D.E. and Williamson, V.M. (1998) The nematode resistance gene *Mi* of tomato confers resistance against the potato aphid. *Proceedings of the National Academy of Sciences of the USA* 95, 9750–9754.

Rosso, M.N., Dubrana, M.P., Cimbolini, N., Jaubert, S. and Abad, P. (2005) Application of RNA interference to root-knot nematode genes encoding esophageal gland proteins. *Molecular Plant–Microbe Interactions* 18, 615–620.

Samac, D.A. and Smigocki, A.C. (2003) Expression of oryzacystatin I and II in alfalfa increases resistance to the root lesion nematode. *Phytopathology* 93, 799–804.

Schaff, J.E., Nielsen, D.M., Smith, C.P., Scholl, E.H. and Bird, D.M. (2007) Comprehensive transcriptome profiling in tomato reveals a role for glycosyltransferase in *Mi*-mediated nematode resistance. *Plant Physiology* 144, 1079–1092.

Schmidt, M. and Bothma, G. (2006) Risk assessment for transgenic sorghum in Africa: crop-to-crop gene flow in Sorghum bicolor (L.) moench. *Crop Science* 46, 790–798.

Seinhorst, J.W. (1967) The relationship between population increase and population density in plant parasitic nematodes. *Nematologica* 13, 157–171.

Seinhorst, J.W. and den Ouden, H. (1971) The relation between density of *Heterodera rostochiensis* and growth and yield of two potato varieties. *Nematologica* 17, 347–369.

Shingles, J., Lilley, C.J., Atkinson, H.J. and Urwin, P.E. (2007) *Meloidogyne incognita*: molecular and biochemical characterization of a cathepsin L cysteine proteinase and the effect on parasitism following RNAi. *Experimental Parasitology* 115, 114–120.

Sick, A.J., Schwab, G.E. and Payne, J. (1994) Genes encoding nematode-active toxins cloned from *Bacillus thuringiensis* isolate PS17. Publication number, Patent number US5281530.

Sijen, T., Steiner, F.A., Thijssen, K.L. and Plasterk, R.H.A. (2007) Secondary siRNAs result from unprimed RNA synthesis and form a distinct class. *Science* 315, 244–247.

Simmer, F., Moorman, C., van der Linden, A.M., Kuijk, E., van den Berghe, P.V.E., Kamath, R.S., Fraser, A.G., Ahringer, J. and Plasterk, R.H.A. (2003) Genome-wide RNAi of *C. elegans* using the hypersensitive *rrf-3* strain reveals novel gene functions. *PLoS Biology* 1(1), e12.

Soberón, M., Pardo-López, L., López I., Gómez I., Tabashink, B.E. and Bravo, A. (2007) Engineering modified Bt toxins to counter insect resistance. *Science* 318, 1640–1642.

Steeves, R.M., Todd, T.C., Essig, J.S. and Trick, H.N. (2006) Transgenic soybeans expressing siRNAs specific to a major sperm protein gene suppress *Heterodera glycines* reproduction. *Functional Plant Biology* 33, 991–999.

Stewart, C.N., Halfhill, M.D. and Warwick, S. I. (2003) Transgene introgression from genetically modified crops to their wild relatives. *Nature Reviews Genetics* 4, 806–817.

Tai, T.H., Dahlbeck, D., Clark, E.T., Gajiwala, P., Pasion, R., Whalen, M.C., Stall, R.E. and Staskawicz, B.J. (1999) Expression of the *Bs2* pepper gene confers resistance to bacterial spot disease in tomato. *Proceedings of the National Academy of Sciences of the USA* 96, 14153–14158.

Tait, J. and Chataway, J. (2007) The governance of corporations, technological change, and risk: examining industrial perspectives on the development of genetically modified crops. *Environment and Planning C – Government and Policy* 25, 21–37.

Timmons, L., Court, D.L. and Fire, A. (2001) Ingestion of bacterially expressed dsRNAs can produce specific and potent genetic interference in *Caenorhabditis elegans*. *Gene* 263, 103–112.

Trett, M.W. and Perry, R.N. (1985) Effects of the carbamoyloxime, aldicarb, on the ultrastructure of the root-lesion nematode, *Pratylenchus penetrans* (Nematoda, Pratylenchidae). *Nematologica* 31, 321–334.

Urwin, P.E. Atkinson, H.J. Waller, D.A. and McPherson, M.J. (1995) Engineered oryzacystatin-I expressed in transgenic hairy roots confers resistance to *Globodera pallida*. *The Plant Journal* 8, 121–131.

Urwin, P.E., Lilley, C.J., McPherson, M.J. and Atkinson, H.J. (1997) Resistance to both cyst- and root-knot nematodes conferred by transgenic *Arabidopsis* expressing a modified plant cystatin. *The Plant Journal* 12, 455–461.

Urwin, P.E., McPherson, M.J. and Atkinson, H.J. (1998) Enhanced transgenic plant resistance to nematodes by dual proteinase inhibitor constructs. *Planta* 204, 472–479.

Urwin, P.E., Levesley, A., McPherson, M.J. and Atkinson, H.J. (2000) Transgenic resistance to the nematode *Rotylenchulus reniformis* conferred by *Arabidopsis thaliana* plants expressing proteinase inhibitors. *Molecular Breeding* 6, 257–264.

Urwin, P.E., Troth, K.M., Zubko, E.I. and Atkinson, H.J. (2001) Effective transgenic resistance to *Globodera pallida* in potato field trials. *Molecular Breeding* 8, 95–101.

Urwin, P.E., Lilley, C.J. and Atkinson, H.J. (2002) Ingestion of double-stranded RNA by pre parasitic juvenile cyst nematodes leads to RNA interference. *Molecular Plant–Microbe Interactions* 15, 747–752.

Urwin, P.E., Green, J. and Atkinson, H.J. (2003) Expression of a plant cystatin confers partial resistance to *Globodera*, full resistance is achieved by pyramiding a cystatin with natural resistance. *Molecular Breeding* 12, 263–269.

Vain, P., Worland, B., Clarke, M.C., Richard, G., Beavis, M., Liu, H., Kohli, A., Leech, M., Snape, J., Christou, P. and Atkinson, H.J. (1998) Expression of an engineered cysteine proteinase inhibitor (OC-IΔD86) for nematode resistance in transgenic rice plants. *Theoretical and Applied Genetics* 96, 266–271.

van der Biezen, E.A. and Jones, J.D.G. (1998) Plant disease resistance proteins and the 'gene-for-gene' concept. *Trends in Biochemical Science* 23, 454–456.

Vishnudasan, D., Tripathi, M.N., Rao, U. and Khurana, P. (2005) Assessment of nematode resistance in wheat transgenic plants expressing potato proteinase inhibitor (*PIN2*) gene. *Transgenic Research* 14, 665–675.

Waterhouse, P.M., Graham, M.W. and Wang, M.B. (1998) Virus resistance and gene silencing in plants can be induced by simultaneous expression of sense and antisense RNA. *Proceedings of the National Academy of Sciences of the USA* 95, 13959–13964.

Wei, J.Z., Hale, K., Carta, L., Platzer, E., Wong, C., Fang, S.C. and Aroian, R.V (2003) *Bacillus thuringiensis* crystal proteins that target nematodes. *Proceedings National Academy Sciences of the USA* 100, 2760–2765.

Williamson, V.M. (1998) Root-knot nematode resistance genes in tomato and their potential for future use. *Annual Review of Phytopathology* 36, 277–293.

Williamson, V.M. and Kumar, A. (2006) Nematode resistance in plants, the battle underground. *Trends in Genetics* 22, 396–403.

Winston, W.M., Sutherlin, M., Wright, A.J., Feinberg, E.H. and Hunter, C.P. (2007) *Caenorhabditis elegans* SID-2 is required for environmental RNA interference. *Proceedings of the National Academy of Sciences of the USA* 104, 10565–10570.

Winter, M.D., McPherson, M.J. and Atkinson, H.J. (2002) Neuronal uptake of pesticides disrupts chemosensory cells of nematodes. *Parasitology* 125, 561–565.

Yadav, B.C., Veluthambi, K. and Subramaniam, K. (2006) Host-generated double stranded RNA induces RNAi in plant-parasitic nematodes and protects the host from infection. *Molecular and Biochemical Parasitology* 148, 219–222.

Zhao, J-Z., Cao, J., Li, Y., Collins, H.L., Roush, R.T., Earle, E.D. and Shelton, A.M. (2003) Transgenic plants expressing two *Bacillus thuringiensis* toxins delay insect resistance evolution. *Nature Biotechnology* 21, 1493–1997.

16 The Complete Sequence of the Genomes of *Meloidogyne incognita* and *Meloidogyne hapla*

Pierre Abad[1,2,3] and Charles H. Opperman[4]
[1]*INRA, Sophia Antipolis, France;* [2]*CNRS, Sophia Antipolis, France;* [3]*Université de Nice Sophia-Antipolis, Sophia Antipolis, France;* [4]*North Carolina State University, Raleigh, North Carolina, USA*

16.1 Introduction	**363**
16.2 *Meloidogyne incognita* Genome	**364**
16.3 *Meloidogyne hapla* Genome	**372**
16.4 Conclusions and Future Directions	**375**
16.5 Acknowledgements	**376**
16.6 References	**376**

16.1 Introduction

The phylum Nematoda comprises over 25,000 described species, including free-living, animal- and plant-parasitic species. Their ability to adapt to severe and changing environmental conditions has made nematodes one of the most successful types of animals on earth (Blaxter, 2003). Despite the profusion of information available for *Caenorhabditis elegans* and its sister species, *C. briggsae*, very little is known about the other members of this diverse phylum. In particular, parasitic nematodes, which constitute half of the earth's nematodes, have remained largely unexplored. Recently, the genome sequence of the human filarial nematode parasite, *Brugia malayi*, has revealed significant differences from *Caenorhabditis*, and underlined the need to obtain additional genome data from representative species to investigate the outstanding diversity of the phylum (Ghedin *et al.*, 2007).

Plant-parasitic nematodes are responsible annually for an estimated €100 billion in crop damage worldwide (Sasser and Freckman, 1987), and species of the root-knot nematodes, genus *Meloidogyne*, are the most damaging. The problem in the subtropics and tropics is particularly severe, and many developing nations are seriously impacted in both food security and economics by root-knot nematodes (see Coyne *et al.*, Chapter 19, this volume). Currently, chemical nematicides are the most used and reliable means of controlling nematodes. However, their active compounds are non-specific and notoriously toxic to human health and the environment, and are consequently being withdrawn (see Nyczepir and Thomas, Chapter 18, this volume).

Many nematode genes have been identified from expressed sequence tag (EST) projects (McCarter *et al.*, 2000). ESTs represent a rapid avenue to gene discovery; because they represent transcripts, each sequence represents an expressed gene. As a first step, this approach begins to pay off immediately and, as such, has been employed to great effect on parasitic nematodes. As recently as 2000, there were only 24,000 ESTs sequenced from nematodes other than *C. elegans*, but that number is now greater than 375,000, with a large emphasis on human parasites (e.g. Daub *et al.*, 2000; Blaxter *et al.*, 2002), animal parasites (e.g. Tetteh *et al.*, 1999) and plant parasites (e.g. Bird *et al.*, 2002; McCarter *et al.*, 2003; Mitreva *et al.*, 2004). An analysis was completed using >250,000 ESTs originating from 30 species, clustered into 93,000 genes and grouped into 60,000 gene families (Parkinson *et al.*, 2004). Using these data, it has been calculated that the diversity of genes within the phylum is great. In nematodes, despite the availability of multiple *Caenorhabditis* species genomes, the addition of new species to the analysis has yielded a rapid increase in discovery of new genes, and it appears that nematodes are more diverse at the molecular level than was previously recognized. The set of ~20,000 genes and 12,000 gene families represented by *C. elegans* provides a baseline for nematodes, and many of the conserved gene families are shared with other eukaryotes, but these findings represent only a portion of the expanding total nematode molecular diversity. It seems very likely that the parasitic species are evolving genes specialized for their niche. The genes that may be critical in the development and evolution of parasitism are the ones that are different from those of the free-living nematodes and other multicellular eukaryotes. However, ESTs do not generally represent full-length sequences and provide no information on promoters or gene structure. Additionally, because ESTs are biased towards abundantly expressed genes, they will generally only identify approximately 50% of the genes in a given organism. That being said, they are absolutely essential for annotation of a full genomic sequence, and have made tremendous contributions to our knowledge and understanding of parasitic nematodes.

The report of the assembled genome of the two main species of root-knot nematodes *M. incognita* (Abad *et al.*, 2008) and *M. hapla* (Opperman *et al.*, 2008) in this review provides new insights into these pests, and should aid the development of environmentally sustainable nematicides and new crops that target several important agricultural pests.

Although belonging to the same genus, these two species are very different in their biology. *Meloidogyne incognita* is asexual and polyphagous, whereas *M. hapla* reproduces sexually and has a host range that does not include most grain or grass crops. These two independent sequencing projects have led to identification of genes reflecting the contrasting biology of these two nematode species, and constitute the first step of an era of comparative and functional genomics in plant-parasitic nematodes.

16.2 *Meloidogyne incognita* Genome

The Southern root-knot nematode, *M. incognita*, is able to infect the roots of almost all cultivated plants, which possibly renders this species among the most damaging crop pathogens in the world (Trudgill and Blok, 2001). This species is an obligatory parasite that reproduces by mitotic parthenogenesis, and cytogenetic analysis has revealed the existence of isolates with chromosome numbers ranging from 32 to 36, as well as isolates with 40 to 48 chromosomes. Assuming that the haploid chromosome number is $n = 18$ (as observed in many meiotic, sexually reproducing *Meloidogyne* species), these isolates can be considered as diploids ($2n$) or hypotriploids ($3n - x$), respectively, both with possible chromosomal/segmental losses leading to observed aneuploidy (Castagnone-Sereno, 2006). Although reproducing by mitotic parthenogenesis, *M. incognita* has the capacity to adapt easily to unfavourable abiotic or biotic environmental conditions.

The *M. incognita* genome has been sequenced in France under the initiative of the nematology group at INRA Sophia Antipolis in close collaboration with the Génoscope at Evry (the French centre for sequencing), and the Bioinformatics platform at INRA Toulouse. The ensemble of predicted and automatically annotated protein-coding genes was manually and carefully annotated by a consortium of laboratories. Each laboratory focused on a particular process or gene family relevant to the different aspects of *M. incognita* biology (Table 16.1). Basically, patterns of presence/absence and expansion/reduction in comparison to *C. elegans* and other nematodes (and other species when appropriate) were examined.

Table 16.1. Teams of the international consortium for annotation of the *Meloidogyne incognita* genome.

INRA Toulouse (F)[a]	Automatic annotation, ncRNA, miRNA[e]
INRA URGI Evry (F)	Repetitive sequences
Lyon Uni. (F)/Lausanne Uni. (CH)[b]	Nuclear receptors
INRA Rennes (F)	Peptidases
INRA Sophia Antipolis (F)	Automatic annotation, peptidases, immune response, CAZymes, detoxification, SL, neuroreceptors[f]
CNR/Bari Uni. (I)[c]	Detoxification compounds
Wageningen Uni. (NL)[d]	Excretory–secretory proteins
Belfast Uni. (UK)	Neuropeptides
Edinburgh Uni. (UK)	Operons
SCRI Dundee (UK)	Sex determination, RNAi pathways[g]
ISU Ames IA (USA)	Kinome
NCSU Raleigh NC (USA)	Pioneer parasitism genes
WHOI Woods Hole MA (USA)	P450

[a]F = France; [b]CH = Switzerland; [c]I = Italy; [d]NL = The Netherlands; [e]ncRNA = non-coding RNA, miRNA = microRNA; [f]CAZymes = carbohydrate-active enzymes, SL = trans-spliced leader; [g]RNAi = RNA interference.

The complete sequence of *M. incognita* constitutes the first genome sequence of an animal that reproduces asexually and provides insights into the adaptations required by metazoans to parasitize plants successfully.

16.2.1 A genome constituted by pairs of homologous but divergent segments

The *M. incognita* genome sequence was established with the use of a whole-genome shotgun and assembly strategy. The sequence reads were assembled with Arachne (Jaffe *et al.*, 2003) in 2817 supercontigs, giving a total coverage of 86 Mb (Table 16.2). This is almost twice the estimated size of 47–51 Mb per haploid genome of this species (Triantaphyllou, 1985; Leroy *et al.*, 2003). Therefore, we suspect that *M. incognita* is a fixed heterozygous organism. An all-against-all comparison of supercontig sequences revealed that the 650 longest supercontigs consist of homologous but diverged segment pairs covering 55 Mb that might represent former alleles. The average sequence divergence between the aligning regions is 7–8%, which is among the highest observed until now for a sequenced heterozygous organism (Fig. 16.1). We also found an additional 3.35 Mb of the assembly, composed of supercontigs that align with two of the previously identified

Table 16.2. Comparison of genome characteristics of *Meloidogyne hapla* and *M. incognita* with those of *Caenorhabditis elegans* and *Brugia malayi*.

Features	*C. elegans*[a]	*M. hapla*	*M. incognita*	*B. malayi*
Genome size, Mb	100	54	47–51	90–95
Scaffolds	N/A	1,523	2,817	8,810
Scaffold N50, bp	N/A	83,645	82,800	93,771
Assembled bp	100,267,623	53,578,246	86,079,672	70,837,048
Gene models	21,193	14,454	19,212	11,515[b]
Gene density	228	270	223	162
Median exon, bp	147	145	136	140
Exon/gene	6	6	7	7
Median intron, bp	68	55	82	219
G+C %[c]	35.4	27.4	31.4	30.5
Predicted peptides	23,662	16,676	20,365	11,500

[a] *C. elegans* wormbase assembly release WS185, November 2007.
[b] Number of genes includes seven predicted pseudogenes. Splice variants were not taken into account. Between 14,500 and 17,800 genes have been estimated after inclusion of genes potentially present in unannotated portion of the genome (Ghedin *et al.*, 2007).
[c] guanine + cytosine %.

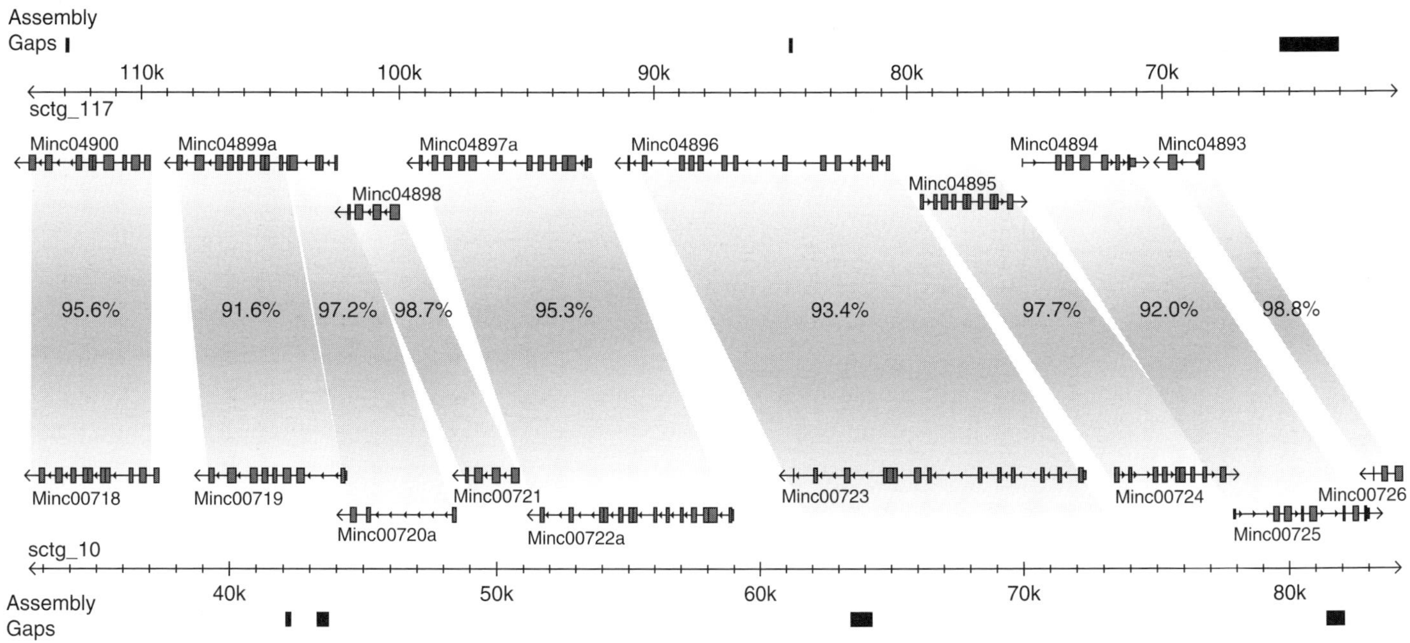

Fig. 16.1. Example of two allelic regions in the *Meloidogyne incognita* assembly. Exons are represented by grey boxes and are linked together to form genes (arrows indicate the strand). The black boxes show the assembly gaps. Homologous genes are linked together using light grey shadows. The gene content appears well conserved between the two alleles, with only local differences in predicted gene structure. Percentages of sequence identity at the protein level between the two allelic regions are indicated.

supercontig pairs, thus corresponding to triplicated genomic regions. Altogether, these observations are consistent with the hypothesis resulting from the cytogenetic studies and with the strictly parthenogenetic lifestyle of *M. incognita*, in which the absence of meiotic recombination may allow alleles to diverge considerably, as hypothesized for bdelloid rotifers (Welch *et al.*, 2004). No DNA attributable to any bacterial endosymbiot genome was identified.

The overall G+C (guanine + cytosine) content (31.4%) is comparable to that of *B. malayi* (30.5%) and lower than that of the free-living nematodes *C. elegans* (35.4%) or *C. briggsae* (37.4%). Repetitive/transposable elements (TEs) comprised 36% of the *M. incognita* genome. In total, 4041 different repeat families have been detected, of which 3066 have no visible TE features. Some of them, organized in tandem arrays, probably correspond to satellite DNA repeats, as previously characterized in the genome of this nematode (Piotte *et al.*, 1994; Mestrovic *et al.*, 2006). Only 690 families have visible TE features, including 252 class I TEs and 430 class II TEs. In terms of dynamics, 111 different families (representing 334 copies and 0.4% of the genome) have a minimum copy-to-family consensus identity percentage greater than 95%, indicating recent families and possible current activity. Although the activity of these TEs remains hypothetical, we can speculate that some may be involved in part in the plasticity of the genome of this parthenogenetic species.

Many *C. elegans* genes possess a trans-spliced leader (SL) that refines the 5' untranslated region of pre-mRNA and enhances translation. Two types of SL have been described: SL1 and SL2. We have identified through the genome of *M. incognita* 283 *Mi*-SL1 genes distributed among 46 contigs, 258 of which were found associated with a satellite DNA. By contrast, we did not identify SL2s in *M. incognita*, strengthening the finding that SL2s are an evolutionary invention of rhabditine (Guiliano and Blaxter, 2006).

One striking characteristic of nematode genomes is the presence of operons, a feature usually associated with prokaryote genomes (Blumenthal and Gleason, 2003). Since operonic genes are trans-spliced to SL and SL-like exons in nematodes, we searched for the existence of operonic structure in the genome of *M. incognita* and identified 1585 candidate operons, containing a total of 3966 *M. incognita* genes (19% of predicted genes). The two longest operons contain ten genes each and are not allelic copies. A comparison with *C. elegans* and *B. malayi* genomes (The *C. elegans* Genome Sequencing Consortium, 1998; Ghedin *et al.*, 2007) shows that these nematodes appear to have the same proportion of genes in operons. An analysis of clusters of orthologous genes between *C. elegans*, *B. malayi* and *M. incognita* revealed that although operonic structures appear to be a common feature of nematodes, different sets of genes compose these operons in each species. Indeed, only one operon was found to be strictly conserved between the three nematodes. Hence, operons are a dynamic structure of the architecture of the nematode genome; however, no clear functional linkage between genes in these structures can be identified.

16.2.2 The gene content of a plant-parasitic nematode

The genome sequence was annotated by using the integrative gene-prediction platform EuGene (Foissac and Schiex, 2005), specifically trained for *M. incognita*. The *M. incognita* genome is predicted to contain 19,212 protein-coding genes (gene models; Table 16.2). The predicted protein-coding genes occupy 25.3 % of the sequence, at an average density of 223 genes per Mb, and 36% are supported by ESTs. Interestingly, 69% of protein sequences are less than 95% identical to another. The high divergence between pseudo-alleles is probably due to the absence of meiotic recombination. For comparison, 21,193 genes were predicted in *C. elegans* and 11,515 were predicted in *B. malayi*. Gene number differences between parasitic nematodes could be explained by the *M. incognita* genome organization.

Gene function assignment was identified based on the identification of InterPro (IPR) domains (release 16.1) (Mulder *et al.*, 2007) using InterproScan (Quevillon *et al.*, 2005). For comparative analysis of Interpro domains found in the predicted *M. incognita* proteome, we ran the program InterproScan on a set of seven different other species: three nematodes (*C. elegans* (WormPep183), *C. briggsae* (rel2) and *B. malayi* (rel1)), one insect (*Drosophila melanogaster* (rel5.4)) and three fungi (*Magnaporthe grisea*, *Gibberella zea* and *Neurospora crassa*). We attributed IPR domains and assigned associated Gene Ontology (Ashburner *et al.*, 2000) terms. Of the predicted *M. incognita* proteins, 55% contain at least one previously identified IPR protein domain and 22% are predicted to be secreted. In all three nematode species, protein kinase domains occur the most frequently. Comparison of domain occurrence in *M. incognita* with *C. elegans* identified an increased abundance of pectin lyase (PL), glycoside hydrolase (GH) of family GH5 and peptidase C48 (SUMO) domains, and a decrease of chemo- and serpentine-receptor domains. Only 52 previously recognized domains were detected uniquely in *M. incognita* (2% of the identified domains). Twenty-seven of these *M. incognita*-unique IPR domains were supported by ESTs, and included protein families involved in plant cell wall degradation and chorismate mutase. We also generated clusters of orthologous protein-coding genes between *M. incognita* and the seven other species listed previously using OrthoMCL (Li *et al.*, 2003). Clusters were constructed on the basis of multidirectional reciprocal best BLAST (Altschul *et al.*, 1997) hits with a MCL clustering algorithm. This method allows clustering of candidate orthologues between different species, but also includes in-paralogues (resulting from species-specific duplications) inside clusters. Thus, OrthoMCL clusters can range from one-species clusters (only composed of species-specific in-paralogues) to eight-species clusters (representing genes shared between all the species considered here). OrthoMCL analysis across these eight proteomes indicated that 52% of the *M. incognita* predicted proteins have no significant orthologues in other organisms. We estimated the core complement of proteins in the Phylum Nematoda, and found 23% of orthologue protein groups with representatives shared by *M. incognita*, *C. elegans* and *B. malayi*.

16.2.3 Identifying plant parasitism genes

Compared with its free-living relatives, *M. incognita*, as a plant-parasitic nematode, has developed specific capabilities and complex biotrophic interactions adapted to its hosts. It is able to penetrate into the plant and navigate between cells to reach its feeding site. Further, the parasite must elude host defence responses for the several weeks during which the feeding site is required to support female development up to hatching of second-stage juveniles (Caillaud et al., 2008). It is thought that nematode proteins produced in, and secreted from, the pharyngeal gland cells into the host via its stylet are the main effectors responsible for these processes (see Abad *et al.*, Chapter 7, this volume).

One of the most remarkable findings that has emerged from this genome study is the identification of an extensive set of plant cell-wall (PCW)-degrading, carbohydrate-active enzymes (CAZymes) in *M. incognita*, which has no equivalent in any animal studied to date. Although a few such individual CAZymes had been identified previously in some plant-parasitic nematodes and in two insect species (Davis *et al.*, 2004; Wei *et al.*, 2006; Caillaud *et al.*, 2008), they are absent from all other metazoans studied to date. This set of plant cell-wall-degrading CAZymes encompasses a total of 61 proteins completely absent in *C. elegans* and *D. melanogaster*, among which are cellulases from family GH5, xylanases from family GH5, polygalacturonases from family GH28 and pectate lyases from family PL3 (Table 16.3). Detailed genome analysis also revealed, for the first time in animals (including other plant-parasitic nematodes), additional CAZymes, including two PCW-degrading enzymes of family GH43 (candidate arabinanases) and two candidate

Table 16.3. Plant cell-wall-degrading and -modifying CAZymes of *Meloidogyne incognita*, *Caenorhabditis elegans* and *Drosophila melanogaster*.

Substrate Family	Cellulose GH5 (eng)[b,c]	Xylan GH5 (xyl)[c]	Arabinan GH43[c]	Pectin GH28[c]	 PL3[c]	Other EXPN[a,b]	Total
M. incognita	21	6	2	2	30	20	81
C. elegans	0	0	0	0	0	0	0
D. melanogaster	0	0	0	0	0	0	0

[a]EXPN: expansin modules (modification of plant cell wall).
[b]cellulose-binding modules of family CBM2 (bacterial type) are found appended to these proteins.
[c]GH = glycoside hydrolase; [d]PL = pectinlyase.

invertases from family GH32. Invertases catalyse the conversion of sucrose (an abundant disaccharide in plants) into glucose and fructose, which can be used by *M. incognita* as a carbon source. This ensemble of plant cell-wall-related CAZymes and the associated invertases was probably acquired via horizontal gene transfer in plant-parasitic nematodes, since most homologues similar to these *M. incognita* proteins are systematically found from other plant-parasitic nematodes (when available), immediately followed by bacterial homologues in which these enzymes are reported. As a supplement to its arsenal of plant cell-wall-degrading CAZymes, we also identified a total of 20 candidate expansins in *M. incognita* (Table 16.3). While the precise biochemical function of these proteins remains unknown, it has been shown (Qin *et al.*, 2004) that plant-parasitic nematodes produce expansin-like proteins, which may disrupt non-covalent bonds, loosening the plant cell wall and making the components more accessible to plant cell-wall-degrading enzymes (Cosgrove, 2000). Examination of the genome of *M. incognita* revealed an unsuspected abundance of genes encoding such enzymes, which illustrates the high specialization of this organism as a plant pathogen. Apart from PCW-related CAZymes, we also identified as plant-parasitic-nematode-specific, a four-member family of secreted chorismate mutases (Lambert *et al.*, 1999; Huang *et al.*, 2005), which most closely resemble bacterial enzymes, suggesting again a critical role of horizontal gene transfer events, which have helped to shape the evolution of plant parasitism within root-knot nematodes.

Apart from genes restricted to *M. incognita* (or plant-parasitic nematodes), we also identified genes or gene families showing marked increases compared with *C. elegans*. For example, a detailed analysis of proteases revealed expansion in some subfamilies in *M. incognita*, compared with *C. elegans*. Proteases are broadly effective factors conferring the capacity for pathogenicity towards hosts from different kingdoms. In nematodes, proteases play essential roles in a broad range of biological processes, as diverse as moulting in *C. elegans* (Frand *et al.*, 2005) and digestive specificity in blood-feeding parasites (Williamson *et al.*, 2003). Among the most important idiosyncrasies in *M. incognita*, we identified more than 20 cysteine proteases of the C48 subfamily, predicted to encode small ubiquitin-like modifier (SUMO) deconjugating enzymes, compared with five in *C. elegans*. Some effectors of phytopathogenic bacteria involved in virulence and activation of plant immunity are SUMO proteases (Hotson and Mudgett, 2004; Xia, 2004), suggesting that the proteolysis of sumoylated host substrates by these enzymes may be a general strategy used by pathogens to manipulate host plant signal transduction. We also noted a profusion of serine proteases (S, signal peptidases) of the S16 subfamily (Lon proteases) in *M. incognita*, known to regulate type III protein secretion in phytopathogenic bacteria (Tang *et al.*, 2006). The relative abundance of these two protease subfamilies in *M. incognita* probably reflects intimate interaction with host plant tissues. Since identified parasitism proteins in root-knot nematodes are secreted (Davis *et al.*, 2004), the observation that 92 proteases in *M. incognita* are predicted to be secreted (~30% of the protease set) reinforces the hypothesis that members of the nematode degradome secretion of proteolytic enzymes play a direct role in the parasitic process. Parasitism genes from this category are likely to have evolved from common ancestral nematode genes that have functions in the nematode itself that are unrelated to parasitism.

Another category of candidate parasitism genes consists of families identified only in *M. incognita* to date. A total of 1819 proteins (338 of which were further validated by ESTs) that appear to be unique to *M. incognita* and for which no predicted function could be attributed on the basis of sequence similarity (i.e. without orthologues and without known IPR domains) were predicted to be secreted by *M. incognita*. Among them, 27 genes had been originally identified as 'pioneer' genes expressed specifically in the pharyngeal glands (Huang *et al.*, 2003). Additional homologues of these 27 genes were identified, all remaining specific to *Meloidogyne* spp. A proportion of these *M. incognita*-specific genes are present in clusters of very similar genes, supporting previous observations that parasitism genes may be present in the gene-duplication region (Yan *et al.*, 2001). Availability of additional genomes of plant-parasitic nematodes will allow determination of whether these secreted proteins of an as-yet-unknown function are further conserved in plant-parasitic nematodes as major determinants of the adaptation of these parasites to their host plants.

16.2.4 A nematode adapted to a privileged plant host environment

Regarding antioxidant enzymes that protect the parasite from cytotoxic oxygen radicals, genes encoding superoxide dismutases and glutathione peroxidases are reduced in *M. incognita* compared with *C. elegans*. More striking is the reduction in glutathione S-transferases (GST) and cytochrome P450s (CYP) involved in protection against peroxidative damage and xenobiotic metabolism. In *C. elegans*, 44 members have been identified that belong to the Omega, Sigma and Zeta classes (Lindblom and Dodd, 2006). *Meloidogyne incognita* possesses only five GST genes, all from the class Sigma, which is involved in protection against oxidants rather than xenobiotics. A similar reduction in GST genes has been observed in the animal-parasitic nematode *B. malayi*. Again, reduction and specialization have been observed in the CYP family in *M. incognita*. Among the 80 different CYP genes divided among the 16 families found in *C. elegans* (Menzel *et al.*, 2001), only 23 full or partial CYP genes were identified in the *M. incognita* genome, divided among at least nine families. In particular, *M. incognita* possesses five CYP13-like genes and nine CYP33-like genes, but lacks CYP35 and other families of xenobiotic-metabolizing P450s. Since plant-parasitic nematodes embedded in root tissues are less exposed to various biotic and abiotic stresses than are free-living nematodes, it is tempting to speculate that such a reduction and specialization of GST and CYP genes result from an active selection exerted by the parasitic lifestyle of the worm on its detoxification processes.

Genes from the signalling pathways of innate immunity were conserved enough to allow us to identify orthologues in *M. incognita* (Ewbank, 2006). By contrast, immune effectors such as lysozymes, C-type lectins and chitinases appear much less abundant in *M. incognita* compared with *C. elegans*. As previously observed in *B. malayi*, entire classes of immune effectors are absent from the genome of *M. incognita*. These included antibacterial genes such as *abf* and *spp* (Alegado and Tan, 2008) and antifungal genes of several classes (*nlp*, *cnc*, *fip*, *fipr*) (Ewbank, 2006). Immune effectors might not be conserved between parasitic nematodes and *C. elegans*. Alternatively, as hypothesized above for GSTs and CYPs, the parasitic nature of *M. incognita* may, in part, be responsible for the lower number of genes involved in response to pathogens, compared with *C. elegans*, as it lives much of its life in a privileged environment, protected from exogenous stresses by the plant tissues.

In *C. elegans*, the *N*-glycosylation modifications result in a broad range of unusual fucosylated structures compared with other metazoans (Haslam and Dell, 2003; Paschinger *et al.*, 2007). This tendency for a complex and rich fucosylation pattern in nematodes is further developed in *M. incognita*, which has almost twice as many candidate fucosyltransferases as *C. elegans*. As suggested for animal-parasitic nematodes, multifucosylated structures in *M. incognita* could help the nematode to evade recognition by its hosts, and thus support the parasitic lifestyle of the worm (Paschinger *et al.*, 2007).

16.2.5 Does the *Caenorhabditis elegans* genome reflect nematode lifestyle diversity?

The ubiquity and conservation of fundamental processes have suggested that studies on *C. elegans* may aid in understanding functions in all organisms. In order to evaluate such conservation within the phylum Nematoda, fundamental processes from *C. elegans*, such as basal metabolism, the ability to respond to a range of physical and environmental stimuli, and sex determination, were targeted for manual expert annotation of the *M. incognita* genome sequence.

Among genes that control pathways in charge of the life of each cell, the kinases are major enzymes that activate and trigger a cascade of reactions. *Meloidogyne incognita* has genes for 499 predicted kinase enzymes, corresponding to 232 orthoMCL clusters. This number compares with a total of 411 kinases in *C. elegans* (Plowman *et al.*, 1999) and 215 in *B. malayi*. Of the predicted 232 *M. incognita* clusters, 158 and 152 clusters are conserved in *C. elegans* and *C. briggsae*, respectively. Similarly, the human parasite *B. malayi* shares 152 kinase gene clusters with *M. incognita* in its genome. Interestingly, orthologues of 24 *M. incognita* kinase clusters are found only in *C. elegans* and *C. briggsae*, as well as in *B. malayi*, suggesting nematode-specific functions for these kinases. Furthermore, orthologues of four *M. incognita* kinase clusters are found only in *B. malayi*, suggesting a potential role for these clusters associated with the parasitic lifestyle of these nematodes. Finally, 66 kinase clusters representing 122 genes appear to be *M. incognita*-specific.

Directly involved in the regulation of gene expression, the superfamily of nuclear receptors (NR) is of widespread relevance to almost all aspects of physiology. The evolutionary history of nematode NRs is known to be quite complex. Many of them, which are of major physiological importance in other animals, were lost in *C. elegans* or but not necessarily in *B. malayi*. Among the 92 predicted NRs in *M. incognita*, we identified clear orthologues to some known nematode NRs, although many of the receptors that were identified in *B. malayi* and not in *C. elegans* were not found. The most striking feature of the *M. incognita* NR set is the presence of a great number of supplementary nuclear receptors (SupNRs). Here, our preliminary analysis indicates that almost 14 of them are one-to-one orthologues between *B. malayi*, *M. incognita* and *C. elegans*, or only between *M. incognita* and *C. elegans*, with secondary losses in *B. malayi*. A large clade of 41 *M. incognita*-specific duplicates was also identified. These findings imply that the duplication event started before the *Brugia–Meloidogyne–Caenorhabditis* split, and proceeded independently in the *C. elegans* and *M. incognita* lineages.

Soil-dwelling nematodes have developed the ability to perceive and respond to the environment using sensory systems such as the cuticle, chemoreception and a huge diversity of neurotransmitters. Collagens are ubiquitous structural proteins that play an essential role, as shown by the range of defects observed after mutations in individual collagen genes. The cuticle collagens are an abundant gene family in *C. elegans*, with over 180 members grouped into six subfamilies according to homology relationships (Page and Johnstone, 2007). The *M. incognita* genome revealed the presence of 122 genes manually annotated as collagen-related. Most of them possess a highly conserved nematode cuticle domain at the N-terminus and a long C-terminal domain, showing a high degree of similarity between members of the same subfamily. Five collagen genes were considered as *M. incognita*-specific as they do not show significant similarity with *C. elegans* and *B. malayi* genes.

The *C. elegans* genome codes for a very large number of G-protein-coupled receptors (GPCRs) with seven transmembrane helices (7% of *C. elegans* genes, 1280 genes), which have been described as playing a crucial role in chemical sensitivity (Bargmann, 2006; Robertson and Thomas, 2006). The *M. incognita* genome shows a great reduction of GPCR genes, with the identification of only 108 genes deriving from serpentine receptor superfamilies. Although *M. incognita* possesses far fewer chemosensory genes than does *C. elegans*, these genes show a similar organizational pattern, i.e. clusters of duplicated genes.

All available data suggest that neuropeptide diversity is uniquely high in the phylum Nematoda, which is particularly remarkable in light of the structural simplicity of the nematode nervous systems. *C. elegans* remains the best yardstick for this diversity, with 28 FMRFamide-like peptide (*flp*) and 35 neuropeptide-like protein (*nlp*) genes, encoding approximately 200 putative neuropeptides (Marks and Maule, 2007). The identified neuropeptide complement of *M. incognita* falls somewhat short of the total in *C. elegans*, with 19 *flp* genes and 22 *nlp* genes readily identifiable in the genome. However, two *Meloidogyne* spp. genes (*flp-30* and *flp-31*) encode peptides that have not been identified in *C. elegans*, suggesting that these peptides could fulfil functions specific to the phytoparasitic lifestyle of *Meloidogyne*.

The pathway of genes responsible for sex determination in *C. elegans* has been studied in detail and is intimately linked to the dosage compensation pathway (Zarkower, 2006). *Meloidogyne incognita* reproduces exclusively by mitotic parthenogenesis (see Chitwood and Perry, Chapter 8, this volume) and males do not contribute genetically to the production of offspring (Triantaphyllou, 1981). However, there is also an environmental influence on sex determination of *M. incognita*, and under less favourable environmental conditions far greater numbers of males are produced. These males can arise due to sex reversal and intersexual forms can be produced (Papadopoulou and Triantaphyllou, 1982). *Meloidogyne incognita* homologues of at least one member of each stage of the sex-determination cascade were identified, including *sdc-1* from the dosage compensation pathway and *tra-1*, *tra-3* and *fem-2* from the sex determination pathway itself, as well as numerous downstream genes such as *mag-1*, which represses male-promoting genes, and *mab-23*, which controls male differentiation and behaviour. In addition, a large family (~35 genes) of *M. incognita*-secreted proteins is similar to *tra-1* from *C. elegans*, which shares C2H2 zinc finger motifs with only one other *C. elegans*-secreted protein. Therefore, it is possible that *M. incognita* uses a similar system for governing sex determination, but with the male developmental pathway also linked to the detection of environmental cues.

Taken together, data from this preliminary comparative analysis of important traits in nematode physiology highlight the fact that the model species *C. elegans* does not represent the whole genomic diversity displayed in the phylum Nematoda.

16.2.6 Exploration of new anti-parasitic drug targets

RNA interference (RNAi) has emerged as one of the most promising technologies for the functional analysis of parasitic nematode genes. RNAi can be induced in *M. incognita* by feeding (Rosso *et al.*, 2005; Huang *et al.*, 2006) Therefore, we expected to find components of the RNAi pathway in the genome. Although many of the components of the RNAi pathway are indeed found in the *M. incognita* genome, *rcd-4* was not found, as in the animal-parasitic nematode *Haemonchus contortus* (Zawadzki *et al.*, 2006). In addition, as reported for *B. malayi* and *H. contortus*, homologues of *sid-1*, *sid-2*, *rsd-2* and *rsd-6* involved in systemic RNAi and dsRNA spreading to surrounding cells were not found either. Novel or poorly conserved factors of spreading could explain the systemic RNAi reported in *M. incognita*.

With the aim of identifying pathways unique to nematode development and parasitism that can serve as new targets for nematicides, we examined the RNAi experiment repository in Wormbase to look for an RNAi-lethal phenotype. We retrieved the 2958 *C. elegans* genes for which RNAi experiments led to a lethal phenotype and searched for orthologous genes in *M. incognita*. Among the 1083 OrthoMCL clusters identified, 148 (corresponding to 344 *M. incognita* genes) were defined as nematode-only clusters. Because of their knockdown lethal phenotype and distinctive sequence properties, those genes provide a source for new and attractive targets for anti-parasitic drugs.

16.3 *Meloidogyne hapla* Genome

The Northern root-knot nematode, *M. hapla*, was chosen for a full genome sequence project because it has a meiotic reproduction lifestyle and genetic crosses are possible. By contrast, other economically important species of root-knot nematodes reproduce by mitotic parthenogenesis, resulting in an inability to make genetic crosses. These traits led to *M. hapla* being developed as a platform for genetic studies (Liu *et al.*, 2007). *Meloidogyne hapla* biology was exploited to develop inbred lines and to construct a genetic map. Combined with the small genome size (54 Mb) and the extensive EST database, this made *M. hapla* an ideal candidate for genome sequencing. *M. hapla* is the smallest multicellular eukaryotic genome sequenced to date. This compares, for example, with the soybean cyst nematode, *Heterodera glycines*, which is 92.5 Mb (Opperman and Bird, 1998), and *C. elegans* at 100 Mb. Further, the *M. hapla* genome is largely composed of unique or moderately repeated sequences, such as the rDNA (ribosomal DNA) cluster and other gene family members. Importantly for genome assembly, less than 5% of the genome represents highly repetitive sequence. The importance of the root-knot nematodes in agriculture, along with the biological features of *M. hapla*, suggested that information from the full genome sequence would provide clues to the evolution of parasitic abilities and phylogenetic relationships.

16.3.1 General characterization of the genome

The strain VW9 (16 chromosomes) was chosen for the primary sequencing owing to the availability of a genetic system, including an AFLP (amplified fragment length polymorphism) linkage map developed by the Williamson laboratory at the University of California, Davis. We have completed a 10.4X draft sequence covering approximately 53.5 Mb, representing >99% of the genome assembled in 1523 scaffolds (Table 16.2). The genome possesses a relatively small percentage of moderately repetitive DNA (~12%), which is made up primarily of low-complexity sequence. Examination of the repetitive sequence revealed that the vast majority occurs as simple repeats. However, 1% of the repetitive sequence encodes characterized repeats, such as DNA transposons. This analysis showed that *M. hapla* carries a similar copy number of transposons like the TC-1 transposons of *C. elegans*. Analysis also showed that both the number of genes and the structure of the SL-1 trans-splice leader in *M. hapla* is equivalent to *C. elegans*, but that the SL-1 loci are dispersed throughout the *M. hapla* genome rather than clustered. Additionally, small groups of satellites were found, as were the rRNA sequences (5S, 16S–5.8S–28S) in clusters. The genome exhibits an extremely low GC content (27%) compared with the free-living nematode *C. elegans* (36%).

16.3.2 Estimation of gene numbers

We used a combination of EST sequence and computational predictions from genomic sequence to identify/predict 14,454 genes in *M. hapla*, fewer than the 21,193 genes of *C. elegans*. In many parasitic and symbiotic species, gene loss due to host dependence is normal. Thus, the difference between *M. hapla* and *C. elegans* may indicate that *M. hapla* does not require as full a gene complement as free-living species due to dependence on the host to provide essential resources. A protein data set (*HapPep3*) was constructed from the predicted genes data set, resulting in 16,259 deduced protein sequences. These predictions were used to perform a comparison between *C. elegans* and *M. hapla*, resulting in 4952 matches in *M. hapla* and 18,062 matches in *C. elegans*. Nine of the top 20 most common pfam (protein family) domains found in *M. hapla* are also in the top 20 for *C. elegans*.

16.3.3 Gene families

Not surprisingly, *M. hapla* carries many of the same gene families as *C. elegans*, and half of the most abundant families in *M. hapla* are also among the most abundant in *C. elegans*. However, in some cases gene family size is very different between the two species. For example, the GPCR family is the largest gene family in *C. elegans* (>1000 genes), yet analysis of the *M. hapla* sequence shows that it has only 18% of the *C. elegans* complement. Other large gene families in *C. elegans* are reduced in *M. hapla*, but the GPCRs are an example of a lowered number that may reflect the lifestyle of the parasite compared with a free-living nematode. This could potentially represent a reduced need of sensory receptors during niche specialization to become an internal parasite of plants, with the life stages outside the plant being restricted to egg and second-stage juvenile dauer (see section 16.3.5). It is also possible that this disparity may reflect gene expansion in *C. elegans* due to its soil environment. Although other gene families in *M. hapla* are reduced compared with *C. elegans*, the disparity seen is not as dramatic. For example, *M. hapla* encodes 81 collagen genes, compared with 165 in *C. elegans*. Collagens are likely to play key roles in basic nematode biology and therefore are more conserved than sensory proteins.

What is abundantly clear is that *M. hapla* carries a significantly smaller complement of genes than the free-living *C. elegans*. It is clear that numerous genes have been acquired by horizontal gene transfer, and all of the genes previously identified have been found in the complete *M. hapla* genome sequence. In addition, genes involved in secretions for parasitism have been identified (Huang *et al.*, 2003), including all those previously reported from EST sequence projects (Mitreva *et al.*, 2005).

16.3.4 Genome organization

As 15% of the *C. elegans* genome is organized into operons, we examined the *M. hapla* sequence for potential similarities. We used a data set of the orthologues of all 4685 *C. elegans* proteins encoded by genes predicted to be found in operons and found 3693 matches in the *M. hapla* assembly. We identified 101 operons from *C. elegans* that are at least partially conserved in *M. hapla*, having at least two genes from an operon within the same genomic locale. The largest operon that is fully conserved consists of three genes, although a larger cassette of four genes from the five-gene operon CEOP3272 has also been conserved (Fig. 16.2). This analysis suggests that the genes co-regulated by virtue of such organization in *C. elegans* are not regulated in a similar manner in *M. hapla*. *Meloidogyne hapla* intergenic regions also tend to be quite short, due to the compact genome size, which makes prediction of operon organization difficult. Comparison of conserved operons has uncovered examples of microsynteny between

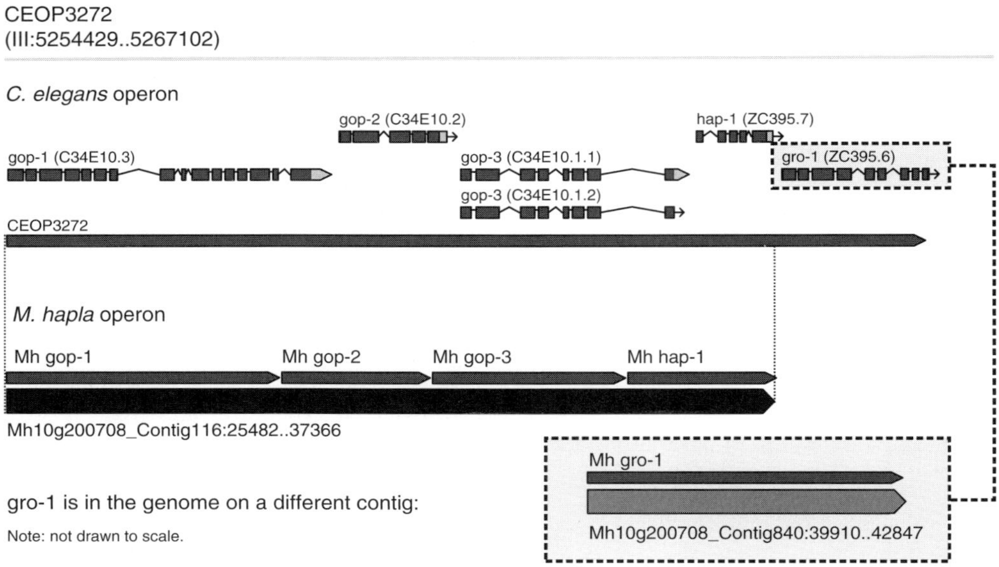

Fig. 16.2. Partial conservation of an operon between *Meloidogyne hapla* and *Caenorhabditis elegans*. This panel depicts a five-gene operon in *C. elegans* that is partially conserved in *M. hapla*. Although all five genes are present in *M. hapla*, only four of five genes are conserved in both orientation and spacing between the two species, but one gene in the *C. elegans* operon is located in a different region of the *M. hapla* genome. This is a common example between these species.

M. hapla and *C. elegans*, but synteny is either broken or non-existent over most of the genome. Lack of operon conservation with *C. elegans* generally results from orthologous genes not being present in *M. hapla*. In most cases the corresponding genes in *C. elegans* do not have an RNAi phenotype (www.wormbase.org), suggesting either redundant or dispensable function. Taken as a whole, these data indicate that *M. hapla* has a minimal genome, due to its obligate biotrophic lifestyle.

16.3.5 Pathway conservation with free-living nematodes

Genes in many important developmental pathways in *C. elegans* are conserved in the *M. hapla* genome, although their roles are not well defined. Sex determination is a key developmental event in all nematodes; however, *M. hapla* carries only a small number of proteins with significant matches to those defined in sex determination in *C. elegans*. Of the major sex-determining genes in *C. elegans*, *tra-1* and *tra-2* alone are highly conserved. Although several dosage-compensation genes are also conserved, as a whole this pathway remains obscure in *M. hapla*. No genes found early in the pathway are conserved, suggesting that the signals that trigger these pathways are substantially diverged. By contrast, many genes in other *C. elegans* pathways have clear orthologues in *M. hapla*. Genes involved in function of small RNAs, including *drsh-1*, *pash-1*, *dcr-1*, *drh-2*, *drh-3*, *agl-1*, *agl-2*, *rrf-3*, *eri-1* and *pir-1*, have strong orthologues in *M. hapla*, with the exception of *rde-4*, which generally is not conserved across phylogenetic distances. Not surprisingly, RNAi is robust in *M. hapla*, where it is persistent over several generations. Similarly, many genes involved in basic nematode development are well conserved, reflecting their primary roles in generalized nematode developmental biology.

The ability to form dauer juveniles is broadly conserved across the phylum Nematoda. Bird and Opperman (1998) previously proposed that the infective stages of *Meloidogyne* function as dauers. Dauers have been best studied in *C. elegans*, where they function as a survival and dispersal stage under adverse conditions. There is no strict definition of a 'dauer' but these survival stages share the properties of being developmentally arrested, motile, non-feeding, non-ageing and hence long lived (Cassada and Russell, 1975; Klass and Hirsh, 1976; Riddle and Albert, 1997). Dauer entry and exit is controlled by the environmental cues of 'food signal' and nematode population density, which is measured based on a secreted pheromone. Although the various arrested parasites functionally resemble *C. elegans* dauers, they clearly respond to a widely different range of cues for entry and exit.

In *C. elegans*, 32 genes have been identified by mutagenesis as affecting dauer formation (*daf*). The molecular identity of 20 genes has been characterized. The related species *C. briggsae* carries 19 of these but does not contain the beta-insulin molecule encoded by *daf-28*. This molecule is key in signal transduction in the dauer pathway. Although *M. hapla* also lacks a *daf-28* orthologue, it does carry 14 orthologues of *C. elegans daf* genes, as well as three further matches that are weak. It is not yet possible to determine whether these genes perform similar roles in *M. hapla* to those in *C. elegans*, but these findings do suggest that basic developmental mechanisms are conserved, although signalling is not. This indicates that developmental response to the environment differs drastically between free-living and parasitic species.

16.4 Conclusions and Future Directions

Although the genomes of *M. hapla* and *M. incognita* differ in numerous important ways, they also share a number of characteristics. For example, both species carry substantially fewer GPCRs than does the free-living nematode, *C. elegans*. Additionally, they carry fewer collagen genes and approximately half the number of nuclear hormone receptors found in *C. elegans*. Both *M. hapla* and *M. incognita* carry a considerable number of carbohydrate active enzymes, a number of which are suspected to be involved in parasitism of plants. Many of the genes encoding these enzymes are thought to have been acquired by horizontal gene transfer, mainly from bacteria. It does appear that, like *C. elegans*, a percentage of genes are organized into operons, although these are not strictly conserved between the three species. Both *M. hapla* and *M. incognita* share some genes that are related to sex determination in common with *C. elegans*, although the pathways are probably not conserved between the parasites and the free-living nematode. *M. hapla* and *M. incognita* carry most of the genes involved in the RNAi pathway. Together, these data indicate that the parasitic species carry similar basal gene suites.

There are also substantial differences between organization of the *M. hapla* and *M. incognita* genomes. *Meloidogyne hapla* VW9 is a diploid with 16 chromosomes and very low levels of repetitive DNA, found mostly as simple, low-complexity repeats. By contrast, *M. incognita* is an aneuploid species with variable chromosome number and carries a higher complement of repetitive DNA, probably due to the increased chromosomal complement. Additionally, *M. incognita* is estimated to possess 19,212 genes, while *M. hapla* appears to carry 14,420 genes. This substantial difference in gene complement may be due to several factors, including both ploidy and duplication of chromosomal segments in the *M. incognita* genome. It may be that these events have enabled *M. incognita* to have an expanded host range compared with *M. hapla*. Gene density in *M. incognita* is very similar to that in *C. elegans*, while *M. hapla* shows substantially greater gene density. This is likely to be a reflection of the small genome size of *M. hapla*. *Meloidogyne hapla* also has smaller introns than either *M. incognita* or *C. elegans*. Both parasitic nematodes have fairly low G+C content compared with *C. elegans*, although that of *M. hapla* is extremely low, at 27.4%. Taken as a whole, these data suggest that *M. hapla* is a basal species and that *M. incognita* has evolved to expand its host range. Deeper comparative analysis of these two genomes is likely to shed light on the evolution of *Meloidogyne* spp. and point to both the basal gene complement and genes involved in host range. Finally, future analysis of transcriptomes and proteomes will provide clues to both the basic parasite biology and parasitism of plants by *Meloidogyne* spp.

Obtaining the complete genome sequence from two species of root-knot nematodes, *M. incognita* and *M. hapla*, has immediate ramifications for both plant nematology and broader biology. The ability to perform comparative genomics using other nematode genomes, such as

C. elegans, *C. briggsae* and *B. malayi*, will provide insight into the evolution of both parasitic ability and general nematode development. In this respect, the comparative analysis of these two root-knot nematode genomes will represent a significant first step towards understanding the genome dynamics and evolution of asexual versus sexual organisms. In addition, these studies will reveal critical junctures in the life cycle of the obligate parasites that may be unique and specific targets for anti-nematode therapies. In particular, events such as sex determination, arrested development and response to host and environmental cues may be examined in a detail that was previously impossible. These events represent key points in the parasites' life cycle, and we expect that the mechanisms will be conserved between species, although the precise machinery may differ. Interestingly, many pathogenic bacteria are evolving to lose genes from pathways for which they rely on their host, and therefore have smaller genomes than their free-living counterparts. The smaller size of the *M. hapla* genome as compared with the free-living *C. elegans* could point to a similar strategy and, as noted above, current evidence is pointing to *M. hapla* possessing fewer genes than its free-living counterpart. If indeed the parasite has a reduced gene repertoire, those lost or disabled pathways may also provide clues to the complex interaction between host and parasite. The situation is more complicated in *M. incognita*, where the genome is composed predominantly of homologous but diverged segment pairs, which are likely to represent ancient allelic regions. Therefore, at this stage of the study it is very difficult to make conclusions about any gene reduction in this parasite. Needless to say, areas such as the identification of pathogenicity islands, virulence operons and horizontally transferred genes will also be amenable to detailed study. Looking to the future, the availability of more free-living, animal- and plant-parasitic nematode genomes will provide an unparalleled opportunity for comparative genomics, with the view to explaining the success of the phylum Nematoda.

16.5 Acknowledgements

Pierre Abad would like to thank the nematology group at INRA Sophia Antipolis (France), the bioinformatics platform at INRA Toulouse (France), the GENOSCOPE at Evry (France) and the teams of the annotation consortium for their contributions in the *M. incognita* genome sequencing project.

Charles Opperman would like to thank the scientists at the Center for the Biology of Nematode Parasitism (both past and present), Professor Valerie Williamson's group at the University of California, Davis, Professor Dan Rokhsar and the DOE-Joint Genome Institute at Walnut Creek, California. Special thanks to Dr Ann Lichens-Park at CSREES, USDA for her support and, of course, to the CSREES for funding this project through the Microbial Genome Sequencing Program.

16.6 References

Abad, P., Gouzy, J., Aury, J.-M. *et al.* (2008) Genome sequence of the metazoan plant-parasitic nematode *Meloidogyne incognita*. *Nature Biotechnology* 8, 909–915.

Alegado, R.A. and Tan, M.W. (2008) Resistance to antimicrobial peptides contributes to persistence of *Salmonella typhimurium* in the *C. elegans* intestine. *Cellular Microbiology* 10, 1259–1273.

Altschul, S.F., Madden, T.L., Schäffer, A.A., Zhang, J., Zhang, Z., Miller, W. and Lipman, D.J. (1997) Gapped BLAST and PSI-BLAST: a new generation of protein database search programs. *Nucleic Acids Research* 25, 3389–3402.

Ashburner, M., Ball, C.A., Blake, J.A. *et al.* (2000) Gene ontology: tool for the unification of biology. The Gene Ontology Consortium. *Nature Genetics* 25, 25–29.

Bargmann, C.I. (2006) Chemosensation in *C. elegans*. *WormBook*, 1–29.

Bird, D.McK. and Opperman, C.H. (1998) *Caenorhabditis elegans*: a genetic guide to parasitic nematode biology. *Journal of Nematology* 30, 299–308.

Bird, D.McK., Clifton, S.W., Kepler, T., Kieber, J.J., Thorne, J. and Opperman, C.H. (2002) Genomic dissection of a nematode–plant interaction: a tool to study plant biology. *Plant Physiology* 129, 394–395.

Blaxter, M.L. (2003) Nematoda: genes, genomes and the evolution of parasitism. *Advances in Parasitology* 54, 101–195.

Blaxter, M., Daub, J., Guiliano, D., Parkinson, J., Whitton, C. and The Filarial Genome Project (2002) The *Brugia malayi* genome project: expressed sequence tags and gene discovery. *Transactions of the Royal Society of Tropical Medicine and Hygiene* 96, 7–17.

Blumenthal, T. and Gleason, K.S. (2003) *Caenorhabditis elegans* operons: form and function. *Nature Review Genetics* 4, 112–120.

Caillaud, M.C., Dubreuil, G., Quentin, M., Perfus-Barbeoch, L., Lecomte, P., de Almeida Engler, J., Abad, P., Rosso, M.-N. and Favery, B. (2008) Root-knot nematodes manipulate plant cell functions during a compatible interaction. *Journal of Plant Physiology* 165, 104–113.

Cassada, R.C. and Russell, R.L. (1975) The dauer larva, a post-embryonic developmental variant of the nematode *Caenorhabditis elegans. Developmental Biology* 46, 326–342.

Castagnone-Sereno, P. (2006) Genetic variability and adaptive evolution in parthenogenetic root-knot nematodes. *Heredity* 96, 282–289.

Cosgrove, D.J. (2000) Loosening of plant cell walls by expansins. *Nature* 407, 321–326.

Daub, J., Loukas, A., Pritchard, D.I. and Blaxter, M. (2000) A survey of genes expressed in adults of the human hookworm, *Necator americanus. Parasitology* 120, 171–184.

Davis, E.L., Hussey, R.S. and Baum, T.J. (2004) Getting to the roots of parasitism by nematodes. *Trends in Parasitology* 20, 134–141.

Ewbank, J.J. (2006) Signaling in the immune response. *WormBook* (online), 1–12.

Foissac, S. and Schiex, T. (2005) Integrating alternative splicing detection into gene prediction. *BMC Bioinformatics* 6, 25.

Frand, A.R., Russel, S. and Ruvkun, G. (2005) Functional genomic analysis of *C. elegans* molting. *PLoS Biology* 3, e312.

Ghedin, E., Wang, S., David Spiro, D. *et al.* (2007) Draft genome of the filarial nematode parasite *Brugia malayi. Science* 317, 1756–1760.

Guiliano, D.B. and Blaxter, M.L. (2006) Operon conservation and the evolution of trans-splicing in the phylum Nematoda. *PLoS Genetics* 2, e198.

Haslam, S.M. and Dell, A. (2003) Hallmarks of *Caenorhabditis elegans* N-glycosylation: complexity and controversy. *Biochimie* 85, 25–32.

Hotson, A. and Mudgett, M.B. (2004) Cysteine proteases in phytopathogenic bacteria: identification of plant targets and activation of innate immunity. *Current Opinion in Plant Biology* 7, 384–390.

Huang, G., Gao, B., Maier, T., Allen, R., Davis, E.L., Baum, T.J. and Hussey, R.S. (2003) A profile of putative parasitism genes expressed in the esophageal gland cells of the root-knot nematode *Meloidogyne incognita. Molecular Plant–Microbe Interactions* 16, 376–381.

Huang, G., Dong, R., Allen, R., Davis, E.L., Baum, T.J. and Hussey, R.S. (2005) Two chorismate mutase genes from the root-knot nematode *Meloidogyne incognita. Molecular Plant Pathology* 61, 23–30.

Huang, G., Allen, R., Davis, E.L., Baum, T.J. and Hussey, R.S. (2006) Engineering broad root-knot resistance in transgenic plants by RNAi silencing of a conserved and essential root-knot nematode parasitism gene. *Proceedings of the National Academy of Sciences of the USA* 103, 14302–14306.

Jaffe, D.B., Butler, J., Gnerre, S., Mauceli, E., Lindblad-Toh, K., Mesirov, J.P., Zody, M.C. and Lander, E.S. (2003) Whole-genome sequence assembly for mammalian genomes: Arachne 2. *Genome Research* 13, 91–96.

Klass, M. and Hirsh, D. (1976) Non-ageing developmental variant of *Caenorhabditis elegans. Nature* 260, 523–525.

Lambert, K.N., Allen, K.D. and Sussex, I.M. (1999) Cloning and characterization of an esophageal-gland-specific chorismate mutase from the phytoparasitic nematode *Meloidogyne javanica. Molecular Plant–Microbe Interactions* 12, 328–336.

Leroy, S., Duperray, C. and Morand, S. (2003) Flow cytometry for parasite nematode genome size measurement. *Molecular and Biochemical Parasitology* 128, 91–93.

Li, L., Stoeckert, C.J Jr and Roos, D.S. (2003) OrthoMCL: identification of ortholog groups for eukaryotic genomes. *Genome Research* 13, 2178–2189.

Lindblom, T.H. and Dodd, A.K. (2006) Xenobiotic detoxification in the nematode *Caenorhabditis elegans. Journal of Experimental Zoology. A Comparative Experimental Biology* 305, 720–730.

Liu, Q.L., Thomas, V.P. and Williamson, V.M. (2007) Meiotic parthenogenesis in a root-knot nematode results in rapid genomic homozygosity. *Genetics* 176, 1483–1490.

Marks, N.J. and Maule, A.G. (2007) Neuropeptides in helminths: occurrence and distribution. In: Geary, T.G. and Maule, A.G. (eds) *Neuropeptide Systems as Targets for Parasite and Pest Control.* Landes Bioscience, Eurekah Bioscience Database, Austin, Texas.

McCarter, J.P., Abad, P., Jones, J. and Bird, D. (2000) Rapid gene discovery in plant-parasitic nematodes via expressed sequence tags. *Nematology* 2, 719–731.

McCarter, J.P., Mitreva, M.D., Martin, J. *et al.* (2003) Analysis and functional classification of transcripts from the root-knot nematode *Meloidogyne incognita. Genome Biology* 4, R26.1–R26.19.

Menzel, R., Bogaert, T. and Achazi, R. (2001) A systematic gene expression screen of *Caenorhabditis elegans* cytochrome P450 genes reveals CYP35 as strongly xenobiotic inducible. *Archives of Biochemistry and Biophysics* 395, 158–168.

Mestrovic, N., Castagnone-Sereno, P. and Plohl, M. (2006) Interplay of selective pressure and stochastic events directs evolution of the MEL172 satellite DNA library in root-knot nematodes. *Molecular Biology and Evolution* 23, 2316–2325.

Mitreva, M., McCarter, J.P., Martin, J., Dante, M., Wylie, T., Chiapelli, B., Pape, D., Clifton, S.W., Nutman, T.B. and Waterston, R.H. (2004) Comparative genomics of gene expression in the parasitic and free-living nematodes *Strongyloides stercoralis* and *Caenorhabditis elegans. Genome Research* 14, 209–220.

Mitreva, M., Blaxter, M.L., Bird, D.M. and McCarter, J.P. (2005) Comparative genomics of nematodes. *Trends in Genetics* 21, 573–581.

Mulder, N.J., Apweiler, R., Attwood, T.K. *et al.* (2007) New developments in the InterPro database. *Nucleic Acids Research* 35 (Database issue), D224–228.

Opperman, C.H. and Bird, D.McK. (1998) The soybean cyst nematode, *Heterodera glycines*: a genetic model system for the study of plant-parasitic nematodes. *Current Opinions in Plant Biology* 1, 342–346.

Opperman, C.H., Bird, D.McK., Williamson, V.M. *et al.* (2008) Sequence and genetic map of *Meloidogyne hapla*: a compact nematode genome for plant parasitism. *Proceedings of the National Academy of Sciences of the USA* 105, 14802–14807.

Page, A.P. and Johnstone, I.L. (2007) The cuticle. *WormBook*, 1–15.

Papadopoulou, J. and Triantaphyllou, A.C. (1982) Sex-determination in *Meloidogyne incognita* and anatomical evidence of sexual reversal. *Journal of Nematology* 14, 549–566.

Parkinson, J., Mitreva, M., Whitton, C. *et al.* (2004) A transcriptomic analysis of the phylum Nematoda. *Nature Genetics* 36, 1259–1267.

Paschinger, K., Gutternigg, M., Rendi , D. and Wilson, I.B.H. (2007) The N-glycosylation pattern of *Caenorhabditis elegans. Carbohydrate Research* 343, 2041–2049.

Piotte, C., Castagnone-Sereno, P., Bongiovanni, M., Dalmasso, A. and Abad, P. (1994) Cloning and characterization of two satellite DNAs in the low-C-value genome of the nematode *Meloidogyne* spp. *Gene* 138, 175–180.

Plowman, G.D., Sudarsanam, S., Bingham, J., Whyte, D. and Hunter, T. (1999) The protein kinases of *Caenorhabditis elegans*: a model for signal transduction in multicellular organisms. *Proceedings of the National Academy of Sciences of the USA* 96, 13603–13610.

Qin, L., Kudlav, U., Roze, E. *et al.* (2004) Plant degradation: a nematode expansin acting on plants. *Nature* 427, 30.

Quevillon, E., Silventoinen, V., Pillai, S., Harte, N., Mulder, N., Apweiler, R. and Lopez, R. (2005) InterProScan: protein domains identifier. *Nucleic Acids Research* 33 (web server issue), W116–W120.

Riddle, D.L. and Albert, P.S. (1997) Genetic and environmental regulation of dauer larva development. In: Riddle, D.L., Blumenthal, T., Meyer, B.J. and Priess, J.R. (eds) *C. elegans II.* Cold Spring Harbor Laboratory Press, Cold Spring Harbor, New York, pp. 739–768.

Robertson, H.M. and Thomas, J.H. (2006) The putative chemoreceptor families of *C. elegans. WormBook*, 1–12.

Rosso, M.N., Dubrana, M.P., Cimbolini, N., Jaubert, S. and Abad, P. (2005) Application of RNA interference to root-knot nematode genes encoding esophageal gland proteins. *Molecular Plant–Microbe Interactions* 18, 615–620.

Sasser, J.N. and Freckman, D.W. (1987) A world perspective on nematology; the role of society. In: Veech, J.A. and Dickson, D.W. (eds) *Vistas on Nematology.* Society of Nematologists, Hyattsville, Maryland, pp. 7–14.

Tang, X., Xiao, Y. and Zhou, J.M. (2006) Regulation of the type III secretion system in phytopathogenic bacteria. *Molecular Plant–Microbe Interactions* 19, 1159–1166.

Tetteh, K.K., Loukas, A., Tripp, C. and Maizels, R.M. (1999) Identification of abundantly expressed novel and conserved genes from the infective larval stage of *Toxocara canis* by an expressed sequence tag strategy. *Infection and Immunity* 67, 4771–4779.

The *C. elegans* Genome Sequencing Consortium (1998) Genome sequence of the nematode *C. elegans*: a platform for investigating biology. *Science* 282, 2012–2018.

Triantaphyllou, A.C. (1981) Oogenesis and the chromosomes of the parthenogenetic root knot nematode *Meloidogyne incognita. Journal of Nematology* 13, 95–104.

Triantaphyllou, A.C. (1985) Cytogenetics, cytotaxonomy and phylogeny of root-knot nematodes. In: Sasser, J.N. and Carter, C.C. (eds) *An Advanced Treatise on Meloidogyne. Volume 1 Biology and Control.* North Carolina State University Graphics, Raleigh, North Carolina, pp. 113–126.

Trudgill, D.L. and Blok, V.C. (2001) Apomictic, polyphagous root-knot nematodes: exceptionally successful and damaging biotrophic root pathogens. *Annual Review of Phytopathology* 39, 53–77.

Wei, Y.D., Lee, K.S., Gui, Z.Z., Yoon, H.J., Kim, I., Zhang, G.Z., Guo, X., Sohn, H.D. and Jin, B.R. (2006) Molecular cloning, expression, and enzymatic activity of a novel endogenous cellulase from the mulberry longicorn beetle, *Apriona germani. Comparative Biochemistry and Physiology B: Biochemistry and Molecular Biology* 145, 220–229.

Welch, M., Cummings, M.P., Hillis, D.M. and Meselson, M. (2004) Divergent gene copies in the asexual class Bdelloidea (Rotifera) separated before the bdelloid radiation or within bdelloid families. *Proceedings of the National Academy of Sciences of the USA* 101, 1622–1625.

Williamson, A.L., Brindley, P.J., Knox, D.P., Hotez, P.J. and Loukas, A. (2003) Digestive proteases of blood-feeding nematodes. *Trends in Parasitology* 19, 417–423.

Xia, Y. (2004) Proteases in pathogenesis and plant defence. *Cellular Microbiology* 6, 905–913.

Yan, Y., Smant, G. and Davis, E.R. (2001) Functional screening yields a new beta-1,4-endoglucanase gene from *Heterodera glycines* that may be the product of recent gene duplication. *Molecular Plant–Microbe Interactions* 14, 63–71.

Zarkower, D. (2006) Somatic sex determination. *WormBook*, 1–12.

Zawadzki, J.L., Presidente, P.J.A., Meeusen, E.N. and De Veer, M.J. (2006) RNAi in *Haemonchus contortus*: a potential method for target validation. *Trends in Parasitology* 22, 495–499.

17 Biological Control Using Microbial Pathogens, Endophytes and Antagonists

Johannes Hallmann[1], Keith G. Davies[2] and Richard Sikora[3]
[1]*Julius Kühn Institute, Münster, Germany;* [2]*Rothamsted Research, Harpenden, Hertfordshire, UK;* [3]*University of Bonn, Bonn, Germany*

17.1 Introduction	**380**
17.2 Bacterial Pathogens and Antagonists	**381**
17.3 Fungal Pathogens and Antagonists	**389**
17.4 Commercialization and Future Directions	**395**
17.5 References	**401**

17.1 Introduction

Microbial pathogens, endophytes and antagonists are extremely important in the regulation of plant-parasitic nematode populations, irrespective of agroecosystems. Many reviews concerning microbial control agents have been published, which have outlined the importance of microbial control agents in nematode management (Mankau, 1981; Jatala, 1986; Kerry, 1987; Sayre, 1988; Stirling, 1991; Sikora, 1992; Siddiqui and Mahmood, 1999). However, none of these reviews was devoted strictly to root-knot nematodes. In this chapter, the present state of the art and future importance of microbial pathogens and antagonists in the biological control of species of *Meloidogyne* will be stressed in the light of recent research progress.

Research on microbial pathogens and antagonists of root-knot nematodes, as well as other economically important species, has progressed over a 50-year developmental process that included: (i) the isolation and identification of the organisms with potential as biological control agents; (ii) ecological manipulation of the soil environment to improve antagonism; (iii) elucidation of mechanisms of parasitism and infection; and (iv) exploration for commercial product development. Therefore, it is surprising that after many years of extensive research the impact of biological control on root-knot management in the field is still marginal. Biological control activity is omnipresent in soil and can range from negligible to complete nematode suppression, with the degree of biological control determined by the diversity and density of communities and/or individual antagonistic microorganisms present in a specific soil.

The potential of microbial pathogens, endophytes and antagonists for biological control of *Meloidogyne* spp. is great when one considers the microbial-based efficacy within a suppressive soil – a soil that totally suppresses nematode multiplication. Nematode-suppressive soils have been detected, and their activity has been shown to be driven by a diverse spectrum of microbes: fungal pathogens of eggs, rhizobacteria, generalized fungal antagonists, mutualistic fungal endophytes

 ©CAB International 2009. *Root-knot Nematodes* (eds R.N. Perry, M. Moens and J.L. Starr)

and obligate parasitic bacteria (Whipps and Davies, 2000). Such suppressive soils are often overlooked because nematicide use masks their existence. Monoculture or short rotations of susceptible crops have also been shown to lead to suppressiveness. Methodologies (Hirsch *et al.*, 2001) have been developed to measure and monitor specific microbial agents comprising the antagonistic potential of a soil suppressive to root-knot nematodes. These techniques can aid in understanding how suppressiveness develops and how it can be manipulated for root-knot nematode control.

The results obtained in the past now allow nematology to enter a new phase of biological control of root-knot nematodes, thus making an impact at the field level. The sections below outline the present status and future direction.

17.2 Bacterial Pathogens and Antagonists

The natural habitat of *Meloidogyne* in the soil and plant is colonized by a broad diversity of bacteria. Quite often bacterial numbers even increase following plant parasitism by *Meloidogyne*, indicating possible interactions. Although most bacteria still remain unidentified, as they are not readily cultured, many of those that can be cultured have been studied for their potential to interfere with nematode behaviour, feeding or reproduction (Table 17.1). Among them the obligate endoparasite *Pasteuria penetrans* represents the best-studied bacterial parasite. However, saprophytic bacteria occurring in the soil, rhizosphere or endorhiza have also been shown to be powerful antagonists with unique modes of action.

17.2.1 Endoparasitic bacteria

There are a number of endoparasitic bacteria that are known to affect nematodes, the most studied of which are from the genus *Wolbachia* (Taylor, 2003). These are Gram-negative bacteria of the family Rickettsiaceae that are non-motile and non-spore forming and are obligate intracellular parasites of invertebrates, including parasitic nematodes. Rickettsia-like organisms were observed in some plant-parasitic nematodes as early as 1973 (Shepherd *et al.*,

Table 17.1. Effect of bacterial pathogens and antagonists on different developmental stages of *Meloidogyne* spp.

Developmental stage	Nematode behaviour intercepted	Mode of action	Place of action	Examples of bacteria	References
Egg or egg mass	Development, hatching	Toxins, lytic enzymes	Soil	*Telluria chitinolytica*	Spiegel *et al.* (1991)
Infective juvenile	Vitality, host attraction, host recognition, penetration	Toxins, lectins, degradation of root exudates, induced resistance, parasitism	Soil, rhizosphere	*Pasteuria penetrans*, *Pseudomonas fluorescens*, *Pseudomonas aeroginosa*, *Rhizobium etli*	Krechel *et al.* (2002); Siddiqui and Shaukat (2004, 2005); Siddiqui *et al.* (2006); Sikora *et al.* (2007); Oliveira *et al.* (2007)
Sedentary juvenile	Formation of feeding site, development	Toxins, induced resistance, parasitism	Endorhiza	*Pasteuria penetrans*, *Rhizobium etli*	Davies *et al.* (1991); Munif *et al.* (2000)
Female	Fecundity	Toxins	Rhizosphere, endorhiza	*Pasteuria penetrans*	Davies *et al.* (2008)

1973). Most recently, a bacterial endosymbiont of *Heterodera glycines* has been identified and described (Noel and Atibalentja, 2006), although the extent to which this organism affects its plant-parasitic nematode host is at present undetermined. So far, there are no reports of these bacteria associated with root-knot nematodes. The most-studied endoparasaitic bacterium of root-knot nematodes is the Gram-positive obligate bacterium *P. penetrans*.

17.2.1.1 Pasteuria penetrans

The *Pasteuria* group of bacteria are hyperparasites of plant-parasitic nematodes and water fleas (*Daphnia* spp.; Cladocera: Anomopoda), and produce highly resistant endospores. The detailed taxonomy of this group of bacteria remains unclear, but the bacterium is a member of the *Bacillus–Closteridium* clade (Charles *et al.*, 2005). All the economically important genera of plant-parasitic nematodes are parasitized by *Pasteuria* spp. To date, five species of *P.* that differ in their host ranges and pathogenicity on nematodes have been described: (i) *P. penetrans* is parasitic on *Meloidogyne* spp. (Sayre and Starr, 1985); (ii) *P. thornei* parasitizes *Pratylenchus brachyurus* (Sayre *et al.*, 1988); (iii) *P. nishizawae* parasitizes cyst nematodes (Sayre *et al.*, 1991); (iv) *P. usgae* is parasitic on the sting nematode *Belonolaimus longicaudatus* (Giblin-Davis *et al.*, 2003); and (v) *P. hartismeri* is parasitic on *Meloidogyne ardenensis* (Bishop *et al.*, 2007). A diagrammatic overview of the life cycle of *P. penetrans* is given in Fig. 17.1. Infection by this bacterium is initiated by endospores adhering to the cuticle of infective second-stage juveniles (J2); endospores then penetrate the body wall to enter the host, usually after the nematode has set up a feeding site (Chen *et al.*, 1996, 1997) (Fig. 17.2). Parasitized females of root-knot nematodes produce only a few eggs (Davies *et al.*, 2008), and the observations that this bacterium has been associated with root-knot nematode suppressive soils (Oostendorp *et al.*, 1991; Weibelzahl-Fulton *et al.*, 1996; Trudgill *et al.*, 2000; Cetintas and Dickson, 2004) has indicated to nematologists that the bacterium has the potential to be developed into a biological control agent of economically important crop pests. However, its restricted host range is a constraint to commercial development.

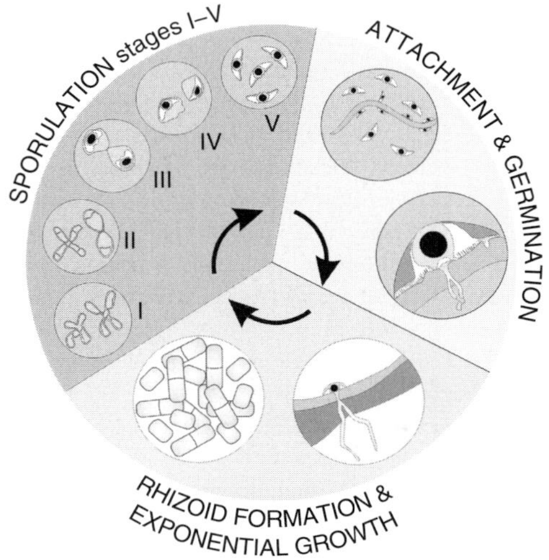

Fig. 17.1. Diagramatic representation of the life cycle of *Pasteuria penetrans*. (Based on Davies, 2009.)

17.2.1.2 Mass production, in vivo *and* in vitro *culturing methods*

A method for the mass production of endospores *in vivo* was first described by Stirling and Wachtel (1980). This method, by which it is possible to produce sufficient endospores for experimental purposes and small-scale pot and field trials, involves encumbering J2 with five to ten endospores per J2 and inoculating tomato plants grown in pots. The time taken for the life cycle of the bacterium to be completed is temperature dependent (Stirling, 1981; Chen and Dickson, 1997; Serracin *et al.*, 1997; Darban *et al.*, 2004) and takes approximately 7–9 weeks at 25 °C. At harvest, the soil is washed from the root systems and the roots dried before they are milled. The endospore yield can be highly variable, but yields of 10^7–10^9/g of milled tomato root powder are not uncommon. The fact that only limited amounts of endospores can be obtained through *in vivo* mass production has stimulated a number of researchers to attempt to develop *in vitro* culturing technology. All the early attempts to produce *P. penetrans in vitro* were unsuccessful (Bishop and Ellar, 1991), although a patent was submitted showing that endospores could be produced in media that contained explanted tissue from nematodes (Previc and Cox, 1992). Other research had shown that vegetative forms of the bacterium could be maintained for short periods of time in one type

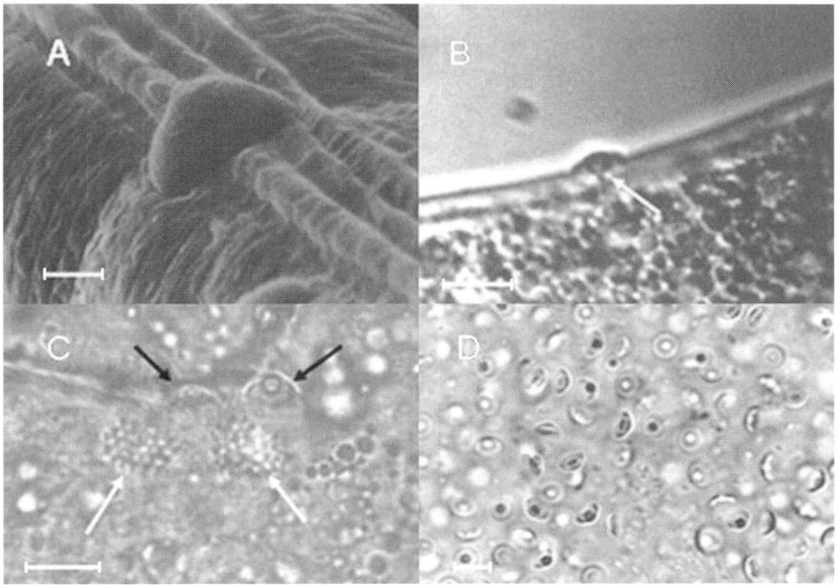

Fig. 17.2. A: Scanning electron micrograph of a *Pasteuria penetrans* endospore adhering to the cuticle of a root-knot nematode (scale bar = 1μm); B: light micrograph of a germinating endospore penetrating the cuticle of a root-knot nematode; C: development of two dichotomously branched microcolonies (lower arrows) associated with two endospores (upper arrows); D: mature endospores (scale bars B–D = 5μm).

of medium and other media could lead to sporulation (Bishop and Ellar, 1991). Interestingly, recent research has suggested that cation concentrations are important in the phosphorylation of Spo0F, a key protein involved in the initiation of sporogenesis in *Bacillus subtilis*. This gene also has a homologue in *P. penetrans* (Kojetin *et al.*, 2005), and it has been suggested that cation concentrations may be prohibiting the vegetative forms of the bacterium from entering sporogenesis and forming endospores. Pasteuria Biosciences LLC, based in Florida, USA, has been intensively pursuing a method of *in vitro* culture, and their proprietary culturing techniques (Gerber and White, 2001; Hewlett *et al.*, 2004) have enabled them to produce sufficient endospores in a fermentation vessel to undertake small trials (Hewlett *et al.*, 2006).

17.2.1.3 Quantification, nematode suppression and the problem of host specificity

There are many microbial pathogens, endophytes and antagonists present in the soil that potentially may reduce root-knot nematode populations but by far the majority are, in fact, neither pathogens nor antagonists. Estimates of total soil microbial biomass are very difficult to determine as many of them cannot be cultured, and *P. penetrans* is a good example of a bacterium whose numbers are difficult to estimate as it does not grow on the traditional media used for quantifying bacteria in soil. However, it has been estimated that soils with around 10^4 *Pasteuria* endospores per g of soil can be suppressive to plant-parasitic nematodes (Davies *et al.*, 1990; Stirling, 1991). But these estimations of endospore numbers occurring in the soil, and those of the suppression of root-knot nematodes should be used with care, as studies have shown that attachment to and subsequent infection of *Pasteuria* endospores in J2 varies both between and within populations of the bacterium (Espanol *et al.*, 1997; Mendoza de Gives *et al.*, 1999; Davies *et al.*, 2001; Wishart *et al.*, 2004). Therefore, the number of endospores that are required in a soil to suppress a particular root-knot nematode population is difficult to measure. Additionally, different species of *Meloidogyne* reproduce using different reproductive strategies (see Chitwood and Perry, Chapter 8, this volume); in standard attachment bioassays, J2 from single egg mass populations of *Meloidogyne* species that reproduce parthenogenetically have been shown to exhibit different numbers of endospores adhering to their cuticles (Davies and Danks, 1992; Fargette *et al.*, 1994; Davies and Redden, 1997). Interestingly, in a comparison of endospore attachment to broods of females derived from single juvenile lines of parthenogenetically reproducing root-knot nematodes, and from single juvenile lines of the amphimictic root-knot nematode *M. hapla*, attachment assays revealed an equally high degree of variability (Davies *et al.*, 2008). This suggests that the variation in cuticles revealed by *Pasteuria* endospore attachment assays is the product of two types of variation: (i) variation in amphimictically reproducing populations, i.e. of genetic origin, in which the numbers of endospores attaching to the nematode cuticle can segregate; and (ii) variation that arises from parthenogenetically reproducing clonal lines, i.e. somaclonal in origin (Davies, 2009). Therefore, and bearing in mind that nematodes usually occur as mixed populations, the inter- and intraspecific variation of root-knot nematode cuticle, as revealed by endopore attachment studies, makes the development of the mathematical relationship between endospore numbers and soil supressiveness very difficult to quantify with any degree of robust predictability.

Both immunological (Fould *et al.*, 2001; Schmidt *et al.*, 2003; Costa *et al.*, 2006) and molecular (Duan *et al.*, 2003) techniques have been employed to quantify *Pasteuria* endospores in soil. The threshold of detection for both immunological techniques and molecular approaches was about 10^2/g endospores soil. However, one of the advantages of immunological techniques over DNA-based approaches is that antibodies can be made to the surface of the endospore, which is immunologically related to nematode cuticle heterogeneity and spore attachment (Davies *et al.*, 1994; Davies and Redden, 1997; Costa *et al.*, 2006). Thus, the specificity of the antibody can also be related to the ability to parasitize a particular nematode host population. At present it is not possible to use molecular techniques to characterize *Pasteuria* in any meaningful way related to host specificity and parasitism. However, as knowledge of the molecules involved in endospore attachment, germination and parasitism becomes understood in terms of genetics, this drawback may only be temporary.

17.2.1.4 Mechanism of endospore attachment

The initial stage of the infection of a nematode by *P. penetrans* commences when endopores of the bacterium adhere to the cuticle of the J2 as it migrates through the soil. To apply *P. penetrans* as a biological control agent there must be compatibility between the bacterial endospore and the nematode cuticle such that endopores will attach, and several studies have examined this aspect (Stirling, 1985; Davies *et al.*, 1988; Sharma and Davies, 1996a,b; Espanol *et al.*, 1997; Mendoza de Gives *et al.*, 1999; Wishart *et al.*, 2004). Biochemical and immunological approaches indicate that this interaction may involve lectin/carbohydrate interactions (Davies *et al.*, 1992; Davies and Danks, 1992, 1993; Afolabi *et al.*, 1995; Spiegel *et al.*, 1996). Monoclonal antibodies raised to the surface of endospores reveal a high degree of heterogeneity, even within an endospore population produced by an individual female (Davies and Redden, 1997), and this heterogeneity has been shown to relate to recognition of a particular 'type' of nematode cuticle (Davies *et al.*, 1994). It has been hypothesized that endospore recognition is mediated through parasporal fibres on the surface of the endospore (Vaid *et al.*, 2002), which interact with a fibronectin-like receptor present in the nematode cuticle through hydrophobic interactions (Davies *et al.*, 1996; Mohan *et al.*, 2001). However, although the nematode cuticle contains collagen, it does not contain fibronectin (Kramer, 1997).

Research on *Bacillus anthracis*, the causative agent of anthrax, suggests that a different mechanism may be operating. Studies have shown that collagen-like fibres are present on the surface of the exosporium of the *B. anthracis* endospore. Further, it has been observed that the genes responsible for these collagen-like fibres exhibit a polymorphism in the number of G-x-y repeats that relate to the length of the collagen fibres (Sylvestre *et al.*, 2003). In a survey of *Pasteuria* genes from a preliminary sequencing project (Bird *et al.*, 2003), similar collagen-like genes have been identified with a predicted filament length of between 56 and 213 nm, suggesting that collagen-like fibres are important structures that may be involved in the attachment process of the endospore to the nematode cuticle (Davies and Opperman, 2006). Interestingly, in two independent studies in which endospores were pretreated, one with collagenase (Davies and Danks, 1993) and another with the collagen-binding domain of fibronectin (Mohan *et al.*, 2001), both pretreatments led to a significant reduction in attachment, suggesting that a collagen-like fibre is present on the surface of the endospore and is involved in adhesion. Whether or not these collagen-like fibres are also responsible for the specificity observed between different isolates of *P. penetrans* remains to be investigated. In *B. anthracis*, the collagen-like fibres are glycosylated and it is likely that glycosylation may be an important determinant in host specificity, but the different lengths of the collagen-like filaments may also play a role.

The other component of this host–parasite interaction relating to host specificity is the molecular structure of the cuticle surface of the J2. The outermost surface of the cortical layer is termed the 'surface coat' and is made up of proteins, carbohydrates and lipids (Lin and McClure, 1996; Blaxter and Robertson, 1998). Immunological studies have shown that this surface is highly dynamic (Spiegel *et al.*, 1997; Gravato-Nobre *et al.*, 1999; Sharon *et al.*, 2002). In animal-parasitic nematodes ethanol is sufficient to extract the surface coat (Page *et al.*, 1992) and therefore it is destroyed in TEM (transmission electron microscope) studies that use ethanol as a dehydration method. Pre-treatments of the J2 cuticle with a series of glycolytic and proteolytic enzymes all reduced the ability of endospores to adhere to the cuticle surface, but such biochemical approaches are unlikely to identify the key molecules that determine host specificity. Genetic approaches using mutants of *Caenorhabditis elegans* have identified genes that alter the surface coat in such a manner as to affect bacterial pathogenesis, and it is likely that the mechanism appears to be linked to the glycan moieties present on the surface coat (Hemmer *et al.*, 1991; Hoflich *et al.*, 2004; Gravato-Nobre *et al.*, 2005; Darby *et al.*, 2007). Similar glycan moieties have been shown to be important in *Pasteuria* endospore attachment (Davies and Danks, 1993; Spiegel *et al.*, 1996). Thus, bacteria appear to be high-resolution probes for genetic identification of nematode components involved in cuticle variability, and it is likely that such approaches will be important in understanding the nature of host specificity in *P. penetrans*–root-knot nematode interactions.

17.2.1.5 Potential for root-knot control

The fact that *Pasteuria* has been associated with nematode-suppressive soils and has been mass produced *in vivo* to add to microplots to control root-knot nematodes successfully (Stirling, 1984; Trudgill *et al.*, 2000) demonstrates that *Pasteuria* clearly has potential to be used as a biological control agent under field conditions. However, the large areas of field soil that will need to be treated will be problematic in the short term. Therefore, it is most likely that markets will be restricted to high-value horticultural crops, the protection of transplants and application to amenity grasses, where good coverage and high spore densities are possible. Host specificity is another problem to be addressed, and it is likely that products will need to contain carefully prepared mixtures of spores with different attachment and infection characteristics to different nematode species and populations to ensure robust control in a range of soils.

17.2.2 Rhizosphere bacteria

In the rhizosphere, bacterial densities are up to 100-fold higher than in bulk soil due to root exudates that serve as nutrients for the bacteria. Rhizosphere bacteria form a complex assemblage of species with many different functions within the soil–plant environment. Among the dominant bacterial genera, *Bacillus* and *Pseudomonas*, there are several species, such as *B. subtilis*, *Bacillus sphaericus* and *Pseudomonas fluorescens*, able to antagonize plant-parasitic nematodes (Sikora, 1992; Tian *et al.*, 2007). Other rhizosphere bacteria expressing antagonistic potential against *Meloidogyne* include, among others, members of the genera *Agrobacterium*, *Alcaligenes*, *Aureobacterium*, *Chryseobacterium*, *Corynebacterium*, *Enterobacter*, *Klebsiella*, *Paenibacillus*, *Phyllobacterium*, *Rhizobium* and *Xanthomonas* (Kloepper *et al.*, 1992; Duponnois *et al.*, 1999; Krechel *et al.*, 2002; Oliveira *et al.*, 2007). Nematode control by rhizosphere bacteria is achieved by mechanisms such as: direct antagonism through the production of toxins, enzymes or other secondary metabolites; interference with plant-nematode recognition; competition for nutrients; plant growth promotion; and induced systemic resistance (reviewed in Tian *et al.*, 2007). Siddiqui *et al.* (2005) suggested that bacterial proteases are involved in nematode antagonism by *P. fluorescens* CHA0; a mutant deficient in the *aprA* gene, encoding a major extracellular protease, resulted in reduced biological control activity against *M. incognita*. Furthermore, culture supernatants of the wild-type strain *P. fluorescens* CHA0 inhibited hatching and induced mortality of J2 of *M. incognita* more strongly than did supernatants of the *aprA*-deficient mutant. By contrast, protease activity does not seem to explain nematode antagonism by *Telluria chitinolytica* (= *Pseudomonas chitinolytica*), a bacterium isolated from chitin-treated soil which has strong biological control activity of *M. javanica* (Spiegel *et al.*, 1991). Although filtrates of *T. chitinolytica* expressed strong chitinolytic and proteolytic activities, nematode eggs pretreated with the *T. chitinolytica* filtrate showed no differences when compared with non-treated eggs of *M. javanica*. The data discussed above were derived from *in vitro* studies and further information is required about the concentrations required for effective control *in vivo*.

Another promising mechanism studied in detail is induced systemic resistance. Using a split-root system, several rhizosphere bacteria, such as *B. sphaericus* B43, *Rhizobium etli* G12 (Plate 32) and *P. fluorescens* Pf1, have been shown to induce systemic resistance towards several plant-parasitic nematodes, including species of *Meloidogyne* (Anita *et al.*, 2004; Siddiqui and Shaukat, 2004; Sikora *et al.*, 2007). For *R. etli* G12, it was shown that resistance is induced by the bacterial lipopolysaccharides (LPS) of the outer cell wall membrane (Reitz *et al.*, 2002), which probably bind to specific receptors on the plant cell surface. Resistance is then expressed in the outer root tissue, as *R. etli* G12-mediated resistance predominantly reduced nematode penetration but had no effect on nematode attraction and only a slight effect on development of those J2 able to bypass the resistance response.

For other rhizosphere bacteria, the identity of the elicitor still awaits identification. The effect of rhizosphere bacteria-mediated induced resistance has been studied more intensively in plants. *R. etli* G12 did not elicit expression of common PR (pathogenesis related)-proteins (acidic/basic chitinase, β-1,3-glucanase, PR-14, PAL (phenylalonine ammonia lyase)) or cause any detectable changes in the lignin content of the root (Reitz *et al.*, 2001), thus supporting the classical rhizosphere bacteria-mediated induced systemic resistance (ISR). As in ISR, salicylic acid (SA) does not seem to be involved in triggering the plant response (Siddiqui and Shaukat, 2004). By contrast, Anita *et al.* (2004) reported that control of *M. incognita* by *P. fluorescens* Pf1 was associated with enhanced PR-protein activity and accumulation of phenolics. Those plant responses are characteristic of systemic acquired resistance (SAR), generally induced in the plant by a local necrotic response such as a pathogen infection. Whatever the pathway, nematode infection may be reduced by >80%, which is sufficient to cause significant yield increase.

Although the control potential of rhizosphere bacteria is generally lower under field conditions, several products based on these bacteria are on the market. Deny is a commercial nematicide based on *Burkholderia cepacia*; BioNem-WP and BioSafe are two biological nematicides based on lyophilized *Bacillus firmus* supplemented with non-toxic additives intended mainly for controlling *Meloidogyne* spp.; and BioYieldTM is a biological inoculant containing *Paenibacillus macerans* and *B. amyloliquefaciens* to be incorporated into glasshouse planting mixes. Although the product's primary effect is early growth promotion and yield increase, a reduction in *Meloidogyne* and other plant-parasitic nematodes on horticultural crops has been reported.

It is important for nematode control that antagonistic bacteria colonize the rhizosphere before non-antagonistic bacteria become established. Seed treatment or soil drenches with antagonistic rhizosphere bacteria immediately after seeding guarantee early root colonization. However, if two or more rhizosphere bacteria are applied simultaneously, they often compete with each other. Competition between rhizosphere bacteria was shown to occur between *B. sphaericus* B43 and *R. etli* G12 (S. Azemoun, 2008, unpublished results). If *R. etli* G12 was applied 3 days before *B. sphaericus* B43, *R. etli* G12 intensively colonized the surface of potato roots. Application of *R. etli* G12 and *B. sphaericus* B43 at the same time resulted in reduced colonization intensity of *R. etli* G12, and if *R. etli* G12 was applied 3 days after *B. sphaericus* B43, the bacterium was not detected on the root surface.

17.2.3 Endophytic bacteria

Endophytic bacteria colonize the internal plant tissue, as do endoparasitic nematodes, which makes them ideal candidates for control of such nematodes. This is especially true for root-knot nematodes, which remain immobile for several weeks during feeding and development to adult stages. The term endophytic bacteria will be used within this context for bacteria that can be isolated from surface-disinfested plant tissue and that do not visibly harm the plant (Hallmann *et al.*, 1997). Most of those bacteria colonize both the rhizosphere and endorhiza simultaneously. This can be an important attribute to enhance disease control and consistency if the control agent is able to avoid unfavourable conditions in one habitat by escaping into the other habitat.

Although the first observations of bacteria residing in healthy plants date back to the 1870s, it was not until the mid-1990s that beneficial effects of endophytic bacteria towards plant-parasitic nematodes were demonstrated (reviewed in Hallmann *et al.*, 1997; Siddiqui and Mahmood, 1999). Research interest in endophytic bacteria was additionally fostered by the fact that several bacteria, originally considered as plant-health-promoting rhizosphere bacteria, turned out to be good endophytic colonizers, such as *R. etli* G12 (Hallmann *et al.*, 2001). Following intensive *in vitro* and *ad planta* screening procedures, several endophytic bacteria have been identified as antagonists of *Meloidogyne* spp.

After application by seed treatment, root dipping or soil drench, endophytic bacteria rapidly colonize the root tissue, where they establish stable population densities of approximately 10^4–10^5 cfu/g root fresh weight (Musson *et al.*, 1995, Hallmann *et al.*, 1997). *Meloidogyne* infection itself supports the establishment of high bacterial densities by increasing root exudation, which serves as a bacterial food source, creating wounds to provide entry avenues for the bacteria, direct transport of adhering bacteria into the plant, and modifying plant physiology (Hallmann *et al.*, 1998). Also, the spectrum of endophytic bacteria changes in response to *Meloidogyne* infection (Hallmann *et al.*, 1998). Even though the dominant species *Agrobacterium radiobacter* reached similar population densities in *M. incognita*-infected and non-infected cotton roots, *Brevundimonas vesicularis* was predominantly isolated from non-infested plants, whereas *Burkholderia pickettii* and *Alcaligenes xylosixydans* were exclusively isolated from *M. incognita*-infested roots (Hallmann *et al.*, 1998). The diversity of indigenous endophytic bacteria was slightly higher in *M. incognita*-infested roots than in non-infested roots (Hallmann *et al.*, 1998).

Bacterial endophytes can, in principle, colonize the plant systemically; however, the bacteria studied so far for biological control have been local colonizers (Hallmann *et al.*, 2001). Colonization studies with GFP (green fluorescent protein)-marked *R. etli* G12 on potato roots showed colonization of the entire root surface but high bacterial densities at the base of emerging lateral roots and root tips, and within epidermal cells (Hallmann *et al.*, 2001). Interestingly, those epidermal cells colonized by *R. etli* G12 were generally packed with bacteria, whereas neighbouring epidermal cells were not colonized. Those observations were confirmed on *Arabidopsis*, which has hyaline roots but, in addition, intracellular colonization of deeper root layers was also demonstrated (Hallmann *et al.*, 2001). Following infection with *M. incognita*, intensive fluorescence caused by the GFP-marked *R. etli* G12 was visible in the proximity of freshly penetrated J2 and within nematode galls. Furthermore, females that disrupted the root tissue at later stages of their development were also intensively colonized by *R. etli* G12 (Hallmann *et al.*, 2001).

17.2.4 Other bacteria

Most bacteria do not have a clear association with either the rhizosphere or endorhiza and primarily occur in the bulk soil. Among these bacteria, strong antagonistic activity towards plant-parasitic nematodes is shown by several species or even strains, such as *B. thuringiensis*, *Streptomyces* spp. and bacterial symbionts of entomopathogenic nematodes (Zuckerman *et al.*, 1993; Samaliev *et al.*, 2000; Krechel *et al.*, 2002). Of these, one of the best studied is *B. thuringienis* (*Bt*), commercially used as a biological insecticide to control Lepidopteran and Coleopteran insects on many crops worldwide. Furthermore, genetically modified plants expressing the *Bt* endotoxin provide control of insects such as *Ostrinia nubilalis* and *Diabrotica virgifera virgifera* on maize or *Helicoverpa armigera* on cotton. The potential of *Bt* for nematode control has also been examined. Commercial products containing *Bt*, such as Dipel and Turex, have been shown to reduce damage caused by root-knot nematodes (Radwan, 2007a). In the case of Dipel, control efficacy was enhanced by the addition of organic amendments such as chicken litter (Radwan 2007b). In insects, toxicity is achieved by crystals of proteinaceous δ-endotoxins causing cell disintegration in the insect gut. Unfortunately, those crystals are too large to be taken up by plant-parasitic nematodes, suggesting that different mechanisms are involved in nematode control. Prasad *et al.* (1972) had previously suggested that the nematicidal fraction is located in the exotoxin of *Bt*; while the purified exotoxin caused 100% mortality of *Meloidogyne* spp., the spore–crystal complex was ineffective. Those results were confirmed by ZhenChuan *et al.* (2004), showing that the β-exotoxin produced by *Bt* B24-14 reduced root-knot nematode infection in repeated experiments. Subsequently, Li *et al.* (2007) expressed the *Bt* crystal protein Cry6A plus the GFP in tomato roots and challenged the plants with *M. incognita*. Their data demonstrated that *M. incognita* was able to ingest the 54-kDa Cry6A protein, as shown by increased fluorescence in the nematode body, and that Cry6A was toxic to the J2, as indicated by a decrease of up to fourfold in progeny production. Thus, transgenic plants expressing the *Bt* Cry6A protein have some potential for suppression of plant-parasitic nematodes.

Another powerful group of antagonists are actinomycetes, i.e. bacteria characterized by forming branching filaments. For example, species of *Streptomyces* are important producers of antibiotics such as streptomycin, tetracycline, cycloheximide and other toxic metabolites. Screening potato-associated bacteria for their antagonistic potential, several *Streptomyces* species were found to control both *M. incognita* and fungal pathogens (Krechel *et al.*, 2002). Species which have shown antagonistic potential towards root-knot nematodes included *Streptomyces avermitilis*, *S. costaricanus*, *S. griseus*, *S. lavendulae* and *S. saraceticus*. Metabolites produced by those species caused either J2 mortality or reduced hatching, or both. For *S. lavendulae* SANK 64297, the minimal effective concentration was below 0.05 ppm (Takatsu *et al.*, 2003). As suggested by the authors, the anti-nematode activity may be derived from the inhibition of RNA synthesis. The best-studied species in terms of mode of action is *S. avermitilis*. This species produces macrocyclic lactones, so-called avermectins, which are the most potent nematicidal compound ever found. For example, the abamectin B1 is now commercialized under the name Avicta® as a seed treatment for cotton and vegetables against a broad spectrum of plant-parasitic nematodes (Plate 45). Avicta® moves from the treated seed alongside the growing roots, thus protecting the young plant from nematode infection.

Entomopathogenic nematodes have been found to reduce populations of root-knot nematodes. Lewis *et al.* (2001) reported a significant decrease in the number of galls and egg masses by *M. incognita* on tomato following soil application with *Steinernema feltiae*. However, Fallon *et al.* (2004) could not find any effect of *S. feltiae* on *M. javanica* root penetration and development in cowpeas, but if the symbiont *Xenorhabdus bovienii* was applied at 2 ml of a 10^{10} cfu/ml suspension per pot, root penetration by *M. javanica* was reduced. This suggests that the adverse effect of entomopathogenic nematodes on root-knot nematodes is probably through the action of the symbiotic bacteria *Xenorhabdus* and *Photorhabdus* that are associated with *Steinernema* and *Heterorhabditis*, respectively. Both exo- and endometabolites of *Xenorhabdus* spp. showed suppressive effects against *M. incognita* and *M. javanica* on tomato (Grewal *et al.*, 1999; Vyas *et al.*, 2006). Cell-free culture filtrates were highly toxic to J2 of *M. incognita*, reducing hatch and nematode penetration in a glasshouse trial. In addition, the bacteria and their culture filtrates showed a repellent effect; *Meloidogyne* J2 were repelled by entomopathogenic nematodes that included their symbiotic bacteria, by *Galleria mellonella* cadavers infected with *Steinernema* spp., and by cell-free culture filtrates of the associated *Xenorhabdus* spp. Grewal *et al.* (1999) considered that the short-term effects of bacterial culture filtrates, namely nematode toxicity and repellency, were almost entirely due to ammonium. In other cases, secondary metabolites of the bacterium contained substances toxic to root-knot nematodes (Hu *et al.*, 1999). For example, the secondary metabolites 3,5-dihydroxy-4-isopropylstilbene and indole from the culture filtrate of *Photorhabdus luminescens* were shown to inhibit hatch of *M. incognita* and, furthermore, indole caused paralysis of J2. In general, symbiotic bacteria of entomopathogenic nematodes seem to affect root-knot nematodes in the soil phase; once in the root tissue, *Meloidogyne* appears to be protected and the nematode can develop normally.

17.3 Fungal Pathogens and Antagonists

17.3.1 Nematophagous fungi

A wide range of fungi have been isolated from both eggs and females of root-knot nematodes (Godoy *et al.*, 1983; Morgan-Jones and Rodriguez-Kabana, 1988; Stirling and West, 1991). Nematophagous fungi can be classified into two broad groups: the obligate parasites and the facultative parasites. Obligate parasites use their spores to infect nematodes, and can initiate infection either by being ingested and then penetrating the gastrointestinal tract, or by adhering to the nematode cuticle and penetrating directly. The facultative parasites can change their trophic state from saprophytes that grow in the soil and rhizosphere into parasites infecting nematodes, either by way of specialized adhesive spores, or by trapping structures that adhere to migrating nematodes, or through specialized hyphae that develop appressoria that can breach the nematode cuticle or eggshell (Barron, 1977). From the numerous fungi that parasitize nematodes, only relatively few have been considered to have the potential to be developed into biological control agents (Siddiqui and Mahmood, 1996), and of these the most commonly isolated fungi that have then been studied further are *Paecilomyces lilacinus* and *Pochonia chlamydosporia* (= *Verticillium chlamydosporium*), which can parasitize both the egg and female stages of the nematode (Morgan-Jones *et al.*, 1982, 1983; Rodriguez-Kabana *et al.*, 1984; Freire and Bridge, 1985; de Leij and Kerry, 1991; Siddiqui and Mahmood, 1996).

There are two distinct barriers to the infection of root-knot nematodes by fungi, either the cuticle of the J2 within the egg or the eggshell from which the J2 emerges. *Pochonia chlamydosporia* is primarily regarded as an egg parasite; observations have shown that during the initial stages of infection, it produces a branched mycelial network that is in close contact with the eggshell (Morgan-Jones *et al.*, 1983; Lopez-Llorca and Duncan, 1988; Lopez-Llorca and Claugher, 1990). The eggshell is composed of three distinct layers – an outer vitelline layer, a chitin layer and an inner lipoprotein layer (see Eisenback and Hunt, Chapter 2, this volume) – and penetration of the eggshell occurs both from a specialized penetration peg, an appressorium, and also from lateral branches of mycelium, and leads to the disintegration of the vitelline layer and dissolution of the chitin and lipid layers (Segers *et al.*, 1996; Morton *et al.*, 2004). Enzymes are thought to be important in the infection process, and experiments indicate that a cocktail of proteases and chitinases are necessary to initiate infection. Fungi differ in their ability to degrade nematode eggshells, and the infection process was found to be affected by the nematode host (Segers *et al.*, 1996, 1998). This suggests that signals from the egg influence fungal growth and development, and penetration of the eggshell (Dackman, 1990). Proteases of nematode parasites have been characterized, including one from *Verticillium suchlasporium* (Lopez-Llorca and Robertson, 1992) and another from *P. chlamydosporia* (Segers *et al.*, 1996).

Studies of different isolates of *P. chlamydosporia* have shown that they produce a range of different proteases, and that the variation in the enzymes perhaps relates to the different ecological niches occupied by each fungus (Segers *et al.*, 1998). Increasing evidence for this comes from fingerprinting studies using Enterobacterial Repetitive Intergenic Consensus PCR (ERIC-PCR); this technique was able to discriminate different isolates of *P. chlamydosporia* (Arora *et al.*, 1996), and subsequent phylogenetic analysis using data generated from ERIC-PCR was able to demonstrate that the different isolates of the fungus were related to the nematode host from which the isolate had been obtained (Morton *et al.*, 2003). Further analysis of sequence similarity between proteases from different nematophagous fungi shows a high level of conservation, with only minor insertions and deletions (Siezen and Leunissen, 1997). Small differences in amino acid sequence can affect substrate utilization and host preference (Segers *et al.*, 1995), and variation in sequence and substrate utilization has been observed in VCP1 proteases from different isolates of *P. chlamydosporia* from different nematode hosts (Morton *et al.*, 2003). These investigations have shown that the replacement of an alanine by a glycine in the S3 substrate-binding region of

VCP1 confers enzymic activity against eggshells of *Meloidogyne* (Morton *et al.*, 2003) (Fig. 17.3). Although other enzymes, such as chitinases and lipases, are also undoubtedly important in the infection processes, these have not as yet been examined to the same extent and detail (Morton *et al.*, 2004) and this is obviously an area in need of further investigation.

To understand and exploit *P. chlamydosporia* as a means to regulate populations of root-knot nematodes, a detailed knowledge is required, not only of the molecular mechanisms involved in infection (as described above) but also of ecological considerations, including the population dynamics of the fungus in the rhizosphere (Kerry, 2000). Knowledge of the dynamics of *P. chlamydosporia* in relation to the nematode populations is essential for the development of a biological control strategy; however, the information that forms the foundation for such an approach is difficult to interpret due to problems quantifying the fungus in the rhizosphere and to the lack of a simple relationship linking abundance to activity. Various methodologies are available for such investigations, and these include selective plating, immunological and PCR-based techniques (Hirsch *et al.*, 2001). However, the quantification of filamentous fungi such as *P. chlamydosporia* is difficult because it is not composed of single, simple-to-quantify units of approximately the same size. *P. chlamydosporia*, as with other fungi, has several life stages, which include multicellular mycelium and chlamydospores mixed together with unicellular conidia. Therefore, calculating a robust relationship between fungal biomass and nematode numbers, which realistically relates to nematode infection and control, is problematic because any unit of fungus quantified could result from any of the life stages, many of which could be resting stages and not necessarily involved with nematode infection at the time of assessment. In a comparison between PCR-based techniques and selective plating, it was concluded that the most accurate interpretation of fungal dynamics in the soil could only be made by combining culture- and PCR-based techniques together rather than using either method alone (Mauchline *et al.*, 2002).

Paecilomyces lilacinus is another facultative parasite that has been used as a biological agent for the control of root-knot nematodes. It is a species that has a wide geographical distribution and was

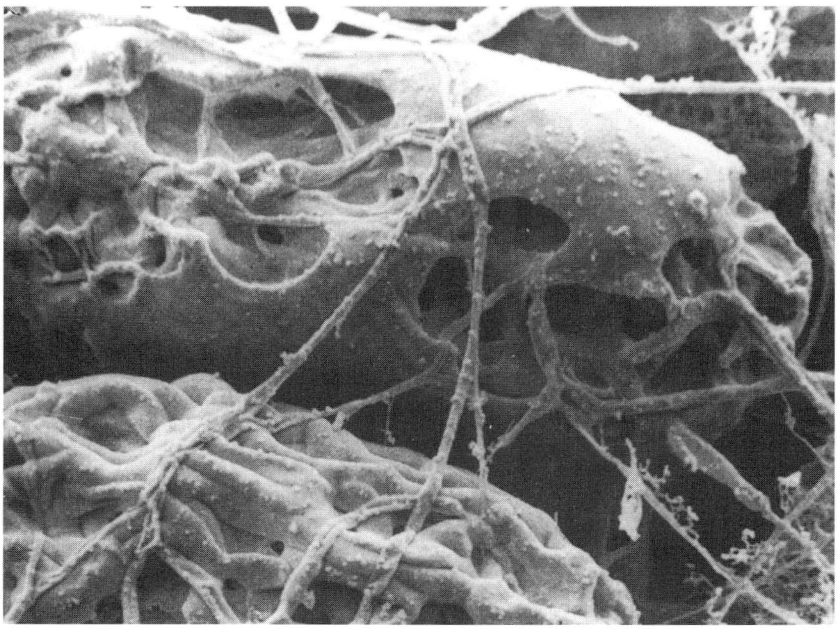

Fig. 17.3. Scanning electron micrograph of an egg of *Meloidogyne incognita* pre-treated with the proteinase VCP1 and colonized by *Pochonia chlamydosporia*. (From Segers *et al.*, 1996.)

first observed in association with nematode eggs (Lysek, 1976), and was found parasitizing *M. incognita* eggs in Peru (Jatala *et al.*, 1979, 1981); like *P. chlamydosporia*, it is primarily regarded as an egg parasite. Early investigations using *P. lilacinus* as a biological control agent were promising (Jatala, 1986); however, isolates known to be parasitic on nematode eggs and present at high population levels were unable to control root-knot nematodes (Rodrigues-Kabana *et al.*, 1984; Hewlett *et al.*, 1988). A number of factors are likely to be responsible for this observed inconsistency. These range from ecological factors pertaining to the ability of the fungus to establish itself in the soil (Hewlett *et al.*, 1988), to genetic components that may be important in determining levels of pathogenic specificity against different nematode populations (Dunn *et al.*, 1982; Stirling and West, 1991). However, as is usual in biological control, there is generally an inconsistency between experiments performed under glasshouse conditions and those undertaken in the field (Kerry and Evans, 1996). Several investigations have used *P. lilacinus* in conjunction with organic materials such as oil cake, leaf residues and seeds (Siddiqui and Mahmood, 1996; Cannayane and Sivakumar, 2001); once again, consistent control of root-knot nematodes has been difficult to reproduce. Production and formulation of filamentous fungal control agents remains problematic (Prabhu *et al.*, 2007) but recent developments in technology have enabled the production of highly concentrated preparations that can be applied and used successfully on a field scale (Kiewnick, 2004, 2006, 2007; Kiewnick and Sikora, 2004, 2006). *Paecilomyces lilacinus* strain 251 has been commercialized and registered for sale under a number of different trade names for the control of nematode pests in several countries (EPA, 2005; Kiewnick and Sikora, 2006).

As with *P. chlamydosporia* mentioned above, *P. lilacinus* also contained a serine protease and several chitinases that were shown to be important in the degradation of the eggshell (Khan *et al.*, 2004). Also as with *P. chlamydosporia*, the major structural changes that occurred in eggs treated with protease and chitinase from *P. lilacinus* strain 251 involved the loss of the lipid layer and disintegration of the vitelline layer, which contains proteins. Damage to these layers caused by the enzymes probably enabled other metabolites to permeate the eggs, causing changes such as swelling, but overall their effect on eggshell structure was drastic (Morgan-Jones *et al.*, 1984; Khan *et al.*, 2004). This seems to indicate that very similar modes of action are present in fungi that are distantly related. Indeed, phylogenetic analysis of a chitinase gene from *P. lilacinus* has shown such similarity to chitinase genes from mycoparasitic, entomopathogenic and nematophagous fungi that it has been hypothesized that the gene is likely to be of bacterial origin and was most likely acquired by gene transfer (Dong *et al.*, 2007).

From the numerous fungi that have been observed to parasitize nematodes, only relatively few have been considered to have potential for development into biological control agents (Siddiqui and Mahmood, 1996) and, as described above, most researchers have focused on facultative parasites that can be cultured easily. However, there are several endoparasitic fungi that have been investigated as they are interesting in terms of their biology and mode of infection, as well as in their potential to be used as biological control agents. For example, *Hirsutella* spp. and *Drechmeria coniospora* are unique because infection is initiated by the adhesion of small conidia to the nematode cuticle, and they are generally difficult to grow in culture (Stirling, 1991). For example, *D. coniospora* produces a teardrop-shaped conidium covered with an adhesive mucous-like layer that contains radiating fibrils (Saikawa, 1982; Jansson, 1993); in experiments investigating attachment of conidia to the nematode cuticle, 70% of the nematodes had spores adhering after 16 h, with young nematodes being preferentially infected (Dijksterhuis, *et al.*, 1993). This is similar to the results with *Pasteuria*, where the ability of endospores to adhere was reduced with increasing age of the nematodes (Davies *et al.*, 1991). These groups of fungi, like *Pasteuria*, are obligate parasites. *Hirsutella rhossiliensis* (Fig. 17.4), which has potential for being developed into a biological control agent, also has an obligate parasitic lifestyle, and it has been shown in microcosm experiments to have a density-dependent relationship with its host nematode (Jaffee *et al.*, 1992), and therefore might be expected to be able to control plant-parasitic nematodes successfully. Studies investigating the population dynamics of *H. rhossiliensis* in relation to *M. javanica* on tomato over two growing seasons failed to demonstrate that the fungus could control the nematode

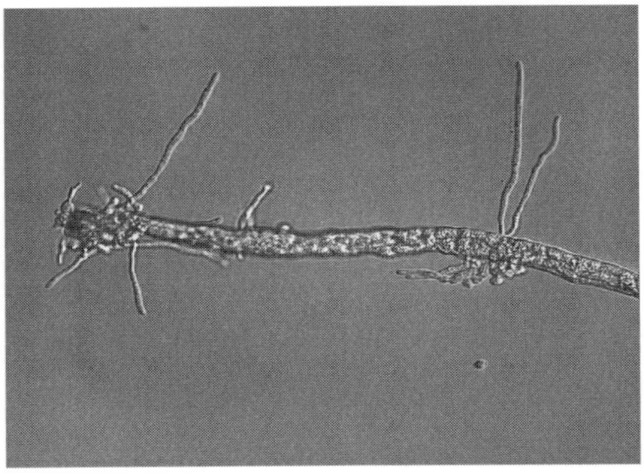

Fig. 17.4. Juvenile of *Meloidogyne incognita* parasitized by *Hirsutella rhossiliensis*. (Courtesy Brigitte Slaats.)

population over the long term (Tedford *et al.*, 1993). However, further recent research using a closely related fungus, *Hirsutella minnesotensis*, found that the fungus reduced populations of *M. hapla* by between 61 and 98%, and was considered to have potential to suppress *M. hapla* populations in vegetable production systems (Mennan *et al.*, 2006, 2007).

17.3.1.1 Predacious fungi

Predacious fungi, also often referred to as nematode-trapping fungi, are specialized forms of nematophagous, soil-borne fungi that form a mycelium able to capture nematodes. Different fungal species produce different trapping structures. The most simple structures are fungal hyphae covered with adhesive secretions (*Stylopage* spp.), followed by adhesive branches such as those produced by *Monacrosporium cionopagum* (Stirling, 1991). Branches consist of one to three cells and even form simple loops or two-dimensional networks. The formation of additional loops leads to three-dimensional adhesive networks, which represent the most common type of fungal traps (e.g. those of *Arthrobotrys oligospora, A. superba, Dactylella pseudoclavata*). Other groups of predacious fungi produce adhesive spores (*Meristacrum* spp.) or adhesive knobs (*A. haptotyla, Nematoctonus* spp.) (Kerry and Jaffee, 1997; Lopez-Llorca *et al.*, 2008). Nematode-trapping structures can vary within a genus, as shown for *Nematoctonus*, where adhesive knobs are found exclusively on hyphae (*Nematoctonus robustus*), exclusively on germinated conidia (*N. leptosporus, N. pachysporus, N. tylosporus*), or on both hyphae and conidia (*N. angustatus, N. geogenius*) (Koziak *et al.*, 2007). Non-constricting rings are formed by lateral branches of vegetative hyphae. The nematode entering the ring becomes ensnared when the ring becomes wedged around its body (Stirling, 1991). The most sophisticated traps are constricting rings (e.g. *Arthrobotrys dactyloides, Monacrosporium doedycoides*), which are formed on the fungal hyphae. If a nematode enters the ring, the three ring cells swell rapidly and hold the nematode tightly. Predacious fungi are not specific in their prey and trap plant-parasitic as well as free-living nematodes. If a nematode is trapped, the fungus penetrates the nematode cuticle with penetration hyphae formed on the trapping organ and thereafter produces infection bulbs inside the nematode, from which trophic hyphae emerge (Jansson and Nordbring-Hertz, 1988). Population density and species composition of predacious fungi varies considerably. Population densities are generally higher in autumn than throughout the rest of the year. Highest densities are generally found in the upper soil layer (0–30 cm), where they range from 2 to 50 propagules/g soil and comprise one to five species (Persmark *et al.*, 1996).

In many cases, trapping structures are induced and/or increased by the presence of nematodes or by organic compounds (amino acids, peptides), but competing microorganisms and nutrient level also play an important role in the transition of the fungi from a saprophytic to a parasitic phase (Quinn, 1987; Persmark and Nordbring-Hertz, 1997). While most predacious fungi colonize the bulk soil and wait for their nematode prey, some fungi, such as *A. superba*, produce secondary compounds that attract J2 of *Meloidogyne*, while others increase their chances of nematode contact by colonizing the rhizosphere. The efficacy of trapping and killing nematodes varies between predacious fungi. For example, *A. dactyloides* is a better pathogen of *M. graminicola* than *Dactylella brochopaga* and *Monacrosporium eudermatum*. Unfortunately, good antagonists are often bad saprophytic colonizers, which limits their potential as biological control agents. Predacious fungi able specifically to colonize the rhizosphere would give superior antagonistic qualities, as they trap plant-parasitic nematodes on their way to the plant roots.

Bordallo *et al.* (2002) showed that *A. oligospora* grew chemotropically towards root tips of tomato and barley, thus explaining its higher abundance in the rhizosphere rather than in the bulk soil. Fungal spread from cell to cell was by means of direct penetration or appressoria. The plant responded to fungal invasion by the formation of cell wall modifications such as papillae, lignotubers and deposits or necrosis. Histochemical labelling indicated that papillae and cell wall deposits consisted of callose and lignin and, in papillae, also of phenolics. Those compounds provide mechanical resistance towards pathogen attack in the cell wall and partly also express antimicrobial properties.

While in most cases trapping structures are formed on fungal mycelium, under certain conditions they can also be formed directly on germinating conidia (Persmark and Nordbring-Hertz, 1997). The ability to form so-called conidial traps is greater for *A. dactyloides* than for *A. superba* and *A. oligospora.* Competition for nutrients may cause conidia to germinate directly into traps. Formation of conidial traps helps the fungi to overcome fungistasis (prevention of fungal growth without killing the fungus). Therefore, in a competitive environment predacious fungi mainly live as parasites.

Genetic comparison of nematode-trapping fungi with other fungi having morphologically similar conidiogenous cells and conidia resulted in two phylogenetically different clusters (Liou and Tzean, 1997), indicating that development of trapping organs may reflect evolutionary relationships, and appear more significant for genus and species delimitation than conidia or conidiogenous cells.

Predacious fungi also produce antimicrobial and nematicidal compounds such as linoleic acid (*A. oligospora*, *A. conoides*) or pleurotin (*N. robustus*, *N. concurrens*) (Anke *et al.*, 1995). The production of linoleic acid was positively correlated with the number of traps formed. Population densities of predacious fungi are probably stimulated by organic amendments. Unfortunately, this conclusion was mainly drawn from glasshouse experiments and in many cases is not supported by field data. Timm *et al.* (2001) found, with few exceptions, similar frequencies and population densities of predacious fungi in plots with and without organic amendments. Jaffee *et al.* (1998) also reported similar densities of predacious fungi in conventional and organically managed field plots; however, *A. dactyloides* and *Nematoctonus leiosporus* reached significantly higher population densities in the organically managed plots. By contrast, *A. haptotyla* and *Arthrobotrys thaumasia* tended to be more numerous in conventional than in organic managed plots. Overall, suppression of *M. javanica* was not related to the management system or population density of predacious fungi.

17.3.2 Saprophagous fungi

Trichoderma is a ubiquitous soil fungus that also colonizes the root surface and root cortex. Several species within the genus *Trichoderma* are well-known antagonists of fungal pathogens, and some species, such as *Trichoderma harzianum*, even provide excellent control of root-knot nematodes (Sharon *et al.*, 2001). Other species within the genus *Trichoderma* with antagonistic activity towards *Meloidogyne* include *T. viride*, *T. atroviride*, and *T. asperellum* (Sharon *et al.*, 2007). Application of *Trichoderma* results in reduced nematode galling and improved plant growth. In the long term, plant growth and yield are enhanced, even when root galling is comparable to that in non-treated plants, suggesting improved plant tolerance (Spiegel and Chet, 1998). The highly branched conidiophores of *Trichoderma* bear conidia that can attach to different nematode stages. Conidial attachment and parasitism varies between fungal species and strains (Sharon *et al.*, 2007). For *T. asperellum*-203, *T. asperellum*-44 and *T. atroviride*, conidia that attached to egg masses led to parasitism of the enclosed eggs and J2. This process was often associated with the formation of fungal coiling and appressorium-like structures. Eggs and J2 free of the gelatinous matrix were parasitized at much lower rates, indicating an important role of the gelatinous matrix in fungal parasitism. These *Trichoderma* species grew well on the gelatinous matrix and, as a result, conidia agglutinated on the matrix and their germination was enhanced. However, other *Trichoderma* species behave differently. For example, *T. harzianum* is not able to grow on gelatinous matrices but colonizes isolated eggs and J2 of *M. javanica* (Sharon *et al.*, 2001).

Successful parasitism of the nematode by *Trichoderma* requires mechanisms to facilitate penetration of the nematode cuticle or eggshell. The involvement of lytic enzymes has long been suggested. Interestingly, biological control activity could be improved by using a proteinase Prb1-transformed line (P-2) carrying multiple copies of this gene. Furthermore, P-2 was also able to penetrate egg masses, suggesting that improved proteolytic activity of the fungus may be important for the biological control of *Meloidogyne*. The involvement of lytic enzymes in *Meloidogyne* parasitism was also demonstrated by using inductive GFP-transformants of *T. asperellum* carrying a fusion of the proteinase or chitinase promoters with the *gfp* gene (Spiegel *et al.*, 2005). GFP expression clearly indicated that both genes were turned on during fungal parasitism of the nematode. However, the role of chitinase during parasitism is not yet fully understood. Chitin is a structural component of the nematode eggshell but is not found in nematodes, which makes it unlikely to be involved in J2 parasitism. This is supported by the fact that if the gene for endochitinase from *T. harzianum* is expressed in transgenic tobacco, it does not provide protection against *M. hapla* (Brants *et al.*, 2000). Besides direct antagonism, other mechanisms involved in *Meloidogyne* control by *Trichoderma* spp. include fungal metabolites and induced resistance (Umamaheswari *et al.*, 2004).

In general, *Trichoderma* should be applied before planting to achieve maximum nematode control (Dababat *et al.*, 2006). There are several methods of application possible, such as seed treatment, dry formulation or soil drench. Combination of *Trichoderma* with organic treatments such as poultry litter has been used successfully to improve overall nematode control (Islam *et al.*, 2005). In all cases, good establishment of the fungus in the rhizosphere seems to be important for nematode control. Although *Trichoderma* is rhizosphere competent, it has not yet been isolated from the endorhiza.

In addition to *Trichoderma*, the bulk soil harbours a diverse spectrum of saprophagous fungi with antagonistic activity towards plant-parasitic nematodes. Antagonistic fungi are found in the genus *Gliocladium*, *Fusarium*, *Acremonium*, *Cylindrocarpon* and many others. They parasitize root-knot nematode eggs and J2 in the soil or release secondary metabolites toxic to the nematode (Freitas *et al.*, 1995, Goswami *et al.*, 2008, Rodríguez-Kábana *et al.*, 1984).

17.3.3 Endophytic fungi

The potential of endophytic fungi to reduce infestation by *Meloidogyne* spp. was first demonstrated for arbuscular mycorrhizal (AM) fungi in vegetable transplants (Sikora and Schönbeck, 1975). Pre-inoculation of tomato seedlings with AM fungi led to high levels of root colonization and, following transplanting into the field, reduced infection by *Meloidogyne* spp. However, if AM fungal spores are directly inoculated into the field soil, nematode control often fails, because endomycorrhizal fungi grow slowly in the endorhiza and do not quickly colonize sufficient root tissue. The latter is of key importance, as nematode control is only achieved when a certain rate of mycorrhization is reached. For cotton, Saleh and Sikora (1984) reported that 38% mycorrhization by *Glomus fasciculatum* was required for control of *M. incognita*; however, that rate might vary between crops and AM fungal species. In general, root tissue colonized by AM fungi will not be parasitized by *Meloidogyne* spp. Another approach to achieve high rates of mycorrhization could be by choosing plants within the crop rotation that are known to promote AM fungal populations in the field, which then allow rapid and extensive root colonization of the following crop. The beneficial effects of AM fungi are manifold. They selectively absorb and accumulate nutrients, such as phosphorus, resulting in improved plant growth and also increased plant tolerance to nematode infection. Mycorrhizal feeder roots are more resistant to infection by plant-parasitic nematodes as well as other soil-born pathogens, and AM fungi compete with plant-parasitic nematodes for nutrient sources and space (Diedhiou *et al.*, 2003). Finally, plant resistance has been reported to be induced by AM fungi (Elsen *et al.*, 2008), but its role in *Meloidogyne* control still awaits further exploitation.

Despite the well-documented beneficial effects of AM fungi in controlling root-knot nematodes (Bagyaraj *et al.*, 1979, Mohanty and Sahoo, 2003), the obligate nature of AM fungi limits commercial production of large quantities. Therefore, facultative saprophytic fungal endophytes with antagonistic properties might be the preferable choice. They can easily be mass-cultured, and commercial formulation and application technology already exist. Saprophytic fungal endophytes are omnipresent in the soil and colonize plant roots immediately after seed germination. As with AM fungi, they colonize the cortex, where they may compete with plant-parasitic nematodes for space and nutrients or even parasitize the nematodes. Interest in saprophytic fungal endophytes for nematode control was stimulated by the observation that several non-pathogenic isolates of *Fusarium oxysporum* reduced the rotting of excised banana roots caused by *Pratylenchus goodeyi* (Speijer, 1993). Hallmann and Sikora (1994) subsequently isolated 142 different isolates of endophytic fungi from tomato roots grown in different agroecozones in Kenya. The frequency of occurrence of fungi in the endorhiza varied among regions from 12 to 50% of the root pieces examined. The most dominant fungal species was *F. oxysporum* at 10–50% of all fungal endophytes recovered. Over 20% of the endophytic fungal isolates tested expressed antagonism towards *M. incognita* in glasshouse trials. Root galling was reduced by up to 75% compared with the non-treated control, and re-isolation studies indicated that 30–50% of the root sections were colonized by the previously applied isolate *F. oxysporum* 162 (FO162). Fungal population densities in the root reached 10^6 cfu/g compared with 10^2 cfu/g in the control. Similar effects have been reported following seed treatments of rice with different isolates of *F. oxysporum*, which led to significant decreases in root galling caused by *M. graminicola* (L. Huong and R.A. Sikora, 2008, unpublished results). The major mechanism by which FO162 reduced nematode parasitism is by reducing J2 penetration (Hallmann and Sikora, 1994). Those J2 that entered the root developed normally and female fecundity was similar to that in endophyte-free control plants. Interestingly, if J2 of *M. incognita* were given the choice between root exudates of a tomato plant treated with FO162 and a non-treated plant, 80% of the J2 moved to the exudates of the non-treated plant (Dababat and Sikora, 2007a). This indicates that plants colonized by FO162 are less attractive or that they exude substances that have repellent activity toward *M. incognita*. An additional mode of action is induced systemic resistance (Dababat and Sikora, 2007b). If tomato roots were split and grown in spatially separated containers and one side of the root system was treated with FO162 while the other side was inoculated with *M. incognita*, roots previously treated with FO162 showed 26–45% fewer nematodes penetrating and 22–26% reduction in the number of egg masses.

The role of fungal nematotoxic metabolites in nematode antagonism is still questionable. Although the potential of endophytic fungi in nutrient-rich artificial media to produce secondary metabolites highly toxic to J2 of *M. incognita* is well documented (Hallmann and Sikora, 1996; Sundararaju *et al.*, 2002), there is no evidence that those metabolites are produced at biologically active concentrations under field conditions.

While AM fungi and saprophytic endophytes differ partly in biology, preferred colonization sites and mode of action, combined application of both types of endophyte might increase overall control efficacy and consistency. Following this approach, Diedhiou *et al.* (2003) applied the AM fungus *Glomus coronatum* and FO162 simultaneously. Although the combined application of *G. coronatum* and FO162 did not result in additive nematode suppression, the experiment indicated some interesting interactions between the two endophytes. First, FO162 stimulated mycorrhization by *G. coronatum* and, second, mycorrhizal roots were not colonized internally by FO162. Nevertheless, synergism of two different fungal antagonists was reported for the AM fungus *Glomus mosseae* when combined with the egg pathogen *P. lilacinus* for the control of *M. javanica* on tomato (Al-Raddad, 1995).

17.4 Commercialization and Future Directions

17.4.1 Commercial products

The ultimate goals of research on biological control are: (i) to develop commercial products that can be applied by growers to manage root-knot nematode problems effectively; (ii) to devise crop management programmes that increase natural soil suppressiveness; and/or (iii) to add to our basic knowledge of the mechanisms existing in the nematode–control agent interrelationship. Different microbial pathogens, endophytes and antagonists offer different control options. However, there is no perfect candidate for biological control since they all have advantages and disadvantages (Table 17.2) when it comes to mass production, formulation, storage, field application and field efficacy.

Stirling (1991) stated that no system utilizing antagonists of nematodes for nematode control

Table 17.2. Advantages and disadvantages of antagonist bacteria and fungi for the biological control of root-knot nematodes.

Type of antagonist	Species	Mechanism	Advantage	Disadvantage
Endoparasitic bacteria	*Pasteuria penetrans*	Parasitism	Highly effective, reduced nematode infection and multiplication, spores survive desiccation, long shelf life	Obligate nature prohibits *in vitro* mass production, bacterial isolates highly specific for few nematode species
Rhizosphere bacteria	*Pseudomonas*, *Bacillus*, *Paenibacillus*, *Rhizobium*	Toxins, reduced nematode attraction, induced systemic resistance	Easy mass production, seed application possible, broad host spectrum, partly plant growth promotion	Seed pelleting technology for some crops missing, short duration of activity limits use on seedlings
Endophytic bacteria	*Rhizobium etli*, *Agrobacterium*	Induced systemic resistance, repellency	Targeted application to seed, intimate contact with nematode, invasion range ensures root coverage	Lack of field data
Predacious fungi	*Arthrobotrys oligospora*, *Arthrobotrys superba*, *Dactylella pseudoclavata*, *Meristacrums* spp., *Arthrobotrys haptotyla*, *Nematoctonus robustus*, *Nematoctonus leptosporus*, *Nematoctonus pachysporus*, *Nematoctonus tylosporus* *Nematoctonus angustatus*, *Nematoctonus geogenius*,	Predation	Applicaton by incorporation after solid state fermentation, marketable products were on market for control in greenhouse vegetables	Costs of production high, traps develop often too late to reduce, infection highly specific, obligate parasitism often prevents large-scale fermentation
Nematophagous fungi	*Pochonia chlamydosporia* *Paecilomyces lilacinus*,	Egg pathogen, enzymes, toxins	Products available, ease in application to soil and drip irrigation, broad spectrum, crop unspecific	Untargeted treatment of soil, only effective between host crops, short-duration survival, multiple applications required on some crops
Saprophytic endophytic fungi	*Fusarium*, *Trichoderma* spp.	Induced systemic resistance, competition, repellency	Targeted application to seed or seedling nursery soil, intimate contact with nematode	Plant cost of fungal colonization often unknown
Symbiotic endomycorrhizae	*Glomus* spp.	Competition, repellency	Targeted application to seedling nursery soil, intimate contact with nematode	Obligate symbiotic nature increases cost of production

Modified after Kerry (1997).

has been adopted widely with obvious success, and this lack of progress towards the commercialization of biological control has left many nematologists sceptical of its potential. The conclusion that Stirling drew at that time was justified. However, significant progress has now been made in biological control and this is clearly outlined in this chapter. A number of commercial biological control products with proven efficacy are on the market for management of root-knot and other nematodes (Table 17.3), and a number of potentially important antagonist systems are in advanced stages of development (Table 17.4). These products further demonstrate the technological advances that nematology has made in developing biological control technology.

However, the small number of products still underscores the need for continued research on biological control. It should be noted that the number of biological control products on the market, even if small in number, outnumbers the number of new commercial nematicides. Both of these control technologies face obstacles for practical use in the field – the former is limited by a lack of broad-spectrum activity and the latter by negative environmental and/or toxicological side

Table 17.3. Commerically available microbial-based biological control products for control of root-knot nematodes.

Product name	Active antagonist	Formulation	Treatment form	Crop	Company/ country
Bioact WG Melocon WG	*Paecilomyes lilacinus*	Water-dispersible granulate, wettable powder	Drench, drip irrigation	Vegetables, banana	Prophyta GmbH, Germany
PIPlus	*Paecilomyces lilacinus*	Wettable powder	Drench, drip irrigation	Vegetables, tobacco, banana, citrus	BCP, South Africa
BioNem-WP BioSafe	*Bacillus firmus*	Wettable powder	Drip irrigation, drench	Vegetables	AgroGreen, Israel
KlamiC	*Pochonia chlamydosporia*	Granulate	Soil incorporation	Vegetables	Cuba
Econem	*Pasteuria penetrans*	Liquid and powder	Drench, drip irrigation	Vegetables, turf, soybean	Pasteuria Bioscience, USA; Nematech, Japan
Deny Blue Circle	*Burkholderia cepacia*				CCT Corp, USA; Stine Microbial Products, USA
Biostart	*Bacillus* spp. mixture	Liquid	Soil drench, irrigation	General use	
BioStart RhizoBoost	*Bacillus chitinosporus* *Bacillus laterosporus* *Bacillus licheniformis*	Liquid	Irrigation, drench	General use, vegetables, fruits	Microbial Solutions, South Africa; Rincon Vilova, USA
Nemix	*Bacillus* spp.	Powder	Drench/drip	Vegetables, fruit trees	AgriLife/Chr Hansen, Brazil

Table 17.4. Pathogens and antagonists with market potential.

Type	Antagonist	Treatment form	Crop	Literature
Rhizobacteria	*Rhizobium* spp. *Bacillus* spp. *Pseudomonas* spp.	Seed coating, drip irrigation, root drench	Sugarbeet, soybean, vegetables, rice	Padgham and Sikora (2007)
Arbuscular mycorrhizae	*Glomus* spp.	Seedling substrate	Vegetables, ornamentals, banana	Saleh and Sikora (1984); Diedhiou *et al.* (2003); Elsen *et al.* (2008)
Mutualistic endophytic fungi	*Fusarium oxysporum, Trichoderma* spp. *F. graminicola*	Seedling substrate, seed coating, root drench	Banana, vegetables, rice, maize	Speijer (1993); Hallmann and Sikora (1994); Dababat *et al.* (2006)
Egg pathogens	*Trichoderma* spp.	Seedling substrate	Vegetables, ornamentals	Sharon *et al.* (2001)

effects. Biological control products are not as widely used as nematicides, because of the overall production costs and the need for development of better control of *Meloidogyne* on high-value crops.

Biological control products are usually applied in granular form by direct incorporation into the soil, or as liquid or wettable powder formulations in the furrow, or through drip irrigation. A major drawback of inundative approaches is the 2000+ t of soil that usually needs to be treated for effective control (Kerry, 1987; Sikora *et al.*, 2007, 2008). Attempts to develop formulations that allow targeted seed and seedling enhancement are in development (Table 17.4) and, if successful, will reduce the amount of product applied and the overall cost to the grower (Sikora and Fernández, 2005; Sikora *et al.*, 2008).

17.4.2 The development of a commercial product

Depending on their mode of action, microbial control agents can be differentiated into protective or curative, and local or systemic. Major industrial considerations for economic development include: (i) high and consistent efficacy; (ii) rapid colonization of the soil or plant in the root zone; (iii) prolonged establishment in the soil or plant; (iv) broad-spectrum control of several plant-parasitic nematodes; (v) inexpensive mass production and formulation; (vi) long shelf life; (vii) lack of environmental side effects; (viii) competitive price; and (ix) grower acceptance (Kerry, 1987; Stirling, 1991). A substantial amount of time, energy and funding was invested in development of biological control in the past, often without a clear understanding of these industrial criteria. Market-driven factors determine whether or not a biological will be successful. The costs of initial product development, toxicology studies, market registration and, finally, promotion are enormous and can only be offset with large financial returns.

17.4.3 Potential markets

Biological control agents have been developed mainly to manage root-knot nematodes of economic importance on high-value crops, which reflects market-driven concerns. The crops that are of interest to the plant-protection industry are the same as those targeted for nematicide development: vegetables, banana, ornamentals and, to a lesser extent, tobacco, sugarbeet, cotton, groundnut and potato. These crops have a higher profit margin than rice, wheat, maize and grain or forage legumes. Biological control of root-knot nematodes in high-value crops grown in glasshouses or under high-intensity protected cultivation systems has been one of the major areas targeted for inundative application of antagonists and pathogens, with application made to the soil, furrow or through irrigation systems (Kerry 1987; Stirling 1991; Sikora and Fernández, 2005).

Many of these glasshouse and field vegetable crops are grown first as seedlings in nurseries before being transplanted into the field and therefore are ideal for biological enhancement technology with rhizosphere and endophytic mircoorganisms (Hallmann *et al.*, 1997; Sikora *et al.*, 2008). Treatment of seedlings of tomato, pepper, melon, cucumber, tobacco or ornamentals in nurseries to ensure proper establishment of antagonists in the rhizosphere or endorhiza before transplanting to the field is considered highly conducive to the control of root-knot nematodes (Hallmann and Sikora, 1994; Sikora and Fernandez 2005; Sikora *et al.*, 2008). The production of hundreds of thousands of vegetable seedlings, e.g. tomato and pepper, by individual companies at central glasshouse complexes for sale to large retail chains is a potent market segment not yet captured by a biological control agent for nematode control. In glasshouses, the enhancement of growth substrates with biological control agents is also a market segment that has been overlooked, and one that is expanding in importance. Large-scale mixing, packaging and shipment of such substrates treated with effective antagonists could be important, for example for home garden growers with root-knot nematode problems. This market requires the treatment of either the planting substrate itself or the seed at planting into the near-sterile planting substrate. The development of effective microbial antagonists that can be applied to planting mixtures or to the seed would be a breakthrough in many countries. However, it should be noted that centralized seedling production is not common in many parts of the world, especially where sustainable production is done by individual growers with little or no capital. In these situations, seed treatment would be a more appropriate technology.

The use of combinations of biological control agents needs to be examined more closely for substrate- and seed-based applications (Dube and Smart, 1987; Siddiqui *et al.*, 2002; Khan *et al.*, 2006; Reimann *et al.*, 2008; Mendoza and Sikora, 2009). Industrial concepts that now stress all-round IPM (integrated pest management) seed-health technology could be advantageous to cocktails of biological control agents combined with modern nematicides. Synergistic effects originating at the seed level could lead to both short- and long-term nematode management.

Also important for development of a biological control agent is the existence of independent plant growth promotion attributes (Reimann *et al.*, 2008). This is important, as growth promotion, even if short lived, is perceived by the grower as a long-lasting effect and relevant to yield. Therefore, screening systems to find antagonists effective against root-knot nematodes should, if possible, also have plant growth promotion activity. In fact, it might be more important to screen organisms first for growth promotion and then second for root-knot nematode control. Good initial seedling growth on some home garden crops is more important to many consumers than control of an unseen nematode over time.

Seed-health technology could lead to breakthroughs for biological control on field crops such as cotton, groundnut, soybean, maize or rice, where profit margins are too small for inundative approaches. The problems caused by root-knot nematodes on these crops are often of widespread importance, but are seldom considered by industrial companies. With increasing prices for these crops, and improved seed-based technology, new markets for microbial control agents could open in the near future for field crops (Padgham and Sikora, 2007).

17.4.4 Enhancement strategies

The stimulation of the natural antagonistic potential in a soil is constantly taking place by rotation with crops that stimulate rhizosphere microorganisms, by the incorporation of organic matter into the soil after harvest and through targeted mulching. In most agricultural production situations, enhancement does not lead to significant levels of natural biological control and seldom to total suppressiveness.

The incorporation of any form of organic substrate will stimulate a broad array of microorganisms and have a negative impact on root-knot nematodes (Singh and Sitaramaiah, 1966, 1967; Muller and Gooch, 1982; Sikora and Fernández, 2005). Reductions in nematode density in organic agriculture, where large quantities of organic matter are incorporated into the soil, are to be expected. In organic agriculture, the combination of an expanded number of diverse crops in a rotation together with high amounts of mulching with organic matter stimulates the antagonistic potential and is an effective strategy for root-knot management. The integrated management of root-knot nematodes in glasshouses using living mulches based on species of *Tagetes* combined with solarization is used effectively in vegetable and fruit production in Morocco (Sikora and Fernández, 2005).

Biofumigation is presently of great interest as a technology for control of root-knot nematodes (Plates 38, 39). The incorporation of cruciferous plants and those that release the nematicidal compound isothiocyanate is considered responsible for control efficacy (Kirkegaard *et al.*, 1993; Stirling and Stirling, 2003). However, it has been stated repeatedly that the amount of nutrient-rich organic matter incorporated into the soil also stimulates microorganisms that stimulate ammonia production, which is also highly nematicidal (Bello *et al.*, 1998).

In a number of cases, attempts have been made to stimulate specific antagonists in the soil by mulching. The best-known method is the use of amendments containing chitin, which stimulate chitin-degrading microorganisms and indirectly destroy the eggshell of nematodes (Mian *et al.*, 1982; Rodríguez-Kábana, 1986; Rodríguez-Kábana and Morgan-Jones, 1987; Galper *et al.*, 1990). Other organic additives have been shown to increase the activity of plant-health-promoting rhizobacteria for root-knot nematode control (Mulawarman *et al.*, 2000, 2001).

Nematode-suppressive soils associated with many crops have been detected worldwide and are a microbial-antagonist-driven phenomenon (Crump and Kerry, 1987; Bird and Brisbane, 1988; Kluepfel *et al.*, 1993; Westphal and Becker, 1999; Pyrowolakis *et al.*, 2002; Sikora *et al.*, 2008). Reductions in plant-parasitic nematode densities exceeding 90% have been reported from suppressive soils harbouring a diverse spectrum of microbial control agents, such as *H. rhossiliensis, Dactylella oviparasitica, P. chlamydosporium, Pseudomonas aureofaciens, P. penetrans* and species of *Trichoderma* and *Fusarium*. Detection of suppressive soils is difficult due to the fact that they are often overlooked because the use of nematicides can mask their existence (Sikora *et al.*, 2008), and monoculture or short rotations that favour development of suppressiveness are not favoured by the growers. Suppressiveness has been detected in the perennial crop banana, and the duplication of suppressiveness in tissue culture plants with selected parts of the microbial community taken from the field, has been demonstrated, and this needs to be examined for other perennial crops.

17.4.5 Transgenic approaches

Transforming biological agents to improve their efficiency has been attempted with entomopathogenic nematodes and some fungi. The release of transgenic microorganisms, which cannot be seen or controlled, into the environment is a controversial issue and will not be discussed here, but more general aspects of the release of GMOs are discussed by Atkinson *et al.*, Chapter 15, this volume. Those organisms that induce systemic resistance in plants to infection by root-knot nematodes probably do so by changing gene expression (Schäfer *et al.*, 2006; Sikora *et al.*, 2007, 2008); therefore, the development of transgenic plants based on plant-based mechanisms that are initiated by antagonists will become an extremely interesting research field to improve crop resilience to combat infection by root-knot nematodes.

17.4.6 Future prospects

Biological control of root-knot would seem to be a straightforward and attainable endeavour; however, in reality it has proven to be an extremely elusive management tool. It is paradoxical that the root-knot nematode, which theoretically is not able to survive one growing season under a non-host break crop, is still extremely important in modern crop production. The reasons include: (i) a high reproductive capacity in the target nematode; (ii) extremely wide host ranges of most *Meloidogyne* species; (iii) population survival on weed hosts; and (iv) the ability to survive over time in deeper soil strata. The ability of root-knot nematodes to survive in the absence of a host crop is examined in detail by Evans and Perry, Chapter 9, this volume.

The future of biological control is closely linked to highly effective microbial pathogens and/or antagonists coupled with modern crop production technology. Therefore, future strategies are needed to improve naturally occurring biological control levels in soils lacking high levels of antagonistic potential (Sikora, 1992). An introduced microbial control agent, or a combination of agents, needs to have a high level of efficacy and consistency. Strategies that use microbial control agents with complementary modes of action, different colonization sites or different biological attributes that favour control need to be developed. Long-term establishment in the soil following inundative applications, or in the rhizosphere and endorhiza following seed or plant treatment, are essential for practical application. Understanding, of the epidemiology, survival and mode of action, even at the gene level where induced resistance is involved, is needed to make biological control an effective tool in nematode management. Research is needed: (i) to investigate mechanisms required to stimulate the naturally occurring antagonistic potential in soils; (ii) to develop molecular methods to increase the detection of antagonists with high control potential; (iii) to develop effective application technologies for treatment of seed or transplants; and (iv) to improve understanding of modes of action for targeted development of new antagonists.

The future of biological control of root-knot nematodes will, of course, be closely related to integration into management programmes that involve reducing pre-plant densities of *Meloidogyne* with conventional management tools, proper weed management and the use of microbial control agents that more effectively target the susceptible stages in the life cycle of the root-knot nematode in the pathozone.

17.5 References

Afolabi, P., Davies, K.G. and O'Shea, P. (1995) The electrostatic nature of the spore surface of *Pasteuria penetrans*, the bacterial parasite of root-knot nematodes. *Journal of Applied Bacteriology* 79, 244–249.

Al-Raddad, A.M. (1995) Interaction of *Glomus mosseae* and *Paecilomyces lilacinus* on *Meloidogyne javanica* on tomato. *Mycorrhiza* 5, 233–236.

Anita, B., Rajendran, G. and Samiyappan, R. (2004) Induction of systemic resistance in tomato against root-knot nematode, *Meloidogyne incognita* by *Pseudomonas fluorescens*. *Nematologia Mediterranea* 32, 47–51.

Anke, H., Stadler, M., Mayer, A. and Sterner, O. (1995) Secondary metabolites with nematicidal and antimicrobial activity from nematophagous fungi and Ascomycetes. *Canadian Journal of Botany* 73, s932–s939.

Arora, D.K., Hirsch, P.R. and Kerry, B.R. (1996) PCR-based molecular discrimination of *Verticillium chlamydosporium* isolates. *Mycological Research* 100, 801–809.

Bagyaraj, D.J., Manjunath, A. and Reddy, D.D.R. (1979) Interaction of vesicular arbuscular mycorrhiza with root knot nematodes in tomato. *Plant and Soil* 51, 397–403.

Barron, G.L. (1977) *The Nematode Destroying Fungi*. Canadian Publications, Ontario, Canada.

Bello, A., González, J.A., Arias, M. and Rodríguez-Kábana, R. (1998) (eds) *Alternatives to Methyl Bromide for the Southern European Countries*. Phytoma-España, DG XI EU, CSIC, Valencia, Spain.

Bird, A.F. and Brisbane, P.G. (1988) The influence of *Pasteuria penetrans* in field soils on the reproduction of root-knot nematodes. *Revue de Nematologie* 11, 75–82.

Bird, D., Opperman, C.H. and Davies, K.G. (2003) Interactions between bacteria and plant-parasitic nematodes: now and then. *International Journal of Parasitiology* 33, 1269–1276.

Bishop, A.H. and Ellar, D.J. (1991) Attempts to culture *Pasteuria penetrans in vitro*. *Biocontrol Science and Technology* 1, 101–104.

Bishop, A.H., Gowen, S.R., Pembroke, B. and Trotter, J.R. (2007) Morphological and molecular characteristics of a new species of *Pasteuria* parasitic on *Meloidogyne ardenensis*. *Journal of Invertebrate Pathology* 96, 28–33.

Blaxter, M.L. and Robertson, W.M. (1998) The cuticle. In: Perry, R.N. and Wright, D.J. (eds) *The Physiology and Biochemistry of Free-living and Plant-parasitic Nematodes*. CAB International, Wallingford, UK, pp. 25–48.

Bordallo, J.J., Lopez-Llorca, L.V. and Jansson, H.B. (2002) Colonization of plant roots by egg-parasitic and nematode-trapping fungi. *New Phytologist* 154, 491–499.

Brants, A., Brown, C.R. and Earle, E.D. (2000) *Trichoderma harzianum* endochitinase does not provide reistance to *Meloidogyne hapla* in transgenic tobacco. *Journal of Nematology* 32, 289–296.

Cannayane, I. and Sivakumar, C.V. (2001) Nematode egg-parasitic fungus 1: *Paecilomyces lilacinus* – a review. *Agricultural Reviews* 22, 79–86.

Cetintas, R. and Dickson, D.W. (2004) Persistence and suppressiveness of *Pasteuria penetrans* to *Meloidogyne arenaria* race 1. *Journal of Nematology* 36 540–549.

Charles, L., Carbonne, I., Davies, K.G., Bird, D., Burke, M., Kerry, B.R. and Opperman, C.H. (2005) Phylogenetic analysis of *Pasteuria penetrans* using multiple genetic loci. *Journal of Bacteriology* 187, 5700–5708.

Chen, Z.X. and Dickson, D.W. (1997) Minimal growth temperature of *Pasteuria penetrans. Journal of Nematology* 29 (Supplement) 635–639.

Chen, Z.X., Dickson, D.W., McSorley, R., Mitchell, D.J. and Hewlett, T.E. (1996) Suppression of *Meloidogyne arenaria* race 1 by soil application of endospores of *Pasteuria penetrans. Journal of Nematology* 28, 159–168.

Chen, Z.X., Dickson, D.W. and Mitchell, D.J. (1997). Suppression mechanisms of *Meloidogyne arenaria* race 1 by *Pasteuria penetrans. Journal of Nematology* 29, 1–8.

Costa, S.R., Kerry, B.R., Bardgett, R. and Davies, K.G. (2006) Exploitation of immunofluorescence for the quantification and characterisation of small numbers of *Pasteuria* endospores *FEMS Microbiology Ecology* 58, 593–600.

Crump, D.H. and Kerry, B.R. (1987) Studies on the population dynamics and fungal parasitism of *Heterodera schachtii* in soil from a sugarbeet monoculture. *Crop Protection* 6, 49–55.

Dababat, A. and Sikora, R.A (2007a) Influence of the mutualistic endophyte *Fusarium oxysporum* 162 on *Meloidogyne incognita* attraction and invasion. *Nematology* 9, 771–776.

Dababat, A. and Sikora, R.A. (2007b) Induced resistance by the mutualistic endophyte, *Fusarium oxysporum* strain 162, toward *Meloidoyne incognita* on tomato. *Biocontrol Science and Technology* 17, 969–975.

Dababat, A., Hauschild, R. and Sikora, R.A. (2006) Use of *Trichoderma harzianum* and *Trichoderma viride* for the biological control of *Meloidogyne incognita* on tomato. *Communications in Agricultural and Applied Biological Sciences* 71, 953–961.

Dackman, C. (1990) Fungal parasites of the potato cyst nematode *Globodera rostochiensis*: isolation and reinfection. *Journal of Nematology* 22, 594–597.

Darban, D.A., Pembroke, B. and Gowen, S.R. (2004) The relationships of time and temperature to body weight and numbers of endospores in *Pasteuria penetrans*-infected *Meloidogyne javanica* females. *Nematology* 6, 33–36.

Darby, C., Chakraborti, A., Politz, S.M., Daniels, C.C., Tan, L. and Drace, K. (2007) *Caenorhabditis elegans* mutants resistant to attachment of *Yersinia biofilms. Genetics* 176, 221–230.

Davies, K.G. (2009) Understanding the interaction between an obligate hyperparasitic bacterium, *Pasteuria penetrans*, and its obligate plant parasitic host, *Meloidogyne* spp. *Advances in Parasitology* 68, 211–245.

Davies, K.G. and Danks, C. (1992) Interspecific differences in the nematode surface coat between *Meloidogyne incognita* and *Meloidogyne arenaria* related to the adhesion of *Pasteuria penetrans. Parasitology* 105, 475–480.

Davies, K.G. and Danks, C. (1993) Carbohydrate/protein interactions between the cuticle of infective juveniles of *Meloidogyne incognita* and spores of the obligate hyperparasite *Pasteuria penetrans. Nematologica* 39, 54–64.

Davies, K.G. and Opperman, C.H. (2006) A potential role for collagen in the attachment of *Pasteuria penetrans* to nematode cuticle. In: Raaijmakers, J.M. and Sikora, R.A. (eds) *Multitrophic Interactions in the Soil and Integrated Control. IOBC/WPRS Bulletin* 29, 11–15.

Davies, K.G. and Redden, M. (1997) Diversity and partial characterisation of putative virulence determinants in *Pasteuria penetrans*, the hyperparasite of root-knot nematodes. *Journal of Applied Microbiology* 83, 227–235.

Davies, K.G., Kerry, B.R. and Flynn, C.A. (1988) Observations on the pathogenicity of *Pasteuria penetrans*, a parasite of root-knot nematodes. *Annals of Applied Biology* 112, 1491–501.

Davies, K.G., Flynn, C.A., Laird, V. and Kerry, B.R. (1990) The life-cycle, population dynamics and host specificity of a parasite of *Heterodera avenae*, similar to *Pasteuria penetrans. Revue de Nématologie* 13, 303–309.

Davies, K.G., Laird, V. and Kerry, B.R. (1991) The motility, development and infection of *Meloidogyne incognita* encumbered with spores of the obligate hyperparasite *Pasteuria penetrans. Revue de Nématologie* 14, 611–618.

Davies, K.G., Robinson, M.P. and Laird, V. (1992) Proteins on the surface of spores of *Pasteuria penetrans* and their involvement in attachment to the cuticle of second-stage juveniles of *Meloidogyne incognita. Journal of Invertebrate Pathology* 59, 18–23.

Davies, K.G., Redden, M. and Pearson, T.K. (1994) Endospore heterogeneity in *Pasteuria penetrans* related to attachment to plant-parasitic nematodes. *Letters in Applied Microbiology* 19, 370–373.

Davies, K.G., Afolabi, P. and O'Shea, P.S. (1996) Adhesion of *Pasteuria penetrans* to the cuticle of root-knot nematodes (*Meloidogyne* spp) inhibited by fibronectin: a study of electrostatic and hydrophobic interactions. *Parasitology* 112, 553–559.

Davies, K.G., Fargette, M., Balla, G., Daudi, A., Duponnois, R., Gowen, S.R., Mateille, T., Phillips, M.S., Sawadogo, A., Trivino, C., Vouyoukalou, E. and Trudgill, D.L. (2001) Cuticle heterogeneity as exhibited by *Pasteuria* spore attachment is not linked to the phylogeny of parthenogenetic root-knot nematodes (*Meloidogyne* spp.) *Parasitology* 122, 111–120.

Davies, K.G., Rowe, J. and Williamson, V.M. (2008) Cuticle variation amongst amphimictic and parthenogenetic populations of nematode (*Meloidogyne* spp.) as exhibited by a bacterial parasite (*Pasteuria penetrans*). *International Journal for Parasitology* 38, 851–860.

de Leij, F.A.A.M. and Kerry, B.R. (1991) The nematophagous fungus, *Verticillium chlamydosporium* Goddard, as a potential biological control agent for *Meloidogyne arenaria* (Neal) Chitwood. *Revue de Nématologie* 14, 157–164.

Diedhiou, P.M., Hallmann, J., Oerke, E.-C. and Dehne, H.-W. (2003) Effects of arbuscular mycorrhizal fungi and a non-pathogenic *Fusarium oxysporum* on *Meloidogyne incognita* infestation of tomato. *Mycorrhiza* 13, 199–204.

Dijksterhuis, J., Marten Veenhuis, M. and Harder, W. (1993) Conidia of the nematophagous fungus *Drechmeria coniospora* adhere to but barely infect *Acrobeloides buetschilii. FEMS Microbiology Letters* 113, 183–188.

Dong, L.Q., Yang, J.K. and Zhang, K.Q. (2007) Cloning and phylogenetic analysis of the chitinase genes from the facultative pathogen *Paeclilomyces lilacinus. Journal of Applied Microbiology* 103, 2476–2488.

Duan, H., Castro, F., Hewlett, T.E., White, J.H. and Ogram, A.V. (2003) Detection and characterization of *Pasteuria* 16S rRNA gene sequences from nematodes and soils. *International Journal of Systematic and Evolutionary Microbiology* 53, 105–112.

Dube, B. and Smart, G.C. Jr. (1987) Biological control of *Meloidogyne* by *Paecilomyces lilacinus* and *Pasteuria penetrans. Journal of Nematology* 19, 222–227.

Dunn, M.T., Sayre, R.M., Carrell, A. and Wergin, W.P. (1982) Colonization of nematode eggs by *Paecilomyces lilacinus* (Thom) Samson as observed with scanning electron microscope. *Scanning Electron Microscopy* 3, 1351–1357.

Duponnois, R., Ba, A.M. and Mateille, T. (1999) Beneficial effects of *Enterobacter cloacae* and *Pseudomonas mendocina* for biocontrol of *Meloidogyne incognita* with the endospore-forming bacterium *Pasteuria penetrans. Nematology* 1, 95–101.

Elsen, A., Gervacio, D., Swennen, R. and de Waele, D. (2008) AMF-induced biocontrol against plant parasitic nematodes in *Musa* sp.: a systemic effect. *Mycorrhiza* 18, 251–256.

EPA, US (Environmental Protection Agency) (2005) *Paecilomyces lilacinus* strain 251; exemption from the requirement of a tolerance. *United States Federal Register* 70, 19278–19283.

Espanol, M., Verdejo-Lucas, S., Davies, K.G. and Kerry, B.R. (1997) Compatibility between *Pasteuria penetrans* and *Meloidogyne* populations from Spain. *Biocontrol Science and Technology* 7, 219–230.

Fallon, D.J., Kaya, H.K., Gaugler, R. and Sipes, B.S. (2004) Effect of *Steinernema feltiae–Xenorhabdus bovienii* insect pathogen complex on *Meloidogyne javanica. Nematology* 6, 671–680.

Fargette, M., Davies, K.G., Robinson, M.P. and Trudgill, D.L. (1994) Characterisation of resistance breaking *Meloidogyne incognita*-like populations, using lectins, monoclonal antibodies and spores of *Pasteuria penetrans. Fundamental and Applied Nematology* 17, 537–542.

Fould, S., Dieng, A.L., Davies, K.G., Normand, P. and Mateille, T. (2001) Immunological quantification of the nematode parasitic bacterium *Pasteuria penetrans* in soil. *FEMS Microbiology Ecology* 37, 187–195.

Freire, F.C.O. and Bridge, J. (1985) Parasitism of eggs, females and juveniles of *Meloidogyne incognita* by *Paecilomyces lilacinus* and *Verticillium chlamydosporium. Fitopatologia Brasileira* 10, 577–596.

Freitas, L.G., Ferraz, S. and Muchovey, J.J. (1995) Effectiveness of different isolates of *Paecilomyces lilacinus* and an isolate of *Cylindorcarpon destructans* on the control of *Meloidogyne javanica. Nematropica* 25, 109–115.

Galper, S., Cohn, E., Spiegel, Y. and Chet, I. (1990) Nematicidal effect of collagen-amended soil and the influence of protease and collagenase. *Revue de Nematologie* 13, 67–71.

Gerber, J.F. and White, J.H. (2001) Materials and methods for the efficient production of *Pasteuria*. International Patent Application WO 01/11017 A2.

Giblin-Davis, R.M., Williams, D.S., Bekal, S., Dickson, D.W., Brito, J.A., Becker, J.O. and Preston, J.F. (2003) *Candidatus Pasteuria usgae* sp nov., an obligate endoparasite of the phytoparasitic nematode *Belonolaimus longicaudatus. International Journal of Systematic and Evolutionary Microbiology* 53, 197–200.

Godoy, G., Rodriguez-Kabana, R. and Morgan-Jones, G. (1983) Fungal parasites of *Meloidogyne arenaria* eggs in an Alabama soil. *Nematropica* 13, 201–213.

Goswami, J., Pandey, R.K., Tewari, J.P. and Goswami, B.K. (2008) Management of root knot nematode on tomato through application of fungal antagonists, *Acremonium strictum* and *Trichoderma harzianum. Journal of Environmental Science and Health* 43, 237–240.

Gravato-Nobre, M.J., McClure, M.A., Dolan, L., Calder, G., Davies, K.G., Mulligan, B., Evans, K. and von Mende, N. (1999) *Meloidogyne incognita* surface antigen epitopes in infected *Arabidopsis* roots. *Journal of Nematology* 31, 212–223.

Gravato-Nobre, M.J., Nicholas, H.R., Nijland, R., O'Rourke, D., Whittington, D.E., Yook, K.J. and Hodgkin, J. (2005) Multiple genes affect sensitivity of *Caenorhabditis elegans* to the bacterial pathogen *Microbacterium nematophilum. Genetics* 171, 1033–1045.

Grewal, P.S., Lewis, E.E. and Venkatachari, S. (1999) Allelopathy: a possible mechanism of suppression of plant-parasitic nematodes by entomopathogenic nematodes. *Nematology* 1, 735–743.

Hallmann, J. and Sikora, R.A. (1994) Influence of *Fusarium oxysporum*, a mutualistic fungal endopyhte on *Meloidogyne incognita* of tomato. *Journal of Plant Diseases and Protection* 101, 475–481.

Hallmann, J. and Sikora, R.A. (1996) Toxicity of fungal endophyte secondary metabolites to plant-parasitic nematodes and soil-borne plant pathogenic fungi. *European Journal of Plant Pathology* 102, 155–162.

Hallmann, J., Quadt-Hallmann, A., Mahaffee, W.F. and Kloepper, J.W. (1997) Bacterial endophytes in agricultural crops. *Canadian Journal of Microbiology* 43, 895–914.

Hallmann, J., Quadt-Hallmann, A., Rodríguez-Kábana, R. and Kloepper, J.W. (1998) Interactions between *Meloidogyne incognita* and endophytic bacteria in cotton and cucumber. *Soil Biology and Biochemistry* 30, 925–937.

Hallmann, J., Quadt-Hallmann, A., Miller, W.G., Sikora, R.A. and Lindow, S.E. (2001) Endophytic colonization of plants by the biocontrol agent *Rhizobium etli* G12 in relation to *Meloidogyne incognita* infestation. *Phytopathology* 91, 415–422.

Hemmer, R.M., Donkin, S.G., Chin, K.J., Grenache, D.G., Bhatt, H. and Politz, S.M. (1991) Altered expression of an l1-specific, O-linked cuticle surface glycoprotein in mutants of the nematode *Caenorhabditis elegans. Journal of Cell Biology* 115, 1237–1247.

Hewlett, T.E., Dickson, D.W., Mitchell, D.J. and Kannwischer-Mitchell, M.E. (1988) Evaluation of *Paecilomyces lilacinus* as a biocontrol agent of *Meloidogyne javanica* on tobacco. *Journal of Nematology* 20, 578–584.

Hewlett, T.E., Gerber, J.F. and Smith, K.S. (2004) *In vitro* culture of *Pasteuria penetrans*. In: Cook, R. and Hunt, D.J. (eds) *Nematology Monographs and Perspectives*, Vol. 2. Koninklijke Brill, Leiden and Boston, pp. 175–185.

Hewlett, T.E., Griswold, S.T. and Smith, K.S. (2006) Biological control of *Meloidogyne incognita* using *in-vitro* produced *Pasteuria penetrans* in a microplot study. *Journal of Nematology* 38, 274 (abstract).

Hirsch, P.R., Atkins, S.D., Mauchline, T.H., Davies, K.G. and Kerry, B.R. (2001) Techniques for studying nematophagous fungi in the root environment. *Plant and Soil* 232, 21–30.

Hoflich, J., Berninsone, P., Gobel, C., Gravato-Nobre, M.J., Libby, B.J., Darby, C., Politz, S.M., Hodgkin, J., Hirschberg, C.B. and Baumeister, R. (2004) Loss of srf-3-encoded nucleotide sugar transporter activity in *Caenorhabditis elegans* alters surface antigenicity and prevents bacterial adherence. *Journal of Biological Chemistry* 279, 30440–30448.

Hu, K., Li, J. and Webster, J.M. (1999) Nematicidal metabolites produced by *Photorhabdus luminescens* (Enterobacteriaceae) bacterial symbiont of entomopathogenic nematodes. *Nematology* 1, 457–469.

Islam, M.N., Ali, M.B., Froz, M.J., Mondol, A.T.M.A.I. and Jahan, M.A.H.S. (2005) Integrated management of root-knot (*Meloidogyne* spp.) disease of tomato using antagonistic isolates of *Trichoderma harzianum* and its combination with organic soil amendments. *Journal of Subtropical Agriultural Research and Development* 3, 78–81.

Jaffee, B.A., Phillips, R., Muldoon, A.E. and Mangel, M. (1992) Density-dependent host–pathogen dynamics in soil microcosms. *Ecology* 73, 495–506.

Jaffee, B.A., Ferris, H. and Scrow, K.M. (1998) Nematode-trapping fungi in organic and conventional cropping systems. *Phytopathology* 88, 344–350.

Jansson, H.-B. (1993) Adhesion to nematodes of conidia from the nematophagous fungus *Drechmeria coniospora. Journal of General Microbiology* 139, 1899–1906.

Jansson, H.B. and Nordbring-Hertz, B. (1988) Infection events in the fungus–nematode system. In: Poinar, G.O. Jr. and Jansson, H.B. (eds) *Diseases of Nematodes*, Vol. II. CRC Press, Boca Raton, Florida, pp. 59–72.

Jatala, P. (1986) Biological control of plant-parasitic nematodes. *Annual Review of Phytopathology* 24, 453–489.

Jatala, P., Kaltenbach, R. and Bocangel, M. (1979) Biological control of *Meloidogyne incognita acrita* and *Globodera pallida* on potatoes. *Journal of Nematology* 11, 303 (abstract).

Jatala, P., Salas, R., Kaltenbach, R. and Bocangel, M. (1981) Multiple application and long-term effect of *Paecilomyces lilacinus* in controlling *Meloidogyne incognita* under field conditions. *Journal of Nematology* 13, 445 (abstract).

Kerry, B.R. (1987) Biological control. In: Brown, R.A. and Kerry, B.R. (eds) *Principles and Practice of Nematode Control in Crops*. Academic Press, New York, pp. 233–263.

Kerry, B.R. (1997) Biological control of nematodes: prospects and opportunities. In: Maqbool, M.A. and Kerry, B. (eds) *Plant Nematode Problems and their Control in the Near East Region. Proceedings of the Expert Consultation on Plant Nematode Problems and their Control in the Near East Region Karachi, Pakistan 22–26 November 1992*. Food and Agriculture Organization of the United Nations, Rome, Italy, pp. 79–92.

Kerry, B.R. (2000) Rhizosphere interactions and exploitation of microbial agents for the biological control of plant-parasitic nematodes. *Annual Review of Phytopathology* 38, 423–441.

Kerry, B.R. and Evans, K. (1996) New strategies for the management of plant-parasitic nematodes. In: Hall, R. (ed.) *Principles and Practice of Managing Soilborne Plant Pathogens*. APS Press, St Paul, Minnesota, pp. 134–152.

Kerry, B.R. and Jaffee, B.A. (1997) Fungi as biological control agents for plant parasitic nematodes. In: Wicklow, D.T. and Soderstrom, B.E. (eds) *The Mycota: a Comprehensive Treatise on Fungi as Experimental Systems for Basic Applied Research: Volume 4 Environmental and Microbial Relationships*. Springer, Berlin, Germany, pp. 203–218.

Khan, A., Williams, K.L. and Nevalainen, H.K.M. (2004) Effects of *Paecilomyces lilacinus* protease and chitinase on the eggshell structures and hatching of *Meloidogyne javanica* juveniles. *Biological Control* 31, 346–352.

Khan, A., Williams, K.L. and Nevalainen, H.K.M. (2006) Control of plant-parasitic nematodes by *Paecilomyces lilacinus* and *Monacrosporium lysipagum* in pot trials. *Biocontrol* 51, 643–658.

Kiewnick, S. (2004) Biological control of plant-parasitic nematodes with *Paecilomyces lilacinus*, strain 251. *IOBC/WPRS Bulletin* 27, 133–136.

Kiewnick, S. (2006) Multitrophic interactions of *Paecilomyces lilacinus* strain 251 in the rhizoshere of host and non-host plants. *IOBC/WPRS Bulletin* 29, 53–61.

Kiewnick, S. (2007) Importance of multitrophic interactions for the efficacy of *Paecilomyces lilacinus* strain 251 to control root-knot nematodes. *Journal of Nematology* 39, 72.

Kiewnick, S. and Sikora, R. (2004) Optimizing the efficacy of *Paecilomyces lilacinus* (strain 251) for the control of root-knot nematodes. *Communications Applied Biological Science* 69, 373–380.

Kiewnick, S. and Sikora, R. (2006) Evaluation of *Paecilomyces lilacinus* strain 251 for the biological control of the northern root-knot nematode *Meloidogyne hapla* Chitwood. *Nematology* 8, 69–78.

Kirkegaard, J.A., Gardner, P.A., Desmarchelier, J.M. and Angus, J.F. (1993) Biofumigation – using *Brassica* species to contol pests and diseases in horticulture and agriculture. In: Wratten, N. and Mailer, R.J. (eds) *Proceedings of the 9th Australian Research Assembly on Brassicas*. Agricultural Research Institute, Wagga Wagga, New South Wales, pp. 77–82.

Kloepper, J.W., Rodríguez-Kábana, R., McInroy, J.A. and Young, R.W. (1992) Rhizosphere bacteria antagonistic to soybean cyst (*Heterodera glycines*) and root-knot (*Meloidogyne incognita*) nematodes: identification by fatty acid analysis and frequency of biological control activity. *Plant and Soil* 139, 75–84.

Kluepfel, D.A., McInnis, T.M. and Zehr, E.I. (1993) Involvement of root-colonizing bacteria in peach orchard soils suppressiveness of the nematode *Criconemella xenoplax. Phytopathology* 83, 1240–1245.

Kojetin, D.J., Thompson, R.J., Benson, L.M., Naylor, S., Waterman, J., Davies, K.G., Opperman, C.H., Stephenson, K., Hoch, J.A. and Cavanah, J. (2005) The structural analysis of divalent metals to the *Bacillus subtilis* response regulator Spo0F: the possibility for *in vitro* metalloregulation in the initiation of sporulation. *Biometals* 18, 449–466.

Koziak, A.T.E., Cheng, K.C. and Thorn, R.G. (2007) Phylogenetic analyses of *Nematoctonus* and *Hohenbuehelia* (Pleurotaceae). *Canadian Journal of Botany* 85, 762–773.

Kramer, J.M. (1997) Extracellular matrix. In: Riddle, D.L., Blumenthal, T., Mayer, B.J. and Priess, J.R. (eds) C. elegans *II*. Cold Spring Harbour Laboratory Press, Cold Spring Harbor, New York, pp. 471–500.

Krechel, A., Faupel, A., Hallmann, J., Ulrich, A. and Berg, G. (2002) Potato-associated bacteria and their antagonistic potential towards plant-pathogenic fungi and the plant-parasitic nematode *Meloidogyne incognita* (Kofoid & White) Chitwood. *Canadian Journal of Microbiology* 48, 772–786.

Lewis, E.E., Grewal, P.S. and Sardanelli, S. (2001) Interactions between the *Steinernema feltiae–Xenorhabdus bovienii* insect pathogen complex and the root-knot nematode *Meloidogyne incognita. Biological Control* 21, 55–62.

Li, X.Q., Wei, J.Z., Tan, A. and Arojan, R.V. (2007) Resistance to root-knot nematode in tomato roots expressing a nematicidal *Bacillus thuringiensis* crystal protein. *Plant Biotechnology Journal* 5, 455–464.

Lin, H.J. and McClure, M.A. (1996) Surface coat of *Meloidogyne incognita. Journal of Nematology* 28, 216–224.

Liou, G.Y. and Tzean, S.S. (1997) Phylogeny of the genus *Arthrobotrys* and allied nematode-trapping fungi based on rDNA sequences. *Mycologia* 89, 876–884.

Lopez-Llorca, L.V. and Claugher, D. (1990) Appressoria of the nematophagous fungus *Verticillium chlamydosporium. Micron and Microscopica Acta*, 21, 125–130.

Lopez-Llorca, L.V. and Duncan, G.H. (1988) A study of fungal endoparasitism of the ceral cyst nematode (*Heterodera avanae*) by scanning electron microscopy. *Canadian Journal of Microbiology* 34, 613–619.

Lopez-Llorca, L.V. and Robertson, W.M. (1992) Immunocytochemical localisation of a 32-kDa protease from the nematophagous fungus *Verticillium suchlasporium. Experimental Mycology* 16, 261–267.

Lopez-Llorca, L.V., Maciá-Vicente, J.G. and Jansson, H.-B. (2008) Mode of action and interactions of nematophagous fungi. In: Ciancio, A. and Mukerji, K.G. (eds) *Integrated Management and Biocontrol of Vegetables and Grain Crops Nematodes.* Springer, Dordrecht, The Netherlands, pp. 51–76.

Lysek, H. (1976) Autodehelminthization of soil in lowland deciduous forests. *Universitatis Palackianae Olomucensis Facultatis Medicae* 41, 73–106.

Mankau, R. (1981) Microbial control of nematodes. In: Zuckerman, B.M. and Rohde, R.A. (eds) *Plant-parasitic Nematodes*, Vol. III. Academic Press, New York, pp. 475–494.

Mauchline, T.H., Kerry, B.R. and Hirsch, P.R. (2002) Quantification in soil and the rhizosphere of the nematophagous fungus *Pochonia chlamydosporia* by competitive PCR and comparison with selective plating. *Applied and Environmental Microbiology*, 68, 1846–1853.

Mendoza, A.R. and Sikora, R.A. (2009) Biological control of *Radopholus similis* in banana by combined application of the mutualistic endophyte *Fusarium oxysporum* strain 162, the egg pathogen *Paecilomyces lilacinus* strain 251 and the antagonistic bacterium *Bacillus firmus. Biocontrol* 54, 263–272.

Mendoza de Gives, P., Davies, K.G., Morgan, M. and Behnke, J.M. (1999) Attachment tests of *Pasteuria penetrans* to the cuticle of plant and animal parasitic nematodes, free living nematodes and *srf* mutants of *Caenorhabditis elegans. Journal of Helminthology* 73, 67–71.

Mennan, S., Chen, S.Y. and Melakeberhan, H. (2006) Suppression of *Meloidogyne hapla* populations by *Hirsutella minnesotensis. Biocontrol Science and Technology* 16, 181–193.

Mennan, S., Chen, S.Y. and Melakeberhan, H. (2007) Effects of *Hirsutella minnesotensis* and N-Viro Soil (R) on populations of *Meloidogyne hapla. Biocontrol Science and Technology* 17, 233–246.

Mian, I.H., Godoy, G., Shelby, R.A., Rodríguez-Kábana, R. and Morgan-Jones, G. (1982) Chitin amendments for control of *Meloidogyne arenaria* in infested soil. *Nematropica* 12, 71–84.

Mohan, S., Fould, S. and Davies, K.G. (2001) The interaction between the gelatine binding domain of fibronectin and the attachment of *Pasteuria penetrans* endospores to nematode cuticle. *Parasitology* 123, 271–276.

Mohanty, K.C. and Sahoo, N.K. (2003) Prospects of mycorrhizae as potential nematode antagonist. In: Trivedi, P.C. (ed.) *Advances in Nematology.* Scientific Publishers, India, pp. 317.

Morgan-Jones, G. and Rodriguez-Kabana, R. (1988) Fungi colonizing cysts and eggs. In: Poinar, G.O. and Janssen, H.-B (eds) *Diseases of Nematodes*, Vol II. CRC Press, Boca Raton, Florida, pp. 39–58.

Morgan-Jones, G., Godoy, G. and Rodríguez-Kábana, R. (1982) *Verticillium chlamydosporium* fungal parasite of *Meloidogyne arenaria* females. *Nematropica* 11, 115–120.

Morgan-Jones, G., White, J.F. and Rodríguez-Kábana, R. (1983) Phytonematode pathology: ultrastructural studies. I. Parasitism of *Meloidogyne arenaria* eggs by *Verticillim chlamydosporium. Nematropica* 13, 245–260.

Morgan-Jones, G., White, J.F. and Rodriguez-Kabana, R. (1984) Phytonematode pathology: ultrastructural studies. II. Parasitism of *Meloidogyne arenaria* eggs and larvae by *Paecilomyces lilacinus. Nematropica* 14, 57–71.

Morton, C.O., Hirsch, P.R., Peberdy, J.P. and Kerry, B.R. (2003) PCR-based DNA fingerprinting indicates host-related genetic variation in the nematophagous fungus *Pochonia chlamydosporia. Mycological Research* 107, 198–205.

Morton, C.O., Hirsch, P.R. and Kerry, B.R. (2004) Infection of plant-parasitic nematodes by nematophagous fungi – a review of the application of molecular biology to understand infection processes and to improve biological control. *Nematology* 6, 161–170.

Mulawarman, Hallmann, J., Schuster, R.-P., Bell, D., Kopp-Holtwiesche, B. and Sikora, R.A. (2000) TerraPy, a natural product with plant growth enhancing effect and nematode control. *Mededelingen Faculteit van de Landbouwkundige. University of Gent*, 65/2b, 537–544.

Mulawarman, Hallmann, J., Bell, D., Kopp-Holtwiesche, B. and Sikora, R.A. (2001) Effects of natural products on soil organisms and plant health enhancement. *Mededelingen Faculteit van de Landbouwkundige. University of Gent*, 66/2b, 609–617.

Muller, R. and Gooch, P.S. (1982) Organic amendments in nematode control: an examination of the literature. *Nematropica* 12, 319–326.

Munif, A., Hallmann, J. and Sikora, R.A. (2000) Evaluation of the biocontrol activity of endophytic bacteria from tomato against *Meloidogyne incognita. Mededelingen Faculteit van de Landbouwkundige. University of Gent*, 65/2b, 471–480.

Musson, G., McInroy, J.A. and Kloepper, J.W. (1995) Development of delivery systems for introducing endophytic bacteria into cotton. *Biocontrol Science and Technology* 5, 407–416.

Noel, R.G. and Atibalentja, N. (2006) *Candidatus paenicardinium endonii*, an endosymbiont of the plant-parasitic nematode *Heterodera glycines* (Nemata: Tylenchida), affiliated to the phylum *Bacteroidetes. International Journal of Systematic and Evolutionary Microbiology* 56, 1697–1702.

Oliveira, D.F., Campos, V.P., Amaral, D.R., Nunes, A.S., Pantaleão, R.A. and Costa, D.A. (2007) Selection of rhizobacteria able to produce metabolites active against *Meloidogyne exigua. European Journal of Plant Pathology* 119, 477–479.

Oostendorp, M., Dickson, D.W. and Mitchell, D.J. (1991) Population development of *Pasteuria penetrans* on *Meloidogyne arenaria. Journal of Nematology* 23, 58–64.

Padgham, J.L. and Sikora, R.A. (2007) Biological control potential and modes of action of *Bacillus megaterium* against *Meloidogyne graminicola* on rice. *Crop Protection* 26, 971–977.

Page, A.P., Rudin, W., Fluri, E., Blaxter, M.L. and Maizels, R.M. (1992) *Toxocara canis* – a labile antigenic surface coat overlying the epicuticle of infective larvae. *Experimental Parasitology* 75, 72–86.

Persmark, L. and Nordbring-Hertz, B. (1997) Conidial trap formation of nematode-trapping fungi in soil and soil extracts. *FEMS Microbiology Ecology* 22, 313–323.

Persmark, L., Banck, A. and Jansson, H.B. (1996) Population dynamics of nematophagous fungi and nematodes in an arable soil: vertical and seasonal fluctuations. *Soil Biology and Biochemistry* 28, 1005–1014.

Prabhu, S., Kumar, S. and Subramanian, S. (2007) Mass production and commercial formulation of *Paecilomyces lilacinus. Annals of Plant Protection Sciences* 14, 444–447.

Prasad, S.S.S.V., Tilak, K.V.B.R. and Gollakota, K.G. (1972) Role of *Bacillus thuringiensis* var. *thuringiensis* on the larval survivability and egg hatching of *Meloidogyne* spp., the causative agent of root knot disease. *Journal of Invertebrate Pathology* 20, 377–378.

Previc, E.P. and Cox, R.J. (1992) Production of endospores from *Pasteuria* by culturing with explanted tissue from nematodes., US Patent Office, US Patent Number 5 084 954, Washington, DC.

Pyrowolakis, A., Westphal, A., Sikora, R.A. and Becker, J.O. (2002) Identification of root-knot nematode suppressive soils. *Applied Soil Ecology* 19, 51–56.

Quinn, M.A. (1987) The influence of saprophytic competition on nematode predation by nematode-trapping fungi. *Journal of Invertebrate Pathology* 49, 170–174.

Radwan, M.A. (2007a) Bioactivity of commercial products of *Bacillus thuringiensis* on *Meloidogyne incognita* infecting tomato. *Indian Journal of Nematology* 37, 30–33.

Radwan, M.A. (2007b) Efficacy of *Bacillus thuringiensis* integrated with other non-chemical materials to control *Meloidogyne incognita* in tomato. *Nematologia Mediterranea* 35, 69–73.

Reimann, S., Hauschild, R., Hildebrandt, U. and Sikora, R.A. (2008) Interrelationships between *Rhizobium etli* G12 and *Glomus intraradices* and multitrophic effects in the biological control of the root-knot nematode *Meloidogyne incongnita* on tomato. *Journal of Plant Diseases and Protection* 115, 108–113.

Reitz, M., Hoffmann-Hergarten, S., Hallmann, J. and Sikora, R.A. (2001) Induction of systemic resistance in potato by rhizobacterium *Rhizobium etli* strain G12 is not associated with accumulation of pathogenesis-related proteins and enhanced lignin biosynthesis. *Journal of Plant Diseases and Protection* 108, 11–20.

Reitz, M., Oger, P., Meyer, A., Niehaus, K., Farrand, S.K., Hallmann, J. and Sikora, R.A. (2002) Importance of the O-antigen, core-region and lipid A of rhizobial lipopolysaccharides for the induction of systemic resistance in potato to *Globodera pallida*. *Nematology* 4, 73–79.

Rodríguez-Kábana, R. (1986) Organic and inorganic nitrogen amendments to soil as nematode suppressants. *Journal of Nematology* 18, 129–135.

Rodríguez-Kábana, R. and Morgan-Jones, G. (1987) Biological control of nematodes: soil amendments and microbial antagonists. *Plant and Soil* 100, 237–247.

Rodríguez-Kábana, R., Morgan-Jones, G., Godroy, G. and Gintis, B.O. (1984) Effectiveness of species of *Gliocladium*, *Paecilomyces* and *Verticillium* for control of *Meloidogyne arenaria* in field soil. *Nematropica* 14, 155–170.

Saikawa, M. (1982) An electron microscope study of *Meria coniospora*, an endozoic nematophagous Hyphomycete. *Canadian Journal of Botany* 60, 2019–2023.

Saleh, H. and Sikora, R.A. (1984) Relationship between *Glomus fasiculatum* root colonization and its effect on *Meloidogyne incognita*. *Nematologica* 30, 230–237.

Samaliev, H.Y., Andreoglou, F.I., Elawad, S.A., Hague, N.G.M. and Gowen, S.R. (2000) The nematicidal effects of the bacteria *Pseudomonas oryzihabitans* and *Xenorhabdus nematophilus* on the root-knot nematode *Meloidogyne javanica*. *Nematology* 2, 507–514.

Sayre, R.M. (1988) Bacterial diseases of nematodes and their role in controlling nematode populations. *Agriculture, Ecosystems and Environment* 24, 263–279.

Sayre, R.M. and Starr, M.P. (1985) *Pasteuria penetrans* (Ex Thorne, 1940) nom-rev, comb-n, sp-n, a mycelial and endospore-forming bacterium parasitic in plant-parasitic nematodes. *Proceedings of the Helminthological Society of Washington* 52, 149–165.

Sayre, R.M., Starr, M.P., Golden, A.M., Wergin, W.P. and Endo, B.Y. (1988) Comparison of *Pasteuria penetrans* from *Meloidogyne incognita* with a related mycelial and endospore-forming bacterial parasite from *Pratylenchus brachyurus*. *Proceedings of the Helminthological Society of Washington* 55, 28–49.

Sayre, R.M., Wergin, W.P., Schmidt, J.M. and Starr, M.P. (1991) *Pasteuria nishizawae* sp-nov, a mycelial and endospore-forming bacterium parasitic on cyst nematodes of genera *Heterodera* and *Globodera*. *Research in Microbiology* 142, 551–564.

Schäfer, K., Fabry, C.S., Sikora, R.A. and Hauschild, R. (2006) Molecular investigations of rhizobacteria-induced systemic resistance toward the root-knot nematode *Meloidogyne incognita* in tomato. In: Raaijmakers, J.A. and Sikora, R.A. (eds) *Multitropic Interactions in Soil and Integrated Control. IOBC/WPRS Bulletin* 29, 135–140.

Schmidt, L.M., Preston, J.F., Dickson, D.W., Rice, J.D. and Hewlett, T.E. (2003) Environmental quantification of *Pasteuria penetrans* endospores using *in situ* antigen extraction and immunodetection with a monoclonal antibody. *FEMS Microbiology Ecology* 44, 17–26

Segers, R., Butt, T.M., Kerry, B.R., Beckett, A. and Peberdy, J.F. (1995) The subtilisins of the invertebrate mycopathogens *Verticillium chlamydosporium* and *Metarhizium anisopiliae* are functionally and serologically related. *FEMS Microbiology Letters* 126, 227–231.

Segers, R., Butt, T.M., Kerry, B.R., Beckett, A. and Peberdy, J.F. (1996) The role of the proteinase VCP1 produced by the nematophagous *Verticillium chlamydpsporium* in the infection process of nematode eggs. *Mycological Research* 100, 421–428.

Segers, R., Butt, T.M., Carder, J.H., Keen, J.N., Kerry, B.R. and Peberdy, J.F. (1998) The subtilisins of fungal pathogens of insects, nematodes and plants: distribution and variation. *Mycological Research* 103, 395–402.

Serracin, M., Schuerger, A.C., Dickson, D.W. and Weingartner, D.P. (1997) Temperature-dependent development of *Pasteuria penetrans* in *Meloidogyne arenaria*. *Journal of Nematology* 29, 228–238.

Sharma, S.B. and Davies, K.G. (1996a) Characterisation of *Pasteuria* isolated from *Heterodera cajani* using morphology, pathology and serology of endospores. *Systematic and Applied Microbiology* 19, 106–112.

Sharma, S.B. and Davies, K.G. (1996b) A comparison of two sympatric species of *Pasteuria* isolated from a tropical vertisol soil. *World Journal of Microbiology and Biotechnology* 12, 361–366.

Sharon, E., Bar-Eyal, M., Chet, I., Herrera-Estrella, A., Kleifeld, O. and Spiegel, Y. (2001) Biological control of the root-knot nematode *Meloidogyne javanica* by *Trichoderma harzianum. Phytopathology* 91, 687–693.

Sharon, E., Spiegel, Y., Salomon, R. and Curtis, R.H.C. (2002) Characterization of *Meloidogyne javanica* surface coat with antibodies and their effect on nematode behaviour. *Parasitology* 125, 177–185.

Sharon, E., Chet, I., Viterbo, A., Bar-Eyal, M., Nagan, H., Samuels, G.J. and Spiegel, Y. (2007) Parasitism of *Trichoderma* on *Meloidogyne javanica* and role of the gelatinous matrix. *European Journal of Plant Pathology* 118, 247–258.

Shepherd, A.M., Clark, S.A. and Kempton, A. (1973) An intracellular microorganism associated with the tissues of *Heterodera* spp. *Nematologica* 19, 31–34.

Siddiqui, Z.A. and Mahmood, I. (1996) Biological control of plant-parasitic nematodes by fungi: a review. *Bioresource Technology* 58, 229–239.

Siddiqui, Z.A. and Mahmood, I. (1999) Role of bacteria in the management of plant parasitic nematodes: a review. *Bioresource Technology* 69, 167–179.

Siddiqui, I.A. and Shaukat, S.S. (2004) Systemic resistance in tomato induced by biocontrol bacteria against the root-knot nematode, *Meloidogyne javanica*, is independent of salicylic acid production. *Journal of Phytopathology* 152, 48–54.

Siddiqui, I.A. and Shaukat, S.S. (2005) *Pseudomonas aeruginosa*-mediated induction of systemic resistance in tomato against root-knot nematode. *Plant Pathology Journal* 4, 21–25.

Siddiqui, I.A., Schaukat, S.S. and Hamid, M. (2002) Combined application of endophytic *Fusarium solani* and *Pseudomonas aeruginosa* for the suppression of *Meloidogyne javanica* in tomato. *Phytopathologia Mediterranea* 41, 138–147.

Siddiqui, I.A., Haas, D. and Heeb, S. (2005) Extracellular protease of *Pseudomonas fluorescens* CHA0, a biocontrol factor with activity against the root-knot nematode *Meloidogyne incognita. Applied and Environmental Microbiology* 71, 5646–5649.

Siddiqui, I.A., Shaukat, S.S., Sheikh, I.H. and Khan, A. (2006) Role of cyanide production by *Pseudomonas fluorescens* CHA0 in the suppression of root-knot nematode, *Meloidogyne javanica* in tomato. *World Journal of Microbiology and Biotechnology* 22, 641–650.

Siezen, R.J. and Leunissen, J.A.M. (1997) Subtilases: the superfamily of subtilisin-like serine proteases. *Protein Science* 6, 501–523.

Sikora, R.A. (1992) Management of the antagonistic potential in agriculture ecosystems for the biocontrol of plant-parasitic nematodes. *Annual Review of Phytopathology* 30, 245–270.

Sikora, R.A. and Fernández, E. (2005) Nematode parasites of vegetables. In: Luc, M., Sikora, R.A. and Bridge, J. (eds) *Plant Parasitic Nematodes in Subtropical and Tropical Agriculture*, 2nd edn. CAB International, Wallingford, UK, pp. 319–392.

Sikora, R.A. and Schönbeck, F. (1975) Effect of vesicular-arbuscular mycorrhizae (*Endogone mosseae*) on the population dynamics of the root-knot nematode *Meloidogyne incognita* and *Meloidogyne hapla*. VIII International Congress of Plant Protection, Moscow, pp. 158–166.

Sikora, R.A., Schäfer, K. and Dababat, A.A. (2007) Mode of action associated with microbially induced *in planta* suppression of plant-parasitic nematodes. *Australasian Plant Pathology* 36, 124–134.

Sikora, R.A., Pocasangre, L., zum Felde, A., Niere, B., Vu, T.T. and Dababat, A.A. (2008) Mutualistic endophytic fungi and *in-planta* suppressiveness to plant parasitic nematodes. *Biological Control* 46, 15–23.

Singh, R.S. and Sitaramaiah, K. (1966) Incidence of root-knot of okra and tomato in oil-cake-amended soil. *Plant Disease Reporter* 50, 668–672.

Singh, R.S. and Sitaramaiah, K. (1967) Effect of decomposing green leaves, sawdust and urea on the incidence of root-knot of okra and tomato. *Indian Phytopathology* 20, 349–355.

Speijer, P.R. (1993) Interrelationships between *Pratylenchus goodeyi* Sher & Allen and strains of non-pathogenic *Fusarium oxysporum* Schl. emd. Snyd. & Hans. in roots of two banana cultivars. Dissertation, University of Bonn, Bonn, Germany.

Spiegel, Y. and Chet, I. (1998) Evaluation of *Trichoderma* spp. as a biocontrol agent against soilborne fungi and plant-parasitic nematodes in Israel. *Integrated Pest Management Reviews* 3, 169–175.

Spiegel, Y., Cohn, E., Galper, S., Sharon, E. and Chet, I. (1991) Evaluation of a newly isolated bacterium, *Pseudomonas chitinolytica* sp. nov., for controlling the root-knot nematode *Meloidogyne javanica. Biocontrol Science and Technology* 1, 115–125.

Spiegel, Y., Mor, M. and Sharon, E. (1996) Attachment of *Pasteuria penetrans* endospores to the surface of *Meloidogyne javanica* second-stage juveniles. *Journal of Nematology* 28, 328–334.

Spiegel, Y., Kahane, I., Cohen, L. and Sharon, E. (1997) *Meloidogyne javanica* surface proteins: characterization and lability. *Parasitology* 115, 513–519.

Spiegel, Y., Sharon, E. and Chet, I. (2005) Mechanisms and improved biocontrol of the root-knot nematodes by *Trichoderma* spp. *Acta Horticulturae* 698, 225–228.

Stirling, G.R. (1981) Effect of temperature on infection of *Meloidogyne javanica* by *Bacillus penetrans. Nematologica* 27, 458–462.

Stirling, G.R. (1984) Biological control of *Meloidogyne javanica* with *Bacillus penetrans. Phytopathology* 74, 55–60.

Stirling, G.R. (1985) Host specificity of *Pasteuria penetrans* within the genus *Meloidogyne. Nematologica* 31, 203–209.

Stirling, G.R. (1991) *Biological Control of Plant-parasitic Nematodes: Progress, Problems and Prospects.* CAB International, Wallingford, UK.

Stirling, G.R. and Stirling, A.M. (2003) The potential of *Brassica* green manure crops for controlling root-knot nematode (*Meloidogyne javanica*) on horticultural crops in a subtropical environment. *Australian Journal of Experimental Agriculture* 43, 623–630.

Stirling, G.R. and Wachtel, M.F. (1980). Mass production of *Bacillus penetrans* for the biological control of root-knot nematodes. *Nematologica* 26, 308–312.

Stirling, G.R. and West, LM. (1991) Fungal parasites of root-knot nematodes eggs from tropical and subtropical regions of Australia. *Australian Plant Pathology* 20, 149–154.

Sundararaju, P., Thangavelu, R. and Cannayane, I. (2002) Management of *Pratylenchus coffeae* and *Meloidogyne incognita* on banana by using endophytic fungi. *Current Nematology* 13, 77–81.

Sylvestre, P., Couture-Tosi, E. and Mock, M. (2003) Polymorphism in the collagen-like region of the *Bacillus anthracis* BclA protein leads to variation in length in the exosporium filament length. *Journal of Bacteriology* 185, 5155–5163.

Takatsu, T., Horiuchi, N., Ishikawa, M., Wanibuchi, K., Moriguchi, T. and Takahashi, S. (2003) 1100-50, a novel nematocide from *Streptomyces lavendulae* SANK 64297. *Journal of Antibiotics* 56, 306–309.

Taylor, M.J. (2003) *Wolbacia* in the inflammatory pathogenesis of human filariasis. *Annals of the New York Academy of Science* 990, 444–449.

Tedford, E.C., Jaffee, B.A., Muldoon, A.E., Anderson, C.E. and Westerdahl, B.B. (1993) Parasitism of *Heterodera schachtii* and *Meloidogyne javanica* by *Hirsutella rhossiliensis* in microplots over two growing seasons. *Journal of Nematology* 25, 427–433.

Tian, B., Yang, J. and Zhang, K.Q. (2007) Bacteria used in the biological control of plant-parasitic nemaodes: populations, mechanisms of action, and future aspects. *FEMS Microbial Ecology* 61, 197–213.

Timm, L., Pearson, D. and Jaffee, B. (2001) Nematode-trapping fungi in conventionally and organically managed corn–tomato rotations. *Mycologia* 93, 25–29.

Trudgill, D.L., Bala, G., Blok, V.C., Daudi, A., Davies, K.G., Fargette, M., Gowen, S.R., Madulu, J.D., Mateille, T., Mwageni, W., Netscher, C., Phillips, M.S., Abdoussalam, S., Trivino, G.C. and Voyoulallou, E. (2000) The importance of tropical root-knot nematodes (*Meloidogyne* spp.) and factors affecting the utility of *Pasteuria penetrans* as a biocontrol agent. *Nematology* 2, 823–845.

Umamaheswari, R., Sivakumar, M., Subramanian, S. and Samiyappan, R. (2004) Induction of systemic resistance by *Trichoderma viride* treatment in greengram (*Vigna radiata*) against root-knot nematode *Meloidogyne incognita. Current Nematology* 15, 1–7.

Vaid, A., Bishop, A.H. and Davies, K.G. (2002) The polypeptide components of the parasporal fibres of *Pasteuria penetrans. World Journal of Microbiology & Biotechnology* 18, 151–157.

Vyas, R.V., Maghodia, A.B., Patel, B.A. and Patel, D.J. (2006) Isolation of native *Xenorhabdus* bacteria from *Steinernema* spp. and role of their exo and endo metabolites for suppression of root-knot nematodes (*Meloidogyne* spp.) on tomato. *Indian Journal of Nematology* 36, 241–246.

Weibelzahl-Fulton, E., Dickson, D.W. and Whitty, E.B. (1996) Suppression of *Meloidogyne incognita* and *M. javanica* by *Pasteuria penetrans* in field soil. *Journal of Nematology* 28, 43–49.

Westphal, A. and Becker, J.O. (1999) Biological suppression and natural population decline of *Heterodera schachtii* in a California field. *Phytopathology* 89, 434–440.

Whipps, J.M. and Davies, K.G. (2000) Success in biological control of plant pathogens and nematodes by microorganisms. In: Gurr, G. and Wratten, S. (eds) *Biological Control: Measures of Success.* Kluwer Academic Publishers, Dortrecht, The Netherlands, pp. 231–269.

Wishart, J., Blok, V.C., Phillips, M.S. and Davies, K.G. (2004) *Pasteuria penetrans* and *P. nischizawae* attachment to *Meloidogyne chitwoodi, M. fallax* and *M. hapla. Nematology* 6, 507–510.

ZhenChuan, M., WenHua, T., RuXian, W. and LiQun, Z. (2004) Action of *Bacillus thuringiensis* B24-14 and its beta-exotoxin on the parasitic nematodes of plant. *Journal of China Agricultural University* 9, 34–37 (English abstract).

Zuckerman, B.M., Dicklow, M.B. and Acosta, N. (1993) A strain of *Bacillus thuringiensis* for the control of plant-parasitic nematodes. *Biocontrol Science and Technology* 3, 41–46.

18 Current and Future Management Strategies in Intensive Crop Production Systems

Andrew P. Nyczepir[1] and Stephen H. Thomas[2]

[1]*USDA-ARS, Southeastern Fruit and Tree Nut Research Laboratory, Byron, Georgia, USA;* [2]*New Mexico State University, Las Cruces, New Mexico, USA*

18.1 Introduction	**412**
18.2 Current Control Practices	**413**
18.3 Current Management Practices	**423**
18.4 Future Opportunities and Challenges	**428**
18.5 Conclusions and Future Directions	**435**
18.6 References	**435**

18.1 Introduction

Species of *Meloidogyne* are recognized as being serious constraints to intensive crop production systems worldwide. Perennial crops such as fruit and nut trees, coffee, banana and grape, as well as many annual vegetable and field crops, all suffer economic losses due to root-knot nematodes. Collectively, *Meloidogyne* spp. are more damaging to such crops than most other plant-parasitic nematodes because: (i) root-knot nematodes are widely distributed throughout the world; (ii) most species complete several generations per growing season and have high fecundity; and (iii) some species have very wide host ranges. Left unmanaged, root-knot nematode populations commonly reach densities that reduce crop yield and vigour. In addition, stressed plants may be predisposed to infection by other pathogens, to pest or environmental injury, and even death.

In this chapter, the terms 'control' and 'management' carry different connotations in regard to suppressing the negative effects of *Meloidogyne* spp. in intensive crop production systems. 'Control' practices refer to specific one-time actions available to growers to reduce root-knot nematode populations below economically damaging density levels, whereas nematode 'management' strategies rely on a combination of control practices to achieve effective suppression of the nematode population density in an effort to minimize the undesirable effects of *Meloidogyne* spp. on agricultural productivity (Bernhard *et al.*, 1985; Norris *et al.*, 2003). Such strategies are governed by the value and characteristics of the crop(s) within a particular system, the various control options that can be utilized, and the level of knowledge that exists pertaining to the biotic and abiotic components affecting the system. As a result of gaps in knowledge and the many uncertainties inherent in such complex systems, growers are far more likely to use control practices rather than management strategies to suppress root-knot nematode damage in most crops.

Successful management of *Meloidogyne* spp. in intensive cropping systems begins with site

 ©CAB International 2009. *Root-knot Nematodes* (eds R.N. Perry, M. Moens and J.L. Starr)

selection. A preferred site is one that is suitable for culture of the desired crop and that does not have a history of disease or root-knot nematode problems. If, as is often the case, a *Meloidogyne*-free site is not available, then either proper control practices or management strategies should be implemented. The decision to use a control practice rather than management strategy occurs most often in crops of higher value, which can support more costly controls. Historically, recommended control practices for *Meloidogyne*-infested sites often included pre-plant and/or post-plant nematicide applications. However, because many of the nematicides that were previously available have been or are being removed from the agricultural market due to environmental and human health awareness issues, effective control practices that employ alternatives to conventional chemical control are being increasingly investigated and implemented. Such practices are likely to become more prevalent in the future for use in controlling root-knot nematodes (Batchelor, 2002). Alternatives to chemical control include the use of cultural practices, host plant resistance and the application of biological control agents, when available. Additional components of root-knot nematode management plans include proper sanitation, quarantine or the use of nematode-free planting stock to prevent reinfestation or the introduction of additional species of *Meloidogyne* into managed sites. This chapter focuses on current control practices and management strategies for root-knot nematodes and on future developments that may affect management in intensive crop production systems.

18.2 Current Control Practices

Producers involved in intensive crop production usually face choices as to the practice(s) best suited to control populations of *Meloidogyne* spp. in the crops they produce. These decisions are heavily influenced by crop characteristics and the species of *Meloidogyne* that are present. Greater levels of control are likely to be necessary in more susceptible high-value crops and in perennials, due to the length of exposure of the host to the parasite. Control practices, such as the use of synthetic chemical nematicides or crop rotation, are more likely to be the exclusive choice in a wider range of crops for the more predominant and more polyphagous root-knot nematode species, such as *M. incognita*, *M. javanica* and *M. hapla*. In general, most control practices must be repeated on a regular basis – often each time a susceptible crop is grown. Most current control practices fall into four categories: chemical, cultural, biological and host plant resistance. This section focuses on chemical and cultural controls; biological control and host plant resistance are discussed by Hallmann *et al.* (Chapter 17, this volume) and Starr and Mercer (Chapter 14, this volume), respectively.

18.2.1 Chemical control

Chemical control was the mainstay for reducing root-knot nematode populations in most moderate-to-high-value crops in intensive production systems throughout much of the 20th century (Johnson, 1985, 1998; Luc *et al.*, 1990; Halbrendt and LaMondia, 2004). Beginning in 1979 (EPA, 2007), environmental and human health concerns have steadily reduced the availability of such control options, starting with the use of 1,2-dibromo-3-chloropropane (DBCP). By 1981, the US Environmental Protection Agency (EPA) suspended registration of DBCP as a soil fumigant for all crops (Johnson and Feldmesser, 1987). Numerous other nematicides have suffered a similar fate or, like methyl bromide, have been voluntarily withdrawn from use in the USA and European Union (EU). In accordance with the 1992 Montreal Protocol, the importation and manufacture of methyl bromide was banned in the USA and Western Europe after January 2005 (Clean Air Act, 1990), due to its role in ozone depletion. It should be noted that methyl bromide can currently be obtained in the USA only for certain exceptions (i.e. Critical Use Exemption (CUE), quarantine and pre-shipment exemptions, and emergency exemptions). In the EU, the use of methyl bromide is restricted to the strict minimum (critical uses which need to be accepted), whereas in developing countries its use will be phased out in 2015. Additionally, within the EU, 1,3-D (1,3-dichloropropene) has also been omitted from the list of authorized nematicides, and its use is expected to diminish during 2009. The chemicals still registered beyond 2009 will mainly be fungicides with a complementary but less marked nematicidal activity (i.e. dazomet, metam sodium). An updated list of authorized nematicides is available at http://e-phy.agriculture.gouv.fr/ (France).

As part of European Union Authorisations Directive 91/414/EEC, the active substances in all new and existing pesticides must undergo a certification process, which has substantially reduced the number of nematicides approved for use by European producers (Haydock *et al.*, 2006). Although still important tools for controlling root-knot nematodes, the predominance of nematicides has decreased in many regions of the world. Despite these reductions, chemical control and host plant resistance remain the predominant means of controlling *Meloidogyne* spp. in several perennial and many high-value annual crops in North America (Roberts, 1993; Starr *et al.*, 2002; Bridge and Starr, 2007). Chemical controls can be categorized according to their method of application as fumigant nematicides, non-fumigant nematicides and those derived from naturally occurring biotic sources (Rich *et al.*, 2004; Haydock *et al.*, 2006). Development of new synthetic nematicides has not kept pace with losses of fumigants and non-fumigants during recent decades. Most recent development efforts pertaining to chemical control options have focused on identifying new efficacious nematotoxic compounds from biotic sources, or on the development of methodologies that mitigate undesirable consequences associated with the use of existing nematicides. Examples of the latter include modifications in the recommended depths of placement of soil fumigants to reduce undesirable effects associated with volatilization and increase crop yield (Lembright, 1990; Thomas, 1994) and development of solenoid-operated shut-offs for application equipment to reduce accidental soil surface contamination at the ends of rows. As of 2008, the only new nematicides that have been widely commercialized in more than three decades are the synthetic non-fumigant fosthiazate and the abamectin seed treatment Avicta® (Rich *et al.*, 2004; Haydock *et al.*, 2006).

18.2.1.1 Fumigant nematicides

The discovery of specific chemicals that were shown to destroy nematodes (i.e. nematicidal chemicals) and the introduction and use of these in agriculture since the late 19th century have increased both yield and quality in crop production systems worldwide. Nematicides used in managing *Meloidogyne* in intensive crop production systems are usually separated into two groups, based on their mode of application: fumigants and non-fumigants. The fumigant nematicides are generally liquids that volatilize to a gaseous phase upon entering the soil. The chemistry of fumigant nematicides currently being used in intensive agricultural production systems includes either compounds containing halogenated hydrocarbons or those that discharge carbon disulfide or methyl isothiocyanate. Examples of halogenated hydrocarbons include methyl bromide, chloropicrin, methyl iodide and 1,3-D. Metam sodium and dazomet release methyl isothiocyanate upon degradation in the soil. Tetrathiocarbonate is classified as an inorganic fumigant, which breaks down in the soil to form carbon disulfide; this compound is not as volatile as the halogenated hydrocarbons and is dependent on soil moisture to move through the soil profile to manage *Meloidogyne*.

Fumigants are further categorized as being either nematicidal or multi-purpose (Lembright, 1990; Dunn and Noling, 1997). Nematicidal fumigants (e.g. 1,3-D) are specific against nematodes at their recommended rates, whereas multipurpose fumigants (e.g. chloropicrin, methyl bromide, methyl iodide and metam sodium) target plant-parasitic nematodes as well as fungi, weeds and insects. Generally, the multipurpose fumigants are more expensive than the nematicidal fumigants and are primarily used on high-value crops.

In California, the soil fumigants most used in treating vineyard acres are sodium tetrathiocarbonate for nematodes and phylloxera and the pre-plant nematicide 1,3-D (California Department of Pesticide Regulation, 2007). Further guidelines for use of soil fumigants in replanting vines or trees in California are summarized elsewhere (McKenry, 2007). In strawberries, 1,3-D or 1,3-D plus chloropicrin are used to control *Meloidogyne*. Until recently, and for many years, methyl bromide was the most widely used multi-purpose fumigant nematicide in Florida vegetable production systems because of its wide spectrum of activity against nematodes, other soil pathogens and weeds (Noling, 1996).

In orchard crops in the south-eastern USA, it is important to get the tree off to a good start to prevent the root-knot nematode from causing major root damage. Trees planted in infested soil with a low initial population density of *Meloidogyne* usually show no sign of stunting or reduced vigour, even though roots eventually become galled (Bertrand and Evert, 1984; Bertrand, 1985). The current pre-plant nematicide recommendations for managing *Meloidogyne* in peach include the soil fumigants, 1,3-D or metam sodium (Horton *et al.*, 2007) (Fig. 18.1, Plate 40). These are the only two soil fumigants available to peach growers since the partial ban of methyl bromide.

As a result of the partial ban on methyl bromide, a number of different alternatives are being evaluated in Europe and the USA against pathogens (e.g. root-knot nematodes) and other pests formerly managed by methyl bromide; these include the use of antagonistic crops. A partial list of chemical alternatives to methyl bromide (see Table 18.1) includes: methyl iodide, propargyl bromide, ozone, formaldehyde, sodium tetrathiocarbonate, carbon disulfide, anhydrous ammonia, inorganic azides, natural compounds, propylene oxide, sulfur dioxide, peroxyacetic acid and acrolein (2-propenal) (MBTOC, 1998; Johnson *et al.*, 2005). In 2007, the EPA approved a 1-year registration for methyl iodide (iodomethane) under highly restrictive provisions regulating its use in the USA (EPA, 2008b). Methyl iodide can be used as a pre-plant soil fumigant to control soil pathogens (including nematodes), insects and weeds. Crop and plant uses include trees, vines, strawberries, tomatoes and peppers, to name a few. In France, dimethyl disulfide has recently been registered as an alternative, with efficacy against fungal pathogens, nematodes and weeds (http://www.fytoweb.fgov.be/indexen.htm). Other potential replacements for methyl bromide are discussed below (section 18.4.2.3).

18.2.1.2 Non-fumigant nematicides

Non-fumigant nematicides do not suppress nematode populations as effectively as fumigant nematicides because they do not have broad-spectrum activity (Luc *et al.*, 2005). These chemicals are formulated as either granular or

Fig. 18.1. Application of 1,3-dichloropropene using a cultipactor to seal the soil surface prior to establishing peach orchard in *Meloidogyne*-infested site. (Photograph by A.P. Nyczepir.)

liquid materials, and include such products as aldicarb, oxamyl, ethoprop/ethoprophos, fenamiphos, carbofuran, fosthiazate and terbufos, all of which are at least moderately effective towards *Meloidogyne* spp. in field and glasshouse conditions. These products have been available globally, but many of the registrations have since been revoked or the products voluntarily withdrawn by the manufacturer in the USA and Europe. Unlike fumigants, non-fumigant nematicides are not volatile and must disperse in the soil water phase (via irrigation and/or rainfall) to be active against nematodes. Non-fumigants are divided into two main chemical classes, which include organophosphates and carbamates. Both classes of non-fumigants are considered nemastatic, meaning that their effects are reversible, and not nematicidal; they act as acetylcholinesterase inhibitors (Opperman and Chang, 1990; Haydock *et al.*, 2006). Therefore, it is important that the non-fumigant remains in contact with the nematode for approximately 4–8 weeks to inhibit nematode infection sufficiently to enable plant growth with minimal impact due to nematode parasitism (Wright, 1981; Rich *et al.*, 2004). Repeated applications are required in orchard and multiple-cropping vegetable systems; these are generally uneconomical and can lead to enhanced biodegradation of the chemical by soil microorganisms (Davis *et al.*, 1993; Smelt *et al.*, 1996; Lawrence *et al.*, 2005). Furthermore, if the chemical moves apoplastically within the plant after being applied to the soil, the residue may persist in the plant, thus limiting crop registration (Rich *et al.*, 2004). In peach, fenamiphos was the only post-plant treatment for suppression of root-knot nematodes in the south-eastern USA, but product registration was voluntarily withdrawn by the manufacturer in May 2007 (Anonymous, 2003a; Horton *et al.*, 2007). However, the effectiveness of fenamiphos in controlling *Meloidogyne* spp. in peach was questionable. In coffee, the reported effectiveness of non-fumigants for control of *Meloidogyne* spp. varied. Poor control of *M. incognita*, *M. coffeicola* and *M. paranaensis* was achieved with contact nematicides, whereas repeated applications (5 years) of nematicides provided increased yields in trees growing in soil infested with *M. exigua* (Campos and Villain, 2005).

18.2.1.3 Other compounds

Numerous other chemical compounds have been tested and found to exhibit nematicidal properties, but most have not been commercialized for a variety of reasons, as reviewed by Chitwood (2003) and Rich *et al.* (2004). However, two such compounds (avermectins and fermentation products from the fungus *Myrothecium verrucaria*) have recently attained commercial status. Avermectins are macrocyclic lactones produced by *Streptomyces avermitilis*. These compounds have been used successfully as mammalian anthelmintics and insecticides for decades (Jansson and Dybas, 1998), but have been considered unlikely choices for controlling soil nematodes due to their poor water solubility and tendency to bind to soil particles. Abamectin, a mixture of avermectins from different fungal strains has been investigated for use as a nematicidal seed treatment, has shown promise for early-season control of *M. incognita* in short-season vegetables and cotton (Becker *et al.*, 2003; Monfort *et al.*, 2006). One difference observed in root-knot nematodes exposed to abamectin compared with other soil-applied non-fumigant nematicides was the failure of juveniles to recover (Faske and Starr, 2006). Abamectin was registered for use as a nematicidal seed treatment for cotton (Avicta®) in the USA in 2006 (Plate 45). Expanded acceptance and efficacy of this material on other crops remains to be determined.

The product label for the second material, the biological nematicide DiTera®, describes the active ingredient as a 'non-viable'/'killed' microbial composition of fermentation solids and solubles from the fungus *Myrothecium verrucaria*. DiTera® has been found to enhance soil antagonism to *M. incognita* but not to inhibit hatch, as has been reported when used against certain *Heterodera* spp. (Fernandez *et al.*, 2001) and *Globodera* spp. (Twomey *et al.*, 2000). Currently, DiTera® is the only registered nematicide listed by the Organic Materials Review Institute in the USA for use with annual and perennial crops and ornamental plants in certified organic production systems.

18.2.2 Cultural control

All human activities designed to reduce populations of *Meloidogyne* can be characterized as cultural control practices, except for nematicide application, the use of biological control agents and resistant plant varieties. Most of these measures are more applicable to annual crops than perennials, with the notable exception of sanitation practices, which are important with both categories of hosts. The benefit/necessity of employing rotation in place of extended monocropping with annual crops has been recognized for millennia (Halbrendt and LaMondia, 2004), and recommended for root-knot nematode control for over a century (Bessey, 1911), whereas our understanding and use of practices such as trap crops, green manures, antagonistic plants and avoidance of infection through adjustments in planting dates have developed only as our knowledge of root-knot nematode biology and identification expanded. The relative efficacy of various cultural control measures differs among *Meloidogyne* species. However, most can be combined with nematicides, host plant resistance or other control practices to enhance the success of root-knot nematode management systems.

18.2.2.1 Crop and fallow rotation

Agricultural growers engaged in intensive annual crop production frequently reduce populations of *Meloidogyne* species by interjecting into their production schemes crops that are poor or non-hosts of the targeted nematode. The greatest advantage of such a strategy compared with most other cultural controls, or the use of a nematicide, is that this control practice produces an economic return on investment. If sufficient markets are available to support profitable production of non-host crops, then host range limitations of many species of *Meloidogyne* can be readily exploited. Key factors in determining economic viability are: (i) accurate species identification and thorough knowledge of the host range for the targeted species of root-knot nematode; (ii) the number of non-host crops that are agronomically suited to the geographic region; and (iii) ensuring that populations of other damaging plant-parasitic nematodes or other pathogens or pests are not significantly enhanced by the rotation crop. Some examples of widely used rotations include growing certain graminaceous crops to control *M. hapla* in vegetables such as carrot and potato, in groundnut, and in soybean (Bélair and Parent, 1996; Bridge and Starr, 2007), and growing cotton or maize to control *M. arenaria* in groundnut (Davis and Timper, 2000).

The use of crop rotation to control species with more extensive host ranges, such as *M. incognita*, *M. javanica* and *M. arenaria*, or to control mixed species of root-knot nematodes remains a serious challenge for growers. Successes in these situations often utilize novel crops such as castor (*Ricinus communis*), velvet bean (*Mucuna deeringiana*), sesame (*Sesamum indicum*), American jointvetch (*Aeschynomene americana*), pangola grass (*Digitaria decumbens*) or *Crotalaria juncea* (McSorley *et al.*, 1994; Whitehead, 1998) or special case scenarios where rotation crops are actually root-knot-resistant varieties of susceptible crops, such as certain cultivars of lucerne (*Medicago sativa*) or cowpea (*Vigna unguiculata*) (Roberts, 1993). Effective weed control is essential to the success of any crop rotation, because the presence of alternative weed hosts in these crops can result in high populations of root-knot nematodes despite the poor host status of the crop (McSorley *et al.*, 1994; Bélair and Parent, 1996). Weeds have also been implicated in the enhancement of *M. chitwoodi* populations in Europe (Kutywayo and Been, 2006), and in negating the suppression of this nematode where resistant potato cultivars were grown in the USA (Boydston *et al.*, 2007; http://www.eppo.org/QUARANTINE/nematodes/Meloidogyne_chitwoodi/MELGCH_ds.pdf). The vertical distribution and potential for migration of the targeted nematode within the soil profile is another consideration when utilizing crop rotation. The within-field vertical distribution of *M. chitwoodi*, a nematode that can exhibit considerable upward migration to infect potato, was relatively unaffected by rotation crops, suggesting that vertical distribution may be more affected by soil characteristics than by host suitability of rotation crops (Mojtahedi *et al.*, 1991; Wesemael and Moens, 2008a).

Most people would agree that controlling *Meloidogyne* spp. in a perennial crop system is more challenging and difficult than in annual crops. Utilizing rotations that have proven successful with annual crops is somewhat impractical when used with perennials. However, there have been some success stories in using pre-plant rotations to control root-knot nematodes with some perennial crops. In coffee, it has been reported that a 1-year pre-plant rotation with sorghum, soybean or *Panicum maximum* in soil infested with *M. exigua* made it possible to plant the site back to susceptible coffee cultivars (Almeida and Campos, 1991). However, it is important to note that when a specific rotation is effective in suppressing populations of *M. exigua* or *M. coffeicola*, it may not be as successful against *M. incognita* or *M. paranaensis*, and the latter two species are considered the more economically important species in Brazil (Carneiro and Carneiro, 1982; Campos and Villain, 2005). In almond, pre-planting a non-host such as cereal grains prior to establishing an orchard reduced the population density of *Meloidogyne* spp. to undetectable levels (McKenry, 1985). The economic benefit derived from a 1-year delay in the planting of a perennial is that it allows the tree to get off to a healthy start by preventing the nematode from causing major root damage (Bertrand and Nyczepir, 1989).

Some crop rotations incorporate an extended period of time where the land is left fallow (uncropped) before replanting an agricultural crop. The general objective in using a fallow rotation is to suppress the population of plant-parasitic nematodes prior to replanting the site. The apparent benefits of fallowing crop land to control species of *Meloidogyne* seem obvious: knowledge of nematode species composition and characteristics are relatively unimportant; nearly all plant-parasitic nematodes are suppressed simultaneously; and there are no expenses associated with the production of a crop. Different types of fallow rotation include: (i) natural vegetation (weedy fallow), in which fields are left undisturbed between crops; (ii) dry or clean fallow, in which weeds are controlled on a regular basis to prevent root-knot nematode reproduction, and water is withheld from irrigated sites; (iii) weed-free wet fallow, in which irrigated production sites are watered to enhance egg development and hatching of *Meloidogyne*, accompanied by regular weed control to prevent reproduction; and (iv) grass fallow, in which a thick stand of grass effectively suppresses populations of *Meloidogyne* that have a narrow host range. Nevertheless, many growers involved in intensive crop production systems do not rely on fallowing as a primary tactic to suppress root-knot nematodes. Lack of income from the fallow land is a primary deterrent, together with the expenses associated with regular weed control (preferably every 3 weeks to prevent root-knot nematode reproduction) and potentially deleterious effects from soil erosion; all of these factors contribute to the limited acceptance of fallowing. In addition, in regions where crops are routinely grown under irrigation, dry-fallowing (irrigation withheld from fallow fields) may produce poor results due to insufficient soil moisture to accommodate egg development, hatching and subsequent starvation during the fallow period.

Some exceptions include crops that are well suited to using grass and clean fallow. Grass fallow has proven effective against some species of *Meloidogyne*. If the rotation grass is a non-host to the *Meloidogyne* species, then the nematode population will starve. The key to a successful pre-plant grass fallow rotation, like any crop rotation system, is that all weed hosts of root-knot nematodes need to be absent so that remnant nematode populations do not survive. Clean fallow has proven to be a useful management tool against *Meloidogyne* in vegetable production. It has been shown to be most effective when used in the dry, hot summer months when weed hosts are not a problem, and when used in conjunction with other convenient management practices (e.g. cultivation and/or root destruction) (Sikora and Fernandez, 2005). In perennials, a rotation grass is sometimes grown during the fallow period before an orchard is re-established. Coastal bermudagrass (*Paspalum* sp.) is an example of a pre-plant grass fallow management strategy that has been used to reclaim peach land infested with *Meloidogyne* (Bertrand and Nyczepir, 1989). Another option that has also been recommended is to use a rye winter cover crop followed by a summer fallow and another rye winter cover crop prior to fumigation (Davis *et al.*, 1996). More detail on fallow rotation types are summarized in reviews published elsewhere (Bridge, 1996; Halbrendt and LaMondia, 2004).

18.2.2.2 Trap crops, cover crops and soil amendments

A number of crops have been identified for use in controlling root-knot nematodes directly (trap crops) or suppressing nematode populations while providing other benefits such as reducing soil erosion, enhancing soil organic matter or providing forage for grazing livestock (cover crops and green manures). These crops differ from rotation crops in that typically there is no expectation of marketability or direct economic return from their production. The economic benefits derived from their use are measured in terms of reduced nematicide costs and/or enhanced productivity in subsequent crops. Roots of trap crops are readily infected by second-stage juveniles (J2) of *Meloidogyne*, which then fail to complete their life cycle due to antagonistic plant responses. Arugula (*Eruca sativa*), a trap crop with some potential for marketability, has been reported to be effective against *M. hapla* (Melakeberhan *et al.*, 2006). However, many good trap crops are plant species with potentially undesirable characteristics, such as toxicity to domestic animals or weedy traits (McSorley, 1998), again limiting grower acceptance.

Cover crops and green manures (cover crops intended for soil incorporation prior to senescence) that are grown during intermittent periods between cash crops, primarily to prevent soil erosion and increase soil quality, may also help suppress populations of *Meloidogyne* if properly selected (Fig. 18.2; Plate 41). The best choices are crops that are poor or non-hosts for problematic species of *Meloidogyne*, so as not to contribute additional inoculum that increases damage in subsequent cash crops (Timper *et al.*, 2006). Upon crop incorporation, populations can be suppressed due to a complex cascade of events triggered by the infusion of organic matter from the cover crop or by the toxic effects of glucosinolates or dhurrin released during the decomposition of green manures of *Brassica* species or sudangrass, respectively (Widmer and Abawi, 2000; Zasada and Ferris, 2004). Cover crops and green manures have provided acceptable suppression of *M. arenaria, M. chitwoodi* and *M. incognita* (Mojtahedi *et al.*, 1993; McSorley, 1999; Zasada and Ferris, 2004). Additional examples and detailed discussions of the role of cover crops and green manures in controlling plant-parasitic nematodes have been presented by Halbrendt and LaMondia (2004) and Widmer *et al.* (2002).

Soil amendments, such as livestock or poultry manure and organic compost, are often applied to fields used for intensive annual crop production, but rarely is control of root-knot nematodes the primary reason for such applications. Rather, amendments are applied as low-cost sources of plant nutrients, acceptable methods for the disposal of animal waste or other agricultural by-products, or to improve soil properties. Application of composted horticultural waste or fresh poultry waste (manure and bedding material) increased vegetable yields and decreased soil populations of *M. incognita*, respectively, in the USA (McSorley and Gallaher, 1995; Riegel and Noe, 2000). More extensive reports of the beneficial effects of soil amendments for suppression of root-knot nematodes exist from other regions of the world (Whitehead, 1998). Suppression of

Fig. 18.2. Incorporation of *Brassica* species (i.e. green manure) as a pre-plant nematode control strategy. (Photograph by A.P. Nyczepir.)

plant-parasitic nematodes due to application of soil amendments is thought to result from the enhanced activity of nematode antagonists responding to elevated levels of organic matter decomposition (Widmer *et al.*, 2002). As organic waste production increases in the future and disposal options become more restricted, the addition of soil amendments to intensive cropping systems is likely to increase (Zasada *et al.*, 2008).

18.2.2.3 Exploitation of phenology

As poikilothermic organisms, ambient temperature has a profound effect on the physiological development of root-knot nematodes and the damage they cause. Temperature delimits the regions in which important species like *M. arenaria*, *M. hapla*, *M. incognita* and *M. javanica* can survive (Taylor and Sasser, 1978), limits hatch of J2 and development (Tyler, 1933; Vrain *et al.*, 1978), and restricts movement in soil and host infection (Roberts *et al.*, 1981). Knowledge of such effects can be coupled with an understanding of crop phenology to avoid or limit damage from *Meloidogyne*. For example, adjusting carrot planting dates greatly reduced the amount of injury resulting from *M. incognita* in infested fields in southern California (Roberts, 1987). Similarly, some vegetable producers in other regions of the southwestern USA refrain from planting autumn lettuce or onion crops in *M. incognita*-infested fields, choosing instead to establish these crops in late winter, when soil temperatures are too cool for nematode infection. Reducing the interval between planting and harvest has been found to lessen the severity of injury to carrot by *M. chitwoodi* in fields with low initial nematode populations (Wesemael and Moens, 2008b). Phenology data for specific root-knot nematode and host associations are becoming increasingly available to the public, as evidenced by the posting of detailed information regarding *M. chitwoodi* development on potato (University of California, Davis, 2003), and demonstration that *M. chitwoodi* and *M. fallax* hatch is affected differently by the onset of host senescence (Wesemael *et al.*, 2006).

18.2.2.4 Sanitation

An important component of cultural control includes approaches that minimize root-knot nematode movement between planting sites or into new production areas (e.g. sanitation). Sanitation practices may include: (i) removal/destruction of infected plant material prior to replanting a site; (ii) inspection and use of certified nematode-free plants from reliable nurseries; (iii) cleaning of farm implements; and (iv) heat treatment of potentially infected plant material (e.g. roots, corms, tubers, etc.).

The use of sanitation to help control root-knot nematodes in annual crops is often limited to one activity – the avoidance of nematode-infected transplants, tubers or other vegetative propagation material. Although producers are typically aware that sanitation can help prevent the dissemination of *Meloidogyne* in soil that adheres to farm implements, few growers clean such equipment between fields, regardless of the level of nematode infestation or their prior use of other nematode control practices. In vegetable-producing regions that rely heavily on the use of contract labour crews or custom mechanized transplanting or harvesting, avoiding field-to-field contamination becomes even more difficult. Where furrow or flood irrigation is practised, the use of tail water from one field to irrigate another field is rarely considered due to concerns about spreading fungal pathogens and weed seed and therefore is less of a factor in the potential dissemination of root-knot nematodes. However, within-field spread of *Meloidogyne* can easily occur due to soil-levelling activities in furrow- or flood-irrigated fields, through movement in irrigation water or with cultivation.

Most annual crops that are established using transplants rely on plant material that is produced in commercial glasshouses using containers filled with growth media that are free of nematodes and plant pathogens. The transplanting process itself is usually mechanized. Under such conditions, the introduction of *Meloidogyne*-infected material in plugs is unlikely. Two crops for which transplants are frequently produced in propagation beds rather than glasshouses are sweet potato and tobacco. Under such conditions, the potential for root-knot nematode infection of transplant stock increases considerably. The use of vine cuttings is the primary sanitation practice used by sweet potato producers to achieve nematode-free planting material. After reaching the eight-leaf stage, sprouts from seed roots in propagation beds are cut above the soil surface, thereby avoiding potentially infected roots (Peet, 2001). Pre-plant fumigation remains the standard practice for production of *Meloidogyne*-free tobacco transplants by producers using traditional seed bed methods (Maksymowicz and Palmer, 1997). The use of potato planting material that has been certified as free of nematodes is a widely recommended and accepted sanitation practice, particularly on land not previously used to produce potato or known to be infested with *M. chitwoodi* or *M. fallax*. Prevention of livestock or other animal movement between infested and potentially non-infested fields is also recommended, to help curtail the spread of *M. chitwoodi* (University of California Pest Management Guidelines, 2007).

In *Prunus*, most species of root-knot nematodes can be controlled with the use of resistant/tolerant rootstocks. However, in California, Nemaguard rootstock has been used for years by *Prunus* growers to manage *M. incognita* and *M. javanica*, but it has been reported that Nemaguard is not necessarily tolerant to feeding by root-knot nematodes under certain field conditions (McKenry *et al.*, 2006). Serious first-year tree damage can occur if trees on Nemaguard are planted into sandy sites having population levels of *Meloidogyne* spp. greater than 100 J2/250 cm^3 soil. One way to combat the high root-knot nematode soil populations associated with California's replant problem without the use of pre-plant fumigation is to start off with proper sanitation management. Briefly, it is recommended to apply herbicide (e.g. glyphosate) to cut stumps, followed by trunk removal 60 days later. The herbicide acts to eliminate a potential food source for nematodes remaining in roots following trunk removal. It is then recommended to wait 1 year, subsoil if necessary, and replant on a rootstock that is resistant to root-knot nematode and does not have Nemaguard in its pedigree.

Hot water dips to control root-knot nematode in *Prunus* seedlings have been shown to provide acceptable nematode control. Submerging *Meloidogyne*-infected dormant seedlings of Lovell peach and cherry (*Prunus mahaleb*) into a hot water bath at 48 °C/30 min, 49 °C/20 min, 50 °C/10 min, 51 °C/5 min or 51 °C/10 min all resulted in killing the nematode. However, it also was determined that dormant trees must be stored for at least 6 weeks after digging prior to treatment, in order to achieve 100% tree survival (Nyland, 1955). In banana, infected corms can be disinfected of *Meloidogyne* spp. first by peeling, then by using hot water treatment (53–55 °C for 20 min) (De Waele and Davide, 1998; Coyne *et al.*, Chapter 19, this volume).

18.2.2.5 Steam heat and solarization

The effectiveness of steam heat and soil solarization in managing root-knot nematodes under glasshouse and field conditions, respectively, is dependent on soil temperature. The lethal soil temperature for fungal or bacterial pathogens occurs at 60–100 °C, whereas 45 °C is considered to be sufficient to control plant-parasitic nematodes (Katan, 1981; Sikora and Fernandez, 2005). Steam heat has been utilized in glasshouses for many years as a means to manage plant-parasitic nematodes and other soil-borne pathogens by way of soil sterilization. However, the use of steam heat in recent years has been limited due to the high cost of heating fuel, possible emission of phytotoxic chemicals into the treated soil, change in soil pH, and death of beneficial soil biota.

Soil solarization is the technique of increasing soil temperatures in both hot, arid climates and humid, temperate regions of the world where high levels of uninterrupted solar radiant energy exist (Gaur and Perry, 1991). Lethal effects on eggs and J2 of *M. incognita* have been achieved below 45 °C when nematodes were exposed to sublethal temperatures for a sufficient period of time (Wang and McSorley, 2008). The length of cumulative exposure necessary to achieve lethality increased with decreasing temperature, down to 39 °C, and was found to be independent of heat unit accumulation and unaffected by temperature oscillation. Moistened soil that is tilled and then covered with a single or doubled-layered polyethylene tarp, with edges buried under the soil, has been shown to control certain soil-borne diseases, insects, weeds and plant-parasitic nematodes (Katan, 1981; Whitehead, 1998). Factors associated with limiting the use of solarization for root-knot nematode suppression are: (i) efficacy decreases with increasing soil depth below 5 cm; (ii) effects are non-specific, affecting pests and beneficial organisms alike (Freitas *et al.*, 1997; McSorley, 1998); (iii) the duration of time needed; and (iv) size of area to be treated. In intensively managed annual crops, the vertical dissemination of *Meloidogyne* below 5 cm within the soil profile can occur through root penetration from preceding crops, cultivation and migration by juveniles, and is largely unaffected by rotation crops (Wesemael and Moens, 2008a). Such movement reduces the effectiveness of solarization and is a possible factor contributing to the inconsistencies in root-knot nematode control that have been reported with use of this practice (Barbercheck and Von Broembsen, 1986; Halbrendt and LaMondia, 2004). Generally, solarization is more suited for annual crops (i.e. organic vegetable production and ornamentals) than perennial crops, and for nursery and raised beds versus large fields (McSorley *et al.*, 1999; McGovern *et al.*, 2002). In Florida, solarization of a fine, sandy soil for 3 months (July–September) suppressed *M. incognita* populations in tomato fields, resulting in increased yields (Overman and Jones, 1986). Furthermore, disposal of large quantities of plastic can be problematic when treating large fields. Solarization offers a potential means for suppressing root-knot nematodes in specialized production systems where other controls are unacceptable or lacking entirely.

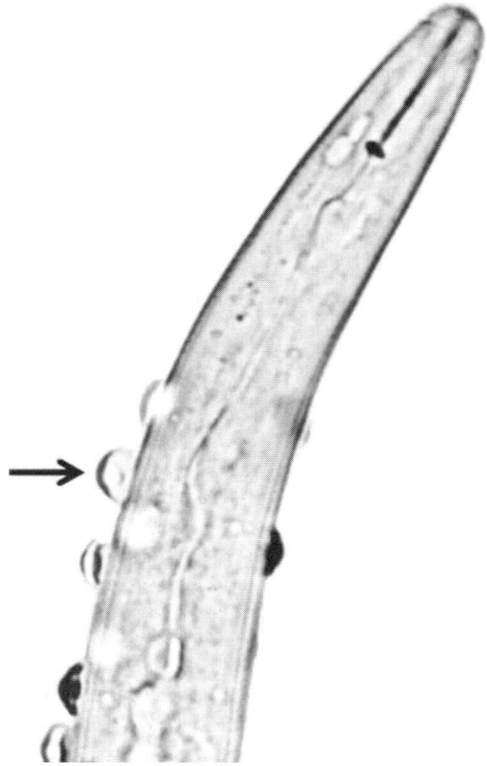

Fig. 18.3. Spores of *Pasteuria penetrans* (see arrow) attached to *Meloidogyne partityla* second-stage juvenile. (Courtesy of J.A. Brito, Gainesville, Florida, USA.)

18.2.3 Biological control and host plant resistance

Biological control of *Meloidogyne* spp. occurs when the action of antagonists maintains the nematode population at a density below the level that would occur in their absence. In *Prunus*, banana and most other intensive cropping systems, biological control success stories have not been too abundant. However, in organic coffee production systems, biological control against *Meloidogyne* spp. appears to be a promising management tactic since chemical control tools are prohibited. One potential biological control agent found infecting 21–65% of J2 of *M. exigua* in a 10-year-old coffee field in Brazil was the bacterium *Pasteuria penetrans* (Maximiniano *et al.*, 2001) (Fig. 18.3). It was also noted that the number of bacterial endospores per J2 varied between 2.1 and 9.2 and generally increased in March and April. A more detailed discussion of natural pathogens and biological control of root-knot nematode is provided by Hallmann *et al.*, Chapter 17, this volume.

Planting certified nematode-free rootstock or planting material, when available, is an important management strategy to avoid any future problems following orchard or field establishment. However, compatibility between scion and rootstock and/or the potential impact of other biological pest problems and edaphic factors on resistant plants must also be considered. Finding resistance to root-knot nematodes in *Prunus* spp., other deciduous fruits and annual crops was recognized as an important consideration for these industries at the beginning of the 20th century. Since then, the development and release of resistant/tolerant germplasm is one of the primary reasons why many *Meloidogyne* spp. are not considered a major threat to the stone-fruit industry today. For example, in the south-eastern USA, Guardian® peach rootstock not only provides protection against *M. incognita* infestations but also imparts longer tree survival on sites conducive to peach tree short life syndrome. The source of root-knot nematode resistance comes from 'Nemaguard' and 'S-37' (Okie *et al.*, 1994), whereas the nature of resistance in Guardian® appears to be the inhibition of nematode development and failure to complete the life cycle (Plate 35) (Nyczepir *et al.*, 1999). In Europe, evaluation of known or new germplasm sources, which include new selections of peach-based hybrids, has been performed to characterize resistance in the parent material. Further, major corresponding genes for root-knot nematode resistance have been identified in several Myrobalan plum accessions, in Nemared peach and in the almond Alnem (Esmenjaud, 2004; Esmenjaud and Dirlewanger, 2007; Nyczepir and Esmenjaud, 2008). In Nemared peach and the GF.557 almond–peach hybrid, the *RMia* gene controls *M. incognita* and *M. arenaria*, whereas in the almond Alnem, the gene *RMja* controls *M. javanica*. In Myrobalan plum, resistance to *M. arenaria*, *M. incognita*, *M. javanica* and *M. floridensis* is conferred by the single major gene *Ma*.

Rootstocks currently available to growers in Europe include Cadaman, Garnem and Felinem peach, which carry *RMia* or alleles from this gene, and the soon-to-be-released Myrobalan plum accessions carrying the complete-spectrum and high-level resistance of the *Ma* gene. Moreover, interspecific hybrids pyramiding the *Ma* and the *RMia* genes are also in a process for registration (Esmenjaud, 2008, unpublished).

Diseases and root-knot nematodes cause major economic losses to vegetable crops worldwide. Current disease- and nematode-control methods rely heavily on fungicides and nematicides. Such pesticides are generally out of favour with consumers due to concerns about food safety and environmental quality, are expensive to growers and must be applied properly to be effective. In many cases, effective pesticides are not available. Therefore, a bio-based, environmentally compatible control strategy that utilizes resistant vegetable cultivars is especially needed. Resistant cultivars present a control strategy that alleviates consumer concerns about food safety and environmental quality, and eliminates or greatly reduces costs associated with pesticides and their application. A few sources of root-knot resistance in vegetables have been reported. Resistance has been found in bean and pepper cultivars and in tomato through the hybridization between a resistant line of *Lycopersicon peruvianum* and tomato by way of embryo culture. Generally, the genetic basis for resistance in vegetables is governed by one major gene. The *Mi-1* gene in tomato confers resistance to *M. arenaria*, *M. incognita* and *M. javanica* but not to *M. hapla* or *M. enterolobii* (= *M. mayaguensis*) (Brito *et al.*, 2004a; Sikora and Fernandez, 2005). *M. enterolobii* is of particular importance to Florida agriculture because of its ability to overcome root-knot-nematode-resistant genes in several crops besides tomato, including the *Mir1* gene in soybean, *N* gene in bell pepper, *Rk* gene in cowpea, and *Tabasco* gene in sweet pepper (Brito *et al.*, 2004a; J.A. Brito, Florida, 2008, personal communication). The *Mi-1* gene in tomato is not heat stable and is inactivated at soil temperatures above 28 °C; however, heat-stable resistance genes have been discovered in tomato (Karssen and Moens, 2006).

The search for improved nematode and disease resistance remains an ongoing process in most intensive cropping systems today. A more detailed discussion regarding the development of resistant and tolerant rootstocks and vegetables is found elsewhere in this book (see Starr and Mercer, Chapter 14, this volume).

18.3 Current Management Practices

The primary goal of any integrated pest management system is to increase crop yield (quality and quantity) by using a combination of the control tools previously discussed, thereby targeting a key pest(s), such as root-knot nematodes. The decision as to which control practices will be part of a management strategy is also governed by economics, technological and societal considerations, perennial versus annual crops, and *Meloidogyne* species present. Management strategies are dynamic, particularly where annual crops are involved, and may require frequent adjustment in response to changes that occur within the production system (e.g. changes in crop cultivars, economic factors, environmental conditions or nematicide availability). The essential components that must be available in order to develop an effective management strategy are: (i) reliable information on nematode densities; (ii) plant damage or crop loss functions for resident *Meloidogyne* species affecting the crop(s) that will be produced; (iii) knowledge of how the root-knot nematode populations are likely to respond with and without the use of control measures; and (iv) the economic consequences associated with using different controls (McSorley and Phillips, 1993).

18.3.1 Significance of diagnostic sampling and government regulation

The cornerstone upon which most successful management strategies rest is accessibility to services that provide growers with accurate nematode identification and quantification. In 1984, approximately one-half of the states in the USA provided such nematode diagnostic services (Barker and Imbriani, 1984). Commercial and governmental laboratories in the USA, Belgium, The Netherlands and other countries also provide nematode diagnostic services. European laboratories typically provide *Meloidogyne* species identification, thereby enabling producers and processors to customize site and crop selection in order to minimize the impact of highly pathogenic species such as *M. chitwoodi* and *M. hapla*. However, a common limitation within the USA when dealing with sedentary endoparasitic genera such as *Meloidogyne* and *Heterodera* is that many laboratories, both governmental and commercial, are not equipped to provide such identification for the J2 recovered from soil samples. Some laboratories within the USA will perform bioassays and species or host race identifications upon request, which by necessity delays results and increases the cost charged to process a sample. Host race identifications are currently not performed in Europe. Species identification remains critical in the development of root-knot nematode management strategies, particularly where more than one species is known to occur within a region or the history of root-knot nematodes in a field is unknown. The recent development of techniques that allow reliable identification of individual J2 of *Meloidogyne* using DNA analysis offers much promise for providing rapid, accurate and readily available (although probably more costly) species identification in the near future (Powers and Harris, 1993; Brito *et al.*, 2004b; Adam *et al.*, 2007).

Quantification of nematode population densities is determined by collecting soil and/or root samples from infested sites or crops and then determining if a nematode action level has been attained, therefore warranting the need for a pre-plant and/or post-plant management strategy. Sampling approaches and nematode distribution patterns, together with damage levels and population dynamics for *Meloidogyne* spp., are discussed in more detail elsewhere (see Greco and Di Vito, Chapter 11 and Duncan and Phillips, Chapter 12, respectively, this volume). Relevant to the management practices discussed in this chapter, root-knot nematode action levels (= action threshold) for *Prunus* in the south-eastern USA have been estimated for Georgia (Davis *et al.*, 1996; Horton *et al.*, 2007) and South Carolina (Dickerson *et al.*, 2000). It is important to note that nematode action levels may differ among states and countries, therefore making it essential that local sources be consulted for updated information. For example, in Georgia, the optimum time to take soil samples for root-knot nematode is in September–October. If one or more J2 per 100 cm^3 soil are detected in a pre-plant sample, or if that site has a history of root-knot problems, it is recommended that prior to orchard establishment a pre-plant fumigant be applied by 1 November and that a root-knot-nematode-resistant rootstock is used. In South Carolina, <50 J2 per 100 cm^3 soil (sandy to sandy loam soil) are considered unlikely to cause a problem, but continuous monitoring of populations is recommended. However, if populations are ≥50 J2 per 100 cm^3 soil, a nematode problem is considered likely to occur and a pre-plant management practice is recommended. It is also important to note that these nematode actions levels differ for clay loam to clay soils. For vegetable production in Belgium, it is highly recommended that vegetable growers obtain soil samples immediately after harvest, especially with crops having long field periods, in order to optimize the detection of *M. chitwoodi* (Wesemael and Moens, 2008a). Furthermore, many factors can affect the relative differences in action levels among regions. These factors include: (i) differences in laboratory extraction techniques and efficiencies; (ii) laboratory systems used to report nematode population densities (i.e. numbers per g, or per 100 or 500 cm^3 soil); (iii) variations in soil type and environmental conditions that may affect nematode population development; and

(iv) differing levels of resistance or tolerance among crops or crop varieties to be grown.

Once a determination has been made as to how the root-knot nematode population is likely to affect the crop, choices can be made as to which, if any, control measures will be applied. For annual crops, the process of selecting a control measure is likely to weigh the efficacy of the control in reducing the nematode population against both the financial cost of the control and its effects on future crop production. After implementing a control practice, additional diagnostic sampling may be necessary to re-evaluate the status of the *Meloidogyne* population at the end of the growing season. This is particularly true in long-season annual crops where nematicides have been selected, because early-season nematode suppression can result in greater root growth, which ultimately leads to pest resurgence and higher root-knot nematode populations at the end of the season (Barker and Imbriani, 1984). If crop loss or damage functions are not available, or insufficient information exists regarding the effects of control measures on *Meloidogyne* populations, then the net economic outcome from any management decision is largely unknown.

The impact of governmental actions, both direct and indirect, on the success of nematode management activities (including those pertaining to *Meloidogyne* of course) is often overlooked in intensive crop production systems. Direct actions (i.e. quarantine) may be associated with any governmental boundaries, but most often occur at the national level or among regional governmental units within nations (e.g. departments, provinces, states, territories), and are typically enacted to reduce the dissemination of pests or pathogens. Europe is a notable exception to this situation, where quarantine decisions apply to all European Union members. Direct governmental actions involve both the enactment of necessary regulations and, perhaps more importantly, the appropriation of public funds to support enforcement of these regulations. The diagnostic capabilities discussed previously contribute to the successes of such actions.

Quarantine strategies are considered a preventive and not a curative approach in stopping the introduction and/or increased dissemination of economically important nematodes into a country, local region or planting site. An excellent review of international legislation as it pertains to phytosanitary approaches against different plant-parasitic nematodes is well documented elsewhere (Hockland *et al.*, 2006), and therefore only a brief discussion on root-knot nematodes will follow. Species of *Meloidogyne* such as *M. incognita*, *M. hapla*, *M. javanica* and *M. arenaria* are generally not regulated, because these species of economic concern are widely distributed throughout the world. One root-knot nematode species that is internationally regulated and occurs on pest-prohibited lists in numerous countries is *M. chitwoodi*. Of lesser international concern, but still considered an economic risk, are *M. fallax and M. citri*, with *M. fallax* on the regulated list in Europe. In mainland USA, *M. floridensis* and *M. enterolobii* are considered important to vegetable production in Florida (Brito *et al.*, 2004b; Kokalis-Burelle and Nyczepir, 2004; J.A. Brito, Florida, 2008, personal communication). Both species are currently limited in distribution to Florida, and therefore necessary inspections and quarantine regulations are in place to prevent introduction of these pests into local counties and other states where they do not occur. Further, *M. floridensis*, formerly characterized as *M. incognita* race 3 (Handoo *et al.*, 2004), is of concern to the south-eastern USA peach industry because it has been shown to reproduce on root-knot-nematode-resistant rootstocks (e.g. Nemaguard and Guardian®) (Sherman *et al.*, 1991; Nyczepir *et al.*, 2006a). In contrast, some of the commonly known peach rootstocks were all rated as either non-hosts (highly resistant) or poor hosts (resistant) for *M. enterolobii* (Nyczepir *et al.*, 2008).

Indirect actions affecting nematode management are also very important and mainly involve the use of government revenue to fund research, including that in the academic and public sectors, that enhances all aspects of knowledge regarding the biology, impact and control of *Meloidogyne* species.

18.3.2 Implementation of management strategies

Most growers are satisfied when a recommendation of one or more control practices achieves effective management of one or more pathogens, resulting in increased crop quality and yield. Within intensive annual crop production systems, the strategies that are applied to manage *Meloidogyne* species vary greatly. Where economically supported, the simplest strategies involve the use of a single pre-season application of a fumigant nematicide in high-value, root-knot-nematode-susceptible crops (Thomas, 1994) (Figs 18.4 and 18.5; Plates 42, 43). The more complex strategies may use nematicides as complementary emergency tools, while emphasizing rotations with non-host crops and/or resistant crops, in combination with judicious weed control to prevent unwanted *Meloidogyne* reproduction, thereby reducing nematode populations and enabling future profitable production of a susceptible crop

Fig. 18.4. Enhanced growth and lint production in cotton following pre-season application of 56 l/ha 1,3-dicholoropropene for suppression of *Meloidogyne incognita* (left) compared with untreated cotton (right). (Photograph by S.H. Thomas.)

Fig. 18.5. Enhanced growth and stand density in pepper (*Capsicum annuum*) following pre-season application of 56 l/ha 1,3-dicholoropropene for suppression of *Meloidogyne incognita* (left) compared with untreated pepper (right). (Photograph by S.H. Thomas.)

(Roberts, 1993; Molendijk and Korthals, 2005). Such strategies require thorough knowledge of the biology and host range of the nematode community for successful reduction of root-knot nematode populations, thus enabling future profitable production of a susceptible crop (Molendijk and Korthals, 2005). When such knowledge is lacking, research has dealt with management of *Meloidogyne* populations in high-value crops by focusing on identifying single, highly effective controls, such as nematicides or host plant resistance. Along this line, most examples of root-knot nematode management strategies fall within the simple, single-control category (Roberts, 1993).

Other less simple 'single-control management strategies' include the use of 1-year rotations involving economically viable non-host crops or the choice of *Meloidogyne*-resistant crop varieties, either of which might also be included as components of more complex management strategies. Rotations are not as simple to implement as a nematicide application because rotation crops often require different farm equipment and production practices, and frequently result in less profit per unit of land, while use of resistant varieties (see Williamson and Roberts, Chapter 13 and Starr and Mercer, Chapter 14, this volume) may carry yield penalties when compared with susceptible varieties grown after nematicide application. Crop rotation and/or use of resistant cultivars are recommended management strategies for such nematode/crop combinations as *M. incognita* and *M. javanica* in soybean, tobacco and tomato, *M. arenaria* in groundnut, soybean and tobacco, *M. hapla* in groundnut and soybean, and *M. naasi* in wheat (Bridge and Starr, 2007). Where root-knot nematode resistance is lacking, susceptible scions from certain vegetable cultivars have been successfully grafted to nematode-resistant rootstocks to reduce injury from *Meloidogyne* spp., particularly under glasshouse conditions (Lee and Oda, 2003; Oka *et al.*, 2004). Duncan and Noling (1998) captured the essence involved with the development of management strategies when they stated 'the complexity of nematode management is generally inversely related to crop value.'

Pressing arguments in favour of the development of more integrated nematode management strategies, which at this point basically involve applying integrated pest management principles to the plant-parasitic nematode component of the crop pest ecosystem, are gaining interest throughout the world and have been thoroughly described in previous works (Roberts, 1993; Duncan and Noling, 1998; Greco and Esmenjaud, 2004). Of foremost concern is the realization that current intensive agricultural management strategies are not likely to be sustainable in the future. Also, public and governmental concerns about human health and environmental quality are likely to curtail further certain uses of the remaining nematicides, which are the basis of many management strategies – particularly where highly damaging parasites such as *Meloidogyne* spp. are concerned. Two major factors are responsible for the slow progress in implementing integrated management of root-knot nematodes: (i) the difficulty in developing such strategies for the predominant highly polyphagous species; and (ii) the reluctance of growers, who must focus on maximizing their return on investment, to accept longer-term and potentially less profitable alternative rotation crops or nematode-resistant cultivars (Thomason and Caswell, 1987; Roberts, 1993).

When multiple control practices are integrated into root-knot nematode management strategies, these controls can be applied simultaneously or sequentially over multiple crops or years (Roberts, 1993). Examples of the use of simultaneous control practices in annual crops include the use of nematode phenology data to determine the timing of nematicide applications for more effective management of *M. chitwoodi* in potato (David, 2007) and the use of compost incorporation and transplants instead of direct seeding to shorten the harvest interval and improve the tolerance of squash to *M. incognita* in organic vegetable production (McSorley and Gallaher, 1995). An example of sequential control practice use involves the successful management of *M. incognita* damage to watermelon in regions where intensive cultural practices and low spring soil temperatures combined to make fumigation with 1,3-D impractical (Westphal, 2007). In this system, successful watermelon production was achieved using a three-step approach, which involved: (i) replacing susceptible soybean varieties with *M. incognita*-resistant varieties during the previous growing season; (ii) replacing typical root-knot-susceptible rye or wheat winter cover crops needed to prevent soil erosion with an *M. incognita*-resistant oilseed radish cultivar (*Raphanus sativus* cv. Boss); and (iii) supplementing seedling transplant plugs with a mycorrhizal fungus amendment to enhance early-season root growth. In Valencia, Spain, the integrated use of biofumigation (i.e. sheep manure and mushroom residue) followed by a trap crop (Chinese vegetables) planted in short cycles proved successful in suppressing the population of *M. incognita* prior to planting tomato containing the *Mi* gene (Bello *et al.*, 2004). In perennial crops, grape root phenology has also been used in determining delivery, placement and timing of nematicide application (McKenry, 1984; Melakeberhan *et al.*, 1989). In North America, some pecan producers have reduced tree decline associated with *M. partityla* by modifying production practices to reduce the impact of added physiological stress to trees from root-knot nematode infection. Profitable yield has been maintained in infested orchards by reducing manageable sources of stress through adjustments in irrigation and nutrient availability, and by using pruning practices that reduce the stress resulting from excessively heavy nut production in alternate years (Wood *et al.*, 2004; Nyczepir *et al.*, 2006b; Nyczepir and Wood, 2008; S.H. Thomas, New Mexico, 2008, personal communication).

For the most part, development of complex integrated root-knot nematode management strategies is left up to the growers themselves. For many major crops, management information is increasingly presented for the entire spectrum of pests that might be encountered, as opposed to only root-knot nematodes (Anonymous, 1986, 1996; MacGuidwin, 1993). For example, in The Netherlands the PPO (Applied Plant Research, Wageningen University and Research Centre) nematode management system is available on the internet (www.digiaal.nl), and provides integrated management strategies at the farm level (Molendijk and Korthals, 2005). Where known, examples of successful and unsuccessful activities against *Meloidogyne* and other species are presented within the different categories of control practices previously presented (e.g. nematicides, rotation crops and resistant varieties). Growers are then left to develop individualized management plans using the different control practices, or combinations thereof, which best fit within the economic, environmental and production confines of their production system. Monitoring and assessment of nematode populations is emphasized as an essential component in the decision-making process. For example, with regard to management of *M. incognita* damage to cotton, the components of control practices that involve rotation crops, cover crops and resistant varieties differ among cotton-producing regions (Anonymous, 1996, 2003b; Koenning *et al.*, 2004).

A few examples of success stories involving intensive perennial cropping systems include the following. In coffee, there is a disease complex known as corky root, which involves *Fusarium oxysporum* and *M. arabicida*. Bertrand *et al.* (2000) showed that no root symptoms occurred in the presence of either pathogen alone, but when both organisms were combined simultaneously they produced corky-root symptoms. Furthermore, field studies have confirmed that use of *M. arabicida*-resistant rootstocks provides an effective management strategy against this fungus–nematode disease complex. A number of different control strategies have also proved effective in managing root-knot nematodes in *Prunus*. In the south-eastern USA and California, pre-plant soil fumigation and the use of root-knot-nematode-resistant/tolerant rootstocks are highly recommended for light, sandy soil where *Meloidogyne* is present, in order to achieve increased orchard productivity (McKenry *et al.*, 2006; Horton *et al.*, 2007). This combined nematode management approach has enabled growers to replant successfully second-, third- and even fourth-generation orchards.

18.4 Future Opportunities and Challenges

Dramatic changes in root-knot nematode management strategies have occurred since the middle of the 20th century. After three decades in which both the number and use of nematicides steadily increased, concerns about risks to the environment and human health abruptly overturned this trend, beginning in 1977 – a reversal that continues today (Rich *et al.*, 2004). At present, substantial gaps exist in our knowledge about how the soil ecosystem, weeds and other pathogens and pests interact with *Meloidogyne* in the post-methyl bromide era. These challenges to high-production agriculture provide opportunities to adapt recent technological and molecular biological advances, together with novel emerging chemical and biological agents for nematode management. These aspects will be addressed in the following sections.

18.4.1 Emerging control options

The success achieved with synthetic chemicals to control nematodes in years past has provided the impetus for industry and university scientists to continue the search for new alternative nematicidal compounds, albeit with a clearer understanding of the detrimental characteristics that must be avoided in order for potential new products to meet regulatory concerns (Kokalis-Burelle *et al.*, 2008). Additionally, since some synthetic nematicides are being removed from the agricultural market due to environmental and human health awareness issues (e.g. methyl bromide), a void in effective nematode management tools for the grower has resulted. Following is a discussion of emerging control options, which include: chemical control, cultural control, and technological advances with potential for use in managing *Meloidogyne* spp. in intensive cropping systems.

18.4.1.1 Chemical control

The limited number of nematicides that are currently available to producers has encouraged the evaluation or re-evaluation of numerous synthesized compounds and naturally occurring products as potential candidates for use in controlling plant-parasitic nematodes. Thorough discussions of the characteristics and future outlook for most of these materials can be found in reviews by Chitwood (2003) and Rich *et al.* (2004). Most of the synthesized candidate compounds lack a clear advantage in both nematicidal efficacy and ease of application over current and past synthetic nematicides, which could affect marketplace acceptance of these materials over existing established products for use as stand-alone root-knot nematode controls.

Two biopesticides have recently been registered for use in controlling a variety of plant pathogens in the USA, including plant-parasitic nematodes. The first is Harpin protein, which is produced naturally by the bacterial plant pathogen *Erwinia amylovora* (EPA, 2002). Harpin protein is believed to activate systemic acquired resistance (SAR), a natural defence mechanism in plants that affords some protection against certain fungal, bacterial, viral and nematode pathogens, including *M. incognita* in cotton. The second material is *Muscodor albus*, a naturally occurring fungus that upon rehydration produces volatile compounds that are reported to kill or inhibit certain other fungi, bacteria and nematodes in the soil (EPA, 2005).

Antagonistic crops and their naturally occurring plant products are two areas being investigated as alternatives to synthetic chemical control of nematodes and have been reviewed elsewhere (Pandey *et al.*, 2003). However, relevant to emerging control options in this chapter, antagonistic crops are plants that produce anthelmintic compounds and that sometimes give the benefit of green manure while suppressing nematode populations (Grainge and Ahmed, 1988; Ferraz and de Freitas, 2004). One important factor associated with the compounds isolated from antagonistic crops is that they are thought to be more acceptable to both public and governmental scrutiny, because: (i) they are derived from plants and not synthesized; (ii) they biodegrade rapidly; (iii) they are less toxic than most synthetic compounds; (iv) they may have multiple modes of action; and (v) they are derived from renewable sources (plants). The goal in developing a nematicide from a natural plant product is to produce the perfect nematicide, which will replace synthetic nematicides and provide similar or better efficacy. An example of a naturally occurring product that is suppressive against *Meloidogyne* spp. is a glycoside (asparagusic acid) that is isolated from the roots of *Asparagus officinalis*. Other examples of naturally occurring compounds and natural nematicides are published elsewhere (Ferraz and de Freitas, 2004; Haydock *et al.*, 2006). Last, it should be emphasized that, like emerging synthetic nematicides, the efficacy and acceptance of biopesticides and antagonistic crops as stand-alone controls for *Meloidogyne* are yet to be determined. However, emerging nematicides, biopesticides and antagonistic crops may provide valuable new components within future integrated management strategies for root-knot nematodes. Biopesticides in particular may play an important role within the confines of organic pest management systems.

18.4.1.2 Cultural control

In intensive production systems, future management strategies for *Meloidogyne* will undoubtedly continue to place greater emphasis on cultural control components than occurred in the last half of the 20th century, particularly where annual crops are involved. As the numbers and uses of chemical controls have declined, coupling cultural controls with chemical controls has offered an alternative method for achieving acceptable root-knot nematode suppression. For example, potato producers who were once reluctant to use sudangrass as a winter cover crop to control *M. chitwoodi*, choosing instead to rely solely on nematicides, are now utilizing this cultural practice on over 12,000 ha (R.E. Ingham, Oregon State University, 2008, personal communication). Other cultural control options, such as the use of a European oilseed radish (*R. sativus*) with resistance to *M. incognita* as a trap crop against this recalcitrant nematode, and the potential selection of brassicaceous amendments based on their glucosinolate profiles to enhance nematicidal efficacy (Zasada and Ferris, 2004), are also promising developments for inclusion in future management strategies.

In the south-eastern USA, pre-plant fumigation, in combination with a nematode-resistant rootstock, is recommended for increased peach tree longevity and maximum protection against root-knot nematode, as previously discussed. However, in recent years growers have been faced with several limitations to continued use of synthetic chemical control as an option to manage plant-parasitic nematodes, including: (i) the high costs associated with pre-plant fumigation; (ii) the worldwide phase-out of methyl bromide due to environmental and human health concerns; and (iii) insufficient time to get the land fumigated at the recommended time of year due to conflicts with managing other crops. The usefulness of a tall fescue crop rotation system as a non-chemical management strategy to reduce the population density of *M. incognita* prior to planting peach may offer a potential alternative. Several tall fescue grass cultivars were found to be poor hosts of *M. incognita*, and the presence of a fungal endophyte in these grasses did not appear to affect nematode reproduction in glasshouse studies (Nyczepir, 2006). Further, *M. incognita* was generally not detected on a subsequent bioassay of the soil previously planted to the tall fescue grass cultivars, possibly indicating that a naturally occurring nematode-suppressive plant product may also be present.

Other important future cultural components of management strategies include increased use of resistant and tolerant crop varieties (see Starr and Mercer, Chapter 14, this volume) and biological control agents (see Hallmann *et al.*, Chapter 17, this volume). It is worth mentioning in this chapter that entomopathogenic nematodes in the families Steinernematidae and Heterorhabditidae are biological control agents against some plant-parasitic nematodes (Stock, 2005). Briefly, these nematodes kill insects with the aid of bacteria carried in the nematode's alimentary canal (Poinar, 1990). Entomopathogenic nematodes have also been reported to suppress populations of *Meloidogyne* spp. (Lewis and Grewal, 2005) and were tested against a recognized root-knot nematode pest of pecan. The pecan root-knot nematode, *M. partityla*, a species previously only reported in South Africa, has been reported in pecan orchards in the USA over the past 12 years, and is associated with tree decline in the orchards or nurseries where it was found (Starr *et al.*, 1996; Thomas *et al.*, 2001; Nyczepir *et al.*, 2002; Crow *et al.*, 2005). However, repeated applications of entomopathogenic nematodes in pecan proved not to be a viable management strategy for suppressing *M. partityla* populations under glasshouse conditions (Shapiro-Ilan *et al.*, 2006).

Genetic engineering (see Atkinson *et al.*, Chapter 15, this volume) will almost certainly impact future management options, particularly considering the recent report that an essential gene required for host parasitism by *M. incognita*, *M. javanica*, *M. arenaria* and *M. hapla* was successfully silenced in transgenic *Arabidopsis*, resulting in disruption of the parasitic process for all four of these major root-knot nematode species (Huang *et al.*, 2006). Additionally, the *Gastrodia* antifungal protein (GAFP) enables the Asiatic orchid (*Gastrodia elata*) to live parasitically off the oak root-rot fungal pathogen *Armillaria mellea*. The GAFP has been shown to confer resistance to *M. incognita* in transgenic tobacco (Cox *et al.*, 2006) and transgenic plum (*Prunus domestica* var. Stanley) rootstock (Nagel *et al.*, 2008). Additional research is needed to determine if root-knot nematode and disease resistance in the transgenic plum lines holds up under orchard conditions.

Another important consideration likely to enhance the future use of cultural control options is increasing consumer demand for organic crops: agricultural products produced without the use of pesticides and, in most cases, synthetic fertilizers. In Germany, it was shown that organic vegetable farming resulted in increased root-knot nematode soil populations, which caused serious crop loss after 5–10 years of farming (Hallmann *et al.*, 2007). The development of suitable practices that prevent such losses is critical to the future success of organic crop production. Current production systems that must adhere to these stringent organic requirements are partly responsible for the growing understanding of the importance of nematodes as components of the soil food web (Phillips *et al.*, 2003; Ferris *et al.*, 2004; Ferris and Bongers, 2006). *Meloidogyne* is not of primary concern in these considerations, which focus mainly on the roles of nematodes that impact soil nutrient-cycling processes through consumption of bacteria and fungi that are involved in primary decomposition of organic matter. However, the reliance on incorporation of organic matter as the main source of plant nutrients will inevitably affect root-knot and other plant-parasitic nematodes through changes in populations of biological control agents (see Hallmann *et al.*, Chapter 17, this volume) and other ecological factors, as discussed previously in section 18.2.2.2. The impacts from increased use of other cultural control options, such as *Meloidogyne*-suppressive cover crops or rotation with resistant crop varieties (Potter and McKeown, 2003), will also differ from the impacts currently encountered in management systems where pesticides and synthetic fertilizers are used.

18.4.1.3 Technological advances

Although vastly different from each other, two emerging technologies are inseparably linked when it comes to their impact on the development of future management strategies for *Meloidogyne*. The first involves recent advances in the rapid, precise identification of individual J2 recovered from soil (see section 18.3.1 of this chapter and Blok and Powers, Chapter 4, this volume. The second involves the application of global positioning system (GPS) technology, to record location of nematode infestations, and global information system (GIS) technology, which can be used to apply nematicides at different rates based on these predetermined nematode populations. Cotton has provided the model system for developing most GIS-based, variable-rate nematicide application, with suppression of *M. incognita* being a major target (Wheeler *et al.*, 1999; Baird *et al.*, 2001; Wrather *et al.*, 2002; Starr *et al.*, 2007). Yield comparisons between conventional fixed-rate application of aldicarb and variable-rate applications were inconsistent among earlier studies, sometimes producing equivalent yields, which exceeded those from untreated control areas, while reducing nematicide usage by an average of 54% over two seasons (Wrather *et al.*, 2002). In one study, yield results from site-specific application of 1,3-D were more favourable than those obtained using similar application methods with aldicarb (Baird *et al.*, 2001).

Despite the potential economic and environmental benefits from variable-rate application of nematicides, the initial impediment to further investigation or widespread use of the technology was the prohibitive cost of collecting and analysing soil samples to characterize root-knot nematode populations (Wrather *et al.*, 2002). Recently developed mobile soil-mapping technology, which measures soil electrical conductivity (EC) as an approximation of soil texture, can be used to rapidly generate precise field maps depicting EC levels at a cost comparable to that required to collect and process two to three traditional nematode soil samples (Mueller *et al.*, 2008; Overstreet *et al.*, 2008). Electrical conductivity has been determined to be the best predictor of populations of *M. incognita* and nematode-induced yield reductions in cotton, and in turn has been used to create nematode management zones (Davis *et al.*, 2008). When such maps were used as the basis for variable-rate application of aldicarb or 1,3-D, nematicide usage was reduced by 34–78%, while cotton yields were equal to or 5% greater than those obtained with single-rate nematicide application (Monfort *et al.*, 2008; Mueller *et al.*, 2008). Data from harvester-mounted yield monitors have also proved useful for generating within-field maps of cotton and grain yields to predict the likely location of damaging populations of *M. incognita*. This information served as the basis for recommending site-specific nematicide application, which resulted in a 58% reduction in 1,3-D use the following year (Overstreet *et al.*,

2008; R. Norton, Arizona, 2008, personal communication). Ultimately the combination of these easy and cost-effective assessment tools may allow annual refinement of estimates for *Meloidogyne* spp. within fields, permitting targeted delivery that achieves the desired pest suppression while substantially reducing the amount of nematicide applied.

It is possible that maps generated from both GPS/GIS technology and yield-monitoring technology, which are available for certain mechanically harvested crops, could be overlaid with maps of soil properties to estimate root-knot nematode populations. However, soil sampling would still be required if the nematode history of the field was unknown, or where other yield-limiting pests or pathogens are likely to be present. Ultimately, easy and cost-effective assessment of the location and density of *Meloidogyne* species within fields must become available before GPS/GIS-directed applications can be implemented effectively.

18.4.2 Emerging management options

As root-knot nematode management strategies shift from relying on highly efficacious synthetic nematicides to more integrated approaches, the number of considerations involved in choosing such strategies increases, along with the need for more thorough knowledge of how different biological components within the production system interact. Without such knowledge, successfully manipulating integrated management systems to suppress root-knot nematode populations will be very difficult.

18.4.2.1 Natural resource availability

Increasing global demand for energy will probably cause many 21st century intensive agricultural producers to alter production practices that directly impact root-knot nematode management. Resulting higher fuel costs could trigger a reduction in mechanical cultivation needed to control weeds that are important alternative hosts for *Meloidogyne* spp. (Bélair and Parent, 1996; Thomas *et al.*, 2005). This is of particular concern with nematode-suppressive rotation crops, where the effects of weed competition on crop growth are of secondary concern, or when fields are fallowed (McSorley *et al.*, 1994). Tillage levels that provide optimal crop residue may also be reduced. While tillage has often been demonstrated to have little effect on densities of *M. incognita* in crop rotation systems (Gallaher *et al.*, 1988; McSorley and Gallaher, 1993), late-season destruction of infected roots is important in maximizing the efficacy of nematicides intended for use the following year. Further, repeated cultivation can reduce carry-over populations by exposing root-knot nematodes to solar radiation and desiccation (Barker and Imbriani, 1984; Thomason and Caswell, 1987).

Global energy demands have resulted in steady increases in the cost of certain nitrogen fertilizers, which may cause some producers to reduce nitrogen inputs to crops. Fewer nutrients could reduce the vigour and ability of plants to partially compensate for root-knot nematode infection through production of additional roots. Along similar lines, the increasing need for water to support human population growth will probably impact the availability of traditional sources of water used for agricultural purposes, requiring more judicious use of that resource. In semi-arid regions, where crop production is dependent upon irrigation, this will probably involve greater use of drip or trickle irrigation to replace flood or furrow irrigation. Drip irrigation offers the advantage of enhancing the efficacy of some control practices by targeting application of nematicides or potential biological control agents directly to the root system of the crop (Roberts, 1993), which is confined within the water-delivery zone. However, where irrigation water is limited, drip irrigation can also be adapted to deliver 'grey water' (partially processed municipal or agricultural waste water). Costs of fuel and nitrogen fertilizer may encourage increased uses of reduced-tillage production systems and alternative sources of plant nutrition, such as greater use of nitrogen-fixing rotation crops, animal waste or other sources of organic matter. Any of these potential changes in crop production practices would most certainly affect root-knot nematode population dynamics and related management strategies.

18.4.2.2 Knowledge gaps

Perhaps the greatest advantage to controlling *Meloidogyne* with a highly effective fumigant nematicide is that all the information required for successful use of the product appears on the accompanying label. The only crop protection products requiring fewer decisions are general biocides (e.g. methyl bromide), which control not only nematodes but also nearly the entire spectrum of soil pests (e.g. fungal and bacterial pathogens, weeds and soil insects). By contrast, multiple-approach, integrated management systems require a high degree of knowledge regarding interactions among *Meloidogyne* species, other pests and the crop. At present, much of this knowledge is incomplete or lacking entirely, especially by growers. Interactions between *Meloidogyne* and other plant-parasitic nematodes, fungal and bacterial pathogens are discussed by Manzanilla-López and Starr, Chapter 10, this volume.

The role of weeds as alternative hosts that maintain and enhance *Meloidogyne* populations is well recognized and documented (Townshend and Davidson, 1962; O'Bannon *et al.*, 1982; Bendixen, 1988; www.ipm.ucdavis.edu/NEMABASE/index.html). Recent work has demonstrated that interactions involving certain weeds extend beyond the simple role of alternative host, affecting other aspects of root-knot nematode development, pest status and related integrated management decisions (Fig. 18.6; Plate 44). An example where interactions can influence root-knot nematode management is evident in the beneficial pest relationship involving yellow nutsedge (*Cyperus esculentus*) and purple nutsedge (*C. rotundus*) and *M. incognita*. These perennial weeds are internationally important pests in crop production, and are both good hosts for *M. incognita* (Bird and Hogger, 1973; Schroeder *et al.*, 1993). It has also been reported that neither weed was injured by nematode populations up to 100-fold greater than the action threshold for chilli pepper (*Capsicum annuum*) (Schroeder *et al.*, 1999), a vegetable crop frequently grown in regions where the pest complex occurs. Subsequent studies revealed that the number and weight of tubers, the primary means by which these nutsedge species propagate, were greater when either weed was infected by *M. incognita* (Schroeder *et al.*, 1994, 1999, 2004). In addition to utilizing the nutsedges as alternative hosts, *M. incognita* invades and overwinters in the tubers, the emergence of which are enhanced by pre-plant irrigation the following season (Fuchs, 2004; Nunez, 2007). Plant competition from nutsedges had no effect on the host susceptibility of chilli pepper for reproduction by *M. incognita*, but pepper competition enhanced the nematode reproductive efficiency on both yellow and purple nutsedge (Fuchs, 2004). Previous crop or nutsedge hosts had a pronounced effect on the subsequent ability of J2 to successfully infect the same or different host species (Thomas *et al.*, 1997; Fuchs, 2004).

Fig. 18.6. Pepper (*Capsicum annuum*) crop, demonstrating successful suppression of the *Meloidogyne incognita*/yellow nutsedge (*Cyperus esculentus*)/purple nutsedge (*Cyperus rotundus*) pest complex following a 3-year rotation with *M. incognita*-resistant lucerne (*Medicago sativa*) (left panel) compared with pepper following a standard 3-year rotation with cotton and pre-season treatment with 56 l/ha 1,3-dichloropropene (right panel). Pepper following lucerne received no pre-season nematicide. (Photograph by S.H. Thomas.)

Currently, information is limited regarding how complex pest interactions may affect established and emerging chemical and transgenic control practices for a variety of pests, including *Meloidogyne*. For example, the use of established fumigants like 1,3-D and metam sodium has been proposed as alternatives to the use of methyl bromide for root-knot nematode control (Table 18.1), but where the previously discussed yellow nutsedge/purple nutsedge/*M. incognita* pest complex occurred 1,3-D had no effect on nematodes within tubers or on the level of resulting crop infection (McSorley, 2004; Thomas *et al.*, 2004). Root-knot nematode infection also reduced the efficacy of herbicides applied to control both of these nutsedges (Schroeder *et al.*, 1994; Norsworthy *et al.*, 2005). Changes in weed composition and the development of herbicide resistance in some weeds have recently been observed in conjunction with the widespread use of the herbicide glyphosate in transgenic crops containing genes for glyphosate tolerance (Patzoldt *et al.*, 2002; Scott and Van Gessel, 2007). Such changes are likely to affect the future dynamics of weeds as alternative hosts for *Meloidogyne*. Recent increases in the use of transgenic *Zea mays* expressing insect-toxic proteins, and shifts away from the use of soil-applied organophosphate and carbamate insecticides in favour of more targeted chemistries have been proposed as two possible explanations for the increases in plant-parasitic nematode injury to maize in the midwestern USA (T.L. Niblack, University of Illinois, 2008, personal communication). Although *Meloidogyne* is not among the nematodes listed, increasing use of transgenic insect resistance and changing insecticide profiles in many crops are likely to impact root-knot nematode populations in the future.

18.4.2.3 Alternatives to methyl bromide

The qualities of the ideal nematicide replacement for methyl bromide include: (i) inexpensive to develop; (ii) high nematode efficacy at low rates; (iii) low water solubility; (iv) low chronic and acute toxicity to humans; (v) non-toxic to non-target organisms; (vi) produced from a living organism or naturally occurring compound; (vii) compatible with other pesticides/biological control agents; and (viii) non-systemic into edible plant parts (Haydock *et al.*, 2006). No synthetic nematicide, past or present, has met all eight criteria. To date, one of the problems encountered in finding a suitable replacement for methyl bromide is the expense associated with the product or combination of products needed to obtain a comparable level of broad-spectrum activity to that typically achieved with methyl bromide. A number of

Table 18.1. Potential chemical alternatives to methyl bromide for soil disinfestations (after Duniway, 2002).

	Requiring further development	
Currently available nematicides[a]	MBTOC[b]	Additions[a,c]
Acrolein (2-propenal)	Propargyl bromide	Other halogenated hydrocarbons
Chloropicrin	Ozone	Propylene oxide
1,3-Dichloropropene (1,3-D)[d]	Formaldehyde	Sulphur dioxide
1,3-D[d] + chloropicrin	Carbon disulfide	Furfural
1,3-D[d] + metam sodium	Anhydrous ammonia	Dimethyl disulfide
1,3-D[d] + chloropicrin + metam sodium	Inorganic azides	Peroxyacetic acid
	Natural compounds	Others to be developed
Methyl iodide		
Methyl isothiocyanate liberators:		
Metam sodium		
Dazomet		
Sodium tetrathiocarbonate		
Muscodor albus strain QST 20799		

[a]Revised according to Kegley *et al.* (2009a,b) and EPA (2008a). Current regional sources of information should be consulted for updated recommendations; [b]Alternatives considered by the 1998 report of the Methyl Bromide Technical Options Committee (MBTOC), United Nations Environmental Programme (Anonymous, 1998); [c]Alternatives added by J.M. Duniway; [d] Withdrawn authorization in some countries.

methyl bromide alternatives are being evaluated, but, as of 2008, only methyl iodide has emerged as a single replacement for methyl bromide (Hutchinson *et al.*, 1999; EPA, 2008b). Despite its registration, cost and availability may limit the acceptance of methyl iodide in large-scale replacement of methyl bromide for soil fumigation. In 2000, the cost of methyl iodide was four times the cost of methyl bromide (Hueth *et al.*, 2000). Of further concern, total world market demand for iodine at that time met only 50% of the agricultural demand for bromine in the form of methyl bromide. Other possible alternatives to methyl bromide are listed in Table 18.1.

18.5 Conclusions and Future Directions

The need for new and better nematicides to supplement the limited arsenal currently available to growers who must manage root-knot nematode problems in intensive cropping systems poses a threat to future production. In recent years, progress has been made in developing additional control practices and management strategies to reduce the impact of root-knot nematode problems. Alternative non-chemical research areas under investigation for *Meloidogyne* suppression include the search for biological control agent(s), biofumigants, rootstock resistance, effective means of soil solarization, soil amendments, cultural practices (i.e. early sowing in spring to avoid nematode attacks and combining crop rotations/cover crops with antagonistic plants) and naturally occurring products that could be used in an integrated root-knot nematode management system. The important phrase here is integrated pest management system (IPM). The term IPM is not new but it appears evident that a combined nematode management system is the best solution and approach for root-knot nematode suppression and needs to be developed and/or improved upon for the benefit of growers in the near future. Promotion of such nematode management systems through basic and applied research and cooperation between extension and regulatory agencies to educate growers will provide the basis for controls and/or strategies that are more effective, economical and less hazardous to man and the environment.

Root-knot nematode resistance in perennial rootstocks and field crops is becoming more available and, when combined with a cultural practice(s) (e.g. rotation) or a chemical nematicide, has provided effective root-knot nematode management. Also, as our understanding increases on how plant resistance mechanisms work at the molecular level and is further combined with the release of transgenic resistant plants to suppress nematode penetration and reproduction, improved plant quality and yield are likely to occur. To date the development of effective biological control agents to manage root-knot nematode infestations in intensive cropping systems has generally been unsuccessful. Perhaps one explanation for this lack of success in effective nematode control may be due to the expectation that a single biological agent will be as efficacious as a nematicide against the targeted pest. However, nematode IPM programmes that seek to combine a biological agent(s) with other control practices may yield more promising results. Better knowledge of the biology of and interactions among agroecosystem components, combined with time, patience and good research will determine the progress made in developing root-knot nematode management alternatives required for the future.

18.6 References

Adam, M.A.M., Phillips, M.S. and Blok, V.C. (2007) Molecular diagnostic key for identification of single juveniles of seven common and economically important species of root-knot nematode (*Meloidogyne* spp.). *Plant Pathology* 56, 190–197.

Almeida, V.F. and Campos, V.P. (1991) Reprodutividade de *Meloidogyne exigua* em plantas antagonistas e em culturas de interesse economico. *Nematologia Brasileira* 15, 24–29.

Anonymous (1986) *Integrated Pest Management for Potatoes in the Western United States.* Publication 3316. Western Regional Integrated Pest Management Project, University of California Division of Agricultural and Natural Resources, Oakland, California.

Anonymous (1996) *Integrated Pest Management for Cotton in the Western Regions of the United States*, 2nd edn. Publication 3305. Western Regional Integrated Pest Management Project, University of California Division of Agricultural and Natural Resources, Oakland, California.

Anonymous (1998) *Assessment Alternatives to Methyl Bromide*. Methyl Bromide Technical Options Committee, United Nations Environmental Programme Ozone Secretariat, Nairobi, Kenya.

Anonymous (2003a) Propanil and fenamiphos; use deletion and product cancellation order. *Federal Register* 68, 68901–68904.

Anonymous (2003b) *Root-knot Nematode* (Meloidogyne incognita). Fact Sheet 46293A. National Cotton Council of America, Memphis, Tennessee.

Baird, R.E., Rich, J.R. and Waters, D. (2001) Evaluation of variable rate nematicide applications using precision farming methods to manage *Meloidogyne incognita* on cotton. *Nematologia Mediterranea* 25, 247–254.

Barbercheck, M.E. and von Broembsen, S.L. (1986) Effects of soil solarization on plant parasitic nematodes and *Phytophthora cinnamomi* in South Africa. *Plant Disease* 70, 945–950.

Barker, K.R. and Imbriani, J.L. (1984) Nematode advisory programs – status and prospects. *Plant Disease* 68, 735–741.

Batchelor, T.A. (2002) International and European community controls on methyl bromide and the status of methyl bromide use and alternatives in the European Community. In: Batchelor, T.A. and Bolivar, J.M. (eds) *Proceedings of International Conference on Alternatives to Methyl Bromide 'The Remaining Challenges'*. European Commission, Brussels, Belgium, pp. 28–32.

Becker, J.O., Morton, H.V. and Hofer, D. (2003) Abamectin seed coating: a new nematicide plant protection tool. *Journal of Nematology* 35, 324.

Bélair, G. and Parent, L.E. (1996) Using crop rotation to control *Meloidogyne hapla* Chitwood and improve marketable carrot yield. *HortScience* 31, 106–108.

Bello, A., Lopez-Perez, J.A., Garcia-Alvarez, A., Sanz, R. and Lacasa, A. (2004) Biofumigation and nematode control in Mediterranean region. In: Cook R. and Hunt D.J. (eds) *Nematology Monographs and Perspectives 2*. IV International Congress of Nematology, Tenerife, Spain, 8–13 June 2002, Brill, Leiden, The Netherlands, pp. 133–149.

Bendixen, L.E. (1988) *A comparative summary of the weed hosts of* Heterodera, Meloidogyne, *and* Pratylenchus *nematodes*. Special Circular 118, Ohio State University Ohio Agricultural Research and Development Center, Wooster, Ohio.

Bernhard, R., Bouquet, A. and Scotto La Massèse, C. (1985) Diversité des problèmes nématologiques en vergers, solutions chimiques et génétiques. Création de variétés résistantes aux nématodes des cultures: intérêt, possibilités et limites. *Compte-Rendus de l'Académie d'Agriculture de France* 71, 705–718.

Bertrand, B., Nuñez, C. and Sarah, J.-L. (2000) Disease complex in coffee involving *Meloidogyne arabicida* and *Fusarium oxysporum*. *Plant Pathology* 49, 383–388.

Bertrand, P.F. (1985) *Peach nematodes and their control in Georgia*. Cooperative Extension Service, University of Georgia, Athens, Georgia.

Bertrand, P.F. and Evert, D.R. (1984) A non-damaging association of root-knot nematode with peach. *Phytopathology* 74, 7.

Bertrand, P.F. and Nyczepir, A.P. (1989) Nematodes. In: Meyers, S.C. (ed.) *Peach Production Handbook*. University of Georgia Cooperative Extension Service, Athens, Georgia, pp. 146–151.

Bessey, E.A. (1911) *Root-knot and its Control*. Bulletin 217. USDA Bureau of Plant Industry, Washington, DC.

Bird, G.W. and Hogger, C.H. (1973) Nutsedges as hosts of plant-parasitic nematodes in Georgia cotton fields. *Plant Disease Reporter* 57, 402.

Boydston, R.A., Mojtahedi, H., Brown, C.R., Anderson, T. and Riga, E. (2007) Hairy nightshade undermines resistance of potato breeding lines to Columbia root-knot nematode. *American Journal of Potato Research* 84, 245–251.

Bridge, J. (1996) Nematode management in sustainable and subsistence agriculture. *Annual Review of Phytopathology* 34, 201–225.

Bridge, J. and Starr, J.L. (2007) *Plant Nematodes of Agricultural Importance*. Academic Press, Burlington, Massachusetts.

Brito, J., Stanley, J., Cetintas, R., Di Vito, M., Thies, J. and Dickson, D. (2004a) *Meloidogyne mayaguensis* reproduction on resistant tomato and pepper. *Proceedings of the Annual International Research*

Conference on Methyl Bromide Alternatives and Emissions Reductions. Methyl Bromide Alternatives Outreach Press, Orlando, Florida, pp. 77/1–77/5.

Brito, J., Powers, T.O., Mullin, P.G., Inserra, R.N. and Dickson, D.W. (2004b) Morphological and molecular characterization of *Meloidogyne mayaguensis* isolates from Florida. *Journal of Nematology* 36, 232–240.

California Department of Pesticide Regulation (2007) Summary of pesticide use report data 2006, http://www.cdpr.ca.gov/docs/pur/pur06rep/comrpt06 (accessed 10 December 2007).

Campos, V.P. and Villain, L. (2005) Nematode parasites of coffee and cocoa. In: Luc, M., Sikora, R.A. and Bridge, J. (eds) *Plant Parasitic Nematodes in Subtropical and Tropical Agriculture*, 2nd edn, CAB International, Wallingford, UK, pp. 529–579.

Carneiro, R.G. and Carneiro, R.M.D.G. (1982) Selecao preliminar de plantas para rotacao de culturas em areas infestadas por *Meloidogyne incognita* nos anos de 1979 e 1980. *Nematologia Brasileira* 6, 141–148.

Chitwood, D.J. (2003) Nematicides. In: Plimmer, J.R. (ed.) *Encyclopedia of Agrochemicals*, Vol. 3. John Wiley and Sons, New York, pp. 1104–1115.

Clean Air Act (1990) Title VI. *Stratospheric Ozone Protection Publication L*. Section 6001. US Congress, Washington, DC, pp. 101–549.

Cox, K.D., Layne, D.R., Scorza, R. and Schnabel, G. (2006) *Gastrodia* anti-fungal protein from the orchid *Gastrodia elata* confers disease resistance to root pathogens in transgenic tobacco. *Planta* 224, 1373–1383.

Crow, W.T., Levin, R., Halsey, L.A. and Rich, J.R. (2005) First report of *Meloidogyne partityla* on pecan in Florida. *Plant Disease* 89, 1128.

David, N.L. (2007) Biology and management of *Meloidogyne chitwoodi* using oxamyl on potato in the western United States. PhD thesis. Oregon State University, Corvallis, Oregon.

Davis, R.F. and Timper, P. (2000) Resistance in selected corn hybrids to *Meloidogyne arenaria* and *M. incognita*. *Journal of Nematology* 32, 633–640.

Davis, R.F., Johnson, A.W. and Wauchope, R.D. (1993) Accelerated degradation of fenamiphos and its metabolites in soil previously treated with fenamiphos. *Journal of Nematology* 25, 679–685.

Davis, R.F., Bertrand, P., Gay, J.D., Baird, R.E., Padgett, G.B., Brown, E.A., Hendrix, F.F. and Balsdon, J.A. (1996) *Guide for interpreting nematode assay results*. Publication Circular 834, The Cooperative Extension Service, The University of Georgia, Athens, Georgia.

Davis, R.F., Ortiz, B.V., Perry, C., Sullivan, D., Kemerait, B., Vellidis, G. and Rucker, K. (2008) Considering field physical characteristics in assessing risk and delineating nematode management zones. *Proceedings of the 5th International Congress of Nematology*, Brisbane, Australia, p. 136.

De Waele, D. and Davide, R.G. (1998) The root-knot nematodes of banana, http://musalit.inibap.org/pdf/IN020258_en.pdf (accessed 11 February 2008).

Dickerson, O.J., Blake, J.H. and Lewis, S.A. (2000) *Nematode guidelines for South Carolina*. Extension Circular EC 703, Clemson University Cooperative Extension Service, Clemson, South Carolina.

Duncan, L.W. and Noling, J.W. (1998) Agricultural sustainability and nematode integrated pest management. In: Barker, K.R., Pederson, G.A. and Windham, G.L. (eds) *Plant and Nematode Interactions*. American Society of Agronomy, Crop Science Society of America, Soil Science Society of America, Madison, Wisconsin, pp. 251–287.

Duniway, J.M. (2002) Status of chemical alternatives to methyl bromide for preplant fumigation of soil. *Phytopathology* 92, 1337–1343.

Dunn, R.A. and Noling, J.W. (1997) *Florida Nematode Management Guide*. Publication SP-54, Florida Cooperative Extension Service, University of Florida, Gainesville, Florida.

EPA (US Environmental Protection Agency) (2002) Harpin protein (006477) fact sheet. www.epa.gov/opp00001/biopesticides/ingredients/factsheets/factsheet_006477.htm (accessed 17 March 2008).

EPA (US Environmental Protection Agency) (2005) *Muscodor albus* QST 20799 (006503) fact sheet. www.epa.gov/pesticides/biopesticides/ingredients/factsheets/factsheet_006503.htm (accessed 17 March 2008).

EPA (US Environmental Protection Agency) (2007) 1,2-Dibromo-3-Chloropropane (DBCP), www.epa.gov/ttn/atw/hlthef/dibromo-.html (accessed 24 January 2008).

EPA (US Environmental Protection Agency) (2008a) Ozone layer depletion – regulatory programs, methyl bromide alternatives. http://www.epa.gov/spdpublc/mbr/alts.html. (accessed 29 January 2009).

EPA (US Environmental Protection Agency) (2008b) Time-limited registration of iodomethane (methyl iodide). Pesticides: Topical & Chemical Fact Sheets, US Environmental Protection Agency, http://www.epa.gov/pesticides/factsheets/iodomethane_fs.htm (accessed 25 January 2008).

Esmenjaud, D. (2004) Breeding for durable resistance to root-knot nematodes in perennials: a European initiative for *Prunus* rootstocks. In: Cook, R. and Hunt, D.J. (eds) *Nematology Monographs and Perspectives 2.* IV International Congress of Nematology, Tenerife, Spain, 8–13 June 2002, Brill Press, Leiden, The Netherlands, pp. 279–287.

Esmenjaud, D. and Dirlewanger, E. (2007) The plum. In: Kole, C. (ed.) *Genome Mapping and Molecular Breeding in Plants. Vol. 4: Fruits and Nuts.* Springer Verlag, Berlin, Heidelberg, Germany, pp. 119–136.

Faske, T.R. and Starr, J.L. (2006) Sensitivity of *Meloidogyne incognita* and *Rotylenchulus reniformis* to abamectin. *Journal of Nematology* 38, 240–244.

Fernandez, C., Rodriguez-Kabana, R., Warrior, P. and Kloepper, J.W. (2001) Induced soil suppressiveness to a root-knot nematode species by a nematicide. *Biological Control* 22, 103–114.

Ferraz, S. and de Freitas, L.G. (2004) Use of antagonistic plants and natural products. In: Chen, Z.X., Chen, S.Y. and Dickson, D.W. (eds) *Nematology Advances and Perspectives*, Vol. 2. CAB International, Wallingford, UK, pp. 931–977.

Ferris, H. and Bongers, T. (2006) Nematode indicators of organic enrichment. *Journal of Nematology* 38, 3–12.

Ferris, H., Venette, R.C. and Scow, K.M. (2004) Soil management to enhance bacterivore and fungivore nematode populations and their nitrogen mineralization function. *Applied Soil Ecology* 25, 19–35.

Freitas, L.G., Mitchell, D.J. and Dickson, D.W. (1997) Temperature effects on the attachment of *Pasteuria penetrans* endospores to *Meloidogyne arenaria* race 1. *Journal of Nematology* 29, 547–555.

Fuchs, J.T. (2004) Characterization of the relationship of *Meloidogyne incognita* in association with chile, yellow nutsedge, and purple nutsedge. MSc thesis. New Mexico State University, Las Cruces, New Mexico.

Gallaher, R.N., Dickson, D.W., Corella, J.F. and Hewlett, T.E. (1988) Tillage and multiple cropping systems and population dynamics of phytoparasitic nematodes. *Annals of Applied Nematology* 2, 90–94.

Gaur, H.S. and Perry, R.N. (1991) The use of soil solarisation for control of plant-parasitic nematodes. *Nematological Abstracts* 60, 153–167.

Grainge, M. and Ahmed, S. (1988) *Handbook of Plants with Pest-control Properties.* John Wiley and Sons, New York.

Greco, N. and Esmenjaud, D. (2004) Management strategies for nematode control in Europe. In: Cook, R. and Hunt, D.J. (eds) *Nematology Monographs and Perspectives 2,*. IV International Congress of Nematology, Tenerife, Spain, 8–13 June 2002, Brill, Leiden, The Netherlands, pp. 33–43.

Halbrendt, J.M. and LaMondia, J.A. (2004) Crop rotation and other cultural practices. In: Chen, Z.X., Chen, S.Y. and Dickson, D.W. (eds) *Nematology Advances and Perspectives*, Vol. 2. CAB International, Wallingford, UK, pp. 909–930.

Hallmann, J., Frankenberg, A., Paffrath, A. and Schmidt, H. (2007) Occurrence and importance of plant-parasitic nematodes in organic farming in Germany. *Nematology* 9, 869–879.

Handoo, Z.A., Nyczepir, A.P., Esmenjaud, D., Van der Beek, J.G., Sereno- Castagnone, P., Carta, L.K., Skantar, A.M. and Higgins, J A. (2004) Morphological, molecular, and differential-host characterization of *Meloidogyne floridensis* n. sp. (Nematoda: Meloidogynidae), a root-knot nematode parasitizing peach in Florida. *Journal of Nematology* 36, 20–35.

Haydock, P.P.J., Woods, S.R., Grove, I.G. and Hare, M.C. (2006) Chemical control of nematodes. In: Perry, R.N. and Moens, M. (eds) *Plant Nematology.* CAB International, Wallingford, UK, pp. 392–410.

Hockland, S., Inserra, R.N., Millar, L. and Lehman, P.S. (2006) International plant health – putting legislation into practice. In: Perry, R.N. and Moens, M. (eds) *Plant Nematology.* CAB International, Wallingford, UK, pp. 327–345.

Horton, D., Brannen, P., Bellinger, B. and Ritchie, D. (2007) *Southeastern Peach, Nectarine and Plum Pest Management and Culture Guide.* University of Georgia Cooperative Extension Services Bulletin No. 1171, University of Georgia, Athens, Georgia.

Huang, G., Allen, R., Davis, E.L., Baum, T.J. and Hussey, R.S. (2006) Engineering broad root-knot resistance in transgenic plants by RNAi silencing of a conserved and essential root-knot nematode parasitism gene. *Proceedings of the National Academy of Sciences of the USA* 103, 14302–14306.

Hueth, B., McWilliams, B., Sunding, D. and Zilberman, D. (2000) Analysis of an emerging market: can methyl iodide substitute for methyl bromide? *Review of Agricultural Economics* 22, 43–54.

Hutchinson, C.M., McGiffen, M.E. Jr., Ohr, H.D., Sims, J.J. and Becker, J.O. (1999) Evaluation of methyl iodide as a soil fumigant for root-knot nematode control in carrot production. *Plant Disease* 83, 33–36.

Jansson, R.K. and Dybas, R.A. (1998) Avermectins: biochemical mode of action, biological activity and agricultural importance. In: Ishaaya, I. and Degheele, D. (eds) *Insecticides with Novel Modes of Action: Mechanisms and Application*. Springer-Verlag, New York, pp. 152–167.

Johnson, A.W. (1985) The role of nematicides in nematode management. In: Sasser, J.N. and Carter, C.C. (eds) *An Advanced Treatise on* Meloidogyne, *Vol. I Biology and Control*. North Carolina State University Graphics, Raleigh, North Carolina, pp. 249–267.

Johnson, A.W. (1998) Vegetable crops. In: Barker, K.R., Pederson, G.A. and Windham, G.L. (eds) *Plant and Nematode Interactions*. American Society of Agronomy, Crop Science Society of America, Soil Science Society of America, Madison, Wisconsin, pp. 595–635.

Johnson, A.W. and Feldmesser, J. (1987) Nematicides – a historical review. In: Veech, J.A and Dickson, D.W. (eds) *Vistas on Nematology*. Society of Nematologists, Hyattsville, Maryland, pp. 448–454.

Johnson, C.S., Way, J. and Barker, K.R. (2005) Nematode parasites of tobacco. In: Luc, M., Sikora, J. and Bridge, J. (eds) *Plant Parasitic Nematodes in Subtropical and Tropical Agriculture*, 2nd edn, CAB International, Wallingford, UK, pp. 675–708.

Karssen, G. and Moens, M. (2006) Root-knot nematodes. In: Perry, R.N. and Moens, M. (eds) *Plant Nematology*. CAB International, Wallingford, UK, pp. 59–90.

Katan, J. (1981) Solar heating (solarization) of soil for control of soilborne pests. *Annual Review of Phytopathology* 19, 211–236.

Kegley, S.E., Hill, B.R., Orme, S. and Choi, A.H. (2009a) PAN pesticide database, Pesticide Action Network, NA, http://www.pesticideinfo.org/List_Chemicals.jsp?ChemUseList=fumigant&dUseList=y&dDPR_Chem_Code=y&dEPA_PCCode=y&dCAS_No=y&dClassList=y&ChemRegList=&ChemUseList=&dS_BA=y&dEPA_Reg=y (accessed 29 January 2009).

Kegley, S.E., Hill, B.R., Orme, S. and Choi, A.H. (2009b) PAN pesticide database, Pesticide Action Network, NA, http://www.pesticideinfo.org/List_Chemicals.jsp?ChemUseList=Nematicide&dUseList=y&dDPR_Chem_Code=y&dEPA_PCCode=y&dCAS_No=y&dClassList=y& ChemRegList=&ChemUseList=&dS_BA=y&dEPA_Reg=y (accessed 29 January 2009).

Koenning, S.R., Kirkpatrick, T.L., Starr, J.L., Wrather, J., Walker, N.R. and Mueller, J.D. (2004) Plant-parasitic nematodes attacking cotton in the United States. *Plant Disease* 88, 100–113.

Kokalis-Burelle, N. and Nyczepir, A.P. (2004) Host-range studies for *Meloidogyne floridensis*. *Journal of Nematology* 36, 328.

Kokalis-Burelle, N., Rosskopf, E.N., Bausher, M., McCollum, G. and Kubota, C. (2008) Alternative fumigants and grafting for tomato and double-cropped muskmelon production in Florida. *Proceedings of the Annual International Research Conference on Methyl Bromide Alternatives and Emissions Reductions*. Methyl Bromide Alternatives Outreach Press, Orlando, Florida, pp. 63/1–63/2.

Kutywayo, V. and Been, T.H. (2006) Host status of six major weeds to *Meloidogyne chitwoodi* and *Pratylenchus penetrans*, including a preliminary field survey concerning other weeds. *Nematology* 8, 647–657.

Lawrence, K.S., Feng, Y., Lawrence, G.L., Burmester, C.H. and Norwood, S.H. (2005) Accelerated degradation of aldicarb and its metabolites in cotton field soils. *Journal of Nematology* 37, 190–197.

Lee, J.M. and Oda, M. (2003) Grafting of herbaceous vegetable and ornamental crops. *Horticultural Reviews* 28, 61–124.

Lembright, H.W. (1990) Soil fumigation: principles and application technology. *Supplement to Journal of Nematology* 22, 632–644.

Lewis, E.E. and Grewal, P.S. (2005) Interactions with plant parasitic nematodes. In: Grewal, P.S., Ehlers, R-U. and Shapiro-Ilan, D.I. (eds) *Nematodes as Biocontrol Agents*. CAB International, Wallingford, UK, pp. 349–362.

Luc, M., Bridge, J. and Sikora, R.A. (1990) Reflections on nematology in subtropical and tropical agriculture. In: Luc, M., Sikora, R.A. and Bridge, J. (eds) *Plant Parasitic Nematodes in Subtropical and Tropical Agriculture*. CAB International, Wallingford, UK, pp. xi–xvii.

Luc, M., Sikora, R.A. and Bridge, J. (2005) *Plant Parasitic Nematodes in Subtropical and Tropical Agriculture*, 2nd edn. CAB International, Wallingford, UK.

MacGuidwin, A.E. (1993) Management of nematodes. In: Rowe, R.C. (ed.) *Potato Health Management*. APS Press, St Paul, Minnesota, pp. 159–166.

Maksymowicz, B. and Palmer, G. (1997) Selecting a tobacco transplant production system. www.ca.uky.edu/agc/pubs/agr/agr155/agr155.htm (accessed 5 March 2008).

Maximiniano, C., Campos, V.P., Souza, R.M. and Almeida, R.A. (2001) Flutuação populacional de *Meloidogyne exigua* em cafezal naturalmente infestado por *Pasteuria penetrans*. *Nematologia Brasileira* 25, 63–69.

MBTOC (1998) Montreal protocol on substances that deplete the ozone layer. United Nations Environmental Programme, http://www.epa.gov/Ozone/mbr/mbtoc98.pdf (accessed 11 February 2008).

McGovern, R.J., McSorley, R. and Bell, M.L. (2002) Reduction of landscape pathogens in Florida by soil solarization. *Plant Disease* 86, 1388–1395.

McKenry, M.V. (1984) Grape root phenology relative to control of parasitic nematodes. *American Journal of Enology and Viticulture* 35, 206–211.

McKenry, M.V. (1985) Nematodes. In: Flint, M.L. (ed.) *Integrated Pest Management for Almonds*. Publication 3308, Division of Natural Resources, University of California Publications, Oakland, California, pp.127–133.

McKenry, M.V. (2007) Nematodes. Kearney Agricultural Center, University of California, Riverside, http://www.uckac.edu/nematode/ (accessed 10 December 2007).

McKenry, M.V., Buzo, T. and Kaku, S. (2006) Replanting stone fruit orchards without soil fumigation. *Proceedings of the Annual International Research Conference on Methyl Bromide Alternatives and Emissions Reductions*. Methyl Bromide Alternatives Outreach Press, Fresno, California, pp. 28/1–28/2.

McSorley, R. (1998) Alternative practices for managing plant-parasitic nematodes. *American Journal of Alternative Agriculture* 13, 98–104.

McSorley, R. (1999) Host suitability of potential cover crops for root-knot nematodes. *Supplement to the Journal of Nematology* 31, 619–623.

McSorley, R. (2004) Research breakthroughs: nutsedge protects root-knot nematodes from soil fumigation. *Phytoprotection* 85, 115–117.

McSorley, R. and Gallaher, R.N. (1993) Effect of crop rotation and tillage on nematode densities in tropical corn. *Supplement to Journal of Nematology* 25, 814–819.

McSorley, R. and Gallaher, R.N. (1995) Cultural practices improve crop tolerance to nematodes. *Nematropica* 25, 53–60.

McSorley, R. and Phillips, M.S. (1993) Modelling population dynamics and yield losses and their use in nematode management. In: Evans, K., Trudgill, D.L. and Webster, J.M. (eds) *Plant Parasitic Nematodes in Temperate Agriculture*. CAB International, Wallingford, UK, pp. 61–85.

McSorley, R., Dickson, D.W., Brito, J.A. and Hochmuth, R.C. (1994) Tropical rotation crops influence nematode densities and vegetable yields. *Journal of Nematology* 26, 308–314.

McSorley, R., Ozores-Hampton, M., Stanley, P.A. and Conner, J.M. (1999) Nematode management, soil fertility, and yield in organic vegetable production. *Nematropica* 29, 205–213.

Melakeberhan, H., Ferris, H., McKenry, M.V. and Gaspard, J.T. (1989) Overwintering stages of *Meloidogyne incognita* in *Vitis vinifera*. *Journal of Nematology* 21, 92–98.

Melakeberhan, H., Xu, A., Kravchenko, A., Mennan, S. and Riga, E. (2006) Potential use of arugula (*Eruca sativa* L.) as a trap crop for *Meloidogyne hapla*. *Nematology* 8, 793–799.

Mojtahedi, H., Ingham, R.E., Santo, G.S., Pinkerton, J.N., Reed, G.L. and Wilson, J.H. (1991) Seasonal migration of *Meloidogyne chitwoodi* and its role in potato production. *Journal of Nematology* 23, 162–169.

Mojtahedi, H., Santo, G.S. and Ingham, R.E. (1993) Suppression of *Meloidogyne chitwoodi* with sudangrass cultivars as green manures. *Journal of Nematology* 25, 303–311.

Molendijk, L.P.G. and Korthals, G.W. (2005) Nematode control strategies in the Netherlands. *Acta Horticulturae* 698, 83–88.

Monfort, W.S., Kirkpatrick, T.L., Long, D.L. and Rideout, S. (2006) Efficacy of a novel nematicidal seed treatment against *Meloidogyne incognita* on cotton. *Journal of Nematology* 38, 245–249.

Monfort, W.S., Kirkpatrick, T.L., Khalilian, A.H. and Mueller, J.D. (2008) Practical site-specific nematicide delivery on cotton farms in the mid-south USA. *Proceedings of the 5th International Congress of Nematology*, Brisbane, Australia, p. 136.

Mueller, J., Khalilain, A. and Henderson, W. (2008) Cost effectiveness of precision nematode management. *Proceedings of the 5th International Congress of Nematology*, Brisbane, Australia, p. 137.

Nagel, A.K., Schnabel, G., Petri, C. and Scorza, R. (2008) Generation and characterization of transgenic plum lines expressing the *Gastrodia* antifungal protein. *HortScience* 43, 1514–1521.

Noling, J.W. (1996) Role of soil fumigants in Florida agriculture. In: Seiber, J.N., Knuteson, J.A., Woodrow, J.E., Wolfe, N.L., Yates, M.V. and Yates, S.R. (eds) *Fumigants: Environmental Fate, Exposure, and Analysis*. American Chemical Society, Symposia Series 652, Oxford University Press, Cary, North Carolina, pp. 14–24.

Norris, R.F., Caswell-Chen, E.P. and Kogan, M. (2003) *Concepts in Integrated Pest Management*. Prentice Hall, Upper Saddle River, New Jersey.

Norsworthy, J.H., Schroeder, J., Thomas, S.H. and Murray, L.W. (2005) Southern root-knot nematode effect on purple nutsedge (*Cyperus rotundus*) and chile pepper response to halosulfuron. *Weed Technology* 19, 1004–1011.

Nunez, S.C. (2007) Effects of early season irrigation and heat unit accumulation on purple nutsedge, yellow nutsedge, and root-knot nematode development. MS thesis. New Mexico State University, Las Cruces, New Mexico.

Nyczepir, A.P. (2006) Evaluation of tall fescue for resistance to root-knot nematode in peach. *Proceedings of the Annual International Research Conference on Methyl Bromide Alternatives and Emissions Reductions*. Methyl Bromide Alternatives Outreach Press, Fresno, California, pp. 136/1–136/2.

Nyczepir, A.P. and Esmenjaud, D. (2008) Nematodes. In: Layne, D.R. and Bassi, D. (eds) *The Peach: Botany, Production and Uses*. CAB International, Wallingford, UK, pp. 505–535.

Nyczepir, A.P. and Wood, B.W. (2008) Interaction of concurrent populations of *Meloidogyne partityla* and *Mesocriconema xenoplax* on pecan. *Journal of Nematology* 40, 221–225.

Nyczepir, A.P., Beckman, T.G. and Reighard, G.L. (1999) Reproduction and development of *Meloidogyne incognita* and *M. javanica* on Guardian peach rootstock. *Journal of Nematology* 31, 334–340.

Nyczepir, A.P., Reilly, C.C., Wood, B.W. and Thomas S.H. (2002) First record of *Meloidogyne partityla* on pecan in Georgia. *Plant Disease* 86, 441.

Nyczepir, A.P., Beckman, T.G. and Reighard, G.L. (2006a) Field evaluation of 'Guardian'™ peach rootstock to different root-knot nematode species. *Acta Horticulturae* 713, 303–309.

Nyczepir, A.P., Wood, B.W. and Reilly, C.C. (2006b) Association of *Meloidogyne partityla* with nickel deficiency and mouse-ear of pecan. *HortScience* 41, 402–404.

Nyczepir, A.P., Brito, J.A., Dickson, D.W. and Beckman, T.G. (2008) Host status of selected peach rootstocks to *Meloidogyne mayaguensis. HortScience* 43, 804–806.

Nyland, G. (1955) Killing root knot nematodes in some stone fruit tree rootstocks. *Plant Disease Reporter* 39, 573–575.

O'Bannon, J.H., Santo, G.S. and Nyczepir, A.P. (1982) Host range of the Columbia root-knot nematode. *Plant Disease* 66, 1045–1048.

Oka, Y., Offenbach, R. and Pivonia, S. (2004) Pepper rootstock graft compatibility and response to *Meloidogyne javanica* and *M. incognita. Journal of Nematology* 36, 137–141.

Okie, W.R., Beckman, T.G., Nyczepir, A.P., Reighard, G.L., Newell, W.C. Jr. and Zehr, E.I. (1994) BY520-9, a peach rootstock for the southeastern United States that increases scion longevity. *HortScience* 29, 705–706.

Opperman, C.H. and Chang, S. (1990) Plant-parasitic nematode acetylcholinesterase inhibition by carbamate and organophosphate nematicides. *Journal of Nematology* 22, 481–488.

Overman, A.J. and Jones, J.P. (1986) Soil solarization, reaction, and fumigation effects on double-cropped tomato under full-bed mulch. *Proceedings of the Florida State Horticultural Society* 99, 315–318.

Overstreet, C., Wolcott, M.C., Burris, G. and Padgett, G.B. (2008) Site-specific technology: an introduction and investigation of its use for management in Louisiana, USA. *Proceedings of the 5th International Congress of Nematology*, Brisbane, Australia, p. 135.

Pandey, R., Sikora, R.A., Kalra, A., Singh, H.B. and Pandey, S. (2003) Plants and their products act as major nematode inhibitory agents. In: Trivedi, P.C. (ed.) *Nematode Management in Plants*. Scientific Publishers, Jodhpur, India, pp. 103–131.

Patzoldt, W.L., Tranel, P.J. and Hager, A.G. (2002) Variable herbicide responses among Illinois waterhemp (*Amaranthus rudis* and *A. tuberculatus*) populations. *Crop Protection* 21, 707–712.

Peet, M. (2001) Sweet potato. Sustainable practices for vegetable production in the south. www.cals.ncsu.edu/sustainable/peet/profiles/c18swpot.html (accessed 5 March 2008).

Phillips, D.A., Ferris, H., Cook, D.R. and Strong, D.R. (2003) Molecular control points in rhizosphere food webs. *Ecology* 84, 816–826.

Poinar, G.O. Jr. (1990) Biology and taxonomy of Steinernematidae and Heterorhabditidae. In: Gaugler, R. and Kaya, H.K. (eds) *Entomopathogenic Nematodes in Biological Control*. CRC Press, Boca Raton, Florida, pp. 23–62.

Potter, J.W. and McKeown, A.W. (2003) Nematode biodiversity in Canadian agricultural soils. *Canadian Journal of Soil Science* 83, 289–302.

Powers, T.O. and Harris, T.S. (1993) A polymerase chain reaction method for identification of five major *Meloidogyne* species. *Journal of Nematology* 25, 1–6.

Rich, J.R., Dunn, R.A. and Noling, J.W. (2004) Nematicides: past and present uses. In: Chen, Z.X., Chen, S.Y. and Dickson, D.W. (eds) *Nematology Advances and Perspectives*, Vol. 2. CAB International, Wallingford, UK, pp. 1179–1200.

Riegel, C. and Noe, J.P. (2000) Chicken litter soil amendment effects on soilborne microbes and *Meloidogyne incognita* on cotton. *Plant Disease* 84, 1275–1281.

Roberts, P.A. (1987) The influence of date of planting of carrot on *Meloidogyne incognita* reproduction and injury to roots. *Nematologica* 33, 335–342.

Roberts, P.A. (1993) The future of nematology: integration of new and improved management strategies. *Journal of Nematology* 25, 383–394.

Roberts, P.A., Van Gundy, S.D. and McKinney, H.E. (1981) Effects of soil temperature and planting date of wheat on *Meloidogyne incognita* reproduction, soil populations, and grain yield. *Journal of Nematology* 13, 338–345.

Schroeder, J., Thomas, S.H. and Murray, L. (1993) Yellow and purple nutsedge and chile peppers host southern root-knot nematode. *Weed Science* 41, 150–156.

Schroeder, J., Kenney, M.J., Thomas, S.H. and Murray, L. (1994) Yellow nutsedge response to southern root-knot nematodes, chile peppers, and metolachlor. *Weed Science* 42, 534–540.

Schroeder, J., Thomas, S.H. and Murray, L.W. (1999) Yellow (*Cyperus esculentus*) and purple nutsedge (*Cyperus rotundus*) are not injured by increasing root-knot nematode (*Meloidogyne incognita*) population density. *Weed Science* 47, 201–207.

Schroeder, J., Thomas, S.H. and Murray, L.W. (2004) Root-knot nematodes affect annual and perennial weed interactions with chile pepper. *Weed Science* 52, 28–46.

Scott, B.A. and Van Gessel, M.J. (2007) Delaware soybean grower survey on glyphosate-resistant horseweed (*Conyza canadensis*). *Weed Technology* 21, 270–274.

Shapiro-Ilan, D.I., Nyczepir, A.P. and Lewis, E.E. (2006) Entomopathogenic nematodes and bacteria applications for control of the pecan root-knot nematode, *Meloidogyne partityla*, in the greenhouse. *Journal of Nematology* 38, 449–454.

Sherman, W.B., Lyrene, P.M. and Sharpe, R.H. (1991) *Flordaguard peach rootstock*. Circular S-376, Institute of Food and Agricultural Sciences, The University of Florida, Gainesville, Florida.

Sikora, R.A. and Fernandez, E. (2005) Nematode parasites of vegetables. In: Luc, M., Sikora, R.A. and Bridge, J. (eds) *Plant Parasitic Nematodes in Subtropical and Tropical Agriculture*, 2nd edn. CAB International, Wallingford, UK, pp. 319–394.

Smelt, J.H., Van De Peppel-Groen, A.E., Van Der Pas, L.J.T. and Dijksterhuis, A. (1996) Development and duration of accelerated degradation of nematicides in different soils. *Soil Biology and Biochemistry* 28, 1757–1765.

Starr, J.L., Tomaszewski, E.K., Mundo-Ocampo, M. and Baldwin, J.G. (1996) *Meloidogyne partityla* on pecan: isozyme phenotypes and other hosts. *Journal of Nematology* 28, 565–568.

Starr, J.L., Bridge, J. and Cook, R. (2002) Resistance to plant-parasitic nematodes: history, current use and future potential. In: Starr, J.L., Cook, R. and Bridge, J. (eds) *Plant Resistance to Parasitic Nematodes*. CAB International, Wallingford, UK, pp. 1–22.

Starr, J.L., Koenning, S.R., Kirkpatrick, T.L., Robinson, A.F., Roberts, P.A. and Nichols, R.L. (2007) The future of nematode management in cotton. *Journal of Nematology* 39, 283–294.

Stock, S.P. (2005) Morphology and systematics of nematodes used in biocontrol. In: Grewal, P.S., Ehlers, R-U. and Shapiro-Ilan, D.I. (eds) *Nematodes as Biocontrol Agents*. CAB International, Wallingford, UK, pp. 3–43.

Taylor, A.L. and Sasser, J.N. (1978) *Biology, Identification and Control of Root-knot Nematodes* (Meloidogyne *Species*). North Carolina State University Graphics, Raleigh, North Carolina.

Thomas, S.H. (1994) Influence of 1,3-dichloropropene, fenamiphos, and carbofuran on *Meloidogyne incognita* populations and yield of chile peppers. *Supplement to the Journal of Nematology* 26, 683–689.

Thomas, S.H., Schroeder, J., Kenney, M.J. and Murray, L.W. (1997) *Meloidogyne incognita* inoculum source affects host suitability and growth of yellow nutsedge and chile pepper. *Journal of Nematology* 29, 404–410.

Thomas, S.H., Fuchs, J.M. and Handoo, Z.A. (2001) First report of *Meloidogyne partityla* in New Mexico. *Plant Disease* 85, 1030.

Thomas, S.H., Schroeder, J. and Murray, L. (2004) *Cyperus* tubers protect *Meloidogyne incognita* from 1,3-dichloropropene. *Journal of Nematology* 36, 131–136.

Thomas, S.H., Schroeder, J. and Murray, L.W. (2005) The role of weeds in nematode management. *Weed Science* 53, 293–298.

Thomason, I.J. and Caswell, E.P. (1987) Principles of nematode control. In: Brown, R.H. and Kerry, B.R. (eds) *Principles and Practices of Nematode Control in Crops.* Academic Press, Orlando, Florida, pp. 87–130.

Timper, P., Davis, R.F. and Tillman, P.G. (2006) Reproduction of *Meloidogyne incognita* on winter cover crops used in cotton production. *Journal of Nematology* 38, 83–89.

Townshend, J.L. and Davidson, T.R. (1962) Some weed hosts of the northern root-knot nematode *Meloidogyne hapla* Chitwood, 1949, in Ontario. *Canadian Journal of Botany* 40, 543–547.

Twomey, U., Warrior, P., Kerry, B.R. and Perry, R.N. (2000) Effects of the biological nematicide, DiTera®, on hatching of *Globodera rostochiensis* and *G. pallida. Nematology* 2, 355–362.

Tyler, J. (1933) Development of the root-knot nematode as affected by temperature. *Hilgardia* 7, 391–415.

University of California, Davis (2003) Phenology model database, Columbia root knot nematode. www.ipm.ucdavis.edu/PHENOLOGY/mn-columbia_root_knot.html (accessed 20 February 2008).

University of California Pest Management Guidelines (2007) Potato nematodes. www.imp.ucdavis.edu/PMG/r607200111.html (accessed 5 March 2008).

Vrain, T.C., Barker, K.R. and Holtzman, G.I. (1978) Influence of low temperature on rate of development of *Meloidogyne incognita* and *M. hapla* larvae. *Journal of Nematology* 10, 166–171.

Wang, K.H. and McSorley, R. (2008) Exposure time to lethal temperatures for *Meloidogyne incognita* suppression and its implication for soil solarization. *Journal of Nematology* 40, 7–12.

Wesemael, W.M.L. and Moens, M. (2008a) Vertical distribution of the plant-parasitic nematode, *Meloidogyne chitwoodi*, under field crops. *European Journal of Plant Pathology* 120, 249–257.

Wesemael, W.M.L. and Moens, M. (2008b) Quality damage on carrots (*Daucus carota* L.) caused by the root-knot nematode *Meloidogyne chitwoodi. Nematology* 10, 261–270.

Wesemael, W.M.L., Perry, R.N. and Moens, M. (2006) The influence of root diffusate and host age on hatching of the root-knot nematodes, *Meloidogyne chitwoodi* and *M. fallax. Nematology* 8, 895–902.

Westphal, A. (2007) Sustainable approaches to the management of plant-parasitic nematodes and disease complexes. *Phytopathology* 97, S155.

Wheeler, T.A., Kaufman, H.W., Baugh, B., Kidd, P., Schuster, G. and Siders, K. (1999) Comparison of variable and single-rate applications of aldicarb on cotton yield in fields infested with *Meloidogyne incognita. Supplement to the Journal of Nematology* 31, 700–708.

Whitehead, A.G. (1998) *Plant Nematode Control.* CAB International, Wallingford, UK.

Widmer, T.L. and Abawi, G.S. (2000) Mechanisms of suppression of *Meloidogyne hapla* and its damage by a green manure of sudengrass. *Plant Disease* 84, 562–568.

Widmer, T.L., Mitkowski, N.A. and Abawi, G.S. (2002) Soil organic matter and management of plant-parasitic nematodes. *Journal of Nematology* 34, 289–295.

Wood, B.W., Reilly, C.C. and Nyczepir, A.P. (2004) Mouse-ear of pecan: a nickel deficiency. *HortScience* 39, 1238–1242.

Wrather, J.A., Stevens, W.E., Kirkpatrick, T.L. and Kitchen, N.R. (2002) Effects of site specific application of aldicarb on cotton in a *Meloidogyne incognita*-infested field. *Journal of Nematology* 34, 115–119.

Wright, D.J. (1981) Nematicides: mode of action and new approaches to chemical control. In: Zuckerman, B.M. and Rhode, R.A. (eds) *Plant Parasitic Nematodes*, Vol. 3. Academic Press, New York, pp. 421–449.

Zasada, I.A. and Ferris, H. (2004) Nematode suppression with brassicaceous amendments: application based upon glucosinolates profiles. *Soil Biology & Biochemistry* 36, 1017–1024.

Zasada, I.A., Avendano, F., Li, Y.C., Logan, T., Melakeberhan, H., Koenning, S.R. and Tylka, G.L. (2008) Potential of an alkaline-stabilized biosolid to manage nematodes: case studies on soybean cyst and root-knot nematodes. *Plant Disease* 92, 4–13.

19 Current and Future Management Strategies in Resource-poor Farming

Danny L. Coyne[1], Hendrika H. Fourie[2] and Maurice Moens[3]
[1]*International Institute of Tropical Agriculture (IITA), Kampala, Uganda;* [2]*Agricultural Research Council-Grain Crops Institute (ARC-GCI), Potchefstroom, South Africa;* [3]*Institute for Agricultural and Fisheries Research, Merelbeke, Belgium and Ghent University, Ghent, Belgium*

19.1	Introduction and Definitions	444
19.2	Options	446
19.3	Correct Diagnosis	446
19.4	Prevention	447
19.5	Cultural Control	451
19.6	Cropping Systems	453
19.7	Resistance	458
19.8	Biological Control	459
19.9	Chemical Control	459
19.10	Conclusions and Future Directions	465
19.11	References	466

19.1 Introduction and Definitions

This chapter is focused on the management of root-knot nematodes, *Meloidogyne* spp. Nematode management aims at the reduction or maintenance of nematode densities at low, sub-damage threshold levels using several strategies, thus enabling sustainable crop production. By contrast, nematode control implies the use of a single measure to reduce or eliminate nematode pests, as outlined by Viaene *et al.* (2006). The chapter is also directed at management in resource-poor regions. The majority of these countries occur under tropical and subtropical climates, with some countries, which lie within the Tropics of Cancer and Capricorn, having relatively cooler or temperate climates at high altitude. Some countries also have economies that are becoming increasingly important on a global scale, while large proportions of their populations remain resource-poor, subsistence smallholders.

Although we have been aware of the existence and importance of root-knot nematodes for over a century, significant knowledge gaps persist. Nowhere is the disparity in knowledge on root-knot nematodes greater than that between developed and developing countries, although much of the progress in understanding nematode biology, epidemiology and management can theoretically be transferred or applied to the less developed areas of the world. Much of the information and technologies can also often be applied across geographical and cropping systems. Thus, the developing world can effectively benefit from advances made elsewhere. Difficulties in the transfer of knowledge and expertise from else-

 ©CAB International 2009. *Root-knot Nematodes* (eds R.N. Perry, M. Moens and J.L. Starr)

where can emerge, especially where knowledge of distribution and identification of certain species which is directly relevant to certain management practices is imprecise or unknown. However, for more generic issues, which can be applied across a range of situations, limited funds in developing countries create obstacles for the use of certain tools or implementation of particular strategies. For example, molecular diagnosis is not possible without the relevant equipment and routine access to costly supplies. On a more basic level, the simple lack of awareness of nematodes as pest problems presents a major hurdle. In North Carolina, USA, the development of a nematode advisory service was catalysed following the discovery of the profound effects of root-knot nematodes on tobacco research (Nusbaum, 1963). Ironically, rice researchers in West Africa were only made aware of nematode problems (including root-knot nematodes) following serious damage to long-term upland rice fertility experiments (Coyne *et al.*, 2004), reflecting similar circumstances to those occurring almost half a century earlier in North Carolina.

In the resource-poor regions of the world, subsistence-style farming predominates, usually on small areas of land. Approximately 90% of the world's poor live in South Asia and Africa, with 75% living in rural areas, where they depend primarily on agriculture and related activities (Hazell and Wood, 2008). Crops that are more usually associated with commercial enterprises (e.g. coffee, cocoa, palm oil, tobacco and cotton) may also be cultivated by smallholders, and these crops are considered within this chapter.

Various definitions have been used to separate what is viewed as subsistence agriculture from commercial, high-value, intensive systems. However, exceptions will always occur. Subsistence agriculture tends to imply the risk-averse cultivation of low-value crops, with the primary aim of attaining food security for home consumption and off-farm sales of excess produce. In terms of nematology, Brown (1987) referred to low-value crops as those for which the conventional use of nematicides could not be economically justified. The resource-poor farmer is generally not dependent on pesticides, but practises organic agriculture by default. However, for smallholder vegetable farmers and Asian rice farmers, the injudicious use of pesticides can be commonplace (e.g. James *et al.*, 2005; Duxbury and Lauren, 2006).

In resource-poor regions, root-knot nematodes are consistently viewed as the economically most important nematode pests (Sasser and Carter, 1985; Luc *et al.*, 2005), whereas in tropical and subtropical regions, such as South Africa (Fourie and McDonald, 2001), they are often viewed as the most widespread nematodes, if not the most important biotic constraint. Undoubtedly, root-knot nematodes are of major economic importance globally, particularly in resource-poor regions (W. Wesemael, 2008, personal communication). The wide host range of most species, persistence, high reproductive capacity and absence of suitable/sustainable management practices secure this status. In a study of peri-urban vegetable systems in Benin (West Africa), originally focused on insect pests, *Meloidogyne* spp. were defined as the most important pests (James *et al.*, 2005). Sikora and Fernández (2005) reported that in tropical and subtropical areas vegetable production is highly dependent on proper nematode control, if not a prerequisite in most cases.

Root-knot nematodes are particularly important in vegetable production. Knowledge and understanding of this problem are scarce and limited, among farmers in particular, but also among agricultural workers and researchers (De Waele and Elsen, 2007). In a study of vegetable production in Ghana (Ntow *et al.*, 2006) and of smallholder cropping problems in general (Dinham, 2003), not a single reference to plant-parasitic nematodes was made, while emphasizing the fundamental importance of pest recognition and pest management training for effective pest management. These shortcomings cannot necessarily be apportioned to the farmer, extension officer or researcher, but rather to the system and its limited support as a whole. This situation, however, can be improved by practices that have the benefit of increasing soil sustainability (Bridge, 1996).

Based on the substantial experience of the authors of this chapter in African farming systems, it is their firm opinion that, as a group, root-knot nematodes pose the greatest single biotic threat to agricultural productivity throughout the continent, and probably across resource-poor areas on a global basis. Therefore, significant and specific attention is needed to even begin addressing the enormity of the issue. It is with this in mind that we present this chapter.

Initially, we first need to determine the current and potential options available for nematode management under resource-poor conditions. However, it is not the aim to report exhaustively on these options here, or to repeat much of the already available information extensively covered by Brown and Kerry (1987), Whitehead (1998) and Luc *et al.* (2005). Subsequently, this chapter aims to identify the shortcomings in the system and how these can best be addressed to improve our ability to manage root-knot nematodes in the future. Much of this is naturally generic to nematology, but here it is specifically focused on root-knot nematodes in resource-poor regions.

19.2 Options

Proper management implies avoiding or preventing the nematode problem in the first instance. However, nematode management strategies are rarely a primary consideration in resource-poor areas, not least because nematode pests are poorly understood. Nematology expertise remains critically low in most developing countries. Thus, nematode management would be improved through problem recognition and through understanding the exact nature of the problem, more specifically through proper knowledge of nematode species (Coyne *et al.*, 2007). Accurate diagnosis is required to enable informed decisions on the most appropriate management measures to be employed (Whitehead, 1998). In most resource-poor situations, such a goal remains an ambitious dream, even though many of the available nematode management measures could be employed with reasonable effectiveness, but essentially require problem recognition, knowledgeable advice, continuous support and ultimately farmer acceptance.

The principal management methods used for plant-parasitic nematodes in general apply also to root-knot nematodes, with the use of resistant or non-host crop plants, fallowing or flooding infested land, disinfestation or protection of planting material, application of amendments or nematicides and, more recently, the use of microbial antagonists and biocontrol agents. The use of any single management tool, perhaps with the exception of nematicides, rarely results in an effective strategy to alleviate nematode problems in resource-poor areas.

19.3 Correct Diagnosis

To employ control strategies such as host plant resistance, biological control and crop rotation, accurate characterization of prevailing nematode populations is essential. The use of resistance is highly dependent on knowledge of the target species against which the resistance is focused. This knowledge is rarely available in the majority of resource-poor areas. The situation is improving (Cook and Starr, 2006; see Starr and Mercer, Chapter 14, this volume), but establishing accurate information on nematode species for many crops in resource-poor areas remains a colossal task. For example, root-knot nematodes from coffee in Central America are often reported without any attempt at species identification (Hernandez *et al.*, 2004). Santos and Triantaphyllou (1992) further implied that where root-knot nematode species are reported from coffee, they were frequently identified incorrectly. Carneiro *et al.* (2004) recently established several root-knot nematodes species associated with coffee, including two unknown species, by using both morphological and molecular diagnostic techniques. With increasing use of molecular diagnostic methods, our understanding of species diversity and distribution will expand. The use of molecular techniques established that *M. fallax* was present in South Africa, although it was previously unrecorded on the African continent (Fourie *et al.*, 2001). In West Africa, identification of root-knot nematodes from peri-urban vegetable systems, based mostly on female morphology, revealed a controversial range of species (Baimey *et al.*, 2007). *Meloidogyne chitwoodi*, otherwise recognized as a temperate species, appears to be established in tropical Africa, as confirmed by its detection in a tropical locality of South Africa (Fourie *et al.*, 1998), and as yet unconfirmed occurrence in Mozambique (Coyne *et al.*, 2005) and Benin (Baimey *et al.*, 2007). Consequently, our use of currently known sources of resistance may be of limited value as our knowledge on nematode diversity broadens (see Whitehead, 1969). Irrespective of diagnostic precision, it is clear that the diversity of species of root-knot nematodes in tropical and subtropical systems is far greater than hitherto has been accepted. The discovery of *Meloidogyne* species that were previously unknown or not perceived as a threat will undoubtedly continue as cropping practices change, develop and adapt to changing needs, such as through the introduction of new cultivars or crops. Add the further complication of intraspecific nematode variation, host range and virulent populations on some sources of resistance (see Moens *et al.*, Chapter 1, this volume; Starr and Mercer, Chapter 14, this volume; Atkinson *et al.*, Chapter 15, this volume) and the complexity of the situation magnifies.

Correct identification is also important, if not critical, when assessing and implementing biological control and rotation strategies. Such strategies are designed as a function of the particular nematode species. A highly specific relationship may be dependent on, or necessary for, successful biological control, which demands accurate knowledge and diagnosis. Crop resistance may also be highly specific, with genus-level identification insufficient to provide a basis on which to advise use as a management option.

19.4 Prevention

Preventing crop infection in the first instance, particularly in resource-poor agriculture, is perhaps the single most important strategy to avoid or limit crop losses in terms of quality and yield. This is particularly true since treatment of nematode-infected crops, or a 'therapeutic' approach, is essentially more complicated and costly for producers.

19.4.1 Healthy planting material

Botanical seeds, the generative means of plant propagation, are not usually infected by root-knot nematodes. However, when sown in infested soil, plants developing from seeds can become infected and be a source of inoculum. In agriculture, the term 'seed' is also used for different forms of vegetative propagation materials (e.g. tubers). This material, when produced in nematode-infested soil, can also become infected. Many tuber and banana and plantain crops rely on planting material derived from the preceding crop, which very often constitutes a primary source of contamination for newly planted fields and plantations, and a major source of crop loss in resource-poor areas.

The use of clean, healthy, nematode-free planting material is a prerequisite for good crop production and cannot be overemphasized. *Meloidogyne*-free seedlings, even when planted into fields infested with these nematode species, will develop better and produce higher yields than infected seedlings. Indeed, the growth stage at which seedlings become infected with root-knot nematodes can be linked directly to performance, with earlier infection leading to increased loss (Bergeson, 1968). The development of sound, healthy seed production systems can result in significant reduction of nematode problems, and effectively represents a long-term strategy towards ensuring reliable and consistent availability of good-quality, pest- and disease-free seed.

For crops that are transplanted after seeding, the management of root-knot nematodes in seedbeds is much easier and more economical than treating larger fields. However, it is imperative that seedbeds are free of root-knot nematodes. In seedbeds, root-knot nematodes can be maintained at sub-threshold levels or eradicated through a number of methods, such as nematicides, heat, biocontrol agents, the use of nematode-free potting material (e.g. sawdust, coconut husk, peat, sand) or commercially available inert material (e.g. vermiculite, rock wool). Seedbeds can also be located at sites that are free of root-knot nematodes, such as locations previously subjected to prolonged flooding (paddy fields, flood plains). Alternatively, sediment (sand or silt) can be relocated from flooded sites or riverbeds for use in a nursery, either in a raised bed (Fig. 19.1), or in containers. A combination of the above strategies can be designed to create a *Meloidogyne*-free nursery bed or facility. However, regular cleaning, sterilization or renewal of potting medium and containers is necessary to maintain a *Meloidogyne*-free zone.

On a local basis, nursery beds and raised constructions can readily be created. Potting or seed trays are often locally available and relatively inexpensive. Alternatively, discarded (plastic) containers, egg trays or other imaginative solutions can provide practical alternatives for seeding individual plants. Soil for use in nurseries can be pasteurized using a relatively simple makeshift method, involving an oil drum semi-filled with

Fig. 19.1. Locally constructed nursery bed raised above the ground for vegetable seedling production in Malawi.

water and heated over a fire. Plastic piping is used to direct steam emerging from the drum to the soil. By covering the soil with plastic sheeting weighted down round the edges, the steam can be contained and provide high standards of pasteurization. Meanwhile, commercial nurseries can supply high-quality, healthy seedlings, often of improved or hybrid cultivars. Seedling production enterprises in resource-poor areas have been stimulated and shown to satisfy such a niche successfully. In Bangladesh, demand was stimulated once the benefits of healthy seedlings became apparent (Duxbury and Lauren, 2006). Nurseries can be readily developed, with scale dependent on needs and capacity. Existing systems can be used as models, such as forestry service tree nurseries, which can provide guidance and can be adapted accordingly (Fig. 19.2). Duxbury and Lauren (2006) also found that, primarily through control of *Meloidogyne graminicola*, rice seedlings produced in improved nurseries produced greater yield and, on average, were 17% less costly to produce than in conventional practice. This unexpected difference was associated with reduced pesticide application costs. In Thailand, yields from healthy vegetable seedlings were between 17 and 20% higher, compared with conventional practices (Duxbury and Lauren, 2006). This practice was also adapted by cut-flower producers in Nepal.

On a more commercial basis, successful implementation of substrates has led to the development of the flotation tray method to produce tobacco seedlings in many countries, such as Brazil, China and Zimbabwe, in addition to cut-flower production in Kenya (UNEP, 2000a; Thomas Seedling Technology Systems, undated). The technique is applicable for both large and small farm operations, has been extremely effective in many regions, and has been adopted in most instances as an alternative to methyl bromide. It has implications for smallholder farmers because high-quality planting material becomes available. Indications are that most tobacco seedlings worldwide could be grown by this method (UNEP, 2000a).

19.4.2 Seed and seedling supply

Using agricultural networks and organizations, rudimentary systems for production and distribution of seed can be introduced on an individual or farmer group basis, which can gradually increase in scale through interested larger-scale providers and non-government organizations (NGOs). Cultivation of seed material for a particular commodity may be undertaken at a locality that is not infested with, and outside the known range of, a particular pest or disease. However, providing healthy planting material through such a seed system production mechanism requires a high standard of knowledge of the biology and distribution of pests or disease in these areas. Some of the best examples of sustainable healthy seed systems are associated with potato, such as in the Philippines (Primavesi, 1989) and Afghanistan (Arif *et al.*, undated).

19.4.3 Heat treatment

Sustainable seed systems can also be employed for tuber crops other than potato. For yam (*Dioscorea*

Fig. 19.2. Forestry Commission tree nursery in Uganda, with a range of horticultural plants and trees for domestic and commercial use.

spp.), root-knot nematodes are an increasingly damaging pest across all yam-growing areas (Bridge *et al.*, 2005). Hot water treatment (HWT) can be used effectively to decontaminate potentially infected material and ensure nematode-free seed stocks. HWT has proved largely impractical in resource-poor areas for treating planting material, such as banana and root crops (e.g. yam), largely due to the bulk of material to be treated, the cost of fuel and the time needed for such treatment (Viaene *et al.*, 2006). However, locally adapted 'improvisations' can often prove suitable, such as the use of halved oil drums for boiling water for immersion of material for a short duration, e.g. 30 s (Tenkuoano *et al.*, 2006; Viaene *et al.*, 2006) (Figs 19.3 and 19.4). Banana corms that have been removed from the field, subjected to HWT and incubated in sawdust or clean potting material in a modified incubator/macropropagator resulted in healthy, *Meloidogyne*-free (and free of other pests and diseases) plantlets, which can be removed, potted and used as a suitable, low-cost alternative to tissue culture (TC) (Tenkuoano *et al.*, 2006). Other crops that similarly bud (e.g. cocoyam) can also be macropropagated in a similar way. In Yemen, severe infection

Fig. 19.3. Freshly clean plantain suckers in Nigeria following paring away of infected roots and corm with a knife, before treating in hot water.

Fig. 19.4. Improvised hot water tank used for disinfesting pared banana suckers in boiling water for 30 s.

by *M. incognita*, associated with heavy banana losses, was reduced through the use of *Meloidogyne*-free propagative stocks (Ibrahim, 1985). However, HWT can result in damage if care is not taken to regulate the temperature or duration of immersion; pre-assessment of crop and cultivar sensitivity may also be necessary (Whitehead, 1998; Viaene *et al.*, 2006). In Nigeria, yam germination was reduced for some cultivars following HWT (D.L. Coyne, 2008, personal observation). The process can also stimulate budding, as observed with bananas in Uganda (D.L. Coyne, 2008, unpublished results). Thus, the use of heat not only provides healthy plantlets but also can stimulate faster budding and more rapid generation of plantlets, probably through a process involving small heat shock proteins produced in treated cells (Sun *et al.*, 2002). Other vegetative planting material, such as root stock, bulbs and vine cuttings, are good candidates for decontamination through use of HWT.

19.4.4 Tissue culture

For resource-poor farmers, the use of TC plants to overcome contamination with root-knot nematodes is yet in its infancy. Success in TC production of pathogen-free planting materials for cassava, yam, banana, plantain, citrus and flowers has been reported from countries such as Kenya and Ghana, and is increasingly attracting private sector investments (Machuka, 2001). However, where farmer awareness of special requirements for handling TC material is poor, high plantlet losses are experienced. With increasing awareness, availability and transport infrastructure in resource-poor areas, the potential of TC offers great promise for reducing the effects of root-knot nematodes and other pests and diseases (Dubois *et al.*, 2006a). Additionally, enhancement of TC plants with mutualistic endophytic fungi can increase plant vigour and provide protection against root-knot nematodes (Pinochet *et al.*, 1997) and pests, increasing TC material durability and potential in resource-poor regions (Sikora *et al.*, 2003; Sikora and Pocasangre, 2004; Dubois *et al.*, 2006b).

19.4.5 Quarantine

Phytosanitary measures are of major importance in reducing the adverse impact of *Meloidogyne* spp. on crops in developing countries. Currently, 25 species of *Meloidogyne* are on the list of exotic nematode plant pests of agricultural and environmental significance to the USA, but the list does not include economically important species, such as *M. chitwoodi* and *M. naasi*, because they are widely distributed in the USA and are not subjected to regulatory controls (APHIS, 2008; see also Moens *et al.*, Chapter 1, this volume). Implementing phytosanitary measures contributes to various regulatory systems designed to minimize the transport and global spread of organisms that are harmful to plants (Hockland *et al.*, 2006). Quarantine and inspection services are often the first to intercept nematode species new to a country, thus assisting in preventing the inadvertent spread of species to new areas. However, the number of nematologists in particular is declining, new nematode species are increasingly being discovered and global trade is intensifying, posing increasing challenges to the interception of new nematode pests. In resource-poor countries, quarantine services face even greater challenges, with significant capacity building necessary for many.

19.5 Cultural Control

19.5.1 Removal of infected material

Within an overall cropping system, the physical removal or destruction of plant material infected with root-knot nematodes, particularly roots, tubers or seeds, should be considered. In tobacco farming in Southern Africa, it is common practice to uproot plants after harvest and expose the roots to the sun (Bridge, 1987) or burn them *in situ* (Shepherd, 1982), thus reducing the root-knot nematode inoculum for the succeeding crop. However, this practice may not always be appropriate for resource-poor farmers, due to labour shortages at critical times. The practice also has limitations because it is impossible to remove all roots. Roots that have deteriorated and roots in dry, hardened soil are difficult to remove properly. The practice should be encouraged and utilized where it is suitable and feasible.

19.5.2 Planting date

Planting crops when temperatures are less favourable for root-knot nematode development and reproduction can suppress nematode problems. In Zimbabwe, early planting of tobacco on ploughed ridges was reported as a key management tactic for root-knot nematodes (Shepherd, 1982; Saka, 1985). Since genetic resistance conferred by the *Mi* gene in tomato is sensitive to temperature and becomes ineffective at soil temperatures above 28 °C (see Williamson and Roberts, Chapter 13, this volume), such resistant cultivars should be planted in areas where soil temperatures remain below 28 °C for at least 6 weeks after planting. Alternatively, as the minimum temperature required for *M. incognita* development in the root is lower than the 18 °C activity threshold for second-stage juveniles (J2) of *M. incognita* (Roberts *et al.*, 1981), it can be exploited for management purposes. Synchronizing date of planting with low soil temperatures was reported to be effective for management of *Meloidogyne* on carrots (Roberts, 1987), and offers potential for management in cooler, higher-altitude areas of the tropics (Sikora and Fernández, 2005).

19.5.3 Flooding

Land that has lain under water for a continuous period of 3 months or more following either natural or artificial flooding will be free of root-knot nematodes (Bridge, 1987). Such soil is almost perfect for use as seedling nursery sites, especially where the areas available are small. Alternatively, the soil may be removed for use in a nursery situated elsewhere or for use later in the season. At least three root-knot nematode species, namely *M. graminicola* (Kinh *et al.*, 1982), *M. triticoryzae* (Garg *et al.*, 1995) and *M. oryzae* (Segeren-v.d. Oever and Sanchit-Becker, 1984), have evolved to survive under flooded conditions. Prolonged periods of flooding and effective water management in paddy rice can, however, successfully control these species. Poor water management or use of intermittent flooding, which is increasingly practised where water is becoming limiting in parts of South-east Asia, aggravates the root-knot nematode problem (De Waele and Elsen, 2007) and, furthermore, can reduce the tolerance of rice to *M. graminicola* (Tandingan *et al.*, 1996).

Following the well-managed flooding of, for example, rice paddies, post-rice crops can benefit from the *Meloidogyne*-free conditions. However, as water management practices become adapted to reduced water availability, *M. graminicola* has become problematic on post-rice, as well as rice, crops (Gergon *et al.*, 2001). Large and sloping areas do not facilitate the effective use of artificial flooding.

19.5.4 Mulching and soil amendments

The effect of soil amendments is generally accepted as an indirect mechanism for promoting nematode suppression through enhanced activity of naturally occurring nematode antagonists such as fungi, bacteria and carnivorous nematodes (Ferraz and de Freitas, 2004). Additionally, some amendments may contain compounds with nematicidal activity (e.g. brassicaceous crop residues). Furthermore, the application of soil amendments, such as fertilizers or organic matter, is a readily accepted practice for improving crop production. This is due primarily to the improvement of soil fertility and structure, which often contribute to a healthier and more robust crop, which is better able to withstand nematode invasion and subsequent damage.

Numerous amendments have been assessed and recommended for nematode management, some of which appear particularly effective. In general, amendments are divided into two broad categories: (i) amendments that have been transported from elsewhere and applied; and (ii) amendments that have been cultivated *in situ* and incorporated as green mulch (manure). Usually, amendments are composed of agricultural by-products or waste products (crop or animal), but they can also be derived from naturally growing vegetation or even human waste. In general, amendments tend to have broad-spectrum activity against root-knot nematodes. Waste crop by-products, such as oilseed waste (cakes, pomace), sawdust, fruit pulp, waste peel, coffee husk, oil palm debris and molasses, are all attractive amendments in this regard. Seed cake applications, such as from castor (*Ricinus communis*), neem (*Azadirachta indica*), cotton (*Gossypium hirsutum*), groundnut (*Arachis hypogaea*) and white mustard (*Sinapis alba*), appear particularly effective at reducing nematode numbers (see section 19.9.2).

Waste products for use as amendments are usually inexpensive but may become unattractive (expensive) through costs of transport to the field, especially if high rates of application are recommended. Amendments originating from animal waste, such as manures, bone meal and chitin, and particularly the addition of crustacean chitin and chicken manure, can also be effective in suppressing populations of root-knot nematodes.

When considering cover crops as green manures, Rodríguez-Kábana and Canullo (1992) referred to either 'passive' or 'active' manures. Passive manures act as a poor or non-host, consequently starving the target nematode species. Active cover crops produce compounds that are nematicidal, either during crop growth (e.g. *Tagetes*, neem, sunn hemp) or upon decomposition (e.g. brassicaceous crops), with the process then being referred to as biofumigation (Kirkegaard *et al.*, 1996, 1998; Tsror *et al.*, 2007). Biofumigation, defined as 'the action of volatile substances produced in the biodecomposition of organic matter for plant pathogen control', is being hailed as a non-chemical alternative to methyl bromide (Bello *et al.*, 2000). Biocidal compounds, such as isothiocyanates, released by brassicaceous crops, and gases, such as ammonia, produced during the decomposition process, act as fumigants. Bello *et al.* (2004) purported that any organic remains can act as a biofumigant against root-knot nematodes, the effect being determined by biochemical characteristics, dosage and method of application, and report numerous examples where biofumigation efficacy compared favourably with conventional nematicides. However, an application rate of 50 t material/ha is generally recommended, and even up to 100 t material/ha where high root-knot nematode and fungal densities are present. 'Active' cover crops incorporated for biofumigation can vary in the duration of crop growth they require before incorporation. Lupins and mustard need only 6–8 weeks' growth, compared with 6–7 months for rapeseed (Riga *et al.*, 2004). Currently, biofumigation as a mode of root-knot nematode management in resource-poor areas would not be construed as a most suitable option, due mainly to the large volumes recommended, the lack of awareness and understanding of the process by farmers, and relative scarcity of suitable material; however, with improved understanding of its applicability and consequent exposure in resource-poor circumstances, it does offer future management potential.

Should a baseline recommendation be made in terms of soil amendments, application whenever and wherever possible, in as high a quantity as is practical, should be practised. Consequently, materials need to be inexpensive and easily accessible. Mulching is beneficial for soil health and crop productivity, but while the nematicidal effects of amendments can often be proven (Ferraz and de Freitas, 2004) they are less than fully understood.

Botanical extracts of amendments also provide a useful aspect of root-knot nematode management and could be used either as a targeted application/treatment (e.g. root dip, seed drench), or through placement of plant parts in planting holes (section 19.9.2).

19.5.5 Physical methods

Thermotherapy or heat treatment has been used widely to disinfest planting material (see section 19.4.3) or treat the soil. The use of steam is possible but expensive and not normally a consideration for resource-poor farmers (Viaene *et al.*, 2006). The use of soil solarization (using plastic or polythene sheeting) to control root-knot nematodes is another strategy but is controversial. Bello *et al.* (2004) claim that soil solarization is ineffective by itself, particularly for mobile organisms such as nematodes. Under low solar energy situations, it is likely to be ineffective, although the technique proved to be effective against root-knot nematodes under suboptimal conditions in Cuban vegetable systems (Fernández and Labrada, 1995). Inconsistency in control from soil solarization can be attributed to variability in biological and physical characteristics at the site, resulting in limited precision for recommending its use. When conducted properly, solarization of soil infested with root-knot nematodes has provided high levels of control (Gaur and Perry, 1991). Soil solarization for at least 4–6 weeks will usually raise soil temperatures to between 35 and 50 °C at depths of up to 30 cm, depending on site/soil conditions. Solarization is more effective on lighter soils that are wet or moist (Stapelton and DeVay, 1986). Effectiveness is reduced with increasing depth and consequent reduction of heat penetration. Efficacy of root-knot nematode suppression can be improved using double-layered, thin (25–30 μm) polyethylene sheeting, transparent as opposed to black sheeting, and during periods of highest solar intensity (Viaene *et al.*, 2006). However, new plastic formulations that increase soil temperature have extended the usefulness of solarization in cool regions (Chase *et al.*, 1998). Although thinner sheeting is more effective, it is less durable and more easily damaged. While it is suitable for use on nursery beds and in glasshouses, relatively larger areas, which help to limit the border effect, can be more effective and practical to treat. Access to, and ultimately disposal of, large quantities of plastic sheeting may also pose a problem. For a simplistic approach, small quantities of soil or compost to raise seedlings or for rooting cuttings, contained in sealed plastic bags, moistened and placed on a suitable surface in direct sunshine for 2 weeks will provide excellent nematode control (Bridge, 1987).

Burning debris on the soil surface is an alternative to soil solarization but is less effective. In traditional slash and burn systems in resource-poor agriculture, burning may contribute to sanitizing the soil in terms of plant-parasitic nematodes. The extended bush and forest fallow period, prior to burning, is likely to be a more effective means of reducing nematode populations, as the burning possibly contributes more to reducing beneficial microorganisms than root-knot nematodes (Tchabi *et al.*, 2008). However, for small areas of land, such as nurseries, burning debris has practical relevance. The burning of rice husks on the soil surface prior to establishing tobacco nurseries has proved effective for root-knot nematode management (Bridge, 1996). In the Philippines, post-harvest burning of rice husks on the soil surface suppressed damage by *M. graminicola* and increased yields of the following onion crop (Gergon *et al.*, 2001). Moist soil can improve the conduction of lethal heat to a greater depth, increasing the efficiency of the process (D.L. Coyne, 2008, unpublished data). As a sanitation exercise, burning can also be used to destroy material potentially contaminated with root-knot nematodes following harvest.

19.6 Cropping Systems

19.6.1 Rotation

The principal of crop rotation lies in distancing susceptible crops in space and time from the target nematode species, in order to maintain nematode populations at levels below damage thresholds. The use of crop rotation to manage root-knot nematodes has adapted and evolved in parallel with agriculture itself and, occurs worldwide. Planting crops that are poor or non-hosts of root-knot nematodes in rotation with susceptible crops remains a highly suitable, yet often neglected, tactic to manage root-knot nematodes in resource-poor areas. In addition to the immediate effect of crop diversity on nematode multiplication, multi-cropping cycles may also facilitate the increase of microbial antagonists of nematodes (Sikora, 1992). Successful crop rotation is therefore dependent on a sufficient diversity of crops within the sequence that are useful for the farmer and that prevent root-knot nematode population increase. Netscher (1978) stated that rotations with non-hosts and resistant cultivars in the tropics should be recommended for use on slightly or non-infested land only, employing their use primarily as a preventive nematode control measure as opposed a cure.

As a consequence of the polyphagous nature of many root-knot nematode species, selecting suitable crop rotation sequences can present quite a challenge (Bridge, 1987). Additionally, sufficient land is necessary to enable the full rotation sequence to be completed, which may be a limitation to smallholder farmers. However, a number of crops (e.g. brassicaceous and graminaceous crops, *Allium* spp. and *Amaranthus* spp.) have been identified as generally useful in managing root-knot nematodes. One rotation that appears quite common involves solanaceous crops with cereals, while rotation with groundnut is generally accepted for *M. incognita* management (Dickson and De Waele, 2005). Although several cultivars of a crop may provide useful resistance against root-knot nematodes, the level of control may differ by geographical site and variation in pathotypes and *Meloidogyne* species (Hussey and Jansen, 2002; Cook and Starr, 2006). It is also worth noting that the recommendation of a particular crop for inclusion in a rotation can be misleading, as susceptibility of individual crop cultivars to root-knot nematode species can differ markedly. For example, sweet potato (*Ipomoea batatas*) cv. Sree Bhadra permits *M. incognita* invasion but not development, and thus is suitable for *M. incognita* management (Mohandas and Ramakrishnan, 1996), whereas most other cultivars appear to be susceptible and therefore unsuitable. Sasser (1954) found sweet potato to have a widely differing reaction to different populations of the same species of *Meloidogyne*, while Struble *et al.* (1966) found that 4343 different sweet potato lines showed extreme variation in host suitability to the same *M. incognita* population. Brassicaceous crops are recommended for management of *M. chitwoodi*, but field mustard (*Brassica rapa*) cv. S94152, proved a good host in South Africa (Fourie *et al.*, 1998). As our knowledge of *Meloidogyne* spp. and their hosts expands, and cropping practices evolve, so do the number of exceptions to the rule. *M. arenaria* has been referred to as the 'peanut root-knot nematode' (Sasser, 1954), although some populations have since been found that fail to reproduce on cv. Florunner (Sasser, 1966, 1979). Groundnut was also first considered a non-host of *M. incognita* and *M. javanica* (Sasser, 1954), but was later found as host for populations of both species in Egypt (Ibrahim and El Saedy, 1976; Taha and Yousif, 1976), South Africa (Fourie *et al.*, 2007) and the USA (Tomaszewski *et al.*, 1994). Therefore, while groundnut is generally susceptible to *M. arenaria*, *M. hapla* and some *M. javanica* populations, it will usually help in controlling *M. incognita* (Dickson and De Waele, 2005). Some crops have also been traditionally viewed as resistant or suppressive to root-knot nematodes, such as cassava (*Manihot esculenta*), but have since been shown to be hosts. This is probably because cassava roots are naturally knobbly, which disguises galling. The introduction of new, higher-yielding cultivars without resistance to local populations/species of root-knot nematodes can be highly vulnerable to local populations, demonstrating the need for local screening. In coastal East Africa, improved lines of cassava suffered heavy damage by *Meloidogyne* spp., including the storage roots (Fig. 19.5; Plate 23), shortly after their release (van den Oever, 1995; Coyne *et al.*, 2004). In the Philippines, the rice root-knot nematode, *M. graminicola*, was discovered in all surveyed rice fields and 74% of onion and garlic fields (Gergon *et al.*, 1998). This discovery in the onion–garlic rotation with paddy rice stimulated the need to identify strategies to reduce *M. graminicola*, including long-term crop rotations (Gergon *et al.*, 1998, 2001).

There exists a myriad of recommended crop rotation sequences for the management of individual species, and for root-knot nematodes as a group. Some may be suitable for general management of *Meloidogyne* spp., while some need to be more specifically focused. Many examples are

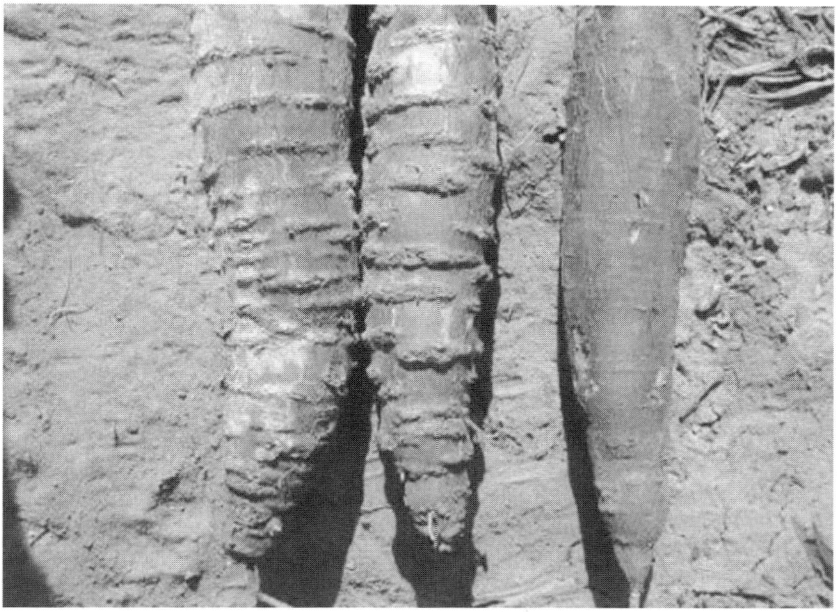

Fig. 19.5. Root-knot-nematode-infected cassava storage roots with ribbed galling.

documented by Sasser and Carter (1985) in addition to more recent publications (Chen *et al.*, 2004; Luc *et al.*, 2005), precluding an exhaustive catalogue for presentation in this volume. Root-knot nematode populations occurring in farmers' fields are often composed of multiple species, whereby rotations with various crops may successively support various species. Presence of multiple *Meloidogyne* spp. at the same location may also affect and interfere with resistance expression against one or more of the species present (Eisenback, 1983). The key, therefore, is not to cultivate the same crop (cultivar) on the same land for too long, while taking into consideration good agricultural practices for using different crop types in the rotation.

19.6.2 Fallow

With few exceptions, land that has lain bare for several seasons or has been cleared from forest or natural vegetation (including weeds and other indigenous plants) rarely has root-knot nematode problems upon initial cultivation. In West African upland rice systems, elevated root-knot nematode densities were observed following no or short fallows, with lower populations present following longer fallows (Coyne *et al.*, 1998). Similarly, root-knot nematodes are an obvious obstacle to production in intensified peri-urban systems, which are characterized by their intensity of production and lack of space for crop rotation. However, in the traditional smallholder cropping systems, fallowing (permitting natural vegetation regrowth) followed by 'slash and burn' can suppress root-knot nematode problems. In addition, reduced weed problems, reduced soil erosion, and the restoration of soil fertility and the natural balance of beneficial soil microorganisms are common additional benefits to fallowing. It should not be overlooked, however, that during a fallow period, the lack of crop production can be a deterrent to the farmer; the loss of productivity during fallow may be greater than the losses due to nematode parasitism.

19.6.3 Cover crops (improved fallow)

An alternative mode of crop rotation is the use of cover crops, which traditionally include leguminous crops, but also refer to grasses, grain crops, etc. Where root-knot nematodes are a problem, the use of poor-host cover crops can provide a useful management tactic, in addition to their soil-erosion-limiting and soil-fertility benefits. Legume cover (or improved fallow) crops can essentially be divided into a number of categories, based on their characteristics and uses, namely creeping annuals that provide good surface cover (e.g. *Mucuna* spp., *Pueraria* spp.), livestock forage legumes (e.g. *Aeschynomene histrix*, *Stylosanthes guianensis*), woody shrubs (e.g. *Crotalaria* spp., *Sesbania rostrata*) and legume food crops (e.g. *Cajanus cajan*, *Vigna unguiculata*). Some cover crops are most useful when incorporated as green mulches, although when used as livestock fodder any mulching benefit would be offset. A balance is therefore required between benefits and uses of such crops. Cover crops may help in reducing root-knot nematode problems, but, as in all rotations, few crop species have impact against a broad spectrum of pests and diseases. Similarly, few cover crops are universally effective against *Meloidogyne* spp., with some being highly susceptible to certain species. For example, sunn hemp is generally known to provide nematode management, although some species readily host *Meloidogyne* spp., such as *Crotalaria pallida* (*M. incognita*) and *Crotalaria juncea* (*M. javanica*) (Silva *et al.*, 1990). Velvet bean (*Mucuna deeringiana*, *M. aterrima*) and sunn hemp are particularly noted for their potential for management of root-knot nematodes and, in general, constitute an excellent cover crop recommendation where root-knot nematodes are problematic (Rodríguez-Kábana *et al.*, 1992; Wang *et al.*, 2007).

Certain crops, such as velvet bean, can also induce soil suppressiveness against nematodes by stimulating build-up of beneficial microorganisms (Vargas-Ayala *et al.*, 2000) through their association with distinctive rhizosphere microflora (Kloepper *et al.*, 1991). The ultimate effect of cover crops in reducing plant-parasitic nematode populations, specifically root-knot nematodes, is due to the presence of bionematicidal compounds present within the roots or other plant parts (see section 19.9.2). However, care is needed in selecting the cover crop in relation to the presence of other plant-parasitic nematodes. Barley (*Hordeum vulgare*), for example, when used in a potato-based cropping system reduced *M. chitwoodi* populations in potato, but led to greater densities of the lesion nematode, *Pratylenchus neglectus* (Ferris *et al.*, 1994).

Difficulties in stand establishment and the length of time required for the suppression of plant-parasitic nematodes appear to be key impediments to farmer adoption of cover crops, although their use as livestock fodder can be attractive.

19.6.4 Antagonistic or trap crops

A number of plants have been identified for their antagonistic (allelopathic) effect on root-knot nematodes (Table 19.1). Some of these crops are planted for their marketable products, while others are used only for reducing damage by a specific nematode species. One of the best studied for management of root-knot nematodes is *Tagetes* spp. (marigold). Although the genus contains 56 species, most reports deal with *Tagetes erecta*, *T. patula* and *T. minuta* (Ferraz and de Freitas, 2004). In principal, *Tagetes* spp. are used in rotation, but can also be effective for root-knot nematode management when intercropped (Khan *et al.*, 1971). *Tagetes* spp. kill root-knot nematodes or prevent their development following root invasion; root exudates can also be strongly nematicidal (Siddiqui and Alam, 1987; section 19.9.2). Reports on the poor host suitability of *Tagetes* spp. to root-knot nematodes are, however, not entirely consistent, suggesting that some species or cultivars are less effective (Chitwood, 2002). Sunn hemp is also considered an effective antagonist of root-knot nematodes when used either in rotation or as an intercrop. It has a similar mode of action against *Meloidogyne* spp. as *Tagetes* spp., since it prevents nematode development after invasion, combined with nematicidal root exudates. However, some of the 350 known *Crotalaria* spp. can act as hosts for *Meloidogyne* spp. (Silva *et al.*, 1990) and so care is required.

Numerous grasses have also been identified as antagonistic to root-knot nematodes (Table 19.1), but some are known hosts to root-knot nematodes, such as *Eragrostis tef* (cv. SA Bruin), *Lolium multiflorum* (cv. Midmar) (Fourie *et al.*, 1998) and *Eragrostis orcuttiana* (O'Bannon and Nyczepir, 1982), which are reported as moderate/good hosts for *M. chitwoodi*. As with most 'alternative' crops, their value to the farmer remains a key feature for their overall acceptance and adoption. In Zimbabwe, use of *Tagetes* spp. by tobacco

Table 19.1. Examples of crops known to be suppressive to root-knot nematode (*Meloidogyne* spp.) populations through antagonistic behaviour in the field.[a]

Plant species	Common name	*Meloidogyne* species
Aeschynomene spp.	Jointvetch	*Meloidogyne* spp.
Allium sativum	Garlic	*M. incognita*
Asparagus officinalis	Asparagus	*Meloidogyne* spp., *M. hapla*, *M. incognita*
Asparagus grayi		*M. incognita*
Bracharia decumbens	Signal grass	*Meloidogyne* spp.
Brassica napus	Rapeseed	*Meloidogyne* spp.
Brassica campestris	Mustard	*Meloidogyne* spp.
Canavalia ensiformis	Horsebean/jack bean	*M. incognita*
Centrosema pubescens	Butterfly pea	*Meloidogyne* spp.
Chrysopogon zizanioides	Vetiver grass	*M. incognita*, *M. javanica*
Crotalaria breviflora	Sunn hemp	*M. incognita*, *M. javanica*
Crotalaria grantiana		*M. incognita*, *M. javanica*
Crotalaria juncea		*M. arenaria*, *M. exigua*, *M. incognita*, *M. javanica*
Crotalaria lanceolata		*M. incognita*, *M. javanica*
Crotalaria longirostrata		*M. arenaria*, *M. incognita*
Crotalaria mucronata		*M. incognita*, *M. javanica*
Crotalaria pallida		*Meloidogyne* spp.
Crotalaria paulina		*M. incognita*, *M. javanica*
Crotalaria retusa		*M. incognita*, *M. javanica*
Crotalaria spectabilis		*M. incognita*
Crotalaria striata		*M. incognita*, *M. javanica*
Cynodon nlemfuensis	Giant star grass	*Meloidogyne* spp.
Cynodon dactylon	Bermuda grass	*M. incognita*
Desmodium spp.	Herbaceous and shrubby legumes	*Meloidogyne* spp.
Digitaria decumbens	Pangola grass	*M. incognita*
Eragrostis curvula	Weeping love grass	*Meloidogyne* spp., *M. javanica*, *M. chitwoodi*
Indigofera spp.	Hairy indigo	*Meloidogyne* spp.
Mucuna deeringiana	Velvet bean	*Meloidogyne* spp.
Mucuna aterrima		*M. incognita*
Panicum maximum	Guinea grass	*M. javanica*
Paspalam notatum	Bahia grass	*M. incognita*
Pennisetum purpureum	Elephant grass	*Meloidogyne* spp.
Ricinus communis	Castor	*M. incognita*
Sesamum indicum	Sesame	*M. incognita*
Sesbania sesban	Egyptian rattle pod, river bean	*Meloidogyne* spp.
Sorghum bicolor	Sorghum	*M. incognita*
Sorghum sudanense	Sudan grass	*M. arenaria* *M. hapla*, *M. incognita*, *M. javanica*
Stylosanthes spp.	Stylo, 'fodder banks'	*Meloidogyne* spp.
Tagetes spp.	Marigold	*Meloidogyne* spp.
Tagetes erecta		*Meloidogyne* spp., *M. incognita*
Tagetes erecta × *Tagetes patula*		*M. arenaria*, *M. incognita*, *M. javanica*
Tagetes jalisciencis		*M. incognita*
Tagetes minuta		*M. incognita*, *M. javanica*
Tagetes patula		*Meloidogyne spp.*, *M. incognita*
Hordeum vulgare	Barley	*Meloidogyne* spp., *M. chitwoodi*

[a]Sources: Murphy *et al.* (1974); Motsinger *et al.* (1977); Silva *et al.* (1990); Villar and Zavaleta (1990); McSorley *et al.* (1994); Fourie *et al.* (1998); Whitehead (1998); Desaeger and Rao (1999); Esparrago *et al.* (1999); Chellemi (2002); Wang *et al.* (2002, 2007); Kandjil *et al.* (2003); Ferraz and de Freitas (2004); Sikora *et al.* (2005); Viaene *et al.* (2006).

farmers has been accepted due to its adverse impact on root-knot nematodes (Shepherd, 1982; Stubbs, 1999). In Malawi, the use of *Tagetes* spp. for management of root-knot nematodes was also promoted because of the crop's value as a food colorant (D.L. Coyne, 2008, personal observation). In South Africa and Egypt, extraction of lucrative essential oils from *T. minuta* is promoted as a useful source of income (Senatore *et al.*, 2004). Woody species such as *Crotalaria* spp. also have additional benefit because of their use as firewood and fencing. Although alternative uses increase the potential of cover crop use in rotations, if it is primarily being employed for the control of the prevalent root-knot nematode species, it is important that it fulfils this role, while additional benefits increase its acceptability or attractiveness to the farmer.

19.7 Resistance

In combination with healthy planting material, host plant resistance, when available, should provide the foundation of any pest management strategy. In most resource-poor areas, nematode resistance breeding programmes pose more than a challenge to any institution, as the elementary information on important *Meloidogyne* species and useful sources of resistance is mostly unavailable or unreliable. Every effort should be made to capitalize on developments made in breeding programmes elsewhere. Although crop cultivars with resistance to root-knot nematodes may not necessarily be suitable or agronomically adapted to conditions outside the target area, the use of such cultivars in breeding programmes to introgress root-knot-nematode-resistant gene(s) could be valuable in developing countries (Starr and Mercer, Chapter 14, this volume). The deployment and use of such resistance in tropical areas in particular could result in a significant increase in specific *Meloidogyne* populations following the high selection pressure exerted on the nematode community. For example, *M. enterolobii* (= *M. mayaguensis*), first described by Yang and Eisenback (1983), presents a substantial threat in tropical and subtropical conditions, where it is a particularly aggressive pest (Rammah and Hirschmann, 1988). It has a wide host range but, importantly, is virulent on tomato with *Mi1*-based resistance. This species remained undetected until recently, most likely due to its morphological variation, which resembles that of *M. incognita*, *M. arenaria* and *M. javanica* (Carneiro *et al.*, 2004). Since its discovery, *M. enterolobii* has been reported from a wide range of countries on various crops (Anonymous, 2008). With increased use of *Mi1*-based resistance, the pest status of *M. enterolobii* could rise dramatically. This questions the extent to which further *Meloidogyne* species remain undiscovered and, consequently, how useful our current sources of host plant resistance are for subtropical and tropical crops. An added complication under tropical conditions involves the breakdown of the *Mi* gene, which is effective against *M. arenaria*, *M. incognita* and *M. javanica*, at soil temperatures exceeding 28 °C (see Williamson and Roberts, Chapter 13, this volume). Resistance may not be a universal tool, but it presents a highly useful component for management of root-knot nematodes, where available.

In resource-poor areas, our knowledge of plant-parasitic nematode communities as well as of resistance sources (crop cultivars and indigenous plants) remains sparse. It is in such areas that we are most likely to discover useful sources of resistance against indigenous species of root-knot nematodes, which should form the basis of future breeding programmes. In Africa, the indigenous rice species, *Oryza glaberrima*, exhibited high levels of resistance against both *M. incognita* and *M. graminicola*. The latter species is not recorded from Africa and has developed independently from *O. glaberrima* (Plowright *et al.*, 1999). Investigating the possible sources of root-knot nematode resistance may yield useful traits. Leafy indigenous vegetables, such as those of the genus *Amaranthus*, which are popular in some peri-urban systems, appear to possess some resistance against root-knot nematodes in Bangladesh (Page, 1979), Uganda (Bafukozara, 1983) and West Africa (James *et al.*, 2005). Grafting of preferred cultivars on to hardier, pest- and disease-resistant rootstocks, an accepted practice with perennial tree, shrub and vine crops in particular, can be used to further exploit root-knot nematode resistance. Commonly used in coffee (*Coffea* spp.), Campos and Villain (2005) imply that the only economic means of producing coffee in Brazil at sites infested with *M. incognita* and *M. paranaensis* is by grafting *C. arabica* on to *C. canephora* cv. Apoatã, which is also immune to *M. exigua*. Of increasing popularity is the innovative use of grafting for control of root-knot nematodes on commercially valuable annual crops (Sikora and Fernández, 2005; Sikora *et al.*, 2005). Production costs associated with such grafting are increased but, through management of root-knot nematodes, grafting is profitable under high infestation levels in high-input systems. Use of resistant rootstocks will depend on the species of root-knot nematode present. Developed and practised in Japan and Korea early in the 20th century, grafting has been applied for disease and root-knot nematode control to avoid the long process of breeding for resistance in popular tomato, aubergine, sweet pepper and cucurbit cultivars. Grafting of such crop plants on to resistant rootstocks or wild *Solanum* spp. can yield good but variable results (Black *et al.*, 2003). The technique has merit and potential for the resource-poor sector, particularly for more valuable crops such as vegetables, but would be dependent upon an organized system using nursery providers.

19.8 Biological Control

Emphasis on the use of biological control agents against root-knot nematodes has increased as our knowledge has progressed, but it has also been catalysed by the increasingly restricted use and removal from the market of effective nematicides. A comprehensive review of biocontrol agents is provided by Hallmann *et al.*, Chapter 17, this volume.

19.9 Chemical Control

Information regarding the use of nematicides in resource-poor agricultural systems remains limited. Although their use offers one of the most reliable control strategies against a wide range of plant-parasitic nematodes, use of these products in subsistence agriculture on low-value crops is more often not recommended (Bridge, 1996), limited or non-existent (Sikora and Fernández, 2005).

Nematicide use in resource-poor agricultural systems is repeatedly stated as low, for the principal reason that farmers can ill afford the high costs. In reality, the simplicity of this assessment undermines the complexity of the issue. The value of a crop is a natural consideration when deciding to use any pest management intervention, especially expensive chemicals. However, relatively inexpensive compounds, such as carbofuran, are often commonly available. The key is whether the resulting gains will provide a profitable cost:benefit ratio following nematicide application. In most cases, the information, as well as the knowledge necessary for making such decisions, simply does not exist. If available for use by resource-poor farmers, such nematicides are often unsuitable, have limited instructions for application, are available in large quantities (and therefore expensive), have been diluted (tampered with) or mixed with other pesticides, are beyond the expiry date, are not always available the next season and may be less effective or have been applied to such an extent at specific sites that they have become ineffective through the development of rapid microbial breakdown (Neuenschwander, 2004; Arbeli and Fuentes, 2007). Vegetable farmers, however, tend to have some limited knowledge of nematicides and their potential impact. They may continue to apply these products as they seemingly provide the only option for nematode management, and vegetables are relatively high-value commodities compared with field crops. Without precise information on the importance and damage incurred by root-knot nematodes on specific crops in specific cropping systems, it remains unethical or unwise to advocate the use of nematicides in most of these cases.

With more intensified systems and cropping of more marginal land, the progressive use of nematicides is likely to rise in resource-poor areas, even on low-value crops. Despite the trend to reduce reliance on nematicides, global pesticide use escalated from 0.49 kg/ha in 1961 to 2 kg/ha in 2004 (Envirostats, 2004) and, consequently, is a factor to consider, even for resource-poor agricultural systems. However, it is equally worth considering that no major synthetic nematicides, with the exception of fosthiozate, have been developed and commercialized since the mid-1970s (see Nyczepir and Thomas, Chapter 18, this volume). Therefore, with the recent phasing out of many nematicides, the identification of alternative nematode management options becomes increasingly urgent and necessary (UNEP, 2000b). Coupled with the loss of effective nematicides, the rise of virulent nematode strains and the detection or spread of nematodes to previously uninfested areas, more complex management programmes are sought (Sikora *et al.*, 2005).

In general, nematicide application follows similar principles whether used in commercial or resource-poor systems, and these are comprehensively discussed by Nyczepir and Thomas (Chapter 18, this volume) as well as in reviews by Johnson (1985), Whitehead (1998), Chitwood (2003) and Haydock *et al.* (2006).

19.9.1 Past and current nematicide use

A recent survey (Haydock *et al.*, 2006) showed that, in terms of global crop production, vegetables attract 38% of the nematicide market, followed by potato (25%), banana (9%), tobacco (8%), sugarbeet (6%) and other crops (14%). Root-knot nematodes are the predominant group, targeted by 48% of global nematicide use across crops, followed by cyst (30%) and other plant-parasitic nematodes (22%). However, to determine nematicide use in developing countries, particularly by resource-poor farmers, is currently a difficult, if not impossible, task. Many of the nematologists from developing countries responding to a recent survey on nematicide use (Table 19.2) emphasized that nematode awareness and control strategies (including chemical treatments) are often limited to larger commercial farms and industrial cropping (plantations) where high-value cash crops are cultivated (W. Wesemael, 2008, personal communication). Additionally, such data relate to plant-parasitic nematodes in general, although root-knot nematodes are the major nematode problem in most cases; this needs to be kept in mind when considering the data in Table 19.2. Nematicide use was reported by 90% of the respondents from developing countries and 100% from least-developed countries that participated in this survey, including the use of both fumigant and non-fumigant nematicides.

While nematicides are being progressively withdrawn from world markets due to increasing environmental and human health concerns, various products remain in use across a wide range of agricultural and horticultural crops, even in the resource-poor sector. In peri-urban vegetable production, for example, significant proportions of farmers are aware of the root-knot nematode problem and will readily apply available nematicides (James *et al.*, 2005). South American potato farmers apply nematicides on a relatively large scale (CIP, undated). Seed treatment or bare-root dips can be effective methods for optimizing nematicide application, and minimizing excess use and environmental and health concerns, particularly in resource-poor areas.

An overview of nematicides used to alleviate, in particular, root-knot nematode problems in developing countries was obtained through the International *Meloidogyne* Project (IMP) during the mid-1980s (Cabanillas, 1985; Davide, 1985; Ferraz, 1985; Ibrahim, 1985; Krishnappa, 1985; Saka, 1985; Sosa-Moss, 1985). Non-

Table 19.2. Relative estimated nematicide use compared with other nematode management strategies in 13 developing and 4 least-developed countries (as indicated by the United Nations), resulting from a global survey (W. Wesemael, Ghent, 2008, personal communication).

Management strategy	Use in developing countries[a] (%)	Use in least-developed countries[b] (%)
Chemical	58	43
Physical	11	11
Biological	4	5
Host plant resistance	4	4
Crop rotation	11	7
Soil amendments/biofumigation	9	26
Others	3	2

[a]Bangladesh, Cameroon, China, Columbia, India, Kenya, Malawi, Nepal, Pakistan, Peru, Thailand, Vietnam, Zambia; [b]Bangladesh, Malawi, Nepal, Zambia.

fumigants and fumigants were used successfully to control root-knot nematodes, such as in Central America, as well as in Caribbean countries (Sosa-Moss, 1985). In some South American countries, such as Chile, the use of carbofuran and aldicarb effectively controlled root-knot nematodes in fruit trees, nurseries, orchards and vineyards, while nematicide application on sugarcane resulted in significant profit margins for farmers in Brazil (Sosa-Moss, 1985). In Asia, *M. incognita* was successfully controlled in sweet potato using 1,3-dichloropropene, ethylene dibromide and products containing chloropicrin, while oxamyl or carbofuran were effective against *Meloidogyne* spp. on tomato in Indonesia (Davide, 1985). In South Korea, *M. hapla* was successfully controlled in groundnut with 1,3-dichloropropene, while carbofuran successfully reduced population levels of root-knot nematodes in tomato in Bangladesh (Davide, 1985). Studies on the chemical control of *Meloidogyne* spp. in the Middle East showed that the mixture of 1,3-dichloropropene and ethylene dibromide was highly successful on a range of crops, while fenamiphos and carbofuran ranked next in their effectiveness against *Meloidogyne* spp. on tomato and tobacco (Stephan, 1978, 1979). In India, aldicarb and carbofuran were the most widely used nematicidal chemicals (Singh and Reddy, 1981; Varma *et al.*, 1981). Farmers on the African continent, particularly in West Africa, applied oxamyl, carbofuran and phorate to increase yields of vegetables (Adesiyan, 1981) and cash crops such as rice (Babatola, 1981). Fumigants were also used in Southern African countries, such as Zimbabwe, particularly against *Meloidogyne* spp. in tobacco (Shepherd, 1982). A wide range of synthetic nematicides is currently available for use on various commercial crops in South Africa (Nel *et al.*, 2007), but are not necessarily used in the resource-poor sector.

Considering the limitations of nematicide use in developing, resource-poor areas, a key question concerns the management of root-knot nematodes by resource-poor farmers without nematicides. How do we foresee these farmers managing these parasites effectively to ensure sustainable food production following removal of many of the available chemical products or the products becoming ineffective? This is of particular relevance when considering production under more intensified systems (e.g. peri-urban and urban agriculture), and on more marginal, infertile land.

19.9.2 Bionematicides

Although not used by resource-poor farmers as such, the phasing out of methyl bromide in developed countries by 2005 and in developing countries by 2015 (UNEP, 2000b; Haydock *et al.*, 2006) has further intensified the search for alternatives that can be used by these farmers, such as phytochemicals with bionematicidal properties (Chitwood, 2002; Ferraz and de Freitas, 2004). A number of alternative fumigants, such as 1,3-dichloropropene, iodemethane and propargyl bromide, have been recommended as alternatives but are unsuitable for subsistence farmers due to their toxicity, high cost (Haydock *et al.*, 2006) and unsuitable package sizes. Since the application of phytochemicals has been used with success to reduce root-knot nematodes across a range of crops (Chitwood, 2002; Ferraz and de Freitas, 2004), there is potential for their use in resource-poor agriculture. Availability and cost-effectiveness of bionematicides will, however, determine their applicability.

Additionally, bionematicides have advantages over synthetic products, in that they: (i) contain novel compounds that plant-parasitic nematodes are not yet able to inactivate; (ii) are less concentrated and thus less toxic than synthetic compounds; (iii) biodegrade relatively rapidly; and (iv) are derived from renewable sources (Chitwood, 2002; Ferraz and de Freitas, 2004). Application of crude phytochemicals by means of cover, green manure or rotation crops, as opposed to synthesized/purified formulations of these products, will most probably be the most viable option for resource-poor farmers to apply against root-knot nematodes. The formulation of synthesized/purified phytochemicals as pre-applied seed/tuber coatings may, however, constitute a significant contribution in assisting resource-poor farmers in the continuous battle against *Meloidogyne* spp.

Chemical compounds with nematicidal properties have been identified from a range of plants (Chitwood, 1992, 1993, 2002; Ferraz and de Freitas, 2004) and other organisms such as algae, bacteria, crustaceans and fungi (Anke *et al.*,

1995; Ehteshamul-Haque, 1997; Warrior *et al.*, 1999; Chitwood, 2002). Various bionematicides of a plant-, microbe- or chitin-based nature continue to be screened and evaluated, but are also beginning to work their way on to the market (Haydock *et al.*, 2006). Some phytochemicals have antagonistic, suppressive or repellent effects on plant-parasitic nematodes, while others are toxic (Viaene *et al.*, 2006).

19.9.2.1 Avermectins

Avermectins, potent macrocyclic lactones produced by the soil-inhabiting bacterium *Streptomyces avermitilis*, have activity against a broad spectrum of helminths (Cayrol *et al.*, 1993; Blackburn *et al.*, 1996, Faske and Starr, 2006), but also against insects (Zufall *et al.*, 1989) and mites (Putter *et al.*, 1981). The chemical has also been investigated for its nematicidal efficacy to control plant-parasitic nematodes in field crops (Sasser *et al.*, 1982; Blackburn *et al.*, 1996; Monfort *et al.*, 2006), and was recently registered as Avicta® (active substance: abamectin, a mixture of avermectins) in the USA as a cotton seed dressing (Anonymous, 2007). For other crops, Avicta® continues to be evaluated to increase its range of application. Abamectin effectively controlled *M. incognita* in vegetables and cotton when applied as a seed dressing comprising several avermectin-producing bacterial strains (Monfort *et al.*, 2006). In contrast, Faske and Starr (2007) found limited effectiveness of Avicta®-treated cotton seed; they reported that protection of the cotton tap root from infection by *M. incognita* extended for only a few centimetres of root length. In terms of nematicidal efficacy, the B group of avermectins are biologically more active than the A group (Lasota and Dybas, 1991). Incorporation of avermectin B_1 into soil (at 0.3, 1.1 and 3.3 kg/ha) was 10–30 times more effective than several organophosphates and carbamates in reducing *M. incognita* populations (Putter *et al.*, 1981). Although not currently being developed as a formulation to be applied in the soil, soil incorporation of granular formulations of avermectin B_1 was also reported to inhibit reproduction of *M. incognita* and root galling on tobacco, at an equivalent efficacy to several synthetic nematicides (Sasser *et al.*, 1982). However, further development and release of products since these early investigations has been slow. The low water solubility and rapid degradation of avermectin means it is unlikely to cause contamination of soil water (Garabedian and Van Gundy, 1983) but, conversely, may limit its potential effectiveness as a seed treatment.

19.9.2.2 Neem products

Neem products, obtained from the tree *Azadirachta indica*, are among the most extensively studied (Akhtar, 2000) and most widely used bionematicides, especially by farmers in India and Pakistan (Guerena, 2006). Neem has insecticidal, antifungal and antifeedant properties for use on a wide range of crops (Guerena, 2006). Various chemical substances in neem (azadirachtin, kaempferol, nimbidin, nimbin, quercetin, salannin, thionemone and others) contribute to its nematicidal properties (Khan *et al.*, 1974; Ferraz and de Freitas, 2004). A range of neem formulations is commercially marketed as nematicides, insecticides, fungicides or miticides. According to Thakur (1995a), optimal root-knot nematode control is obtained within 3 weeks after incorporation of neem, since polyphenols are released in the highest concentrations during this period. *In vitro* studies showed that products from neem seed resulted in significant mortality, immobility and reduction of hatching of J2 of *Meloidogyne* spp. (Paruthi *et al.*, 1996; Javed *et al.*, 2008). Incorporation of neem oilcakes, leaves or leaf powder in soil reduced penetration by J2 of *Meloidogyne* spp., gall formation and final population densities on a wide range of crops (Sharma, 1987; Haseeb, 1991; Thakur, 1995b).

Coating of tomato seed with Suneem or neem oil reduced *M. incognita* infection and population development substantially (Dash, 1990; Akhtar, 1997). Similarly, a root dip with neem substantially delayed the development of *M. incognita* (Akthar, 1996) and *M. javanica* (Vats, 1993) on tomato seedlings. On pea, populations of *M. incognita* were reduced and yields increased following seed coating with neem products (Mojunder *et al.*, 2002). Numerous examples have further demonstrated the effective management of root-knot nematodes when neem-based products were combined with other products, including biocontrol agents, even though the effects of neem on biocontrol agents could be detrimental. Combining neem products with *Paecilomyces lilacinus* spores (Rao, 1997a), *Pasteuria penetrans* and

Pasteuria lilacinus (Reddy, 1997), *Trichoderma harzianum* (Rao, 1997b) or arbuscular mycorrhizal fungi (*Glomus mosseae*) (Rao, 1997c) all resulted in substantial root-knot nematode reduction on a range of crops.

However, results are not always consistent between studies. Variation may arise from inconsistency of product formulation, or especially from preparations made *in situ* from fresh material, which can vary in content and quality of active compounds between locations and plant parts. Although most reports indicate that neem-based products successfully reduce root-knot nematodes, neem cake did not reduce *M. javanica* galling on tobacco when applied at 100 and 200 g/m^2 (Krishnamurthy, 1990), for example. Agbakli (1992) also reported a lack of nematode control following application of foliar neem extracts on jute (*Corchorus olitorius*), lettuce (*Lactuca sativa*) and celosia (*Celosia argentea*) in Benin. Phytotoxicity has also been recorded, such as on tomato after application of neem oil (Akthar, 1997).

Neem products and locally processed formulations do, however, offer great cost-efficient potential for management of root-knot nematodes. Neem products are reputedly safe for humans (Schmutterer, 1997) and, due to their relative selectivity, are ideal for use in integrated pest management programmes without causing environmental disturbance.

19.9.2.3 Glucosinolates in Brassica *spp.*

Research on brassicaceous (*Brassica* spp.) crops as 'natural' nematicides commenced as early as the 1930s (Smedley, 1939). Successful reduction of *Meloidogyne* spp. following brassicaceous crop biofumigation is now recorded across a wide geographical spectrum (Stirling and Stirling, 2003; Monfort *et al.*, 2007; Qing *et al.*, 2007). Brassicaceous plant material contains volatile sulfur-containing compounds (glucosinolates), which are hydrolysed to active fungicidal, bactericidal and nematicidal isothiocyanates (Kirkegaard *et al.*, 1996; Brown and Morra, 1997). Stapleton *et al.* (1998) demonstrated the benefit of biofumigation in reducing multiple soil-borne pathogens such as *M. incognita*, *Sclerotium rolfsii* and *Pythium ultimum* 7 days after incorporating brassicaceous residues into the soil. Rapeseed (*Brassica napus*) green manure grown prior to potatoes in the USA was also shown to significantly reduce populations of *Meloidogyne* spp. on potato (Stark, 1995). Recent work has also shown that exposure to sublethal concentrations of isothiocyanates can play a role in nematode suppression by affecting root-knot nematode behaviour. Exposure of J2 of *M. incognita* to sublethal concentrations of benzyl isothiocyanate reduced infectivity and virtually eliminated egg production (Zasada *et al.*, 2009).

While almost all brassicaceous crops produce glucosinolates, several are good hosts for *Meloidogyne* spp. (McSorley and Frederick, 1995; Sikora and Fernández, 2005; Pattison *et al.*, 2006), e.g. field mustard cv. Norfolk (Liebanas and Castillo, 2004). This is generally explained by the variation in glucosinolate content present, as well as by environmental effects (Stirling and Stirling, 2003). Brassicaceous crops with high glucosinolate concentrations should therefore be selected to obtain optimal control of root-knot nematodes. During a screening exercise, Pattison *et al.* (2006) identified a number of fodder radishes (*Raphanus sativus*) that combined relatively high levels of resistance with good biofumigant activity. Farmers should also be made aware that adverse effects on crop growth and yield, as observed in vegetables by Monfort *et al.* (2007), can occur as a result of biofumigation. In dryland conditions, insufficient disruption of crop tissue and incorporation of residues during periods of low temperatures are also factors that can contribute to the lack of a biofumigation effect (Stirling and Stirling, 2003).

19.9.2.4 Polythienyls in Tagetes *spp.*

Goff (1936) first observed resistance to plant-parasitic nematodes in *Tagetes* spp., reporting that both *T. patula* and *T. erecta* were poor hosts to *Meloidogyne* spp. Polythienyls in the roots of *Tagetes* spp. are the nematicidal active ingredient (Chitwood, 2002), particularly against root-knot and lesion nematodes (Ferraz and de Freitas, 2004). The formation of singlet oxygen by photo-activated α-terthienyl is probably the mechanism present in *Tagetes* spp. and responsible for nematode mortality (Ferraz and de Freitas, 2004). Inhibition of hatching, as well as a reduction in gall formation, number of egg masses and final population of *M. incognita*, were recorded in tomato and aubergine when undiluted extracts and chopped leaves of *Tagetes* spp. were applied as a combination treatment (Walia, 1997). Intercropping *T. erecta* with aubergine was also superior to carbofuran application in reducing final *M. javanica* densities (Dhanger *et al.*, 1996), and when intercropped with tomato resulted in fewer *M. javanica* root galls and increased growth, compared with monocropped tomato (Abid and Maqbool, 1990). It also provided successful management of root-knot nematodes when alley cropped in 'annually' replaced banana plantations (UNEP, 2000b).

19.9.2.5 Ricin in Ricinus communis

Ricin, the active substance in castor (*Ricinus communis*), a fast-growing tropical shrub, has been identified as nematicidal (Ferraz and de Freitas, 2004), with numerous examples attesting its effect. On tomato, furrow and spot application of castor bean and mustard oilcake effectively reduced *M. incognita* populations, with spot applications leading to a substantial increase in yield (Deka, 1997). Incorporation of castor cake in soil resulted in a substantial decrease in *M. incognita* populations in davana (*Artemisia pallens*) (Pandey, 1994); when combined with karanj (*Pongamia pinnata*) and mahua (*Madhuca longifolia*) seed cake effectively prevented penetration of J2 of *M. incognita* and gall formation on tomato (Poornima, 1997). When castor, mahua and groundnut oilcakes combined with arbuscular mycorrhizal fungi (*Glomus fasciculatum*) were incorporated into soil prior to sowing blackgram (*Vigna mungo*), population levels of *M. incognita* were reduced substantially (Sankaranarayanan, 1997).

19.9.2.6 Velvet bean compounds

Velvet bean (*Mucuna* spp.) contains several compounds with reported nematicidal activity, such as alcohols, fatty acids, allantion, daucosterol + stigmasterol, D-glycoside and L-dopa (Barbosa *et al.*, 1999; Chitwood, 2002). Although their mode of action is yet to be determined, velvet bean appears particularly effective at reducing populations of *Meloidogyne* spp. In Brazil, for example, *M. javanica* was reduced by 65% following 100 days of cultivation of *M. aterrima* before incorporation into the soil, compared with a 200% increase in *M. javanica* on adjacent tomato (Asmus and Ferraz, 1988). Quénéhervé *et al.* (1998), meanwhile, demonstrated the positive value of *Mucuna pruriens* in reducing *M. incognita* populations when planted 3 months prior to a vegetable crop. Use of *Mucuna* spp. can also have an adverse affect on pathogenic fungi, such as *Fusarium oxysporum*, and therefore offers the possibility of providing multiple-purpose pest management (Ferraz *et al.*, 1977).

19.9.2.7 Monocrotaline in Crotalaria *spp.*

The active substance monocrotaline in sunn hemp has been reported to exhibit nematicidal properties (Mori *et al.*, 2000). Incorporation of *Crotalaria spectabilis* residue in soil resulted in reduced galling by *M. incognita* and *M. javanica* in tomato (Villar and Zavaleta, 1990), while a similar response was observed for *M. incognita* on okra (Wang *et al.*, 2007). Villar and Zavaleta (1990) indicated that successful reduction of *M. incognita* and *M. arenaria* galling of tomato after incorporation of *C. longirostrata* residues was due to toxic products of microbial degradation, and not to the toxic exudates from the plant. It must be noted, however, that some alkaloids contained in *Crotalaria* spp. have proved hepatotoxic to livestock, with monocrotaline one of the most toxic (Ferraz and de Freitas, 2004).

19.9.2.8 Glucoside in cassava

Applications of the cassava (*Manihot esculenta*) flour by-product known as manipueira or cassareep have been reported to provide some level of control of *Meloidogyne* spp. (Whitehead, 1998). The cyanogenic glucoside linamarin present in *Manihot* spp. roots is responsible for the nematicidal effect and has been used for management of root-knot nematodes in Brazil (Sena and Ponte, 1982; Ponte *et al.*, 1996). Incorporation of manipueira as a soil amendment at rates of 20–80 m^3/ha resulted in substantial reductions of both *M. incognita* and *M. javanica* populations in okra (Ponte *et al.*, 1987) and cassava (Ponte and Franco, 1981).

19.9.2.9 Other sources of phytochemicals with nematicidal properties

In addition to various bionematicides derived from plants (Table 19.2), a number of products or compounds based on algae, fungi and bacteria

(Goswami, 1993; Whitehead, 1998; Chitwood, 2002; Haydock *et al.*, 2006) and crustacean chitin (Rodriguez-Kábana, 1990; Ehteshamul-Haque, 1997; Chitwood, 2002; Ferraz and de Freitas, 2004) are also antagonistic, suppressive or detrimental to root-knot nematodes (see Hallmann *et al.*, Chapter 17, this volume). The class of plant secondary metabolites 1,2-dehydropyrrolizidine alkaloids (PAs) may have potential for management of *Meloidogyne*. In pot tests, Thoden *et al.* (2009) found that, although *M. hapla* was not repelled by commercially available PA-containing plants, the development of J2 was completely suppressed on floss flower (*Ageratum houstonianum*) and silver ragwort (*Senecio bicolour*). Other plant by-products to note, such as furfural (Al-Hamdany, 1999; Ferraz and de Freitas, 2004; Ismail, 2007) and molasses (Bettiol, 1996; Vawdrey, 1997), have also been highlighted for their nematicidal properties. Furfural, a by-product of sugarcane, is currently registered for use against plant-parasitic nematodes in a number of countries for a range of crops (Haydock *et al.*, 2006; Nel *et al.*, 2007).

While the extracts of many plants often show potential in the laboratory or in glasshouse studies, the practicality of preparing such extractions, ensuring quality control and maintaining their efficacy under field conditions is very often not realized, leading to contradictory reports, which question the suitability and usefulness of the product concerned.

19.10 Conclusions and Future Directions

In order to achieve the improved productivity necessary to maintain a sustainable food supply in developing countries and resource-poor areas, farmers need to be cognizant of plant-parasitic nematodes and constantly update and maintain appropriate pest management systems. Marginal areas of poorer-quality land with limited water availability and/or heavy pest pressures will be increasingly required for food production as prime land becomes scarcer. Paradoxically, expansion on to such land will challenge pest management systems further and add to the cost per unit food production. In such situations, root-knot nematodes will become increasingly prominent. We have outlined a variety of options possible for resource-poor farming conditions, to aid the agriculturist and field nematologist. However, without the expertise to understand the problem in the first instance, the various management options will be of limited value. A crucial underlying premise that requires imminent attention is the scarcity of expertise and awareness of nematode problems in resource-poor situations. For many years there has been continuous and gradual erosion globally of nematology expertise (Coomans, 2000; Luc *et al.*, 2005). Resource-poor areas have traditionally been deficient of nematological expertise, with complete absence in many cases. A key objective of the International *Meloidogyne* Project (IMP) (1975–1985) was to address this shortage and the limited awareness, with input from approximately 200 nematologists based in 70 countries (Sasser *et al.*, 1983). Since then, no other project or consolidated effort has come close to sustaining the progress made during this commendable effort, with the all-too-inevitable loss of momentum on the one hand and a consequent decay of the nematological infrastructure on the other. Thus, some of the most-wanting places remain the most in need of such support.

During its 10 years of activity, the accomplishments of the IMP included the promotion of nematological awareness, improved knowledge on species distribution, identification of new species, improved taxonomic methods and enhanced research capability in developing nations. Upon conclusion of the IMP in 1985, key priority areas for future investment included geographical distribution and species identification records (surveys), information on economic importance, resistance identification, crop systems management (including chemical and biological), training of nematologists and creating awareness. Ironically, these 'needs' reflect very closely those identified in a recent synthesis of tropical nematology (De Waele and Elsen, 2007), with the possible exception of an additional priority to attain greater understanding of the role of nematodes in disease complexes. Consequently, this begs the question as to how nematology for resource-poor countries is progressing and developing. How is support ultimately being attracted from international aid, national programmes or the private sector? How can the continuity of the likes of the IMP be maintained? Perhaps, more importantly, from where will the next generation of nematologists and soil health specialists emerge? Support remains meagre, notwithstanding some truly outstanding efforts of nematology support for developing countries, such as the Postgraduate International Nematology Course, now supplemented with the European Master of Science in Nematology (EUMAINE), both based at Ghent University, with support from the Belgian Government and the EU, respectively; the Nematology Initiative in East and Southern Africa (NIESA), with support from The Gatsby Charitable Trust; and the Flemish-Interuniversity Project (V.L.I.R.) 'Mobilising IPM for sustainable nematode management in household and community gardens of resource-poor farmers in South Africa', in association with relevant South African universities and national Institutes. Even in the Consultative Group for International Agricultural Research (CGIAR) system, which provides support and underscores capacity building for national programmes, nematologists are scarce and declining (Sharma *et al.*, 1997; Coyne *et al.*, 2008). There is real concern across the nematological world for the future development and support of nematological expertise, which now constitutes a major limiting factor in agricultural research and services, particularly for resource-poor areas. A scarcity of nematologists has obvious consequences, and impacts adversely on research efforts aimed at problem diagnosis and developing solutions. Furthermore, it also has a negative impact in transfer of crucial information, while the lack of expert nematologists involved in quarantine services reduces the likelihood of nematode pests being detected in cross-border trade and commerce.

19.11 References

Abid, M. and Maqbool, M.A. (1990) Effects of inter-cropping of *Tagetes erecta* on root knot disease and growth of tomato. *International Nematology Network Newsletter* 7, 41–42.

Adesiyan, S.O. (1981) Comparison of methods and rates of application of some short residual toxicants for the control of *Meloidogyne incognita* attacking some vegetable crops. In: *Proceedings of the Third International* Meloidogyne *Project Research Planning Conference on Root-knot Nematodes,* Meloidogyne *spp., Regions IV & V,* 16–20 November 1981, Ibadan, Nigeria.

Agbakli, R. (1992) Use of neem extracts (*Azadirachta indica*) to control root knot nematodes (*Meloidogyne* spp.) in leafy vegetable crops in South Benin. *Mededelingen van de Faculteit Landbouwetenschappen, Rijksuniversiteit, Gent,* 57, 913–918.

Akthar, M. (1996) Evaluation of nematicidal properties of a biopesticide against root-knot nematode *Meloidogyne incognita. Tests of Agrochemicals and Cultivars* 17, 104–105.

Akthar, M. (1997) Control of root-knot nematode *Meloidogyne incognita* in tomato plants by seed coating with suneem and neem oil. *Journal of Pesticide Science* 22, 37–38.

Akhtar, M. (2000) Nematicidal potential of the neem tree *Azadirachta indica* (A. Juss.). *Integrated Pest Management Reviews* 5, 57–66.

Al-Hamdany, M.A. (1999) Use of furfural for control of the root-knot nematode *Meloidogyne javanica* on cucumber and eggplant under greenhouse conditions. *Arab Journal of Plant Protection* 17, 84–87.

Anke, H., Stadler, M., Mayer, A. and Sterner, O. (1995) Secondary metabolites with nematicidal and antimicrobial activity from nematophagous fungi and Ascomycetes. *Canadian Journal of Botany* 73, 932–939.

Anonymous (2007) Planting corn on cotton acres may open a whole new can of worms. *Corn South,* 13–16.

Anonymous (2008) Safe Food International: Plant Health Issues. http://www.regional news.safefoodinternational.org (accessed 5 November 2008).

APHIS (USDA Animal and Plant Inspection Service) (2008) Exotic nematode plant pests of agricultural and environmental significance to the USA. http://nematode.unl.edu/projectpest.htm (accessed 15 November 2008).

Arbeli, Z. and Fuentes, C.L. (2007) Accelerated biodegradation of pesticides: an overview of the phenomenon, its basis and possible solutions; and a discussion on the tropical dimension. *Crop Protection* 12, 1733–1746.

Arif, M., Moeen-Uddin, Kadian, M.S., Ilangantileke, S. and Wassimi, N. (undated) Increasing potato production in Afghanistan. Press release: International Potato Center, Lima, Peru. www.cipotato.org/pressroom/press_releases_detail.asp (accessed 19 November 2008).

Asmus, R.M.F. and Ferraz, S. (1988) Antafonismo de algumas espésies vegetais, principalmente leguminosas, *Meloidogyne javanica. Fitopatologia Brasileira* 13, 20–24.

Babatola, J.O. (1981) Root-knot nematode problems in rice in Nigeria. In: *Proceedings of the Third International* Meloidogyne *Project Research Planning Conference on Root-knot Nematodes,* Meloidogyne *spp., Regions IV & V,* 16–20 November 1981, Ibadan, Nigeria.

Bafokuzara, N.D. (1983) Reaction of *Amaranthus hybridus* subsp. *incurvatus* to varying levels of *Meloidogyne javanica* inoculum. *Nematropica* 13, 17–26.

Baimey, H., Coyne, D.L., Al Banna, L. and James, B. (2007) Diversity of *Meloidogyne* species affecting vegetable crops in Benin, West Africa. In: *Proceedings of the Eighteenth Symposium of the Nematological Society of Southern Africa. South African Journal of Plant and Soil* 24, 255–256.

Barbosa, L.C.A., Barcelos, F.F., Demuner, A.J. and dos Santos, M.A. (1999) Chemical constituents from *Mucuna aterrima* with activity against *Meloidogyne incognita* and *Heterodera glycines. Nematropica* 29, 81–88.

Bello, A., López-Pérez, J.A., Sanz, R., Escuer, M. and Herrero, J. (2000) Biofumigation and organic amendments. *Regional Workshop on Methyl Bromide Alternatives for North Africa and Southern European Countries.* UNEP, Paris, France, pp. 113–141.

Bello, A., López-Pérez, J.A., García-Álverez, A., Sanz, R. and Lacasa, A. (2004) Biofumigation and nematode control in the Mediterranean region. In: Cook, R.C. and Hunt, D.J. (eds) *Nematology Monographs and Perspectives* 2, Brill, Leiden, The Netherlands, pp. 133–149.

Bergeson, G.B. (1968) Evaluation of factors contributing to the pathogenicity of *Meloidogyne incognita. Phytopathology* 58, 49–53.

Bettiol, W. (1996) Soil solarization for controlling the root-knot nematode in okra crop. *Horticultura Brasileira* 14, 158–160.

Black, L.L., Wu, D.L., Wang, J.F., Kalb, T., Abbass, D. and Chen, J.H. (2003) *Grafting tomatoes for production in the hot-wet season.* AVRDC International Cooperators' Guide No. 03–551. Shanhua, Taiwan.

Blackburn, K., Alm, S.R. and Yeh, T.S. (1996) Avermectin B_1, isazofos, and fenamiphos for control of *Hoplolaimus galeatus* and *Tylenchorhynchus dubius* infesting *Poa annua. Journal of Nematology* (S) 28, 687–694.

Bridge, J. (1987) Control strategies in subsistence agriculture. In: Brown, R.H. and Kerry, B.R (eds) *Principles and Practice of Nematode Control in Crops.* Academic Press, Melbourne, Australia, pp. 389–420.

Bridge, J. (1996) Nematode management in sustainable and subsistence agriculture. *Annual Review of Phytopathology* 34, 201–225.

Bridge, J., Coyne, D. and Kwoseh, C.K. (2005) Nematode parasites of tropical root and tuber crops. In: Luc, M., Sikora, R.A. and Bridge, J. (eds) *Plant Parasitic Nematodes in Subtropical and Tropical Agriculture,* 2nd edn. CAB International, Wallingford, UK, pp. 221–258.

Brown, P.D. and Morra, M.J. (1997) Control of soil-borne plant pests using glucosinolate-containing plants. *Advances in Agronomy* 61, 167–231.

Brown, R.H. (1987) Control strategies in low value crops. In: Brown, R.H. and Kerry, B.R (eds) *Principles and Practice of Nematode Control in Crops.* Academic Press, Melbourne, Australia, pp. 351–388.

Brown, R.H. and Kerry, B.R. (eds) (1987) *Principles and Practice of Nematode Control in Crops.* Academic Press, Melbourne, Australia.

Cabanillas, E. (1985) Summary report on the current status, progress and needs for *Meloidogyne* research in South America. In: Sasser, J.N. and Carter, C.C. (eds) *An Advanced Treatise on* Meloidogyne. *Volume I. Biology and Control.* North Carolina State University Graphics, Raleigh, North Carolina, pp. 347–350.

Campos, V.P. and Villain, L. (2005) Nematode parasites of coffee and cacao. In: Luc, M., Sikora, R.A. and Bridge, J. (eds) *Plant Parasitic Nematodes in Subtropical and Tropical Agriculture,* 2nd edn. CAB International, Wallingford, UK, pp. 529–580.

Carneiro, R.M.D.G., Tigano, M.S., Randig, O., Almeida, M.R.A. and Sarah, J.L. (2004) Identification and genetic diversity of *Meloidogyne* spp. (Tylenchida: Meloidogynidae) on coffee from Brazil, Central America and Hawaii. *Nematology* 6, 287–298.

Cayrol, J.-C., Dijan, C. and Frankowski, J.P. (1993) Efficacy of abamectin B_1 for the control of *Meloiodygne arenaria. Fundamentals of Applied Nematology* 16, 239–246.

Chase, S.A., Sinclair, T.R., Chellemi, D.O., Olson, S.D., Gilreath, J.P. and Locascio, S.J. (1998) Heat sensitive films for increasing soil temperatures during solarization in a humid cloudy environment. *Hortscience*, 34, 1085–1089.

Chellemi, D.O. (2002) Nonchemical management of soilborne pests in fresh market vegetable production systems. *Phytopathology* 92, 1367–1372.

Chen, Z.X., Chen, S.Y. and Dickson, D.W. (2004) *Nematology – Advances and Perspectives*, Vol. 2. CAB International, Wallingford, UK.

Chitwood, D.J. (1992) Nematicidal compounds from plants. In: Nigg, H.N. and Siegler, D. (eds) *Phytochemical Resources for Medicine and Agriculture*. Plenum Press, New York, pp. 185–204.

Chitwood, D.J. (1993) Naturally occurring nematicides. In: Chitwood, D.J., Duke, S.O., Menn, J.J. and Plimmer, J.R. (eds) *Pest Control with Enhanced Environmental Safety*. ACS Symposium Series No. 524, American Chemical Society, Washington, DC, pp. 300–315.

Chitwood, D.J. (2002) Phytochemical based strategies for nematode control. *Annual Review of Phytopathology* 40, 221–249.

Chitwood, D.J. (2003) Nematicides. In: Plimmer, J.R. (ed.) *Encyclopedia of Agrochemicals*, Vol. 3. John Wiley and Sons, New York, pp. 1104–1115.

CIP (International Potato Center) (Undated) Hazardous pesticides commonly used in the Andes. www.cipotato.org/pressroom/press_releases_detail.asp?cod=30 (accessed 15 November 2008).

Cook, R. and Starr, J.L. (2006) Resistant cultivars. In: Perry, R.N. and Moens, M. (eds) *Plant Nematology*. CAB International, Wallingford, UK, pp. 370–391.

Coomans, A. (2000) Nematode systematics: past, present and future. *Nematology* 2, 3–7.

Coyne, D.L., Plowright, R.A., Thio, B. and Hunt, D. (1998) Plant parasitic nematode diversity and prevalence in upland rice based cropping systems in Ivory Coast: preliminary observations on the effects of cropping intensification. *Fundamental and Applied Nematology* 21, 723–732.

Coyne, D.L., Khizzah, W. and Whyte, J. (2004) Root knot nematode damage to cassava in Kenya. *Roots* 9, 3–5.

Coyne, D.L., Toko, M., Andrade, M., Hanna, R., Sitole, A., Kagoda, F., Al Banna, L. and Marais, M. (2005) *Meloidogyne* spp. and associated galling and damage on cassava in Kenya and Mozambique. In: *Proceedings of the Seventeenth Symposium of the Nematological Society of Southern Africa. African Plant Protection* 12, 35–36.

Coyne, D.L., Nicol, J. and Claudius-Cole, A. (2007) *Practical Plant Nematology: Field and Laboratory Guide*. International Institute of Tropical Agriculture, Ibadan, Nigeria.

Coyne, D.L., Nicol, J. and Sibanda, M. (2008) Support for nematology in developing world agriculture: how is the future looking? In: *Proceedings 5th International Congress of Nematology, Brisbane, Queensland, Australia, 13–18 July 2008*, p. 85 (abstract).

Dash, B.K. (1990) Response of root-knot nematode infecting tomato to seed treatment by some neem products. *Journal of Agricultural Research* 2, 148–150.

Davide, R.G. (1985) Summary report on the current status, progress and needs for *Meloidogyne* research in Region VI. In: Sasser, J.N. and Carter, C.C. (eds). *An Advanced Treatise on* Meloidogyne. *Volume I. Biology and Control*. North Carolina State University Graphics, Raleigh, North Carolina, pp. 369–372.

Deka, U. (1997) Management of root-knot nematode, *Meloidogyne incognita* on tomato through soil amendments. *Journal of the Agricultural Science Society of North East India* 10, 91–94.

Desaeger, J. and Rao, M.R. (1999) The root-knot nematode problem in sesbania fallows and scope for managing it in western Kenya. *Agroforestry Systems* 47, 273–288.

De Waele, D. and Elsen, A. (2007) Challenges in tropical plant nematology. *Annual Review of Phytopathology* 45, 457–485.

Dhanger, D.S., Gupta, D.C. and Jain, R.K. (1996) Effect of marigold (*Tagetes erecta*) intercropped with brinjal in different soil types on disease intensity of root-knot nematode (*Meloidogyne javanica*). *Indian Journal of Nematology* 25, 181–186.

Dickson, D. and De Waele, D. (2005) Nematode parasites of groundnut. In: Luc, M., Sikora, R.A. and Bridge, J. (eds) *Plant Parasitic Nematodes in Subtropical and Tropical Agriculture*, 2nd edn. CAB International, Wallingford, UK, pp. 393–436.

Dinham, B. (2003) Growing vegetables in developing countries for local urban populations and export markets: problems confronting small-scale producers. *Pest Management Science* 59, 575–582.

Dubois, T., Coyne, D., Kahangi, E., Turoop, L. and Nsubuga, E.W.N. (2006a) Endophyte-enhanced banana tissue culture: an example of public–private partnerships in Kenya and Uganda to transfer technology. *African Technology Development Forum* 3, 18–24.

Dubois, T., Gold, C.S., Paparu, P., Athman, S. and Kapindu, S. (2006b) Tissue culture and the *in vitro* environment. Enhancing plants with endophytes: potential for ornamentals? In: Teixeira Da Silva, J. (ed.)

Floriculture, Ornamental and Plant Biotechnology: Advances and Topical Issues. Global Science Books, London, pp. 397–409.

Duxbury, J.M. and Lauren, J.G. (2006) *Enhancing Technology Adoption for the Rice–Wheat Cropping System of the Indo-Gangetic Plains Project Report*. Soil Management CRSP Management Entity, University of Hawaii, Hawaii.

Ehteshamul-Haque, S. (1997) Use of crustacean chitin and plant growth promoting bacteria for the control of *Meloidogyne javanica* root knot nematode in chickpea. *Pakistan Journal of Nematology* 15, 89–93.

Eisenback, J.D. (1983) Loss of resistance in tobacco cultivars 'NC 95' by infection of *Meloidogyne arenaria* and *M. hapla*. *Journal of Nematology* 15, 478 (abstract).

Envirostats (2004) Environmental statistics and other topics. http:/envirostats.info/ (accessed 2 April 2008).

Esparrago, G., Barreiro, J.M., Ruales, C. and Bieche, B.J. (1999) Comparison of the reproduction of *Meloidogyne* populations on roots of *Crotalaria spectabilis* and processing tomato varieties with the *Mi* gene. *Acta Horticulturae* 487, 267–270.

Faske, T.R. and Starr, J.L. (2006) Sensitivity of *Meloidogyne incognita* and *Rotylenchulus reniformis* to abamectin. *Journal of Nematology* 38, 240–244.

Faske, T.R. and Starr, J.L. (2007) Cotton root protection from plant-parasitic nematodes by abamectin-treated seed. *Journal of Nematology* 39, 27–30.

Fernández, E. and Labrada, R. (1995) Experiencia en el uso de la solarizacíon en Cuba. In: *Memorias del taller 'Solarizacíon del Suelo' Escuela Agrícola Panamericana 'El Zamorano'*, Honduras, 18–21 September, pp. 5–6.

Ferraz, C.A.M., Cia, E. and Sabino, N.P. (1977) Effect of Bengal beans and groundnuts in rotation with cotton. *Bragantia* 36, 1–9.

Ferraz, S. (1985) Summary report on the current status, progress and needs for *Meloidogyne* research in Brazil (Region III). In: Sasser, J.N. and Carter, C.C. (eds) *An Advanced Treatise on* Meloidogyne. *Volume I. Biology and Control*. North Carolina State University Graphics, Raleigh, North Carolina, pp. 351–360.

Ferraz, S. and de Freitas, L.G. (2004) Use of antagonistic plants and natural products. In: Chen, Z.X., Chen, S.Y. and Dickson, D.W. (eds) *Nematology – Advances and Perspectives. Vol 2: Nematode Management and Utilization*. CAB International, Wallingford, UK, pp. 931–977.

Ferris, H., Carlson, H.L. and Westerdahl, B.B. (1994) Nematode population changes under crop rotation sequences: consequences for potato production. *Agronomy Journal* 86, 340–348.

Fourie, H. and McDonald, A.H. (2001) *Nematode pests of maize and sorghum*. Research report: Project M151/10 for 2000/2001. ARC-Grain Crops Institute, Potchefstroom, South Africa.

Fourie, H., Zijlstra, C. and McDonald, A.H. (1998) ITS-PCR sequence-based identification of *Meloidogyne chitwoodi* from Mooi River, South Africa, and screening of crops for host suitability. *African Plant Protection* 4, 107–111.

Fourie, H., Zijlstra, C. and McDonald, A.H. (2001) Identification of root-knot nematode species occurring in South Africa using the SCAR-PCR technique. *Nematology* 3, 675–680.

Fourie, H., Leswifi, C., McDonald, A.H. and De Waele, D. (2007) Host suitability of vetiver grass to *Meloidogyne incognita* and *M. javanica*. *Nematology* 9, 49–52.

Garabedian, S. and Van Gundy, S.D. (1983) Use of avermectins for the control of *Meloidogyne incognita* on tomatoes. *Journal of Nematology* 15, 503–510.

Garg, R.N., Gaur, H.S. and Singh, G. (1995) Effect of puddling and water regimes on soil physical properties, plant parasitic nematodes and performance of rice crop. *Annals of Plant Protection Sciences* 3, 121–126.

Gaur, H.S. and Perry, R.N. (1991) The use of soil solarization for control of plant parasitic nematodes. *Nematological Abstracts* 60, 153–167.

Gergon, E.B., Gapasin, R., Castillo, M., Judal, V., Pile, C., Ravina, C., Prot, J.C. and Halbrendt, J. (1998) The rice root-knot nematode and its management in rice–vegetable systems. IPM CRSP Symposium abstracts www.oired.vt.edu/ipmcrsp/meetings/symp98/abstracts1.html. (accessed 3 April 2008).

Gergon, E.B., Miller, S.A., Davide, R.G., Opina, O.S. and Obien, S.R. (2001) Evaluation of cultural practices (surface burning, deep ploughing, organic amendments) for the management of rice root-knot nematode in rice–onion cropping system and their effect on onion (*Allium cepa* L.) yield. *International Journal of Pest Management* 47, 265–272.

Goff, C.C. (1936) Relative susceptibility of some annual ornamentals to root-knot. *Agricultural Experimental Station Bulletin* 291, University of. Florida, USA, pp. 15.

Goswami, B.K. (1993) Effect of Algan – a product of brown alga in the management of root-knot nematodes, *Meloidogyne incognita* in brinjal. *Annals of Agricultural Research* 13, 391–393.

Guerena, M. (2006) Nematodes: alternative controls. http://attra.ncat.org/attra-pub/PDF/nematode.pdf. (accessed 15 April 2008).

Haseeb, A. (1991) Evaluation of different pesticides and oil-cake on the control of root-knot nematodes (*Meloidogyne incognita*) infesting on davana (*Artemisia pallens*). *Current Nematology* 2, 201–204.

Haydock, P.P.J., Woods, S.R., Grove, I.G. and Hare, M. (2006) Chemical control of nematodes. In: Perry, R.N. and Moens, M. (eds) *Plant Nematology.* CAB International, Wallingford, UK, pp. 392–410.

Hazell, P. and Wood, S. (2008) Drivers of change in global agriculture. *Philosophical Transactions of the Royal Society B. Biological Science* 363, 495–515.

Hernandez, A., Fargette, M. and Sarah, J.-L. (2004) Characterisation of *Meloidogyne* spp. (Tylenchida: Meloidogynidae) from coffee plantations in Central America and Brazil. *Nematology* 6, 193–204.

Hockland, S., Inserra, R.N., Millar, L. and Lehman, P.S. (2006) International plant health – putting legislation into practice. In: Perry, R.N. and Moens, M. (eds) *Plant Nematology.* CAB International, Wallingford, UK, pp. 327–345.

Hussey, R.S. and Jansen, G.J.W. (2002) Root-knot nematodes: *Meloidogyne* species. In: Starr, J.L., Cook, R. and Bridge, J. (eds) *Plant Resistance to Parasitic Nematodes.* Biddles, Guildford and King's Lynn, UK, pp. 43–71.

Ibrahim, A. (1985) The status of root-knot nematodes in the Middle East, Region VII of the International *Meloidogyne* Project. In: Sasser, J.N. and Carter, C.C. (eds) *An Advanced Treatise on* Meloidogyne. *Volume I. Biology and Control.* North Carolina State University Graphics, Raleigh, North Carolina, pp. 373–378.

Ibrahim, I.K.A. and El Saedy, M.A. (1976) Development and pathogenisis of *Meloidogyne javanica* in peanut roots. *Nematologica Mediterranea* 4, 231–234.

Ismail, A.E. (2007) Effect of different concentrations of furfural as a botanical nematicide and the application methods in controlling *Meloidogyne incognita* and *Rotylenchulus reniformis* infecting sunflower. *Pakistan Journal of Nematology* 25, 45–52.

James, B., Atcha, C., Godonou, I. and Baimey, H. (2005) *Summary of Activities and Achievements, 2003–2005, Healthy Vegetables through Participatory IPM in Peri-urban Areas of Benin.* International Institute of Tropical Agriculture, Ibadan, Nigeria.

Javed, N., Gowen, S.R., El-Hassan, S.A., Inam-ul-Haq, M., Shahina, F. and Pembroke, B. (2008) Efficacy of neem (*Azadirachta indica*) formulations on biology of root-knot nematodes (*Meloidogyne javanica*) on tomato. *Crop Protection* 27, 36–43.

Johnson, A.W. (1985) The role of nematicides in nematode management. In: Sasser, J.N. and Carter, C.C. (eds). *An Advanced Treatise on* Meloidogyne. *Volume I. Biology and Control.* North Carolina State University Graphics, Raleigh, North Carolina, pp. 249–268.

Kandjil, S.T., Ogol, C.K.P.O. and Albrecht, A. (2003) Crop damage by nematodes in improved-fallow fields in western Kenya. *Agroforestry Systems* 57, 51–57.

Khan, A.M., Saxena, S.K. and Siddiqi, Z.A. (1971) Efficacy of *Tagetes erecta* in reducing root infesting nematodes in tomato and okra. *Indian Phytopathology* 24, 166–169.

Khan, M.W., Alam, M.M., Khan, A.M. and Saxena, S.K. (1974) Effect of water soluble fractions of oilcakes and bitter principles of neem on some fungi and nematodes. *Acta Botanica Indica* 2, 120–128.

Kinh, D.-N., Huong, N.M. and Ut, N.U. (1982) Root knot disease of rice in the Mekong Delta, Vietnam. *International Rice Research Newsletter* 6, 6–7.

Kirkegaard, J.A., Wong, P.T.W. and Desmarchelier, J.M. (1996) *In vitro* suppression of fungal root pathogens of cereals by brassica tissues. *Plant Pathology* 45, 593–603.

Kirkegaard, J.A., Sarwar, M., Mathiessen, J.N., Thomas, G. and Monteiro, A.A. (1998) Assessing the biofumigation potential of crucifers. *Acta Horticulturae* 459, 105–111.

Kloepper, J.W., Rodríguez-Kábana, R., McInroy, J.A. and Collins, D.J. (1991) Analysis of populations and physiological characterization of microorganisms in rhizospheres of plants with antagonistic properties to phytopathogenic nematodes. *Plant and Soil* 136, 95–102.

Krishnamurthy, G.V.G. (1990) Effect of oil seed cakes on the control of root-knot nematode, *Meloidogyne javanica*, in tobacco nursery. In: Chari, M.S. and Ramaprasad, G. (eds) *Botanical Pesticides in Integrated Pest Management: Proceedings of National Symposium.* Indian Society of Tobacco, Rajahmundry, India, pp. 434–437.

Krishnappa, K. (1985) Nematology in developing countries: India – IMP Region VIII. In: Sasser, J.N. and Carter, C.C. (eds) *An Advanced Treatise on* Meloidogyne. *Volume I. Biology and Control.* North Carolina State University Graphics, Raleigh, North Carolina, pp. 379–398.

Lasota, J.A. and Dybas, R.A. (1991) Avermectins, a novel class of compounds: implications for use in arthropod pest control. *Annual Review of Entomology* 36, 91–117.

Liebanas, G. and Castillo, P. (2004) Host suitability of some crucifers for root-knot nematodes in southern Spain. *Nematology* 6, 125–128.

Luc, M., Sikora, R.A. and Bridge, J. (eds) (2005) *Plant Parasitic Nematodes in Subtropical and Tropical Agriculture*, 2nd edn. CAB International, Wallingford, UK.

Machuka, M. (2001) Agricultural biotechnology for Africa. African scientists and farmers must feed their own people. *Plant Physiology* 126, 16–19.

McSorley, R. and Frederick, J.J. (1995) Responses of some common cruciferae to root-knot nematodes. *Journal of Nematology* 27(S), 550–554.

McSorley, R., Dickson, D.W. and de Brito, J.A. (1994) Host status of selected tropical rotation crops to four populations of root-knot nematodes. *Nematropica* 24, 45–53.

Mohandas, C. and Ramakrishnan, S. (1996) Sweet potato variety – *Sree Bahadra* – a trap crop of root knot nematode. *Journal of Root Crops* 22, 135–136.

Mojumder, V., Pankaj and Singh, J. (2002) Comparative efficacy of seed coating of pea with neem based pesticides and carbosulfan against root-knot, reniform and other plant parasitic nematodes. *Annals of Plant Protection Sciences* (India) 10, 404–406.

Monfort, W.S., Kirkpatrick, T.L., Long, D.L. and Rideout, S. (2006) Efficacy of a novel nematicidal seed treatment against *Meloidogyne incognita* on cotton. *Journal of Nematology* 38, 245–249.

Monfort, W.S., Csinos, A.S., Desaeger, J., Seebold, K., Webster, T.M. and Diaz-Perez, J.C. (2007) Evaluating *Brassica* species as an alternative control measure for root-knot nematode (*Meloidogyne incognita*) in Georgia vegetable plasticulture. *Crop Protection* 26, 1359–1368.

Mori, T., Mohammed, M. and Yamasaki, T. (2000) Ellagitannin toxicity in the free-living soil-inhabiting nematode, *Caenorhabditis elegans. Journal of Pesticide Science* 25, 405–409.

Motsinger, R.E., Moody, E.H. and Gay, C.M. (1977) Reaction of certain French marigold (*Tagetes patula*) cultivars to three *Meloidogyne* spp. *Journal of Nematology* 9, 278.

Murphy, W.S., Brodie, B.B. and Good, J.M. (1974) Population dynamics of plant parasitic nematodes in cultivated soil: effects of combinations of cropping systems and nematicides. *Journal of Nematology* 6, 103–107.

Nel, A., Krause, M. and Khelawanlall, N. (2007) *A Guide for the Control of Plant Pests*, 39th edn. Department of Agriculture, Republic of South Africa. Government Printer, Pretoria, South Africa.

Netscher, C. (1978) Morphological and physiological variability of species of *Meloidogyne* in West Africa and implications for their control. *Mededelingen Landbouwhogeschool Wageningen* 43, 78–83.

Neuenschwander, P. (2004) Harnessing nature in Africa: biological pest control can benefit the pocket, health and the environment. *Nature* 432, 801–802.

Ntow, W.J., Huub, J.G., Kelderman, P. and Drechsel, P. (2006) Farmers perceptions and pesticide use in practices in vegetable production in Ghana. *Pest Management Science* 62, 356–365.

Nusbaum, C.J. (1963) Nematode assays – a program to aid agricultural workers. *Down to Earth* 19, 2.

O'Bannon, J.H. and Nyczepir, A.P. (1982) Host range of the Columbia root-knot nematode. *Plant Disease* 66, 1045–1048.

Page, S.L.J. (1979) *An Assessment of the Importance and Control of Plant Parasitic Nematodes of Vegetable Crops in Bangladesh*. ODM Report of a visit to Bangladesh. Ministry of Overseas Development, Ascot, UK.

Pandey, R. (1994) Comparative performance of oil seed cakes and pesticides in the management of root-knot disease of davana. *Nematologica Mediterranea* 22, 17–19.

Paruthi, I.J., Kanwar, R.S. and Gupta, D.C. (1996) *In vitro* evaluation of achook – a neem based pesticide against root-knot nematode, *Meloidogyne incognita. Indian Journal of Nematology* 26, 270–272.

Pattison, A.B., Versteeg, C., Akiew, S. and Kirkegaard, J. (2006) Resistance of Brassicaceae plants to root-knot nematodes (*Meloidogyne* spp.) in northern Australia. *International Journal of Pest Management* 52, 53–62.

Pinochet, J., Fernández, C., de Carmen Jaizme, M. and Tenoury, P. (1997) Micropropagated banana infected with *Meloidogyne javanica* responds to *Glomus intraradices* and phosphorus. *HortScience* 32, 101–103.

Plowright, R.A., Coyne, D.L., Nash, P. and Jones, M.P. (1999) Resistance to the rice nematodes *Heterodera sacchari, Meloidogyne graminicola* and *M. incognita* in *Oryza glaberrima* and *O. glaberrima* × *O. sativa* interspecific hybrids. *Nematology* 1, 745–751.

Ponte, J.J. da and Franco, A. (1981) Manipueira, un nematicida nao convencional de comprovada potencialidade. In: *Trabalhos Apresentados Reuniae de Sociedade Brasileira de Nematologia, Piracicaba, Brazil* 5, 25–33.

Ponte, J.J. da, Franko, A. and Ponte, A.E.L. (1987) Study of the use of Manipueira as a nematicide under field conditions. *Nematologia Brasileira* 11, 42–47.

Ponte, J.J. da, Cavada, B.S. and Silveira-Filho, J. (1996) Teste com lectina no controle de *Meloidogyne incognita* em tomateiro. *Fitopatologia Brasileira* 21, 489–491.

Poornima, K. (1997) Effect of some organic oils and their cakes on the effectivity of root-knot nematode, *Meloidogyne incognita* affecting tomato. *Indian Journal of Nematology* 27, 70–73.

Primavesi, A. (1989) *Seed Potato Systems in the Philippines: a Case Study.* International Potato Center, Lima, Peru.

Putter, I., MacConnell, F.A., Preiser, A.A., Haidri, S.S. and Dybas, R.A. (1981) Avermectins: novel insecticides, acaricides, and nematicides from a soil microorganism. *Experienta* 37, 963–964.

Qing, Y., Tsao, R., Chiba, M. and Potter, J. (2007) Elucidation of the nematicidal activity of bran and seed meal of oriental mustard (*Brassica juncea*) under controlled conditions. *Food, Agriculture and Environment* 5, 374–379.

Quénéhervé, P., Topart, P. and Martiny, B. (1998) *Mucuna pruriens* and other rotational crops for control of *Meloidogyne incognita* and *Rotylenchulus reniformis* in vegetables in polytunnels in Martinique. *Nematropica* 28, 19–30.

Rammah, A. and Hirschmann, H. (1988) *Meloidogyne mayaguensis* n. sp. (Meloidogynidae), a root-knot nematode from Puerto Rico. *Journal of Nematology* 20, 58–69.

Rao, M.S. (1997a) Integration of *Paecilomyces lilacinus* with neem leaf suspension for the management of root-knot nematodes on egg plant. *Nematologica Mediterranea* 25, 249–252.

Rao, M.S. (1997b) Management of root-knot nematode, *Meloidogyne incognita* on tomato by integration of *Trichoderma harzianum* with neem cake. *Zeitschrift für Pflanzenkrankheiten und Pflanzenschutz* 104, 423–425.

Rao, M.S. (1997c) Management of root-knot nematodes, *Meloidogyne incognita* in tomato nursery by integration of endomycorrhiza, *Glomus fasciculatum* with castor cake. *Pest Management in Horticultural Ecosystems* 3, 31–35.

Reddy, P.P. (1997) Effect of integration of *Pasteuria penetrans*, *Paecilomyces lilacinus* and neem cake for the management of root-knot nematodes infecting tomato. *Pest Management in Horticultural Ecosystems* 3, 100–104.

Riga, E., Mojtahedi, H., Ingham, R. and McGuire, A. (2004) Green manure amendments and management of root knot nematodes on potato in the Pacific North West of USA. In: Cook, R.C. and Hunt, D.J. (eds) *Nematology Monographs and Perspectives* 2, 151–158.

Roberts, P.A. (1987) The influence of planting date of carrot on *Meloidogyne incognita* reproduction and injury to roots. *Nematologica* 33, 335–342.

Roberts, P.A., Van Gundy, S.D. and McKinney, H.E. (1981) The effects of soil temperature and planting date of wheat on *Meloidogyne incognita* reproduction, soil populations and grain yield. *Journal of Nematology* 13, 338–345.

Rodríguez-Kábana, R. (1990) Chitinous materials from blue crab for control of root-knot nematode. II. Effect of soybean meal. *Nematropica* 20, 153–168.

Rodríguez-Kábana, R. and Canullo, G.H. (1992) Cropping systems for the management of phytonematodes. *Phytoparasitica* 20, 211–224.

Rodríguez-Kábana, R., Pinochet, J., Robertson, D.G. and Wells, L. (1992) Crop rotation studies with velvet bean (*Mucuna deeringiana*) for the management of *Meloidogyne* spp. *Journal of Nematology* 24 (4S), 662–668.

Saka, V.W. (1985) *Meloidogyne* spp. research in region V of the International *Meloidogyne* Project (IMP). In: Sasser, J.N. and Carter, C.C. (eds) *An Advanced Treatise on* Meloidogyne. *Volume I. Biology and Control.* North Carolina State University Graphics, Raleigh, North Carolina, pp. 361–368.

Sankaranarayanan, C. (1997) Effect of oil cakes and nematicides on the growth of blackgram (*Vigna mungo*) inoculated with VAM fungus (*Glomus fasciculatum*) and root-knot nematode (*Meloidogyne incognita*). *Indian Journal of Nematology* 27, 128–130.

Santos, J.M. and Triantaphyllou, H.H. (1992) Determinação dos fenótipos isoenzimáticos e estudos comparativos da morfologia de 88 populações de *Meloidogyne* spp., parasitas do cafeeiro. *Nematologia Brasileira* 16, 88 (abstract).

Sasser, J.N. (1954) *Identification and host–parasite relationships of certain root-knot nematodes* (Meloidogyne *spp.*). Maryland Agricultural Experimental Station Technical Bulletin A-77, 31 pp.

Sasser, J.N. (1966) Behaviour of *Meloidogyne* spp. from various geographical locations on ten host differentials. *Nematologica* 12, 97–98.

Sasser, J.N. (1979) Pathogenicity, host ranges, and variability in *Meloidogyne* spp. In: Lamberti, F. and Taylor, C.E. (eds) *Root-knot Nematodes* (Meloidogyne *Species*) *Systematics, Biology and Control.* Academic Press, New York, pp. 257–268.

Sasser, J.N. and Carter, C.C. (eds) (1985) *An Advanced Treatise on* Meloidogyne. *Vol. 1. Biology and Control.* North Carolina State University, Raleigh, North Carolina.

Sasser, J.N., Kirkpatrick, T.L. and Dybas, R.A. (1982) Efficacy of avermectins for root-knot control in tobacco. *Plant Disease* 66, 691–693.

Sasser, J.N., Eisenback, J.D., Carter, C.C. and Triantaphyllou, A.C. (1983) The International *Meloidogyne* Project – its goals and accomplishments. *Annual Review of Phytopathology* 21, 271–288.

Schmutterer, H. (1997) Side-effects of neem (*Azadirachta indica*) products on insect pathogens and natural enemies of spider, mites and insects. *Journal of Applied Entomology* 121, 121–128.

Segeren-v.d. Oever, H.A. and Sanchit-Becker, M.L. (1984) Observations on *Meloidogyne oryzae* Maas, Sanders and Dede 1978 in irrigated rice in Suriname. *De Surinaamse Landbouw* 32, 51–59.

Sena, E.S. and Ponte, J.J. da (1982) A manipueira no controle da meloidoginosa da cenoura. *Revista Sociedade Brasileira Nematologia* 6, 95–98.

Senatore, F., Napolitano, F., Mohamed, M.A-H., Harris, P.J.C., Mnkeni, P.N.S. and Henderson, J. (2004) Antibacterial activity of *Tagetes minuta* L. (Asteraceae) essential oil with different chemical composition. *Flavour and Fragrance Journal* 16, 574–578.

Sharma, R.K. (1987) Effect of nematicides and organic amendments on root-knot nematodes infecting potato. *International Nematology Network Newsletter* 4, 8–10.

Sharma, S.B., Price, N. and Bridge, J. (1997) The past, present and future of plant nematology in International Agricultural Research Centres. *Nematological Abstracts* 66, 119–142.

Shepherd, J.A. (1982) Nematode pests of tobacco. In: Keetch, D.P. and Heyns, J. (eds) *Nematology in Southern Africa.* Science Bulletin No. 400, Department of Agriculture and Fisheries. Government Printer, Pretoria, South Africa, pp. 58–68.

Siddiqui, M.A. and Alam, M.M. (1987) Control of plant parasitic nematodes by intercropping with *Tagetes minuta. Nematologia Mediterranea* 15, 205–211.

Sikora, R.A. (1992) Management of the antagonistic potential in agricultural ecosystems for the biological control of plant parasitic nematodes. *Annual Review of Phytopathology* 30, 245–270.

Sikora, R.A. and Fernández, E. (2005) Nematode parasites of vegetables. In: Luc, M., Sikora, R.A. and Bridge, J. (eds) *Plant Parasitic Nematodes in Subtropical and Tropical Agriculture*, 2nd edn. CAB International, Wallingford, UK, pp. 319–392.

Sikora, R.A. and Pocasangre, L.E. (2004) New technologies to increase root health and crop production. *InfoMusa* 13, 25–29.

Sikora, R.A., Niere, B. and Kimenju, J. (2003) Endophytic microbial biodiversity and plant nematode management in African agriculture. In: Neuenschwander, P., Borgemeister, C. and Langewald, U. (eds) *Biological Control in IPM Systems in Africa.* CAB International, Wallingford, UK, pp. 179–192.

Sikora, R.A., Bridge, J. and Starr, J.L. (2005) Management practices: an overview of integrated nematode management technologies. In: Luc, M., Sikora, R.A. and Bridge, J. (eds) *Plant Parasitic Nematodes in Subtropical and Tropical Agriculture*, 2nd edn. CAB International, Wallingford, UK, pp. 793–826.

Silva, G.S. da, Ferraz, S. and dos Santos, J.M. (1990) Histopatologia de raízes de *Crotalaria* parasitadas por *Meloidogyne javanica. Fitopatologia Brasileira* 15, 46–48.

Singh, D.B. and Reddy, P.P. (1981) Note on the chemical control of root-knot nematodes infecting French bean. *Indian Journal of Nematology* 7, 127.

Smedley, E.M. (1939) Experiments on the use of isothiocynates in the control of the potato strain of *Heterodera schachtii* (Schmidt). *Journal of Helminthology* 17, 31–38.

Sosa-Moss, C. (1985) Report on the status of *Meloidogyne* research in Mexico, Central America and the Caribbean countries (Region I). In: Sasser, J.N. and Carter, C.C. (eds) *An Advanced Treatise on* Meloidogyne. *Volume I. Biology and Control.* North Carolina State University Graphics, Raleigh, North Carolina, pp. 327–346.

Stapleton, J.J. and DeVay, J.E. (1986) Soil solarization: a non-chemical approach for management of plant pathogens and pests. *Crop Protection* 5, 190–198.

Stapleton, J., Duncan, R.A. and Johnson, R. (1998) Soil disinfestation with cruciferous amendments and sub-lethal heating: effects on *Meloidogyne incognita*, *Sclerotium rolfsii* and *Pythium ultimum*. *Plant Pathology* 47, 737–742.

Stark, J.C. (1995) *Development of Sustainable Potato Production Systems for the Pacific Northwest.* SARE Final Report, Washington, DC.

Stephan, Z.A. (1978) Report on the status of root-knot nematodes, *Meloidogyne* spp., in Iraq. In: *Proceedings of the First International* Meloidogyne *Project Research Planning Conference on Root-knot Nematodes,* Meloidogyne *spp., Region VII*, Cairo, Egypt, pp. 42–43.

Stephan, Z.A. (1979) The survey program of IMP project in Iraq. In: *Proceedings of the Second International* Meloidogyne *Project Research Planning Conference on Root-knot Nematodes,* Meloidogyne *spp., Region VII*, Athens, Greece, pp. 64–66.

Stirling, G.R. and Stirling, A.M. (2003) The potential of *Brassica* green manure crops for controlling root-knot nematode (*Meloidogyne javanica*) on horticultural crops in a subtropical environment. *Australian Journal of Experimental Agriculture* 43, 623–630.

Struble, F.B., Morrison, L.S. and Cordner, H.B. (1966) Inheritance of resistance to stem rot and root-knot nematode in sweet potato. *Phytopathology* 56, 1217–1219.

Stubbs, V. (1999) UNEP/NTO *Workshop on Alternatives to Methyl Bromide for Eastern and Southern African Countries*, 6–10 September 1999. Tobacco Research Board, Lilongwe, Malawi,.

Sun, W., Montagu, M.V. and Verbruggen, N. (2002) Small heat shock proteins and stress tolerance in plants. *Biochimica et Biophysica Acta* 1577, 1–9.

Taha, A.H.Y. and Yousif, G.M. (1976) Histology of peanut underground parts infected with *Meloidogyne incognita. Nematologica Mediterranea* 4, 175–181.

Tandingan, I.C., Prot, J.-C. and Davide, R.G. (1996) Influence of water management on tolerance of rice cultivars for *Meloidogyne graminicola. Fundamental and Applied Nematology* 19, 189–192.

Tchabi, A., Coyne, D., Hountondji, F., Lawouin, L., Wiemken, A. and Oehl, F. (2008) Arbuscular mycorrhizal fungal communities in sub-Saharan savannahs of Benin, West Africa, as affected by agricultural land use intensity and ecological zone. *Mycorrhiza* 18, 181–195.

Tenkuoano, A., Hauser, S., Coyne, D. and Coulibaly, O. (2006) Clean planting materials and management practices for sustained production of banana and plantain in Africa. *Chronica Horticulturae* 46, 14–18.

Thakur, S.G. (1995a) Determination of the waiting period (period of decomposition) of some non-edible oilseed cakes for the control of root-knot nematode. *Current Nematology* 6, 1–4.

Thakur, S.G. (1995b) Effect of some non-edible oilseed cakes against root-knot nematode, *Meloidogyne incognita* and growth parameters of brinjal. *Current Nematology* 6, 21–26.

Thoden, T.C., Hallmann, J. and Boppré, M. (2009) Effects of plants containing pyrrolizidine alkaloids on the northern root-knot nematode *Meloidogyne hapla. European Journal of Plant Pathology* 123, 27–36.

Thomas Seedling Technology Systems (undated) http://www.berlthomas.com (accessed on 7 January 2009).

Tomaszewski, E.K., Khalil, M.A.M., El-Deeb, A.A., Powers, T.O and Starr, J.L. (1994) *Meloidogne javanica* parasitic on peanut. *Journal of Nematology* 26, 436–441.

Tsror (Lahkim), L., Lebiush, S., Meshulam, M., Erlich, O., Hazanovsky, M., Aharon, M., Matan, E., Tregerman, M. and Gamliel, A. (2007) Biofumigation for the control of soil borne diseases. *Acta Horticulturae* 747, 389–394.

UNEP (2000a) *Montreal Protocol on Substances that Deplete the Ozone Layer.* UNEP Technology and Economic Assessment Panel. April 2000 Report. http://www.teap.org (accessed 7 January 2009).

UNEP (2000b) Methyl bromide alternatives for North African and southern European countries. In: *Proceedings from the Workshop on Methyl Bromide Alternatives for North African and Southern European Countries.* Rome, Italy.

van den Oever, R. (1995) Nematological surveys and testing of varieties in Mozambique in 1993/94. In: Botha-Greeff, M., Le Roux, H., Jones, R., Huisman, L. and McDonald, A.H. (eds) *Proceedings of the Twelfth Symposium of the Nematological Society of South Africa*, Kruger Park, South Africa (abstract).

Vargas-Ayala, R., Rodríguez-Kabána, R., Morgan-Jones, G., McInroy, J.A. and Kloepper, J.W. (2000) Shifts in soil microflora induced by velvetbean (*Mucuna deeringiana*) in cropping systems to control root-knot nematodes. *Biological Control* 17, 11–22.

Varma, M.K., Sharma, R.C. and Mathur, J.R. (1981) Efficacy of granular nematicides against *Meloidogyne incognita* in moth bean (*Vigna acontifolia*). *Nematology Mediterranea* 9, 109–110.

Vats, R. (1993) Optimization of dipping periods of tomato seedlings in plant extracts for the management of root-knot nematode, *Meloidogyne javanica*. *Current Nematology* 4, 161–164.

Vawdrey, L.L. (1997) Control of root-knot nematode (*Meloidogyne javanica*) on tomato with molasses and other organic amendments. *Australasian Plant Pathology* 26, 179–187.

Viaene, N., Coyne, D.L. and Kerry, B.R. (2006) Biological and cultural control. In: Perry, R.N. and Moens, M. (eds) *Plant Nematology*. CAB International, Wallingford, UK, pp. 346–369.

Villar, E.M.J. and Zavaleta, M.E. (1990) Effecto de *Crotalaria longirostrata* Hook y Arnott sobre nematodes agalladores (*Meloidogyne* spp.) revista. *Mexicana de Fitopatologiea* 8, 166–172.

Walia, (1997) Management of root-knot nematode, *Meloidogyne javanica* on vegetable crops with *Tagetes* sp. *Indian Journal of Nematology* 27, 18–23.

Wang, K.-H., Sipes, B.S. and Schmitt, D.P. (2002) *Crotalaria* as a cover crop for nematode management: a review. *Nematropica* 32, 35–57.

Wang, Q.R., Li, Y., Handoo, Z.A. and Klassen, W. (2007) Influence of cover crops in rotation on populations of soil nematodes. *Nematropica* 37, 79–92.

Warrior, P., Rehberger, L.A., Beach, M., Grau, P.A., Kirfman, G.W. and Conley, J.M. (1999) Commercial development and introduction of DiTera™, a new nematicide. *Pesticide Science* 55, 76–79.

Whitehead, A.G. (1969) The distribution of root-knot nematodes (*Meloidogyne* spp.) in tropical Africa. *Nematologica* 15, 315–333.

Whitehead, A.G. (1998) *Plant Nematode Control*. CAB International, Wallingford, UK.

Yang, B. and Eisenback, J.D. (1983) *Meloidogyne enterolobii* n. sp. (Meloidogynidae), a root-knot nematode parasitizing pacara earpod tree in China. *Journal of Nematology* 15, 381–391.

Zasada, I.A., Masler, E.P., Rogers, S.T. and Halbrendt, J.M. (2009) Behavioural response of *Meloidogyne incognita* to benzyl isothiocyanate. *Nematology*, 603–611.

Zufall, F., Franke, C.H. and Hatt, H. (1989) The insecticide avermectin Bl activates chloride channel in crayfish muscle membrane. *Journal of Experimental Biology* 142, 191–205.

Gene Index

abf 370
ace-1 193
ACT2 172
ACT7 172
agl-1 374
agl-2 374
aprA 386
AtAAP6 176
AtAUX4/LAX3 175
AtCAT6 175
AtFH1 172
AtFH6 172
AtFH10 172
AtPIP1.5 175
AtPIP2.5 175
AtTIP1.1 175
AUX1 175
Avr genes 168–169, 307
AvrPi-ta 168
AvrXv4 168

Bt 340–341, 350, 387–382

CCS52 173
CCS52a 176
Cg-1 313–314, 318
cnc 370
collagen-like genes 384
cry 341, 348
CYCD3 176

16D10 167, 343–345, 349
daf-2 375
dcr-1 374
drh-2 374
drh-3 374
drsh-1 374

EDS1 310
EIN3 174
ENOD11 176
ENOD40 176
ERF1 174
eri-1 374

fem-2 371
fip 370
fipr 370
flp 342, 371
flp-30 371
flp-31 371

γTIP1;1 175

H1 74, 100
HasHSP17.7 174
Hsp 90, 310, 346

ICS1 174

KNOX 175, 334

Ma 304, 344, 423
mab-23 371
mag-1 371
map-1 168, 193, 313
map65-3 172–173
Me1 6, 303, 305, 309–310, 346
Me3 6, 303, 305
Me4 305
Me7 303, 307
Mech1 303, 305
Mech2 303, 305
Mex-1 303, 311
Mi-1 9, 149, 151, 302, 304–306, 308–311, 313–316, 318, 346, 423
Mi-1.1 308–309
Mi-1.2 308–309, 346
Mi-1.3 308
Mi-3 305, 316
Mi-9 304–305, 309–310, 316
Mi-ace-2 193
Mi-cpl-1 168
Mig-1 303, 306–307
Mi genes 316, 326–328, 428, 451, 458
Mir-1 303, 306–307
mjap-1 313
mjap-2 313
Mjg-1 303, 306–307
Msp 125, 345

N. 224
NHL3 174
nlp 370–371
NOD26 175–176
NPR1 174

PAD4 174
pash-1 374
PHAN 175, 334
pir-1 374
PR4 174
PR5 174

Rar1 310
rde-4 374
red4 372
R-genes 307–311, 346, 352
Ribulose-5-phosphate 3-epimerase (RPE) 171
Rk 303–306, 311–318, 423
Rk^2 305, 316–317
rk^3 303
rkn1 305–306
RKN2 303, 306
$R_{mc1(blb)}$ 305, 316
Rme1 309–310, 346
RMia 423
RMja 423
RNAi genes 169, 342–345, 348–349, 372, 374
Rpi-blb2 309
rrf-3 342, 374
rsd-2 342, 374
rsd-6 372

sdc-1 371
Sgt1 310
SID1 174
sid-1 372
sid-2 372

Tabasco 423
TobRB7 175, 349
tra-1 371–372, 374
tra-2 374
tra-3 371

UidA

vip3a 340

WHY1 174
WRKY 174

XopD 168

Nematode Genus and Species Index

Anguillula arenaria 58, 63
Anguillula javanica 64
Anguillula marioni 56, 66
Anguillula vialae 58, 66
Aphelenchus avenae 232
Ascaris suum 106

Belonolaimus longicaudatus 232, 382
Brugia malayi 363, 365

Caenorhabditis briggsae 363, 367, 370, 375–376
Caenorhabditis elegans 106, 126–127, 148, 150, 152, 154, 182, 191–193, 218–219, 341–342, 345, 354, 363–365, 367–376, 385

Dactylella oviparasitica 400
Ditylenchus dipsaci 147, 223, 232, 235, 238

Globodera 141, 327, 416
Globodera pallida 125, 140, 152–153, 224, 233, 276, 280, 339–340, 342, 345, 349
Globodera rostochiensis 139–141, 153–154

Haemonchus contortus 372
Helicotylenchus dihystera 288
Helicotylenchus multicinctus 223, 232
Hemicycliophora arenaria 233
Heterodera 416
Heterodera arenaria 63
Heterodera avenae 218, 339
Heterodera cajani 232
Heterodera exigua 63
Heterodera glycines 152, 223, 232, 234–237, 345, 372, 382
Heterodera javanica 56, 59, 64
Heterodera marioni 2, 65, 207
Heterodera oryzae 146
Heterodera oryzicola 232, 353
Heterodera schachtii 224, 232, 234–236, 238, 339
Heterodera vialae 66
Heterodera zeae 232, 235
Hoplolaimus 234
Hoplolaimus columbus 223, 232
Hoplolaimus galeatus 232
Hypsoperine acronea 63
Hypsoperine graminis 64
Hypsoperine megriensis 66
Hypsoperine ottersoni 65
Hypsoperine propora 65–66
Hypsoperine spartinae 66

Meloidogyne (species of *Meloidogyne* are given below, except for *M. arenaria, M. hapla, M. incognita* and *M. javanica* where citations are too numerous to list)
Meloidogyne acrita 64
Meloidogyne acronea 63, 66, 77–78, 213
Meloidogyne acronea apud 66
Meloidogyne actinidiae 63
Meloidogyne africana 63
Meloidogyne aquatilis 63
Meloidogyne arabicida 63, 101, 107, 122, 124, 127, 130–133, 428
Meloidogyne ardenensis 63, 101, 105, 124, 129, 131–133, 382
Meloidogyne arenaria arenaria 63, 216
Meloidogyne arenaria thamesi 63, 216

Meloidogyne artiellia 63, 121–124, 129–133, 140, 190, 228, 248, 251, 253–254, 256–257, 261–262, 264, 342
Meloidogyne baetica 63, 101, 122–124, 129–133
Meloidogyne bauruensis 65
Meloidogyne brasilensis 63
Meloidogyne brevicauda 23, 31, 63
Meloidogyne californiensis 63, 66
Meloidogyne camelliae 63, 105
Meloidogyne caraganae 63
Meloidogyne carolinensis 5, 33, 63, 66, 101, 165, 184, 186
Meloidogyne chitwoodi 2, 5, 7, 10–12, 48, 63, 69, 72, 77, 79–80, 84–85, 98, 101, 103–111, 121–128, 130–133, 140, 143–144, 147, 184, 186–187, 201, 207, 251, 256–257, 259, 261, 277, 283, 289, 291, 301, 303–302, 315–316, 417, 419–421, 424–425, 427, 430, 446, 451, 454, 456–457
Meloidogyne chosenia 64
Meloidogyne christiei 64
Meloidogyne cirricauda 64
Meloidogyne citri 2, 64, 425
Meloidogyne coffeicola 64, 101, 416, 418
Meloidogyne cruciani 64, 101, 184
Meloidogyne cynariensis 64
Meloidogyne decalineata 64
Meloidogyne deconincki 63
Meloidogyne donghaiensis 64
Meloidogyne dunensis 64, 101, 122, 124, 129–133
Meloidogyne duytsi 64, 101, 105, 121–122, 124, 130–133
Meloidogyne elegans 64
Meloidogyne enterolobii 3, 5, 7, 9, 64, 69, 80–81, 98–99, 101, 108, 111, 122, 124–127, 130–133, 184, 187, 301, 308, 423, 425, 458
Meloidogyne ethiopica 58, 64, 69, 72, 80, 82, 101, 106–107, 122, 124, 127, 130–133, 184
Meloidogyne exigua 2, 5, 7, 24, 33, 56–59, 63, 69, 72, 74, 80–84, 101–102, 104, 109, 120–124, 130–133, 140, 184, 207, 249, 257, 303, 311, 416, 418, 422, 457
Meloidogyne fallax 2, 10–11, 64, 69, 72, 84–85, 101, 103–111, 124–126, 130–133, 143–144, 184, 186–187, 301, 303–305, 315, 420–421, 425, 446
Meloidogyne fanzhiensis 64
Meloidogyne floridensis 64, 98, 101, 107–108, 111, 122, 124, 127, 129–133, 183–184, 423, 425
Meloidogyne fujianensis 64, 66*n*2
Meloidogyne goeldii 66
Meloidogyne grahami 64
Meloidogyne graminicola 5, 33, 64, 69, 72, 84, 86, 101, 105, 107–108, 121–124, 129–133, 184–185, 232, 250, 353, 393, 395, 448, 451–454, 458
Meloidogyne graminis 64, 101, 107–108, 124, 129–133, 184
Meloidogyne gyulai 63
Meloidogyne hainanensis 64
Meloidogyne haplanaria 64, 101, 107
Meloidogyne hispanica 64, 101, 106, 122, 124, 130–133, 184
Meloidogyne ichinohei 64, 121–122, 129–133
Meloidogyne incognita acrita 64, 233, 237
Meloidogyne incognita grahami 64
Meloidogyne incognita incognita 64
Meloidogyne incognita inornata 64
Meloidogyne incognita wartellei 64
Meloidogyne indica 64
Meloidogyne inornata 64, 101, 184
Meloidogyne izalcoensis 64, 98, 101, 113, 184
Meloidogyne javanica bauruensis 64
Meloidogyne javanica javanica 64
Meloidogyne jianyangensis 64–65, 101
Meloidogyne jinanensis 64
Meloidogyne kikuyensis 5, 65, 97*n*3, 184
Meloidogyne kirjanovae 64
Meloidogyne konaensis 65, 101, 122–123, 130–133, 184
Meloidogyne kongi 65
Meloidogyne kralli 65, 101
Meloidogyne lini 65
Meloidogyne litoralis 63
Meloidogyne lordelloi 65
Meloidogyne lucknowica 65
Meloidogyne lusitanica 65, 101
Meloidogyne mali 65, 105, 107–108
Meloidogyne marioni 59, 66
Meloidogyne maritima 65, 101, 105, 121–122, 124, 130–133
Meloidogyne marylandi 65, 101, 107–108
Meloidogyne mayaguensis 3, 9, 64, 80, 98–99, 107–109, 125, 187, 301, 423, 458
Meloidogyne megadora 65
Meloidogyne megatyla 5, 7, 24, 33, 65, 184, 186
Meloidogyne megriensis 66
Meloidogyne mersa 65
Meloidogyne microcephalus 65, 183–184
Meloidogyne microtyla 5, 65, 101, 105, 107–108, 121–122, 124, 129–133, 164, 184, 186
Meloidogyne mingnanica 65
Meloidogyne minor 2, 5, 11, 65, 101, 105, 110, 124, 130–133, 184
Meloidogyne morocciensis 65, 101, 107, 122, 124, 127, 130–133, 184
Meloidogyne naasi 7, 33, 65, 82, 101, 105, 107–109, 124, 129–133, 140, 149, 184, 202, 207, 212–213, 218, 231, 233, 248–249, 254, 427, 451
Meloidogyne nataliae 23, 31, 33, 65
Meloidogyne oryzae 5, 65, 91, 101, 105, 107–108, 122, 129–133, 142–143, 451
Meloidogyne oteifai 65

Meloidogyne ottersoni 65, 185
Meloidogyne ovalis 65
Meloidogyne panyuensis 65–67, 101, 124, 129–133
Meloidogyne paranaensis 9–10, 65, 69, 84–88, 99, 101, 107, 109, 122–124, 127, 130–133, 416, 418, 458
Meloidogyne partityla 3, 6, 65, 101, 107, 124, 130–133, 185, 422, 428, 430
Meloidogyne petuniae 65, 101, 185
Meloidogyne phaseoli 65, 230
Meloidogyne pini 5, 7, 39, 65, 165, 185
Meloidogyne piperi 65
Meloidogyne pisi 65
Meloidogyne platani 65, 185
Meloidogyne poghossianae 66
Meloidogyne propora 65–66
Meloidogyne querciana 66, 101, 185
Meloidogyne salasi 66, 84, 185
Meloidogyne sasseri 66, 72
Meloidogyne sewelli 66
Meloidogyne sinensis 66
Meloidogyne spartinae 5, 7, 66, 97*n*3, 185–186
Meloidogyne subarctica 66, 185
Meloidogyne suginamiensis 66, 105, 107–108
Meloidogyne tadshikistanica 66
Meloidogyne thailandica 66, 107, 124, 130–133
Meloidogyne thamesi 63, 216
Meloidogyne thamesi gyulai 63
Meloidogyne trifoliophila 3, 5, 66, 107, 122–124, 129–133, 185, 303, 331–332
Meloidogyne triticoryzae 66, 142–143, 451
Meloidogyne turkestanica 66
Meloidogyne ulmi 66
Meloidogyne vandervegtei 66
Meloidogyne vialae 66
Meloidogyne wartellei 64
Meloidogyne zhanjiangensis 65, 66n3

Nacobbus aberrans 108, 224

Oxyuris incognita 59, 64

Panagrellus redivivus 207
Paratrichodorus minor 231, 233, 286
Pratylenchus 123, 226, 230–231, 234, 286, 340
Pratylenchus brachyurus 223, 232, 235, 237, 288, 382
Pratylenchus coffeae 223, 232, 235
Pratylenchus goodeyi 395
Pratylenchus neglectus 456
Pratylenchus penetrans 147, 152, 231–234, 237, 275, 288, 291, 339, 382–385, 400
Pratylenchus vulnus 233

Radopholus 234
Radopholus similis 150, 223, 233, 235, 339
Rotylenchulus reniformis 145, 224, 233–234, 236–239, 276, 293, 305, 339

Scutellonema brachyurus 223, 231, 233, 235, 237
Spartonema kikuyense 64
Spartonema spartinae 65
Steinernema 388
Subanguina 63
Subanguina radicicola 97*n*1

Tylenchorhynchus agri 231, 233
Tylenchorhynchus brassicae 233, 236–238
Tylenchorhynchus claytoni 147
Tylenchorhynchus nudus 233
Tylenchorhynchus vulgaris 233, 235
Tylenchulus semipenetrans 276–277, 280
Tylenchus arenarius 63
Tylenchus javanicus 64

Xiphinema americanum 231–232

General Index

Abamectin 388, 414, 416, 462
Actin 34, 172
Agrobacterium 229, 342, 385, 387, 396
Agrobacterium radiobacter 387
Alcaligene 385, 387
Alcaligenes xylosixydans 387
Algae 461, 464
Allelochemicals 148, 150
Almond 418, 423
Amino acids 103, 127, 166–167, 175, 190–191, 193, 238, 308, 341, 343, 349, 389, 392
Amixis 183
Amphid 21, 23, 28–29, 36–37, 39–40, 62, 141, 152–153, 165, 168, 189, 192–193, 313, 345
Amphimixis 5, 21, 119, 165, 183–185
Amplified fragment length polymorphism (AFLP) 168, 186, 372
Aneuploidy 5, 119, 186, 364
Anhydrobiosis *See* Survival
Antagonistic plant 417, 419, 435
Antagonists 11–12, 152, 205, 230, 235, 262, 380–401, 415, 417, 419–420, 422, 429, 435, 446, 452, 454, 456–458, 462, 465
Antioxidants 6, 165–166, 312, 370
Anus 20, 22, 29, 31, 34–35, 45, 60, 62–63, 68, 71, 74, 140
Apomixis 5, 183
Arbuscular mycorrhizal (AM) fungi 394–395, 463–464
Arthrobotrys 392–393, 396
Arugula 419
Ascorbic acid 6, 151
Associations 5, 13, 175, 223, 226, 229–230, 232, 352, 387, 391, 420, 456, 466
Aureobacterium 385
Automixis 5, 183
Auxin 153–154, 166, 175, 254
Avermectin 388, 416, 462

Bacillus sphaericus 385–386
Bacillus subtilis 383, 385
Bacillus thuringiensis (Bt) 340–341, 350, 387–388
Beta-1,4-endoglucanase 166
Biological control 11, 284, 291, 334, 380–401, 413, 417, 422–423, 430–432, 434–435, 446–447, 459
Biosynthesis
    amino acids 166
    nucleic acids 171
    proteins 174, 347
Body wall 21, 23–36, 41–44, 189, 382
Brevundimonas vesicularis 387
Burkholderia pickettii 387

Cabalase 6, 189
Calreticulin 164, 167
Carbohydrates 150, 152–153, 171, 191–192, 340, 368, 375, 385
Carrot 2, 10, 12, 60, 80, 84, 259, 264, 277, 283–284, 293, 303, 333, 417, 420, 451
Cassava 257, 450, 454–455, 464
Castor 417, 452, 457, 464
Cellulase 166, 368
Cephalic framework 33, 41
Cherry 421
Chilli pepper 333

Chitin 44, 140, 190–191, 342, 389, 394, 400, 452, 462, 465
Chitinase 141, 370, 386, 389–391, 394
Chord 34, 36–37, 46
Chorismate mutase 166, 190, 334, 343, 368–369
Chromosome number 5, 47, 74, 183–186, 313, 364, 375
Chryseobacterium 385
Clade 121–127, 129–131, 134, 371, 382
Classification 2, 67, 147, 254, 389, 414
Clover 5, 210, 233–234, 286, 303, 331–333
Collagen 141, 190, 371, 373, 375, 384–385
Competition 110–111, 217, 224, 230–231, 235–236, 247, 251–252, 385–386, 393, 396, 432–433
Control
 biological 11, 284, 290, 334, 380–401, 413, 417, 422–423, 430–432, 434–435, 446–447, 459
 chemical 294, 413–416, 422, 429–430, 459–465
Corky root 428
Corn 147, 254, 258, 286
Corynebacterium 385
Cotton 6, 8, 77, 140, 201, 210, 211, 223, 250, 276, 288, 290, 293, 294, 398, 399, 445, 452
 interactions 223, 224, 226,–229, 232, 234, 235, 236–239, 387, 394
 nematode control 416, 417, 426, 428, 429, 431, 433
 resistance 302, 303, 305, 306, 307, 328, 331, 333, 334, 348, 355
 seed treatment 388, 462
Cover crop 201, 286, 418–420, 427–428, 430–431, 435, 452, 455–456
Cowpea 168, 228, 232–233, 254, 258, 285–286, 303, 305–306, 311–312, 314–318, 328, 339, 388, 417, 423
Crop rotation 2, 250, 263, 266, 288, 334, 394, 413, 417–418, 427, 430, 432, 435, 446, 453–455, 460
Cry6A protein 388
Cryophil 7, 140, 207
Cultural control 413, 416–422, 429–431, 451–453
Cuticle
 basal layer 23, 26, 40, 187
 cortical layer 23–24, 26, 187, 385
 epicuticle 153
 median layer 23, 26, 187
Cysteine protease 168, 369
Cyst nematode 2, 59, 139, 141, 143, 153, 165–168, 234, 269, 280–281, 284, 314, 339, 342, 345, 349, 354, 372, 382
Cytochrome P450 (CYP) 370
Cytogenetics 2, 5, 47, 119–120, 182–187, 364, 367
Cytokinesis 170, 172–173
Cytokinins 169, 175, 309–310, 334

Damage levels 246–269, 424
Dauer 191–192, 218, 373–375
1,2-dehydropyrrolizidine alkaloids 465
Desiccation 12, 148, 192, 201, 203–205, 396, 432
Diagnostics 13, 46, 49, 60–61, 67–68, 70, 72, 88, 98–100, 102–108, 110–112, 119–120, 124, 126, 285, 294, 314, 424–425, 446
Diapause 3, 19, 140, 143–144, 202–203, 205–206, 212–213, 216, 218–219, 249
Differential test 8
Digestive system 21, 40–44
Dispersal 2, 98–99, 170, 342, 374
Distribution
 spatial patterns 236, 276, 278
 vertical 147, 291, 417
Dormancy 143, 202–203, 205, 212–213, 218
Drechmeria coniospora 391
Dystrophin 120, 125

Eclosion *See* Hatching
Egg mass 3, 5, 6, 11, 19, 62, 67, 70, 71, 103, 104, 140, 142–144, 187, 188, 189, 201–211, 216, 218, 237, 239, 247, 248, 254, 261, 262, 265, 276, 281, 282, 289, 290, 311, 314, 329, 332, 341, 346, 384, 388, 393, 394, 463
Egg morphology 22–23
Eggshell 22, 44, 140–141, 190–191, 202–203, 213, 342, 389–391, 394, 400
Elongation factor 120, 125–126
Enterobacter 385, 389
Entomopathogenic nematodos 387–388, 400, 430
Environmental impact 350
Enzymes
 cell-wall degrading 164, 369
Epidermis 290, 301, 311
Excretion 188–189
Expansin 166, 168, 173, 369
Expressed sequence tag (EST) 100, 102, 127, 190, 363, 372–373
Extraction
 from plant material 103, 145, 265, 288, 294, 333, 465
 from soil 98, 103–104, 263, 281–283, 285, 287–288, 292

Fallow
 dry 418
 grass 418
 weedy 277, 292, 418
 wet 211, 418
Fatty acids 150, 191–192, 214, 351, 464
Fibronectin 384–385
Flavonoids 150–151, 166
Flooding 12, 446–447, 451–452
FMRFamide-like peptides (FLPs) 191

Food reserves 21, 187, 191, 211–214
Fucosyltransferase 370
Fusarium wilt 223, 226–229, 305

Gall index 207, 247, 254, 283–284
Gastrodia anti-fungal protein (GAFP) 430
Gelatinous matrix 3, 22, 40, 62, 140, 142, 165, 189, 191, 201–202, 204–205, 342, 393
Gene flow 351
Gene for gene hypothesis 239, 335, 346, 369–370, 391
Genes *See* Gene index
Genetic engineering 334, 430
Giant cell
  development 4–6, 19, 170–176, 311, 312
  general 12, 21, 34, 37, 58, 70, 226, 227, 236, 238, 239, 306, 334, 343, 347, 349
  mini cell plate 165, 170, 172–173
Global information system (GIS) 431–432
Global positioning system (GPS) 46, 293, 431–432
Glomus fasciculatum 394, 464
Glucoside 464
Glucosinolates 419, 430, 463
Glutathione peroxidase 370
Glutathione S-transferase (GST) 166, 370
Glycoproteins 3, 140, 152, 189, 192
Glycosyltransferase 310, 346
Glycoside hydrolase 368
Green manure 417–419, 429, 452, 461, 463
Groundnut 6–8, 229, 250, 256–257, 260, 291–292, 326, 328, 331, 333, 398–399, 417, 427, 452, 454, 461, 464
Growth reduction 237–238, 254–255, 263, 266
Guard hypothesis 346

Harpin protein 429
Hatching
  hatching mechanism 140–142
Heat shock proteins 174, 346, 450
Heat treatment 420, 448–450, 453
Hemicellulose 166
Hemizonid 20, 27, 37, 43, 46, 62, 69, 73–74, 77, 80, 84, 88
Hirsutella minnesotensisi 392
Hirsutella rhossiliensis 391–392, 400
Horizontal gene transfer 166, 169, 369, 373, 375–376
Host efficiency 347
Host races 7–8, 13, 70, 88, 237, 424
Hot water treatment *See* Heat treatment
Hydrostatic pressure *See* Pseudocoelom
Hypersensitive response (HR) 6, 12, 167–168, 237, 302, 307–311, 318
Hypodermis 20, 23, 26, 28, 33–34, 40, 42–43, 153–154, 165–166, 187

Interactions
  between *Meloidogyne* species 225–226, 230, 236–237, 433
  with microbial pathogens 226–230, 239
  with other plant-parasitic nematodes 230–239
  with root-rot pathogens 228–229
  with vascular wilt pathogens 226–229
Intestine 20–22, 27, 30, 35, 40–42, 44, 61–62, 189, 212, 340, 342
Invasion 6, 139, 146–151, 153, 163–176, 188, 204, 210, 212–213, 216–218, 231, 234–235, 237, 251, 259, 338, 345, 393, 452, 454, 456
Ionic regulation 188–189, 213
Isothiocyanates 188, 400, 414, 434, 452, 463
Isozymes 2–3, 9, 44, 47, 61, 67, 72–74, 77, 80, 84, 88, 99–102, 112, 119, 122, 308

Jasmonates 6

Kinase 166, 173, 341, 346, 349, 368, 370–371
Kinesis 170, 172–173
Klebsiella 385
K-strategist 234

Labial disc 21, 29, 36, 40, 46, 61–62, 72, 74, 77, 80, 82, 84
Lactate dehydrogenase 190, 205
Lateral field 25–26, 29–32, 34, 46, 48–49, 61–62, 68–69, 72, 74–78, 80, 82–84, 86, 152
Lectin 152, 191–192, 204, 342, 370, 381, 384
Linoleic acid 393
Lipid 3, 7, 22, 34, 36, 42, 44, 140, 144, 153, 190–191, 203, 207, 209, 211–212, 341, 351, 385, 389, 391
Lucerne 149, 223–224, 232, 234–235, 238, 257, 284, 303, 311, 315–316, 318, 328, 339, 417, 433

Major sperm protein (msp) 120, 125–126, 342
Matric potential 145, 203
Metabolomics 194
Methyl bromide 188, 327, 355, 413–415, 428–430, 433–435, 448, 452, 461
Methyl iodide 414–415, 434–435
Microfilaments 172
Microtubules (MTs) 44, 172–173
Migration 47, 100, 165–166, 189, 212, 311, 343, 417, 422
Mitochondrial DNA (mtDNA) 99, 104, 106–108, 112, 120–121, 125–129, 134
Monacrosporium 392
Monocrotaline 464

Morphology
egg 22–23
female 19, 21–22, 39, 41–42, 44–46, 49, 66–67, 71–72, 74, 77, 79–80, 84
juvenile 19–23, 43, 49
male 19, 21, 23–49
Moulting 3–4, 19, 21, 26, 40, 62, 164–165, 182–188, 191, 202–203, 369
Movement 144–147, 148, 149, 152
mtDNA. *See* Mitochondrial DNA
Mulching 216–217, 399–400, 452–453, 456
Multiplex PCR 110
Muscles 20–21, 23, 27–28, 33–37, 41–44, 125, 193, 345
Muscodor albus 429, 434
Myrothecium verrucaria 416

Neem 452, 462–463
Nematicides
application 259, 275, 431–432
contact nematicides 416
formulation 462–463
fumigant 188, 259, 414–415, 426, 432
granular 12, 415
non-fumigant 12, 414–416, 460
Nematoctonus 392–393, 396
Nematode management *See* Management
Nematophagous fungi 389–393, 396
Nervous system 21, 23, 37, 39–40, 371
Neuron 152, 192, 342, 345
Neuropeptide 191, 342, 365, 371
Neurotransmission 192–193
Neurotransmitters 193, 371
Nodulation factor 334
Nuclear receptor (NR) 365, 371
Nursery management 259, 422, 447–448, 451, 453, 459
Nutsedge 433–434

Oilseed radish 427–428, 430
Operon 365, 367, 373–376
Orientation 26, 34, 145–147, 149, 152, 192–193, 266, 286, 345, 374
Osmotic regulation 189
Overwintering 19, 208–209, 211–212, 217, 263, 289, 433
Oxidative burst (OB) (oxidative response) 6, 308, 311, 318

Paecilomyces lilacinus 389–391, 395–397, 462
Paenibacillus 385–386, 396
Parthenogenesis
evolution 125, 186–187
meiotic 47, 120–121, 131, 134, 183–185, 187, 315
mitotic 5, 119–121, 123, 125, 130, 134, 183–187, 216, 313, 315, 364, 371–372
origin 186
Pasteuria penetrans 187, 232–233, 288, 290–292, 381–385, 396–397, 400, 422, 462
Pathotypes 13, 315, 329, 454
Peach 58, 98, 257, 327–328, 415–416, 418, 421–423, 425, 430
Pecan 6, 428, 430
Pectate lyase 166, 173, 190, 343, 368
Pectin 166, 173, 204, 368
Pepper 6, 8–9, 80, 233, 250, 257–258, 264–265, 302–303, 305, 311, 313, 316, 328, 331, 333, 399, 415, 423, 426, 433, 459
Peptidase 365, 368
Peptide 103, 153, 167–168, 172, 175, 191, 334, 342–345, 349, 365, 371, 392
Perineal pattern/region 2, 19, 24, 31, 35, 38–40, 44, 46–49, 58, 60–62, 67–69, 73–87, 99, 285
Peroxidase 6, 190, 370
Pest risk analysis (PRA) 11
Pharyngeal glands 20–21, 27, 35–36, 41, 46, 61–62, 110, 141, 153, 163, 165–166, –168, 189, 203, 313, 343, 368–369
Pharynx 20, 22–23, 27, 30, 32, 36–37, 40–42, 163
Phasmid 23, 31–32, 35, 37, 40, 45, 62, 68, 152–153, 165, 193
Phenology 420, 427–428
Pheromone 148, 191, 218, 375
Phospholipid 191, 351
Phyllobacterium 385
Phylogenetic analysis/tree 108, 119–125, 127–128, 134, 389, 391
Phylogeny 61, 72, 88, 119–134
Phytoalexin 166
Phytosanitary measures 450–451
Plant breeding 301–302, 316, 318, 330–331, 333–335
Plant cell-wall-degrading enzyme 368–369
Plant hormones
cytokinins 169–170, 175, 309–310, 334
Plantibodies 102
Pleurotin 393
Ploidy 5, 98, 119–120, 173, 186, 313, 315, 364, 375
Plum 423, 430
Pochonia chlamydosporia 389–391, 396–397
Polygalacturonase 166, 173, 190, 368
Polymerase chain reaction (PCR) 3, 67, 72, 103–106, 108, 110–111, 171, 175, 284, 294, 308, 371, 373, 389–390
Polythienyls 463–464
Population dynamics 108, 231, 240, 246–269, 286, 289, 291, 390–391, 424, 432
Prestoma 21, 27–29, 36, 39–40
Primers 3, 72, 105–111, 125–126, 344
Promoter 171–172, 174–175, 339, 342, 345–347, 349–351, 364, 394

Protease 168–169, 190, 369, 386, 389, 391
Proteinase inhibitors 339–340, 353–354
Protein electrophoresis 3, 103, 189, 205
Proteomics 194
PR-proteins 386
Prunus 304, 316, 333, 421–422, 424, 428, 430
Pseudocoelom 23, 34, 188–189
Pseudomonas fluorescens 381, 385–386

Quantitative trait loci (QTL) 302, 304–305
Quarantine 2, 11–12, 67, 80, 84, 99, 103, 110, 143, 201, 254, 283, 413, 417, 425, 450–451, 466
Quiescence 3, 11, 140, 202, 206, 211, 213, 216, 218

Race 7–8, 13, 70, 82, 84, 103, 121, 153, 184, 217, 228, 237, 250, 288, 315–316, 326, 329, 424–425
Random amplified polymorphic DNA (RAPD) 47, 72, 84, 99, 106, 108–110
Reactive oxygen 6, 166, 174, 310–311, 346
Real-time PCR 110, 284, 294
Reporter genes 171, 349
Reproductive system
  female system 21, 44, 342
  male system 21, 43–44
  ovary 22, 25, 35, 37, 43–44, 62, 183, 312
  sperm 21, 27, 43, 120, 125–126, 183, 342
  spicules 43
  uterus 35, 37, 42, 44
Resistance
  breeding 301–302, 305–307, 318, 328–331, 333–335, 458–459
  genes (*See* Genes index)
Respiration 145–146, 188, 209, 212, 214
Restriction fragment length polymorphism (RFLP) 72, 104–105
Rhizobium 165, 174–176, 334, 381, 385–386, 396, 398
Rhizoctonia 229
Rhizosphere bacteria 385–387, 396
Ribosomal DNA (rDNA) 60, 72, 105–106, 108, 120–125, 127, 134, 372
Rice 72, 84, 168, 185, 188, 232, 250, 257–258, 339, 352–353, 395, 398–399, 445, 448, 451–455, 458, 461
Ricin 464
RNA interference (RNAi) 169, 191, 309, 315, 335, 338, 342–345, 348–350, 354, 365, 372, 374–375
RNA polymerase 2 120, 126
r-strategist 234

Salicylic acid 151, 166, 174, 308–310, 386
Sampling
  accuracy 276, 278–281
  precision 268, 275, 278, 280, 284, 294
  processing 278–281
  programmes 280–281, 283–293
Sanitation 413, 417, 420–421, 453
Satellite bioimagery 284
Secretions 29, 42, 62, 102, 141, 151–153, 163–170, 189–190, 192–193, 345, 369, 373, 392
Secretory-excretory products 189
Secretory-excretory system 42–43, 141, 189
Seed treatment 388, 462
Semiochemicals 148
Sensilla/sense organs 21, 28–29, 36–37, 39, 61, 147, 152, 155, 192
Sensory perception 152–153, 165, 192–193
Sequencing 103, 133, 193, 318, 364, 367, 372, 376, 384
Sequential control 427
Sesame pangola grass 417, 457
Sex determination 215, 365, 370–372, 374–376
Soil amendment 418–420, 435, 452–453, 460, 464
Soil moisture 21, 140, 145, 148, 203–204, 211, 251, 261, 264, 414, 418
Soil type 2, 149, 209–211, 240, 253, 262, 281, 284, 291, 424
Solarization 12, 208, 215, 259, 400, 421–422, 435, 453
Soybean 8–9, 80, 88, 175, 223, 232–233, 235–237, 240, 279–280, 285–286, 288, 304, 307, 311, 333, 355, 372, 397–399, 417, 423, 427
Species concepts 61, 67, 99
Spermatozoa 44
Spicules 27, 32, 37, 40, 43, 45–46, 49, 62, 69
18S rDNA 60, 72, 120
28S rDNA 120
Stacked defence 348
Steam heat 421–422
Steroids 6, 188, 191–192
Stoma 21, 27–29, 33, 36, 39–42
Streptomyces 387–388, 416, 462
Streptomyces avermitilis 388, 416, 462
Striae 26, 31, 35, 46, 68, 72, 74, 77, 79–80, 82, 84
Stylet
  protractor muscles 20, 28, 34, 41
Sudan grass 209, 419, 430, 457
Sugars 56, 80, 190–191, 230, 232, 234–235, 249, 253–254, 257–258, 264, 282, 285–286, 304, 339, 398, 460–461, 465
Sunn hemp 452, 456–457, 464
Superoxide dismutase (SOD) 3, 6, 100, 190, 370
Supertree 127–130, 134
Suppressive soil 380–382, 385, 400
Surface coat 153, 165, 193, 385
Survival 7, 11, 42–43, 140, 142–144, 189, 201–219, 246, 248, 262, 316, 374–375, 396, 400–401, 421–422
Syncytium 33, 349

Tall fescue 430
Taxis 192
Telluria chitinolytica 381, 386
Temperature
- base 144, 214–215, 237
- and diapause 212–213
- lethal 208, 421
- thermal time 214–216, 218

Thermophil 7, 140, 206–207, 213, 215
Tobacco 8–10, 80, 88, 167, 174–175, 217, 223, 227, 229, 232, 235–238, 250, 256, 258, 260, 278, 290–291, 304, 327, 331, 333, 343, 345–346, 394, 397–399, 420–421, 427, 430, 445, 448, 451, 453, 456, 460–463
Tolerance
- tolerance limit 254, 256–260, 263, 266, 269, 280

Transgenic plants 153, 169, 308, 339, 343–345, 350–351 354–355, 388, 400
Transgressive segregation 302, 306
Transplant 12, 249, 259, 261, 286, 385, 394, 398–399, 401, 420–421, 427–428, 447
Trans-spliced leader 365, 367, 373
Transposable element 313, 367
Trap crop 12, 417–420, 428, 430, 456–458
Trichoderma 229–230, 393–394, 396, 398, 400, 463
Tubulin 172, 349

Ubiquitin 334, 369

Velvet bean 286, 417, 456–457, 464
Verticillium wilt 226, 305
Virulence 8, 13, 16, 99, 167–169, 302–304, 307, 309, 311–317, 327, 329, 346, 348, 369, 376
Vulva 22–23, 31, 35, 42, 44–46, 60, 62, 67, 77

Water channel 175
Watermelon 9–10, 77, 88, 427
White mustard 452
Wolbachia 381

Xanthomonas 168, 385
Xylanase 166, 190, 368

Yam 283, 448–450
Yield
- drag 327
- minimum yield 226, 254, 257–260, 263, 269, 334
- reduction 237, 254, 431